Wireless Network Traffic and Quality of Service Support:
Trends and Standards

Thomas Lagkas
University of Western Macedonia, Greece

Pantelis Angelidis
University of Western Macedonia, Greece

Loukas Georgiadis
University of Western Macedonia, Greece

Information Science REFERENCE

INFORMATION SCIENCE REFERENCE

Hershey · New York

Director of Editorial Content: Kristin Klinger
Director of Book Publications: Julia Mosemann
Acquisitions Editor: Lindsay Johnston
Development Editor: Joel Gamon
Publishing Assistant: Keith Glazewski
Typesetter: Keith Glazewski
Production Editor: Jamie Snavely
Cover Design: Lisa Tosheff
Printed at: Yurchak Printing Inc.

Published in the United States of America by
Information Science Reference (an imprint of IGI Global)
701 E. Chocolate Avenue
Hershey PA 17033
Tel: 717-533-8845
Fax: 717-533-8661
E-mail: cust@igi-global.com
Web site: http://www.igi-global.com/reference

Library of Congress Cataloging-in-Publication Data

Wireless network traffic and quality of service support : trends and standards / Thomas Lagkas, Pantelis Angelidis, and Loukas Georgiadis, editors. p. cm.
Includes bibliographical references.
Summary: "This book offers cutting edge approaches for the provision of quality of service in wireless local area networks"--Provided by publisher. ISBN 978-1-61520-771-8 (hardcover) -- ISBN 978-1-61520-772-5 (ebook) 1. Wireless LANs--Standards. 2. Wireless LANs--Quality control. I. Lagkas, Thomas, 1980- II. Angelidis, Pantelis, 1967- III. Georgiadis, Loukas, 1977- TK5105.78.W5685 2010
621.384--dc22

2009036914

British Cataloguing in Publication Data
A Cataloguing in Publication record for this book is available from the British Library.

Editorial Advisory Board

List of Reviewers

Table of Contents

Detailed Table of Contents

Section 1
QoS Provision in Wireless Local Area Networks

This section discusses QoS provision techniques for wireless networks that mainly target local area networks (WLANs). State of the art solutions for QoS support in WLANs, like the well-known IEEE 802.11e standard, are examined. Moreover, modern proposals which improve the efficiency of medium access control and routing algorithms in small-scale wireless networks are presented. Furthermore, latest traffic differentiation and prediction techniques for wireless networks are studied. This section, also, examines cross-layer approaches for the efficient support of multimedia applications inside a wireless network, while power saving methodology is discussed, too. Lastly, the reader is introduced to cutting edge management techniques of multi-hop mobile networks and wireless sensor networks.

This chapter discusses the hybrid concept of providing QoS in wireless networks via medium access control by presenting the Priority Oriented Hybrid Access (POHA) scheme. POHA implements a hybrid access scheme via combining two modern medium access control protocols for WLANs: POAP (Priority Oriented Adaptive Polling) and POAC-QG (Priority Oriented Adaptive Control with QoS Guarantee). The comparison of POHA with another well known hybrid protocol, the HCF (Hybrid Control Function) access scheme of the IEEE 802.11e standard, shows that in a mixed traffic-type network environment POHA performs exhibits superior performance.

This chapter presents a high performance distributed MAC protocol, named DQCA (Distributed Queueing Collision Avoidance), suitable for infrastructure wireless LANs. In addition, four scheduling algorithms are proposed, that modify the transmission order of DQCA in order to increase efficiency under heterogeneous traffic conditions. Depending on the network configuration and objectives, the more suitable algorithm can be applied over the DQCA MAC mechanism to ensure that the service requirements are satisfied.

This chapter proposes a strictly guaranteed QoS for CBR and VBR traffic in IEEE 802.11e wireless LANs. The performance of the proposed scheme is analyzed and compared with the original IEEE 802.11e protocol under standard MAC and physical layer parameters proposed by the IEEE standard for wireless LANs. Through extensive simulations, important performance metrics such as average access delay, achievable throughput, dropping probability, and delay variation performance are thoroughly investigated.

This chapter presents a dynamic queue length scheduling for downlink multiuser and multi-antenna WLAN systems with heterogeneous traffic. A dynamic DLC queue length control is proposed, so that the maximum length is allowed to obtain the highest average system throughput, but restricted to the satisfaction of the users QoS. Several alternative QoS measures are presented along the chapter and in closed form expressions, so that the wireless operator can choose among them for the most suitable ones for each scenario and QoS requirements. Besides the dynamic queue proposal, another important outcome of this chapter is on how applications and link layers (or in general higher layers) take profit of the advances introduced by multiple antennas and signal processing techniques in the physical layer.

representative schemes from each with respect to major distinguishing characteristics of the WiMAX MAC layer and PHY layer as specified in the IEEE 802.16d standard. They also discuss and highlight research trends on QoS support issues in WiMAX networks.

This chapter looks at what might be the next generation of wireless networks not only in terms of QoS provision and the mechanisms related to, but also with regard to network architectures and system design. QoS requirements related to 4G networks are examined. Heterogeneous networking challenges are discussed focusing on the interoperability between different access technologies (e.g. 3G/UMTS-WLAN). Various resource management techniques are presented including resource reservation, QoS renegotiation, game theoretic resource allocation and dynamic pricing.

This chapter examines the QoS-predictions service, providing predictions for QoS of networks available at a given user's geographical location and time. In a case study, the authors show the feasibility of deriving predictions from historical data collected by a mobile service user. Specifically, this mobile user is assumed to be a patient, using his health tele-monitoring service in his daily environments for a period of one month. The QoS-predictions service is considered as a novel support for mobile services operational in 4G heterogeneous network environments.

This chapter collects trends on efficient resource allocation through cooperation among entities in a wireless heterogeneous environment and selection of the best solution among a set of alternatives. Initially, the problem of network selection is presented. Then, the authors described the basic trends and ideas on user cooperation in future networks, emphasizing on cooperation on both physical and application layers. Next, it is briefly demonstrated how game theory can be applied to wireless networking.

This chapter provides a thorough study of the basic concepts considering Call Admission Control (CAC) design and a comprehensive analysis of the fundamental CAC schemes employed in wireless networks. The basic performance criterion considering CAC schemes is the probability of denying the access to the network for an arriving call, which is extensively studied in this chapter. Moreover, additional performance criteria are presented and discussed, which may help to provide an overall efficiency estimation of the available CAC schemes.

This chapter introduces the key concepts of MBMS (Multimedia Broadcast / Multicast Service) services. The main target is to highlight the importance of power control and its commanding role during the delivery of MBMS multicast content. The power profiles of several transport channels are investigated. Moreover, the reader is introduced to certain problems that MBMS current specifications are facing and becomes familiar with techniques/solutions proposed to overcome such limitations. Finally, the authors discuss several radio bearer selection mechanisms that focus on conceiving and adapting to continuous changes that occur in dynamic wireless environments.

Foreword

Wireless networks have been rapidly expanded in business and home environments making the wireless market a huge market. PDAs are widely spread, mobile phones with computer capabilities are almost everywhere and laptops with wireless connections are selling in great numbers. Wireless networks allow somebody to stay connected and be productive no matter where he is. Driven by the mobility of society, convenience and ease of use of wireless products, users require the elimination of distinction between services offered from wireless and wireline networks. Interactive applications such as IP telephony, streaming video, interactive video games and video telephony have become very popular and pose certain tight constraints at the underline network.

Wireless communication poses new challenges in Wireless Network design. Efficient Medium Access techniques, transmission errors at the wireless medium, variable bit rate adjusting to wireless medium conditions, connection to the most suitable base station as the mobile user moves around, energy saving techniques due to limited battery power of mobile devices, best path selection in multi-hop mesh networks and admission control of new stations are some of the critical issues that should be addressed in efficient Network Design. These new and challenging problems of mobile and wireless network design, combined with the tight constraints and QoS required by interactive applications is a new fruitful area of research.

In short, this book is a valuable reference to researchers and developers in the fields of wireless network design considering quality of service support; in addition, it will be very useful for new entrants in this field as it includes contributions in the most interesting research topics in the field of providing interactive applications the required Quality of Service support in modern wireless networks. The variety of topics discussed in this book makes it o valuable reference to active researchers. This book presents to the wireless network community the current research trends for efficient MAC design to provide QoS in wireless networks and inspire future research in this area.

Vasileios Vitsas
Technological Educational Institution of Thessaloniki, Greece

***Vasileios Vitsas**, B.Sc Electrical Engineering (Aristotelian University of Thessaloniki, GR), M.Sc. Computer Science, (University of California, Santa Barbara, USA), Ph.D. Wireless Communications (Bournemouth University, UK), is a professor in Computer Networks at the Information Technology Department at Technological Educational Institution of Thessaloniki, Greece. He has worked for several years as a Senior Telecommunications Engineer at the Greek Telecommunications Organization (OTE). Vasileios has served as Head of Department and Head of Division at the Information Technology Department at Technological Educational Institution of Thessaloniki. His current research interests lie in wireless and multimedia communications. Vasileios has served as Chair, Technical Program Committee member and Referee for several international IEEE journals, conferences and workshops.*

Preface

Nowadays, wireless networks are considered to be a significant and emerging part of the communication networks field. The current prevalent opinion is that in some years a considerable percentage of the data networks will be wireless networks. The same time, the network applications are evolving and become more resource demanding. Specifically, multimedia traffic tends to be a significant portion of the total network load, which leads to stricter transmission requirements. The modern time-bounded network applications need to be provided with high throughput, low delay, low jitter, and low loss/drop rate. Today, users expect that this type of high quality services can be offered by all kinds of communication networks. However, meeting strict transmission requirements under the harsh wireless environment, which is characterized by scarce bandwidth, unreliable links, and limited range, is quite challenging and it has actually formed a very active research field in the last years. A straightforward approach to this whole issue is the effort to physically maximize the available data rate and link quality. It is certain that there has been lately great development in the respective area, which has led to efficient signal modulation techniques and has definitely improved the overall performance of wireless data transmissions. However, the increasing demands necessitate global solutions that involve cross-layer approaches. Specifically, modern wireless networks need to provide total Quality of Service (QoS), which in practice means efficient differentiation of the offered traffic flows based on their nature and serving them according to their specific needs. This concept includes adaptive control of the physical layer, optimized medium access control for serving mixed-type traffic load, efficient routing algorithms that can provide end-to-end transport guarantees, and QoS-aware higher layer protocols which can be automatically adjusted to the user needs and the network limitations. Furthermore, part of the provided Quality of Experience (QoE) in a wireless network environment has become the available energetic autonomy of the mobile devices, thus, power conservation techniques are also considered to participate in the total QoS provision concept.

Cutting edge approaches for the provision of QoS in wireless local area networks are examined in this book. The Enhanced Distributed Channel Access (EDCA) and the Hybrid Control Channel Access (HCCA) protocols, which constitute the Hybrid Coordination Function (HCF), that is the Medium Access Control (MAC) scheme of the IEEE 802.11e standard, are discussed. The authors present latest solutions towards the optimization of the channel management and routing in infrastructure and ad hoc wireless networks. This book also analyzes traffic categorization issues and methods that can predict the available QoS. Moreover, cross-layer designs (especially PHY-MAC) are presented, which can efficiently serve multimedia traffic over wireless links. Energy conservation techniques for the maximization of the mobile devices' battery lifetime are also examined. Furthermore, the book introduces the reader to modern methods of managing wireless mesh networks and controlling wireless sensor networks.

State of the art mechanisms that ensure QoS support in wireless wide area networks are presented, too. Topology control issues for extended point-to-point wireless networks are examined. The book also discusses the new IEEE 802.16 standard, known as WiMAX, focusing on traffic scheduling techniques. The reader can thoroughly study future network cooperation under the promising 4G architecture by examining issues related with allocating resources, managing new connections, and predicting the QoS level that can be offered. Additionally, the role of power control in supporting MBMS (Multimedia Broadcast / Multicast Service) services over 4G networks is analyzed.

The target audience of this book includes students on computer science and communications engineering that need a good background and understanding of the subject area, scholars and researchers whose area of interests is wireless networks – QoS and need a reliable reference for their study, and people working in the wireless communications industry and require a modern book that can support their effort to enhance their current services and develop new ones. The book can be proved valuable as a library reference, useful in computer science, informatics, electronic-electrical engineering, and communications engineering departments. It can also serve as a course supplement for graduate studies on computer networks, wireless communications, and multimedia applications. Instructors in the above mentioned scientific areas would find this book useful as a resource, when teaching (among others) about the evolution of wireless networks, the nature and characteristics of network traffic, resource management algorithms, medium access control protocols, multihop network control, network simulation techniques, and power saving trends.

The book is organized into two sections. Section 1 discusses QoS provision techniques for wireless networks that mainly target local area networks (WLANs). Chapter 1 presents the Priority Oriented Hybrid Access (POHA) scheme, which is formed by the combination of two different MAC protocols for WLANs. This scheme is examined in comparison to the hybrid MAC protocol defined in the IEEE 802.11e standard, HCF. Chapter 2 presents the Distributed Queueing Collision Avoidance (DQCA) MAC protocol for infrastructure wireless networks. Based on a cross-layer approach, four traffic scheduling algorithms are also proposed. Chapter 3 introduces a mechanism for strict QoS guarantee in WLANs. Its performance is examined compared to the defined IEEE 802.11e HCCA scheme. Chapter 4 proposes a dynamic queue length scheduling technique for wireless networks with heterogeneous traffic. It also examines the impact of modern multi-antenna solutions to the higher network layers. Chapter 5 describes traffic prediction models, including time series models, artificial neural network models, wavelet-based models, and support vector machine-based models. The authors study the application of a support vector machine in a WLAN and examine its behavior along with other models. Chapter 6 analyzes power consumption minimization of video transmission over wireless links. The authors adopt a cross-layer approach, which takes into account the video coding and the wireless communication process. Chapter 7 examines power saving issues related with medium access control and routing algorithms in wireless networks. The respective protocols proposed in the chapter are finally evaluated in comparison to other well-known protocols. Chapter 8 discusses modern end-to-end QoS support solutions. The authors study the issues raised by the coexistence of mobility management and MAC protocols. Chapter 9 describes the Control and Provisioning Wireless Access Protocol (CAPWAP), which is related with the management of centralized WLANs. The authors examine the possibility to use this protocol for QoS monitoring and adjusting. Chapter 10 provides a detailed overview of QoS provision in multi-hop ad-hoc networks. It presents related issues in each one of the physical, data link, and routing layers. Cross-layer approaches are also examined. Chapter 11 discusses heuristic geographic routing techniques for wireless networks. The authors introduce a new performance measure and compare the presented algorithms using simulation.

Section 2 discusses QoS provision techniques for wireless networks that mainly target wide area networks (WWANs). Chapter 12 presents topology control solutions in wireless mesh networks, in terms of employing power control or not. The authors present the Path Reduction (PR) algorithm that is described via simulation. Chapter 13 is about QoS provision techniques in WiMAX networks. Various resource management schemes proposed for IEEE 802.16 networks are examined and evaluated. Chapter 14 provides a complete overview of the next generation wireless networks area. The 4G concept is examined and various techniques for QoS provision in a heterogeneous network environment are provided. Chapter 15 examines the process of predicting the QoS characteristics provided by the locally available networks in a 4G architecture. The specific case study involves a health tele-monitoring service. Chapter 16 analyzes the network selection problem inside a heterogeneous network environment. The authors present the latest approaches and introduce the application of the game theory concept in the specific subject area. Chapter 17 thoroughly examines the call admission control in wireless networks. The authors analytically study the performance of the available relative schemes focusing on the admission denying probability. Lastly, Chapter 18 presents the basic approaches in delivering MBMS content inside a 4G network architecture. The authors focus on power control adaptation issues when operating under variable network conditions.

In overall, this book aggregates technologies related to QoS support in wireless networks. This publication targets in explaining all related issues, the problems that arise when trying to provide different and demanding services over wireless networks, the theoretical background, the evolution of traditional related technologies and the solutions given so far, and moreover, the book presents the latest research concepts and relevant ideas regarding near-future implementations. The selected chapters cover most of the aspects of this modern and promising area that till now was usually presented in a segmental way in literature. Finally, this book could potentially become: a) a guidebook for readers entering the area of QoS support in wireless networks, b) a reference for scientists who need to gain up-to-date knowledge of the latest related technologies, and c) a source of ideas and new trends for researchers who work on the area of QoS provision in wireless networks.

Acknowledgment

The editors would like to thank all people involved in the editing process, without whose efforts the completion of this project would not be possible. Special thanks to all chapter reviewers for their dedicated and consistent work. We would also like to acknowledge the valuable help offered by the editorial content department of IGI Global. The editors are of course thankful to all chapter authors for contributing their qualitative work.

Finally, special thanks to our families and close friends for their steady encouragement and understanding throughout this project.

Thomas Lagkas
University of Western Macedonia, Greece

Pantelis Angelidis
University of Western Macedonia, Greece

Loukas Georgiadis
Hewlett-Packard Co, USA

Section 1
QoS Provision in Wireless Local Area Networks

Chapter 1
Hybrid Approach to Integrated QoS Capable Protocols for Wireless LANs

Thomas D. Lagkas
University of Western Macedonia, Greece

ABSTRACT

The recent evolution of wireless networking has led the market to increased service demands. Thus, the emerged necessity is to develop specialized mechanisms that provide efficient QoS (Quality of Service) for both traditional and modern network applications in the challenging wireless environment. The respective IEEE proposal comes from the 802.11e workgroup which has developed the Hybrid Coordination Function (HCF). HCF is definitely capable of providing QoS, however, it exhibits significant limitations. This work presents an alternative protocol with improved behavior and performance. The Priority Oriented Hybrid Access (POHA) is a complete channel access mechanism able to provide integrated QoS for all types of traffic and network applications. POHA combines a polling based and a TDMA access scheme, adapts to the dynamic conditions of modern WLANs, improves channel utilization and station feedback, provides medium contention fairness, eliminates collisions, differentiates traffic based on priorities, supports dynamic resource assignment, and instantly negotiates the quality levels of the offered traffic streams trying to support multiple streams with best possible quality. POHA, compared with HCF, exhibits generally superior performance.

INTRODUCTION

Nowadays, users expect the wireless networks to provide qualitative services similarly to the wired ones. The initial role of the WLANs gradually changes, so they are no longer considered to be simple extensions of the wired networks with limited capabilities. The modern needs for extended mobility, real-time communications, and multimedia traffic transmissions have significantly increased the network requirements. Lately, great development has been observed in the area of the wireless physical layer. In a few years, the maximum throughput has risen from 1 Mbps, in legacy IEEE

DOI: 10.4018/978-1-61520-771-8.ch001

802.11 (IEEE 802.11 WG, 1999) to 100 Mbps or more, as measured at the MAC data service access point in IEEE 802.11n (IEEE 802.11n/D11.0, 2009). This development has provided the opportunity to employ WLANs for demanding network applications. However, the wireless environment exhibits special characteristics that make it rather unreliable. Thus, specialized access mechanisms are needed to optimize the wireless network performance.

Efficiently serving different types of simultaneous traffic flows under the harsh wireless conditions is challenging. Modern networks have to be able to provide extended QoS, so that the final user experiences no deficiencies in any kind of communication. The medium access control mechanism plays a crucial role in QoS provision. However, it is definitely inefficient to adopt a wired LAN access control protocol in a wireless LAN. The wired LANs are characterized of fixed links, while, nowadays, the use of inexpensive switching devices has eliminated their collision domains. On the other hand, in a wireless LAN the links are variable and the bandwidth is shared to all the nodes through the common transmission medium (the air), that comprises a large collision domain. Furthermore, there are some phenomena which are only met in the wireless environment, like the hidden station and the exposed station problems.

The IEEE proposal regarding QoS in WLANs comes from the 802.11e workgroup and is called HCF (IEEE 802.11e WG, 2005). Of course, there are several relative access mechanisms proposed in literature, however, HCF is the most promising one, since it belongs to the well known 802.11 family of standards, it is backward compatible with the dominant 802.11 network, it seems to be supported by the industry (although there are hardly yet any network products adopting it), and it is generally a high-quality approach. The HCF scheme considers a contention based (Enhanced Distributed Channel Access - EDCA) and a contention free protocol (Hybrid Control Channel Access - HCCA). Besides their qualitative characteristics, EDCA and HCCA exhibit some significant limitations regarding the provided QoS, while efficiently supporting integrated traffic in a wireless environment is rather challenging.

This chapter discusses a hybrid access scheme capable of efficiently supporting all types of network communications under the harsh conditions of the wireless medium. The Priority Oriented Hybrid Access (POHA) protocol (Lagkas et al., 2007b) is a combination of two different channel access control mechanisms, POAP (Priority Oriented Adaptive Polling) and POAC-QG (Priority Oriented Adaptive Control with QoS Guarantee). POAP (Lagkas et al., 2008) is a polling-based access scheme which uses an efficient network feedback method in order to improve decision making regarding medium allocation. It is a flexible random access control protocol able to provide QoS by exploiting packet priorities and stations' status information without necessitating QoS requests. At this point, it should be clarified that in this chapter we call "packets" all the data units that require transmission without differentiating between frames (lower layer data units) and packets (higher layer data units). There is no need for this kind of differentiation in this work, thus, we intend to keep terms simple and clear. POAC-QG (Lagkas et al., 2007a) is a resource reservation access scheme which dynamically allocates time slots to the QoS requesting stations. It is capable of guarantying QoS to real-time Traffic Streams (TSs) of variable bandwidth demands, while instantly negotiating the best combination of the served applications' quality levels.

The chapter initially provides some background information about QoS supportive MAC protocols. Then, the IEEE 802.11e HCF protocol is presented and the behavior of the EDCA and the HCCA access schemes is discussed. The description of the POHA protocol follows next, where the combined operation of POAP and POAC-QG is analyzed. After that, we present our simulation environment and we comment on the comparison results of POHA and HCF. The chapter completes with some future research directions and a section of conclusions.

BACKGROUND

Medium Access Control (MAC) protocols are responsible for ensuring efficient and fair sharing of the available bandwidth. In wireless networks, the role of the MAC protocol is crucial. The available resources are limited, so there is a great need for efficient control of the transmissions. QoS support is also strongly related with the access control mechanism. A QoS supportive MAC protocol is able to distinguish different types of traffic and treat them accordingly. Usually, traffic is prioritized and high priority data is favored by the access control mechanism. A classification of the traffic considered by this work can be found in (Chandra et al., 2000; Akyildiz et al., 1999) and is presented in Table 1. Furthermore, the employed channel access mechanism can involve power saving methods which help the mobile stations to conserve battery energy.

Regarding QoS support in wireless LANs, the ad hoc network topology is the most ineffective one, mainly due to the absence of central control. However, the use of packet priorities can partially provide QoS, thus, there are some distributed MAC protocols that favor high priority packets. In decentralized WLANs, the level of QoS support depends on the network characteristics, such as load and number of stations. Specifically, distributed access mechanisms are contention based, thus, high load and increased number of stations cause high collision rate and low channel utilization. Under these conditions, packet delay and loss rate are increased. Representative distributed protocols which support QoS are EY-NPMA (Elimination Yield – Non Preemptive Multiple Access) and EDCA. The EY-NPMA protocol (ETSI, 1998; Papadimitriou et al., 2003) is used in HIPERLAN (HIgh PERformance Local Area Network), which is standardized by ETSI (European Telecommunications Standards Institute), and it is based on active signaling in order to avert simultaneous data transmissions. The EDCA protocol is used by IEEE 802.11e and it is described in more detail later in the chapter.

Centralized infrastructure wireless networks are more suitable for supporting QoS. The access control and the schedule mechanism are implemented in the AP, which is responsible for giving transmission permissions to the mobile stations. One of the centralized access methods that provide QoS involves station polling according to the previous or following packet priorities. This method does not include bandwidth reservation. The AP analyzes the feedback that gets from the stations and decides which one should be allowed to transmit, taking into account packet priorities. These polling schemes usually ensure low collision rate and high channel utilization, and they can provide QoS, while supporting background traffic transmissions. The QAP (QoS supportive Adap-

Table 1. Classification of traffic

Traffic Type	Examples	Characteristics
CBR (Constant Bit Rate)	real-time voice-video	efficient bandwidth reservation fast digital encoding increased produced data
nrt-VBR (non-real-time Variable Bit Rate)	background data transmission	high reliability required delay-jitter tolerant
rt-VBR (real-time Variable Bit Rate)	real-time audio-video	changeable bandwidth requirements increased encoding delay compressed produced data

tive Polling), the SQAP (Simple QoS supportive Adaptive Polling), the GRAP (Group Randomly Addressed Polling) [12] and the POAP protocols belong to this class of access mechanisms. QAP (Lagkas et al., 2006a) exploits the known priority of the last sent packet to implement polling. SQAP (Lagkas et al., 2006b) is based on a similar technique, but using a much simpler decision mechanism. GRAP (Chen & Lee, 1994) adopts the concept of stations group forming according to their priorities. Lastly, POAP is presented in detail as part of the POHA protocol.

Guaranteed QoS in a WLAN can be provided by the reservation centralized protocols. The respective access mechanisms give the ability to the different traffic streams to reserve bandwidth. According to this model, the stations send transmission requests to the AP asking for transmission intervals, usually using a contention based scheme. The scheduling algorithm implemented in the AP decides the bandwidth distribution in the contention free period according to the stations' requests, the priorities, the available resources etc. This type of channel access method guarantees QoS by ensuring that the packet delay of a traffic stream will not exceed an agreed maximum limit, however the value of the actual packet delay and jitter vary and depend on the specific protocol. The usual drawbacks of this model include the waste of bandwidth at the contention based period, because of the high collision probability, and the inability to efficiently support all types of real-time traffic. Specifically, if the assigned transmission periods remain constant for the whole duration of the communication, then VBR traffic cannot be efficiently supported. Moreover, this type of protocols cannot operate independently, since they are not able to serve best-effort transmissions. Typical examples of reservation centralized protocols are: DQRUMA (Distributed-Queuing Request Update Multiple Access), MASCARA (Mobile Access Scheme based on Contention and Reservation for ATM), DSA++ (Dynamic Slot Assignment), PRMA (Packet Reservation Multiple Access), HCCA, and POAC-QG. The general concept of the previously mentioned protocols is focusing on the real-time traffic and the use of a contention based scheme, like EDCA, for the transmission of the requests and the non-real-time data. For example, in DQRUMA (Karol et al., 1995), the uplink consists of a request channel, used to send contention requests, and a data channel, used to send data. The downlink slot is responsible for acknowledging the contention requests, granting transmit-permission and carrying data to the nodes. MASCARA (Bauchot et al., 1996) uses variable-length time frame, which consist of three periods: broadcast, reservation, and contention. The broadcast period contains control information from the AP to the nodes, the reservation period consist of the uplink and downlink data transmissions, and the contention period is used to send requests to the AP. DSA++ (Petras & Cramling, 1996) schedules the data transmissions using a heuristic algorithm. This algorithm prioritizes the requests and assigns the next slot to the node with the highest priority. In PRMA (Kim & Widjaja, 1996; Dyson & Haas, 1999; Bianchi et al., 1997), the transmission of voice packets requires reservation of uplink slots, while no reservation is made for a data transmission. Different versions of the PRMA protocol have been proposed to improve its performance. Lastly, HCCA and POAC-QG are thoroughly described in the following sections. This classification of the QoS supportive MAC protocols is presented in Table 2.

THE IEEE 802.11E HYBRID COORDINATION FUNCTION

The need for QoS support in the wireless local area networks has led IEEE to form the 802.11e workgroup. The objective was to enhance the existing 802.11 MAC protocol with QoS capable characteristics. The new medium access control

Table 2. Classification of QoS Supportive MAC Protocols

Protocol Type		Examples	Characteristics
Distributed		EY-NPMA EDCA	no infrastructure required low performance poor QoS support
Centralized	Random Access	QAP SQAP GRAP POAP	high performance not guaranteed QoS support low feedback requirements
	Reserved Access	DQRUMA MASCARA DSA++ PRMA HCCA POAC-QG	increased QoS guarantee not optimal channel utilization high feedback requirements

had to retain backward compatibility with the legacy 802.11 scheme, so that it would be possible for the old and new wireless stations to cooperate under the same base station.

The super-frame of the legacy 802.11 MAC protocol involves the operation of a distributed and a centralized access scheme. The first one is the Distributed Coordination Function (DCF), while the latter is the Point Coordination Function (PCF). DCF is based on a CSMA/CA algorithm and is able to implement virtual carrier sensing using a control packets' handshake. PCF is optionally employed in the legacy 802.11, it can operate along with DCF, and it adopts polling in order to give the wireless stations the permission to transmit.

The above schemes have no QoS capabilities, however, some modifications that enhance partial QoS support have been proposed (Ni et al., 2004). IEEE decided to enhance them with advanced features, thus, the new version that corresponds to DCF is the EDCA protocol, while the one corresponding to PCF is the HCCA protocol. EDCA is able to operate in ad hoc and infrastructure mode, it is based on the same CSMA/CA algorithm employed by DCF, however, it provides QoS by differentiating traffic using packet priorities. HCCA adopts resource reservation to guarantee QoS, it adopts station polling similarly to PCF, and it is designed to serve exclusively real-time traffic in infrastructure mode.

EDCA and HCCA operate in consecutive periods of the HCF super-frame. The periods during which EDCA takes place are called Contention Periods (CPs), while the periods during which HCCA takes place are called Contention Free Periods (CFPs) or Controlled Access Periods (CAPs). Every super-frame starts with a CFP followed by a CP, however, the base station (called Hybrid Coordinator – HC by the HCF protocol) is able to initiate several CAPs during a CP. The time interval during which a station is allowed to transmit data is called Transmission Opportunity (TXOP). The interval between two consecutive HCCA TXOPs assigned to a station defines a Service Interval (SI). The EDCA and HCCA protocols are described in more detail in the following two subsections.

Enhanced Distributed Channel Access

As DCF is necessary for the operation of the legacy IEEE 802.11 MAC protocol, EDCA is necessary for the operation of the IEEE 802.11e MAC protocol. It is already mentioned that EDCA enhances

DCF with QoS capabilities. This is achieved via the support of prioritized traffic based on the DiffServ model.

The medium access algorithm adopted by EDCA is CSMA/CA. According to it, each wireless station that needs to transmit a packet senses the carrier. If the carrier is found idle for an interval equal to an Arbitrary Distributed Interframe Space (AIFS), then the station can proceed to the transmission of the packet. When the channel is sensed busy, the station waits until it becomes idle and defers for an AIFS interval. After this interval, if the channel has remained idle, the station backs off selecting a uniformly distributed random number of time-slots from a contention window (CW). When the channel is sensed idle, the backoff counter decreases. When it reaches zero, the packet is transmitted and an acknowledgement (ACK) is expected. Since a wireless network interface is not able to listen and transmit simultaneously, a wireless station cannot directly detect collisions, but it assumes transmission failure by the ACK absence. In this case, it starts the whole backoff procedure again, using a larger contention window. The reason is to avoid a possible collision the next time.

The collision avoidance mechanism of the CSMA/CA algorithm is based on the use of the RTS/CTS (Request To Send/Clear To Send) handshake. This mechanism is able to limit the hidden station problem, which is present in the wireless environment and the cause of collisions' increase and general performance degradation. This problem is due to the fact that each wireless station may have a different view of the medium status. Thus, a station that has won medium contention and is ready to send data initially transmits a RTS control packet in order to inform the listening stations of the coming data transmissions. This way, these stations will not interfere with the new communication. The destination station which receives the RTS replies with a CTS control packet. The neighboring stations detect the CTS transmission and will remain silent during the communication that is going to take place. It should be noticed that the stations close to the receiver are mainly responsible for data collisions. The drawback of this scheme is the significant overhead increase and the fact that it does not completely eliminate collisions.

In order to differentiate traffic, EDCA divides it to Access Categories (ACs). Each AC contends independently for the medium, thus, in every station multiple backoff instances take place, each one parameterized specific for each AC. There is a different packet buffer corresponding to each one of the four ACs. According to the DiffServ model, there are eight possible User Priorities (UPs) assigned to the generated traffic. The eight UPs are mapped to the four ACs. This way traffic is differentiated. Each AC within the stations starts an independent backoff procedure after detecting the channel idle for an AIFS. The high priority traffic should have better chances of occupying the medium, for this reason, the higher ACs are assigned lower AIFS values. Moreover, the high ACs select backoff slots from smaller contention windows. Thus, a high priority packet will probably adopt a smaller waiting interval, so it will have increased chances of winning the contention. This scheme completes the EDCA QoS support.

There are some drawbacks related to the operation of EDCA. First of all, it is only able to provide limited QoS. The backoff intervals waste valuable bandwidth and the hidden station problem increases collisions which lead to performance degradation. The use of the RTS/CTS handshake limits this problem, however, it increases overhead. Furthermore, the use of exponential backoff does not favor QoS provision. According to the respective algorithm, the colliding packets have less chance to gain medium access than the newly contending packets. This is negative for the QoS support, because the colliding packets experience longer waiting times while the newly contending packets are transmitted sooner. Thus, time-critical packets may exhibit increased drop rates under increased load and bad link conditions. Generally, exponential backoff can cause high variation of throughput and delay in overloaded situations. Moreover, it has been shown that EDCA can be unfair when wireless stations experience different conditions (Pong & Moors, 2004). Some approaches that enhance EDCA can be found in (Hamidian & Komer, 2006; Ge et al., 2007; Shankar & Schaar, 2007). An analysis on the performance limitations caused by EDCA overhead can be found in (Wang & Helmy, 2006). Conclusively, EDCA definitely enhances the legacy DCF access mechanism with QoS support, however, it is shown that it can actually serve only limited traffic of low QoS requirements. It becomes obvious that an alternative protocol with higher channel utilization, more efficient QoS support, and fair resource assignment could be probably used.

Hybrid Control Channel Access

HCCA is the optional part of HCF. According to it, the HC decides the TXOPs granted to the QoS enhanced wireless stations. The HC, which is responsible for the central control, is actually located at the AP, so this work never refers particularly to the HC, but generally to the AP.

As it is already mentioned, the HCF superframe (or beacon interval according to the IEEE 802.11e terminology) is composed of alternated CFP, CP, and CAP modes. HCCA operates both during CPs and CAPs. At the start of the superframe, the AP transmits a beacon signal containing control information. A CFP-End message broadcasted by the AP signals the end of the CFP. The AP can initiate a CAP by occupying the channel using a CF-Poll message, which grants HCCA TXOPs to the stations.

According to the HCCA operation, every TS, which has its own packet buffer, sends a QoS request to the AP. This request contains the Traffic Specifications (TSPECs) of the specific TS, such as the mean data rate, the MAC Service Data Unit (MSDU) size and the maximum Required Service Interval (RSI). The scheduler calculates first the minimum value of all the RSIs, and then chooses the highest submultiple value of the beacon interval duration as the selected Service Interval (SI), which is less than the minimum of all the maximum RSIs.

The basis of the scheduler that calculates the TXOPs is the fact that a TXOP should be long enough to allow the transmission of all the packets generated during a SI in a TS buffer. The mean number of packets (N_{ij}) generated in the TS buffer (i) for a station (j) during a SI is:

$$N_{ij} = \left\lceil \frac{\bar{r}_{ij} SI}{M_{ij}} \right\rceil \tag{1}$$

where $\bar{r}_{ij}$ is the application mean data rate and M_{ij} is the nominal MSDU size. The TXOP (T_{ij}) is finally calculated as follows:

$$T_{ij} = max(\frac{N_{ij} M_{ij}}{R} + 2SIFS + T_{ACK}, \frac{M_{max}}{R} + 2SIFS + T_{ACK}) \tag{2}$$

where R is the transmission rate supported by the physical layer and M_{max} is the maximum MSDU size. The time interval $2SIFS+T_{ACK}$ corresponds to the overhead during a TXOP. Equation (2) guarantees that the TXOP will be long enough for the transmission of at least one packet with maximum size. The total TXOP assigned to a station is the sum of the TXOPs assigned to the different TSs of this station, that is:

$$TXOP_j = \sum_{i=1}^{Fi} T_{ij} \tag{3}$$

where F_j is the number of TSs in station j. The admission control algorithm checks for available bandwidth before assigning TXOP to a new TS. The fraction of total time assigned to a station j

Figure 1. The POHA superframe

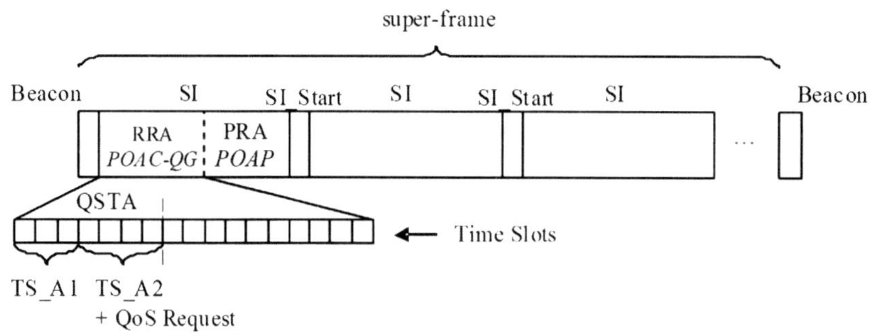

is: $TXOP_j/SI$. If the total number of QoS stations that are assigned TXOPs is K, then the scheduler needs to check if the new request of $TXOP_{K+1}$ will keep the fraction of time allocated for TXOPs lower than the maximum fraction of time that can be used by HCCA:

$$\frac{TXOP_{K+1}}{SI} + \sum_{j=1}^{K} \frac{TXOP_j}{SI} \leq \frac{T_{CAPLimit}}{T_{Beacon}} \tag{4}$$

where $T_{CAPLimit}$ is the maximum duration of HCCA in a beacon interval (T_{Beacon}).

The HCCA protocol exhibits some significant weaknesses. For example, the use of polling packets wastes valuable bandwidth. A major drawback of HCCA is related to the scheduler. Specifically, the allocation of fixed TXOPs leads to inefficient support of VBR traffic, because possible sudden increases of the generated bit rate would cause increased delays and packet drops. Moreover, the scheduling algorithm does not take into account prioritized TSs. It just uses the quality requirements in order to assign TXOPs. This means that the traffic is not efficiently differentiated according to the demands for QoS. It can be seen that a new efficient protocol that could always guarantee QoS would be useful.

PRIORITY ORIENTED HYBRID ACCESS

The necessity for a proposal alternative to the HCF protocol, which would offer improved QoS support and exhibit generally superior performance led to the development of the POHA protocol. POHA comprises two modern different access mechanisms; POAP and POAC-QG. The concept is to combine a QoS supportive flexible general purpose access scheme, like POAP, with a scheme of strict QoS provision specified for real-time communications, like POAC-QG. Thus, the POHA super-frame is divided into POAP and POAC-QG operation periods. The periods during which POAP takes place are named Polling-based Random Access (PRA) periods and they are essential for the operation of the POHA protocol. POAC-QG operates during the Resource Reservation Access (RRA) periods, which can dynamically change duration having a predefined maximum limit. The POHA super-frame is depicted in Figure 1.

Priority Oriented Adaptive Polling

POAP is built in order to exploit the usual presence of an AP in a WLAN and support qualitative services. Its purpose is to serve all types of traffic by improving the fundamental network feedback. It targets on providing QoS by taking dynamically into account the traffic characteristics. Fairness is

Figure 2. The polling model employed in POAP

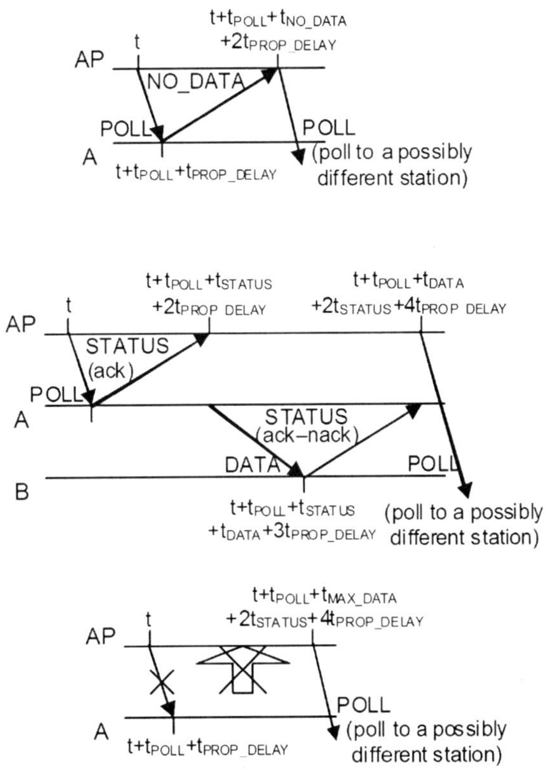

provided be preventing any station from dominating the channel.

POAP assumes a cellular topology where the AP polls the stations to give transmission opportunities. The introduced model nullifies collisions and decreases overhead while ensuring efficient feedback. The employed control packets are: POLL, NO_DATA, and STATUS, with transmission duration t_{POLL}, t_{NO_DATA}, and t_{STATUS}, respectively. A STATUS packet is marked as ACK or NACK according to the specific case. The time needed to transmit a DATA packet is t_{DATA} and the propagation delay is t_{PROP_DELAY}. Figure 2 presents the polling events.

- Polling a station that has no buffered packets to transmit.

The AP sends POLL to the wireless station at time t and waits for feedback. The station responds with NO_DATA, which is received by the AP at $t+t_{POLL}+t_{NO\_DATA}+2t_{PROP\_DELAY}$. Then, the latter initiates new polling.

- Polling a station that has buffered packets to transmit.

The AP sends POLL to the wireless station at time t and waits for feedback. The station replies with a STATUS packet marked as ACK, which carries the destination address and the size of the following DATA packet. Then, the polled station starts transmitting the DATA packet directly to the destination station. Upon successful reception of the DATA packet, the destination station broadcasts a STATUS packet marked as ACK. Otherwise, if the reception fails but the station has realized that the specific packet is destined to it, it responds with a STATUS packet marked as NACK. The transmission of a NACK is not wasted time, since either way the stations had to wait for a possible ACK. As we will see later, each STATUS packet contains valuable feedback for the AP. The latter can proceed to a new poll at time $t+t_{POLL}+t_{DATA}+2t_{STATUS}+4t_{PROP\_DELAY}$. It should be noticed that we consider variable DATA packet size, thus, t_{DATA} is not constant.

- Polling fails or the receiver fails to receive any feedback after polling.

In case the corresponding station does not successfully receive the POLL packet from the AP, the polling procedure fails. The AP has to wait for the maximum polling cycle before proceeding to a new poll, since it has to be certain that it will not collide with a possible ongoing transmission which is not detectable by the AP. When the POLL packet is received successfully by the polled station, but then the AP fails to receive any feedback, that is it cannot detect the following control and data packets, it waits for the maximum polling cycle similarly to the previous case. The duration of the maximum polling cycle is $t_{POLL}+t_{MAX\_DATA}+2t_{STATUS}+4t_{PROP\_DELAY}$, where t_{MAX_DATA} is the transmission duration of the largest allowed DATA packet. At the end of the maximum polling cycle, it is certain that the medium is idle in any event. When such a communication failure occurs, the AP lowers the probability to choose this station in the new polling procedure assuming a bad link between them. It also has to be mentioned that it is most likely that the AP will eventually receive some feedback either from the polled or the destination station.

Furthermore, it is possible to have multiple successive data packets destined to the same station with total duration no longer than t_{MAX_DATA} and a single block acknowledgement. This way, bursty traffic with strict QoS requirements could be more effectively supported.

The presented polling model provides efficient feedback, low overhead, and high channel utilization, while it eliminates collisions. The purpose of the control packets is to keep the stations informed of the network's status and minimize the idle intervals. The AP needs to monitor transmissions so that it can proceed to the next poll right after the completion of a communication. For this reason, it has to be aware of the actual duration of the specific polling cycle. In order to gain this knowledge, the AP just has to successfully detect the NO_DATA packet or the STATUS ACK packet, which contains the duration of the following data transmission, or the DATA packet from the polled station or the STATUS ACK-NACK packet from the destination station.

Figure 3 shows the packet choice mechanism introduced in POAP. First of all, if there are no buffered packets, the polled station replies with NO_DATA. The normalized priority (P_{PR}) and normalized number of buffered packets (P_B) are then calculated for each buffer. Specifically, assuming that the priority of buffer i is equal to $p[i]=i+1$, so that it is not null for AC[0], then:

$$P_{PR}[i] = p[i] \Big/ \sum_{k=0}^{3} p[k] \qquad (5)$$

Also, if $b[i]$ is the number of packets carried by buffer i, then it holds:

$$P_B[i] = b[i] \Big/ \sum_{k=0}^{3} b[k] \qquad (6)$$

In order to control the contribution of the buffer priority and the buffer load to the final buffer choice probability (P), we use the weights W_{PR} (default value 6) and W_B (default value 2) for P_{PR} and P_B, respectively. Specifically, when in the network configuration the purpose is to extendedly favor high priority traffic, then W_{PR} is set to a high value compared to W_B, otherwise, if the configuration should be able to efficiently serve highly loaded stations, then the value of W_B is raised. The default values have resulted from the actual meaning of

Figure 3. The packet choice mechanism of POAP

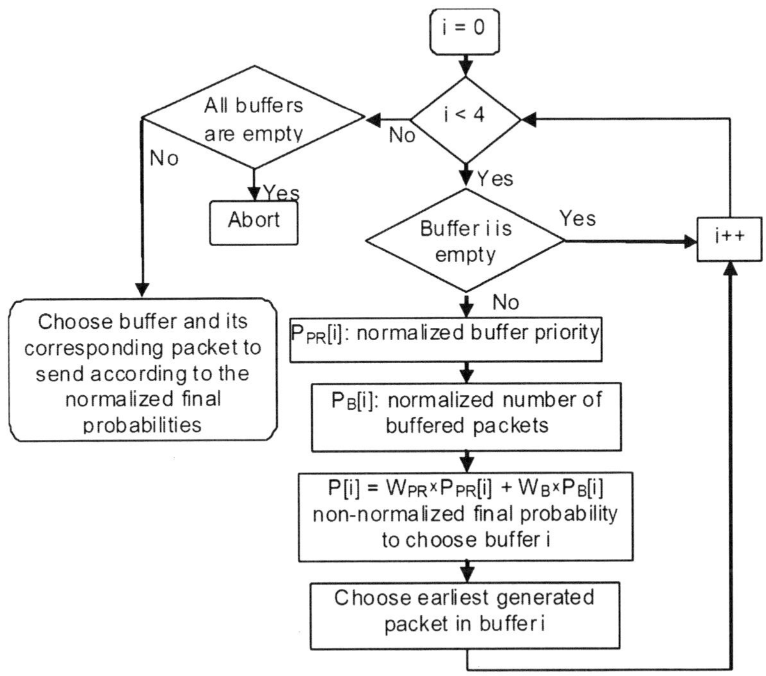

the parameters, simulation analysis, and tests which have shown that when the priority weight is three times higher than the buffer load weight, then the resulted buffer choice probability ensures the combination of efficient traffic differentiation and relatively low packet delays for all the buffers in most network conditions. We use the values *6* and *2* rather than *3* and *1*, because value *1* is assigned to the weight W_T which is introduced later. The non-normalized final probability of choosing a packet from buffer *i* is:

$$P[i] = W_{PR} \times P_{PR}[i] + W_B \times P_B[i] \quad (7)$$

At this point, it should be clarified that when we use the term "non-normalized probability" (like in Equation 7), we imply a quantity that does not necessarily range between *0* and *1*, but when normalized it represents some probability. Apparently, when W_{PR} is higher than W_B, probably a high priority buffer will be chosen for transmission. On the other hand, if W_B is increased, it is more probable to choose a packet from a highly loaded buffer. The normalized choice probability is finally equal to:

$$P_N[i] = P[i] \Big/ \sum_{k=0}^{3} P[k] \quad (8)$$

The earliest generated packet in the selected buffer is finally chosen for transmission.

In order to keep the AP informed of the stations' status, we exploit the use of the ACK and NACK messages, which are already useful in the polling scheme. Specifically, apart from the use of the STATUS control packet in acknowledging the packet receptions, it also carries its source station's priority score. Specifically, according to this modern feedback engine, when a station broadcasts a STATUS packet, it includes its priority score, which is introduced as an indication of the station's buffered traffic status. It depends on the priority of each buffer and the number of packets it contains. The priority score of station *j* is:

Figure 4. The station choice mechanism of POAP, where N are all the stations including the AP

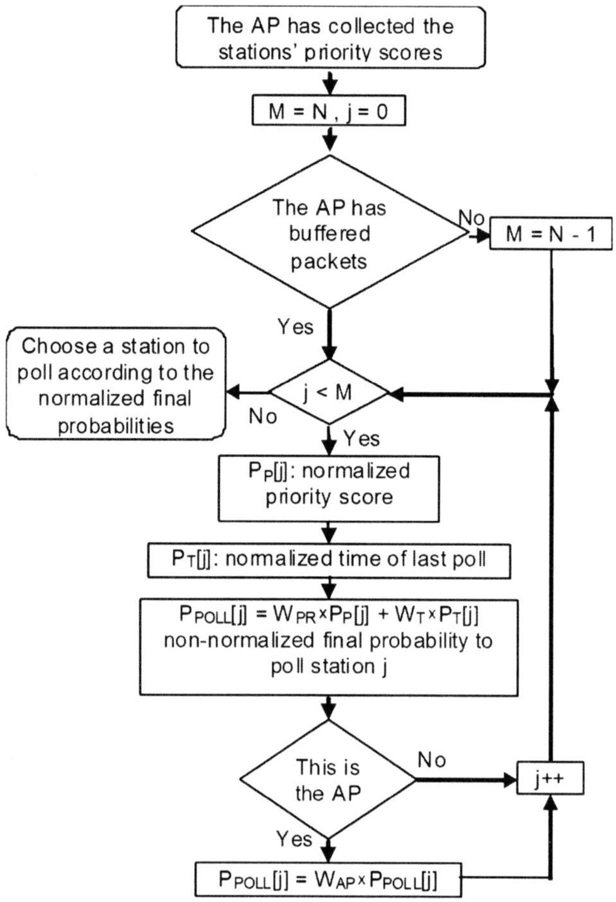

$$P_S[j] = \sum_{k=0}^{3} p[k] \times b[k] \qquad (9)$$

The AP examines every STATUS packet in order to update the stored priority score of the transmitting station. The priority score formula is based on the concept that a station with more packets in high priority buffers should have higher transmission probability. Furthermore, the presented mechanism ensures frequent feedback, which describes the status of every station causing minimal overhead. Finally, the AP implements an efficient new station choice mechanism based on the gathered priority scores, as it is shown in Figure 4.

The main concern of the station choice algorithm is to favor stations that carry high priority traffic and have a large number of buffered packets, thus, the priority score is initially taken into account. The second factor affecting the AP's decision is the time elapsed since the last poll of each station (τ). Specifically, in order to provide fairness and avoid the total exclusion of stations that are inactive for quite long, the stations that have not been polled for a long time are favored to some degree. Moreover, the AP is assigned a higher probability of getting access, since it usual plays a central role in the network communications. Lastly, a wireless station's priority score is halved, when no feedback is received after polling, assuming the existence of a bad link.

$$P_P[j] = P_S[j] \Big/ \sum_{l=0}^{M-1} P_S[l] \quad (10)$$

where M is the number of stations considered by the algorithm. The time elapsed since its last polling is also normalized:

$$P_T[j] = \tau[j] \Big/ \sum_{l=0}^{M-1} \tau[l] \quad (11)$$

The non-normalized final probability of polling station j is:

$$P_{POLL}[j] = W_{PR} \times P_P[j] + W_T \times P_T[j] \quad (12)$$

where W_T (default value *1*) is the weight of the contribution of the P_T factor. Obviously, the priority score plays a significant role in choosing which station to poll, since our primary objective is to provide priority based QoS. Therefore, according to our analysis, when the value of the weight W_T is six times lower than the value of W_{PR}, then fairness is provided without degrading QoS. As far as fairness is concerned, a station that at some point carried low priority score should not be totally excluded. The proposed method gives this station the chance to be polled, as it might have generated new traffic since its last poll. Obviously, a station that has not been polled for a long time has a high P_T value, so its polling probability increases. A high W_T value would provide extended fairness among stations, however, this way traffic differentiation would fade. If the examined station j is the AP, then its non-normalized final probability of getting channel access is multiplied by the factor W_{AP} (default value *10*):

$$P_{POLL}^{AP}[j] = W_{AP}(W_{PR} \times P_P[j] + W_T \times P_T[j]) \quad (13)$$

This way, we do not allow the AP to dominate the channel whenever it needs to transmit, so that it does not monopolize it and QoS is provided to all stations. Nevertheless, because of the AP's central role in the network, we give it the significant advantage of accessing the channel with ten times higher chances. Finally, the AP decides which station will be given access according to each one's normalized polling probability, which is for station j equal to:

$$P_{POLL\_N}[j] = P_{POLL}[j] \Big/ \sum_{l=0}^{M-1} P_{POLL}[l] \quad (14)$$

Priority Oriented Adaptive Control with QoS Guarantee

The purpose of POAC-QG is to efficiently serve demanding real-time network applications providing guaranteed QoS with efficient VBR traffic support, and traffic type distinction. The POAC-QG protocol operates during the RRA periods, which are contention free. It is not based on polling, but on a TDMA scheme. The basic concept is to assist station synchronization by dividing the RRA period into time slots, and keep the stations informed of the time interval, source and destination of the coming transmissions. The AP uses the beacon signal to inform the stations of the assigned slots for real-time traffic transmissions and the SI duration for the current super-frame. In the beginning of every SI, except from the first one in the super-frame, the AP broadcasts a SI_Start message which carries the same information with the initial beacon signal. If a station fails to receive the beacon signal, it defers, until it successfully receives a SI_Start (or a new beacon). The stations send their QoS requests for every TS during the PRA periods or the last RRA slots assigned to them (Figure 1).

Different quality levels can be usually supported by a multimedia network application

Figure 5. The operation cycle of POHA

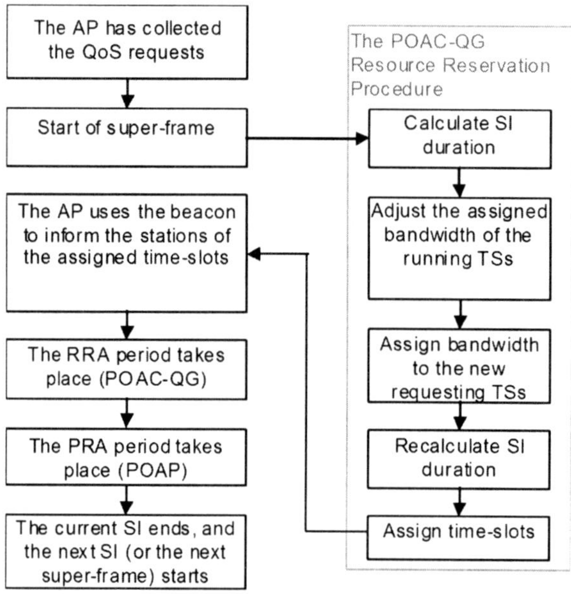

(depending on the codec, the audio-video quality etc). The admission control of POAC-QG negotiates instantly multiple quality levels that can be supported by the requesting TS. The objective is to serve the higher priority TSs with maximum quality level, however, the provided quality levels can be lowered in order to allocate slots for lower priority TSs, as well. Of course, we assume that the higher the quality level is, the higher are the resource requirements (bandwidth, delay). The main purpose of POAC-QG is to serve as many TSs as possible, favor the higher priority TSs, and provide the higher possible quality levels. A station that needs slot allocation for its TSs sends a QoS request which includes the traffic specifications of the different supported quality levels.

While the TSs are in progress, they are able to request a different number of RRA slots, according to their current traffic rate and the total size of their buffered data. Thus, every QoS request can include traffic specifications for both running and new TSs. This way VBR traffic can be efficiently supported. The respective algorithm calculates first the minimum value of all the maximum inter-transmission intervals required by the running and the new TSs, and then chooses the highest submultiple value of the beacon interval duration as the selected SI, which is less than the minimum of all the maximum inter-transmission intervals. Next, the AP assigns slots to the running TSs based on their QoS requests. The TSs that are in progress are examined first in order to keep the quality of the existing communications steady. After all, a new requested voice call can wait for admission, but it is unacceptable for a running call to be suddenly terminated or experience increased delays. The remaining resources are then assigned to the new TSs, using an admission control algorithm. The new SI duration is then calculated, based on the requests of the accepted TSs and the slots are finally assigned. The resource reservation procedure of POAC-QG can be detected as part of the POHA engine in Figure 5.

According to the presented TS admission control, the new TSs that request bandwidth allocation are sorted according to their priorities (highest priority first). The algorithm starts with the highest priority TS and checks if there is enough

Figure 6. A quality levels negotiation example according to the POAC-QG traffic streams admission control

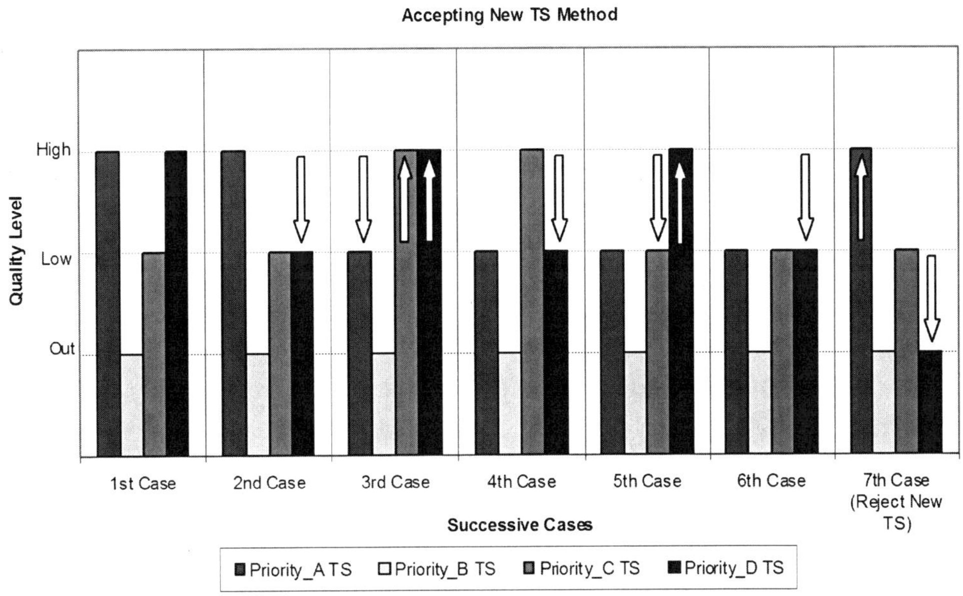

available bandwidth to serve the specific TS with maximum (highest) quality level. Otherwise, the QoS requirements of the next lower quality level are checked. If neither the minimum (lowest) quality level can be supported, then the TS is rejected and the next priority TS is examined. When not enough bandwidth is left to serve a TS with minimum quality, then the quality levels of the previously examined higher priority TSs are lowered in order to save some bandwidth for the new TS. In that case, when the quality levels of the high priority TSs are lowered, then we also check if it becomes possible to increase the quality of the low priority TSs. This way, the best combination of supported quality levels is provided. In Figure 6, an example of this procedure is presented. We assume two available Quality Levels (High QL, Low QL) and four new TSs with different priorities (Piority_A is the highest, while Priority_D is the lowest). The first three TSs are considered examined. Let us assume that, so far, Priority_A TS has been accepted with High QL, Priority_B TS has been rejected, Priority_C TS has been accepted with Low QL, and Priority_D TS is now examined for admission. Thus, we are looking for the best quality levels combination of these TSs, which can be served using the available bandwidth. There can be seven cases in our example. This method checks at each step if there is enough available bandwidth to serve the TSs providing the corresponding quality levels combination. If the available bandwidth is not enough, then we proceed to the next best combination (case). The last (and worst) possible case is the rejection of the examined TS (quality level: OUT).

A dynamic traffic control mechanism is built in the POAC-QG protocol, which is able to adapt to the changing requirements of the running TSs, thus, it efficiently supports VBR real-time traffic. Before a station sends a QoS request, it calculates the current traffic rate of all the TSs that are in progress by counting the generated bits for a short time interval (default value is *2* sec). The size of the corresponding packet buffer is included in the request packet. The AP assigns slots to the running TSs according to their latest QoS requests.

The remaining RRA slots are then assigned to the new TSs as it has already been discussed. The quality level of a TS cannot change while it is in progress, so that transmissions remain steady and reliable.

This dynamic traffic control tries to adapt to the variable traffic rate without sudden alterations of the allocated bandwidth. If there are not enough available slots, the algorithm assigns a proportion of the requested bandwidth to each TS according to its priority. All the generated and buffered packets of a TS can be transmitted during a SI, if the allocated bandwidth corresponds to the theoretical traffic rate:

$$TheoreticalTR = CurrentTR + BufferedBits / SI \tag{15}$$

where *CurrentTR* is the current traffic rate included in the QoS request. Our effort to avert sudden and continuous alterations of the allocated bandwidth has led us to consider as the target a proportion of the requested bandwidth accession or reduction. Specifically, the considered target traffic rate is:

$$TargetTR = PreviousTR + BW\_DifPercent \times (TheoreticalTR - PreviousTR) \tag{16}$$

where *PreviousTR* is the traffic rate corresponding to the last assigned bandwidth and *BW_DifPercent* (default value is *0.8*) is the percentage of the requested bandwidth accession or reduction which is considered to be the target. Furthermore, this algorithm employs a down limit for the target traffic rate related to the initial traffic rate requested, so that sudden increases of the generated packets after long idle intervals do not cause excessive packet drops.

In case a TS requests to free some of its allocated resources because they are not necessary anymore, it is of course immediately allowed to do so. However, resource allocation is challenging when there are not enough available slots to cover all the extra requests of the running TSs. In order to handle this situation, we have developed a method that distributes the available bandwidth taking into account the traffic priorities. This algorithm initially calculates the percentage of the available bandwidth that each requesting TS deserves (eligible bandwidth). The available bandwidth corresponds to the free slots of the maximum RRA period, after assigning to all the running TSs the previously occupied slots and freeing the returned slots. The eligible bandwidth percentage is calculated based on the traffic priority and the amount of extra bandwidth requested by the TS. The weights rW_{PR} (default value is *5*) and rW_{BW} (default value is *1*) are introduced at this point in order to control the contribution of the traffic priority and the extra bandwidth requested, respectively, to the eligible extra bandwidth. Clearly, the traffic priority is considered the most significant factor. The non-normalized eligible bandwidth percentage for the TS *k* is given by the equation:

$$Per[k] = rW_{PR} \times PerPR[k] + rW_{BW} \times PerBW[k] \tag{17}$$

where *PerPR* is the normalized traffic priority:

$$PerPR[k] = \frac{PriorityWeight[k]}{\sum_{l=0}^{NumberOfRequestingTSs-1} PriorityWeight[l]} \tag{18}$$

and *PerBW* is the normalized extra bandwidth requested:

Table 3. An example of the POAC-QG dynamic resource reservation method, where extra requested bandwidth is assigned to three running TSs

Step	TS	Priority	Requested Bandwidth	Available Bandwidth	Eligible Bandwidth	Assigned Bandwidth
1	A	6	5 Mbps	10 Mbps	5.6 Mbps	5 Mbps
	B	3	3 Mbps		2.9 Mbps	-
	C	1	4 Mbps		1.5 Mbps	-
2	B	3	3 Mbps	5 Mbps	3.3 Mbps	3 Mbps
	C	1	4 Mbps		1.7 Mbps	-
3	C	1	4 Mbps	2 Mbps	2 Mbps	2 Mbps

$$PerBW[k] = \frac{ExtraBW\_Requested[k]}{\sum_{l=0}^{NumberOfRequestingTSs-1} ExtraBW\_Requested[l]} \quad (19)$$

The term "priority weight" is used instead of "priority", because the weight of a traffic priority might be considered different than its index. We assume that $PriorityWeight = PriorityIndex + 1$. Because of the AP's central role, which often interconnects the WLAN with the backbone wired network, any traffic coming from the AP should be served with definitely higher priority. The weight rW_{AP} (default value is 5) is introduced in order to favor the AP's TSs. So, for every TS k transmitted by the AP it stands:

$$Per[k] = rW_{AP} \times (rW_{PR} \times PerPR[k] + rW_{BW} \times PerBW[k]) \quad (20)$$

The normalized eligible bandwidth for TS k is finally equal to:

$$nPer[k] = \frac{Per[k]}{\sum_{l=0}^{NumberOfRequestingTSs-1} Per[l]} \quad (21)$$

In case a TS requests for less bandwidth than its eligible bandwidth, then the former is immediately assigned to it. The TSs that cannot be provided with all the bandwidth they request, they get a part of it. Table 3 provides an example of the algorithm which calculates the extra bandwidth that would be assigned to every requesting TS. This mechanism completes the POAC-QG dynamic resource reservation.

SIMULATION ANALYSIS

It is better to clarify at this point that it is not feasible to model channel access in POHA based on the concept of the classical Bianchi two-state Markov chain (Bianchi, 2000). The POHA access scheme does not actually involve any idle time. However, we do use a three-state Markov process to simulate the link status between each pair of stations, as it is explained later. Generally, it is a common practice in the related literature to adopt a simulation approach, when the algorithmic complexity and the heuristic nature of the access control mechanism make further theoretical analysis impossible and actually unnecessary. Thus, no asymptotic analysis is performed to validate the behavior of the mechanism, instead, we implement a thorough and precise simulation comparison based on real traffic.

Simulation Characteristics

POHA and HCF were compared by using a C++ simulator that was specially developed for this study. The physical layer protocol adopted for the simulated WLAN is the IEEE 802.11g (IEEE 802.11g WG, 2003). In order to simulate realistic wireless links, we assumed that all stations use the ERP-OFDM (Extended Rate PHY – Orthogonal Frequency Division Multiplex) transmission mechanism at *36* Mbps based on the *16* QAM (Quadrate Amplitude Modulation) technique.

The condition of any wireless link was modeled using a finite-state machine with three states (good, bad, and hidden) based on the work of Zorzi et al. (1995):

- State *G* denotes that the wireless link is in a relatively "clean" condition and is characterized by a small BER, which is given by the parameter *G_BER*.
- State *B* denotes that the wireless link is in a condition characterized by increased BER, which is given by the parameter *B_BER*.
- State *H* denotes that the pair of communication stations is out of range (hidden stations).

The time spent by a link in states *G*, *B* and *H* is exponentially distributed, but with different average values, given by the parameters *TG*, *TB*, *TH*, respectively. The status of a link probabilistically changes between the three states. When a link is in state *G* and its status is about to change, the link transits either to state *H*, with probability given by the parameter P_h, or to state *B*, with transition probability P_h. When a link is in state *B* and its status is about to change, the link transits either to state *H*, with probability given by the parameter P_h, or to state *G*, with transition probability $1–P_h$. Finally, when a link spent its time in state *H*, it transits either to state *G* or *B*, with the same probability (*0.5*). Apparently, when setting the parameter P_h to zero, a fully connected network topology can be assumed, whereas for values of P_h greater than zero, the effect of the well-known "hidden station" problem on protocol performance can be studied.

For our simulation network, it stands for the inter-station links: *TG=3* sec, *TB=1* sec, *TH=0.5* sec, *G_BER=0, B_BER=0.00001*, P_h*=0.05*. For the AP-station links are characterized by the following parameters: *TG_AP=6* sec, *TB_AP=0.5* sec, *TH_AP=0.25* sec, *G_BER_AP=0, B_BER_AP=0.000001*, P_h*_AP=0.01*. We assume that the links among the AP and the stations are more reliable and "clear" than the inter-station links, because the range of the AP is usually greater than the stations' range, its emitted signal is usually stronger, and its default position is the center of the cell. Moreover, the BERs are assumed to be resulted after the application of the standard's predefined coding and error handling techniques.

Regarding some other parameters of the simulation network, the signal propagation delay is *0.0005* msec corresponding to distances among the stations of *150 m* and the maximum duration of the RRA period is *30%* of the super-frame. The considered control packet sizes agree with the 802.11e specifications. Furthermore, the simulator takes into account any of the PHY, MAC, RTP, UDP, IP, or SNAP headers that may be added to a data packet.

As far as the simulation engine is concerned, the random number generator used by our simulator is a classic multiplicative congruential random number generator with period 2^{32}. The simulation results presented in this section are produced by a statistical analysis based on the "sequential simulation" method (Pawlikowski et al., 2002). For this statistical analysis, we used *95%* confidence intervals. The relative statistical error threshold varies depending on the meaning of the statistic and the magnitude of its value. However, this threshold was usually assumed lower than *2%* and never exceeded *5%*.

Figure 7. Transmissions defined in the simulation scenario (FT: File Transfer, RDB: Remote DB)

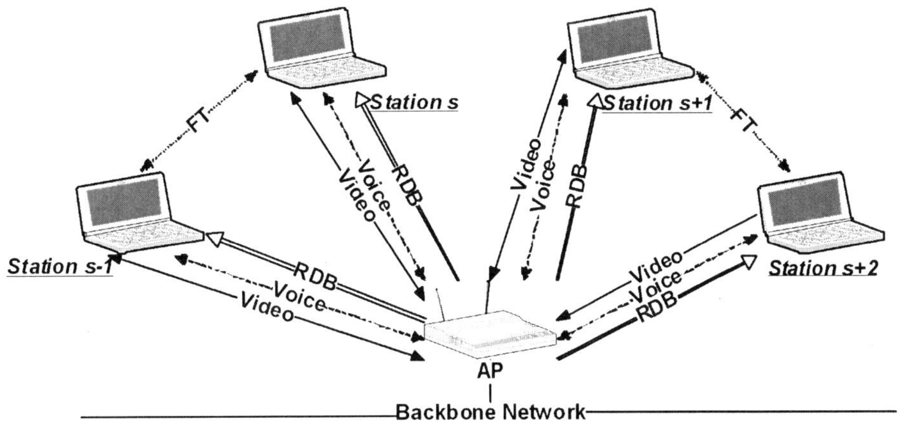

The Simulation Scenario

The objective of the specific simulation scenario is to examine the network behavior when there is a combination of different priority traditional data flows and different priority multimedia flows. Specifically, the scenario involves one bidirectional file transfer transmission (two PRA flows) between any two adjacent wireless stations (that is station *s-1* communicates with station *s* and station *s+1* communicates with station *s+2*), one unidirectional remote database transmission (one PRA flow) from the AP to each wireless station, one bidirectional video communication (two RRA streams) between the AP and each wireless station, and one bidirectional voice communication (two RRA streams) between the AP and each wireless station, as it is depicted in Figure 7. The traffic offered in the PRA and RRA periods has characteristics that are shown in Table 4 and Table 5, respectively. It can be seen that RRA traffic supports two quality levels (QL: MIN, MAX). The traffic characteristics have derived from real network traffic properties. The simulation duration is *60* sec, every communication lasts for *30* sec, and a new set of flows is added every *1.5* sec. We simulated networks consisting of *2* to *22* wireless stations, creating eleven topologies.

Simulation Evaluation of POHA Compared with HCF

In this subsection, the simulation results are presented. The simulation measurements that better reveal the general network behavior are the average packet delay, the achieved throughput, and the packet loss rate. We assume that packet delay is the end-to-end delay. A packet is considered lost, when its lifetime expires or it cannot enter the corresponding buffer due to overflow. Each

Table 4. Characteristics of the PRA traffic

Type	Packet Data Size (bytes)	Packet Interarrival Time (msec)	Data Bit Rate	Packet Delay Bound (msec)
File Transfer (User Priority: 0)	1500	Expo. Mean: 12	~1000 Kbps (VBR)	60000
Remote DB (User Priority: 3)	1500	Expo. Mean: 50	~240 Kbps (VBR)	1000

Table 5. Characteristics of the RRA traffic

Type	QL	Codec	Packet Data Size (bytes)	Packet Inter-arrival Time (msec)	On/Off Periods (sec)	Data Bit Rate	Packet Delay Bound (msec)
Voice (User Priority: 6)	MAX	G. 711 (PCM)	160	20	Expo. (mean) On: 1.5 Off: 1.8	64 Kbps (CBR)	50
	MIN	G.729A CS-ACELP	20			8 Kbps (CBR)	
Video (User Priority: 5)	MAX	H.261 [CIF]	Expo. [40-2048] Mean: 1320	Expo. Mean: 13	Always On	~800 Kbps (VBR)	100
	MIN	H.261 [QCIF]	Expo. [20-1024] Mean: 660	Expo. Mean: 26		~200 Kbps (VBR)	

time we simulate a different topology where the number of stations increases, there is a proportional increase of the offered traffic flows and the network load. However, it should be noticed that the RRA streams, that request service from the POAC-QG or the HCCA protocol, although they are admeasured as offered flows, they do not increase load, until they are admitted and get the requested resources.

Figure 8 presents the simulation results regarding file transfers. It is obvious that in this mixed traffic type environment, POHA manages to exhibit significantly low packet delays and high throughput (Figure 8a). Moreover, POAP, which serves file transfers, achieves almost null packet losses, whereas EDCA increasingly starts to lose packets for file transfer load higher than 7 Mbps, as shown in Figure 8b. The reason is that POHA utilizes the medium more efficiently, thus, its buffers start overflowing due to saturation later than HCF. The latter suffers from increased overhead, collisions, and waiting intervals, especially when more stations participate in the network, because of the adopted contention scheme. Notice that if the generated file transfer data bits are equal to *g*, the produced file transfer control bits are equal to *c*, and the total simulation time is *s*, then the considered file transfer load (in terms of rate) is equal to *(g+c)/s*. The same way, we calculate the file transfer throughput using the successfully transmitted bits. Similar calculations are performed for the other types of traffic.

The simulation results that concern remote database traffic are plotted in Figure 9. In Figure 9a, it can be seen that POHA ensures particularly low packet delays for this high priority non-multimedia traffic. On the other hand, HCF exhibits increasing packet delays as the throughput increases. Similarly to the file transfer traffic, POHA ensures almost null remote database packet losses, whereas HCF suffers from increasing packet losses when load is over *1.25* Mbps (Figure 9b). These results reveal that POHA efficiently manages to differentiate traffic and improve channel utilization, whereas HCF saturates at "lighter" load conditions (*~1.5%* packet loss rate at *~1.5* Mbps load). It is reminded that remote database traffic is handled by POAP and EDCA.

Figure 10 presents the results regarding video traffic, which is handled by POAC-QG and HCCA. POHA ensures about *25%* lower video packet delays than HCF, as it can be seen in Figure 10a.

Figure 8. File Transfer (FT): a) Average packet delay versus FT throughput and b) Packet loss rate versus FT load, in POHA (POAP) and HCF (EDCA)

a)

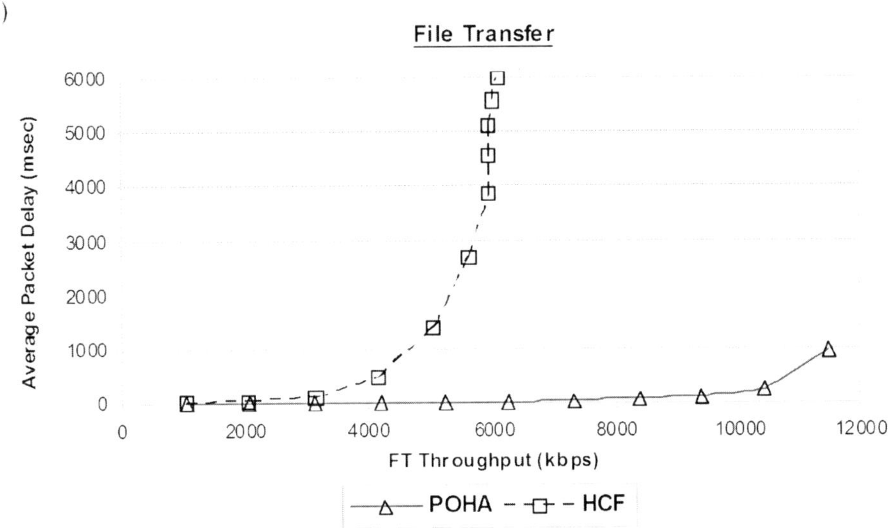

b)

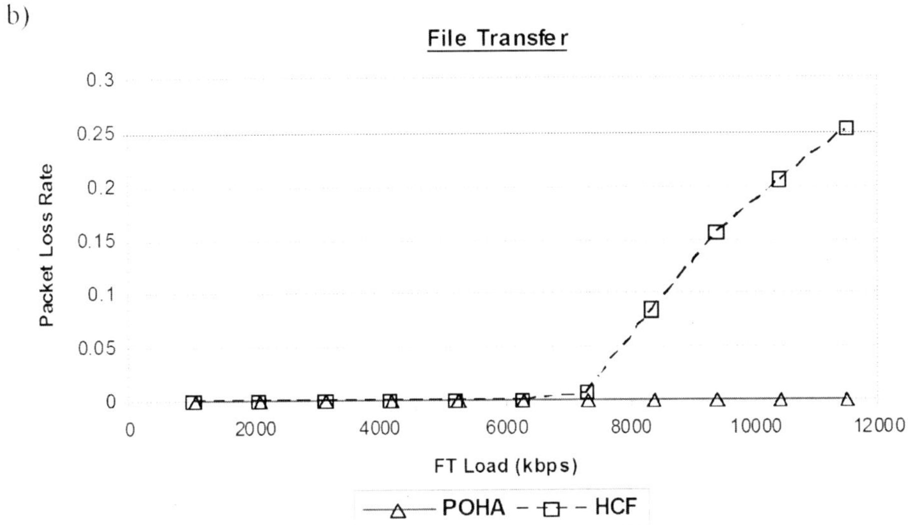

Both access schemes manage to keep delay steady while throughput increases. Furthermore, Figure 10b reveals that POHA manages to keep the video packet loss rate at low values while the load increases, whereas HCF exhibits about *5* times more packet losses than POHA at high video load (~*5* Mbps). According to these results, the dynamic resource reservation of POAC-QG successfully adapts to the demanding VBR video traffic.

The voice traffic streams simulated in the examined network are also treated as RRA traffic, so they are handled by the POAC-QG and HCCA protocols. As it is shown in Figure 11a, POHA keeps voice packet delays at steady low values (not higher than *24* msec), whereas HCF exhibits varying voice packet delays at the range of *30* msec to *35* msec while it achieves about *25%* lower maximum voice throughput than POHA. In Figure 11b, it can be seen that POHA exhibits

Figure 9. Remote DB (RDB): a) Average packet delay versus RDB throughput and b) Packet loss rate versus RDB load, in POHA (POAP) and HCF (EDCA)

a)

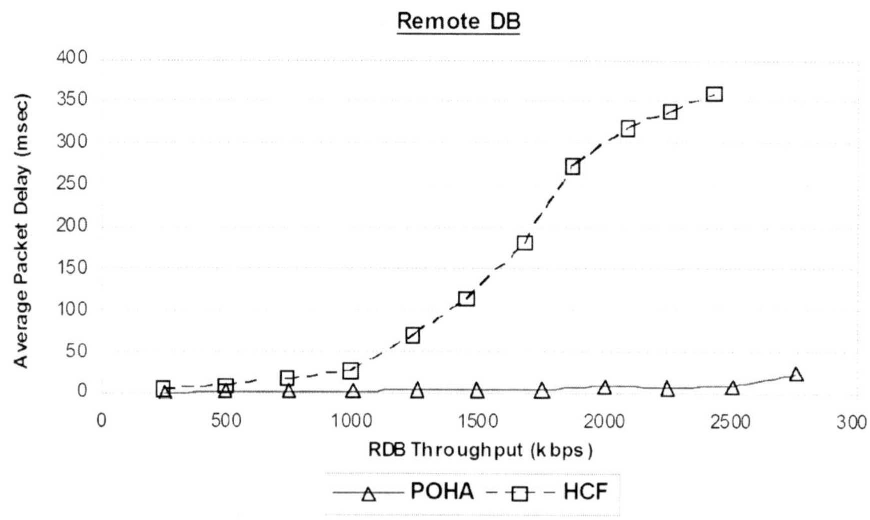

b)

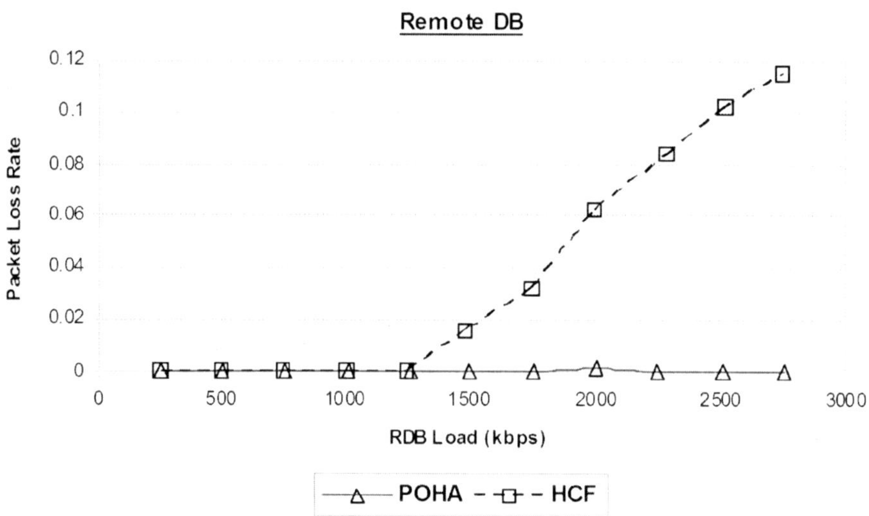

a negligible packet loss rate despite of the load increase, whereas HCF suffers from increased voice packet losses (reaching *5.5%*) caused by the high voice load. It should be noticed that the maximum voice load of POHA is higher than HCF, because more voice TSs were admitted by POAC-QG compared to HCCA. Another reason is the fact that POAC-QG was able to offer the MAX quality level to more TSs than HCCA. The MAX quality level voice TSs are characterized by *8* times higher data bit rate than the MIN quality level voice TSs (Table 5).

Figure 12 provides an indication of the way POHA and HCF assigned the two different quality levels to the TSs and served them. POHA directly negotiates the quality levels employing a new algorithm that was presented in the POAC-QG subsection. On the other hand, in HCF, we assume that a stream which is not assigned the required resources when requesting the MAX quality level, the next time tries again requesting to be served

Figure 10. Video: a) Average packet delay versus video throughput and b) Packet loss rate versus video load, in POHA (POAC-QG) and HCF (HCCA)

a)

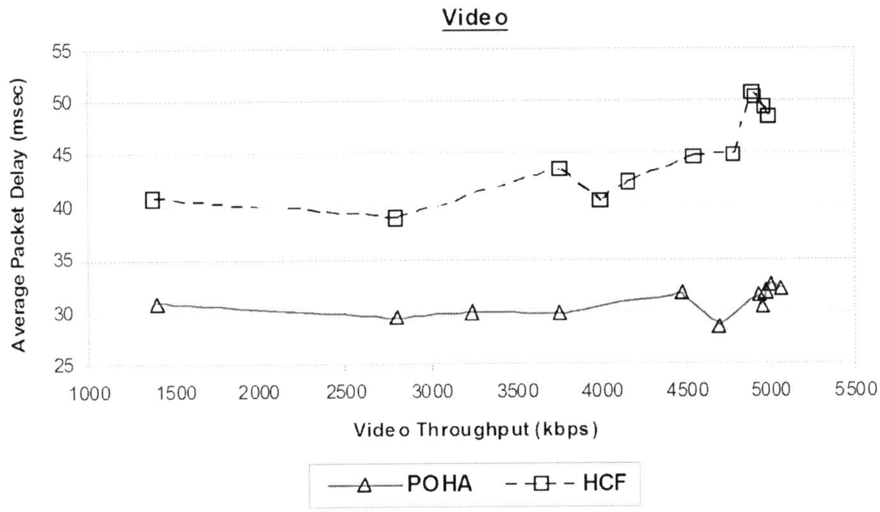

b)

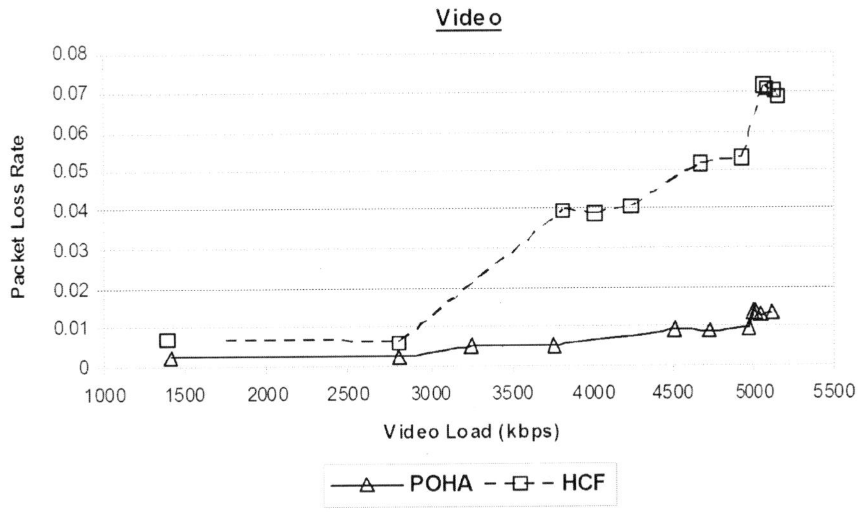

with the MIN quality level. The curves present the aggregate time the TSs were served by the two protocols as a function of the number of the offered TSs. It is obvious that while the offered TSs increase, the number of the MIN quality level TSs increases and the number of the MAX quality level TSs decreases. This graph shows that POHA manages to provide the TSs with more service time, while it succeeds to devote more time to MAX quality level TSs than HCF. These results reveal that the POAC-QG protocol achieves an improved quality level combination for the admitted RRA streams.

FUTURE RESEARCH DIRECTIONS

The research area of resource allocation in wireless networks is definitely emerging and quite promising. There are a lot of challenges related with this work and researchers all over the world try hard to come up with optimized decision making

Figure 11. Voice: a) Average packet delay versus voice throughput and b) Packet loss rate versus voice load, in POHA (POAC-QG) and HCF (HCCA)

a)

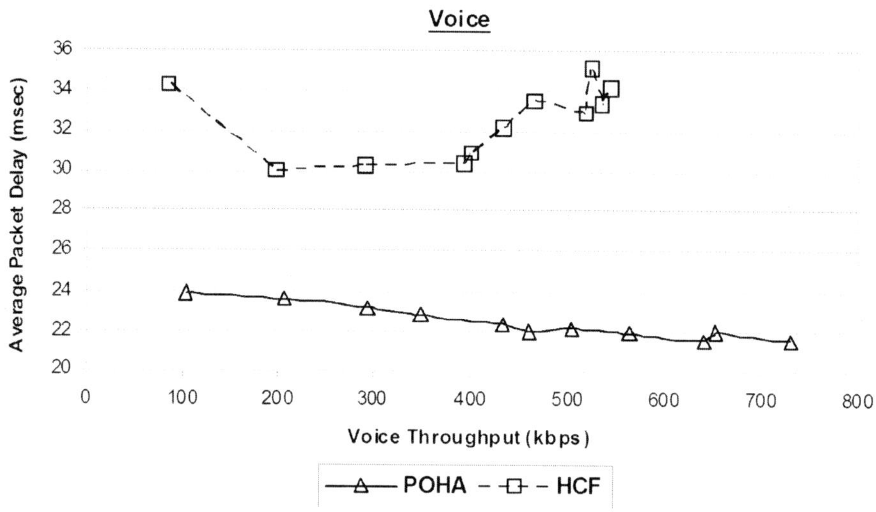

b)

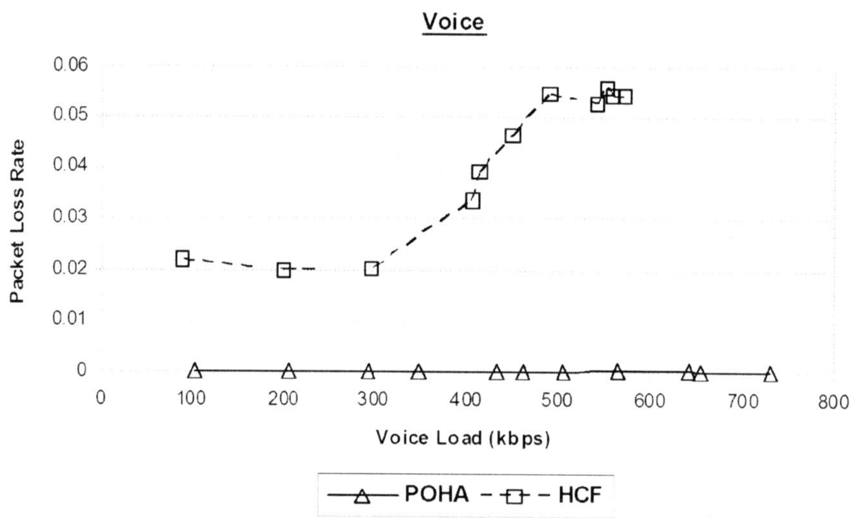

algorithms, in order to allow most efficient use of the wireless medium according to the modern users' needs.

Today, significant effort is devoted on developing efficient resource allocation methods based on mathematical analysis. The HCF protocol parameters are thoroughly analyzed and are still tuned. This type of analysis could take place for the POHA scheme as well, in order to configure it optimally according to the network characteristics. Another feature which is expected to be necessary for the near future mobile networks is power saving. Hybrid medium access control protocols that provide QoS in wireless networks should also combine an efficient energy conservation mechanism to effectively support mobile devices. An additional important research direction is towards nomadic networks. Providing QoS in dynamic WLANs of no infrastructure is quite challenging, thus, it seems essential to develop specialized access schemes capable of serving this difficult cause. Moreover, we witness a mod-

Figure 12. The aggregate time the different quality level traffic streams were served versus the number of the offered traffic streams in POHA (POAC-QG) and HCF (HCCA)

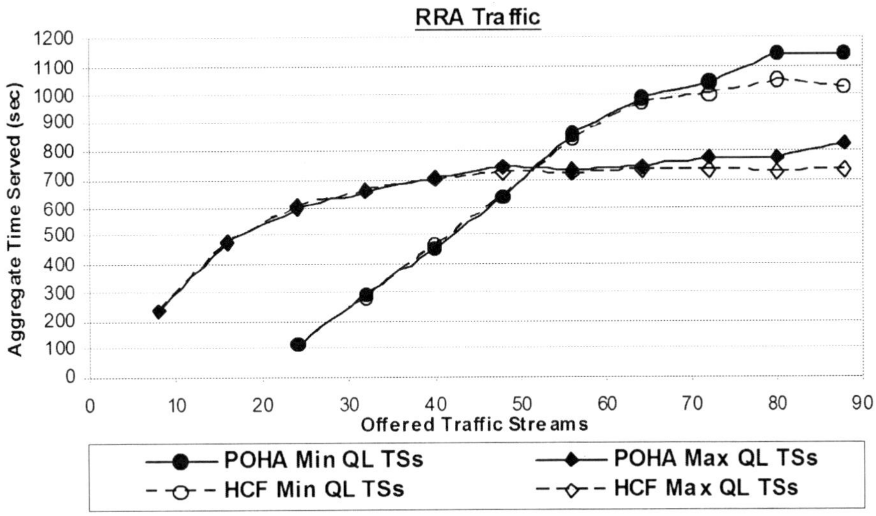

ern trend towards wireless broadband networks. WiMAX (Worldwide Interoperability for Microwave Access) is a promising technology that could probably bridge the gap between wireless local area networks and mobile cell networks. IEEE is standardizing 802.16 for broadband wireless access and a lot of research needs to be carried out on its QoS provision capabilities, especially because no specific mechanism is defined by the standard. Lastly, hybrid medium access control solutions need not to be limited in the data link layer, since the objective of optimized QoS provision concerns the whole network architecture. For this reason, researchers have lately started and are expected to continue to develop mechanisms that combine elements of different layers (physical – data link – network), in order to achieve better global results.

CONCLUSION

This chapter discussed the hybrid concept of providing QoS in wireless networks via medium access control by presenting the Priority Oriented Hybrid Access scheme. The motivation for the introduction of POHA was the necessity for a complete WLAN MAC protocol, which would be able to efficiently serve all types of traffic simultaneously, differentiate them accordingly, improve channel utilization, and provide the needed QoS by successfully adapting to the dynamic resource requirements. For these reasons, POHA implements a hybrid access scheme via combining two modern medium access control protocols for WLANs: POAP and POAC-QG. The former performs polling-based random access and provides extensive QoS by employing packet priorities and adopting an efficient network feedback mechanism without the need for resource reservation, whereas the latter is a resource reservation access scheme that serves exclusively real-time traffic providing strict QoS by supporting direct quality levels negotiation for the traffic streams, adopting traffic priorities, and dynamically allocating resources. The comparison of POHA with another well known hybrid protocol, the HCF access scheme of the IEEE 802.11e standard, has shown that in a mixed traffic-type network environment POHA performs clearly better, provides superior

QoS support, and generally gets saturated more difficultly. This type of hybrid approaches seem to be, in general, very promising solutions for the demanded modern wireless networks.

ACKNOWLEDGMENT

This work was supported by the State Scholarships Foundation of Greece.

REFERENCES

Akyildiz, I. F., McNair, J., Carrasco, L., & Puigjaner, R. (1999). Medium Access Control Protocols for Multimedia Traffic in Wireless Networks. *IEEE Network Magazine*, *13*(4), 39–47. doi:10.1109/65.777440

Bauchot, F., Decrauzat, S., Marmigere, G., Merakos, L., & Passas, N. (1996). MASCARA, a MAC Protocol for Wireless ATM. In *Proceedings of ACTS Mobile Summit 1996.*

Bianchi, G. (2000). Performance Analysis of the IEEE 802.11 Distributed Coordination Function. *IEEE Journal on Selected Areas in Communications*, *18*(3), 535–547. doi:10.1109/49.840210

Bianchi, G., Borgonovo, F., Fratta, L., Musumeci, L., & Zorzi, M. (1997). C-PRMA: A centralized packet multiple access for local wireless communications. *IEEE Transactions on Vehicular Technology*, *46*(2), 422–436. doi:10.1109/25.580781

Chandra, A., Gumalla, V., & Limb, J. O. (2000). Wireless Medium Access Control Protocols. *IEEE Communications Surveys and Tutorials*, *3*(2), 2–15.

Chen, K.-C. (1994). Lee, & C.-Hua. (1994). Group randomly addressed polling for wireless data networks. *Proceedings of the IEEE, ICC*, 1713–1717.

Dyson, D. A., & Haas, Z. J. (1999). A dynamic packet reservation multiple access scheme for wireless ATM. *ACM/Baltzer Journal of Mobile Networks & Applications, 4*, 87-99.

ETSI. (1998), EN 300 652 V1.2.1, Broadband Radio Access Network (BRAN); HIgh PErformance Radio Local Area Network (HIPERLAN) Type 1. Functional Specification.

Ge, Y., Hou, J. C., & Choi, S. (2007). An analytic study of tuning systems parameters in IEEE 802.11e enhanced distributed channel access. *Elsevier Computer Networks*, *51*(8), 1955–1980. doi:10.1016/j.comnet.2006.07.018

Hamidian, A., & Korner, U. (2006). An Enhancement to the IEEE 802.11e EDCA Providing QoS Guarantees. *Springer Telecommunication Systems*, *31*(2-3), 195–212. doi:10.1007/s11235-006-6520-z

IEEE 802.11 WG. (1999). International Standard [for] Information Technology – Telecommunications and Information Exchange between systems-Local and metropolitan area networks-Specific Requirements - *Part 11:Wireless LAN Medium Access Control (MAC) and Physical Layer (PHY) specifications*, Reference number ISO/IEC 8802-11:1999(E).

IEEE 802.11g WG. (2003). International Standard for Information Technology – Telecommunications and Information Exchange between systems-Local and metropolitan area networks-Specific Requirements - *Part 11:Wireless LAN Medium Access Control (MAC) and Physical Layer (PHY) specifications, Amendment 4: Further Higher Data Rate Extension in the 2.4GHz Band.*

IEEE 802.11e WG. (2005). IEEE Standard for Information Technology--Telecommunications and Information Exchange Between Systems--LAN/MAN Specific Requirements--*Part 11 Wireless Medium Access Control and Physical Layer specifications, Amendment 8: Medium Access Control Quality of Service Enhancements*.

IEEE 802.11n/D11.0. (2009). Unapproved Draft Standard for Information Technology-Telecommunications and information exchange between systems-Local and metropolitan area networks-Specific requirements-*Part 11: Wireless LAN Medium Access Control (MAC) and Physical Layer (PHY) specifications Amendment: Enhancements for Higher Throughput.*

Karol, M. J., Liu, Z., & Eng, K. Y. (1995). An Efficient Demand- Assignment Multiple Access Protocol for Wireless Packet (ATM) Networks. *ACM/Baltzer Journal of Wireless Networks, 1*(3), 267-279.

Kim, J. G., & Widjaja, I. (1996). PRMA/DA: A New Media Access Control Protocol for Wireless ATM. In *Proceedings of ICC 1996*, (pp. 1-19).

Lagkas, T. D., Papadimitriou, G. I., Nicopolitidis, P., & Pomportsis, A. S. (2007a). Priority-Oriented Adaptive Control With QoS Guarantee for Wireless LANs. *IEEE Transactions on Vehicular Technology, 56*(4), 1761–1772. doi:10.1109/TVT.2007.897223

Lagkas, T. D., Papadimitriou, G. I., Nicopolitidis, P., & Pomportsis, A. S. (2007b). A New Hybrid Medium Access Mechanism with Integrated QoS Support for Wireless LANs. In *Proceedings of IEEE/CVT '07*, Delft, the Netherlands.

Lagkas, T. D., Papadimitriou, G. I., Nicopolitidis, P., & Pomportsis, A. S. (2008). A Novel Method of Serving Multimedia and Background Traffic in Wireless LANs. *IEEE Transactions on Vehicular Technology, 57*(5), 3263–3267. doi:10.1109/TVT.2007.912341

Lagkas, T. D., Papadimitriou, G. I., & Pomportsis, A. S. (2006a). QAP: A QoS supportive Adaptive Polling Protocol for Wireless LANs. *Elsevier Computer Communications, 29*(5), 618–633.

Lagkas, T. D., Papadimitriou, G. I., & Pomportsis, A. S. (2006b). SQAP: A Simple QoS supportive Adaptive Polling Protocol for Wireless LANs. *Elsevier Computer Communications, 29*(8), 934–937.

Ni, Q., Romdhani, L., & Turletti, T. (2004). A Survey of QoS Enhancements for IEEE 802.11 Wireless LAN. *Journal of Wireless Communication and Mobile Computing, 4*(5), 547–566. doi:10.1002/wcm.196

Papadimitriou, G. I., Lagkas, T. D., & Pomportsis, A. S. (2003). HIPERSIM: A Sense Range Distinctive Simulation Environment for HiperLAN Systems. *Simulation. Transactions of The Society for Modelling and Simulation International, 79*(8), 462–481. doi:10.1177/0037549703040235

Pawlikowski, K., Jeong, H. D. J., & Lee, J. S. R. (2002). On Credibility of Simulation Studies of Telecommunication Networks. *IEEE Communications Magazine, 40*(1), 132–139. doi:10.1109/35.978060

Petras, D., & Kramling, A. (1996). MAC protocol with polling and fast collision resolution for an ATM air interface. In *Proceedings of IEEE ATM Workshop 1996.*

Pong, D., & Moors, T. (2004). Fairness and capacity trade-off in IEEE 802.11 WLANs. In *Proceedings of 29th Annual IEEE International Conference on Local Computer Networks 2004*, (pp. 310-317).

Shankar, S., & van der Schaar, M. (2007). Performance analysis of video transmission over IEEE 802.11a/e WLANs. *IEEE Transactions on Vehicular Technology, 56*(4), 2346–2362. doi:10.1109/TVT.2007.897646

Wang, S. C., & Helmy, A. (2006). Performance Limits and Analysis of Contention-based IEEE 802.11 MAC. In *Proceedings of 31st Annual IEEE Conference on Local Computer Networks 2006*, (pp. 418-425).

Zorzi, M., Rao, R. R., & Milstein, L. B. (1995). On the accuracy of a first-order Markov model for data transmission on fading channels. In. *Proceedings of ICUPC, 1995*, 211–215.

ADDITIONAL READING

Akyildiz, I. F., McNair, J., Martorell, L. C., Puigjaner, R., & Yesha, Y. (1999). Medium access control protocols for multimedia traffic in wireless networks. *IEEE Network, 13*(4), 39–47. doi:10.1109/65.777440

Boggia, G., Camarda, P., Grieco, L.A., & Mascolo, S. (2007). Feedback-Based Control for Providing Real-Time Services With the 802.11e MAC. *IEEE/ACM Transactions on Networking, 15(2)*, 323-333.

Bohge, M., Gross, J., Wolisz, A., & Meyer, M. (2007). Dynamic resource allocation in OFDM systems: an overview of cross-layer optimization principles and techniques. *IEEE Network, 21*(1), 53–59. doi:10.1109/MNET.2007.314539

Chlamtac, I., Conti, M., & Liu, J. J. N. (2003). Mobile ad hoc networking: imperatives and challenges. *ELSEVIER Ad Hoc Networks, 1*(1), 13–64. doi:10.1016/S1570-8705(03)00013-1

Dangerfield, I., Malone, D., & Leith, D. J. (2006). Experimental Evaluation of 802.11e EDCA for Enhanced Voice over WLAN Performance. *Proceedings of 4th International Symposium on Modeling and Optimization in Mobile, Ad Hoc and Wireless Networks 2006*, 1-7.

Fallah, Y. P., & Alnuweiri, H. (2005). A controlled-access scheduling mechanism for QoS provisioning in IEEE 802.11e wireless LAN. *Proceedings of 1st ACM workshop on QoS and Security in Wireless and Mobile Networks 2005*, 120-129.

Gilbert, E. (1960). Capacity of a burst noise channel. *The Bell System Technical Journal, 39*, 1253–1265.

Haines, R. J., Lewis, T., Coon, J., & Fanning, N. (2006). Non-linear optimization of IEEE 802.11e super-frame configuration. *Proceedings of 63rd IEEE Vehicular Technology Conference 2006, 3*, 1211–1215.

Iera, A., Ruggeri, G., & Tripodi, D. (2005). Providing throughput guarantees in 802.11e WLAN through a dynamic priority assignment mechanism. *Springer Wireless Personal Communications, 34*(1-2), 109–125. doi:10.1007/s11277-005-8729-7

Issariyakul, T., Hossain, E., & Kim, D. I. (2003). Medium access control protocols for wireless mobile ad hoc networks: issues and approaches. *Wiley Journal of Wireless Communication and Mobile Computing, 3*(8), 935–958. doi:10.1002/wcm.118

Lagkas, T. D., Papadimitriou, G. I., Nicopolitidis, P., & Pomportsis, A. S. (2006). A New Approach to the Design of MAC protocols for Wireless LANs: Combining QoS Guarantee with Power Saving. *IEEE Communications Letters, 10*(7), 537–539.

Lagkas, T. D., Papadimitriou, G. I., Nicopolitidis, P., & Pomportsis, A. S. (2006). Low Energy Priority Oriented Adaptive Polling for Wireless Networks. *Proceedings of IEEE CSNDSP, 06*, 139–143.

Li, T., Ni, Q., Turletti, T., & Xiao, Y. (2005). Performance analysis of the IEEE 802.11e block ACK scheme in a noisy channel. *Proceedings of 2nd International Conference on Broadband Networks 2005*, 511-517.

Mowlaei, A., Lee, B. S., Lim, T. M., & Yeo, C. K. (2007). Service Differentiation in Wireless Networks: Adaptive Ad Hoc Qs Versus IEEE 802.11e. *Proceedings of 15th IEEE International Conference on Networks 2007*, 306-311.

Nicopolitidis, P., Obaidat, M. S., Papadimitriou, G. I., & Pomportsis, A. S. (2003). Wireless Networks. Wiley.

Nicopolitidis, P., Papadimitriou, G. I., Lagkas, T. D., & Pomportsis, A. S. (2007). A Power Efficient Approach to Adaptive Polling Protocols for Wireless LANs. *IEEE Communications Letters*, *11*(6), 483–485. doi:10.1109/LCOMM.2007.070184

Nicopolitidis, P., Papadimitriou, G. I., & Pomportsis, A. S. (2002). Using Learning Automata for Adaptive Push-Based Data Broadcasting in Asymmetric Wireless Environments. *IEEE Transactions on Vehicular Technology*, *51*(6), 1652–1660. doi:10.1109/TVT.2002.802978

Nicopolitidis, P., Papadimitriou, G. I., & Pomportsis, A. S. (2003). Learning-Automata-Based Polling Protocols for Wireless LANs. *IEEE Transactions on Communications*, *51*(3), 453–463. doi:10.1109/TCOMM.2003.809788

Nicopolitidis, P., Papadimitriou, G. I., & Pomportsis, A. S. (2004). Distributed Protocols for Ad-Hoc Wireless LANs: A Learning-Automata-Based Approach. *Elsevier Ad Hoc Networks*, *2*(4), 419–431. doi:10.1016/j.adhoc.2003.09.004

Pahalawatta, P., Berry, R., Pappas, T., & Katsaggelos, A. (2007). Content-Aware Resource Allocation and Packet Scheduling for Video Transmission over Wireless Networks. *IEEE Journal on Selected Areas in Communications*, *25*(4), 749–759. doi:10.1109/JSAC.2007.070511

Papadimitriou, G. I., Lagkas, T. D., & Pomportsis, A. S. (2003). A New Approach to the Simulation of HIPERLAN Wireless Networks. *Proceedings of ESS*, *03*, 459–468.

Raychaudhuri, D., French, L. J., Siracusa, J., Biswas, S. K., Yuan, R., Narasimhan, P., & Johnston, C. A. (1997). WATMnet: A Prototype Wireless ATM System for Multimedia Personal Communication. *IEEE Journal on Selected Areas in Communications*, *15*(1), 83–95. doi:10.1109/49.553680

Sanchez, J., Martinez, R., & Marcellin, M. W. (1997). A survey of MAC protocols proposed for Wireless ATM. *IEEE Network*, *11*(6), 52–62. doi:10.1109/65.642360

Wang, P.-C., Wang, K., & Lee, L. S. (2005). A QoS scheme for digital home applications in IEEE 802.11e wireless LANs. *Proceedings of 16th IEEE International Symposium on Personal, Indoor and Mobile Radio Communications 2005*, *3*, 1845-1849.

Xiao, Y., Li, F. H., Wu, K., Leung, K. K., & Ni, Q. (2006). On optimizing backoff counter reservation and classifying stations for the IEEE 802.11 distributed wireless LANs. *IEEE Transactions on Parallel and Distributed Systems*, *17*(7), 713–722. doi:10.1109/TPDS.2006.92

Zhao, Q., & Tsang, D. H. K. (2008). An Equal-Spacing-Based Design for QoS Guarantee in IEEE 802.11e HCCA Wireless Networks. *IEEE Transactions on Mobile Computing*, *7*(12), 1474–1490. doi:10.1109/TMC.2008.71

KEY TERMS AND DEFINITIONS

EDCA: Enhanced Distributed Channel Access

HCCA: Hybrid Control Channel Access

HCF: Hybrid Coordination Function

POAC-QG: Priority Oriented Adaptive Control with QoS Guarantee

POAP: Priority Oriented Adaptive Polling.

POHA: Priority Oriented Hybrid Access.

QoS Provision: Quality of Service provision

WLAN MAC: Medium Access Control for Wireless Local Area Networks.

Chapter 2
Cross-Layer Scheduling with QoS Support over a Near-Optimum Distributed Queueing Protocol for Wireless LANS

E. Kartsakli
Technical University of Catalonia (UPC), Spain

J. Alonso-Zárate
Telecommunications Technological Centre of Catalonia (CTTC), Spain

L. Alonso
Technical University of Catalonia (UPC), Spain

C. Verikoukis
Telecommunications Technological Centre of Catalonia (CTTC), Spain

ABSTRACT

Distributed Queueing Collision Avoidance (DQCA) is an efficient MAC protocol designed for infrastructure Wireless LANs. In this chapter, a thorough description of the protocol is given, along with a set of protocol rules and an example of its operation. In continuation, four algorithms are proposed that alter the FIFO scheduling order of DQCA in order to meet specific network requirements. The proposed schemes combine the efficiency of opportunistic scheduling with the QoS provisioning through service differentiation. The opportunistic policy encourages transmissions at higher rates when the channel condition is good and is implemented through a cross-layer dialogue between the PHY and the MAC layers. The key idea of service differentiation is to assign priorities to traffic flows with different requirements in order to provide QoS guarantees. The throughput, delay and jitter performance of the proposed schemes have been evaluated through simulations for a scenario with heterogeneous traffic of voice, video, best-effort and background data traffic flows.

DOI: 10.4018/978-1-61520-771-8.ch002

INTRODUCTION

Wireless Local Area Network (WLAN) technology has become a very appealing solution to network connectivity, providing user mobility, flexibility and easy deployment at a relatively low cost. As the popularity of WLANs grows, the need for higher transmission rates and Quality of Service (QoS) guarantees becomes imperative, especially for real-time multimedia applications.

Considerable amount of research has been directed towards the improvement of the Physical layer (PHY), resulting in more sophisticated transmission techniques and advanced modulation and coding schemes. However, even though the increase in the available transmission rates and the robustness against channel errors is significant, the network capacity is greatly dependent on the Medium Access Control (MAC) protocol employed at the Data Link Control layer. Especially in wireless communication systems where the medium is unpredictable, the need for an efficient MAC protocol to handle channel access and scheduling issues is imperative.

The most prevailing MAC protocol specification for WLANs is the IEEE 802.11 standard and its amendments (IEEE Std. 802.11, 2007) which are based on a Carrier Sensing Multiple Access protocol with Collision Avoidance (CSMA/CA) and employ an exponential backoff mechanism to resolve collisions. The standard, although widely deployed, suffers from inefficiency induced by the long idle periods due to the backoff mechanism and the high presence of collisions as the offered traffic load grows (Bianchi, 2000). Therefore, there is a need to develop novel MAC schemes that can offer high performance in terms of throughput, delay and jitter and which can also adapt dynamically to changing conditions, such as the traffic load or the number of users.

In this context, an innovative and efficient scheme for WLAN systems, called Distributed Queueing with Collision Avoidance Protocol (DQCA) is presented (Alonso-Zárate, Verikoukis, Kartsakli, Cateura, & Alonso, 2008). DQCA behaves as a random access mechanism under low traffic conditions, thus reducing transmission delays, and switches smoothly and automatically to a reservation scheme when traffic load grows. Collisions are handled by a blocked-access splitting algorithm that works in parallel with the data transmission process. DQCA is a fair protocol that yields a performance that approaches the theoretical system capacity. In addition, it is a flexible scheme that can be easily adapted to include more advanced scheduling algorithms.

The objective of this chapter is twofold. First, the DQCA protocol will be described in detail, along with an explicit set of rules and an operation example. In the second place, a number of scheduling algorithms that enhance the performance of DQCA under heterogeneous traffic will be presented. These algorithms combine the efficiency of opportunistic scheduling with the QoS provisioning through service differentiation. The opportunistic policy encourages transmissions at higher rates when the channel condition is good and is implemented through a cross-layer dialogue between the PHY and the MAC layers. The key idea of service differentiation is to assign priorities to traffic flows with different requirements in order to provide QoS guarantees.

The remainder of this chapter is organized in seven sections. In the background section a literature review on cross- layer design and scheduling is provided. The two subsequent sections are dedicated to the description of DQCA and four cross-layer scheduling algorithms respectively, along with implementation details and operation examples. In continuation, the proposed schemes are evaluated and compared through simulations for a specific scenario that includes mixed real and non-real time applications. The chapter proceeds by indicating possible future research directions and is completed with some concluding remarks.

BACKGROUND

Advances in the physical layer have led to a significant capacity increase in WLANs. The widely deployed IEEE 802.11g standard (IEEE Std. 802.11g, 2003), for example, defines a set of eight rates ranging from 6 to 54 Mbps and the emerging IEEE 802.11n standard promises rates beyond 100Mbps (IEEE Std 802.11n, 2008). Lower rates are more robust to channel noise whereas higher ones require larger signal-to-noise ratio (SNR) values over the wireless link.

The process of selecting the optimal transmission rate depending on the channel condition is known as link adaptation and is usually performed at the MAC layer. Several link adaptation algorithms have been proposed for 802.11-based WLANs. One approach is based on modifying the Request-to-Send (RTS)–Clear-to-Send (CTS) handshake (Holland, Vaidya, & Bahl, 2001). The signal strength of a received RTS is used to estimate the channel condition and select the most appropriate rate, which is returned to the transmitter via the CTS packet. In another case the link estimation is performed at the transmitter, based on the history of SNR measurements of previously received frames (Pavon & Choi, 2003).

The time-varying nature of the wireless channel and the multirate PHY capability can be further exploited by the MAC layer to achieve more efficient scheduling. The key idea is to encourage communication between nodes with high link quality, as reflected by the perceived SNR at the receiver. High SNR values mean that a higher transmission rate can be supported while maintaining the error probability at an acceptable level. Although the consideration of a PHY layer metric, such as the SNR, in MAC layer scheduling decisions is a violation of the OSI architecture, it can be exploited to optimize the performance of the system.

The exchange of information between different OSI layers is known as cross-layer design and is a very promising field of investigation (Shakkottai, Rappaport, & Karlsson, 2003). Many related schemes can be found in the literature, especially regarding opportunistic algorithms based on a PHY/MAC layer dialogue. In one case, nodes with a good link are permitted to transmit multiple packets back-to-back, resulting to more high-rate transmissions (Sadeghi, Kanodia, Sabharwal, & Knightly, 2002). A contrary approach is to discourage nodes from transmitting if their available rate fails to satisfy their QoS demands (Bouam & Othman, 2004).

There are many other paradigms of cross-layer design that do not focus only on opportunistic scheduling. An example of a more complex scheme that exploits multiuser diversity is the Opportunistic packet Scheduling and Media Access control (OSMA) (Wang, Zhai, & Fang, 2004). A multicast RTS is transmitted to a group of nodes who evaluate their link quality and reply with a CTS if their channel is better than a certain level. The transmitter then establishes communication with the node with the best channel and consequently the highest available rate. Another scheme employs the air-time usage as a metric of the channel condition and distributes resources in order to accomplish a proportional fairness objective, through a token-based scheduling policy (Garroppo, Giordano, Lucetti, & Tavanti, 2007).

Although opportunistic scheduling can increase the system capacity, it cannot provide the QoS guarantees required by real-time multimedia traffic. One way to alleviate this problem is to incorporate QoS demands into the MAC layer scheduling decisions. This approach has been adopted in the relatively new IEEE 802.11e standard through service differentiation (IEEE Std. 802.11e, 2005). Traffic flows are mapped to four Access Categories (AC), with each category assigned a different level of priority; MAC layer parameters are set accordingly so that higher priority ACs have smaller medium access delays and may hold the channel for a longer time.

All these concepts of link-adaptation, opportunistic scheduling and service differentiation are considered in the design of the DQCA protocol and the incorporated cross-layer scheduling algorithms. DQCA is part of an extended protocol family that originates from the Distributed Queueing Random Access Protocol (DQRAP) (Xu & Campbell, 1992). DQRAP was initially designed for the distribution of cable TV but since then the distributed queue concept has been successfully adapted to diverse scenarios, including ad hoc cluster (Alonso-Zárate, Kartsakli, Skianis, Verikoukis, & Alonso, 2008) and body sensor networks (Otal, Alonso, & Verikoukis, 2009). Currently, an Ethernet-based distributed protocol and a DQCA-based wireless air-interface for WLANs are being implemented by Ether2 for commercial release (Ether2, 2005).

A very good reference on the distributed queue protocol design is given by Alonso, Agusti and Sallent (2000) and a description of DQCA can be also found in the literature (Alonso-Zárate, Verikoukis, Kartsakli, Cateura, & Alonso, 2008). Some preliminary work on cross-layer scheduling on DQCA has been focused on a heterogeneous voice and data traffic scenario (Kartsakli, Cateura, Alonso-Zárate, Verikoukis, & Alonso, 2007a). A more complete study on QoS-aware scheduling with extensive simulations emphasizes the flexibility of the DQCA protocol which, through cross-layer optimization, can adapt to different network requirements, entailing a tradeoff between efficiency in link utilization and fairness among data flows (Kartsakli, Alonso-Zárate, Verikoukis, & Alonso, 2009).

THE DQCA PROTOCOL

Protocol Overview

DQCA is a distributed high-performance medium access protocol that behaves as a random access mechanism under low traffic and switches smoothly and automatically to a reservation scheme when traffic load grows. DQCA eliminates back-off periods and minimizes data packet collisions. In addition, unlike slotted Aloha, the maximum achieved throughput is maintained even when the traffic load exceeds the channel capacity.

The main idea of DQCA is that the nodes may ask for channel access in a reserved time interval, thus confining collisions almost exclusively to a part of the frame. Any collisions are resolved by a blocked access *m*-ary splitting tree collision resolution algorihtm in a FIFO order, managed by a distributed queue. Once having successfully sent an access request, a node enters in another distributed queue (devoted to the scheduling of collision-free data transmission) and waits for its turn to transmit. The next section contains a thorough description of the DQCA.

Protocol Description

Consider an infrastructure wireless network with N nodes communicating with an Access Point (AP) through a shared wireless channel. The time axis is divided into continuous DQCA frames which consist of three parts of different duration.

The first part, also referred to as Contention Window (CW), is divided into m control minislots. Provided that there are no pending collisions for resolution, nodes that want to gain access to the channel randomly select one minislot and transmit an Access-Request Sequence (ARS). It has been demonstrated that with three or more minislots the collision resolution works faster than the data transmission process and therefore three minislots (m=3) are sufficient to ensure near-optimum performance for any traffic load (Xu & Campbell, 1992). Higher values of m attain slightly lower delay at the cost of increased overhead.

The operation of DQCA is based on the m-ternary feedback information on the state of each of the minislots. The AP must be able to distinguish between idle, success and collision state for each minislot, since this information is crucial for the application of the protocol rules at the end of each frame. Adopting a pattented technology (Campbell

& Xu, 2001) each node is assigned a unique bit pattern that has the property that when two or more ARS collide, the pattern of the overlapping signal is distinguishable from the original pattern of any single ARS; hence, the AP can detect the collision. It should be noted that the ARS does not need to contain any additional information, such as the transmitter address, thanks to the distributed execution of DQCA.

The second part of the frame is reserved for the almost collision-free transmission of a data packet by a single user. Without losing generality, the length of the packets is fixed and their transmission time is determined by the transmission rate. Large messages are fragmented into packets and are transmitted within consecutive DQCA frames.

In the third part of the DQCA frame, the AP broadcasts a Feedback Packet (FBP) that contains ternary feedback information on the state of all access minislots; an acknowledgement (ACK) to verify the correct reception of data packets; and a final-message bit which is set to 1 when the last packet of a message has been received and to 0 when more packets of the same message are expected to follow. In certain configurations, the FBP may contain additional information, as it will be explained later. Before and after the FBP a Short InterFrame Space (SIFS) is introduced in order to compensate for propagation delays, turn around times (i.e., time required for a node to switch from receive to transmit mode) and for processing purposes.

The protocol is based on two concatenated distributed queues, the Collision Resolution Queue (CRQ) and the Data Transmission Queue (DTQ). The CRQ is responsible for the resolution of collisions among ARS whereas the DTQ handles the data transmission. The number of occupied positions in each queue is indicated by an integer counter (*RQ* and *TQ* for the CRQ and the DTQ, respectively). Since these counters represent the state of the distributed queues at any time instance, they hold the same value for all the nodes in the system. Each node must also maintain and update another set of counters that reveal its position in the queue (*pRQ* and *pTQ* for the CRQ and the DTQ, respectively). By the term position, it is meant the relative order of arrival (or age) of the node in the respective queue. In particular, these values are set to zero if a node does not belong to any queue; otherwise, their values range from 1 to *TQ* or *RQ*, respectively, with the value 1 indicating the head of the queues. In the CRQ, each position (or element) is occupied by a set of nodes that suffered an ARS collision (i.e., attempted an ARS transmission in the same control slot of the same CW). The DTQ contains the nodes that successfully reserved the channel through an ARS and therefore each queue element corresponds to exactly one node.

In principle, every node, including those that do not have a message to transmit, must update these counters upon the reception of the FBP at the end of each frame by executing the set of rules described below. However, it is also convenient to periodically send the *TQ* and *RQ* values within the FBP to allow inactive nodes to leave or reenter the system, thus increasing robustness against possible miscalculation of the counters by the nodes.

Three sets of rules are defined and executed by each node upon decoding of the FBP. They are, in order of execution, the data transmission rules (DTR), that indicate who can transmit data in the following frame, the request transmission rules (RTR), that manage the collision resolution process and the queueing discipline rules (QDR), that handle the state update of the queues. The prococol rules are described in detail next and can be better understood with the help of the flow chart depicted in Figure 1.

DTR (Data Transmission Rules)

1. If the DTQ is empty (*TQ*=0), meaning that there are no nodes waiting to transmit and there are no pending collisions to be resolved (*RQ*=0), any node that wants to gain channel

Figure 1. DQCA operation flow chart

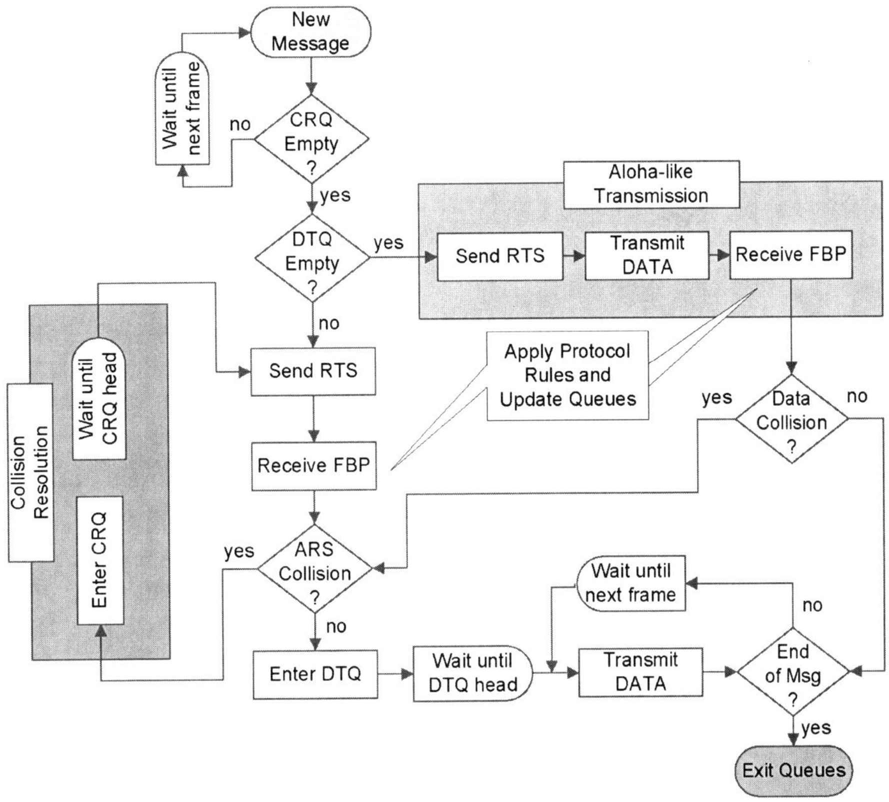

access transmits an ARS in a randomly selected minislot. The node is also enabled to transmit a data packet in the data part of the same frame, so that an empty data slot is avoided. This rule is referred to as the immediate access rule. When applied, DQCA behaves as a random access protocol and it is the only occasion in which a data collision may occur. In practice, the rule is only executed under light traffic (reflected by the empty state of the both queues), when the probability of having a collision is low.

2. If a node is at the head of the data transmission queue (*pTQ*=1), it is enabled to transmit a packet in the following frame. The final-message-bit in the MAC header of the packet must be set to 1, if this packet is the last of a message, otherwise is set to 0.

RTR (Request Transmission Rules)

1. If there are no pending collisions (*RQ*=0) and there are transmissions scheduled (*TQ*>0), any node that is in neither queue (*pTQ*=0 and *pRQ*=0) and wants to gain channel access randomly selects one of the *m* control minislots and transmits an ARS in the next frame.
2. If a node is at the head of the CRQ (*pRQ*=1), it transmits an ARS in a randomly selected control minislot in an attempt to resolve the collision in the following frame.

QDR (Queueing Discipline Rules)

1. Each node increases the value of *TQ* by one unit for each control minislot with a successful state (indicating that a new node has entered the DTQ).
2. The value of *TQ* is reduced by one unit if a data packet has been successfully transmitted and the final-message-bit was set to 1, since the transmitting node will leave the queue.
3. If there are collisions pending to be resolved (*RQ*>0) the value of *RQ* is reduced by one unit, since those nodes at the head of CRQ will attempt to resolve their collision within the following frame.
4. The value of *RQ* is incremented by one unit for each control minislot where an ARS collision occurred.

The protocol operates in a distributed manner, in the sense that each node executes the above rules to update the queue counters individually, with the use of the information in the FBP. The protocol operation can be better understood with the help of the example that follows.

Example of the DQCA Operation

An example of the DQCA operation for two consecutive frames and three control minislots is shown in Figure 2. Nodes are denoted by ni and their respective pointer values by pTQi and pRQi. It is assumed that in the exactly previous frame (i.e., the (k–1)th) the CRQ is empty (RQ=0), the DTQ contains two nodes (n1, n2, with TQ=2), and n1 has transmitted a packet (pTQ1=1). Three new messages have also arrived at nodes n3, n4 and n5 and these nodes will contend for channel access in the next frame. Indeed, in the contention window of the *k*th frame, n_4 randomly selects the first control minislot and n_3 and n_5 the third one and transmit an ARS. In the data part of the same frame, n_1 continues and completes its message transmission (the end of message is signaled in the MAC header of the data packet). The AP detects the state of all the control minislots (Success – Idle – Collision respectively) and incorporates this information into the FBP. It also acknowledges the correct reception of the data packet and sets the final-message-bit to 1, indicating the exit of n_1 from the DTQ. The successful state of the first control minislot causes all nodes to increase *TQ* by one unit while the collision of the third minislot leads to an increase of *RQ* (*RQ*=1). *TQ* is then reduced by one for the successful data transmission. The nodes set their *pTQ* and *pRQ* values appropriately, with n_4 entering the DTQ in the second position (pTQ_4=2) and n_3 and n_5 entering the CRQ (pRQ_3=pRQ_5=1).

In the CW of the second (i.e the *(k+1)*th) frame, the two nodes in the head of the CRQ send an ARS to a randomly selected control minislot in an attempt to resolve their previous collision. In the meantime, n_2, who is now at the head of DTQ, transmits the first packet of its message (with more to follow). The FBP indicates that the ARS and the data transmission have been successful and that the final-message-bit is 0. All nodes decrease *RQ* by one unit, since one collision has been resolved, and increase *TQ* by 2 (*TQ*=4, *RQ*=0) for the two successful ARS. The nodes n_3 and n_5 enter the DTQ by the order in which the ARS had been sent within the CW (first n_5 and then *n3*). This process will continue for the next frames.

CROSS-LAYER SCHEDULING ALGORITHMS

In DQCA, the data transmissions take place in a FIFO (First-In First-Out) order, thus ensuring fairness among nodes. In this section, four cross-layer algorithms that modify the scheduling discipline of the DTQ in order to increase the system efficiency are presented. The main idea is

Figure 2. Frame structure and operation example

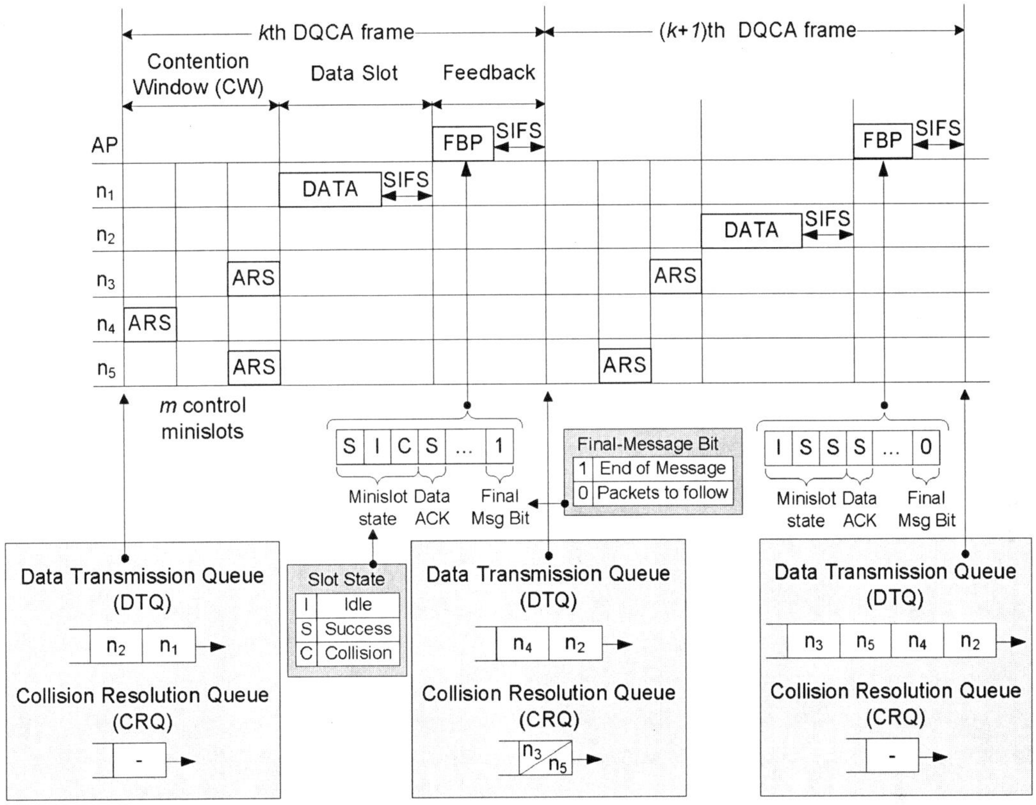

on one hand to exploit the multirate capability of the PHY layer by opportunistically transmitting when higher rates are available and on the other hand to meet QoS requirements through service differentiation.

To implement the proposed algorithms some minor modifications need to be made to the DQCA protocol. First, a link adaptation mechanism must be incorporated to adapt the transmission rate to the channel condition. Hence, the AP measures the SNR of the link whenever an ARS is successfully received and selects the maximum rate that can be supported for upcoming data transmissions from the respective node. This rate is included in the FBP so that the node can appropriately adjust its modulation and coding scheme at the time of data transmission. Obviously, when the link condition is good (i.e., the measured SNR level is sufficiently high), a modulation and coding set of higher order can be employed to yield higher transmission rates.

In the context of this chapter, an indoor quasi-static channel has been assumed so that the channel state remains unchanged from the moment of the SNR measurement until the data transmission. A link adaptation scheme with a periodic update mechanism, suitable for rapid-changing channels, has been proposed and evaluated by the authors but is beyond the scope of this work (Kartsakli, Cateura, Alonso-Zárate, Verikoukis, & Alonso, 2007b).

Another issue is that in order to perform QoS oriented scheduling, service differentiation capabilities should be introduced to the DQCA protocol. To this end, all traffic flows are classified into P service classes ($P \geq 1$), with class P having the most demanding requirements, usually translated to most stringent delay constraints. It has been assumed that it is possible to form P distinguishable types of ARS, one for each service class. Each node that wants to access the channel must transmit an ARS of the appropriate format, so as to make the service class of its traffic load known to the other nodes of the network, including the AP.

Cross-Layer Opportunistic Scheduling Algorithm (CL-algorithm)

The first scheme, named CL-algorithm, implements an opportunistic policy that prioritizes the node with the best channel quality, irrespectively of its service class. Since transmissions generally take place at a higher rate, the overall throughput of the system is increased.

The CL-algorithm is implemented in a distributed manner. The AP keeps a table with the available rates of all the nodes waiting to transmit in the DTQ. These rates have been acquired through link estimation for every received ARS. The entries of this table are fed back to the nodes in the form of a vector that is included in the FBP. This vector has a specific structure; it contains the rates of all the nodes in the DTQ placed by ascending order of their respective *pTQ* values. Note that the *pTQ*s coincide with the order of arrival of the nodes in the DTQ (which is also an indication of their waiting time in the queue). Upon reception of the FPB, the nodes sort the vector entries from higher to lower rate and the node with the highest rate gains chnnel access. If more than one node has the same rate, the one with the smallest *pTQ* value has priority.

Strict Service Priority Scheduling Algorithm (SP-algorithm)

The second scheme, called SP-algorithm, implements a strict service priority policy. In particular, the nodes of the highest-priority service class (P) are always scheduled to transmit first. As a result, delay-sensitive applications are more likely to meet their QoS demands.

This scheme could be better understood by visualizing P transmission queues, one for each service class, instead of a single DTQ, as was the case in DQCA. Hence, P counters are defined, denoted by TQ_i ($0<i\leq P$), to keep the number of the nodes waiting in each queue. A node can belong to only one queue at a time and its position in that queue is expressed, as before, by the *pTQ* value. Within each queue the scheduling is FIFO, whereas among queues there is a strict priority discipline. Consequently, a node that belongs to the ith queue can transmit only if all the queues with a higher priority are empty (i.e., $TQ_{i+1}=\ldots=TQ_P=0$) and if it has the smallest *pTQ* value (i.e., $pTQ=1$) among the other nodes of the same queue. Note that the position of a node can be completely described by the pair (i, *pTQ*), where i denotes the service class and consequently the queue in which the node belongs.

It is also worth noting that there is no need for additional overhead in order to implement this algorithm. The only difference with respect to DQCA is the need to transmit distinguishable types of ARS to reveal the service class of each message. This is essential, since at the end of the frame when the protocol rules are executed, the nodes must know the service class of any successful ARS in order to update the corresponding TQ_i counter.

Cross-Layer with Strict Service Priority Scheduling Algorithm (CLSP-algorithm)

The CLSP-algorithm is a combination of the two previously described schemes. It implements a strict priority scheme among service classes and an opportunistic scheduling among the nodes of the same service class. For the implementation of this algorithm, a rate vector must be transmitted, as in the case of the CL-algorithm. In this case, however, the vector entries are sorted by order of the service class and the position of the nodes in each service class queue.

Virtual Priority Function Scheduling Algorithm (VPF-algorithm)

The fourth technique defines a virtual priority function (VPF) that determines the transmission order of the nodes waiting in the DTQ and has the general form

$$\text{VPF} = f\,(\textit{PHY parameters, MAC parameters, service class}) \qquad (1)$$

The VPF is known to all nodes and the AP is responsible for providing the required feedback so that at the end of each frame every node is capable of calculating the VPF values of all the nodes in the DTQ, including its own. The node with the highest VPF value is then enabled to transmit. If more than one node has the same VPF value, then priority is given to the one with the longest waiting time in the DTQ (i.e., smallest *pTQ* value). Note that in this scheme, the data transmission system consists of a single DTQ.

The VPF can be selected in various ways according to the available parameters and the scheduling objective, which is usually a trade-off between throughput maximization and fairness. To evaluate the VPF-algorithm, the following definition has been chosen:

$$\text{VPF} = \alpha\left(\frac{SC}{P}\right) + \left(1-\alpha\right)\left(\frac{R}{R_{max}}\right)\left(\frac{1}{pTQ}\right) \qquad (2)$$

where α is a tunable weighting factor ($0 \leq \alpha \leq 1$), P is the total number of service classes and SC is an integer within $[1, P]$ that represents the respective service class (with P being the highest priority class). The available rate of a node is denoted by R and R_{max} is the maximum rate defined at the rate set. Finally, as mentioned before, pTQ represents the position of the node in the queue and is an integer within $[1,TQ]$, where TQ is the total number of nodes in the queue.

Note that the VPF has two parts. The first part depends solely on the service class of the node. The second part is proportional to the (normalized) transmission rate divided by the pTQ value, meaning that preference is given to nodes that have higher rates and smaller *pTQ*s (i.e., longer waiting time in the queue). By tuning the parameter α, the system behavior can vary from a strict service priority scheme (for α=1) to an opportunistic scheme (for α=0). The overhead required for the implementation of this scheme is the same as in the case of the CL-algorithm.

Example of the Cross-Layer Scheduling Operation

An example that stresses the difference between DQCA and the four cross-layer algorithms is depicted in Figure 3. In the example, six nodes ($n_1 - n_6$) are waiting in the DTQ and a rate set of R={6, 9, 12, 18, 24, 36, 48, 54}Mbps, as defined in IEEE 802.11g, is considered. There are four service classes, denoted by SC, with SC=4 having the higher priority. The available rate of each node and its pTQ value, as calculated by the DQCA rules, have been also marked on the figure.

In DQCA, the nodes transmit by order of their pTQ, regardless of the available rate or the service class. When the CL-algorithm is executed, the nodes that have higher rates are scheduled to transmit first. Since there are two nodes that have

Figure 3. Operation example of the cross-layer scheduling algorithm

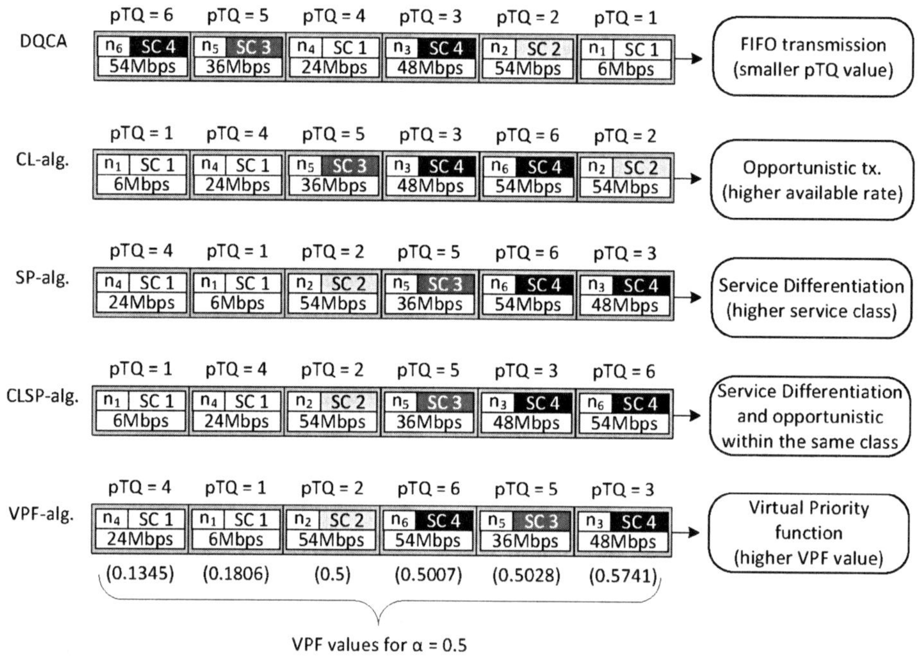

the same maximum rate of 54 Mbps, the one with the smaller *pTQ* (*pTQ*=2) is given priority.

According to the SP-algorithm the nodes transmit by priority order that corresponds to their service class, whereas among the nodes of the same class, a FIFO order is maintained. In the example, there are two nodes (n_3 and n_6) that belong to the highest priority service class (*SC*=4) and they transmit by order of their pTQ (i.e., first n_3 and then n_6).

In the CLSP-algorithm, the service class again determines the transmission order of the nodes, with the difference that among the nodes of the same class opportunistic scheduling is employed. Consequently, the first node to transmit is n_6 since it is the node with the highest transmission rate among the nodes that belong to the highest priority class.

Finally, in the VPF-algorithm the transmission order is determined by the VPF value. In the example parameter α is set to 0.5. It can be observed that nodes with priority service classes and higher rates tend to transmit first. However, the scheme also displays some amount of fairness. For example n_1 gets to transmit before n_4 because it has a longer waiting time in the queue (i.e., *pTQ*=1), even though its service class and rate are lower.

CASE STUDY AND PERFORMANCE EVALUATION

The proposed algorithms have been evaluated through simulations performed by a C++ link-layer simulation tool. The basic features of the simulated scenario and a detailed performace evaluation are given in this section.

Channel Model

Without losing generality, the IEEE 802.11g specification(IEEE Std. 802.11g, 2003) has been selected as the underlying PHY layer for the simulations, although any other multi-rate PHY could also have been considered. The nodes can transmit data at a set of eight basic rates $R=\{6, 9, 12, 18, 24, 36, 48, 54\}$ Mbps, depending on their channel condition.

The wireless channel has been modeled as a eight-state discrete Markov chain similar to the one described in (Konrad, Zhao, Joseph, & Ludwig, 2001), where each state represents one of the eight available rates. The idea is based on the fact that although wireless channels are characterized by fast-fading, some correlation exists between the current transmission state and the immediately previous one. The Markov chain is represented by a transition matrix T which expresses the probability for each node to select a certain rate every time the coherence time has elapsed. Each scenario environment (basically defined by a radio channel model and a mobility model for the transmitting nodes) will have its own matrix values, therefore any matrix can be used in the study. For evaluation purposes, and without losing generality, the following matrix T has been used in the simulated scenario:

$$T = \overbrace{\begin{array}{cccccccc} 6 & 9 & 12 & 18 & 24 & 36 & 48 & 54 \end{array}}^{\text{future state}} \atop \left[\begin{array}{cccccccc} 0.4 & 0.5 & 0.1 & 0 & 0 & 0 & 0 & 0 \\ 0.1 & 0.4 & 0.5 & 0 & 0 & 0 & 0 & 0 \\ 0 & 0.1 & 0.4 & 0.4 & 0.1 & 0 & 0 & 0 \\ 0 & 0 & 0.1 & 0.4 & 0.4 & 0.1 & 0 & 0 \\ 0 & 0 & 0 & 0.1 & 0.5 & 0.4 & 0 & 0 \\ 0 & 0 & 0 & 0 & 0.3 & 0.5 & 0.2 & 0 \\ 0 & 0 & 0 & 0 & 0.1 & 0.2 & 0.5 & 0.2 \\ 0 & 0 & 0 & 0 & 0 & 0.1 & 0.4 & 0.5 \end{array}\right] \left.\begin{array}{c} 6 \\ 9 \\ 12 \\ 18 \\ 24 \\ 36 \\ 48 \\ 54 \end{array}\right\} \text{current state} \qquad (3)$$

The above channel model corresponds to an average transmission rate of approximately 35 Mbps. The coherence time of the channel is supposed to be long enough to ensure that the channel state remains unchanged from the moment of the SNR measurement until the data transmission. This is a realistic assumption for indoor quasi-static WLAN situations. In the evaluated scenarios a coherence time of 150 ms has been considered. It has also been assumed that, regardless of the channel conditions, the appropriate rate selection guarantees that no errors are introduced during transmissions.

Traffic Models

Three types of data sources have been used in the study case and will be described next.

Data Traffic Sources

The data traffic sources have a Poisson distributed generation process with arrival rate λ. The size of the generated messages is exponentially distributed with a mean value equal to $10 \cdot L$, with L set to 1000 bytes. This results to an offered load of $C=\lambda \cdot (10 \cdot L) \cdot 8 \cdot 10^{-6}$ Mbps. Since data messages are relatively large, their transmission takes place in consecutive DQCA frames with L being the maximum number of bytes transmitted per each frame.

Voice Traffic Sources

The voice traffic sources are based on the Brady's ON-OFF model (Brady, 1969). The time spent at the ON and OFF state is exponentially distributed with mean values of 1 s and 1.35 s respectively. While at the ON state, packets of 160 bytes are generated every 20 ms resulting to a constant bit rate (CBR) of 64 kbps and a mean offered load of 27.23 kbps, as defined in the G.711 voice codec (ITU-T Recommendation G.711, 1998). The packets generated within a single ON interval will be referred to as a voice burst. Since voice packets are relatively small, voice nodes transmit all the buffered packets of the same burst in a single DQCA frame.

Video Traffic Sources

A near real-time video model defined in (3GPP2, 2004) has been used for the generation of streaming video traffic. Each video frame arrives at a regular rate of 10 frames per second and consists of 25 packets whose size follows a truncated Pareto distribution (a=1.2, min=50 bytes and max=200 bytes). The packet arrival process is also modelled by a truncated Pareto distribution (a=1.2, min=2.5 ms and max=4 ms), to account for the video encoding delays. The mean offered load for video traffic is 180 kbps and video nodes transmit all the buffered packets that belong to the same frame in a single DQCA frame.

Service Differentiation and QoS Requirements

The service differentiation paradigm used in IEEE 802.11e (IEEE Std. 802.11e, 2005) has been adopted. In particular, four classes are defined for Background, Best-Effort, Video and Voice traffic, with service class ids from 1 to 4, respectively. Voice traffic is assigned the highest priority due to stringent delay constraints. In particular, a maximum delay of 150 ms per voice packet can be tolerated, after which the packet is dropped. For the Video class, the maximum delay has been set to 300 ms per packet. Best-effort data can tolerate a delay up to 5 s and finally background data has no delay constraints.

MAC Layer Parameters

The DQCA protocol parameters are summarized in Table 1. All control transmissions (i.e., the FBP) are sent at the minimum rate of 6 Mbps in order to ensure reliable transmission.

The FBP has a length of 13 bytes, including 2 bytes for the Frame Control (FC) field, 6 bytes for the feedback information, 1 byte for the ACK and 4 bytes for the FCS (Frame Control Sequence). Some of the proposed algorithms require a rate

Table 1. PHY/MAC layer parameters

Parameter	**Value**
MAC Header	34 bytes
PHY Header	20 μs
SIFS	10 μs
Propagation Time	1 μs
Control Minislots m	3
ARS	10 μs
FBP	13bytes+CL_Overhead
CL_Overhead	
DQCA and SP-alg.	0 bits
CL,CLSP,VPF-alg.	$3 \cdot TQ$ bits

vector with the available rate of all the nodes in the DTQ to be incorporated in the FBP. Given a set of eight rates, 3 bits are sufficient to represent all possible rate values. If TQ is the number of nodes in the DTQ, then $TQ \cdot 3$ bits must be appended to the FBP. When multiple transmission queues are defined, as in the case of the CLSP-algorithm, the number of nodes is calculated as $TQ = \sum_{i=1}^{P} TQ_i$.

PERFORMANCE EVALUATION

The throughput and delay performance of the proposed algorithms have been evaluated through simulations. Throughput is defined as the total number of successfully transmitted data bits to the unit of time. The relative throughput is defined as the ratio of the successfully transmitted bits to the number of generated bits and is evaluated for each class as well as for the whole system. Obviously, packets that do not satisfy the previously defined QoS constraints are dropped and not included in the throughput calculation. The mean delay is defined as the average time from the generation of a packet until its successful transmission and is evaluated separately for each service class. In addition, for the voice and video service classes

Figure 4.

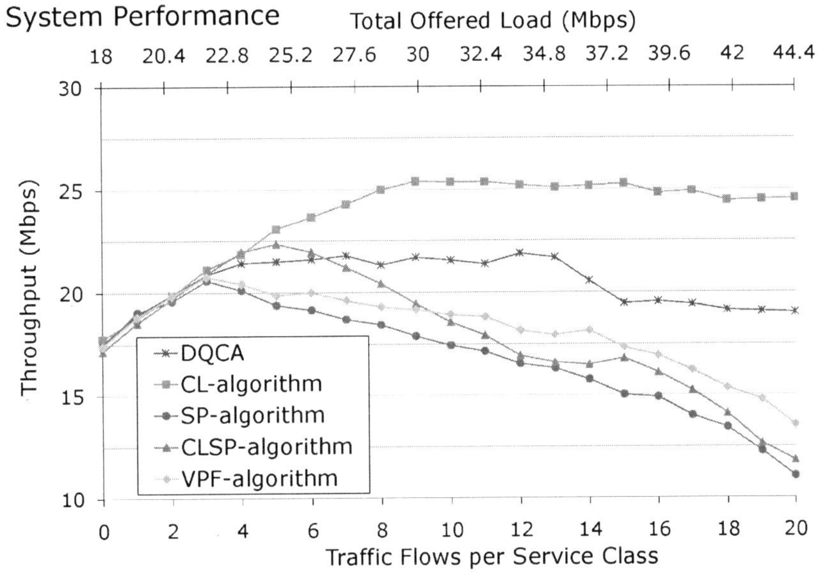

the delay jitter has also been measured. Jitter $J_{(i)}$ is defined as the mean deviation (smoothed absolute value) of the difference $D_{(i-1,\, i)}$ between the delays of two consecutive packets (i-1) and i and is calculated by the formula presented in (Schulzrinne, Casner, Frederick, & Jacobson, 1996)

$$J_{(i)} = J_{(i-1)} + (|D_{(i-1,\, i)}|-J_{(i-1)})/16,\ i\geq 1 \qquad (4)$$

and $J_{(0)}$=0, $D_{(0,1)}$=0.

In the considered scenario each node generates a single type of traffic, so the terms node and traffic flow (tr. flow) can be used interchangeably. Initially, a constant number of 10 background tr. flows that generate 1.8 Mbps each, has been considered. Then, tr. flows of voice, video and best-effort service classes are gradually added. The number of additional tr. flows is marked on the x-axis of all the figures in this section. For example, at x=0 there are only 10 background tr. flows. At x=1, a voice, a video and a best-effort tr. flow are added, raising the total number of flows to 13. At the final evaluation point (x=20) there are 70 nodes, 10 with background traffic and 20 with each of the other three service classes. Note that the generated traffic is approximately 27.23 kbps per voice node, 180 kbps per video node and 1 Mbps per best-effort node. Hence, at each step the offered load increases by about 1.2 Mbps. The approximate value of the total offered load is marked at the upper part of the figures. Note also that the parameter α of the VPF-algorithm has been set to 0.5.

The total system throughput is illustrated in Figure 4. The best performance is achieved by the CL-algorithm since, due to the opportunistic scheduling, higher transmission rates are used. A maximum throughput of 25 Mbps is reached and maintained even when the traffic load increases. The performance of DQCA comes next with a maximum throughput of approximately 21.5 Mbps, which drops after point 14 and stabilizes at 19 Mbps. The other three proposed algorithms yield a lower throughput that decreases as the traffic load grows. The CLSP-algorithm performs better than the SP-algorithm especially under light traffic (e.g., for 5 tr. flows CLSP achieves 22.3

Figure 5.

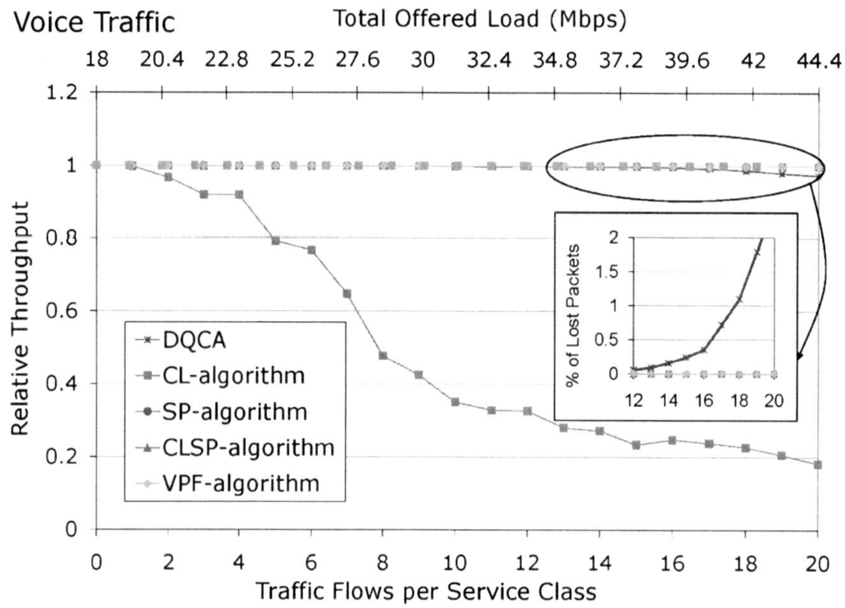

Mbps which is about 3 Mbps higher than the SP-algorithm and 1Mbps higher than DQCA). The VPF-algorithm performs slightly better than the SP-algorithm for light traffic and has a milder decrease rate. As a result it eventually outperforms both SL and CLSP algorithms for higher traffic loads.

In order to appreciate the contribution of the proposed schemes, the relative throughput for the three upper service classes has been plotted. The SP, CLSP and VPF algorithms achieve a relative throughput of 1 for the high-priority voice class since they employ service differentiation (Figure 5). DQCA has a close to one throughput that slightly drops as the traffic grows. The percentage of lost packets due to excess delay is shown at the rightmost part of the figure and for DQCA it remains under 1% for up to 17 voice tr. flows. On the contrary, the CL-algorithm totally fails to satisfy the voice QoS demands. In the case of video traffic (Figure 6), SP, CLSP and VPF algorithms again guarantee maximum throughput. DQCA can support up to 13 video nodes, but after that point the video performance drops dramatically. The CL-algorithm yields the highest throughput for the best-effort traffic that has very relaxed QoS constraints (Figure 7). The strict priority schemes (SP and CLSP) have a throughput of 1 for up to 13 best-effort nodes and then deteriorate rapidly. The mild-priority VPF-algorithm can fully support 9 best-effort flows but the drop in performance is not so steep.

The above observations make sense since all algorithms have the same bandwidth but allocate resources in a different way. Hence, there is a trade-off between fairness among all service classes and QoS provisioning. As the traffic load grows, the majority of packets transmitted under the service priority schemes belong to the voice and video classes. Since those packets are generally small, the impact of the PHY/MAC overhead is more noticeable and is reflected as a decrease on the total system throughput.

As far as the delay performance is concerned, for the algorithms that prioritize voice (SP, CLSP and VPF algorithms) the mean voice delay is kept

Figure 6.

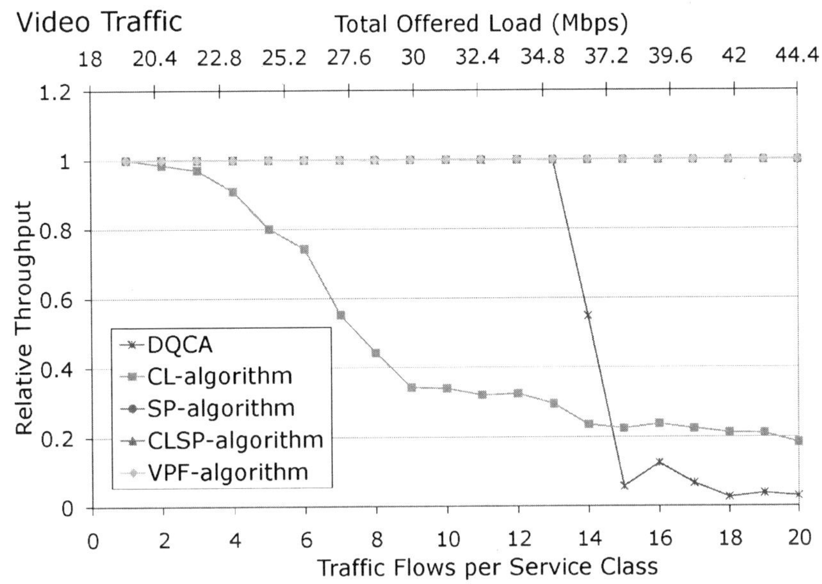

below 5 ms, as shown in Figure 8. In DQCA, the mean voice delay gradually increases as the traffic grows but it remains below 75ms. On the contrary, the delay for the CL-algorithm quickly exceeds 150 ms, which is the maximum tolerated value by the voice service. Similar behavior can be observed in Figure 9 for the video service where the maximum tolerated delay is 300 ms. Even though the delay

Figure 7.

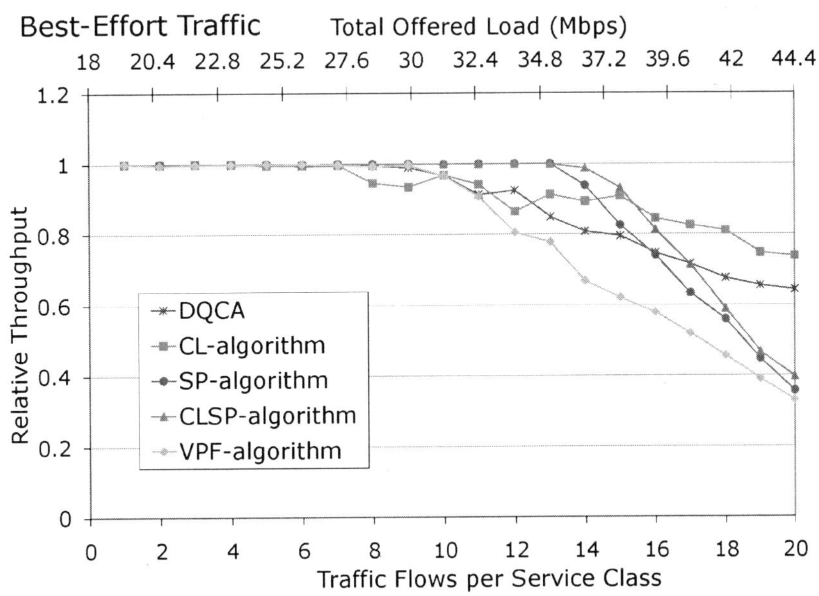

Figure 8.

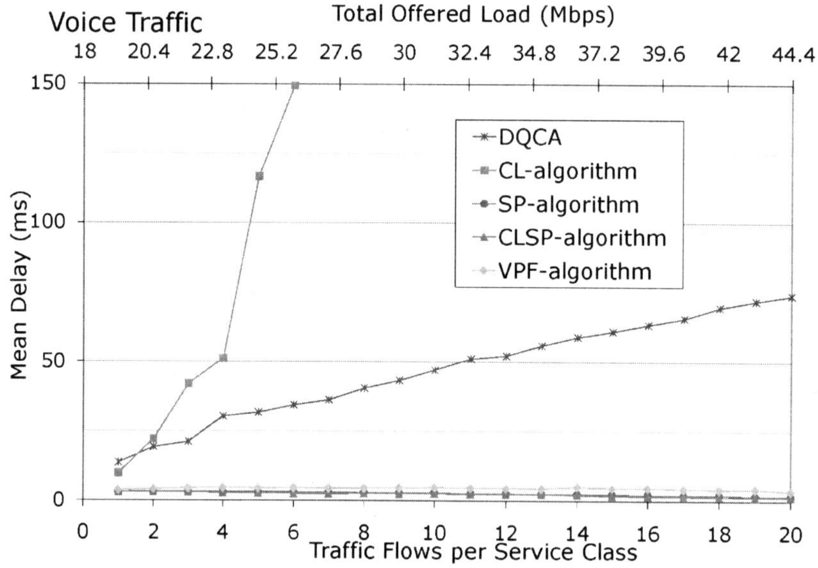

Figure 9.

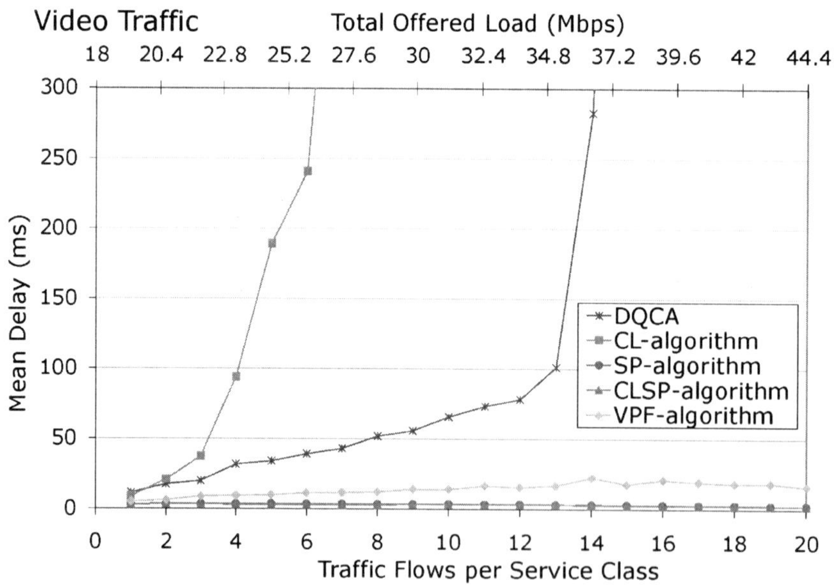

performance for the data services has not been plotted due to space constraints, higher delays are anticipated for these classes, as a trade-off for the priorities assigned to the delay-sensitive multimedia services.

To complete the delay analysis, the delay jitter for the voice and the video service classes has been plotted in Figure 10 and Figure 11 respectively. The jitter for the SP and the CLSP algorithms is almost zero, meaning that all the delivered voice

Figure 10.

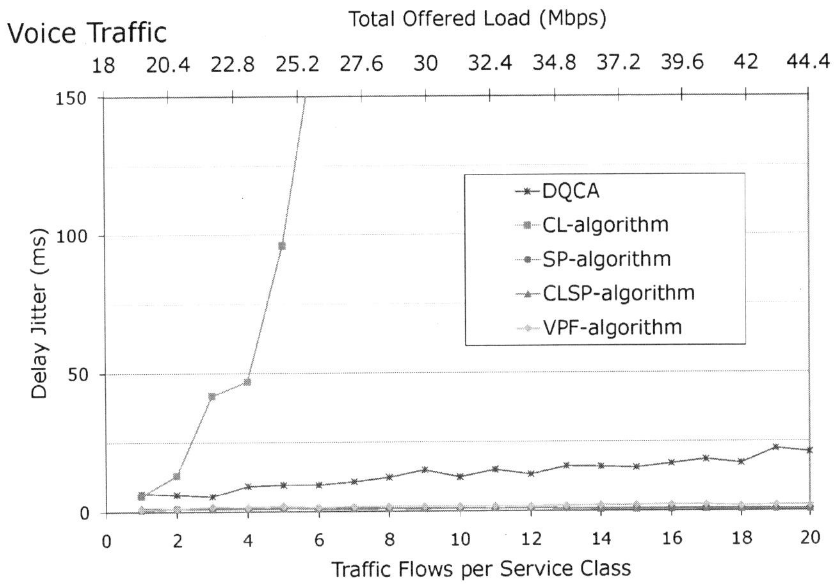

and video packets suffer the same amount of delay. The jitter for the VPF algorithm is also very small, and does not exceed 3 ms for voice and 10 ms for video traffic. For DQCA the voice jitter increases with a slight slope and is kept below 25 ms whereas the video jitter rapidly rises when there are more than 13 tr. flows (which is exactly when its relative throughput performance drops dramatically). Clearly, the jitter is considerably large for the CL-algorithm.

Finally, Figure 12 evaluates the throughput performance of the VPF-algorithm for three values of the parameter α, in particular for α=0.3, 0.5 and 0.7. Smaller values of α correspond to a more opportunistic scheduling scheme whereas larger values mean that the service differentiation plays a more important role. The throughput for the voice, video and best-effort class has been plotted in the figure, as well as the total throughput of the system. It is evident that more opportunistic schemes have a higher total throughput but cannot always guarantee QoS.

The performance evaluation in terms of throughput, mean delay and delay jitter for the four proposed algorithms has been presented in this section. It is difficult to compare the proposed schemes, since each one of them has a different objective. The DQCA is a balanced, fair protocol that does not take into consideration QoS demands. As a result, it cannot provide QoS guarantees, which is mostly evident in the case of the video service class. The CL-algorithm is focused on maximizing the system throughput but also ignores QoS demands. Therefore, it enhances significantly the performance for classes that can tolerate delay (such as the data best-effort and background traffic) but is not suitable for networks with multimedia traffic.

On the other hand, the main objective of the SP-algorithm is to give priority to delay-sensitive service classes and therefore has a near-optimum throughput performance and very low delays for voice and video applications. This, however, has an impact on the best-effort and background classes, whose performance drops significantly

Figure 11.

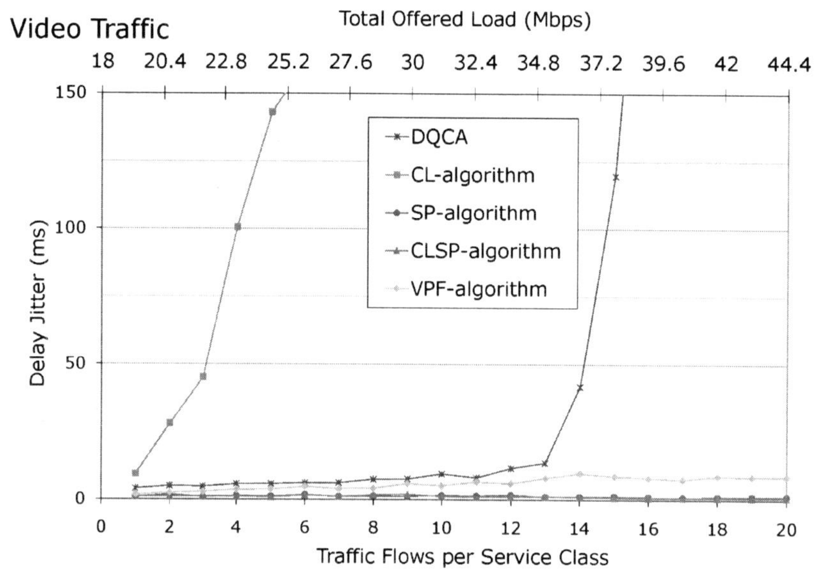

and eventually reaches zero. The CLSP-algorithm tackles this problem by employing opportunistic scheduling while maintaining the service priority scheme. As a result, higher throughput and lower delays with respect to the SP-algorithm is achieved for all classes. However, when the traffic load is very high, the background traffic class is also led to starvation (zero background throughput).

Figure 12.

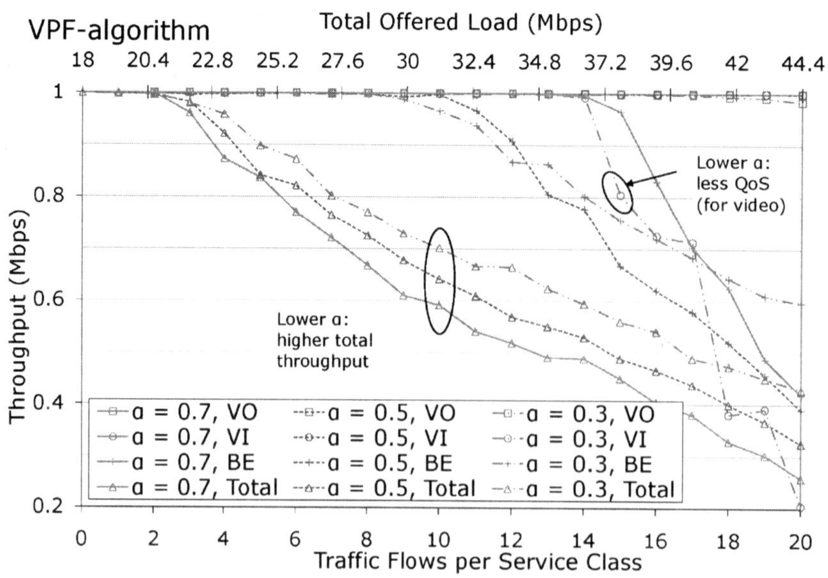

Finally, the VPF-algorithm implements a trade-off between service differentiation and a mild opportunistic scheme that takes into account the age of the nodes in the transmission queue (indicated by the *pTQ* value). Its performance depends on the selection of the tunable parameter α. For α=1 the VPF-algorithm performance becomes identical to the SP-algorithm and for smaller values of α it tends (but cannot reach) towards the performance of the CL-algorithm. For α=0.5, which is the case evaluated in this section, the VPF-algorithm ensures QoS performance for the voice and the video traffic and in addition overcomes the starvation problem. The improvement of the background traffic performance is achieved at the cost of best-effort throughput degradation (with respect to the other algorithms). Nevertheless, the system fairness is increased since all classes get access to the channel (non-zero throughput).

FUTURE RESEARCH DIRECTIONS

There are several interesting open research lines in the field of MAC layer design based on DQCA. Some currently conducted work is focused on the development of a power consumption model that will facilitate the assessment of the suitability of DQCA for energy constraint networks, such as sensor networks.

Moreover, since DQCA is a very efficient and flexible protocol, the identification and performance evaluation of realistic applications also constitutes an important line for future research. Among many examples, disaster rescue operations or inter-vehicular communications can be identified as innovative topics that are getting more relevance in the scientific community and which match with the characteristics of DQCA. In fact, they are infrastructureless networks established in an ad hoc fashion to satisfy a spontaneous need and where a high number of users, which tend to be clustered, compose a highly loaded network. A variation of DQCA, named DQMAN (Distributed Queueing Mac protocol for Ad hoc Networks) and specially adapted for this kind of networks, has been already presented in the literature (Alonso-Zárate, Kartsakli, Skianis, Verikoukis, & Alonso, 2008).

Finally, the most interesting and important future goal would be the actual hardware implementation of the protocol. Most of the efforts have been focused on developing a testbed wherein DQCA can be implemented and thus the ideas presented in this chapter can be finally validated in a real environment. An American company called Ether2 (Ether2, 2005) is already implementing a prototype of wireless network cards based on DQCA, and hopefully some real equipment will be available in the near future and a commercial application could be possible.

REFERENCES

Alonso, L., Agusti, R., & Sallent, O. (2000). A near-optimum MAC protocol based on the distributed queueing random access protocol (DQRAP) for a CDMA mobile communication system. *Journal on Selected Areas in Communications*, *18*(9), 1701–1718. doi:10.1109/49.872957

Alonso-Zárate, J., Kartsakli, E., Skianis, C., Verikoukis, C., & Alonso, L. (2008). Saturation Throughput Analysis of a Cluster-based Medium Access Control Protocol for Ad Hoc Wireless Networks. *Simulation: Transactions of the Society for Modeling and Simulation International*, *84*(12), 619–633. doi:10.1177/0037549708099232

Alonso-Zárate, J., Verikoukis, C., Kartsakli, E., Cateura, A., & Alonso, L. (2008). A near-optimum cross-layered distributed queuing protocol for wireless LAN [medium access control protocols for wireless LANs]. *Wireless Communications*, *15*(1), 48–55. doi:10.1109/MWC.2008.4454704

Bianchi, G. (2000). Performance analysis of the IEEE 802.11 distributed coordination function. *Journal on Selected Areas in Communications*, *18*(3), 535–547. doi:10.1109/49.840210

Bouam, S., & Othman, J. (2004). A 802.11 multiservices cross-layer approach for QoS management. In *Vehicular Technology Conference, 2004. VTC2004-Fall. 2004 IEEE 60th*, (Vol. 4, pp. 2698-2702).

Brady, P. T. (1969). A model for generating on-off speech patterns in two-way conversation. *The Bell System Technical Journal*, *48*(7), 2445–2472.

Campbell, G., & Xu, W. (2001). *Patent No. 6292493.* USA.

Ether2. (2005). *Ether 2 - Distributing the Queue*. Retrieved from www.ether2.com

Garroppo, R. G., Giordano, S., Lucetti, S., & Tavanti, L. (2007). Providing air-time usage fairness in IEEE 802.11 networks with the deficit transmission time (dtt) scheduler. *Wireless Networks*, *13*(4), 481–495. doi:10.1007/s11276-006-9201-7

3GPP2. (2004). *cdma2000 Evaluation Methodology Revision 0.* TSG-C.R1002-0.

Holland, G., Vaidya, N., & Bahl, P. (2001). A rate-adaptive MAC protocol for multi-hop wireless networks. In *Proceedings of the 7th annual international conference on Mobile computing and networking (MobiCom '01)* (pp. 236–251). New York: ACM.

ITU-T Recommendation G.711. (1998). Pulse code modulation (PCM) of voice frequencies.

Kartsakli, E., Alonso-Zárate, J., Verikoukis, C., & Alonso, L. (2009). Cross-Layer Scheduling with QoS Support over a Distributed Queuing MAC for Wireless LANs. Mobile Networks and Applications (MONET) Special Issue on Recent Advances in IEEE 802.11 WLANs: Protocols, Solutions and Future Directions.

Kartsakli, E., Cateura, A., Alonso-Zárate, J., Verikoukis, C., & Alonso, L. (2007a). Cross-layer enhancement for wlan systems with heterogeneous traffic based on DQCA. In *IEEE International Conference onCommunications, 2007. ICC '07.*, (pp. 5708–5713).

Kartsakli, E., Cateura, A., Alonso-Zárate, J., Verikoukis, C., & Alonso, L. (2007b). Opportunistic Scheduling using an Enhanced Channel State Information Update Scheme for WLAN Systems with DQCA. In *Proc. of the IEEE Vehicular Technology Conference Spring (VTC Spring 2007).* Dublin, Ireland.

Konrad, A., Zhao, B. Y., Joseph, A. D., & Ludwig, R. (2001). A markov-based channel model algorithm for wireless networks. In *Proceedings of the 4th ACM international workshop on Modeling, analysis and simulation of wireless and mobile systems (MSWIM '01)* (pp. 28–36). New York: ACM.

Otal, B., Alonso, L., & Verikoukis, C. (2009). Highly Reliable Energy-Saving MAC for Wireless Body Sensor Networks in healthcare systems. *Journal Selected Area of Communications special issue on "Wireless and Pervasive Communications for Healthcare*.

Pavon, J. d., & Choi, S. (2003). Link adaptation strategy for IEEE 802.11 WLAN via received signal strength measurement. In *IEEE International Conference on Communications, 2003. ICC '03*, (Vol. 2, pp. 1108–1113).

Sadeghi, B., Kanodia, V., Sabharwal, A., & Knightly, E. (2002). Opportunistic media access for multirate ad hoc networks. In *Proceedings of the 8th annual international conference on Mobile computing and networking (MobiCom '02)* (pp. 24–35). New York: ACM.

Schulzrinne, H., Casner, S., Frederick, R., & Jacobson, V. (1996). RTP: A Transport Protocol for Real-Time Applications. RFC 1889, IETF Audio-Video Transport Working Group.

Shakkottai, S., Rappaport, T., & Karlsson, P. (2003). Cross-layer design for wireless networks. *Communications Magazine, IEEE*, *41*(10), 74–80. doi:10.1109/MCOM.2003.1235598

IEEE Std. 802.11g. (2003). IEEE Standard for Wireless LAN Medium Access Control (MAC) and Physical Layer (PHY) Specifications - Amendment 4: Further Higher Data Rate Extension in the 2.4 GHz Band. *IEEE Std. 802.11g*.

IEEE Std. 802.11e. (2005). IEEE Standard for Wireless LAN Medium Access Control (MAC) and Physical Layer (PHY) Specifications - Amendment 8: Medium Access Control (MAC) Quality of Service Enhancements. *IEEE Std. 802.11e*.

IEEE Std. 802.11. (2007). IEEE Standard for Wireless LAN Medium Access Control (MAC) and Physical Layer (PHY) Specifications. *IEEE Std 802.11-2007 (Revision of IEEE Std 802.11-1999)*.

IEEE Std 802.11n. (2008). IEEE Draft Standard for Wireless LAN Medium Access Control (MAC) and Physical Layer (PHY) Specifications - Amendment 5: Enhancements for Higher Throughput. *IEEE Unapproved Draft Std. P803.22n/D5.0*.

Wang, J., Zhai, H., & Fang, Y. (2004). Opportunistic packet scheduling and media access control for wireless LANs and multi-hop ad hoc networks. *Wireless Communications and Networking Conference, 2004. WCNC. 2004 IEEE*, (pp. 1234–1239 Vol.2).

Xu, W., & Campbell, G. (1992). A Near Perfect Stable Random Access Protocol for a Broadcast Channel. In *Proc. of the IEEE International Conference on Communications 1992 (ICC'92)*, (Vol. 1, pp. 370-374).

ADDITIONAL READING

Babu, A., & Jacob, L. (2007). Fairness Analysis of IEEE 802.11 Multirate Wireless LANs. *Vehicular Technology, IEEE Transactions on*, 3073-3088.

Bredel, M., & Fidler, M. (2009). Understanding Fairness and its Impact on Quality of Service in IEEE 802.11. *INFOCOM 2009. The 28th Conference on Computer Communications. IEEE*, (pp. 1098-1106).

Cheng, Y., Ling, X., Song, W., Cai, L. X., Zhuang, W., & Shen, X. (2007). A Cross-Layer Approach for WLAN Voice Capacity Planning. *Selected Areas in Communications, IEEE Journal on*, 678-688.

Choi, J., Park, K., & Kim, C.-k. (2009). Analysis of Cross-Layer Interaction in Multirate 802.11 WLANs. *Mobile Computing, IEEE Transactions on*, 682-693.

Deng, D.-J., & Yen, H.-C. (2005). Quality-of-service provisioning system for multimedia transmission in IEEE 802.11 wireless LANs. *Selected Areas in Communications, IEEE Journal on*, 1240-1252.

Dimic, G., Sidiropoulos, N., & Zhang, R. (2004). Medium access control - physical cross-layer design. Signal Processing Magazine, IEEE, 40-50.

Foukalas, F., Gazis, V., & Alonistioti, N. (2008). Cross-layer design proposals for wireless mobile networks: a survey and taxonomy. Communications Surveys & Tutorials, IEEE, 70-85.

Jiang, H., & Weihua, Z. (2006). Resource allocation with service differentiation for wireless video transmission. *Wireless Communications, IEEE Transactions on*, 1456-1468.

Jiang, H., & Zhuang, W. (2006). Cross-layer resource allocation for integrated Voice/Data traffic in wireless cellular networks. *Wireless Communications, IEEE Transactions on*, 457-468.

Kawadia, V., & Kumar, P. (2005). A cautionary perspective on cross-layer design. Wireless Communications, IEEE, 3-11.

Park, E.-C., Kim, D.-Y., Choi, C.-H., & So, J. (2007). Improving Quality of Service and Assuring Fairness in WLAN Access Networks. *Mobile Computing, IEEE Transactions on*, 337-350.

Ramos, N., Panigrahi, D., & Dey, S. (2005). Quality of service provisioning in 802.11e networks: challenges, approaches, and future directions. Network, IEEE, 14-20.

Shiang, H.-P., & van der Schaar, M. (2007). Multi-user video streaming over multi-hop wireless networks: a distributed, cross-layer approach based on priority queuing. *Selected Areas in Communications, IEEE Journal on*, 770-785.

Srivastava, V., & Motani, M. (2005). Cross-layer design: a survey and the road ahead. Communications Magazine, IEEE, 112-119.

Villalon, J., Cuenca, P., Orozco-Barbosa, L., Seok, Y., & Turletti, T. (2007). Cross-Layer Architecture for Adaptive Video Multicast Streaming Over Multirate Wireless LANs. *Selected Areas in Communications, IEEE Journal on*, 699-711.

Wang, X., Yu, Y., & Giannakis, G. (2008). Design and Analysis of Cross-Layer Tree Algorithms for Wireless Random Access. *Wireless Communications, IEEE Transactions on*, 909-919.

Xia, Q., & Hamdi, M. (2005). Cross Layer Design for IEEE 802.11 WLANs: Joint Rate Control and Packet Scheduling. *Local Computer Networks, 2005. 30th Anniversary. The IEEE Conference on*, (pp. 624-631).

Xiao, Y., & Pan, Y. (2005). Differentiation, QoS guarantee, and optimization for real-time traffic over one-hop ad hoc networks. *Parallel and Distributed Systems, IEEE Transactions on*, 538-549.

Zhai, H., Chen, X., & Fang, Y. (2005). How well can the IEEE 802.11 wireless LAN support quality of service? *Wireless Communications, IEEE Transactions on*, 3084-3094.

Zhao, Q., & Tong, L. (2003). A multiqueue service room MAC protocol for wireless networks with multipacket reception. IEEE/ACM Trans. Netw., 125-137.

Chapter 3
Delay Constrained Admission Control and Scheduling Policy for IEEE 802.11e HCCA Method

Der-Jiunn Deng
National Changhua University of Education, Taiwan

Yueh-Ming Huang
National Cheng Kung University, Taiwan

Hsiao-Hwa Chen
National Cheng Kung University, Taiwan

ABSTRACT

In order to obtain better Quality-of-Service (QoS) requirements for multimedia traffic, the 802.11TGe has proposed HCF Controlled Channel Access (HCCA) method for the Controlled Access Period (CAP) in the HCF (Hybrid Coordination Function) to enhance the original IEEE 802.11 Medium Access Control (MAC) protocol, and it is expected to provide integrated traffic service to realize mobile multimedia communications. However, the reference design of admission control and scheduling policy in HCCA still can not provide stringent delay requirements to fulfill a hard QoS guarantee, a necessary feature for most multimedia applications. In this chapter, the authors propose a pragmatic admission control scheme with a novel polling based packet scheduling policy for multimedia transmission, such as Constant Bit Rate (CBR) and Variable Bit Rate (VBR) traffic, for IEEE 802.11e HCCA method. Our design is simple, it is compatible with the standard, it can guarantee delay constraint, and it utilizes bandwidth efficiently. A simple and accurate analytical model is carried out to study the average queueing delay estimation of the proposed scheme. In addition to theoretical analysis, simulations are conducted in NS2 network simulator to verify our analysis and to validate the promising performance of the proposed scheme.

DOI: 10.4018/978-1-61520-771-8.ch003

INTRODUCTION

In the past, network designers had to contend with only one form of traffic, voice or data. Today, most new applications have become much richer in content than simply text, or strictly telephony, and require transmission of multitude of media. Hence, there is a strong economic and practical push to integrate all of these applications onto a single network. However, data, video and audio media have different transmission requirements, in terms of sustained data rates, acceptable latency, and tolerated error rates. Hence, the presence of QoS support in wireless networks is crucial since a global, ubiquitous wireless network will play a vital role in creating new user-centric communication service in the next generation Internet.

In Wireless Local Area Networks (WLANs), the MAC protocol is the key component that provides the efficiency in sharing the common radio channel while satisfying the QoS requirements for multimedia applications. That is, MAC protocols that aim to carry multimedia traffic must be able to meet the differing requirements of each of the different traffic classes. Time-bounded data are useless unless arrived in time. Such data usually have stringent delay constrains and in many cases they should be delivered exactly as they were generated. Examples of such traffic include voice and video. On the other hand, asynchronous data, such as email or file transfer, can be delayed without causing any inconvenience.

In general, there are two methods in wireless MAC protocols to facilitate the transmission of time-bounded data, reservation schemes and priority schemes. Reservation schemes allow time-bounded traffic to reserve a periodic time slot on the channel that they alone can access. When the time-bounded traffic being transmitted is voice or video, the reserved time is a periodic slot in time within a larger time frame. Once a slot is assigned, no other stations may contend for that portion of resources. All reservation schemes suffer from one drawback. When reserved and unused, the resource is simply wasted. This is where priority schemes come in. Priority schemes share resources and at the same time allow some stations to have a larger share of the pie. They assign higher priority to the time-bounded traffic and high priority traffic has precedence for using network resources. However, depending on the protocol design (for example, whether the resource usage is preemptive), performance can not be absolutely guaranteed.

The original IEEE 802.11 standard (Chen, Zhai, Tain & Fang, 2006) is designed for best effort service only. The lack of a built-in mechanism for support of real-time services makes it very difficult to provide quality-of-service guarantees for throughput-sensitive and delay-sensitive multimedia applications. Therefore, modification of existing standards is necessary. To expand support for applications with QoS requirements, the 802.11e Task Group (IEEE Draft Std. 802.11e/D8.0, 2004) was formed to enhance the original IEEE 802.11 MAC protocol and it is expected to provide integrated traffic service to realize mobile multimedia communications.

A simple admission control scheme and packet scheduling policy for IEEE 802.11e HCCA method has been developed as a reference in the IEEE 802.11e standard, where the mean data rate and the mean packet size are used to calculate the resource need by stream. However, a guaranteed stringent delay constrain for every single MAC Service Data Unit (MSDU) to provide multimedia traffic with their pledged QoS requirements still cannot be satisfied in WLANs since the instantaneous and fluctuating data rate generated by multimedia applications are usually quite different from the corresponding mean values.

This chapter aims to provide a comprehensive study of the limitations and merits of mechanisms that have been proposed toward the provision of hard QoS support to WLANs at MAC layer. First, we will give a brief introduction of the IEEE 802.11e HCCA method. We then explore various proposed admission control schemes and packet scheduling policies in WLANs. These schemes significantly enrich the enhancement of QoS guaranteed from different aspects.

Moreover, as a result from the study of these strategies, in this chapter, we propose a pragmatic admission control scheme with a novel polling based packet scheduling policy for multimedia transmission, such as CBR and VBR traffic, for IEEE 802.11e HCCA method. Under such a scheme, the proposed admission control scheme considers the expected delay of new connection and its effect on existing sessions upon arrival of new connection for admission decision to ensure that the channel is not overloaded and the delay constrains are not violated. Besides, the proposed polling based packet scheduling policy can derive sufficient conditions such that all accepted streams satisfy their delay constraints to provide hard QoS guarantees for every single MSDU in IEEE 802.11e WLANs. Compared with previous works, our design is simpler and provides better performance over a wider range of scenarios. Furthermore, the proposed scheme not only can guarantee the delay constrains, but also it is compatible with the IEEE 802.11e HCCA method.

An analytical analysis is carried out to study the average queueing delay estimation of proposed scheme under equilibrium state. In addition to the analytical analysis, we have also carried out comprehensive simulations implemented by network simulator NS2 to evaluate the performance of the proposed scheme. The results confirmed that the proposed scheme not only guarantee delay constrains but also achieves lower packet dropping probability, lower average packet access delay, and higher system throughput compare to the reference design in HCCA method.

Finally, we will take a look at emerging technologies that are on horizon and provide some thoughts on how these technologies might address particular problems that may exist today and tomorrow.

The remainder of this chapter is organized as follows. Section 2 gives an overview of IEEE 802.11e HCCA method along with the reference scheduler design. Section 3 discusses the related QoS studies for transmitting multimedia traffic over HCCA method. Section 4 introduces the proposed scheme. Mean queueing delay analysis of the proposed scheme is presented in Section 5. Simulation and experimental results are given in Section 6, followed by Section 7 which draws conclusions of this chapter.

THE REFERENCE DESIGN IN IEEE 802.11E HCCA METHOD

The IEEE 802.11e HCCA method, which improves over the Point Coordination Function (PCF) of legacy IEEE 802.11 MAC protocol, is a polling based mechanism where the access to the channel is arbitrated centrally. For HCCA mechanism, admission control is an important component for the QoS provision. This section presents the reference scheduler design in HCCA method.

IEEE 802.11e introduces the concept of Traffic Stream (TS) which can be thought of as a set of MSDU that has to be delivered conforming to a corresponding Traffic Specification (TSPEC). A TSPEC characterizes the traffic streams and its QoS requirements. A TSPEC negotiation takes place between a station with real-time traffic (QSTA) and the HC collocated with the QoS Access Point (QAP) before a TS can be served through the HCCA. The TSPEC parameters which are considered during the negotiation include nominal MSDU size in octets, mean data rate in bps, Maximum Service Interval (MaxSI) which is the maximum interval in micro seconds between two successive polls for the stream, minimum PHY rate, and delay bound. Once the TSPEC negotiation is successful, the TS is admitted and offered Transmission Opportunities (TXOPs) by the HC in each polling cycle. The length of the TXOPs offered to the stream is decided through a scheduling scheme. The periods in which the HC has the exclusive control of the channel for

Figure 1. Timing diagram of the reference HCCA access method

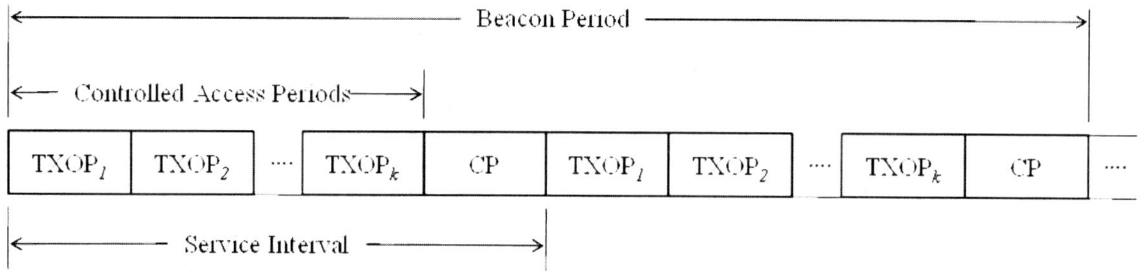

data transmission are called Controlled Access Periods (CAPs). The HC polls the QSTAs during these CAPs and offers TXOPs to the admitted streams.

A reference admission control algorithm was developed in the IEEE 802.11e draft. Its admission control is based on a simple scheduler that uses the mandatory set of TSPEC parameters to generate a schedule. In particular, when a new stream requests for admission, the admission control process performs the following tasks.

1) The admission control unit calculates the number of MSDUs that arrive at the mean data rate during the scheduled service interval SI, ς,as

$$N_i = \left\lceil \frac{\varsigma \times \gamma_i}{\omega} \right\rceil$$

where γ_i is the mean data rate and ω is the nominal MSDU size.

2) For stream *i*, the $TXOP_i$ is calculated as

$$TXOP_i = max(\frac{N_i \times \omega}{\phi} + X, \frac{\omega^*}{\phi} + X)$$

where ϕ is the minimum physical transmission rate, ω^* is the maximum size of an MSDU, and *X* denotes the overhead in time unit.

3) Assuming that there are *k* admitted streams, a new stream $k+1$ is accepted if it satisfies the following criterion:

$$\frac{\text{TXOP}_{k+1}}{\varsigma} + \sum_{i=1}^{k} \frac{\text{TXOP}_i}{\varsigma} \leq \frac{T_{CAP}}{T_{CAP} + T_{CP}}$$

where T_{CAP} is the duration of Controlled Access Period and T_{CP} is the duration of Contention Period (CP). Figure 1 shows the channel timing diagram encompassing a beacon period in an HCCA enabled IEEE 802.11e WLAN with the reference HCCA scheduler operated in Round-Robin (RR) fashion, where *k* in the figure denotes the number of scheduled streams.

BACKGROUND

The need for providing QoS for multimedia traffic in 802.11e WLANs has been driving research activities and standardization efforts for some time. In the past few years, we have seen many research works focusing on providing QoS for real-time traffic in WLANs (David, Jose, Daniel & Javier, 2007), (Stefan, Sunghyun, Gudio, Ole

& Bernhard, 2003), (Chen, Zhai, Tian & Fang, 2006) (Deng & Yen, 2005). For example, a media optimization network architecture (MONA) is proposed in (Koga, Kashihara, Fukida, Lida & Oie, 2006) to improve voice performance. Since voice communication is bidirectional, the authors improved downlink voice throughput from access point to individual wireless stations. In (Wang, Liew & Li, 2005), the authors have used black-burst contention for high priority access category. A voice station waits for designated arbitrary inter frame space period when the channel is idle and then sends a sequence of energy pulses called black burst. The station with longest burst will send the voice packet. In (Park, Kim, Choi & So, 2007), the authors proposed a fair QoS agent (FQA) to simultaneously provide per-class QoS enhancement and per-station fair channel sharing in WLANs.

The concept of PIFS waiting period is introduced at the access point for sending voice packets to wireless nodes instead of regular backoff process as stipulated by EDCA protocol. Handover is performed based on layer 2 retransmissions. For similar reasons, multiplexing and multicasting (MM) scheme is proposed in (Wang, Jian & Zhuang, 2006). The scheme combines voice packets destined to several wireless stations into a single large packet at the central access point. The large multiplexed packet is then multicasted on to the channel so that all wireless stations can receive it. In another work, the performance of EDCA protocol is compared with European standard HIPERLAN/1 media access control protocol namely "Elimination Yield Non-Preemptive Medium Access (EY-NPMA)" (Dimitriadis & Pavlidou, 2004).

Most of the above approaches are based on priorities, giving access preference to those stations that have been assigned higher priority. However, service differentiation mechanism is helpful in providing better QoS for multimedia data traffic under low to medium traffic load conditions, but service differentiation mechanism does not perform well under high traffic load conditions (Hua, Ming, Imrich & Prabhakaran, 2004). Besides, with the contention-based Enhanced Distributed Channel Access (EDCA) mechanism, bandwidth provisioning is almost impossible, leading to only soft QoS guarantee.

In HCCA method, the QAP polls the stations and allocates TXOPs to a QSTA. However, there are some drawbacks concerning the operation of HCCA. To begin with, some valuable bandwidth is spent because of the polling packets sent to the stations. Besides, the use of acknowledgments is bandwidth costly, too. Furthermore, extra TXOPs may be allocated to a station (Qinglin & Tsang, 2008). Finally, the reference design in HCCA method is not suitable for VBR traffic, where the instantaneous and fluctuating data rate is usually quite different from the corresponding mean values (Mihaela, Yiannis & Zhiping, 2006).

To remedy these problems, some researchers proposed novel methods that use additional information such as the queue size and the transmission status of stations to determine how long and how often should TXOPs be allocated. In the method proposed in (Antonio, Mario & Mario, 2003), the AP allocates the TXOPs with different lengths to stations and sends the polling messages to the stations with service intervals of varying lengths. The FHCF (Efficient HCF) mechanism presented in (Ansel, Ni & Turletti, 2006) uses the statistics of estimated queue size to evaluate the proper queue size allocated in the next polling procedure. In (Deyun, Jianfei & King Ngi, 2005), the authors proposed an admission control algorithm named Physical Rate-Based Admission Control (PRBAC), which considers physical rate variance due to station mobility and wireless channel characteristics. The basic idea of PRBAC is to use the long-term average physical rates for admission control, and at the same time use the instantaneous physical rates to distribute TXOPs to individual stations. In (Deyun, Jianfei

Table 1. Notations and variables used in analytical analysis

Notations and variables	Meaning and explanation
m	number of CBR sessions
n	number of VBR sessions
γ	mean data rate
ω	nominal MSDU size
β	maximum burstiness of VBR traffic
d	maximum tolerant delay of multimedia traffic
ϕ	minimum physical transmission rate
ς	maximum service interval
η	maximum delay constrain for admitted inactivated VBR source
T_{SIFS}	time needed for SIFS
T_{poll}	time needed to poll a QSTA
T_{ACK}	time needed for transmitting ACK frame
N_Q^i	average number of packets in queue for i^{th} multimedia session
W_i	packet average waiting time in queue for i^{th} multimedia session
ρ_i	system utilization factor for i^{th} multimedia session

& Chen, 2008), the authors proposed a Rate Variance envelop based Admission Control (RVAC) for VBR traffic over HCCA. The RVAC uses the dual token bucket shaper to guarantee that each traffic flow conforms to a certain QoS specification. In (Mohammad, Ekram & Vijay, 2008), the authors proposed a new scheduling scheme, namely, the Prediction and Optimization-based HCCA (PRO-HCCA), based on the insights gained from the queueing analysis since queue management is one of the fundamental mechanisms for providing the underlying QoS and one of the fundamental mechanisms for differentiating service levels.

THE PROPOSED SCHEME

In this section, we present the proposed scheme in detail. Our method involves two basic components: (a) admission control scheme for access decision (b) packet scheduling policy for HCCA method.

Before we start to discuss the issues of interest, important notations and variables are defined in Table 1, and they will be used throughout this chapter.

Admission Control Scheme for Access Decision

Admission control is a crucial part of any QoS implementation. Within the context of the IETF Integrated Services architecture, admission control determines whether a network node has sufficiently available resources to supply the requested QoS (Paul & Geoff, 1998). In the proposed scheme, if the originator of the traffic stream requests QoS levels from the network using parameters within the TSPEC specification, the admission control module in the QAP checks the requester's TSPEC to determine whether the resources are available under the current network status. That is, upon arrival of a new flow, the admission control scheme decides to grant or deny admission permit. The decision is made based on the current information of the system and the analytical model for estimating of the expected delay.

Admission Control for CBR Traffic

Assume there are m CBR sources, and we denote (γ_i, d_i) as the traffic parameters of the *ith* CBR source.

Let $t_p = 2T_{SIFS} + T_{poll} + \frac{\omega}{\phi} + T_{ACK}$ and $d_i^* = t_p + \sum_{k=1}^{i-1}\left\lceil\frac{\gamma_k}{\gamma_i}\right\rceil \times t_p$, if $d_i^* < \omega/\gamma_i$ and $d_i^* \le d_i$ for all i =1, 2…m, then all the packets generated by the *ith* CBR source meet their delay constraints.

Proof

Suppose that the first token generated from the *ith* CBR source has the maximum delay time $\overline{d_i}$. We want to prove that $\overline{d_i} \le d_i$ for $1 \le i \le m$.

Considering the first CBR source, since the maximum delay is to wait the channel to be cleared, thus we have $\overline{d_1} \le t_p = d_1^* \le d_i$, which establishes the induction basis.

Suppose our induction hypotheses holds up to the *(i-1)st* CBR sources. Now we consider the *ith* CBR source. Suppose $\overline{d_i} > d_i^*$. Then it means that up to the time d_i^* the channel must be serving all the CBR sources from 1 to i-1. Since we assume $d_i^* < \frac{\omega}{\gamma_i}$, we have $(\sum_{k=1}^{i-1}\left\lceil\frac{\gamma_k}{\omega} \times d_i^*\right\rceil + 1) \times t_p < (\sum_{k=1}^{i-1}\left\lceil\frac{\gamma_k}{\gamma_i}\right\rceil + 1) \times t_p = d_i^*$. Hence, the claim itself is proved false, so opposite of the claim is proved true. This shows $\overline{d_i} \le d_i^* \le d_i$ (Q.E.D.)

Admission Control for VBR Traffic

Assume there are n VBR sources, and we denote (γ_j, β_j, d_j) as the traffic parameters of the *jth* VBR source.

Let $\overline{\beta_0} = (m+1) \times t_p$, $\overline{\gamma_0} = t_p \times \sum_{i=1}^{m}\frac{\gamma_i}{\omega}$

$\overline{\beta_j} = (\beta_j + 1) \times t_p$, $\overline{\gamma_j} = t_p \times \frac{\gamma_j}{\omega}$,

$d_1^* = \eta_1 + \frac{\overline{\beta_0} + \overline{\beta_1}}{1 - \overline{\gamma_0}}$,

$$d_j^* = \eta_j + \frac{\sum_{k=0}^{j}\overline{\beta_k} + t_p \times \sum_{k=1}^{j-1}\frac{\gamma_k}{\omega} \times d_k^*}{1 - \sum_{k=0}^{j-1}\overline{\gamma_k}},$$

j=2,3…n.

If $\sum_{k=0}^{n}\overline{\gamma_k} \le 1$ and $d_j^* \le d_j$ for all j =1, 2…n, then all the packets generated by the *jth* VBR source meet their delay constraints.

Proof

Consider a nonnegative, left limited, and right continuous stochastic process $A \equiv \{a(t), t \ge 0\}$.

Let $A(t_1,t_2) = \int_{t_1}^{t_2} a(t)dt$. We say that A is (β, r)-upper constrained if $A(s, t+s) \le rt+\beta$ for all s, $t\ge 0$. Similarly, A is (β, r)-lower constrained if $A(s, t+s) \ge rt+\beta$ for all s, $t\ge 0$. Since the number of departures in (t_1 , t_2] from a (β, r)-leaky bucket is bounded above by $\beta + \lceil r(t_2 - t_1)\rceil$, the departure process from a (β, r)-leaky bucket is (β+1, r)-upper constrained.

Now consider the first VBR source. Let $C1\equiv\{c1(t), t\ge 0\}$ be the stochastic process that denotes the available bandwidth to the first VBR source at time t. If the channel is available to the first VBR source at time t, then $c1(t)$=1. Otherwise, $c1(t)$=0.

As mentioned above, the maximum number of packets from the m CBR sources that can be served in (t_1 , t_2] is at most $\sum_{i=1}^{m}\left\lceil\frac{\gamma_i}{\omega}\times(t_2-t_1)\right\rceil$. Hence, the bandwidth that is available to the first VBR source in (t_1 , t_2] is at least $t_2 - t_1 - t_p\times\{1+\sum_{i=1}^{m}\left\lceil\frac{\gamma_i}{\omega}\times(t_2-t_1)+1\right\rceil\}$. Thus, $C_1(t_1,t_2)\ge(1-t_p\times\sum_{i=1}^{m}\frac{\gamma_i}{\omega})\times(t_2-t_1)-t_p\times(m+1)$. That is, C_1 is ($\overline{\beta_0}$, $1-\overline{\gamma_0}$)-lower constrained.

Let $A1\equiv\{a1(t), t\ge 0\}$ be the amount of workload of VBR source 1 that arrives at the channel at time t. Since the number of departures in ($t1$, $t2$] from the first VBR traffic is (β_1 +1, γ_1)-upper constrained. We have $A_1(t_1,t_2)\le t_p[\gamma_1(t_2-t_1)+\beta_1+1]$. This shows $A1$ is ($\overline{\beta_0}$, $1-\overline{\gamma_0}$)-upper constrained.

Consider an instant after the last packet was sent (but not the end-of-file packet) by the first VBR source. Mark the instant as time 0. By letting $q1(t)$ be the amount of backlogged workload from the first VBR source in the channel at time t, we have $q1(t)$=0. Since the next token for the first VBR source will be generated after time $\eta 1$ at the latest. We have $q1(t)=A1(0,t)-C1(\eta 1,t)$. Note that the delay for an arrival at time t is bounded by the amount of time needed to deplete $q1(t)$, and the time to deplete $q1(t)$ is bounded by $\inf\{d\ge 0: A1(0, t)-C1(\eta 1, t+d)\le 0\}$. Maximizing over t, we have the following upper bound for the maximum delay: $d_1^* = \sup_t \inf\{d\ge 0 : A_1(0,t) - C_1(\eta_1, t+d)\le 0\}$.

Since $\sum_{k=0}^{n}\overline{\gamma_k}\le 1$, we have $\overline{r_0} + t_p\times\frac{\gamma_1}{\omega}\le 1$. Applying the upper constraint for A_1 and the lower constraint for C_1, we have $d_1^* = \eta_1 + \frac{\overline{\beta_0}+\overline{\beta_1}}{1-\overline{\gamma_0}}$. This completes the argument for the first VBR source.

Since the maximum delay of the kth VBR source is bounded above by d_k^*, k=1,2...j-1, the number of packets from the kth source that can be served in $(t_1,t_2]$ is bounded above by $\beta_k + \left\lceil\frac{\gamma_k}{\omega}\times(t_2-t_1+d_k^*)\right\rceil$. Hence, the amount of workload from the kth source that can be served in $(t_1,t_2]$ is bounded above by $t_p\times[\frac{\gamma_k}{\omega}(t_2-t_1)+\beta_k+1+\frac{\gamma_k}{\omega}\times d_k^*]$. Parallel to the argument for the first VBR source, the maximum delay of the jth VBR source is bounded above by:

$$\eta_j + \frac{\sum_{k=0}^{j}\overline{\beta_k} + t_p\times\sum_{k=1}^{j-1}\frac{\gamma_k}{\omega}\times d_k^*}{1-\sum_{k=0}^{j-1}\overline{\gamma_k}}.$$

Finally, we still need to engineer η_i to complete this scheme. In order to maximize the bandwidth utilization, one should have as large as possible. The largest η_i can be obtained by solving $d_j^* = d_j$

(Q.E.D.)

Figure 2. System diagram for the proposed packet scheduling policy

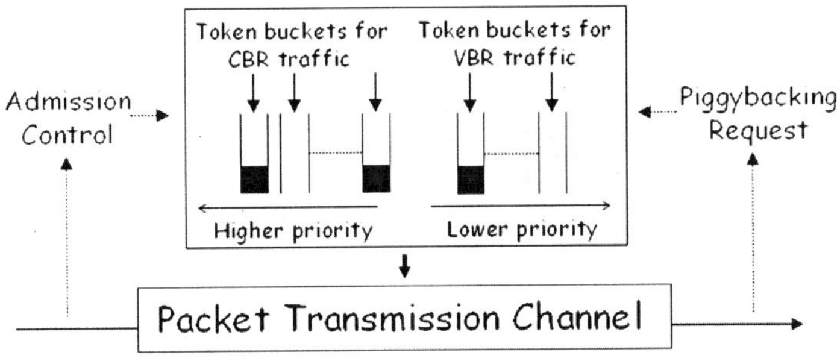

Packet Scheduling Policy for HCCA Method

In the proposed scheme, we use token buckets to provide traffic shaping and ingress rate control. The token buckets hold tokens, with each token representing the capability to send a packet. The QAP implements a token bucket for each multimedia stream, and the stream with smaller tolerated delay constraint will get higher priority. The token bucket contains tokens. For CBR traffic, each of which can represent the capability to send a packet. For VBR traffic, the token bucket allows bursty traffic to continue transmitting while there is a token in the bucket thereby accommodating traffic flows with bursty characteristics. We depict the packet scheduling policy in Figure 2.

Since QAP acts as a bandwidth coordinator, it will perform the following algorithm once it has the control of the medium.

```
Function Packet Scheduling Policy
do {
repeat Contention Period
until any token found in token bucket
repeat Controlled Access Period
scan token buckets of CBR streams according to present priority order
if a token found in CBR flow
ithen {
  remove one token from this token bucket
  poll corresponding CBR source and transmit a packet
  if not the last packet then {
     generate next token after
ω/γ_i − t_p sec }else {remove this token bucket}}
else {
scan token bucket of VBR streams according to present priority order
if a token found in VBR flow
jthen {
  poll corresponding VBR source and transmit a packet
  if not the last packet thenif piggyback bit was not set then {
       remove the token from this token bucket
       generate next token after η sec }else remove this token bucket }}until no token found in token buffer
} while(1)
```

The QAP first scans the token buckets of CBR sources according to the preset priority order. If a token is found, it removes one from this token bucket and polls this CBR terminal. On receiving a poll the station transmits its packet after a SIFS interval. Then, the QAP generates the next token for this CBR source after $\frac{\omega}{\gamma_i} - t_p$ second if this is not the last packet.

If no tokens are found in the token buckets of CBR sources, the QAP continues to scan the token buckets for VBR sources according to the preset priority order. If a token is found, it polls this VBR terminal, and it will not remove the token if the piggyback was set while this VBR source transmit it packet. If the piggyback was not set and it is not the last packet (End-of-File) either, the QAP removes the token, and then generates the next token for this VBR source after η seconds.

If there is no token found in all token buckets, the QAP will not know which, if any, of the stations have packets to transmit, then, it can end the Controlled Access Period, and, for assuring the time constraint of admitted real-time traffic, the QAP shall announce the beginning of the next Controlled Access Period by consistently observing the token buckets among its polling list.

MEAN QUEUEING DELAY ANALYSIS

A packet sent from an upper-layer protocol over MAC layer will first be placed in a transmission queue. The packet delay caused by waiting here is referred to as the queueing delay. In this section, we focus on the evaluation of average queueing delay for the proposed scheme.

Denote

N_Q^i = Average number of packets in queue for i^{th} multimedia session

W_i = Packet average waiting time in the ready-to-transmit buffer for i^{th} multimedia session.

$\rho_i = \frac{\gamma_i}{\phi}$ = System utilization factor for i^{th} multimedia session

Firstly, we assume that the overall system utilization is less than 1, that is $\rho_1 + \rho_2 + ... + \rho_n < 1$

Considering the first multimedia flow, i.e. $i = 1$, its average waiting time in the ready-to-transmit buffer is $W_1 = {t_p}/{2} + N_Q^1 \cdot t_p$. Eliminating N_Q^1 from this equation using Little's Theorem, $N_Q^1 = r_1 \cdot W_1$, we obtain $W_1 = {t_p}/{2} + \rho_1 \cdot W_1$, and finally we have $W_1 = \frac{t_p}{2(1-\rho_1)} = \frac{t_p}{2(1-r_1 \cdot t_p)}$.

For the second multimedia flow, i.e. i=2, we have a similar expression for W_2 except that we have to count the additional delay due to packets of higher priority that arrive while a packet is waiting in ready-to-transmit buffer. This is the meaning of the last term in the following formula:

$$W_2 = {t_p}/{2} + N_Q^1 \cdot t_p + N_Q^2 \cdot t_p + r_1 W_2 \cdot t_p$$

Using Little's Theorem, we obtain:

$$W_2 = {t_p}/{2} + \rho_1 \cdot W_1 + \rho_2 \cdot W_2 + \rho_1 \cdot W_2$$

Which yields $W_2 = \frac{{t_p}/{2} + \rho_1 \cdot W_1}{1 - \rho_1 - \rho_2}$. Using the expression of W_1 obtained earlier, we have $W_2 = \frac{t_p}{2(1 - r_1 \cdot t_p)(1 - r_1 \cdot t_p - r_2 \cdot t_p)}$. Since the derivation is similar for all priority classes, the formula for W_i is $\frac{t_p}{2(1-\rho_1 - ... - \rho_{i-1})(1 - \rho_1 - ... - \rho_i)}$.

Hence, the average queueing delay of each packet for i^{th} multimedia flow is:

$$\frac{t_p}{2(1 - r_1 \cdot t_p - ... - r_{i-1} \cdot t_p)(1 - r_1 \cdot t_p - ... - r_i \cdot t_p)}$$

PERFORMANCE EVALUATION

In this section, we evaluate the performance of the proposed scheme.

Simulation Environment

Our simulation model is built using the network simulator NS2 (The Network Simulator, n.d.). The model represents a BSS in the IEEE 802.11e standard WLANs with all stations in the BSS (Basic Service Set) capable of directly communicating with the remaining parties. In other words, the transmitting power used for each station is assumed to be high enough to cover a 250 meters transmission range. To focus on the access control issue and to reduce the complexity of the simulation, what follows are the basic assumptions in our simulation environment. First, the "hidden terminal" and the "exposed terminal" problems are not addressed in the simulation model. Second, no stations operate in the "power-saving" mode. Third, no interference is considered from nearby BSSs. Finally, the transmitted propagation delay is 1 microsecond. This is a realistic assumption if the transmission distances between stations are of tens meters. The default parameter values used in the simulation are listed in Table 2. The values for the simulation parameters are chosen carefully in order to closely reflect the realistic scenarios as well as to make the simulation feasible and reasonable.

Simulation Results

In what follows, the performances of the proposed scheme and the reference design in HCCA are compared based on simulations. Our aim is to compare the QoS provided by these two protocols.

In reference design, first come first serve (FCFS) is adopted as the random access protocol, and the round-robin discipline is chosen as the scheduling policy for QAP in the CAPs. It is noteworthy that the duration of the CAPs are dynamically allocated according to the current channel and traffic status by the QAP using the

Table 2. Default parameter values used in the simulation

Parameter	Value
maximum data rate (WaveLAN DSSS)	11 Mb/s
time needed for each time slot	20 *us*
duration of short interframe space (SIFS)	10 *us*
duration of PCF interframe space (PIFS)	30 *us*
duration of DCF interframe space (DIFS)	50 *us*
ACK frame size	112 bits
MAC overhead	224 bits
duration of a PLCP preamble	144 *us*
duration of a PLCP header	48 *us*
propagation delay	1 *us*
minimum contention window size	31 slots
maximum backoff stages	5
simulation topology	250m × 250m
sustained traffic rate for multimedia traffic	64~256Kb/s

Figure 3. Packet delay variation measurement concerning high priority multimedia traffic for the proposed scheme

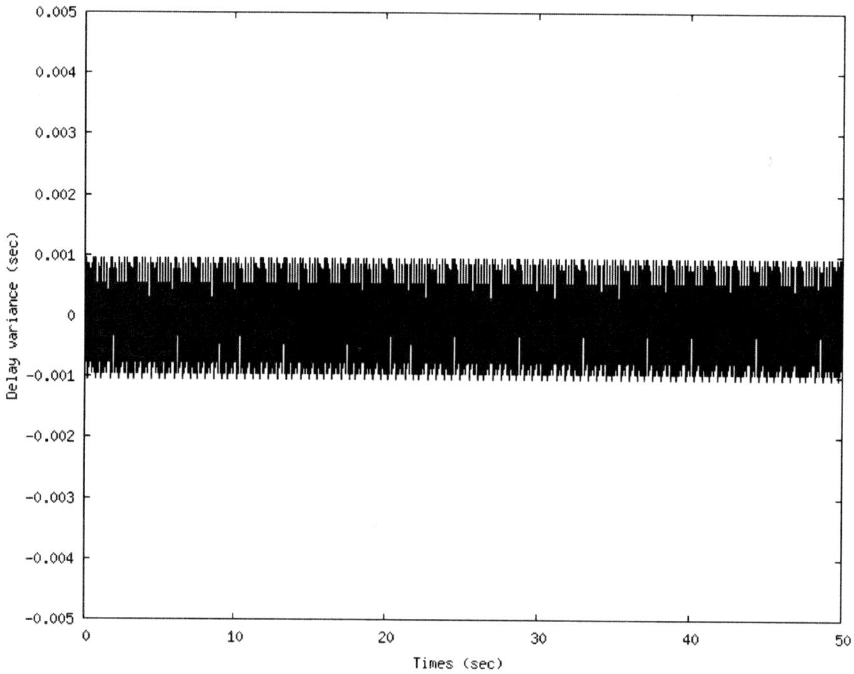

proposed packet scheduling policy in the proposed scheme.

Simulation results are shown below in the form of plots. Firstly, we measured the packet delay variation for the proposed scheme and reference design respectively. This metric is the major representative capability of the MAC protocol to efficiently provide QoS. In Figure 3 to Figure 6, we have plotted the results regarding packet delay variation. Figure 3 and Figure 4 report this major performance metric, gathered in a 50 seconds experiment, for high priority streams. It is obvious that the proposed scheme exhibits much lower delay variation than the reference design, as the delay variation of the packets is always kept below the required delay constraint. This result shows that the proposed scheme can efficiently support multimedia traffic by providing significantly low delay variations.

Figure 5 and Figure 6 report the packet delay variation, gathered in a 50 seconds experiment, for low priority multimedia sessions. As shown in the figures, the variation for the proposed scheme seems a slightly larger compared to the reference design. This phenomenon may be attributed to that we allow the high priority traffic to use bandwidth exclusively with priority over low priority traffic for the proposed scheme, and in the reference design some of the streams are simply dropped since these streams exceed their delay constraints. However, in the proposed scheme the packet variations for the accepted streams are found to be well below the corresponding QoS limits.

Figure 7 illustrates the dropping probability of accepted multimedia sessions for the proposed scheme and reference design respectively. It should be noticed that in both scenarios, packets will be dropped if they exceed their maximum tolerated delay constraint. As illustrated in the figure, we can see that although there is not much difference in the values of the performance measures when load is light, the proposed scheme provides significantly

Figure 4. Packet delay variation measurement concerning high priority multimedia traffic for the reference design

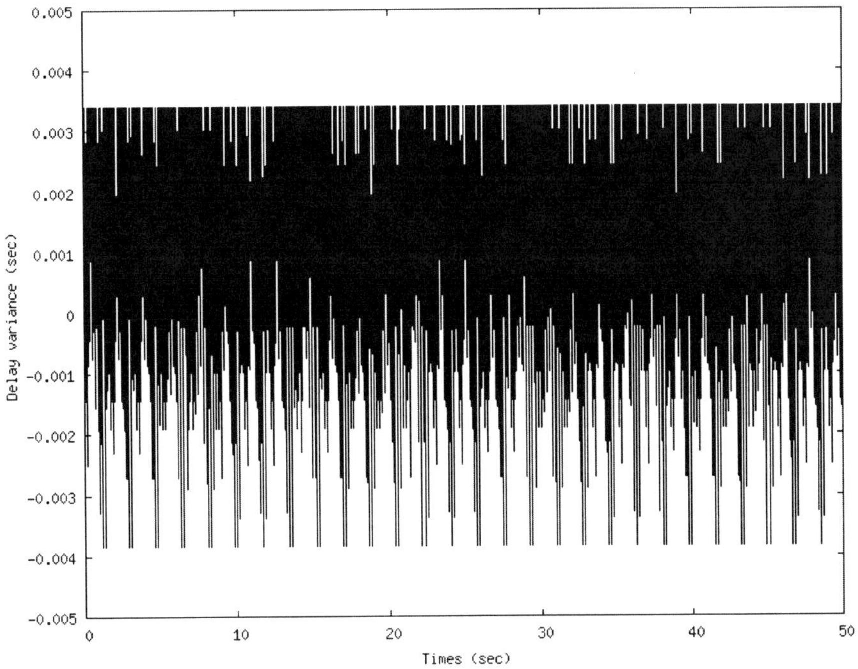

Figure 5. Packet delay variation measurement concerning low priority multimedia traffic for the proposed scheme

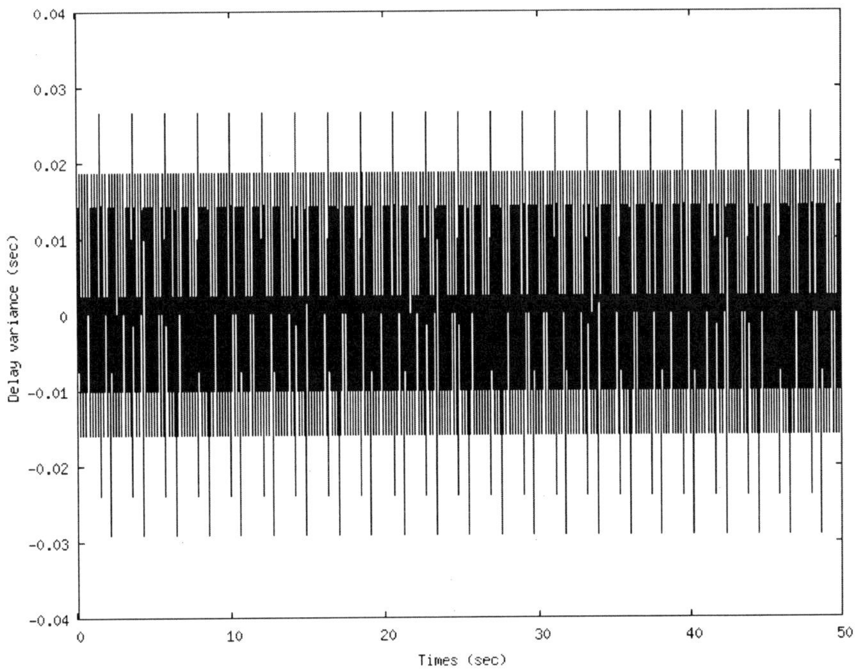

Figure 6. Packet delay variation measurement concerning low priority multimedia traffic for the reference design

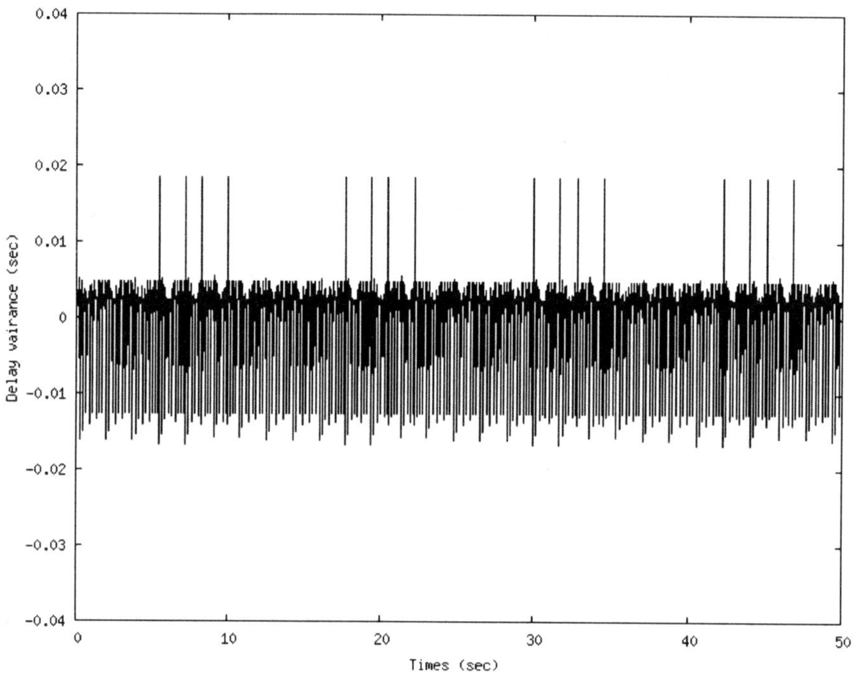

Figure 7. Dropping probability versus number of flows for multimedia traffic

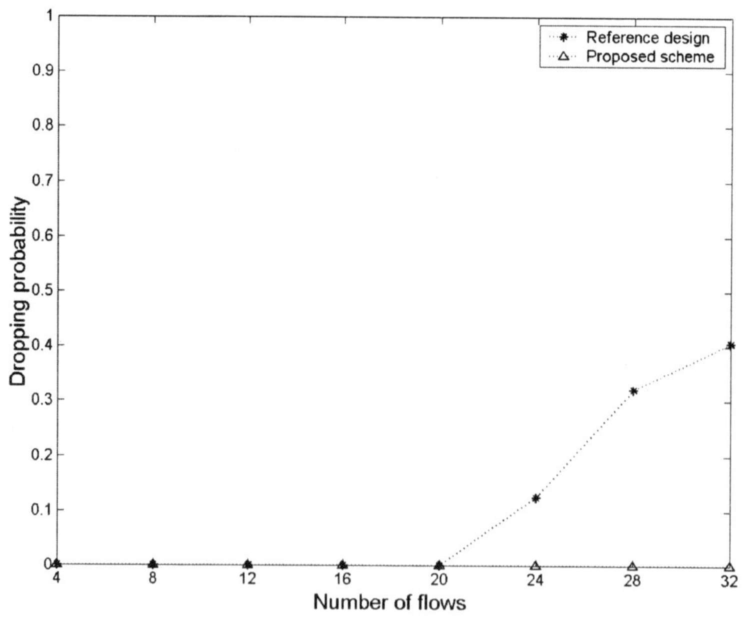

better performance than the reference design at heavy load. In other words, the average access delay of the proposed scheme remains zero when the offered load is high; in contrast, the reference design shows a sharp rise as the load increases since it lacks appropriate admission control scheme and packet scheduling policy. The simulation results imply that the proposed scheme is appropriate for transmitting multimedia traffic.

Figure 8 compares the average access delays of multimedia traffic from the proposed scheme and the reference design. We can see that although there is not much difference in the values of the performance measures when traffic load is low. The proposed scheme provides a much better performance than the reference design as we increase the traffic load. Since we allow the high priority multimedia traffic to use bandwidth exclusively with priority over other traffic, the average access delay of the proposed scheme will be kept lower than the reference design.

Figure 9 depicts the achievable throughput as the traffic load increases under both scenarios. As shown in the figure, the throughput improvement can be as much as 15% in congested wireless environments. It reveals that our proposed scheme could reduce the dropping probability without sacrificing the overall system performance, which means that our proposed scheme gets up to 15% more throughput from the channel while still meeting the QoS constraints.

FUTURE RESEARCH DIRECTIONS

There are several interesting problems arise from this research. First, the proposed scheme provides hard QoS for real-time traffic. This strategy might work well in the environment where resource is abundant. Unfortunately, wireless transmission links are noisy and highly unreliable. Path loss, terminal noise, fading, and interference may cause significant bit errors. Hence, to improve efficiency,

Figure 8. Average access delay versus number of flows for multimedia traffic

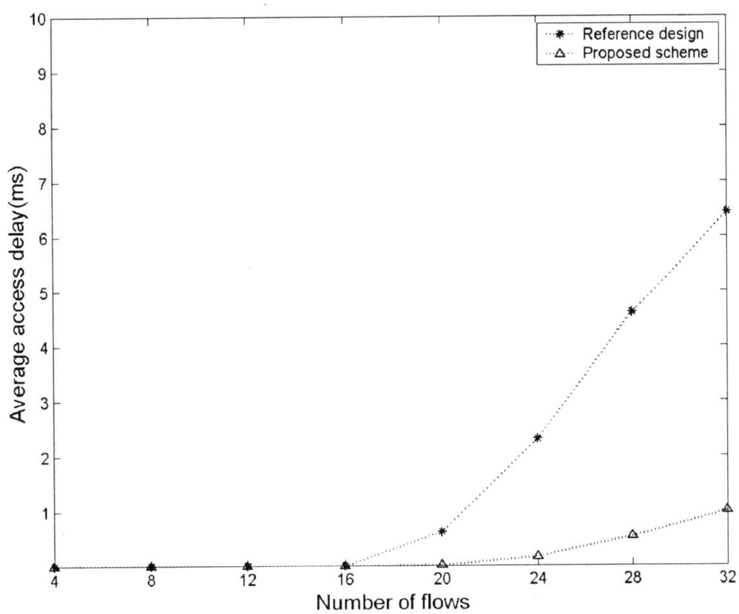

Figure 9. Achievable throughput under different traffic load

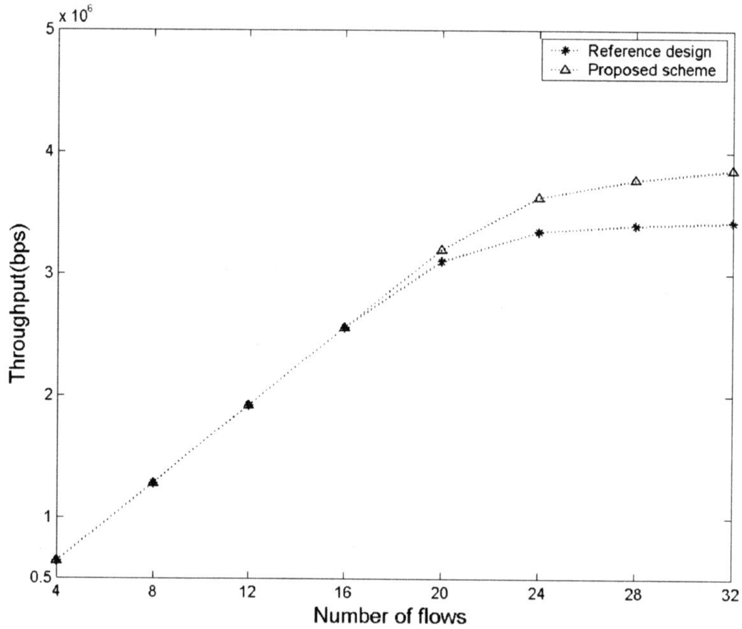

researchers could consider soft QoS approaches. Besides, whether retransmission for a corrupted real-time packet is worthwhile is another problem that needs further investigation. Second, the proposed admission control scheme might require extensive computing if the number of real-time sources is large. This could be a serious problem for handoff or roaming mobile station. Finally, providing QoS in wireless ad hoc networks is another interesting task for future research due to node mobility, distributed channel access, and fading radio signal effects.

CONCLUSION

Currently, many wireless applications and devices are emerging, and this trend is expected to continue in the foreseeable future. Many problems remain, require further intensive investigation and need to be solved. So far ongoing efforts to provide perfect solutions have illustrated that attempts to solve all possible problems result in technologies that are far too complex, have poor scaling properties, or simply do not integrate well into the diversity of the Internet. In this chapter, we have proposed a strictly guaranteed QoS for CBR and VBR traffic in IEEE 802.11e wireless LANs. The performance of proposed scheme is analyzed and compared with original IEEE 802.11e protocol under standard MAC and physical layer parameters proposed by IEEE standard for wireless LANs. Through extensive simulations, important performance metrics such as average access delay, achievable throughput, dropping probability, and delay variation performance are thoroughly investigated. We have demonstrated a satisfactory performance of our proposed scheme in a quantitative way. The proposed scheme is significant in that it paves the way for providing hard QoS guaranteed service in the future wireless networks.

REFERENCES

Ansel, P., Ni, Q., & Turletti, T. (2006). A simple and efficient scheduling scheme for IEEE 802.11e wireless LAN. *Journal of Mobile Networks and Applications*, 391-403.

Antonio, G., Mario, M., & Mario, N. (2003). A scheduling algorithm for QoS support in IEEE802.11 networks. *IEEE Wireless Communications*, 36-43.

Chen, X., Zhai, H., Tian, X., & Fang, Y. (2006). Supporting QoS in IEEE 802.11e Wireless LANs. *IEEE Transaction on Wireless Communications*, 2217-2227.

David, G., Jose F., M., Daniel, C., & Javier, G. (2007). Policy-based channel access mechanism selection for QoS provision in IEEE 802.11e. *IEEE Vehicular Technology Magazine*, 29-34.

Deng, D. J., & Yen, H. C. (2005). Quality-of-Service Provision System for Multimedia Transmission in IEEE 802.11 Wireless LANs. *IEEE Journal on Selected Areas in Communications*, 23*(6)*, 1240–1252. doi:10.1109/JSAC.2005.845632

Deyun, G., Jianfei, C., & Chen, C. W. (2008). Admission Control Based on Rate-Variance Envelop for VBR Traffic Over IEEE 802.11e HCCA WLANs. *IEEE Transactions on Vehicular Technology*, 1778–1788.

Deyun, G., Jianfei, C., & King Ngi, N. (2005). Admission control in IEEE 802.11e wireless LANs. *IEEE Network*, 19*(4)*, 6–13. doi:10.1109/MNET.2005.1470677

Dimitriadis, G., & Pavlidou, F. N. (2004). Comparative performance evaluation of EDCF and EY-NPMA protocols. *IEEE Communications Letters*, *8(1)*, 42–44. doi:10.1109/LCOMM.2003.822501

IEEE Draft Std. 802.11e/D8.0. (2004). *Medium Access Control (MAC) Quality of Service (QoS) Enhancements*.

Hua, Z., Ming, L., Imrich, C., & Prabhakaran, B. (2004). A survey of quality of service in IEEE 802.11 networks. *IEEE Wireless Communications*, 6-14.

Koga, H., Kashihara, S., Fukuda, Y., Lida, K., & Oie, Y. (2006). A quality-aware VoWLAN architecture and its quantitative evaluations. *IEEE Wireless Communications*, 52-59.

Mihaela, V. D. S., Yiannis, A., & Zhiping, H. (2006). Optimized scalable video streaming over IEEE 802.11 a/e HCCA wireless networks under delay constraints. *IEEE Transactions on Mobile Computing*, 755–768.

Mohammad, M. R., Ekram, H., & Vijay, K. B. (2008). Controlled Channel Access Scheduling for Guaranteed QoS in 802.11e-Based WLANs. *IEEE Transactions on Wireless Communications*, 1287–1297.

Park, E. C., Kim, D. Y., Choi, C. H., & So, J. M. (2007). Improving Quality of Service and Assuring Fairness in WLAN Access Networks. *IEEE Transactions on Mobile Computing*, 6(4), 337–350. doi:10.1109/TMC.2007.53

Paul, F., & Geoff, H. (1998) Quality of Service. New York: John Wiley & Sons, Inc.

Qinglin, Z., & Tsang, H. K. (2008). An Equal-Spacing-Based Design for QoS Guarantee in IEEE 802.11e HCCA Wireless Networks. *IEEE Transactions on Mobile Computing*, 1474–1490.

Stefan, M., Sunghyun, C., & Gudio, R. H., Ole, K., & Bernhard, W. (2003). Analysis of IEEE 802.11e for QoS support in wireless LANs. *IEEE Wireless Communications*, 40-50. *The Network Simulator ns-2*. (n.d.). Retrieved from http://www.isi.edu/nsnam/ns/

Wang, P., Jiang, H. & Zhuang, W. (2006) IEEE 802.11e enhancements for voice service. *IEEE Journal for Wireless Communications*, 30-35.

Wang, W., Liew, S. C., & Li, V. O. K. (2005). Solutions to performance problems in VoIP over a 802.11 wireless LAN. *IEEE Transactions on Vehicular Technology*, 54(1), 366–384. doi:10.1109/TVT.2004.838890

Wireless LAN Medium Access Control (MAC) and Physical Layer(PHY) specifications. (1999). ANSI/IEEE Std 802.11: 1999 (E) Part 11, ISO/IEC 8802-11.

Chapter 4
Quality of Service in Heterogeneous Traffic Wireless Systems

Nizar Zorba
University of Jordan, Jordan

Christos V. Verikoukis
Telecommunications Technological Centre of Catalonia (CTTC), Spain

ABSTRACT

The use of real-time delay-sensitive applications in wireless systems has significantly increased during the last years. Consequently, the demand to guarantee certain Quality of Service (QoS) is a challenging issue for the system's designers. On the other hand, the use of multiple antennas has already been included in several commercial standards, where the multibeam opportunistic transmission beamforming strategies have been proposed to improve the performance of the wireless systems. A cross-layer based dynamically tuned queue length scheduler is presented in this chapter, for the Downlink of multiuser and multiantenna WLAN systems with heterogeneous traffic requirements. An opportunistic scheduling algorithm is applied, while users from higher priority traffic classes are prioritized. A trade-off between the throughput maximization of the system and the guarantee of the QoS requirements is obtained. Therefore the length of the queue is dynamically adjusted to select the appropriate conditions based on the operator requirements.

INTRODUCTION

The demand for using in-home Wireless Local Area Networks (WLANs) to support real-time delay-sensitive applications such as voice, video streaming or online-gaming has been remarkably growing during the last years. However, current IEEE 802.11 WLAN systems fail to fulfil the strict Quality of Service (QoS) requirements in terms of maximum allowed delay and/or delay jitter for such applications. The fact that the wireless environments are characterized by a harsh scenario for communications adds certain difficulties to guarantee QoS in WLAN based systems, where the wireless channel suffers from multiple undesired effects such as deep fades and multipath that produce errors to the original information. Therefore, providing QoS by

DOI: 10.4018/978-1-61520-771-8.ch004

using the scarce resources in the home wireless medium is a challenging aspect for such system objective.

Different notions of QoS are available at different communication layers (Zorba & Perez-Neira, 2007). At the physical layer, QoS means an acceptable signal strength level and/or Bit Error Rate at the receiver, while at the Data Link Control (DLC) or higher layers, QoS is usually expressed in terms of minimum guaranteed throughput, maximum allowed delay and/or delay jitter. The fulfilment of QoS requirements depends on procedures that follow each layer. At the DLC layer, QoS guarantees can be provided by appropriate scheduling and resource allocation algorithms, while at the physical layer, adaptation of transmission power, modulation level or symbol rate are employed to maintain the link quality.

The consideration of the physical layer transmission characteristics from the higher layers can significantly improve the efficiency of the wireless systems. The vertical coupling among layers is well-known as Cross-Layer (Shakkottai, Rappaport & Karlsson, 2003). Cross-layer between the physical layer and the higher layers seems to be unavoidable in wireless environments in order to exploit the physical layer instantaneous conditions. Such kind of schemes is needed to guarantee the QoS requirements in heterogeneous traffic systems, where the network includes users with different applications and different QoS restrictions. Cross-Layer further advantages can include improvements in terms of link throughput, reduction of the network latency, energy savings in the mobile nodes or minimization of transmitted power (Shakkottai, Rappaport & Karlsson, 2003).

One of the system resources that can be employed to improve the system performance in terms of both rate and QoS is the spatial resource. The Multiple-Input-Multiple-Output (MIMO) technology in multiuser scenarios shows very interesting results as several users can be simultaneously serviced within the same frequency, time and code. Its employment has been already standardized in IEEE 802.11n and IEEE 802.16e, while it is expected to be part of 4th Generation Long Term Evolution (LTE) Standard. Among all the techniques within the MIMO technology, the Multibeam Opportunistic Beamforming (MOB) strategy has been suggested in (Sharif & Hassibi, 2005) to boost the wireless link capabilities, showing high performance, low complexity design and only partial channel information is rçequired at the transmitter side. MOB can be operated and adopted to fulfil the QoS requirements demanded by the users for their correct operation (Zorba & Perez-Neira, 2007).

An interesting remark concerning the QoS compliance in commercial wireless systems refers to the outage concept (Chalise & Czylwik, 2004), where due to the wireless channel characteristics, the 100% satisfaction of the strict QoS demands is impossible, for what is known as outage in the QoS requirements (Chalise & Czylwik, 2004). The notion of outage is widely used in cellular systems where several commercial systems (e.g. GSM and WCDMA) allow for 2-5% outage. Therefore, the extension of the QoS outage to WLAN based systems, when running delay-sensitive applications, seems to be the most tractable approach to asset their efficiency.

Taking into account the aforementioned concepts, the main contribution of this chapter is to propose a Dynamic Queue Length in the Data Link Control Layer, in order to guarantee certain QoS, in the Downlink of multiuser and multiantenna WLAN systems with heterogeneous traffic. Therefore, the proposed solution considers both the physical and application layers characteristics of the system. To be more precise, the length of the queue depends on the QoS system requirements, in terms of the system throughput and the maximum allowed delay (and jitter) of the delay-sensitive applications, where some outage is considered in the QoS requirements of these applications.

Chapter Organization

As a summary, the contributions of this chapter are in the field of Dynamic queues management under QoS demands as follows

- The paper tackles a multiantenna scenario and chooses the MOB scheme for its transmission strategy.
- Through the use of the outage concept, this chapter is able to formalize the service distribution characteristics of the MOB scheme; allowing to obtain the minimum rate and maximum scheduling delay in closed form expressions.
- An approach to obtain the opportunistic multiuser gain, while providing the system QoS constraints in terms of minimum guaranteed rate and maximum allowed delay, is presented.
- This paper presents a Cross-Layer Dynamic queues management strategy, and studies its performance. A Cross-Layer design is required in order to consider the instantaneous channel conditions and QoS demands.

The remainder of this chapter is organized as follows: while section II deals with the chapter background, the system model is presented in section III. Section IV makes a review of the MOB procedure and section V exposes the QoS performance under outage specifications to enable the discussion about the Dynamic queues management in section VI. The numerical results and simulations are in section VII, followed by the future research directions and conclusions in sections VIII and IX, respectively.

BACKGROUND

With respect to the aforementioned concepts in a Downlink system with heterogeneous traffic, several proposals in literature tackle the dynamic queue consideration, but with different objectives and requirements.

The authors in (Zhao & Tong, 2000) propose a Media Access Control (MAC) protocol for a finite-user slotted channel with multipacket reception (MPR) capability. By adaptively changing the size of the contention class (defined as a subset of users who can access the channel at the same time) according to the traffic load and the channel MPR capability. However, this protocol is dynamic in terms of traffic load queue and does not deal with the problem of having different users with different QoS demands.

An admission control problem for a multi-class single-server queue is considered in (Ata, 2006). The system serves multiple demand streams, each having a rigid due-date lead time. To meet the due-date constraints, a system manager may reject orders when a backlog of work is judged to be excessive, thereby incurring lost revenues. Nevertheless, in this chapter, service classes are turned-away based on pre-defined load (packets in the queue) thresholds and only the average mean delay is guaranteed.

A dynamically queuing feature for service enhancement is proposed in (Choi, Igm, An & Nam, 1999), according to the increment of service subscribers and their mobility. In addition, it presents a dynamic queue manager that handles the queue size to increase call completion rates for service enhancements in wireless intelligent network environments. In spite of this, other QoS demands are not possible and the problem of having different users with different QoS demands is not dealt with.

A Cross-layer resource allocation technique is proposed in (Xue, Nguyen & Nahrstedt, 2007) where the prioritization over the resources is given to the user with the best channel conditions. The presented technique accomplishes a joint flow control and active queue management solution, but there is not any consideration of the heterogeneity of the users and their QoS demands.

Various QoS requirements of bursty traffic and a dynamic priority queue with two types of traffic are proposed and analyzed in (Choi & Lee, 2001). The system has two separate buffers to accommodate two types of customers, the capacities of the buffers being assumed to be finite. But the service order is only determined by the queue length of the first buffer, so that only average QoS demands can be satisfied.

The scheduler gives some buffers and bandwidth to every priority class at every port in (Hahne & Choudhury, 2002). The scheme adapts to changes in traffic conditions, so that when the load changes the system goes through a transient. Therefore, each queue individually carries out its blocking process, which does not provide any tight control on the QoS demands.

SYSTEM MODEL

We focus on the single cell Downlink channel where N receivers, each one of them equipped with a single receiving antenna, are being served by a transmitter at the Base Station (BS) provided with n_t transmitting antennas, and supposing that N is greater than n_t. The considered scenario is actually a multiuser Multiple Input Single Output (MISO) but its results can be immediately applied to multiuser MIMO with any receiver processing. This scenario is considered for easiness, as the receiver processing is out of this chapter scope, and all the chapter's objectives and conclusions are independent of the processing carried out at the receiver. The scenario is identified to be a heterogeneous scenario where users run any of the four different classes of applications. Class 1 represents voice users (the most delay-sensitive application) and has the highest priority, while Class 4 is the lowest priority best-effort class.

It is worth mentioning that the demand of real-time services, such as Voice over IP, for strict QoS delay demands, leads to the re-consideration of the ring scattering model (Pavon & Choi, 2003), which is widely used in the evaluation of **WLAN** systems with non-real time (e.g. data traffic) applications. This is because the QoS requirements have to be satisfied in a tighter time scale, which requires for detailed models to account for the instantaneous channel random fluctuations.

A wireless multiantenna channel $\mathbf{h}_{[1\times n_t]}$ is considered between each of the users and the BS, where a quasi-static block fading model is assumed, which keeps constant through the coherence time, and independently changes between consecutive time intervals with independent and identically distributed (i.i.d.) complex Gaussian entries $\sim CN(0,1)$. Therefore, the channel for each user is assumed to be fixed within each fading block (i.e. scenario coherence time) and i.i.d from block to block, so that for the QoS objective, this model captures the instantaneous channel fluctuations in a better approach than the circular rings model. Let $x(t)$ be the (n_t $x1$) transmitted vector (as we are in a Downlink scenario), while denote $y_i(t)$ as the ith user received signal, given by

$$y_i(t) = \mathbf{h}_i(t)\mathbf{x}(t) + z_i(t) \qquad (1)$$

where $z_i(t)$ is an additive Gaussian complex noise component with zero mean and $E\{|z_i|^2\} = \sigma^2$. The transmitted signal $x(t)$ encloses the independent data symbols $s_i(t)$ to all the selected users with $E\{|s_i|^2\} = 1$. A total transmitted power constraint $P_t=1$ is considered, and for ease of notation, time index is dropped whenever possible.

MULTIBEAM OPPORTUNISTIC BEAMFORMING (MOB)

One of the main transmission techniques in multiuser multiantenna scenarios is the MOB scheme (Sharif & Hassibi, 2005), where random beams are generated at the BS to simultaneously serve more than one user. The beam generation follows an orthogonal manner to decrease the interference among the served users, where n_t beams are generated. Within the acquisition step, a known training sequence is transmitted for all the users in the system. Therefore, each user sequentially calculates the Signal-to-Noise-Interference-Ratio (SNIR) related to each beam, and feeds back to the BS only the best SNIR value together with an integer number indicating the index of the selected beam. To increase the system throughput, the BS scheduler chooses the user with the highest SNIR value for each one of the beams, it enters the transmission stage and simultaneously transmits to each one of the n_t selected users, where no user can obtain more than one beam at a time.

Since the users with the best channel conditions are selected for transmission, the scheduler is called *Opportunistic Scheduler*. Therefore, the low complexity MOB strategy achieves high throughput by spatial multiplexing the n_t users with the best channel conditions, making the transmitted signal to enclose the data symbols for the n_t selected users as

$$\mathbf{x} = \sqrt{\frac{1}{n_t}} \sum_{m=1}^{n_t} \mathbf{b}_m s_m \qquad (2)$$

with $\boldsymbol{b}_m$ as the unit-power beam assigned to the mth user, where the square root term is due to a total power constraint of $P_t = 1$.

This scheme is characterized by its SNIR term due to the interference that each beam generates to its non-intended users, representing a major drawback of this system. Even though the beams are orthogonally generated, some of the orthogonality is lost due to the propagation channel (Sharif & Hassibi, 2005), stating the SNIR formulation for the ith user through the mth beam as

$$SNIR_{i,m} = \frac{\frac{1}{n_t}\left|\mathbf{h}_i \mathbf{b}_m\right|^2}{\sigma^2 + \sum_{u \neq m}^{n_t} \frac{1}{n_t}\left|\mathbf{h}_i \mathbf{b}_u\right|^2} \qquad (3)$$

with $\boldsymbol{b}_u$ as the unit-power beam assigned to the uth user, and where a uniform power allocation is considered. As the user with highest SNIR value is selected for each transmitting beam, then the average system throughput of MOB can be written (Sharif & Hassibi, 2005) as

$$TH = E\left\{\sum_{m=1}^{n_t} \log_2(1 + \max_{1 \leq i \leq N} SNIR_{i,m})\right\} \qquad (4)$$

where $E\{.\}$ is the expectation operator to denote the average value. Notice that the value of $\max_{1 \leq i \leq N} SNIR_{i,m}$ reflects the serving SNIR (i.e. the SNIR that the selected user i receives when serviced through the mth beam).

The MOB scheme is shown to improve the system average throughput (Sharif & Hassibi, 2005), but the main target of this work is in providing a precise and guaranteed QoS control for all the users, mainly in terms of the maximum allowed delay and minimum guaranteed throughput. It has to be noted that the minimum allowed rate, the maximum allowed delay and the minimum guaranteed throughput stand as QoS realistic constraints for both real and non-real time applications, providing the commercial operator with a wider view than the fairness concept, as the QoS is stated in terms of per user exact requirements.

QoS SYSTEM PERFORMANCE

For the consideration of any transmission scheme in commercial standards that run real-time applications, the QoS of the users is a very important aspect that can be characterized by several metrics or indicators based on the design objectives. So, QoS can be expressed in terms of rate, reflecting the minimum required rate per user, or in terms of delay, showing the maximum delay that a user can tolerate for its packets. This chapter considers both of the aforementioned QoS concepts, where the proposed transmission scheme guarantees a minimum rate R per user, which is presented by a minimum SNIR restriction ($snir_{th}$), through the classical relation ($R = \log_2\left(1+\text{snir}_{th}\right)$), and delivered to it within a maximum tolerable time delay K.

As this work deals with real-time applications in WLAN systems, then the QoS demands can not be satisfied for the 100% of cases due to the channel characteristics. Therefore, some outage ξ_{out} in the QoS is accepted (Chalise & Czylwik, 2004). Similar situation currently happens in cellular systems as GSM and UMTS, and expected in WLAN systems when running real-time applications. As an example, VoIP can accept erroneous packets up to 10^{-3} of the total number of packets.

The chapter defines two concepts for outage (Zorba & Perez-Neira, 2007): the scheduling delay outage and the rate outage. The first one is related to the opportunistic access policy and the time instant when the ith user is provided service. The second one accounts for the received data rate once the ith user is selected for transmission, and whether its rate requirement is satisfied or not.

Access Delay Outage

In TDMA systems (e.g. GSM) each user knows, in advance, its exact access slot; but in an opportunistic scheduler, as a continuous monitorization of the users' channel quality is performed to select the best ones in each slot, then the access to the wireless medium is not guaranteed. Therefore, the study of the access to the channel in the MOB scheme offers several challenges that must be solved in practical systems.

This section calculates the maximum access delay from the time that a user's packet is available for transmission at the scheduler until the user is serviced through any of the n_t beams of the BS. If an active user is in the system, but it is not scheduled within its maximum allowed delay (e.g. because its channel conditions are not good enough to be selected by the MOB scheduler), then that user is declared as being in access delay, with an outage probability ξ_{access} given by

$$\xi_{access} = 1 - V(K) \quad (5)$$

with $V(K)$ as the probability that a maximum of K time slots are required to select a user i from a group of N i.i.d. users, where this probability follows a Geometric Distribution (Spiegel, 1992) as

$$V(K) = 1 - (1 - \overline{P}_{access})^K \quad (6)$$

Along this chapter, all the users are assumed to have the same average channel characteristics, and showing the same distribution for the maximum SNIR value, so that each user has the same probability to be selected. If this is not the case (e.g. heterogeneous users distribution in the cell, with some users far from the BS), then a channel normalization (e.g. division by the path loss) can be accomplished for such a scenario. In the MOB scheme, each one of the N independent users attempts to be serviced by one of the n_t generated beams with $\overline{P}_{access} = \frac{n_t}{N}$, therefore from previous equation, the maximum number of time slots K until the ith user is selected for transmission, with a probability of delay outage ξ_{access}, is given by

$$K = \frac{\log_2(1-V)}{\log_2(1-\overline{P}_{access})} = \frac{\log_2(\xi_{access})}{\log_2(1-n_t/N)} \quad (7)$$

showing the effects of the number of active users N and the number of serving beams n_t.

Minimum Rate Outage

If the BS scheduler selects a user for Downlink transmission, it means that he/she has the maximum SNIR among the users for a specific beam. But the instantaneous channel conditions (i.e. the instantaneous SNIR) may correspond to a transmission rate that does not satisfy its current application rate requirements (e.g. for a predefined Packet Error Rate, the channel can only provide 6 Mbps while the application asks for 24 Mbps). As a consequence, the user is unable to correctly decode the received packets during the current time unit and suffers a rate outage.

Based on the MOB philosophy to deliver service to the users, the serving SNIR value is the maximum SNIR over the active users in the system, corresponding to each generated beam. Using the SNIR equation in (3), note that the numerator follows a Chi-square $\chi^2(2)$ distribution while the interference terms in the denominator are modelled as $\chi^2(2(n_t-1))$, which allows to obtain the SNIR probability distribution function (pdf) as (Sharif & Hassibi, 2005; Zorba & Perez-Neira, 2007).

$$f(x) = \frac{e^{-(x \cdot n_t \sigma^2)}}{(1+x)^{n_t}} \left(n_t \sigma^2 (1+x) + n_t - 1 \right) \tag{8}$$

and the cumulative distribution function (cdf) is then formulated as

$$F(x) = 1 - \frac{e^{-(x \cdot n_t \sigma^2)}}{(1+x)^{n_t - 1}} \tag{9}$$

and since the serving SNIR is the maximum over all the users' SNIR values (i.e. the opportunistic philosophy), then it's cdf is stated as

$$FF(x) = (F(x))^N = \left(1 - \frac{e^{-(x \cdot n_t \sigma^2)}}{(1+x)^{n_t - 1}} \right)^N \tag{10}$$

Therefore the minimum required SNIR ($snir_{th}$) for each user is achieved with a probability U as

$$U(snir_{th}) = 1 - \left(1 - \frac{e^{-(snir_{th} \cdot n_t \sigma^2)}}{(1+snir_{th})^{n_t - 1}} \right)^N \tag{11}$$

which relates to the predefined rate outage ξ_{rate} as

$$\xi_{rate} = \left(1 - \frac{e^{-(snir_{th} \cdot n_t \sigma^2)}}{(1+snir_{th})^{n_t - 1}} \right)^N \tag{12}$$

where the values of $snir_{th}$ and ξ_{rate} can be computed on the basis on any system objectives, under the number of users N. With further manipulations, the expression (12) can be re-formulated as

$$\log_2\left(1 + snir_{th}\right) = \frac{\log_2\left(\frac{1}{1 - \sqrt[N]{\xi_{rate}}} \right) - \lambda snir_{th} \cdot n_t \sigma^2}{n_t - 1} \tag{13}$$

obtaining the minimum guaranteed-rate, and where $\lambda = \log_2(e) = 1.4427$ is adopted.

Eqn.(13) shows the rate limits of the system, indicating that high $snir_{th}$ requirements induce high outage ξ_{rate} in the system. Negative values in the right hand term indicate infeasibility of the requested rate. We assume in this chapter that the minimum SNIR guarantees successful decoding of packets. Therefore, the following unit step function defines the Packet Success Rate (PSR) related to the $snir_{th}$ as

$$PSR = \begin{cases} 1 & \text{if} \quad serving\ SNIR \geq snir_{th} \\ 0 & \text{if} \quad serving\ SNIR < snir_{th} \end{cases} \tag{14}$$

where a direct relation to ξ_{rate} is obtained from Eqn.(12).

Outage of the System

As previously explained, the MOB scheme comes controlled by two different outage measures, but the total system performance has to be defined through a single parameter. Notice that the two discussed kinds of outage are totally independent, as the user's access to the channel happens when its SNIR is the maximum over all the other users, with respect to a given beam, but being the user with largest SNIR does not guarantee that this SNIR is larger than an application predefined threshold $snir_{th}$. Therefore, the **total outage** ξ_{out} is defined as

$$\xi_{out} = 1 - (1 - \xi_{access}) \cdot (1 - \xi_{rate}) \quad (15)$$

standing as the global measure of system outage.

Maximum Scheduling Delay

In point to point scenarios, the queuing delay is the dominant factor in the system delay (Neely, Modiano & Rohrs, 2005) while in multiuser systems an additional delay factor is introduced, because the system resources are not all the time available to the same user. We name this additional delay factor as the scheduling delay in multiuser systems. In the round robin systems (e.g. TDMA) the user access to the channel is known in advance, so that its scheduling delay can be easily calculated. However, in opportunistic multiuser systems where the users with the best channel conditions are selected for transmission based on their instantaneous SNIR, a user does not have any guarantee for being scheduled in a specific time, which increases its scheduling delay.

In the context of this chapter, we define the maximum scheduling delay as the time period from the instant that a user's packet is available for transmission at the scheduler until the packet is correctly received at its destination. The difference with the access delay definition is the requirement of a rate threshold in order to guarantee the decoding without errors, as in Eqn. (14). Notice that this definition includes both the delay resulting from the scheduling process (i.e. the opportunistic selection) and the delay caused by the requirements to get a rate above to a minimum required threshold to be correctly received. Therefore, the maximum number of time slots to select a user with a total outage ξ_{out} is equal to the K access slots (7), defining the maximum allowed scheduling delay.

As we consider the scheduling delay, both the buffer management and source statistics for arriving packets are not addressed (Issariyakul & Hossain, 2006); and the queues stability target (Neely, Modiano & Rohrs, 2005) is neither considered. Therefore, we assume a saturated system and only consider the delay resulting from the scheduling process. The total delay (scheduling + queuing) will be tackled as a future work.

Minimum Guaranteed Throughput per User and per Slot

Obtaining the system throughput formulation is difficult as several processes are included in the communication procedure. The receiver decoding through the unit step function in Eqn.(14) simplifies the throughput formulation, as the effects of several steps in the communication process (e.g. coding) are avoided.

In opportunistic multiuser scenarios, the user in not always served by the system, so that its throughput is zero for several time units. Therefore, a normalized minimum-guaranteed throughput per user over the time is required. Notice that such definition of throughput per user and per slot accounts for the user's waiting time and hence, for its corresponding scheduling delay expression. Considering that the bandwidth of the system is B_w, then the minimum-guaranteed throughput per user and per slot is denoted as T, in bits, and given as

$$T = \frac{B_w \log_2\left(1 - n_t / N\right)\left(\log_2\left(\frac{1}{1 - \sqrt[N]{\xi_{rate}}}\right) - \lambda snir_{th} \cdot n_t \sigma^2\right)}{(n_t - 1)\log_2(\xi_{access})} \quad (16)$$

where the expression in (13) is used to provide a closed form solution for the minimum-guaranteed throughput per user, with all the operating variables. Notice that by increasing the number of users N, the minimum guaranteed rate R goes up and as a consequence higher throughput is obtained. On the other hand, larger N induces larger scheduling delay, increasing in this way the value of K, that drives lower throughput values. This shows a tradeoff on the number of available users in the systems, motivating a control over the N value to achieve the system QoS requirements, as will be shown in the next section.

Note that the minimum-guaranteed throughput is the worst case awarded throughput to the users, but it actually defines the throughput value that an operator can guarantee to its customers, obviously, with a given outage ξ_{out}; where the guaranteed throughput per user is different from the concept of average throughput in Eqn.(4), previously presented. A very common example in commercial systems for average throughput and the minimum guaranteed throughput is seen in the ADSL service, where for example, an operator can provide its costumers 20Mbps (which is the value that appears in its advertisements), while the minimum guaranteed value for the user is 2Mbps (National regulatory telecommunication agencies often ask for a guaranteed value of at least *10%* of the average value).

DATA LINK CONTROL WITH DYNAMIC QUEUE LENGTH

Two important aspects to achieve QoS for the serviced users are extracted from the analytical study in the previous section: the impact of the number of available users and their exact QoS demands. To control the different user requirements and their sensitivity to delay and rate, a control on the DLC queue length L is proposed in this chapter. The aim of this section is to provide a description of this proposal, performed through a cross-layer scheduling algorithm at the DLC layer of WLAN systems. The main idea of the proposed scheme is depicted in Figure 1. It can be seen that each IP packet is stored at the corresponding priority queue in the IP layer, before moving down to the DLC layer queue. Users from higher priority IP queues are placed at the beginning of the DLC queue following by users with lower priorities traffic.

At the Physical layer, the WLAN systems use different modulation levels, so that variable transmission rates depending on the channel conditions (measured through the received SNIR) are obtained. The MOB scheme is applied to select the users with the best channel conditions per beam in order to maximize the system average throughput.

Regarding the dynamic queue length mechanism, when the maximum allowed delay (or minimum allowed rate) in the delivery of the most delay sensitive application is smoothly satisfied, then the length of the queue can be increased so that more users can be placed in the DLC layer queue. As a consequence, the MOB scheduler can select the user per beam with the best channel conditions in a bigger pool of choices, increasing in this way the performance of the system in terms of the average throughput in Eqn.(4). On the other hand, when the maximum allowed delay requirements are hardly satisfied, then the length of the DLC queue is decreased. Therefore, only packets form users within the higher priority classes can be available in the DLC layer queue, so that the MOB scheduler can only select, for each one of the beams, among these users. Likewise, the same procedure can be applied when the minimum guaranteed throughput per user is the considered QoS indicator.

Note that the proposed dynamic adjustment in the size of the queue shows the tradeoff between the real-time users' QoS demands and the system average throughput in the network, where the best operating point depends on the network operator requirements. It has to be noted that very delay sensitive applications are in general characterized

Figure 1. Dynamic queue length scheme

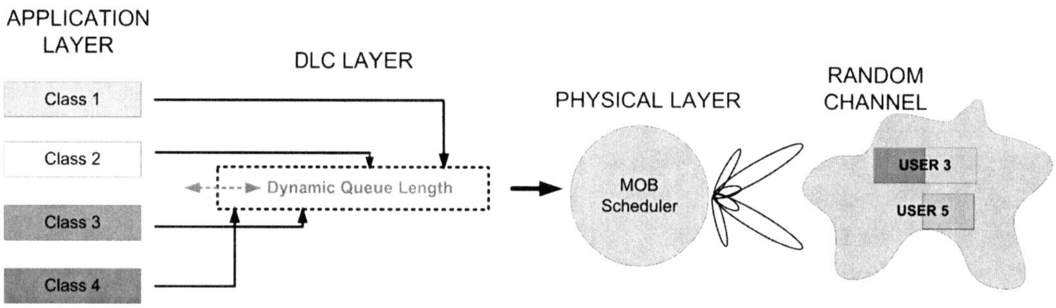

by short packets lengths, such as VoIP, that do not extract all the benefit from the throughput of the system. To find the best operating point, the dynamic queue length L is maximized, subject to some system requirements in terms of the users' QoS demands. Taking into consideration the existence of outage in the QoS satisfaction, a proposed optimization procedure for the system performance can be stated as

$$\max \ L$$

$$s.t._1 \ Prob\left\{SNIR_i < snir_{th}\right\} \leq \xi_{rate} \ \forall i \in L$$

$$s.t._2 \ Prob\left\{D_{max} < K_i\right\} \leq \xi_{delay} \ \forall i \in L$$

$$s.t._3 \ Prob\left\{T_i \geq T_{min}\right\} \geq 1 - \xi_{out} \ \forall i \in L \qquad (17)$$

where D_{max} is the maximum allowed delay and Tmin is the minimum required throughput per user and per slot. It is has to be noted that the previous scheme presents the dynamic queue length adjustment together with the QoS concepts (minimum allowed rate, maximum allowed delay and minimum guaranteed throughput), where the operator can choose among the QoS demands for the most appropriate ones for each scenario.

PERFORMANCE EVALUATION

To evaluate the performance of the proposed dynamic DLC queue mechanism, an heterogeneous scenario is set up where users with four types of applications coexist in the system. Two transmitting antennas nt=2 are available, so that two beams are generated and two users in the Downlink can be simultaneously serviced through the same frequency, code and time. A total of $N=20$ users are available in the scenario with 5 users for each service traffic class. The length of the packets for the classes 1, 2, 3 and 4 are 100, 512, 1024 and 2312 bytes respectively. Class 1 has the highest priority, while class 4 is the lowest priority class. A saturated system is considered, where all users have at least one packet available for transmission. A total system bandwidth of *10 MHz* and a slot service time of *1 ms* are assumed in the simulations. An Indoor complex i.i.d. Gaussian channel with $\sim CN(0,1)$ entries is considered. A time scale of 10^6 channel visualizations is employed to display the channel continuous variations. Results for an opportunistic scheduler that only transmits to a single user (Knopp & Humblet, 1995) are also shown in the figures to realize the benefits of MIMO from a higher layers perspective. Obviously, the same total power constraint is imposed on both systems in order to have a fair comparison. Table 1 shows how the SNIR values for IEEE 802.11 legacy systems are mapped to the transmission rate per beam, as stated in (Pubill & Perez-Neira, 2006).

Table 1. SNIR values mapping to rate

Rate (Mbps)	SNIR value (dB)
0	<-8
6	-8 to 12.5
9	12.5 to 14
12	14 to 16.5
18	16.5 to 19
24	19 to 22.5
36	22.5 to 26
48	26 to 28
54	>28

The efficiency of our dynamic queue length scheme is compared with a Round Robin based scheme (Fu & Kim, 2006) where the channel conditions are not taken into consideration in the scheduling process, and the users access to the channel are guaranteed at fixed intervals. This technique is implemented in TDMA-based systems and it has been proved to provide the lowest possible scheduling delay, but the obtained throughput is very low as the channel conditions are not regarded in the scheduling process. Moreover, it can not be combined with the MIMO Multiuser capability, since the latter needs for the users' selection principle to choose n_t users that show the least interference among themselves (Sharif & Hassibi, 2005).

In Figure 2, the percentage of the outage in the maximum delay satisfaction for Class 1 users versus the length of the queue is presented. A maximum allowed delay of *20ms* is assumed for the class 1 users. It can be seen from Figure 2 that when the length of the queue is $L=5$ (so that only users of the class 1 exist in the DLC queue), the maximum allowed delay is guaranteed for almost *100%* of the cases (with an outage of *0.049%*). Notice that increasing the queue length to *20*, so that all users are eligible to be selected, the outage reaches a value of *12%*. The single user service (indicated as ``no-MIMO" in the figures) provides an outage value of *2%* for a DLC queue length of 5 and when the DLC queue length is 20, the outage value boosts to *36%*, which is an unacceptable value for any communications system. The results show the great benefit of providing QoS delay guarantees with the MOB technique as the users are provided service more frequently, thus the probability to violate the maximum delay restriction is lower. Note the exact match between the theory and the simulations results, as approximations were not employed in the equations derivation. From Figure 2 we can also see that in a scenario of 20 users with a maximum of *20 ms* maximum allowed delay (remind that the service slot time is *1 ms*), then all users are serviced through the Round Robin strategy, delivering a *0%* in the outage delay.

From Figure 2 we saw that increasing the DLC queue length increases the outage probability which is harmful for the performance of the system. On the other hand, in order to increase the system average throughput a longer length of the DLC queue is required, so that more users are eligible for scheduling selection in the system. This means that class 1 users have lower chance to be serviced by the BS scheduler, which has a direct impact on the time delivery of their packets. Figure 3 shows the performance of the average throughput (from Eqn.(4)) for a variable DLC queue length, where as expected, increasing the queue length (i.e. the number of available users

Figure 2. Outage probability (%) in the maximum delay satisfaction for Class 1 users, with a maximum allowed delay threshold=20ms

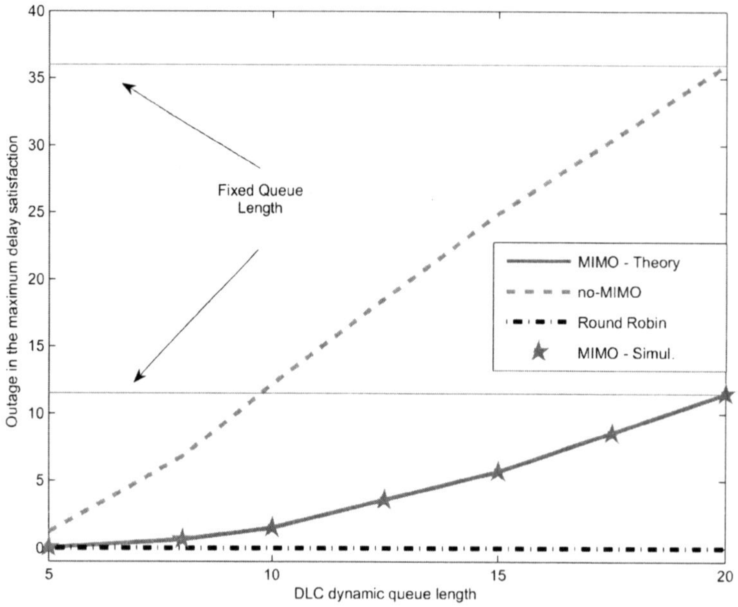

for scheduling), the average throughput values go up due to the opportunistic way of user/s selection in both of MOB and single user selection in (Knopp & Humblet, 1995). Once again, it can be seen the exact match between simulations and the theoretical analysis. Notice the deficient performance of the round robin strategy, as the scheduler does not tackle the channel conditions, thus delivering very low system throughput performance, which handicaps its implementation in current broadband wireless systems, even of its outstanding delay performance.

The figure shows how the gap between the two schedulers enlarges as the DLC queue length increases, which is motivated by the MOB performance, where a larger number of users enables a better search for a set of users (2 users in our simulations) that do not interfere between them a lot (i.e. better SNIR value). Also realize that the average throughput gain of MOB is not as amazing as the MOB gain in the outage of the QoS satisfaction, as seen in Figure 2. The explanation for this matter is due to the MOB technique where more users can be serviced (2 users in our study case), so that the users have almost twice the probability to be serviced in comparison with the single user scheduling approach in (Knopp & Humblet, 1995). But on the other hand, the throughput average gain is not twice due to the interference that the users generate between them. Therefore, we can claim that MOB scheme is more suitable for QoS demands than average throughput performance. This conclusion have not been stated previously in the literature (up to the authors' knowledge), where this result is very interesting for the implementation of MOB (and most probably, for any other MIMO multiuser technique).

FUTURE RESEARCH DIRECTIONS

This work developed a QoS optimization over the system metrics to guarantee the QoS for the users, but as a future work, a joint optimization

Figure 3. System average throughput for a variable DLC queue length

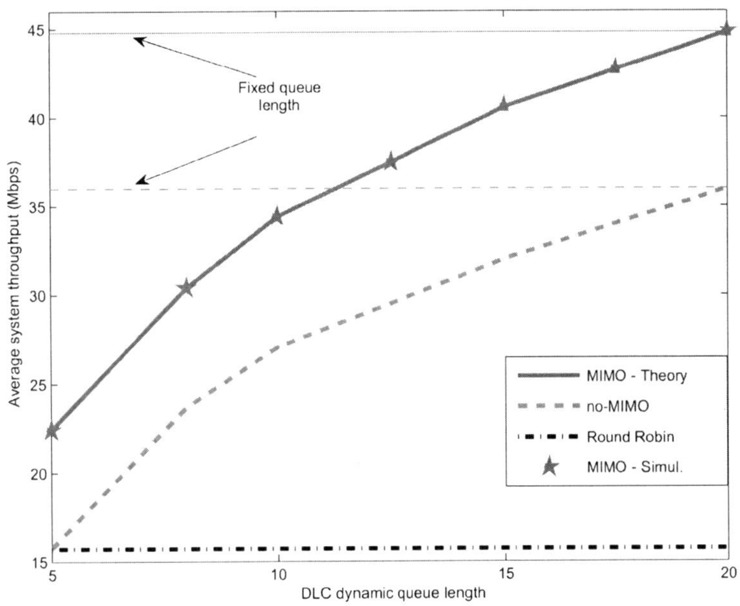

over the QoS metrics is also required to avoid any controversial results among them. Moreover, as all current broadband wireless systems are based on the OFDM Access (OFDMA) scheme, a resource management based on the subcarriers allocation is also required to align with current standards.

Another future work is related to the Hour-Aware Resource Management (HA-RMM). As it is defined in the literature, applications running over the different hours have different QoS requirements (e.g. during the night background traffic is the dominant one while during the morning, real time traffic is needed; where each application has its own QoS demands). Therefore, a smart resource management strategy over the different day hours is required to achieve a further optimization of the system resources.

CONCLUSION

A dynamic queue length scheduling has been presented in this work for Downlink multiuser and multiantenna WLAN systems with heterogeneous traffic. In the considered system, the users with the best channel conditions are selected for transmission. Through the MOB scheme, the length of the queue defines the maximum achievable average throughput of the system. On the other hand, the QoS requirements of the delay sensitive applications are guaranteed with short DLC queue lengths. A tradeoff appears between the system average throughput and the QoS demands of the users.

The chapter proposed a dynamic DLC queue length control, so that the maximum length is allowed to obtain the highest average system throughput, but restricted to the satisfaction of the users QoS. Several alternative QoS measures are presented along the chapter and in closed form expressions, so that the wireless operator can choose among them for the most suitable ones for each scenario and QoS requirements.

Besides the dynamic queue proposal, another important outcome of this chapter is on how applications and link layers (or in general higher layers) take profit of the advances introduced by multiple antennas and signal processing techniques in the physical layer. A challenge faced by this chapter is on how to deal with several aspects from the different layers of the communication process, so that we tried to make the physical layer concepts to be clear for high layers researchers, and viceversa.

ACKNOWLEDGMENT

This work has been funded by the Research Projects PERSEO and R2D2 .

REFERENCES

Ata, B. (2006). Dynamic Control of a Multiclass Queue with Thin Arrival Streams. *Operations Research*, 54.

Chalise, B. K., & Czylwik, A. (2004). Robust Downlink Beamforming based upon Outage Probability Criterion. In *IEEE Vehicular Technology Conference (VTC)*.

Choi, D.I., & Lee, Y. (2001). Performance Analysis of a Dynamic Priority Queue for Traffic Control of Bursty Traffic in ATM Networks. *IEEE Communications Proceedings, 148.*

Choi, H., Igm, Y., An, S., & Nam, C. (1999). Dynamic Queue Management Mechanism for Service Enhancement in Wireless Intelligent Network Environment. In *IEEE Global Communications Conference (Globecom).*

Fu, H., & Kim, D. I. (2006). Analysis of Throughput and Fairness with Downlink Scheduling in WCDMA Networks. *IEEE Transactions on Wireless Communications*, 5.

Hahne, E. L., & Choudhury, A. K. (2002). Dynamic Queue Length Thresholds for Multiple Loss Priorities. IEEE/ACM Transactions on Networking, 10.

Issariyakul, T., & Hossain, E. (2006). Channel-Quality-Based Opportunistic Scheduling with ARQ in Multi-Rate Wireless Networks: Modeling and Analysis. *IEEE Transactions on Wireless Communications*, 5.

Knopp, R., & Humblet, P. (1995). Information Capacity and Power Control in Single Cell Multiuser Communications. *IEEE International Conference on Communications (ICC).*

Kountouris, M., & Gesbert, D. (2005). Robust Multi-user Opportunistic Beamforming for Sparse Networks. In IEEE Signal Processing Advances for Wireless Communications (SPAWC) workshop.

Neely, M.J., Modiano, E., & Rohrs, C.E. (2005). Dynamic Power Allocation and Routing for Time Varying Wireless Networks. *IEEE Journal Selected Area of Communications, 23.*

Pavon, J., & Choi, S. (2003). Link Adaptation Strategy for IEEE 802.11 WLAN Via Received Signal Strength Measurement. *IEEE International Conference on Communications (ICC).*

Pubill, D., & Perez-Neira, A. I. (2006). Handoff Optimization with Fuzzy Logic in 802.11 Networks. *Information Processing and Management of Uncertainty (IPMU) Conference.*

Shakkottai, G. S., Rappaport, T. S., & Karlsson, P. C. (2003). Cross-Layer Design for Wireless Networks. *IEEE Communications Magazine*, 41.

Shakkottai, G. S., & Stolyar, A. (2001). Scheduling algorithms for a mixture of real time and non-real-time data in HDR. *International Teletraffic Conference (ITC).*

Sharif, M., & Hassibi, B. (2005). On the Capacity of MIMO Broadcast Channel with Partial Side Information. *IEEE Transactions on Information Theory*, 51.

Spiegel, M. R. (1992). Theory and Problems of Probability and Statistics. New York: McGraw-Hill.

Xue, Y., Nguyen, H., & Nahrstedt, K. (2007). CA-AQM: Channel-Aware Active Queue Management for Wireless Networks. *IEEE International Conference on Communications (ICC)*.

Zhao, Q., & Tong, L. (2000). A Dynamic Queue MAC Protocol for Random Access Channels with Multipacket Reception. *Asilomar Conference*.

Zorba, N., & Perez-Neira, A. I. (2007). CAC for Multibeam Opportunistic Schemes in Heterogeneous WiMax Systems under QoS Constraints. *IEEE Global Communications Conference (Globecom)*.

ADDITIONAL READING

Abu-Ali, N., Ekram, E., Eljasmy, A., & Shuaib, K. (2008). Delay Distribution in WMN: Test-bed experiments. *ACS/IEEE International Conference on Computer Systems and Applications (AICCSA-08)*.

Al-Manthari, B., Abu-Ali, N., Nasser, N., & Hassanein, H. (2008). Dynamic Bandwidth Provisioning with Fairness and Revenue Considerations for Broadband Wireless Communication. *IEEE International Conference on Communications (ICC)*.

Andrews, M. (2001). Providing quality of service over a shared wireless link. IEEE Communications Magazine, vol. 39.

Bae, Y. H., Lyakhov, A. I., Vishnevsky, V. M., Kim, K. J., & Choi, B. D. (2008). Matrix method to study IEEE 802.11 network. Automation and Remote Control, v. 69.

Bender, P., Black, P., Grob, M., Padovani, R., Sindhushayana, N., & Viterbi, A. (2000). CDMA/HDR: A bandwidth-efficient high-speed wireless data service for nomadic users. IEEE Communications Magazine, vol. 38.

Chakrabarti, G., & Kulkarni, S. (2006). Load Balancing and resource reservation in mobile ad hoc networks. Ad Hoc Networks, vol. 4.

Foschini, G., Karakayali, K., & Valenzuela, R. (2006). Coordinating multiple antenna cellular networks to achieve enormous spectral efficiency. IEE Proceedings on Communications, vol. 4.

Gesbert, D., & Alouini, M. S. (2004). How Much Feedback is Multi-user Diversity Really Worth? *IEEE International Conference on Communications (ICC)*.

Gupta, P., Cerruti, I., & Fumagalli, A. (2004). Three transmission scheduling policies for a cooperative ARQ protocol in radio networks. *WNGG conference*.

Jalali, A., Padovani, R., & Pankaj, R. (2000). Data throughput of CDMA-HDR a high efficiency-high data rate personal communication wireless system. *IEEE Vehicular Technology Conference (VTC)*.

Kafetzakis, E., Kontovasilis, K., & Sarakis, L. (2009). A Distributed Simulator Coordination Platform and its Application for Integrating an IEEE 802.11 Effective Capacity-based Admission Control Algorithm. *Simutools Conference*.

Liu, Q., Wang, X., & Giannakis, G. B. (2006). A Cross-Layer Scheduling Algorithm with QoS Support in Wireless Networks. IEEE Transactions on Vehicular Technology, vol. 55.

Malone, D., Duffy, K., & Leith, D. (2007). Modeling the 802.11 Distributed Coordination Function in Nonsaturated Heterogeneous Conditions. IEEE/ACM Transactions on Networking, vol. 15.

Nazari, V., & Ziarati, K. (2006). Performance Comparison of Routing Protocols for Mobile Ad Hoc Networks. *IEEE Asia-Pacific Conference on Communications.*

Wang, X., Giannakis, G. B., & Marques, A. G. (2007). A Unified Approach to QoS-Guaranteed Scheduling for Channel-Adaptive Wireless Networks. *Proceedings of the IEEE, vol. 95.*

Woo, K., & Lee, B. (2001). Non blocking localized routing algorithm for balanced energy consumption in mobile ad hoc networks. IEEE MASCOTS.

Wu, D., & Negi, R. (2003). Effective Capacity: A Wireless Link Model for Support of Quality of Service. IEEE Transactions on Wireless Communications, vol. 2.

Yashkov, S. F., & Yashkova, A. S. (2007). Processor Sharing: A survey of the mathematical theory. Automation and Remote Control, vol. 68.

Zhang, X., Tang, J., Chen, H., Ci, S., & Guizani, M. (2006). Cross-Layer-Based Modeling for Quality of Service Guarantees in Mobile Wireless Networks. IEEE Communications Magazine, vol. 44.

Chapter 5
Traffic Prediction over Wireless Networks

Huifang Feng
Northwest Normal University, China

Maode Ma
Nanyang Technological University, Singapore

ABSTRACT

The predictability of network traffic is an important and widely studied topic because it can lead to the solutions to get more efficient dynamic bandwidth allocation, admission control, congestion control and better performance of wireless networks. In this chapter, firstly, the authors briefly describe a number of traffic models that include time series models, artificial neural networks models, wavelet-based models, and support vector machine-based models. Secondly, they give the prediction method and metrics of measuring the accuracy of a prediction. Finally, they examine the feasibility of applying support vector machine into the prediction of actual traffic in WLANs and evaluate the performance of different prediction models such as ARIMA, FARIMA, and artificial neural network, wavelet-based and support vector machine-based models for the prediction of the real WLANs traffic.

INTRODUCTION

The predictability of network traffic is of significant interest in many areas including network design, management, control, and optimization (Stemm & Seshan & Katz, 2000; Kim & Noble, 2001). Prediction on long time scale is the base of long-term planning of Internet topology (Ostring, 2000; Shah, 2003). Prediction on short time scale is necessary for congestion control, admission control, bandwidth allocation and buffer management (Yi, 2004; Qiu, 2004; Chisci, 2006). The goal is to forecast future traffic variations as precisely as possible based on the measurements of the traffic history. An accurate traffic prediction model should have the ability to capture the prominent traffic characteristics. Therefore, highly accurate traffic prediction can help to make the maximum usage of bandwidth while guaranteeing the quality of service (QoS) for real-time applications which have stringent requirements on delay or packet rate

DOI: 10.4018/978-1-61520-771-8.ch005

loss. An accurate traffic prediction is, therefore, critical in preventive control scheme in order to utilize its resources efficiently.

The widespread deployment of IEEE 802.11 wireless local area networks (WLANs) as a fundamental approach for enabling ubiquitous wireless networking is mainly due to their easy installation, flexibility and robustness against failures. WLANs traffic prediction plays a fundamental role in the design and analysis of WLANs (Papadopouli, Shen, Raftopoulos, Ploumidis & Campos, 2005). For example, the bandwidth utilization at an access point (AP) can affect the performance of the wireless clients in terms of throughput, delay, and energy consumption. The traffic prediction can help the AP to perform better load balancing, admission control, and QoS provisioning. Specifically, an AP can use the expected traffic estimations to decide whether to accept a new association request or advise a client to associate with a neighboring AP. Thus, it can improve wireless network throughput and ensure the QoS requirements from different traffic types.

Recently, there has been a significant change in the understanding of network traffic. A number of studies of traffic measurements in wired and wireless networks have convincingly demonstrated that the Internet traffic is self-similar or long-range dependent (LRD) in nature (Park & Willinger, 2000; Jiang, Nikolic, Hardy & Trajkovic, 2001). It implies the existence of concentrated periods of low activity and high activity (i.e., burstiness) at a wide range of time scales. Self-similarity traffic cannot be captured by classical models. Predicting the future traffic from the past observations is an important way to obtain traffic control under self-similar traffic loads. Under the long-range dependent network traffic and with their slow convergence properties, it becomes more difficult to design rigorously effective predictors with desirable properties. Effective optimal prediction remains a technical challenge (Beran, 1994). Hence the new fundamental problem in the network traffic prediction is to find efficient self-similar models to predict future traffic variations precisely and produce good predictability.

BACKGROUND

In the past several decades, many models have been proposed for the prediction of traffic in wired networks. In the earliest work, some researchers have proposed autoregressive moving average (ARMA) and autoregressive integrated moving average (ARIMA) models to predict network traffic (Sang & Li, 2002; Adas, 1997; Krithikaivasan, Dekal & Medhi, 2004). All these models are linear time series models that can capture short-range dependence. Although ARIMA models are proved to be quite powerful to model a class of non-stationary data traffic, it can't capture long-range dependent characteristics.

Currently, self-similarity and LRD features have been observed in the Internet traffic over the wired and wireless networks. Traditional linear time series models cannot explain and capture self-similarity and LRD features of the traffic. To deal with the self-similar nature of the network traffic, some researchers (Shu, Jin, & Zhang, 1999; Corradi, Garroppo, Giordano, & Pagano, 2001) have proposed using fractional autoregressive integrated moving average (FARIMA) to model and predict the traffic in the networks. Since FARIMA is a self-similar model with the capability to capture both the short-range dependent (SRD) and LRD characteristics, the FARIMA model has shown its prediction ability in admission control and dynamic bandwidth allocation (Shu, Jin, Wang & Yang, 2000; Sadek, Khotanzad & Chen, 2003). Another non-stationary and non-linear threshold autoregressive (TAR) model has also been introduced to model and predict network traffic (You & Chandra, 1999). The TAR model is also non-linear and behaves as long memory. It can capture self-similarity and LRD well. Both FARIMA and TAR models do improve the performance of prediction for self-similar time series with the cost of computational complexity. Recently, a related model combining ARIMA with generalized autoregressive conditional heteroscedasticity (GARCH) has been proposed to provide even better prediction performance (Zhou, He & Sun, 2006).

The above mentioned models can capture the self-similarity (mono-fractal) and LRD features, which have been found in the network traffic in large time scale. But they cannot capture the multi-fractal in small time scale. For this reason, another multi-fractal wavelet model (MWM) has been introduced to address the issue (Feldmann, Gilbert & Willinger, 1998). The general prediction model combined with wavelet theory has also been proposed to predict the network traffic (Papagiannaki, Taft, Zhang & Diot, 2003; Qiao, Skicewicz & Dinda, 2004).

Meanwhile, lots of researches have also tried to apply artificial neural network (ANN) to improve the traffic prediction accuracy. ANN is a non-linear, non-parametric and data driven modeling approach. It allows one to fully utilize the available information to determine the structure and parameters of a model without any restrictive parametric modeling assumptions. An approach using ANN (Tarraf, Habib, Saadawi & Ahmed, 1993) has been proposed to characterize and predict multimedia traffic in ATM networks. Chang and Hu have suggested employ a pipelined recurrent neural network (PRNN) model to produce an adaptive traffic prediction of MPEG video traffic via dynamic ATM networks (Chang & Hu, 1997). Bhattacharya, et al. have provided a recurrent neural networks (RNN) model to develop MPEG-coded real-time video traffic predictors for use in one-step and multi-step prediction (Bhattacharya, Parlos & Atiya, 2003). ANN has been proved to capture any kind of relationship between the output and the input theoretically (Khotanzad & Sadek, 2003). However, it might suffer from over-fitting (Doulamis, Doulamis & Kollias, 2003).

Support vector machine (SVM) is a new and valid machine-learning algorithm with many advantages that has been put forward by Vapnik (Vapnik, 1995). Its architecture does not have to be determined beforehand. And the input data of any arbitrary dimension can be treated with only a linear cost at the number of inputs. Moreover, the training is able to produce a unique solution. SVM has been well used for pattern recognition (Schmidt, 1996), regression estimation, and prediction, etc. (Liang & Du, 2007; Li, Zhu, Cao, Sheng & Hu, 2008; Sakiyama, Yuki & Moriya et al. 2008). It has been proved that it is a successful approach used in regression and time series prediction.

The use of SVM for classification has been relatively more popular in networking research, especially, in the context of anomaly and intrusion detection (Haffner, Tur & Wright, 2003; Sung & Mukkamala, 2003; Khan, Awad & Thuraisingham, 2007). The bandwidth reservation algorithm based on SVM has been used to predict the moving direction of a mobile host by retrieving the records kept in the base station (Huang, Lai, Luo & Yan, 2005.). Beverly et al. (2006) has used the SVM to predict the round-trip latency to an unknown IP address based on the latency knowledge of other previously contacted IP addresses. Mirza et al. (2007) have proposed to explore a combination of queueing delays and available bandwidth for a TCP throughput prediction based on the SVM model. The authors (Bermolen & Rossi, 2008) have explored to use SVM for the purpose of traffic load forecast.

While there is a rich literature traffic prediction in wired networks, there are only a few studies of wireless traffic prediction. Liang (2002) has applied a fuzzy logic model to forecast the traffic in ad hoc wireless networks. And the simulation results have shown that the proposed model can perform much better than a least mean square (LMS) adaptive filter. The researchers (Papadopouli, Shen, Raftopoulos, Ploumidis & Campos, 2005) have studied the ARIMA models based traffic forecasting at each wireless access point in a long timescale (hourly timescale). The researchers (Ghaderi, Capka & Boutaba, 2003) have investigated the application of traffic prediction to address the admission control issue in wireless mobile Internet. The authors (Feng, Shu, Wang & Ma, 2006) have examined the feasibility of applying SVM to perform one-step-ahead prediction and multi-step-ahead prediction without any assumption on the statistical property of actual WLANs traffic. They have also evaluated the performance of different prediction models such as ARIMA, FARIMA, neural network, and wavelet-based models using three types of real WLANs traffic. The wireless network traffic for short time step has also been modeled as the time series and future values of time series can be predicted by radial basis function network, echo state network (ESN) and FARIMA models (Gowrishankar & Satyanarayana, 2008).

TRAFFIC PREDICTION MODELS

In this section, a brief introduction on the general network traffic prediction models will be presented, including time series models, artificial neural network models, wavelet-based models, and support vector machine-based models, etc.

Time Series Models

AR Model

Autoregressive (AR) model is widely used stochastic model that is extremely useful in the representation of certain practically occurring series (Box, Jenkins& Reinsel, 1994). By this model, the current value of the process is expressed as a finite, linear of previous values and a white noise series $\{a_t\}$ with zero mean and variance σ^2. Denote the values of a process at equally spaced times $t, t-1, t-2....$ by $x_t, x_{t-1}, x_{t-2}, \cdots$, then the AR model of order p can be shown in Eq.1.

$$x_t = \phi_1 x_{t-1} + \phi_2 x_{t-2} + \cdots + \phi_p x_{t-p} + a_t \qquad (1)$$

where $\phi_1, \phi_2, \ldots, \phi_p$ are autoregressive coefficients of AR model, p is the order of AR model.

By employing the backshift operator B that defines $x_{t-1} = Bx_t$, and hence $x_{t-m} = B^m x_t$, equation (1) can be written in the form:

$$\phi(B)x_t = a_t \qquad (2)$$

where,

$$\varphi(B) = 1 - \varphi_1 B - \varphi_2 B^2 - \ldots - \varphi_p B^p \qquad (3)$$

The parameters ϕ_i should be calculated for the prediction. There are two methods for the calculation. One is the least squares estimation while another is the maximum likelihood estimation (MLE).

MA Model

Another kind of model of great practical importance in the representation of observed time-series is the finite moving average (MA) process. The MA model of order q is defined as

$$x_t = a_t - \theta_1 a_{t-1} - \theta_2 a_{t-2} - \cdots - \theta_q a_{t-q} \qquad (4)$$

A similar application of the backshift operator on the white noise series would allow equation (4) to be written as:

$$x_t = \theta(B)a_t \qquad (5)$$

where $\theta(B) = 1 - \theta_1 B - \theta_2 B^2 - \ldots - \theta_q B^q$.

ARMA Model

To achieve greater flexibility in fitting of actual time-series, it is advantageous to include both autoregressive and moving average model. It leads to the mixed autoregressive–moving average (ARMA) model

$$x_t = \phi_1 x_{t-1} + \phi_2 x_{t-2} + \cdots + \phi_p x_{t-p} + a_t - \theta_1 a_{t-1} - \theta_2 a_{t-2} - \cdots - \theta_q a_{t-q} \quad (6)$$

or

$$\phi(B) x_t = \theta(B) a_t \quad (7)$$

where, φ_i and θ_j are the *autoregressive* and *moving average* parameters, respectively. And in this case, it is an ARMA (*p*, *q*) model. In practice, it is frequently true that adequate representation of actual occurring stationary time series can be obtained with autoregressive, moving average, or mixed models, in which p and q are less than or equal to 2.

ARIMA Model

The time series defined previously are stationary processes. It implies that the mean value of the series of any of these processes and the covariances among its observations do not change with time. A generalization of ARMA models, named autoregressive integrated moving average model (ARIMA), which incorporates a wide class of non-stationary time-series, can be obtained by introducing the differencing into the model. ARIMA time series models form a general class of linear models that are widely used in modeling and forecasting time series.

Let $\nabla = 1 - B$ be the differencing operator, a differenced time series of order 1 can be written as:

$$\nabla x_t = x_t - x_{t-1} = (1 - B) x_t \quad (8)$$

Consequently, an order d differenced time series is written as

$$\nabla^d x_t = (1 - B)^d x_t \quad (9)$$

where $\nabla^d = (1 - B)^d$ is defined by means of binomial expansion,

$$(1 - B)^d = \sum_{k=0}^{\infty} \binom{d}{k} (-B)^k \quad (10)$$

An ARIMA(p, d, q) process is a stochastic time-series process where d is the degree of differencing, p is the autoregressive order, and q is the moving average order. All three parameters are non-negative integers. Then the ARIMA (p, d, q) process can be described by the following relationship:

$$\phi(B) \nabla^d x_t = \theta(B) a_t \quad (11)$$

where

$$\phi(B) = 1 - \phi_1 B - \phi_2 B^2 - \ldots - \phi_p B^p \quad (12)$$

$$\theta(B) = 1 - \theta_1 B - \theta_2 B^2 - \ldots - \theta_q B^q \quad (13)$$

When $d = 0$, we have the usual ARMA model, that is ARIMA$(p, 0, q)$=ARMA(p, q).

Generally, construction of an ARIMA prediction model entails four steps:

- **Step 1: Model identification.** 1) To detect the time series is stationary. If it is not, it involves a determination of the degree of differencing d needed to ensure stationarity; 2) To identify the order p of the AR aspect of the time series model; 3) To identify the order q of the MA aspect of the model.
- **Step 2: Model parameters estimation.** The main approaches to fitting ARIMA model are non-linear least squares and maximum likelihood estimation. Maximum likelihood estimation is generally the preferred technique (Box, Jenkins & Reinsel, 1994).
- **Step 3: Model validation (or diagnostics).** It involves the examination of diagnostic statistics that measure goodness-of-fit of the data to the model and the asasses

of the extent to which the data conform to the necessary statistical assumptions.

- **Step 4: Prediction.** The best model selected is used for prediction.

FARIMA Model

Fractionally integrated ARMA (FARIMA) model has been introduced independently by Granger and Joyeux (1980) and Hosking (1981) in order to model processes with long-range dependence. FARIMA(p,d,q) is the natural extension of the ARIAM model when the degree of differencing d is allowed to real values. The time series $\{x_t\}$ is a stationary invertible FARIMA(p,d,q) process, if

$$\phi(B)\nabla^d x_t = \theta(B)a_t \tag{14}$$

where d is a real number, $d \in (-0.5, 0.5)$, $\phi(B)$ and $\theta(\mathrm{B})$ are stationary AR and invertible MA polynomials. The parameter d can be calculated by $d = H - 1/2$, where H is the Hurst parameter.

Let $\{x_t\}$ be the causal invertible FARIMA (p,d,q) process, we can write

$$x_t = \sum_{j=0}^{\infty} \psi_j a_{t-j} \tag{15}$$

and

$$a_t = \sum_{j=0}^{\infty} \pi_j x_{t-j} \tag{16}$$

where $\sum_{j=0}^{\infty} \psi_j B^j = \theta(B)\phi^{-1}(B)(1-B)^{-d}$ and $\sum_{i=0}^{\infty} \pi_j B^j = \theta(B)\phi^{-1}(B)(1-B)^{d}$.

Denote $\hat{x}_{t+h}$ as the best linear predictor of x_{t+h} in terms of $x_1, x_2, \ldots, x_t$, which is also called the h-step-ahead prediction of x_t. From the theorems on linear prediction (11), the h-step-ahead prediction $\hat{x}_{t+h}$ of a FARIMA process is

$$\hat{x}_{t+h} = \sum_{j=1}^{t+h-1} \pi_j \hat{x}_{t+h-j} \tag{17}$$

We can obtain their minimum mean square error of the h-step-ahead prediction

$$\hat{\sigma}_{t+h}^2 = E(x_{t+h} - \hat{x}_{t+h})^2 = \sigma^2 \sum_{j=0}^{h-1} \psi_j^2 \tag{18}$$

GARCH Model

If an autoregressive moving average model (ARMA model) is assumed for the error variance, the model is a generalized autoregressive conditional heteroskedasticity (GARCH) model (Bollerslev, 1986). The GARCH (p, q) model, where p is the order of autoregressive terms and q is the order of moving average, is defined by

$$x_t = z_t \sigma_t \tag{19}$$

where $\{z_t\}$ are independent and identically distributed standard normal random variables, i.e. $z_t \sim iid\ N(0,1)$

The conditional variance can be expressed as

$$\sigma_t^2 = a_0 + \sum_{i=1}^{q} \alpha_i x_{t-i}^2 + \sum_{i=1}^{p} \beta_i \sigma_{t-i}^2 \tag{20}$$

where $\alpha_0 > 0, \alpha_i \geq 0, \beta_i \geq 0$ and

$$\sum_{i=1}^{p} \beta_i + \sum_{i=1}^{q} \alpha_i < 1$$

When $\beta_i = 0, i = 1, 2, \ldots, p$, the above model reduces to ARCH(q) which was proposed by Engle (1982).

The general scheme to obtain the proposed GARCH model is as follows:

- **Step 1:** A general GARCH formulation is selected to model time series. This selection can be carried out by careful inspection of the main characteristics of the time series.
- **Step 2:** A model is identified for the observed data. Firstly, the observation of the autocorrelation and partial autocorrelation plots of the time series can help to make this selection. Secondly, the same observation of the residuals obtained in Step 3 can refine the structure of the functions in the model.
- **Step 3:** The model parameters are estimated. The maximal likelihood method will be used for estimating the parameters in GARCH (p, q).
- **Step 4:** A diagnosis check is needed to validate the model assumption of the GARCH model. To validate the assumptions of the GARCH model selection is a careful inspection of the ACF and PACF plots of the residuals. If the hypotheses of the model are validated, go to Step 5, otherwise go to Step 1 to refine the model.
- **Step 5:** The model can be used for prediction.

Artificial Neural Network Models

An ANN model is a model whose architecture essentially mimics the learning capability of the human brain. The processing elements of an ANN resemble the biological structure of neurons and the internal operation of a human brain. ANN is a non-linear, non-parametric and data driven modeling approach. It allows one to fully utilize the available information to determine the structure and parameters of a model without any restrictive parametric modeling assumptions. Artificial neural networks are important alternatives to the traditional methods of data analysis and modeling.

There are different types of neural networks, which can be distinguished on the basis of their structure and directions of information flow. Each kind of neural network has its own method of training. Generally, neural networks could be differentiated as follows.

Multi-Layer Feed-Forward Neural Networks

A multi-layer feed-forward neural network (FNN) is an artificial neural network where connections between the units do not form a directed cycle. In this network, the information moves in only one direction, forward, from the input nodes, through the hidden nodes, if any, and to the output nodes. There are no cycles or loops in the network. Each of input layer or output layer contains m input nodes or n output nodes such that each unit corresponds to a particular input or output variable. In the hidden layer, there are q hidden nodes connected to all input and output nodes. The strengths of such connections are labeled by unknown parameters known as the network connection weights. Figure 1 illustrates the architecture of a 3-layer feed-forward network.

The training of a multi-layer feed-forward neural network is usually accomplished by a back-propagation algorithm (BP) (Werbos, 1974; Rumelhart et al., 1986). When each entry of the sample set is presented to the network, the network examines its output response to the sample input pattern. The output response is then compared to the known and desired output and the error value will be calculated. Based on the error, the connection weights will be adjusted. The set of these sample patterns are repeatedly presented to the network until the error value is minimized.

Another popular multi-layer feed-forward network is the radial-basis function (RBF) network which has been introduced into the neural network literature by Broomhead & Lowe (1988), and its structure is shown in Figure 2.

Figure 1. A Feed-forward network

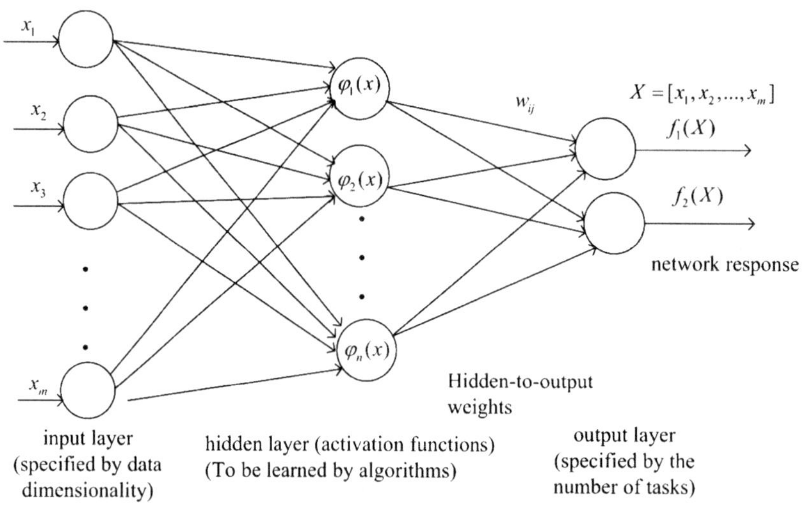

The radial basis function as implemented has a single hidden layer of units which can use one of three basis functions: Gaussian, thin plate spline, or multi-quadratic.

RBF networks differ from multi-layer feed-forward neural network with the back-propagation learning algorithm in some fundamental respects (Haykin, 2001):

- RBF networks are local approximations, whereas back-propagation FNN are global approximators.
- RBF networks have a single hidden layer, whereas back-propagation FNN can have any number of hidden layers.
- The activation function of the hidden layer in a RBF network computes the Euclidean

Figure 2. A radial-basis function network

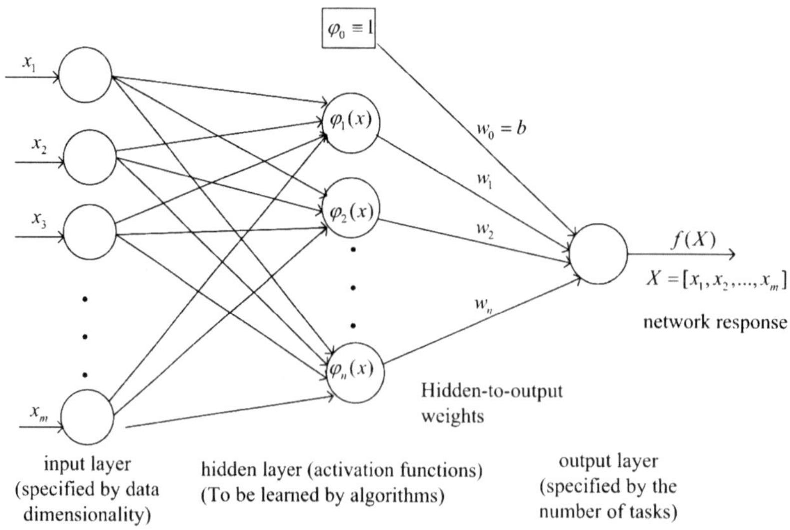

distance between the input signal vector and parameter vector of the network, whereas the activation function of a back-propagation FNN computes the inner product between the input signal vector and the pertinent synaptic weight vector.

- The adjustment of the output layer is simple and RBF has a guaranteed learning procedure for convergence. However, in back-propagation FNN, the parameters of transfer functions both in hidden and output layers should be adjusted by using the Sigmoid transfer functions and generally it is time-consuming (Akhlaghi & Kompany-Zareh, 2005).

Recurrent Neural Network

A recurrent neural network (RNN) is a class of neural network where connections between units form a directed cycle. This creates an internal state of the network which allows it to exhibit dynamic temporal behavior. There are many types of formal RNN models. The structure proposed by Elman (1990) is an illustration of this kind of architecture. Figure 3 illustrates the architecture of an Elman recurrent neural network. An Elman network structure is composed with an input layer, a hidden layer, a context layer and an output layer. The input and output units interact with the outside environment, while the hidden and context units do not. Figure shows the architecture of an Elman RNN. The nodes in the context layer (context neurons) hold a copy of the output of the hidden nodes. The output of each hidden node is copied into a specific node in the context layer. The value of the context neuron is used as an extra input signal for all the neurons in the hidden layer one time step later. Therefore, the Elman network has an explicit memory of one time lag (Elman, 1990).

The feed-forward back-propagation algorithm cannot be directly transferred to RNN, because the back-propagation pass presupposes that the connections between the units induce a cycle-free ordering. The RNN is trained by an algorithm based on a gradient such as back-propagation through time (BPTT) algorithm. More details about the BPTT networks and training algorithms can be found in (Werbos, 1990; Ahmad & Abdul Manan Ahmad, 2004).

The general procedures for developing the artificial neural network model are as follows (Su, Tong & Leou, 1997):

- **Step 1:** Normalize the learning set.
- **Step 2:** Decide the architecture and parameters: i.e., learning rate, momentum, and architecture.
- **Step 3:** Initialize all weights randomly.
- **Step 4:** Training. The stopping criterion is either the number of iterations reached or when the total sum of squares of error is lower than a pre-determined value.
- **Step 5:** Choose the network with the minimum error.
- **Step 6:** The best network is used for prediction.

Wavelet-Based Models

The discrete wavelet transform is based on the concept of multi-resolution analysis (MRA) introduced by Mallat (Mallat, 1989; Daubechies, 1992). It is a linear transformation with a special property of time and frequency localization at the same time.

The goal of discrete wavelet transform is to decompose arbitrary signal *f(t)* into a finite summation of wavelets at different scales or levels according to the expansion

$$f(t) = \sum_j \sum_k c_{j,k} \psi_{j,k}(t) \tag{21}$$

where $c_{j,k}$ are called the discrete wavelet transform coefficients of $f(t)$ and the function $\psi_{j,k}(t)$ are the wavelet expansion functions.

Figure 3. Elman recurrent neural network

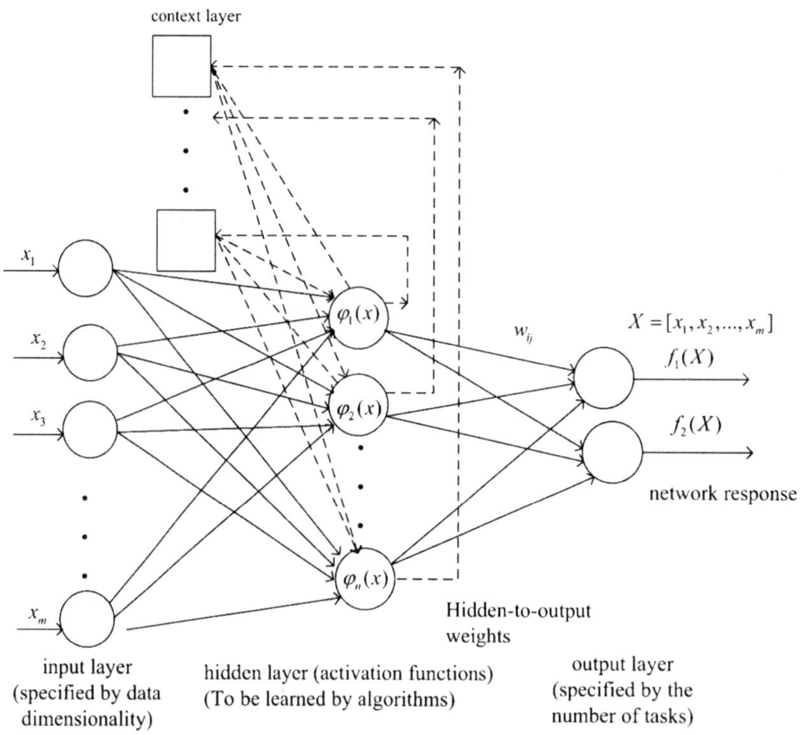

The coefficients are given by:

$$c_{j,k} = \int_{-\infty}^{+\infty} f(t)\psi_{j,k}(t)\,dt = \langle f, \psi_{j,k} \rangle \tag{22}$$

The wavelet basis functions can be computed form a function $\psi(t)$ called the mother wavelet through translation and dilation:

$$\psi_{j,k}(t) = 2^{-j/2}\psi(2^{-j}t - k), \quad j,k \in Z \tag{23}$$

where j is the translation and i is the dilation parameter. For some choices of $\psi(t)$, this is an orthonormal basis. Examples are the Haar wavelets and the Daubechies wavelets. The dilation equations

$$\begin{aligned} \phi(t) &= \sqrt{2}\sum_k g_k \phi(2t - k) \\ \psi(t) &= \sqrt{2}\sum_k h_k \phi(2t - k) \end{aligned} \tag{24}$$

are used for this purpose, where $\phi(t)$ is scaling function or father wavelet. The coefficients g_k and h_k are low-pass and high-pass filter coefficients, respectively, satisfying $h_k = (-)^k g_{1-k}$.

A function f can be represented as

$$f_0(t) = \sum_k c_{0,k}\phi_{0,k}(t) = \sum_k (c_{J,k}\phi_{J,k}(t) + \sum_{j=1}^{J} d_{j,k}\psi_{j,k}(t)) \tag{25}$$

where coefficient $c_{0,k}$ k are given and coefficients $c_{j+1,n}$ and $d_{j+1,n}$ at scale $j+1$. They can be obtained if coefficient at scale j is available:

$$\begin{aligned} c_{j+1,n} &= \sum_k c_{j,k} h(k - 2n) \\ d_{j+1,n} &= \sum_k c_{j,k} g(k - 2n) \end{aligned} \tag{26}$$

The smooth coefficients $c_{J,k}$ mainly capture the underlying smooth behavior of the data at the coarsest scale, while the details coefficients $d_{1,k}, d_{2,k}, \ldots, d_{J,k}$ representing deviations from the smooth behavior, and provide progressively finer scale deviations.

The multi-resolution decomposition of the original signal $f(t)$ is given by the following expression

$$f(t) = C_J + D_J + D_{J-1} + \ldots. + D_1 \quad (27)$$

where $C_F = \sum_k c_{J,k}\phi_{J,k}(t)$, $D_j = \sum_k d_{j,k}\psi_{j,k}(t)$ with $j = 1, 2, .., J$. The sequence of terms $C_J, D_J, D_{J-1}, \ldots.., D_1$ represents a set of signals components that provide representations of the signal at the different resolution levels 1 to *J*.

In practical applications, such decomposition can be obtained by using a specific wavelet. Several families of wavelets have been proven to be special useful in various applications. They differ with respect to orthogonality, smoothness and other related properties such as vanishing moments or size of the support. Readers interested in mathematical and practical aspects of wavelets are directed to the monographs of Chui (1992), Daubechies (1992), and Meyer (1992).

The basic idea of wavelet-based methods for prediction is to first decompose the original signal into components and then applies some predicting method to theses individual components. High frequency components can be used to predict the near future while low frequency components can usually tell the long-term trend. In other words, the lower level of the decomposition can capture the long-range dependencies, while the higher levels capture the usual short-term dependencies. The procedure of this algorithm can be described below:

- **Step 1:** Choose number of data points such that 2^L can be maximized within the time series interval.
- **Step 2:** Do the wavelet transform for time series, and then obtain the approximation coefficients and detail coefficients.
- **Step 3:** Perform prediction of wavelets coefficients for each level of scale.
- **Step 4:** Reconstruct the signal based on the coefficients predicted at each level. The forecast can be obtained based on the reconstructed signal. Then the final prediction for the original data can be obtained.

Support Vector Machine

SVM was developed by Vapnik and his coworkers in 1995 (Vapnik, 1995). It works based on the principle of structural risk minimization (SRM) which seeks to minimize an upper bound of the generalization error consisting of the sum of the training error and a confidence interval. This induction principle is different from the commonly used empirical risk minimization (ERM) principle which only minimizes the training error. Their major advantage over ANN is that they formulate the regression problem as a quadratic optimization problem. SVM performs by nonlinearly mapping the input data into a high dimensional feature space by means of a kernel function. Then the linear regression in the transformed space is performed. The whole process results in nonlinear regression in the low-dimensional space.

By the SVM, the basic idea is to map the data into a higher-dimensional feature space via a nonlinear mapping and then to do linear regression in this space (Cristianini & Shawe, 2000; Suykens, Gestel,et al. 2002; Cao & Tay, 2003). It is assumed that a set of a training data set $G = \{(x_i; d_i) \in R^n \times R, i = 1, 2, \ldots, l\}$, whichconsists of l pairs of training data $(x_i, d_i), i = 1, 2, \ldots, l$ are given. x_i is an input vector, d_i is the desired value. The issue of the regression is to determine a function that can approximate future values accurately. SVM approximates the function using the following form:

$$f(x) = w\phi(x) + b \tag{28}$$

where $w \in R^n, b \in R$ and ϕ represents a non-linear transformation form R^n to high-dimensional feature spaces. The coefficients w and b are estimated by minimizing the regularized risk function (2)

$$R = \frac{1}{2}\|w\|^2 + C\frac{1}{l}\sum_{i=1}^{l} L_\varepsilon(d_i, f(x_i)) \tag{29}$$

where

$$L_\varepsilon(d_i, f(x_i)) = \begin{cases} | d_i - f(x_i) | - \varepsilon, & | d_i - f(x_i) | \geq \varepsilon \\ 0 & \text{otherwise} \end{cases} \tag{30}$$

and ε is a prescribed parameter.

In equation (29), the first term $\|w\|^2$ is called the regularized term and used as a measure of function flatness. Minimizing $\|w\|^2$ will make a function as flat as possible, thus playing the role of controlling the function capacity. The second term $(1/l)\sum_{i=1}^{l} L_\varepsilon(d_i, f(x_i)$ is a so-called empirical error, which is measured by the $\varepsilon -$ insensitive loss function (30). The constant $C > 0$ is a regularized constant determining the tradeoff between the training error and model flatness. C and ε are both user-prescribed parameters and determined empirically.

The slack variables ξ_i, ξ_i^* $i = 1, \ldots n$ will be used to measure the deviation of training samples outside ε-insensitive zone. Thus, SVM regression is formulated as minimization of the following functional:

$$\min \{\frac{1}{2} \| w \|^2 + C\sum_{i=1}^{n}(\xi_i + \xi_i^*)\} \tag{31}$$

subject to

$$\begin{cases} y_i - f(\mathbf{x}_i, w) \leq \varepsilon + \xi_i^* \\ f(\mathbf{x}_i, w) - y_i \leq \varepsilon + \xi_i \\ \xi_i, \xi_i^* \geq 0, i = 1, \ldots, n \end{cases}$$

Finally, by introducing Lagrange multipliers and exploiting the optimality constraints, the decision function (1) has the following explicit form:

$$f(x) = \sum_{i=1}^{l}(a_i - a_i^*)K(x_i, x) + b \tag{32}$$

where a_i, a_i^* are the so-called Lagrange multipliers. They satisfy the equalities $a_i a_i^* = 0, a_i \geq 0, a_i^* \geq 0; i = 1, 2, \ldots, l$ and can be obtained by maximizing the dual form of Eq. (31), which has the following form:

$$W(a_i^*) = \sum_{i=1}^{l} d_i(a_i - a_i^*) - \varepsilon\sum_{i=1}^{l}(a_i + a_i^*) - \frac{1}{2}\sum_{i=1}^{l}\sum_{j=1}^{l}(a_i - a_i^*)(a_j - a_j^*)K(x_i, x_j) \tag{33}$$

with the following constrains

$$\begin{cases} \sum_{i=1}^{l}(a_i - a_i^*) = 0 \\ 0 \leq a_i \leq C \qquad , i = 1, 2, \ldots, l \\ 0 \leq a_i^* \leq C \end{cases}$$

$K(x_i, x_j)$ is defined as the kernel function. The value of the kernel is equal to the inner product of two vectors x_i and x_j in the feature space $\phi(x_i)$ and $\phi(x_j)$. That is $K(x_i, x_j) = \phi(x_i) \cdot \phi(x_j)$. The elegance of using the kernel function is that one can deal with feature spaces of arbitrary dimensionality without having to compute the map $\phi(x)$ explicitly. Any function that satisfies Mercer's condition can be used as the kernel function. Four commonly-used kernel functions are:

1. Linear: $K(x_i, x_j) = < x_i \cdot x_j >$
2. Polynomial:
   $K(x_i, x_j) = (< x_i \cdot x_j > + 1)^d, d = \text{degree}$

3. Radial Basis:
$$K(x_i, x_j) = \exp(-\left\|x_i - x_j\right\|^2 / 2\sigma^2), \sigma = \text{width}$$

The general prediction procedure can be described as follows:

- **Step 1:** Construct the SVM input vector
- **Step 2:** Choose the kernel function and the parameters used for the training SVM.
- **Step 3:** Train the SVM using the selected training data sets according to equation (6)
- **Step 4:** Repeat steps (2) to (3) by changing the parameter values until prediction error is reached
- **Step 5:** Train the training data sets by the best SVM model with the best parameters.
- **Step 6:** Use the best SVM model to predict

PREDICTION METHOD AND PERFORMANCE METRICS

Prediction Methods

One-Step-Ahead Prediction

One-step-ahead prediction (OSP) supposes that at the current time t, to predict $\hat{x}_{t+1}$ for the future time $t+1$ with the knowledge of the value $x_t, x_{t-1}, \ldots, x_1$ for past time t, t - 1, ..., 1, respectively. The prediction function is expressed as:

$$\hat{x}_{t+1} = g(x_t, x_{t-1}, \cdots, x_1) \quad (34)$$

Multi-Step-Ahead Prediction

By the multi-step-ahead prediction (MSP), the value of the time series at the time step $t+k$ is predicted recursively, using time series measurements up time t, where $k > 1$. MSP includes the direct prediction and the iterated prediction. Direct prediction is a method by which a model is built to directly predict $\hat{x}_{t+k}$. The direct prediction equation is expressed as:

$$\hat{x}_{t+k} = g(x_t, x_{t-1}, \cdots x_1) \quad (35)$$

The Iterated prediction is a method by which a model is built to predict one step ahead. To predict more than one step, earlier predictions are used as model of inputs. That is to say, the l -step-ahead prediction is performed by iterating a one-step-ahead estimator. The Iterated prediction equation is expressed as:

$$\hat{x}_{t+k} = g(\hat{x}_{t+k-1}, \hat{x}_{t+k-2}, \cdots \hat{x}_{t+1}, x_t, x_{t-1}, \cdots x_1) \quad (36)$$

It is clear that the multi-step-ahead forecasting uses the recent predicted values instead of the actual ones. This makes the performance of multi-step prediction become more deteriorative than that of one-step-ahead forecasting.

There have been many debates on two methods to find the better one. Although most researchers have found that the iterated prediction is more accurate than direct prediction (Casdagli, Jardins,et al., 1992), the iterated prediction is questionable because it has not taken into account the accumulated errors in the input vector (Stock& Watson, 1999). Clements and Hendry found in a Monte Carlo study that direct prediction can outperform iterated prediction in the cases when the negative moving average component is large (Clements & Hendry, 1996). Contributions to the theory of iterated vs. direct prediction can be found in the paper by Chevillon and Hendry and its references (Chevillon & Hendry, 2005). The authors (Chevillon & Hendry, 2005) have shown that direct prediction exhibited some robustness compared to iterated prediction. Their analysis has shown that direct prediction could prove either more efficient asymptotically or more precise than iterated prediction when the model is well specified regardless of stationary or non-stationary. The results have also indicated that even if a model is mis-specified for a non-stationary time series, direct prediction could predict more accurately in finite samples.

Performance Metrics

There are several metrics of measuring the performance of a prediction model. These statistics are given as follows:

The mean absolute error (MAE) is given by:

$$MAE = \frac{1}{M}\sum_{t=1}^{M} | x_t - \hat{x}_t | \quad (37)$$

where x_t is the observed value of the time series at time t, $\hat{x}_t$ is the predicted value of x_t, and M is the total number of the predicted values. If the MAE is zero, it indicates that the forecast is perfect and the value increases proportionally with the discrepancies in the forecast.

The mean squared error (MSE) is given by:

$$MSE = \frac{1}{M}\sum_{t=1}^{M}\left(x_t - \hat{x}_t\right)^2 \quad (38)$$

The MSE will be more sensitive to larger errors than the MAE. The MSE indicates a perfect forecast when it exhibits a zero with errors increasing with larger MSE values.

The root mean squared error (RMSE) is given by:

$$RMSE = \sqrt{\frac{1}{M}\sum_{t=1}^{M}\left(x_t - \hat{x}_t\right)^2} \quad (39)$$

The RMSE is simply the square root of the MSE and has the same physical dimensions as the forecasts and observations and is useful since it is considered as a typical magnitude for forecast errors.

The normalized mean squared error (NMSE) is given by:

$$NMSE = \frac{1}{\sigma^2}\frac{1}{M}\sum_{t=1}^{M}\left(x_t - \hat{x}_t\right)^2 \quad (40)$$

where σ^2 is the variance of the time series over the prediction duration. The NMSE is widely used for evaluating prediction performance. It can be seen that, for a perfect predictor, $NMSE = 0$ and for a trivial predictor, which statistically predicts the mean of the actual time series, $NMSE = 1$. If $NMSE > 1$, it means that the prediction performance is worse than that of trivial predictor.

TRAFFIC PREDICTION IN WLANS

The prediction of traffic in WLANs plays a fundamental role in the design and analysis of WLANs. In this section, we will examine the feasibility of applying SVM to predict actual WLANs traffic. We will also evaluate the performance of different prediction models such as ARIMA, FARIMA, artificial neural network, and wavelet-based model using three types of actual WLANs traffic.

WLANs Traffic Traces

Three traffic traces have been used to verify the proposed model. One trace (t030801) has been collected from the WLANs at the Network Research Laboratory at Tianjin University. The collection has been conducted at an AP in the laboratory, which carried all WLANs traffic to access the Internet.

Another one (final.anon) has been collected from the Mobile Computing Group at Stanford University (http1). The wireless network is a WaveLAN network with WavePoint II access points acting as bridges between the wireless and wired networks. Tang & Baker (2000) have described the network, methodology, and the characteristics of the trace of wireless traffic in detail.

The last one (trace.pcap) has been collected from the intranet traffic at the ACM SIGCOMM'01 conference held in the U.C. San Diego in August 2001(http2). The trace consists of two parts. The first part is the record of performance monitoring data sampled from wireless APs serving the conference. And the second part is the record of anonymized packet headers of all wireless traffic. The trace has been obtained by monitoring the traffic at the APs for three days of the conference capturing the workload of 300,000 flows from 195 users. The auditorium was covered by four ORiNOCO™ AP-1000 wireless APs. The subnet of APs was connected to a Cisco Catalyst 2924 switch over a 100BaseT link, which connected to the venue's intranet, then the campus gigabit backbone, and finally to the Internet. Balachandran et al. (2002) has described the configuration of the wireless network, methodology for trace collection and analysis (See Table 1).

One-Step-Ahead Prediction

In the prediction, we have divided each traffic trace into two parts. The first part is used to predict the future traffic. The predicted traffic has been compared with the second part (real traffic) to evaluate the performance of our prediction algorithm. We have trained 489, 293 and 294 samples, respectively. And we have constructed the SVM classifier with RBF kernel function. Then, we have predicted the next 32, 75 and 50 examples, respectively. Figure 4, Figure 5 and Figure 6 show the results of one-step-ahead prediction in comparison with the three actual WLANs traffic traces, respectively.

To evaluate the performance of WLANs traffic prediction with supporting vector regression. Some common baseline traffic prediction models such as ARIMA, FARIMA ANN and wavelet-based models have been exploited for performance comparison.

For all WLANs traffic traces, 1000 samples have been trained by the baseline prediction models and then next samples have been predicted. The OSP performances for various prediction models have been shown in Table 2.

From the statistics summarized in Table 2, it can be obviously observed as follows.

One can conclude that the accurate ratio of SVM predictor is more precise than that of other predictors. For each WLANs traffic trace, both the MSE and NMSE produced by the SVM model are smaller than those obtained by the other prediction models.

Compared with other predictors, ARIMA model delivers worse forecasting performance. Although ARIMA models are proved to be quite powerful to model a class of non-stationary data, it can't capture long-range dependent characteristics.

The ANN prediction shows better superiority to the FARIMA model on the same testing set because the ANN model can capture the nonlinear nature of network traffic. However, ANN might suffer from over-fitting. Although FARIMA models was considered to be preferable to forecasting network traffic (Sadek, Khotanzad & Chen, 2003; Doulamis, Doulamis & Kollias, 2003), its prediction performance is worse than that of ANN and SVM models in our experiments.

The wavelet-based forecasting can't give remarkable performance as in the work (Qiao, Skicewicz & Dinda, 2004). This is due to the ef-

Table 1. WLANs traffic statistical parameters

Trace name	Max (pkts/sec)	Min (pkts/sec)	Mean (pkts/sec)	Standard variance
t030801	614	0	278.3411	96.0897
final.anon	187	0	64.9151	42.6428
sigcomm'01	676	16	120.4476	123.0416

Figure 4. One-step-ahead prediction of t030801

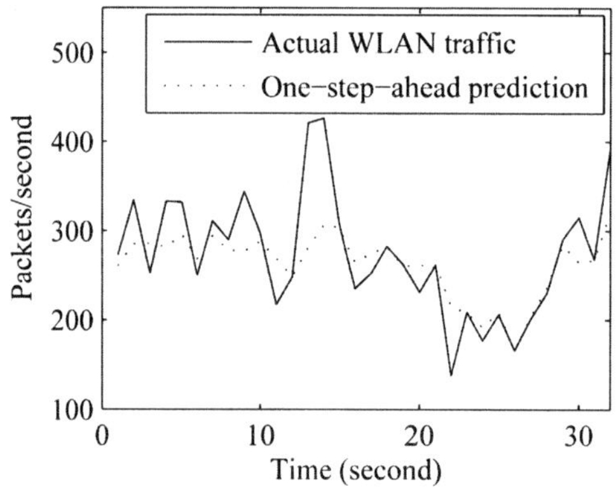

fect of boundary conditions in wavelet transform. Although it has worse prediction performance compared to other predictors, wavelet-based prediction is attractive due to its lower computational complexity. Compared with baseline prediction models, there are only a few training samples to achieve desired prediction accuracies using SVM.

The OSP performance for various prediction models is given in Table 3 in unit of bytes/second. The same conclusion, that the SVM model possesses better prediction performance than those above-mentioned models, can be obtained.

Multi-Step-Ahead Prediction

Figure 7, Figure 8, and Figure 9 show the results of four-step-ahead prediction in comparison with the three actual WLANs traffic traces respectively. Performance metrics of four- and five-step-ahead prediction for all the WLANs traffic traces have also be provided in Table 4 and 5. The performance metrics shown in Table 2, Table 3, Table 4, and Table 5 indicate deterioration in the performance of the MSP prediction compared with that of the OSP. From Table 2, Table 3, and Table 4, it can be seen that, for trace t030801 both the NMSE and MSE of four-step-ahead produced by the SVM

Figure 5. One-step-ahead prediction of final.anon

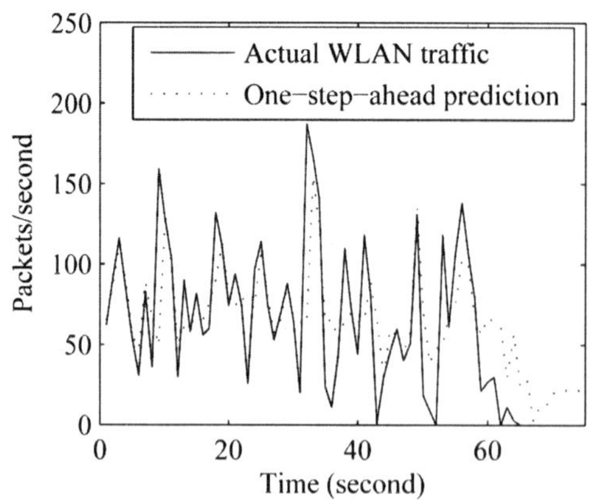

Figure 6. One-step-ahead prediction of sigcomm'01

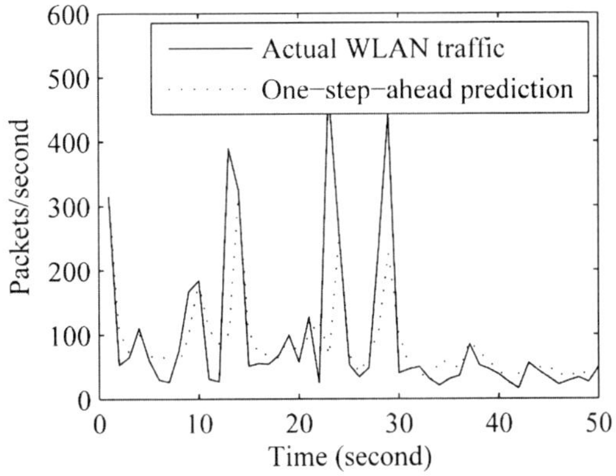

model are even smaller than that of OSP obtained by other prediction models. For trace final.anon in the unit of packets/sec, both the NMSE and MSE of four-step-ahead produced by SVM model are almost equal to that of OSP provided by ARIMA model. For trace final.anon in the unit of bytes/sec, these performance metrics are smaller than that of OSP obtained by other prediction models. For trace sigcomm'01, its four-step-ahead performance of the SVM predictor is better than that of OSP obtained by ARIMA and FARIMA models.

Computational Complexity

The computational complexity for FARIMA and ARIMA predictors are $O(n^2)$ (Krunz & Makowski, 1998). When training back-propagation neural networks, there are several factors that determine the computational complexity. It is assumed that the network is fully connected. The variable i, n, o represents the number of input nodes, hidden nodes and output nodes, respectively. The total order of complexity is $O(i * o * n + n * o)$ for training a single epoch (Istook& Martinez, 2002). The algorithm, which computes the orthogonal wavelet coefficients of a time-series of length, n, is very

Table 2. Performance metrics of the one-step-ahead prediction in unit of packets/second

Models	Metrics	t030801	final.anon	sigcomm'01
ARIMA	MSE	4081	1881	11632
	NMSE	0.87	0.8	0.9
FARIMA	MSE	3954	1720	10829
	NMSE	0.84	0.72	0.83
ANN	MSE	3411	1601	9819
	NMSE	0.73	0.68	0.76
Wavelet	MSE	3446	1836	11794
	NMSE	0.74	0.79	0.91
SVM	MSE	2025	1095	7194
	NMSE	0.44	0.47	0.55

Table 3. Performance metrics of the one-step-ahead prediction for various WLANs traffic in unit of bytes/second

Models	Metrics	t030801	final.anon	sigcomm'01
ARIMA	MSE	1233039616	1537012021	371469993
	NMSE	0.81	0.87	0.77
FARIMA	MSE	1186539169	1472094166	369381802
	NMSE	0.78	0.83	0.77
ANN	MSE	1204150178	1472188921	351671999
	NMSE	0.79	0.83	0.73
Wavelet	MSE	1292082697	1539463459	384604427
	NMSE	0.85	0.87	0.80
SVM	MSE	418839085	723103523	160830775
	NMSE	0.27	0.41	0.33

fast and its complexity is of the order of n, e.g. $O(n)$ (Daubechies, 1992).

The optimization problems for SVM can be formulated as standard quadratic programming (QP) problems with complexity of $O(n^3)$. It is too time consuming in the case of a large number of samples. Many attempts have been made to reduce the computational complexity of SVM algorithms. A sequential minimal optimization (SMO) training algorithm for the SVM was suggested by Platt (1998). It has been considered as the state-of-the-art SVM implementation, with training time complexity empirically observed to be between $O(n)$ and $O(n^{2.3})$ (Keerthi, Shevade, et al., 2001). Moreover, computationally complexity of SMO is linear in the number of support vectors (Salomon, King & Osborne, 2002). In our experiment, an implementation of a SMO training algorithm has been taken to achieve low complexity. With recent developments in efficient algorithms, coupled with the rapid improvement of computer performance, real-time prediction of SVM is now feasible (Zhang & Shen, 2005; Wang, Men & Lu, 2008; Wang, Pi, & Sun,. 2007; Alenezi, Moses & Trafalis, 2008).

Figure 7. Four-step-ahead prediction of t030801

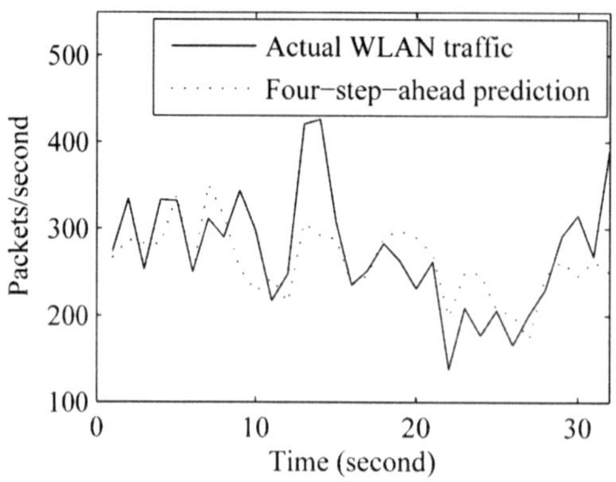

Figure 8. Four-step-ahead prediction of final.anon

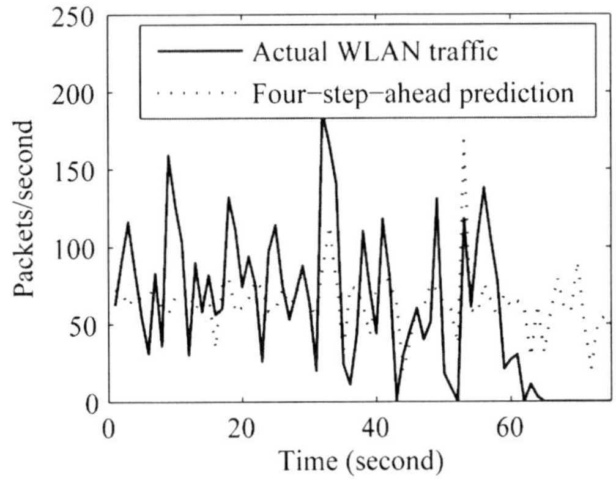

FUTURE RESEARCH DIRECTIONS

Several prediction models have been proposed in the literature to deal with the traffic prediction over various wired networks. But there are only a few studies on the prediction of the traffic over wireless networks. SVM is a successful model for the prediction of wireless traffic. Compared to ARIMA, FARIMA, ANN and wavelet-based predictors, the SVM predictor possesses better prediction performance due to its features of noise-tolerance, high stability, adaptive properties and better generalization performance. The trace collection is still ongoing, recently, some lager wireless traces have been made available at CRAWDAD (CRAWDAD). In the future, it is necessary to analyze significantly large and broad traffic traces over WLANs with hope to achieve moer accurate predictions. Moreover, it is the fact that performance of SVM is sensitive to the proper choice of a kernel function and the parameters. The optimal choice of kernel function and SVM parameters affect the SVM prediction performance in wireless networks.

Seeking a perfect prediction models is not the final goal. The final goal is to make the maximum usage of wireless network resources. We plan to provide an in-depth study on the performance of different prediction models with various trace, seek a perfect prediction models for wireless traffic,

Figure 9. Four-step-ahead prediction of sigcomm'01

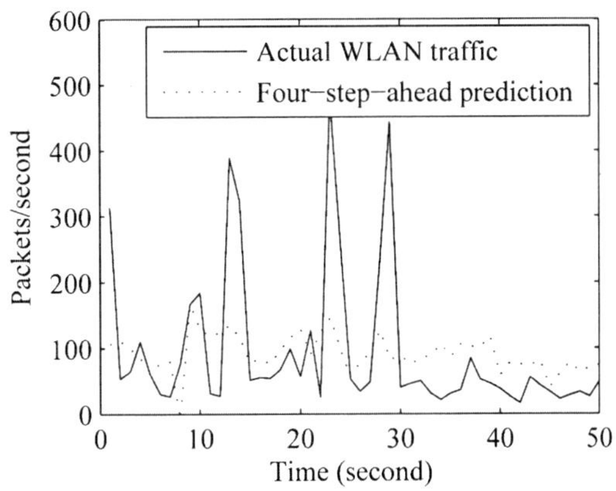

Table 4. Performance metrics of the four-step-ahead prediction for various WLANs traffic

Models	Metrics	t030801	final.anon	sigcomm'01
packets/second	MSE	3056	1889	10792
	NMSE	0.66	0.81	0.82
bytes/second	MSE	447584308	811438475	351351775
	NMSE	0.29	0.46	0.73

Table 5. Performance metrics of the five-step-ahead prediction for various WLANs traffic

Models	Metrics	t030801	final.anon	sigcomm'01
packets/second	MSE	4220	2667	11831
	NMSE	0.90	0.97	0.91
bytes/second	MSE	1104243984	1600349537	410965388
	NMSE	0.72	0.90	0.86

and investigate prediction in various time-scales. High accurate traffic prediction on large time scale is the base of long term planning of the resources of wireless networks. Prediction on small time scale is necessary for dynamic bandwidth allocation, admission control. This is very important for bandwidth constrained wireless networks. The prediction-based dynamic bandwidth allocation and admission control methods are suitable to real-time applications over the wireless networks and can improve the network utilization. Providing real-time traffic prediction is essential and necessary for wireless network resource management. Therefore, providing adaptive models that can be used for real-time prediction of traffic over wireless networks will be the future work.

CONCLUSION

In this section, the performance of SVM has been evaluated by using three actual WLANs traffic traces. Firstly, a SVM-based one-step-ahead prediction has been performed. From the results, it can be concluded that, compared to ARIMA, FARIMA, ANN and wavelet-based predictors, the SVM model can predict WLANs traffic with higher estimation accuracy. Secondly, a multi-step-ahead prediction has been developed for WLANs traffic. As expected, although the results of multi-step-ahead prediction have exhibited a deterioration behavior compared to that of the one-step-ahead prediction, the results have indicated that a lower horizon forecast can be obtained with little degradation in the performance. That is to say, the SVM-based multi-step-ahead prediction is able to provide a good prediction performance within a finite future horizon.

A better prediction model should be adaptive to the change of the network traffic. The choice of a prediction model is a tradeoff between the prediction interval, prediction error and computational cost. In the future, more traffic information will be available and more characteristics will be known. More adaptive traffic prediction models could be built up, which can improve and update the parameters to capture new available information and behave new characteristics.

ACKNOWLEDGMENT

I would like to thank my supervisor, Prof. Yantai Shu of Tianjin University. Thanks for instructing me through all the steps in my study. I have benefited a lot and will still benefit in the future of my life from his enthusiastic, optimistic and critical view to both research and life. This research was supported in part by the National Natural Science Foundation of China (NSFC) under grant No. 90604013, No. 60702038, the National 863 Program of China under grant No. 2007AA01Z220, the Cultivation Fund of the Key Scientific and Technical Innovation Project Ministry of Education of China under grant No. 708024, the Science and Research Project of the Education Department of Gansu Province (0901-03), NWNU-KJCXGC-03-52, NWNU-KJCXGC (network computing 2006-2010).

REFERENCES

Adas, A. (1997). Traffic models in broadband networks. *IEEE Communications Magazine, 35*, 82–89. doi:10.1109/35.601746

Ahmad, A. M., & Abdul Manan Ahmad. (2004). Recurrent neural network with backpropagation through time algorithm for arabic recognition. In *proceedings of 18th European Simulation Multiconference Graham Horton(c) SCS Europe*.

Akhlaghi, Y., & Kompany-Zareh, M. (2005). Comparing radial basis function and feed-forward neural networks assisted by linear discriminant or principal component analysis for simultaneous spectrophotometric quantification of mercury and copper. *Analytica Chimica Acta, 537*, 331–338. doi:10.1016/j.aca.2004.12.079

Alenezi, A., Moses, S., & Trafalis, T. B. (2008). Real-time prediction of order flow times using support vector regression. *Computers & Operations Research, 35*, 3489–3503. doi:10.1016/j.cor.2007.01.026

Balachandran, A., Voelker, G. M., Bahl, P., et al. (2002). Characterizing user behavior and network performance in a public wireless LAN. In proceedings of ACM SIGMETRICS Performance Evaluation Review, 195-205.

Beran, J. (1994). Statistics for long-Memory Processes. New York: Chapman and Hall.

Bermolen, P., & Rossi, D. (2008). Support vector regression for link load forecast. In *Proceedings of Telecommunication Networking Workshop on QoS in Multiservice IP Networks*, 268-273.

Beverly, R., Sollins, K., & Berger, A. (2006). SVM Learning of IP address structure for latency prediction. In *Proceedings of SIGCOMM Workshop on Mining Network Data*, 299-304.

Bhattacharya, A., Parlos, A. G., & Atiya, A. F. (2003). Prediction of MPEG-coded video source traffic using recurrent neural networks. *IEEE Transactions on Signal Processing, 51*(8), 2177–2190. doi:10.1109/TSP.2003.814470

Bollerslev, T. (1986). Generalized autoregressive conditional heteroscedasticity. *Journal of Econometrics, 31*, 307–327. doi:10.1016/0304-4076(86)90063-1

Box, G., Jenkins, G. M., & Reinsel, G. (1994). Time series analysis: Forecasting and control, (3rd ed.). Upper Saddle River, NJ: Prentice-Hall.

Broomhead, D., & Lowe, D. (1988). Multivariable functional interpolation and adaptive Networks. *Complex Systems, 2*, 321–355.

Cao, L. J., & Tay, F. E. H. (2003). Support vector machine with adaptive parameters in financial time series forecasting. *IEEE Transactions on Neural Networks, 14*(6), 1506–1518. doi:10.1109/TNN.2003.820556

Casdagli, M., Jardins, D. D., Eubank, S., et al. (1992). Nonlinear modeling of chaotic time series: Theory and applications. In J. H. Kim & J. Stringer (Eds.), Applied Chaos (pp. 335-380). New York: John Wiley Sons, Inc.

Chang; P. R., & Hu, J, T, (1997). Optimal nonlinear adaptive prediction and modeling of MPEG video in ATM networks using pipelined recurrent neural networks, *Selected Areas in Communications, 15*(6), 1087-1100.

Chevillon, G., & Hendry, D. (2005). Non-parametric direct multi-step estimation for forecasting economic processes. *International Journal of Forecasting, 21*, 201–218. doi:10.1016/j.ijforecast.2004.08.004

Chisci, L., Pecorella, T., & Fantacci, R. (2006). Dynamic bandwidth allocation in GEO satellite networks: A predictive control approach. *Control Engineering Practice, 14*, 1057–1067. doi:10.1016/j.conengprac.2005.06.010

Chui, K. (1992). An introduction to wavelets. New York: Academic Press.

Clements, M. P., & Hendry, D. F. (1996). Multi-step estimation for forecasting. *Oxford Bulletin of Economics and Statistics, 58*(4), 657–684.

Corradi, M. Garroppo, R.G., Giordano, S. & Pagano, M. (2001). Analysis of f-ARIMA processes in the modeling of broadband traffic. *Proceedings of IEEE International Conference on Communications, 3*, 964-968.

CRAWDAD. (n.d.). Retrieved from http://crawdad.cs.dartmouth.edu

Cristianini, N., & Shawe-Taylor, J. (2000). An Introduction to Support Vector Machines and Other Kernel-Based Learning Methods. Cambridge, UK: Cambridge Univ. Press.

Daubechies, I. (1992). Ten lectures on wavelets. SIAM, Philadelphia, PA.

Doulamis, A. D., Doulamis, N. D., & Kollias, S. D. (2003). An adaptable neural-network model for recursive nonlinear traffic prediction and modeling of MPEG video sources. *IEEE Transactions on Neural Networks, 14*(1), 150–166. doi:10.1109/TNN.2002.806645

Elman, J. L. (1990). Finding structure in time. *Cognitive Science, 14*(2), 179–211.

Engle, R. F. (1982). Autoregressive conditional heteroscedasticity with estimates of the variance of united kingdom inflation. *Econometrica, 50*, 987–1008. doi:10.2307/1912773

Feldmann, A., Gilbert, A. C., & Willinger, W. (1998). Data networks as cascades: investigating the multifractal nature of Internet WAN traffic. *Computer Communication Review, 28*(4), 42–55. doi:10.1145/285243.285256

Feng, H. F., & Shu, Y. T. Wang, S. Y. & Ma, M. D. (2006). SVM-Based models for predicting WLAN traffic. In *Proceedings of IEEE International Conference on Communications*, (pp. 597-602).

Ghaderi, M., Capka, J., & Boutaba, R. (2003). Prediction-based admission control for DiffServ wireless internet. *Proceedings of IEEE Vehicular Technology Conference, 3*, 1974–1978.

Gowrishankar & Satyanarayana, P.S. (2008). Neural network based traffic prediction for wireless data networks. *International Journal of Computational Intelligence Systems, 1*(4), 379–389. doi:10.2991/ijcis.2008.1.4.9

Granger, C. W. J., Joyeux. R. (1980). An introduction to long memory time series models and fractional differencing. *Journal of Time Series Analysis, 1*, 15–39. doi:10.1111/j.1467-9892.1980.tb00297.x

Haffner, P., Tur, G., & Wright, J. H. (2003). Optimizing SVMs for complex call classification. *Proceedings of International Acoustics, Speech, and Signal Processing, 1*, 632–635.

Haykin, S. (2001). *Feedforward: Neural Networks: An Introduction*. Retrieved from media.wiley.com/product_data/excerpt/19/04713491/0471349119.pdf

Hosking, J. R. M. (1981). Fractional differencing. *Biometrika, 68*(1), 165–176. doi:10.1093/biomet/68.1.165

http://mosquitonet.stanford.edu/software.html. (n.d.).

http://ramp.ucsd.edu/pawn/sigcomm-trace/. (n.d.).

Huang, C., Lai, W., Luo, R., & Yan, Y. (2005). Application of support vector machines to bandwidth reservation in sectored cellular communications. *Engineering Applications of Artificial Intelligence, 18*, 585–594. doi:10.1016/j.engappai.2004.11.009

IEEE. (1999). *IEEE Std 802.11b-Wireless LAN Medium Access Control (MAC) and Physical layer (PHY) Specifications: High Speed Physical Layer (PHY) in the 2.4 GHZz Band, 1999.*

Istook, E., & Martinez, T. (2002). Improved backpropagation learning in neural networks with windowed momentum. *International Journal of Neural Systems, 12*(3-4), 303–318. doi:10.1142/S0129065702001114

Jiang, M., Nikolic, M., Hardy, S., & Trajkovic, L. (2001). Impact of self-similarity on wireless data Network performance. In *Proceedings of IEEE International Conference on Communications*, (pp. 477-481).

Keerthi, S., Shevade, S. K., Bhattacharyya, C., & Murthy, K. (2001). Improvements to Platt's SMO algorithm for SVM classifier design. *Neural Computation, 13*(3), 637–649. doi:10.1162/089976601300014493

Khan, L., Awad, M., & Thuraisingham, B. (2007). A new intrusion detection system using support vector machines and hierarchical clustering. *The Very Large Data Bases Journal, 16*(4), 507–521. doi:10.1007/s00778-006-0002-5

Khotanzad, A., & Sadek, N. (2003). Multi-scale high-speed network traffic prediction using combination of neural network. *Proceedings of the International Joint Conference on Neural Networks, 2*, 1071–1075.

Kim, M., & Noble, B. (2001). Mobile network estimation. In *Proceedings of the Seventh Annual International Conference on Mobile Computing and Networking*, (pp. 298-309).

Krithikaivasan, B. Dekal K., & Medhi, D. (2004). Adaptive bandwidth provisioning envelope based on discrete temporal network measurements. In *Proceedings of IEEE Computer and Communications Societies*, (pp. 1786-1796).

Krunz, M., & Makowski, A. (1998). A source model for VBR video traffic based on M/G/∞ input processes. In *Proceedings of IEEE Computer and Communications Societies*, (pp. 1441-1449).

Li, C. H., Zhu, X. J., Cao, G. Y., Sheng, S., & Hu, M. R. (2008). Identification of the Hammerstein model of a PEMFC stack based on least squares support vector machines. *Journal of Power Sources*, (1): 303–316. doi:10.1016/j.jpowsour.2007.09.049

Liang, J., & Du, R. (2007). Model-based fault detection and diagnosis of HVAC systems using support vector machine method. *International Journal of Refrigeration, 30*(6), 1104–1114. doi:10.1016/j.ijrefrig.2006.12.012

Liang, Q. (2002). Ad Hoc wireless network traffic—self-similarity and forecasting. *IEEE Communications Letters, 6*(7), 297–299. doi:10.1109/LCOMM.2002.801327

Mallat, S. (1989). A theory for multiresolution signal decomposition: the wavelet representation. *IEEE Transactions on Pattern Analysis and Machine Intelligence, 11*, 674–693. doi:10.1109/34.192463

Meyer, Y. (1992). Wavelets and Operators. Cambridge, UK: Cambridge University Press.

Mirza, M., Sommers, J., Bardford, P., & Zhu, X. (2007). A machine learning approach to TCP throughput prediction. In *Proceedings of the 2007 ACM SIGMETRICS international conference on Measurement and modeling of computer systems*, (pp. 97-108).

Ostring, S., Sirisena, H., & Hudson, I. (2000). Designing and managing networks for self-similarity. In Proceedings of Engineering Mathematics and Applications, (pp. 235-238).

Papadopouli, M., Shen, H., Raftopoulos, E., Ploumidis, M., & Campos, F. (2005). Short-term traffic forecasting in a campus-wide wireless network. In *Proceedings of 16th Annual IEEE International Symposium on Personal Indoor and Mobile Radio Communications,* (pp. 1446-1452).

Papagiannaki, K. Taft, N. Zhang, Z., & Diot, C. (2003). Long-term forecasting of internet backbone traffic: Observations and initial models. In *Proceedings of IEEE Computer and Communications Societies*, (pp. 1178-1188).

Park, K., & Willinger, W. (Eds.). (2000). Self-Similar Network Traffic and Performance Evaluation. New York: John Wiley & Sons, Inc.

Platt, J. (1998). *Sequential minimal optimization: a fast algorithm for training support vector machines,* (Tech. Rep. MSR-TR-98-14). Microsoft Research; 1998.

Qiao, Y. Skicewicz, J., & Dinda, P. (2004). An empirical study of the multiscale predictability of network traffic. In *Proceedings of the 13th IEEE International Symposium on High Performance Distributed Computing*, (pp. 66-76).

Qiu, D., & Shroff, N. B. (2004). A predictive flow control scheme for efficient network utilization and QoS. *IEEE/ACM Transactions on Networking, 12*(1), 161-172.

Rumelhart, D. E., Hinton, G. E., & Williams, R. J. (1986). Learning internal representations by error propagation. In D. E. Rumelhart & J. L. McCleland (Eds.), Parallel Distributed Processing, (vol. 1). Cambridge, MA: MIT Press.

Sadek, N., Khotanzad, A., & Chen, T. (2003). ATM dynamic bandwidth allocation using F-ARIMA prediction model, In *Proceedings of International Conference on Computer Communications and Networks*, (359-363).

Sakiyama, Y., Yuki, H., Moriya, T., Hattori, K., Suzuki, M., Shimada, K., & Honma, T. (2008). Predicting human liver microsomal stability with machine learning techniques. *Journal of Molecular Graphics & Modelling*, (6): 907–915. doi:10.1016/j.jmgm.2007.06.005

Salomon, J., King, S., & Osborne, M. (2002). Framewise phone classification using support vector machines. In Proceedings of International Spoken Language Processing, (2645-2648).

Sang, A., & Li, S. (2002). A predictability analysis of network traffic. *Computer Networks, 39*, 329–345. doi:10.1016/S1389-1286(01)00304-8

Schmidt, M. (1996). Identifying speaker with support vector networks. In Proceedings of Interface, (pp. 305-308).

Shah, K., Bohacek, S., & Jonckheere, E. (2003). On the predictability of data network traffic. In *Proceedings of the 2003 American Control Conference, 2*, 1619-1624.

Shu, Y. T., Feng, H. F., Wang, H., & Ma, M. D. (2006). Dynamic admission control in hybrid QoS networks with WiFi access. *Science in China Series F-. Information Sciences*, *49*(4), 409–420.

Shu, Y. T., Jin, Z. G., Wang, J. D., & Yang, O. W. (2000). Prediction-based admission control using FARIMA models. *Proceedings of IEEE International Conference on Communications, 3*, 1325–1329.

Shu, Y. T., Jin, Z. G., Zhang, L. F., et al. (1999). Traffic prediction using FARIMA models. In *Proceedings of IEEE International Conference on Communications*, (pp. 891-895).

Stemm, M., Katz, R., & Seshan, S. (2000). A network measurement architecture for adaptive applications. *Proceedings of IEEE Computer and Communications Societies*, *1*, 285–294.

Stock, J. H., & Watson, M. W. (1999). A comparison of linear and nonlinear univariate models for forecasting macroeconomic time-series. In R. F. Engle & H. White, (Eds.), Cointegration, causality and forecasting: A Festschrift in honour of Clive W. J. Granger, (pp. 1-44). Oxford, UK: Oxford University Press.

Su, C. T., Tong, L. I., & Leou, C. M. (1997). Combination of time series and neural network for reliability forecasting modeling. *Journal of Chinese Institute of Industrial Engineers*, *14*(4), 419–429.

Sung, A. H., & Mukkamala, S. (2003). Identifying important features for intrusion detection using SVM and neural networks. In *Proceedings of the 2003 International Symposium on Applications and the Internet Technology*, (pp. 209-216).

Suykens, J. A. K., Gestel, T. V., Brabanter, J. D., Moor, B. D., & Vandewalle, J. (2002). Least squares support vector machines. Hoboken, NJ: World Scientific.

Tang, D., & Baker, M. (2000). Analysis of a local-area wireless network. In *Proceedings of ACM Mobile Computing and Networking*, (pp. 1-10).

Tarraf, A. A., Habib, I. W., Saadawi, T. N., & Ahmed, S. A. (1993). ATM multimedia traffic prediction using neural networks. In *Proceedings of IEEE Global Data Networking*, (pp. 77–84).

Vapnik, V. N. (1995). The Nature of Statistical Learning Theory. New York: Springer-Verlag.

Wang, H., Pi, D., & Sun, Y. (2007). Online SVM regression algorithm-based adaptive inverse control. *Neurocomputing*, *70*, 952–959.

Wang, W., Men, C., & Lu, W. (2008). Online prediction model based on support vector machine. *Neurocomputing*, *71*, 550–558. doi:10.1016/j.neucom.2007.07.020

Werbos, P. (1990). Backpropagation trough time, what it does and how do it. *Proceedings of IEEE on Special Issue on Neural Networks*, *78*, 1550–1560.

Werbos, P. J. (1974). *Beyond regression: New tools for prediction and analysis in the behavioral sciences*. Ph.D. Thesis, Harvard University, Cambridge, MA.

You, C., & Chandra, K. (1999). Time series models for Internet data traffic. In *Proceedings of the 24th Conference on Local Computer Networks*, (pp. 164-171).

Zhang, Z., & Shen, H. (2005). Application of online-training SVMs for real-time intrusion detection with different considerations. *Computer Communications*, *28*, 1428–1442. doi:10.1016/j.comcom.2005.01.014

Zhou, B. He, D., & Sun, Z. (2006). Traffic modeling and prediction using ARIMA/GARCH model. In *Proceedings of 2nd EuroNGI Conference on Next Generation Internet Design and Engineering*.

ADDITIONAL READING

Alenezi, A., Moses, S., & Trafalis, T. B. (2008). Real-time prediction of order flow times using support vector regression. *Computers & Operations Research*, *35*, 3489–3503. doi:10.1016/j.cor.2007.01.026

Balachandran, A., Voelker, G. M., & Bahl, P. (2002). Characterizing user behavior and network performance in a public wireless LAN, *proceedings of ACM SIGMETRICS. Performance Evaluation Review*, 195–205. doi:10.1145/511399.511359

Bermolen, P., & Rossi, D. (2008). Support vector regression for link load forecast, *Proceedings of Telecommunication Networking Workshop on QoS in Multiservice IP Networks*, 268-273.

Clements, M. P., & Hendry, D. F. (1996). Multi-step estimation for forecasting. *Oxford Bulletin of Economics and Statistics*, *58*(4), 657–684.

Farkas, K., Hossmann, T., Legendre, F., Plattner, B., & Das, S. K. (2008). Link quality prediction in mesh networks. *Computer Communications*, *31*, 1497–1512. doi:10.1016/j.comcom.2008.01.047

Ghaderi, M. (2003). On the relevance of self-similarity in network traffic prediction (Tech. Rep., CS-2003-28). School of Computer Science, University of Waterloo, Waterloo.

Gowrishankar & Satyanarayana, P.S. (2008). Neural network based traffic prediction for wireless data networks. *International Journal of Computational Intelligence Systems*, *1*(4), 379–389. doi:10.2991/ijcis.2008.1.4.9

Hernández-Campos, F. & Maria Papadopouli (2005), Assessing the real impact of 802.11 WLANs: a large-scale comparison of wired and wireless traffic. *Proceedings of the 14th IEEE Workshop on Local and Metropolitan Area Networks (LANMAN)*, Chania, Crete, Greece, September 18-21.

Katsaros, D., & Manolopoulos, Y. (2009). Prediction in wireless networks by Markov chains. *IEEE Wireless Communications*, *16*(2), 56–63. doi:10.1109/MWC.2009.4907561

Papadopouli, M., Shen, H., Raftopoulos, E., Ploumidis, M., & Campos, F. (2005). Short-term traffic forecasting in a campus-wide wireless network. *Proceedings of 16th Annual IEEE International Symposium on Personal Indoor and Mobile Radio Communications*, 1446-1452.

Park, K., & Willinger, W. (Eds.). (2000). Self-Similar Network Traffic and Performance Evaluation, New York, John Wiley & Sons, Inc.

Qiao, Y. Skicewicz, J., & Dinda, P. (2004). An empirical study of the multiscale predictability of network traffic. *Proceedings of the 13th IEEE International Symposium on High Performance Distributed Computing*, 66-76.

Sun, J. (2004). *Wireless video: Complexity, adaptability and predictability.* Unpublished Lic. Thesis, Umea University, Sweden.

Wang, H., Pi, D., & Sun, Y. (2007). Online SVM regression algorithm-based adaptive inverse control. *Neurocomputing*, *70*, 952–959.

Wang, W., Men, C., & Lu, W. (2008). Online prediction model based on support vector machine. *Neurocomputing*, *71*, 550–558. doi:10.1016/j.neucom.2007.07.020

Wu, C. F., Lee, L. T., & Tao, D. F. (2007). An HMM prediction and throttling-based call admission control scheme for wireless multimedia networks. *Computers & Mathematics with Applications (Oxford, England)*, *54*, 364–378. doi:10.1016/j.camwa.2007.01.035

Chapter 6
Cross-Layer Optimization for Energy-Efficient QoS Support of Multimedia Streams

Carolina Blanch Perez del Notario
Interuniversity MicroElectronics Center (IMEC), Belgium

Sofie Pollin
Interuniversity MicroElectronics Center (IMEC), Belgium

Tong Gan
Interuniversity MicroElectronics Center (IMEC), Belgium

Claude Desset
Interuniversity MicroElectronics Center (IMEC), Belgium

Antoine Dejonghe
Interuniversity MicroElectronics Center (IMEC), Belgium

Bart Masschelein
Interuniversity MicroElectronics Center (IMEC), Belgium

ABSTRACT

A major limitation for wireless video communication on portable devices is the limited energy budget. For this reason, efficient usage of the scarce energy becomes a critical design constraint, in addition to meeting the Quality of Service constraints related to the video quality. In this chapter the authors focus on minimizing the energy cost of the two main energy consumers in the handheld wireless video device: the video encoding and wireless communication tasks. For this purpose, they present a cross-layer approach that explores the tradeoff between coding and communication energies. They then exploit the Rate-Distortion-Complexity tradeoffs and flexibility of the Scalable Video Codec. The results show that by adapting the codec configuration at runtime to the specific scenarios up to 50% of the total energy can be saved with marginal video quality loss. Moreover, the approach presented is of low complexity and easily deployable in practical systems.

DOI: 10.4018/978-1-61520-771-8.ch006

INTRODUCTION

Wireless video communication with portable devices has become the driving technology of many important applications such as personal communication, gaming and security. In this respect, both mobile communication protocols and video coding technologies have experienced rapid advances.

On the one hand, in the context of video coding technologies, several video compression standards have been developed within the past years such as MPEG-4 Part 10, H.264/Advanced Video Codec (AVC) or the most recent Scalable Video Codec (SVC). The purpose of the H.264/AVC project was to create a standard capable of providing good video quality at substantially lower bit rates than previous standards (e.g. half or less the bit rate of MPEG-2, H.263, or MPEG-4 Part 2), without increasing the complexity of design so much that it would be impractical or excessively expensive to implement. In a similar way, the Scalable Video Codec (SVC) has been developed as an extension of the H.264/AVC providing scalability aspects in the encoded video bitstream relying on a wide range of spatio-temporal and quality scalability. This higher flexibility comes also at the cost of an increased complexity or energy consumption.

In general, we can see that the trend in the evolution of video codecs aims at achieving higher compression degree and higher flexibility in terms of scalability and adaptability. Both features are highly valuable in mobile communications. First, bandwidth is a scarce resource and a high compression degree lowers the required bandwidth. Secondly, the variability of the available bandwidth together with the heterogeneity of devices and processing capabilities makes the bitstream scalability and adaptability a desired feature.

One obvious way to compress the video information is to reduce the target video quality. By allowing a coarser quantization the output rate is reduced at the cost of an increased video distortion. This exploits the inherent Rate – Distortion tradeoffs of video codecs. However, for a fixed target video quality, obtaining a higher compression degree generally requires the use of more complex coding tools, which increases the complexity of the coding process and with it its coding energy. This extends the Rate – Distortion tradeoff to a three-dimensional Rate – Distortion – Energy tradeoff, as shown in Figure 1.

On top of this, to avoid severe degradation of the end video quality when transmission errors occur, the video codec should provide a certain degree of error resilience or robustness. This is achieved at the cost of an increased redundancy, or in other words, increased output rate, which is translated again in terms of Rate-Distortion tradeoff.

On the other hand, the trend in the development of new wireless communication standards is to provide, among other aspects, higher bandwidth/data rate to the end users. This way, the new WLAN standard, 802.11n, aims to provide up to 600 Mbits/s with respect to the 54 Mbits/s provided by previous WLAN standard such as 802.11g. Achieving high data rate is indeed one of the key aspects of wireless communications. The available bandwidth is variable as it is dependent on the channel conditions during transmission and the distance between mobile terminal and base station. Given the error-prone nature of wireless channels, providing a certain Quality of Service (QoS) to the end user is another challenging task. This QoS is generally measured in terms of latency and Packet Error Rate (PER) as this can have a dramatic impact on the end video quality. Finally, when addressing wireless communications from battery-powered devices, energy consumption becomes also a critical issue. The required energy consumption is directly linked to the amount of transmitted data: the more bits transmitted, the higher the energy consumption is. On top of this, the required transmission energy depends on both distance between transmitter and receiver and experienced channel conditions. This way, guaranteeing a specific QoS under

Figure 1. Tradeoffs at communication and codec sides

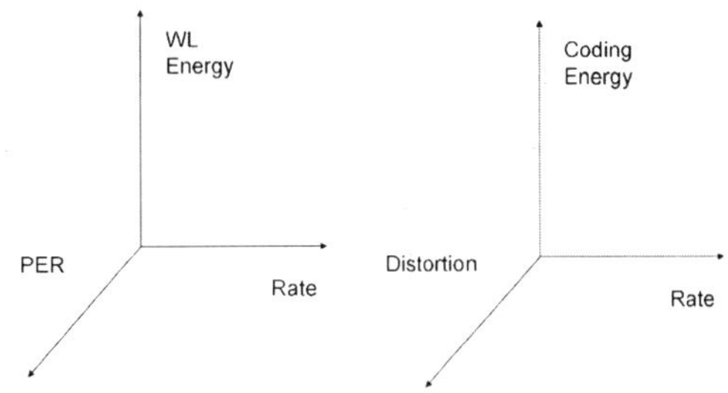

higher transmission distance or worse channel conditions usually involves higher energy consumption. In general, we could summarize the available tradeoffs in wireless communications as Rate – PER – Energy tradeoffs. Most of the existing work focuses on capacity-energy tradeoffs, by combining the wireless rate and PER dimensions into a general capacity dimension (Schurgers, 2002), (Raghunattan & Ganeriwal, 2002), (Uysal & Prabhakar, 2002). For cross-layer coupling with video coding, it is however more practical to treat rate and PER separately. In (Pollin et al, 2007), (Ji, 2008), (Blanch & Gan, 2009), a detailed wireless communication model is given relating total wireless communication energy (including transmission power and static power), rate and PER. It is shown that a relevant trade-off is achieved between wireless energy, rate and PER, in the sense that a higher rate or lower PER requirements correspond to a higher energy cost.

Nevertheless, to reduce the total energy consumption at the mobile terminal a more holistic approach is needed where not only communication energy is considered. Indeed, the two dominant power consumers at a mobile terminal are shown to be wireless transmission and video encoding processes as illustrated in (He & Cheng, 2008), (Mohapatra & Dut, 2007). Therefore, in order to minimize the total energy consumption at the terminal, both energy components need to be addressed. For this purpose, the available tradeoffs at both video codec and network side (see Figure 1) can be combined into one global tradeoff (Figure 2), where the Total Energy at the device is the sum of coding and wireless energy which depends also on the rate or number of bits transmitted. In a similar way, the video distortion takes into account the impact of both coding and transmission distortion or PER. For simplicity, we can assume that a certain QoS is imposed at the network side guaranteeing a very low or marginal PER.

In order to analyze the global tradeoffs given in Figure 2, a cross-layer approach is required as the interdependencies and tradeoffs between different layers (such as coding and transmission) are considered.

The remainder of this chapter describes an optimization technique for the transmission over a WLAN of the latest video compression standard, the Scalable Video Codec. The purpose is to minimize the total energy subject to distortion constraints by exploring the link between the video rate axis and the wireless and coding energy components. The chapter is organized as follows. First, some related work is presented where some

Figure 2. Combined tradeoffs in cross-layer approach

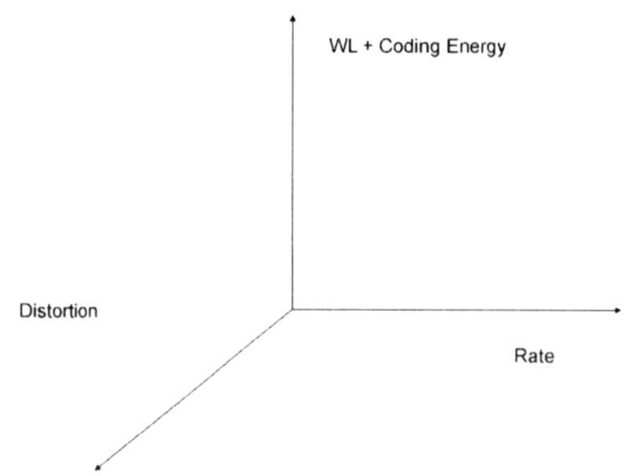

approaches to consider the global tradeoffs and total energy consumption are given. Then, the Scalable Video Codec is introduced together with its Complexity – Performance tradeoffs. The network model is presented next. This is followed by the proposed cross-layer approach to minimize the total energy consumption and the achieved results. Finally we conclude this chapter and propose some future research directions.

BACKGROUND

A main challenge in wireless communications is the limited battery capacity of the portable devices involved. Hence, achieving low energy consumption becomes a critical issue. To this end, most of the existing work focuses on minimizing the energy consumption of either the video codec or the wireless transmitter, which are considered the main energy consumers in the device. However, few authors consider the joint minimization of the codec and wireless energy by taking their interdependencies into account. In (Katsaggelos & Zhai, 2005) the impact of source coding on the wireless transmission energy is considered but the authors do not take into account the coding energy. In (Gan et al, 2007) a cross-layer optimization scheme is proposed to minimize coding and wireless energy for MPEG-4 video. In (Mohapatra & Dut, 2007) several cross-layer adaptation techniques are proposed at application, OS, network and hardware level to reduce the energy of video streaming on mobile handheld systems. In (He & Cheng, 2008) the authors propose energy minimization based on power-rate-distortion optimization. The energy tradeoff between video encoding and wireless communication is analyzed and the MPEG-4 encoding process is adapted based on the video content variations. However, none of the mentioned works consider the user mobility and the impact of the path loss scenario.

Other interesting works that considers the minimization of the total energy by jointly considering transmission and source coding energy are found in (Lu & Wang, 2002), (Lu & Wang, 2003), (Lu & Wang, 2007). In (Lu & Wang, 2002) and (Lu & Wang, 2003) a mathematical model of H.263 video codec, Reed Solomon channel codec and wireless transmitter is presented with the purpose of minimizing the total energy consumption of an uplink transmission from mobile to Base Station at varying transmission distances. In (Lu & Wang, 2007) the work is extended to a multi-user context over CDMA cellular networks for the H.263 video codec where the sum of compression and transmission power for all users is minimized. This is done in a centralized context at the cost of significant communication overhead.

In (Lan & Tewfik, 2003) the coding and transmission models of mobile users in a cellular network are configured by the Base Station in order to reduce the total energy consumption. Energy savings from 10% to 30% are reported.

All mentioned work in literature addresses video codecs such as MPEG-4 or H.263 that provide very limited scalability. This chapter extends the analysis to transmission over 802.11e WLANS of the more advanced and challenging Scalable Video Codec (SVC). This upcoming video standard provides high flexibility and compression efficiency. But this occurs at the cost of an increased coding complexity, which makes energy minimization a critical requirement. The approach presented is however general and can be applied to all video and wireless communication standards.

As in (He & Cheng, 2008) we consider the dependencies and tradeoffs between video encoding and wireless energy. Based on this energy tradeoff and codec **complexity-rate-distortion** modeling we adapt the SVC codec configuration to reduce the total energy consumption, as already presented in our earlier works (Blanch & Gan, 2009). The impact of user mobility and path loss scenario is considered, similarly to the approach we presented in (Blanch & Gan, 2009). On top of this an approach for optimal target quality selection is presented.

THE SCALABLE VIDEO CODEC

The Scalable Video Codec (SVC) is an extension of the H.264/Advanced Video Codec (AVC) and it is the latest standardized video codec (Reichel & Schwarz, 2005) developed in the Joint Video Team (JVT), which is a collaboration between MPEG and ITU-T. The purpose of SVC is to provide scalability aspects in the encoded video bitstream in a way that a wide range of spatio-temporal and quality scalability can be achieved.

There are three axes of scalability provided in SVC:

- Temporal scalability enables the adaptation of the frame rate,
- Spatial scalability enables the adaptation of different resolutions,
- Quality scalability enables the adaptation of different qualities,
- Combined scalability is any combination of the above.

These scalability aspects are further explained in the following subsections.

Temporal Scalability

Temporal scalability is achieved by means of hierarchical prediction structures as illustrated in Figure 3 and Figure 4. The types of video frames are explained as follows. I frames are intra-predicted, and do not require any previously coded frame. P frames are uni-directional predicted from previous I frames or P frames. B frames are bi-directional predicted from previous I, P or B frames, in a hierarchical manner. A special type of I pictures are called IDR pictures (Instantaneous Decoder Refresh). At the occurrence of an IDR pictures, following P and B pictures cannot be predicted from any picture preceding the IDR picture. This enables the refresh and flushing of the decoder. Both Intra and IDR pictures are coded periodically.

The basic unit for hierarchical prediction structure is called a Group of Pictures (GOP). The pictures at the boundaries of the GOP are either intra-coded or inter-coded by using previous GOP boundaries as reference. The remaining pictures of a GOP, are hierarchically predicted, which inherently provides temporal scalability. This way, when this hierarchical structure is used to encode

Figure 3. High delay GOP structure

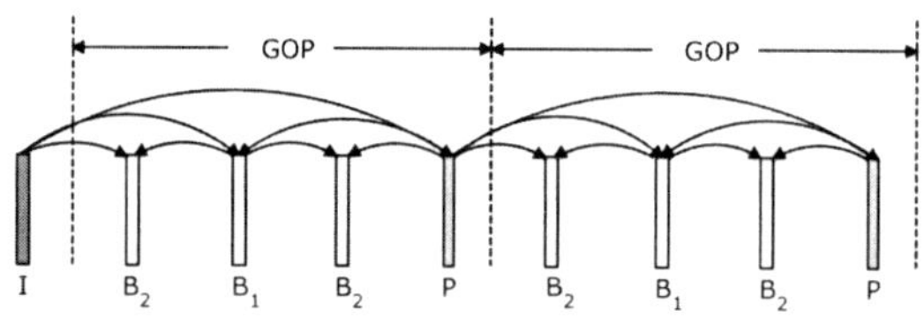

a sequence, the total number of temporal layers $L_{temporal}$ is given by:

$$L_{temporal} = \log_2 GOPSize + 1 \quad (1)$$

Therefore, a prediction structure with a bigger GOP size provides a higher degree of temporal scalability as a higher number of temporal layers can be extracted from the encoded bit stream.

Note that it is possible to arbitrarily adjust the structural delay between encoding and decoding of a picture by restricting the prediction to motion-compensated prediction from pictures that follow the picture to be predicted in display order. This is illustrated in Figure 4, where a hierarchical prediction structure is shown which does not employ motion-compensated prediction from pictures in the future. Although this structure provides the same degree of temporal scalability as the one of Figure 3, its structural delay is equal to zero in contrast to 4 pictures for the prediction structure in Figure 3. However, such a low-delay structure typically decreases coding efficiency.

Spatial Scalability

For supporting spatial scalable coding, SVC follows the conventional approach of multilayer

Figure 4. Low delay GOP structure

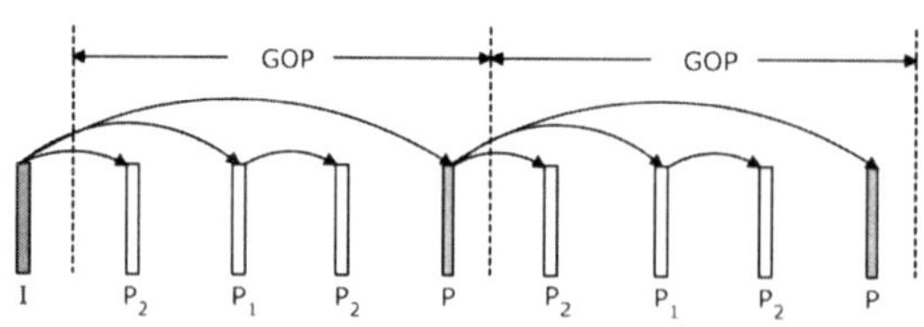

coding, where the spatial resolution can be changed by adding or dropping spatial enhancement layers. In each spatial layer, motion-compensated prediction and intra-prediction are employed as for single-layer coding. But in order to improve coding efficiency in comparison to simulcasting different spatial resolutions, the high resolution layers are predictive coded from low resolution layers using inter-layer prediction mechanisms. In particular, the pixels in lower resolution layer have to be up-sampled to match the resolution of higher layer.

SNR Quality Scalability

With SNR (Signal-to-Noise-Ratio) scalability, also called quality scalability, the reconstructed video quality (often measured in mean square error, MSE or peak signal-to-noise ratio, PSNR) can be tuned by adding/dropping quality enhancement layers (EL). Therefore, when quality scalability is used, each video frame in the sequence consists of a base layer (BL) encoded at lower quality and one or more quality enhancement layers (EL) on top.

The latest SVC standard supports two types of quality scalability: coarse granular scalability (CGS) and medium granular scalability (MGS). Both of them employ the same inter-layer prediction technique as that in spatial scalability. The main difference between CGS and MGS lies in the capability of rate adaptation. Similar to spatial scalability, CGS layers can only be added/dropped at IDR pictures. In contrast, MGS layers can be added/dropped for each picture regardless of the picture type, hence providing much more flexibility than CGS.

An important concept in MGS is the key picture (KP), which aims at making a trade-off between error resilience and coding efficiency. More specifically, for a picture that is not a KP (i.e., non-KP or NKP), the EL of the reference pictures are employed for motion compensation. In contrast, for a KP, only the BL of the reference pictures are used for motion compensation. In practice, the KP concept can be efficiently combined with the GOP structure, where I and P pictures on the GOP boundaries are coded as KPs, and all the other pictures within a GOP are coded as NKPs (see Figure 5). Each frame in the sequence consists of a base layer (BL) encoded at low quality and one quality enhancement layer (EL) on top.

In this way, if an EL of a picture is lost due either to packet loss or rate adaptation, the error propagation can be restricted within a GOP. In other words, the KPs are serving as resynchronization points. However, such error robustness comes at a cost of coding efficiency, as only the BL of reference pictures with coarse quality are used for motion-compensation of KPs.

There exists a third type of scalability called Fine Granular Scalability (FGS). FGS provides higher flexibility of adaptation than either CGS or MGS. However, this type of quality scalability is not included in the current profile definition of the SVC standard.

The rate of both Base Layer (BL) and Enhancement Quality Layer (EL) is highly dependent on the Quantization Parameter (QP) chosen to encode the specific layer, where a lower QP corresponds to a higher encoded quality and higher rate. This way, the control of the rate distribution between Base Layer and Enhancement Layer (EL) is done via the ΔQP parameter given as the difference between QP's of BL and EL:

$$\Delta QP = QP_{BL} - QP_{EL} \qquad (2)$$

The choice of ΔQP causes a specific rate distribution between BL and EL. This determines the percentage of discardable (EL) information in the bitstream and therefore its adaptation flexibility.

The following section analyzes the impact of the chosen SVC codec configurations on the rate-distortion-complexity tradeoffs. For this purpose, the analysis is focused on configurations

Figure 5. Key Picture prediction

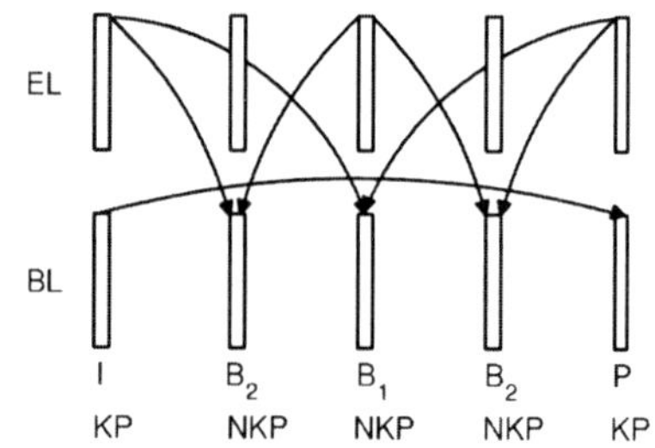

providing temporal and quality scalability. In terms of quality scalability, Medium Grain SNR scalability (MGS) will be used as this provides higher flexibility than CGS.

Rate – Distortion – Complexity Tradeoffs in SVC

The selection of the SVC codec configuration has a clear impact on the Rate-Distortion-Complexity tradeoffs. This way, the support of quality scalability usually incurs a loss in coding efficiency when compared to single-layer coding. On the contrary, the use of bigger GOP sizes, providing increased temporal scalability, generally increases the coding efficiency while incurring a higher coding complexity. The purpose of this section is to analyze all these Rate-Distortion-Complexity tradeoffs provided by different SVC configurations.

To perform the analysis the JSVM version 9.10 (JSVM, 2008) implementation of the SVC codec is used and its complexity is measured as number of cycles on a Pentium 4 PC. In terms of temporal scalability we analyze configurations with GOP sizes ranging from 1 to 16, which correspond to using from 1 up to 5 temporal layers. Both prediction structures with high coding delay (Figure 3) and low coding delay (see Figure 4) are considered. Moreover, to guarantee a degree of error robustness configurations with one Base Layer and one MGS quality layer are considered. This is combined with a prioritization mechanism, where packets from the quality layer are dropped first in case of congestion. In addition, the Key Picture functionality is always enabled to avoid drift problems when an MGS layer of a frame is lost or dropped. Finally, to analyze different rate distributions between base layer (BL) and MGS layer, ΔQP parameter values of 2 and 6 are considered.

Table 1 shows the rate and coding complexity tradeoffs corresponding to encoding the Mobile & Calendar sequence with different SVC configurations. Each codec configuration is given by a combination of parameters such as GOP size, ΔQP and Delay structure that determine the bitstream spatio-temporal and quality scalability. The QP for each configuration is selected such that the resulted video quality is approximately 37.5 dB for all configurations. The analysis is performed in our previous work in (Blanch, 2009) and the results are shown in Table 1 and Figure 6. Codec configurations shown in Table 1 are ordered by coding complexity from the highest to the lowest. Only Pareto optimal configurations, in terms of normalized coding complexity versus output rate,

Table 1. Complexity – Rate tradeoffs per SVC codec configuration (Blanch, 2009)

Point	GOP	QP_BL	ΔQP	Delay	Coding Complexity	Rate (Mbps)
1	16	26	2	High	1.76	2.37
2	8	26	2	High	1.68	2.39
3	16	30	6	High	1.66	2.47
4	8	30	6	High	1.58	2.53
5	4	31	6	High	1.41	2.58
6	2	31	6	High	1.12	3.17
7	4	26	2	Low	0.658	3.29
8	16	30	6	Low	0.635	3.31
9	8	30	6	Low	0.63	3.32
10	4	30	6	Low	0.62	3.57
11	2	30	6	Low	0.59	3.99
12	1	32	6	Low	0.54	4.47

are shown in Table 1. This is, all other possible configurations require higher coding complexity and rate and are therefore suboptimal.

Figure 6 illustrates the complexity-rate tradeoffs given by the configurations in Table 1.

From the analysis of both Figure 6 and Table 1 the following observations can be made:

- Achieving higher coding efficiency requires an increase of coding complexity
- Configurations 1 to 5 form one cluster of points. These configurations have a high coding efficiency (low rate) and a high complexity, and use a High Delay structure (bi-directionally predicted frames).
- Configurations 7 to 12 form another cluster, with lower complexity and higher rates. They use the Low Delay structure (forward only prediction).
- Within each cluster of configurations, increased GOP sizes yield higher compression, in particular when low ΔQP are used.
- The configuration with the lowest compression efficiency and lowest coding energy corresponds to a GOP size of 1.

This analysis is repeated for other types of video content, providing similar conclusions in terms of Pareto optimal configurations. In general we observe that the **scalable video codec** provides a wide range of tradeoffs in terms of rate and coding complexity. By selecting different codec configurations we can cover, for a fixed video quality, a range of factor 2 to 4 (depending on the video content) in terms of rate and a range of more than factor 3 in coding complexity.

Codec Energy Modeling

The presented complexity measurements give us an indication of the relative complexity cost between codec configurations. However, to compare the cost of video encoding and wireless transmission processes, it is necessary to map the coding complexity measurements onto energy consumption values. To do so, the SVC codec energy is estimated by taking as reference our previous work on MPEG-4 (Gan et al, 2007) and

Figure 6. Complexity – Rate tradeoffs for the mobile &calendar sequence

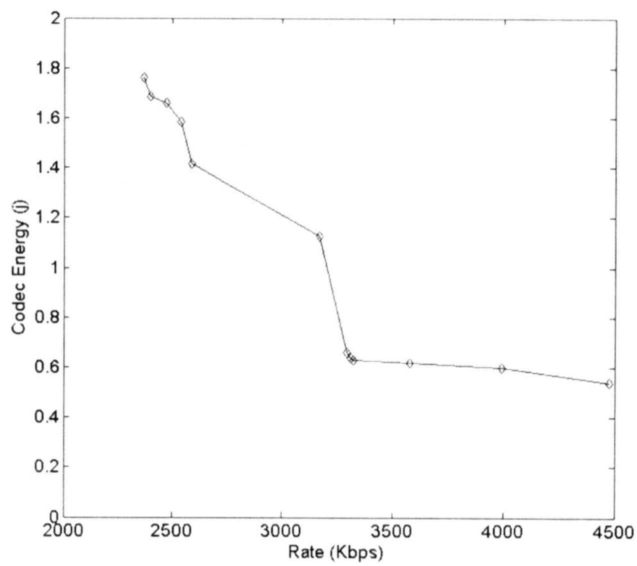

the Advanced Video Codec (AVC) (Saponara & Blanch, 2003). Then the following assumptions are taken:

- Despite some restrictions (like constrained intra prediction), the Base Layer of an SVC stream is fully compliant with AVC, and hence it is assumed that its coding complexity is very similar.
- Then the complexity of the AVC codec can be related to that one of the MPEG-4 codec. By coding an AVC bit-stream with IPPP structure and disabling complex AVC tools similar rate-distortion characteristics are achieved as with MPEG-4 at very comparable coding complexity. This is the outcome of the analysis and observations performed in (Saponara & Blanch, 2003).

After combining the two above assumptions, the coding energy of the SVC Base Layer with a configuration of GOP 1 (IPPP structure) can be considered comparable to the coding energy of an MPEG-4 bitstream, obtained in our previous work in (Gan et al, 2007). In this work, wireless energy and MPEG-4 coding on ASIC was shown to be comparable for certain path losses and video rate scenarios. Next, to estimate the coding energy for all other SVC configurations the relative complexity factors between configurations (given in Table 1) are applied.

The reader should note, however, that the presented assumptions only aim at obtaining approximate values of the coding energy that can be compared with the wireless communication energy. The real coding energy depends on the specific implementation chosen and therefore may differ. However, it is out of the scope of this chapter to obtain very accurate coding energy numbers.

Figure 7. Probability density function of channel attenuation (Blanch & Gan, 2009)

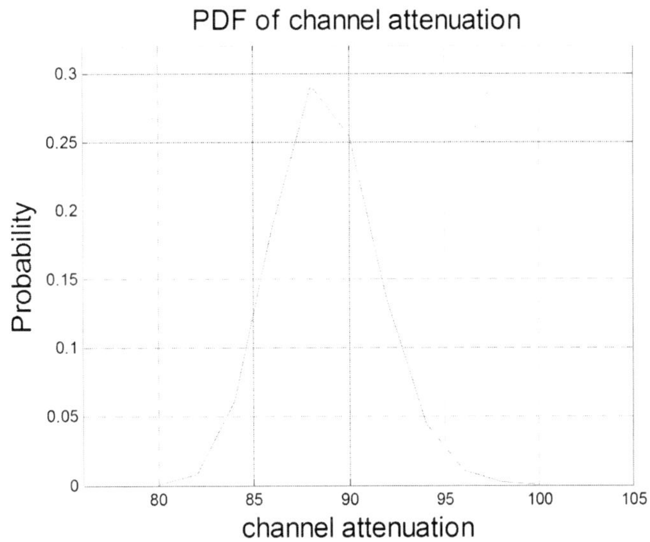

Its purpose is to obtain estimated values, which are sufficient to show the available tradeoffs and the added value of cross-layer optimizations. However, as will be shown later, the optimization approach remains valid under different codec energy assumptions arising from different codec implementations.

WIRELESS ENERGY MODELING IN IEEE 802.11E

The efficient transmission of video over wireless local area networks (WLANs) is a challenging goal, especially when considering multiple mobile users on an error-prone channel and sharing the same channel resources. To address this challenge and provide Quality of Service the WLAN IEEE 802.11e standard (WLAN, 2004) proposes the Hybrid Coordination Function (HCF) with two different access schemes, namely HCF Controlled Channel Access (HCCA) and Enhanced Distributed Channel Access (EDCA). Both schemes support user mobility and provide high data rates but face the limitation of the high energy consumption. As wireless stations are battery-powered, achieving the required performance at minimal energy consumption becomes a critical issue. Further details on these WLAN standards can be found in other chapters throughout this book.

In this chapter we focus our analysis on the HCF functionality of the 802.11e standard. However, the cross-layer approach presented can be generalized and implemented on the distributed EDCA functionality or any other wireless standard such as cellular systems.

The ns-2 network simulator (NS) is used to simulate the transmission of scalable video over the Hybrid Coordinated Function of the 802.11. To model the energy consumption the power and performance models from our previous work in (Mangharam et al, 2005) are used. At the MAC and PHY layer the behavior of state-of-the-art wireless systems such as 802.11a devices is as-

Figure 8. Power-performance tradeoffs according to path loss (Blanch & Gan, 2009)

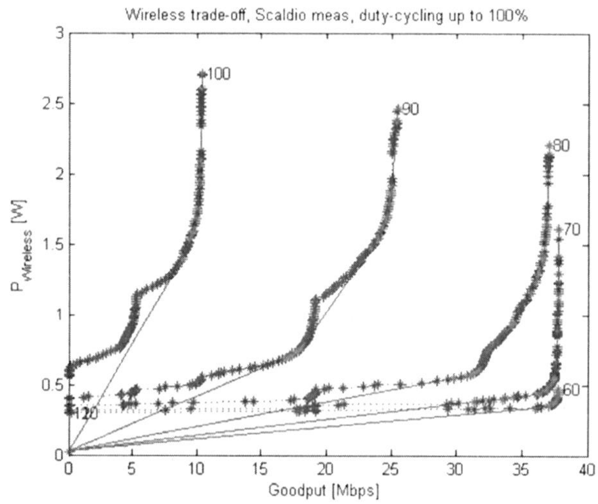

sumed where the highest feasible physical rate is always used and the power amplifier operates at the maximum transmit power (Atheros, 2003). To consider the impact of sleep energy in the total wireless energy, the assumptions from (Timmers et al, 2007) are taken where the sleep state power corresponds to 26 mW.

The channel conditions are modeled so that the effect of path loss attenuation and fast fading is combined. The Probability Density Function, shown in Figure 7, is considered. This models the total channel attenuation: the fast fading and the average path loss value. The experiments consider average path loss values ranging from 70 to 105 dB as these will lead to different tradeoffs between coding and wireless energies. Figure 8 shows typical power-performance tradeoffs for the average path loss values considered.

STACK-WIDE CROSS-LAYER OPTIMIZATION

The energy modeling and resulting energy-performance tradeoffs at both video codec and wireless side has been presented in previous sections. The tradeoff between wireless and coding energies is now introduced and the optimization cross-layer approach developed in our previous work (Blanch & Gan, 2009) is presented. In addition, a method for optimal target video quality selection is proposed.

Coding Energy and Wireless Energy Tradeoffs

In the previous section the inherent compression-complexity tradeoffs in the video codec have been analyzed showing that achieving high compression (lower rates) requires configurations with higher encoding energy. In addition, the codec output rate, linked to the compression achieved, also has

Figure 9. Impact of path loss on energy breakdown

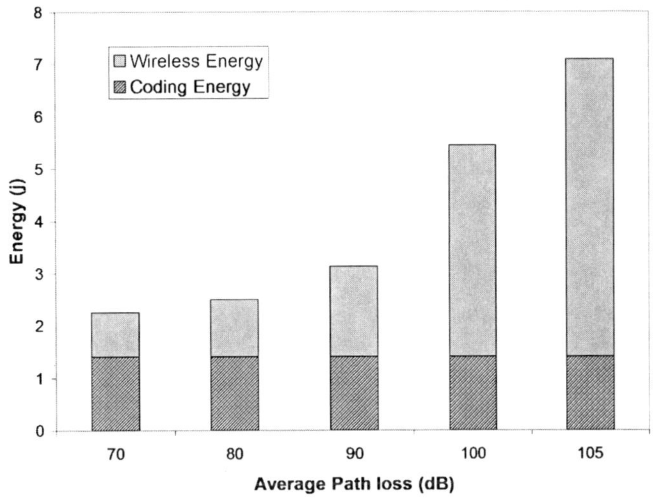

a clear impact on the required communication energy as the more bits transmitted, the higher the wireless energy consumption. This is illustrated as well in Figure 8 where higher good-put requires increased wireless power consumption.

In addition to the rate, the channel attenuation (or path loss) highly influences the wireless energy consumption. Indeed, a high path loss between transmitter and receiver requires higher communication energy to guarantee a successful reception. To illustrate this Figure 9 shows the breakdown of the total energy in wireless and video coding when transmitting a video rate of 3 Mbps at an increasing average path loss. We observe that for an average path loss below 90 dB the coding energy is dominant, while for path loss over 90 dB the wireless energy increases dramatically and dominates over the coding energy.

From the analysis performed in both Figure 6 and Figure 9 the following conclusions can be extracted:

- Communication energy and encoding energy follow opposite trends with respect to the video rate. In order to reduce the coding energy the compression efficiency can be reduced allowing higher output video rate energy. However, this increases the required communication energy. Similarly, by applying higher compression the bit rate is reduced and with it the communication energy, but the encoding energy increases. Hence, decreasing in this way the coding energy increases the communication energy and viceversa. This makes it necessary to trade-off both types of energy consumption.
- The video rate is the parameter that links coding energy and wireless energy and has a noticeable impact on both. This rate is controllable through the quantization parameter and the selected codec configuration.

Figure 10. Energy-rate tradeoffs at 70 dB path loss

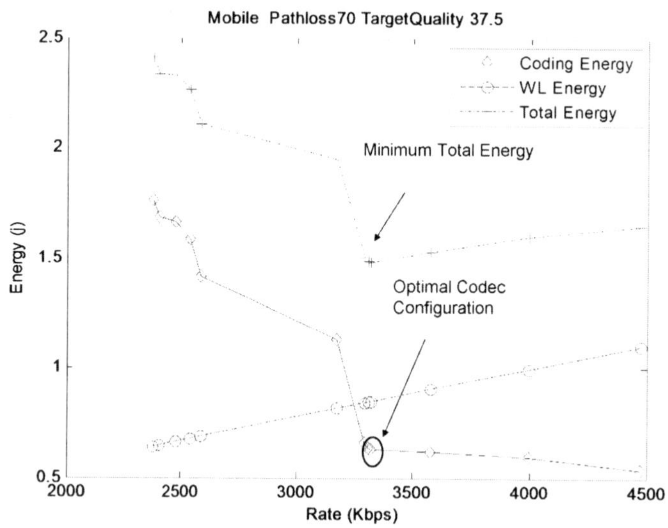

This is reflected in Figure 10 and Figure 11, which show the coding energy–rate tradeoffs of the SVC codec together with the associated wireless communication energy. Each point on the coding energy curve corresponds to a specific codec configuration. For the wireless energy consumption we assume an average path loss of 70 dB in Figure 10 and 100 dB in Figure 11. The global energy curve (sum of coding and wireless energy) is shown as well.

The proposed cross-layer approach jointly considers the tradeoff between codec and wireless energy and selects the optimal SVC codec configuration that minimizes the total energy consumption. These configurations providing the lowest energy consumption are circled in Figure 10 and Figure 11. When the codec energy is dominant (see Figure 10 at 70 dB), we should reduce the coding energy by allowing lower compression ratio (higher rate). On the other hand, when the wireless energy is high and dominates over the codec energy (see Figure 11 at Path loss of 100 dB), to minimize the total energy it is imperative to reduce the wireless energy. To do so we need a high compression ratio at the codec side, regardless the coding energy increase, that reduces the video transmission rate.

As explained previously, the actual coding energy values will vary according to the specific codec implementation assumed. However, the presented optimization approach remains valid under any other coding -wireless energy tradeoff. Let us assume that the coding energy in Figure 10 is halved for a particular codec implementation while the relative complexity between codec configurations still holds. This would simply increase the relative contribution of wireless energy to the total energy and would move the optimal codec configuration to a lower rate configuration. In a similar way, if the coding energy was doubled with respect to Figure 11, this would just lead to a different tradeoff where lowering the coding energy would be preferred. This way, the optimal codec configuration would move to the right in

Figure 11. Energy-rate tradeoffs at 100 dB path loss

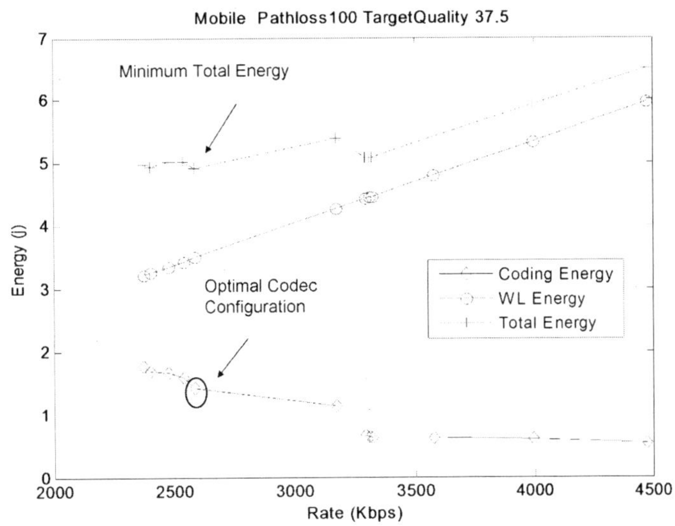

Figure 11, corresponding to one of higher rate and lower complexity.

Practical Implementation: Sensing of Average Wireless Energy

The channel attenuation (path loss) highly determines the existing tradeoff between wireless and coding energy and therefore the optimal codec configuration. However, other factors such as network load or underlying communication technology also influence the required wireless communication energy. Therefore, since an estimate of the wireless energy from the current path loss value does not consider the impact of such factors, it is better to track the required average wireless energy per bit. From this energy-per-bit value the wireless energy cost for transmitting different SVC configurations can be estimated, based on the known output rate for each configuration.

The selection of the optimal configuration is summarized as follows:

- **Step 1:** At design-time databases of SVC Pareto optimal configurations in terms of energy-rate tradeoffs $\{E_{enc}\left(k_{enc}\right), R_{enc}\left(k_{enc}\right)\}$ are built for the desired video quality Q_{target}. These codec configurations are defined by the combination of the following codec parameters:

$$k_{enc} = \left\langle GOPsize, QP, \#\, ofMGSlayers, \Delta QP, CodingDelay, KP \right\rangle$$

- **Step 2:** for each optimal codec setting k_{enc}, its corresponding wireless energy $E_{wl}(k_{enc})$ is estimated as:

$$E_{wl}(k_{enc}) = R(k_{enc}) * E_{measured} \tag{3}$$

where $R(k_{enc})$ is the average rate generated by k_{enc} and $E_{measured}$ is the measured wireless energy per bit. Note that it is not necessary to know the underlying tradeoffs and wireless configurations

Figure 12. Rate – Distortion tradeoffs for the mobile sequence

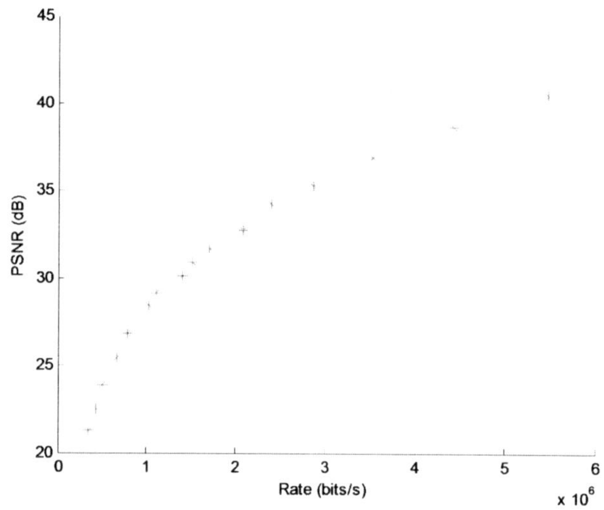

k_{wl} at the MAC layer. The total energy is then obtained by:

$$E_{tot}(k_{enc}) = E_{wl}(k_{enc}) + E_{enc}(k_{enc}) \quad (4)$$

- **Step 3**: the SVC codec configuration that minimizes the total energy is selected:

$$k^*_{enc} = \arg\min_{\{k_{enc}\}} \left\{ E_{tot}\left(k_{enc}\right) : Q(k_{enc}) \geq Q_{target} \right\} \quad (5)$$

Optimal Target Quality for Energy Minimization

Previously presented Figure 9 illustrated how the wireless energy consumption rapidly increases with a higher path loss. When this happens, the wireless energy becomes the main component of the total energy. Hence, to minimize energy consumption the cross-layer controller chooses the SVC configuration with highest compression efficiency to minimize the rate and therefore the wireless energy. The only mean to further reduce the rate, and with it the associated wireless energy, is to allow an increased video distortion. Figure 12 shows the rate-distortion tradeoffs for the Mobile sequence at a fixed SVC configuration of high coding efficiency (GOP 16 with short delay). This way, at the highest compression efficiency the only way to reduce the rate is by targeting a lower PSNR.

We exploit the SVC rate-distortion tradeoffs to further reduce the total energy consumption. The basic idea behind this approach is that the video quality can be increased when the cost of the wireless energy per bit is low while it is decreased when the energy per bit becomes high. For this purpose we implement an algorithm that maintains the average target video quality constant but trades off video quality with rate depending on the wireless scenario.

At design-time the optimal target quality selection is computed as follows:

- **Step 1:** we first identify N possible path loss scenarios $PL_i \in \{PL_1....PL_N\}$, or in other words, scenarios of typical wireless energy consumption per transmitted bit. From the modeling performed in (Mangharam et al, 2005) we can extract then typical values E_bit_i for $i=1..N$ where N is the number of PL scenarios considered. From statistical analysis we can also assume a certain probability of occurrence for each PL_i: $prob_i \in \{prob_1...prob_N\}$.
- **Step 2:** The values for the desired average quality $\overline{Q}$, minimum required quality $Q_{\min}$, and maximum quality $Q_{\max}$ are determined. Then the set of possible target quality values can be defined as: $Q \in \{Q_{\min}...Q_{\max}\}$ where steps of 1 dB are chosen between consecutive Q values
- **Step 3:** Based on the set of possible Path loss PL and possible target quality Q we will assign an optimal target quality TQ_i per Path loss scenario PL_i. The optimization algorithm is summarized as follows:

```
Initialize TQ_i = Q_max ∀i = 1..N While
Σ_{i=N} TQ_i / N ≥ Q̄   and  TQ_i ≥ Q_min ∀i = 1..N Do
For j=1:N
   TQ_j_temp = TQ_j − 1dB  /* Propose
a change in target Quality */
   For i=1:N
   Global_j = (Σ_{i=1}^{i=N} (TotE_i * prob_i))
   end
end
/* Accept the change in target
quality at that Path loss where
energy reduction is highest */
j* = arg min(Global_j) TQ_j = TQ_j_temp
end
```

Note that $CodingE_i$ and Rate corresponds to the optimal SVC configuration k^* for the specific path loss PL_i.

The presented algorithm can be described as follows:

- The maximum target quality is initially assigned to each path loss scenario.
- Then the algorithm identifies at which path loss a quality reduction of 1dB leads to the highest reduction of total energy.
- This process is iterated while the desired average quality and minimum quality requirements are still met. As a result we obtain a set of $\{TQ_i, PL_i\}\ \forall i = 1..N$

The target of the algorithm is to assign lower target qualities when the Path loss and wireless energy consumption is high and higher target qualities when the Path loss is low. This Path loss is a function of the distance between mobile terminal and access point. In practice, statistics about the path loss occurrence probability can be gathered at runtime but for simplicity we assume equal probability of occurrence for different average Path loss.

We next show a practical example of the application of the algorithm for optimal target quality selection described above. Let us assume that the set of possible Target Qualities is {38.5, 37.5, 36.5, 35.5, 34.5, 33.5}, this is $Q_{\min}$ =33.5 dB, $Q_{\max}$ =38.5 dB. Moreover, at most 1 dB degradation of the initial average video quality (37.5 dB) is allowed, this is, $\overline{Q}$=36.5 dB. The Path loss takes the values in the set {PL_1=70, PL_2=80, PL_3=90, PL_4=100, PL_5=105} with identical probability of occurrence, this is, $Prob_i = \frac{1}{N}$, where N= 5.

The output of the optimal target quality selection is {TQ_1=38.5, TQ_2=38.5, TQ_3=37.5, TQ_4 =35.5, TQ_5=33.5} resulting in an average quality of $\overline{Q}$=36.7 dB

Figure 13. Highest energy reduction achieved by MM-WL-XL with quality control

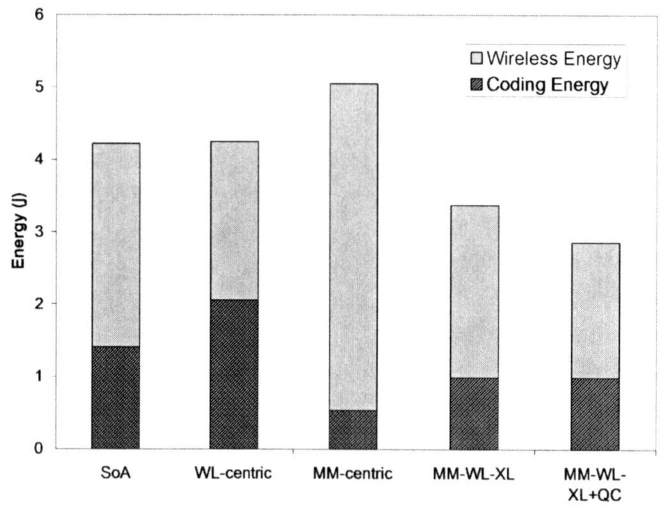

Note that reducing the quality requirements to 33.5 dB during a path loss of 105 dB (*i=5*) does not have a very noticeable impact on the visually perceived quality, as the visual degradation is not occurring due to packet errors but only to a coarser quantization parameter.

At run-time, based on the measured wireless energy per bit consumption and the estimated consumption per path loss scenario E_bit_i it is possible to predict the current Path loss scenario PL_i and set the corresponding target quality TQ_i.

Once the new target quality is selected the optimal SVC codec configuration is recomputed. This is due to the fact that targeting a different video quality modifies the video rate range and with it the tradeoff between coding and wireless energy. Depending on this energy tradeoff a different codec configuration will be optimal for energy minimization.

Energy Savings

This section presents the results obtained from ns-2 simulations where a single mobile user transmits 150 frames of the high rate *Calendar & Mobile* sequence over a WLAN link targeting an average video quality of 37.5 dB. The wireless energy per bit is tracked and averaged every 6 video frames. At the codec side, changes on the codec configuration are allowed at IDR pictures, which are coded every 64 frames. As the video sequence is encoded at 30 frames per second, this is equivalent to a codec configuration every two seconds. Finally, a certain degree of user mobility is considered where the user is moving away from the Access Point at walking speed. To mimic the user's mobility, it is assumed that the average path loss experienced by the user changes every two seconds and takes the following values: 70, 80, 90, 100, and 105 dB. The results presented are averaged over the different path loss scenarios.

This section compares different transmission approaches: the first one being the state-of-the-art approach where no energy minimization is sought. Secondly, approaches that target the minimization of either wireless or multimedia energy consumption are considered. Finally, the proposed joint multimedia-wireless cross-layer approach is also shown, where joint minimization of multimedia and wireless energy is targeted by exploiting the interdependencies and tradeoffs between multimedia and wireless tasks.

This way, Figure 13 shows the wireless and coding energy consumption for the transmission of the Mobile sequence under the mentioned approaches:

- State-of-the-art (SoA): no optimization seeking energy minimization is performed. A fixed SVC configuration (GOP = 4, 1 MGS layer, $\Delta QP = 6$, KP, High Delay) with a reasonable compression-complexity tradeoff is selected, and this is fixed regardless of the transmission scenario.
- Wireless-centric optimization (WL): seeks to minimize the wireless energy. To this end, at the codec side the most coding efficient configuration is selected in order to reduce the rate and hence the wireless energy.
- Multimedia-centric optimization (MM): targets the reduction of the coding energy. This way, the codec configuration that minimizes the encoding energy is selected, regardless of its impact on the wireless energy.
- Stack-wide Multimedia-Wireless Cross-Layer (MM-WL-XL): aims at minimizing the sum of coding and wireless energy. To do so, according to the approach presented previously, the SVC coding configuration that minimizes the total energy is selected.
- Stack-wide Multimedia-Wireless Cross-Layer with Quality Control (MM-WL-XL + QC): in addition to the previous approach, the target video quality is modified in order to further reduce the total energy consumption.

The corresponding savings on the total energy with respect to the SoA approach are shown in Table 2. Results for other video sequences such as Foreman and the slow motion *Mother & Daughter* are also given.

We can see from Figure 13 and Table 2 that the approach that minimizes coding energy disregarding the impact on wireless energy (or vice versa), fails to minimize the total energy. This way, the MM-centric and WL-centric approaches provide at most minor savings or even result in energy increase. In contrast, the cross-layer approach, which jointly considers the wireless and coding energy and their interdependencies, succeeds in minimizing the total energy. The savings range from 20% to 46% depending on the video content. On top of this, if the optimal target quality is selected based on the wireless consumption scenario an additional energy savings is achieved reaching savings from 33% to 50.5% on the total energy.

To conclude, Table 3 shows the SVC configurations that are selected in the MM-WL-XL

Table 2. Total energy savings per approach

	Sequence	MM-centric	WL-centric	MM-WL-XL	MM-WL-XL + QC
Saving (%)	Mobile	-19%	- 1%	20%	33%
Saving (%)	Foreman	-3.5%	-10%	38%	46%
Saving (%)	Mother	3%	- 11%	46%	50.5%

Table 3. Optimal SVC configurations and target quality for mobile sequence

Path Loss(dB)	GOP size	QP	ΔQP	Delay	Rate (Mbps)	Coding Energy (j)	Target Quality
70	8	24	2	Low	3.95	0.69	38.5 dB
80	8	24	2	Low	3.95	0.69	38.5 dB
90	16	30	6	Low	3.31	0.63	37.5 dB
100	4	30	2	High	1.65	1.43	35.5 dB
105	16	36	2	High	1.51	1.00	33.5 dB

optimization for the transmission of the *Mobile* sequence targeting an average of 37.5 dB for all path loss scenarios. Confirming the findings in previous section, at a low path loss (70 dB), a low complexity codec configuration is selected, while at 105 dB, a configuration with high coding efficiency is preferred. In addition, the target quality is adapted to the wireless energy consumption scenario, which is closely related to the path loss scenario.

FUTURE RESEARCH DIRECTIONS

One of the many possible extensions of the presented optimization approach would be to extend it to the multi-user context. In a multi-user scenario different users may experience different path loss, different video content and require a different target quality. All this causes different tradeoffs between wireless and coding energy per user that have to be considered when aiming at a global energy minimization. Moreover, the available bandwidth would be shared among users and the selected user's codec configuration would influence the transmission conditions of other users. In this sense, a distributed context would pose bigger challenges than a centralized one.

In addition, most of the work presented in (Katsaggelos & Zhai, 2005), (Gan & Dejonghe, 2008), (Lu & Wang, 2002), (Lu & Wang, 2003), (Lu & Wang, 2007) and (Lan & Tewfid, 2003) focuses on uplink scenarios whereas it would be interesting to address also downlink scenarios or even optimization of uplink (coding and transmission) with downlink (decoding and receiving) jointly in peer-to-peer scenarios. Moreover, other aspects of video encoding, such as error resilience coding techniques and adaptive error control coding should also be integrated in the optimization.

To successfully implement the presented cross-layer approach a model of the available tradeoffs at the video codec side is required. Such a model should be able to predict the energy-rate tradeoffs offered by different codec configurations and different target video qualities for a variety of video content. Obtaining these Rate-Distortion-Complexity tradeoffs at design-time could be achieved by extensive profiling of a variety of codec configurations and video content. However, this would be time consuming and the actual video content may differ from the modeled ones. To address this issue the required model should measure at run-time some video features that give an indication of the type of video content and based on this and the selected configuration predict the Rate-Distortion-Complexity tradeoffs. Moreover, the profiling performed at design-time should be as simple and less time consuming as possible. Finally, at run-time some model re-calibration should be taken into account.

On top of this, to be able to tradeoff between wireless and coding energy consumption a practical implementation should allow the measurement of the average energy consumption on the wireless card.

To conclude, the presented approach could be combined with many other energy reduction techniques such as Dynamic Voltage Scaling (DVS) (Mohapatra & Dut, 2007) and Frequency Scaling (FS). The exploration of holistic approaches that combine these diverse optimization techniques remains an open and interesting research path.

CONCLUSION

This chapter has presented a cross-layer approach that focuses on the joint minimization of coding and wireless energy. To do so, the flexibility and power–rate-distortion tradeoffs of the Scalable Video Codec (SVC) are exploited. By configuring the SVC codec settings in order to optimally trade-off coding and wireless energies, savings of up to 40% of the total energy consumption can be achieved without any video quality loss. If on top of this, the video target quality is adapted to the wireless energy scenario, savings of up to 50% on the total energy can be reached. In practice, both the joint cross-layer approach and the target quality control approach are implementable on any video codec that provides configurations with energy-rate tradeoffs. The SVC codec has been selected as it provides a wider range in terms of energy-rate tradeoffs than other video codecs such as MPEG-4 or the AVC. Moreover, the cross-layer approach is based on the sensing of the wireless energy consumption and of low complexity. This makes it easily deployable and suitable for real-time implementations.

In addition, it has been shown that approaches that seek to minimize separately coding or wireless energy fail to minimize the total energy consumption.

REFERENCES

Blanch, C., Gan, T., Dejonghe, A., & Masschelein, B. (2009). Cross-Layer Optimization for the Scalable Video Codec over WLAN. Packet Video Workshop (PV), Seattle, USA.

Blanch, C., Lenoir, G., Pollin, S., & Dejonghe, A. (2009). Energy-efficient transmission of H.264 scalable video over IEEE 802.11e. In *International Conference on Acoustics, Speech and Signal Processing (ICASSP), Taiwan.*

Consumption, P. & *Energy Efficiency Comparisons of WLAN Products*. (2003). Atheros White Paper.

Gan, T., Dejonghe, A., Lenoir, G., et al. (2008). Cross-Layer optimization for multi-user video streaming over IEEE 802.11e HCCA wireless networks. In *International Conference of Multimedia and Expo (ICME).*

Gan, T., Denolf, K., Lafruit, G., et al. (2007). Modelling Energy Consumption of an ASIC MPEG-4 Simple Profile Encoder. In *International Conference of Multimedia and Expo (ICME)*.

He, Z., Cheng, W., & Chen, X. (2008). Energy Minimization of Portable Video Communication Devices based on Power-Rate-Distortion Optimization. In IEEE Transactions on Circuits and Systems for Video Technology, (Vol. 18).

IEEEP802.11e/D11.0, Part 11: Wireless LAN Medium Access Control (MAC) and Physical Layer (PHY) specifications: Amendment 7 - Medium Access Control Quality of Service Enhancements. (2004 October). Washington, DC: IEEE.

Ji, X. (2008). *Energy-efficient video transmission over multi-user wireless networks*. Ph D. Thesis in Electrical Engineering, Katholieke Universiteit Leuven, Belgium.

Joint Scalable Video Model JSVM-10. (2008, April).

Katsaggelos, A. K., Zhai, F., Eisenberg, Y., & Berry, R. (2005). Energy-Efficient Wireless Video Coding and Delivery. IEEE Wireless Communications.

Lan, T.-H., & Tewfik, A.-H. (2003). A Resource Management Strategy in Wireless Multimedia Communications – Total Power Saving in Mobile Terminal with a Guaranteed QoS. *IEEE Transactions on Multimedia*, *5*(2).

Lu, X., Wang, Y., & Erkip, E. (2002). Power Efficient H.263 Video Transmission over Wireless Channels. In *International Conference on Image Processing (ICIP)*.

Lu, X., Wang, Y., & Erkip, E. (2007). Total Power Minimization for Multiuser Video Communications Over CDMA Networks. *IEEE Transactions on Circuits and Systems for Video Technology*, *17*(6).

Lu, X., Wang, Y., Erkip, E., & Goodman, D. (2003). Power Efficient Mutlimedia Communication over Wireless Channels. *IEEE Journal on Selected Areas in Communications*, *21*(10).

Mangharam, R., et al. (2005). Optimal Fixed and Scalable Energy Management for Wireless Networks. In INFOCOM, USA.

Mohapatra, S., Dut, N., Nicolau, A., & Venkatasubramanian, N. (2007). DYNAMO: A Cross-Layer Framework for End-to-End QoS and Energy Optimization in Mobile Handheld Devices. *Journal on Selected areas in Communications*.

NS-2. (n.d.). Retrieved from http://www.isi.edu/nsnam/ns

Pollin, S., Mangharam, R., Bougard, B., Van der Perre, L., Moerman, I., Rajkumar, R., & Catthoor, F. (2007). MEERA: cross-layer methodology for energy efficient resource allocation in wireless networks. *IEEE Transactions on Wireless Communications*, *6*(2), 617–628. doi:10.1109/TWC.2007.05356

Raghunathan, V., Ganeriwal, S., Schurgers, C., & Srivastava, M. B. (2002). E2WFQ: An energy efficient fair scheduling policy for wireless systems. In *Proc. ISLPED*, (pp. 30-35).

Reichel, J., Schwarz, H., & Wien, M. (2005). Scalable Video Coding – Working Draft 1, Joint Video Team, JVT-N020, China.

Saponara, S., Blanch, C., Denolf, K., & Bormans, J. (2003). The JVT Advanced Video Coding Standard:Complexity and Performance Analysis on a Tool-by-tool Basis. In *Packet Video Conference, Nantes, France*.

Schurgers, C. (2002). *Energy-Aware Communication Systems*. Ph.D. Thesis in Electrical Engineering, Univ. of California, L.A, Los Angeles, CA.

Timmers, M., et al. (2007). A Distributed Multichannel MAC Protocol for Cognitive Radio Networks with Primary User Recognition. In CROWNCOM.

Uysal-Biyikogly, E., Prabhakar, B., & El Gamal, A. (2002, Aug.). Energy-efficient packet transmission over a wireless link. *ACM/IEEE Transactions on Networking*, *10*(4), 487-499.

Chapter 7
QoS and Energy Saving Routing and MAC Mechanisms for Wireless Networks

Vasileios Gkamas
University of Patras, Greece

Panagiotis Kokkinos
University of Patras, Greece

Emmanouel Varvarigos
University of Patras, Greece

ABSTRACT

This chapter describes routing and medium access control (MAC) mechanisms for providing Quality of Service (QoS) together with energy savings in wireless ad hoc networks. The proposed mechanisms operate in a cross-layer optimization logic, in the direction of either minimizing total energy consumption in the network or maximizing network lifetime, while at the same time providing QoS to the end users. The authors present a multi-cost routing approach, where various cost parameters and optimization functions are defined and used for selecting the paths to be followed. Also, routing and MAC protocols are investigated for the case where nodes have variable transmission power capabilities. Finally, the performance of the proposed protocols is evaluated and compared to that of other well-known routing and MAC protocols.

INTRODUCTION

Quality of Service (QoS) mechanisms are important for ad hoc wireless networks, especially since they determine their ability to support delay-sensitive or high-bandwidth applications. Traditional distributed algorithms for QoS routing in wired networks may not work well for wireless ad hoc networks, given that these networks have mobility characteristics, and energy and capacity constraints that are not present in wired networks. Since stations in wireless networks are usually battery-operated and require long operating lifetimes, energy is a crucial resource limiting the performance and range of applicability of such networks. However, the advances in battery lifetime during recent years have not kept in pace

DOI: 10.4018/978-1-61520-771-8.ch007

with the significant decline achieved in computation and communication costs in wireless ad hoc networks. As a result, the design of routing and MAC protocols that make best use of the nodes' energy resources is of outmost importance.

A node in a wireless network consumes energy when transmitting, receiving, or simply listening to the channel. In the transmitting mode, energy is spent in two main ways. The first is in the front-end amplifier that supplies the power for the actual RF transmission (transmission energy), while the second is in the node processor that implements the various signal processing functions (processing energy). In the receiving mode, energy is consumed entirely by the processor. Finally, in the listening mode, the energy consumed is again due to processing operations. However, even in the listening mode, the network protocol may require a device to emit periodic beacon signals. All of the aforementioned modes of operation involve all the components (hardware and software) that constitute a wireless node.

Moreover, energy consumption in a wireless network relates to the higher-level protocols used. At the network layer, extensive collaboration is required among network nodes, in order to establish the routes and to secure the resources that are necessary to provide to the users the required QoS. Most routing protocols proposed to date are based on the single-cost (shortest path) approach, where a single metric is used to represent the cost of using a link. This link metric can be a function of several network parameters, but it is still a scalar. Routing algorithms of this kind calculate the path that has the minimum cost for each source-destination pair, but they cannot optimize the performance with respect to general cost functions, and they do not easily support QoS differentiation. Also, they usually yield only one path per source-destination pair, leading to non-uniform traffic and energy distribution, and to possible instability problems in the network. At the MAC layer, the proposed protocols for ad hoc networks have solved several problems occurring mainly due to the broadcast nature of the wireless medium (for example, the hidden terminal problem). However, many of these MAC protocols do not take into account energy and QoS considerations. At the physical layer, the transmission power used by the nodes obviously affects the energy consumed in the network, since the use of large transmission power results in more energy expenditure. When the wireless nodes have fixed transmission power, independently of the receiving node they are trying to reach, they may expend an unnecessarily large amount of energy and cause unwarranted interference to other nodes (when the desired recipient is at a smaller distance than the static transmission range used), resulting in packet re-transmissions, and consequently more energy expenditure.

At this chapter we describe cross-layer protocols to reduce energy consumption and support QoS differentiation in wireless ad hoc networks. To achieve energy savings in the routing procedure, the multi-cost routing approach is presented where routing decisions are made based on several cost parameters, including information provided by lower levels (e.g., MAC layer). In particular, a cost vector consisting of several cost parameters is assigned to each link. The cost parameters of a link can include many parameters of interest (e.g., the hop count, the energy expended by the transmitting node, its residual energy, etc). These cost parameters are combined in a number of different ways, each corresponding to a different routing algorithm, in order to select the optimal route. Another interesting approach considered is the joint multi-cost routing and power control in wireless networks, where various cost parameters of interest are considered, such as the interference caused, the residual energy and the variable transmission power of the nodes. This joint multi-cost routing and power control approach obtains energy/interference efficient routes and optimizes network performance. Finally, a power control MAC protocol is presented where every node by using information exchanged during the RTS/CTS frames exchange, computes the minimum required power for coherent packets' transmission.

The rest of the chapter is organized as follows. In Section 2 we review related routing and MAC energy-aware protocols for wireless ad hoc networks. In Section 3, new cross-layer energy-aware protocols for wireless ad hoc networks are presented. In particular, Section 3.1, describes the multi-cost routing approach, Section 3.2, describes various interference/energy aware routing protocols that are based on the multi-cost approach, and Section 3.3 describes a new energy-aware MAC-layer power control protocol. The performance of the proposed energy-aware protocols is evaluated in Section 4. Finally, the conclusions and future research directions are pointed out in Section 5.

BACKGROUND

In this section several works are presented, investigating energy and Quality of Service (QoS) issues for wireless ad hoc networks. Generally, it is accepted that in ad hoc networks there is a strong coupling between the different layers of the ISO/OSI network model (Ephremides, 2002), and the decisions that are made at one layer affect the decisions and performance of the protocols at the other layers, making cross-layer optimization necessary. For this reason in this section we describe both routing and MAC protocols for wireless networks. The various works presented however are not in any way exhaustive, but indicative of the different approaches that can be followed.

Energy-efficient routing has been considered from the perspective of either minimizing the total energy consumption or maximizing the network lifetime. The work from Singh et al. (1998) is one of the first, proposing energy-aware routing protocols for ad hoc networks, investigating these two optimization goals. These goals are contradicting, since the selection of routes that minimize energy consumption, tends to create nodes with quite different energy profiles. In this way the lifetime of individual nodes is prolonged instead of the lifetime of the entire network. On the other hand, protocols that increase the network's lifetime tend to overlook performance related parameters, such as packet delay, or delay jitter and other. For example, these protocols may select a route with underutilized nodes than a shortest route. There are various definitions of the network's lifetime; it can be defined as the time period until the first packet drop due to energy issues occurs, as the time until certain percentage of nodes fails, or as the time until network partitioning occurs. Independently of the optimization goal followed, most energy related protocols search for a path that minimizes a energy related cost metric. Different works define differently this cost metric. Moreover, the various wireless link characteristics (e.g., interference) should also be included into the links' cost metrics, since they affect the energy consumption, e.g., by leading to exhaustive retransmission of packets.

In general, the algorithms that attempt to maximize the network's lifetime, actually perform load balancing. For example, by distributing the network load to all the nodes, all of them consume energy at the same rate and eventually run out of energy at approximately the same time. This may result in longer network lifetime, but also may result on less performance efficient routes. Chang et al., (2004) propose a protocol where the link costs are defined based on the initial and residual energy at the transmitting nodes. Toh et al. (2001) propose a Min-Max Battery Capacity Routing (MMBC) protocol to exclude the energy starving nodes (those having least battery capacity) from route selection. Energy-aware algorithms that also take into account link related parameters have also been proposed. Gupta & Das (2002) suggest a new cost metric for routing, which is a function of the remaining battery level and the number of neighbors of a node. Other works have focused on the discovery of energy efficient routes under the constraint of a fixed end-to-end bit error rate, or by considering the expected number of retransmissions for reliable packet delivery (Younis et al., 2003; Banerjee & Misra, 2002). Woo et al. (2001) present the LEAR protocol, where a node decides to forward or not traffic based on its residual energy.

The most common technique for minimizing energy consumption in the network is minimizing the energy consumed when nodes are in transmission mode or when some nodes are inactive. The former is achieved though transmission power control, where a node's transmission power is adjusted to the minimum required in order for a packet to be successfully received by a neighbour node. The latter goal is achieved by putting nodes into sleep mode (e.g., turn-off, hibernate etc) when they do not have any data to transmit or receive. Both approaches have advantages and disadvantages. By putting nodes into sleep mode large energy savings can be achieved. However, this approach is not so efficient when the network is under high load, since data delivery can be hindered by nodes in sleep mode. This situation requires efficient protocols for routing packets and disabling (putting into sleep mode) and/or enabling (waking up) nodes. Also, by using transmission power control a protocol may choose a path with many short range links that actually perform worse than a path with fewer long-range links, both in terms of latency as well as energy consumption. For example, a path with many short-range links may cause more link errors which result in more retransmissions. In the same context, when the transmission power is controllable, the optimal adjustment of the power level is essential not only for energy conservation but also for the interference control. A minimum energy consumption protocol is presented in Rodoplu et al. (1998), where the link costs are defined based on the energy expenditure for unit flow transmission. Ramanathan & Rosales (2000) propose two algorithms for selecting the node transmission power. At another work (Chang & Tassiulas, 2000), a distributed algorithm is presented, that incorporates power control in the routing of packets, and tries to increase energy consumption at nodes with plenty of energy, while reducing consumption at nodes with small energy reserves. In ElBatt et al. (2000), Li et al. (2007), joint power control, scheduling and routing algorithms are presented. Also, Stojmenovic & Lin (2001) propose a localized, distributed energy-aware routing protocol, which assumes that each node has the location information of its neighbours and the destination. Using this information the protocol selects the next hop, through which the total transmission power to the destination is minimized. Also, a term usually encountered in this context is topology control (Burkhart et al. (2004)), where the goal is to maintain a connected topology using the minimum power. Span (Chen et al. 2001) is a distributed, randomized algorithm where nodes make local decisions on whether to sleep, or to join a backbone infrastructure.

As mentioned, cross-layer optimization is necessary for achieving energy efficiency in wireless ad hoc networks (Ephremides, 2002). In particular, the transmission power control and the nodes' sleep mode capabilities, relate to lower layers of the ISO/OSI model and the corresponding routing protocols use MAC layer mechanisms for implementing these operations. Also, several energy efficient MAC protocols have been proposed. These MAC protocols alter the transmission powers of the nodes as a means to reduce the energy consumption. Agarwal et al. (2001) propose a scheme where power control is incorporated in the MAC layer by using the RTS-CTS-DATA-ACK sequence to reach an agreement on the transmission power to be used. Also, Jung & Vaidya (2005) proposed a power control MAC protocol for mobile ad hoc networks that allows nodes to vary their transmission power on a per-frame basis. The main idea in this scheme is to use different power levels for RTS/CTS and DATA/ACK frames transmissions so as to save energy. A Power Controlled Dual Channel MAC protocol was proposed by Muqattash & Krunz (2003), where the MAC layer indirectly influences the routing decisions at the network layer by controlling the power level of the broadcasted route request packets. The potential benefits of adjusting the transmission power are also considered by ElBatt et al. (2000), where the authors investigate the effects of power adjustment on the average power consumption and on the end-to-end throughput in a wireless ad hoc environment. Finally, Ye et al. (2003) propose a MAC protocol, where nodes form virtual clusters based on common sleep schedules to reduce the control overhead and enable traffic-adaptive wake-up capability. Each node bases its decision on an estimation of how many of its neighbours will benefit from it being awake and the amount of energy available to it.

Different energy related approaches have also been proposed. In Gupta & Das (2002) network clustering is used and the manager of each cluster is chosen as the least energy constrained node. Network models that assume energy recharging capabilities have also been considered in Rahimi et al. (2003). The selection of multiple energy efficient paths for a given source-destination pair has been proposed by Shah & Rabaey (2002).

Also, very few of the previously described algorithms include (at least directly) performance related parameters in their formulations, like delay, number of hops and other. In this way Quality of Service (QoS) cannot be achieved. For some energy related algorithms (Singh et al. (1998) and Toh et al. (2001)), it has been observed that it may not be necessary for the energy related algorithm to be used all the time. Instead, other routing protocols (e.g., a shortest path algorithm) can initially be used. After some time or after the nodes consume a predefined amount of their energy reserves, an energy-aware algorithm can be used. Another approach would be to route some packets (e.g., based on their priority) with energy-aware algorithms and some other packets with a shortest path algorithm. Incorporating performance related parameters in the links' cost metrics is also another approach.

In general, most of the proposed (energy efficient or not) routing algorithms are single-cost and incorporate into a single link metric various different and completely diverse parameters. In single-cost routing one or more parameters (link characteristics) are combined in a single-cost metric characterizing the link. However, in this case the path's final cost value does not indicate almost anything about the values of the path's constituent parameters. For example, in the end we cannot be certain whether the values of particular parameters (e.g., load, delay) are above a predefined threshold, except if we rerun the algorithm once for each parameter. In this way it is not possible to provide Quality of Service (QoS) and this is the reason the multi-constraint and multi-cost approaches have been considered. These approaches better captures the meaningfulness of each cost parameter by considering it for the whole path than for a single link. Also, operators like minimization (restrictive parameters) and maximization (maximum representative parameters) can be used, something that is not possible in the single-cost approach. Multi-constrained routing algorithms have also been investigated, especially for wired networks in order to provide QoS (Chen & Nahrstedt, 1998; Jafee 1984; Korkmaz & Krunz, 2001; Mieghem & Kuipers, 2004; Wang & Crowcroft, 1996). Finding paths subject to two or more cost parameters/constraints is in most cases an NP-complete problem (Yuan & Liu, 2002; Garey & Johnson, 1979). As a result, most algorithms proposed in this area concentrate on solving the Multi-Constrained Path (MCP) problem or the Multi-Constrained Optimal Path (MCOP) problem in a heuristic and approximate way with polynomial and pseudo-polynomial-time complexities, paying little attention to the parameters/costs used and their effects on network performance. The multi-constrained problem has been less studied in the context of wireless ad hoc networks, even though these networks have important reliability, energy and capacity constraints that are not present in wired networks. Wang &

Crowcroft (1996) propose a probabilistic modelling of the link state for wireless sensor networks, and propose an approximation of a local multipath routing algorithm to provide soft-QoS under delay and reliability constraints. At another work (Huang & Fang, 2006) a multi-constrained QoS routing algorithm for mobile ad hoc networks is proposed that uses simulated annealing. Liu & Feng (2007) present an algorithm based on depth-first-search that solves the general *k*-constrained MCP problem with pseudo-polynomial time complexity. At other works (Badis & Agha, 2007; Perkins et al, 2000), well-known routing algorithms for ad hoc networks, are extended to support QoS through the usage of multiple constraints. These algorithms focus on the bandwidth and delay constrained routing problem. A QoS routing scheme for ad hoc networks that uses flooding is proposed by Perkins et al., (2000). The idea of multi-cost routing, which is a generalization of multi-constrained routing, was presented by Gutierrez et al. (2000) where it was applied to wireline max-min fair share networks. Some of the ideas concerning multi-cost routing in wireless ad hoc networks are scattered over our previous works (Kokkinos et al., 2005; Papageorgiou et al., 2006) and are presented here coherently and more structured.

It is self-evident that it is difficult to compare directly the various presented approaches, since each approach has different goals and may work better under different conditions. However, we can draw some basic remarks. Energy preservation should not be performed without considering also the performance requirements of the packets and of the corresponding applications. The single-cost approach is not the solution to this problem, since in this way the significance of each parameter is lost in the final cost value of a candidate path. Instead, the multi-constrained and the multi-cost approaches seem the most appropriate ones. Finally, the energy and performance/QoS related issues must be handled in all the layers of the ISO/OSI network model and in particular in the routing and MAC layers, either jointly or disjointly. The algorithms presented next, are consistent with these basic remarks.

ENERGY AWARE PROTOCOLS FOR WIRELESS AD HOC NETWORKS

We are interested in the design of energy aware MAC and routing protocols for wireless ad hoc networks that also provide Quality of Service (QoS) differentiation to the users. As it was stated in Section 2, cross-layer optimization is a principle which should be followed during the design of energy efficient and QoS protocols for wireless networks.

In the routing procedure, the wireless nodes forward, in addition to their own packets, data or control packets generated by other nodes. Many routing protocols compute the paths using as selection criterion the minimization of the hop-count. In this case some nodes may end up getting considerably more traffic than other nodes and may soon run out of energy. As a result the network may become disconnected, even though many nodes with sufficient energy still exist. Even when the network does not become disconnected, the lack of sufficient energy at some nodes may force future packets to take very long paths. A better strategy is to select the paths to be followed taking into account, in addition to the number of hops, various energy parameters, such as the transmission power and residual energy of the nodes.. Furthermore, the traffic should be routed so that the energy consumption is balanced among the nodes, that is, so that the variance of the nodes' energy reserves is minimized. The goal is to make the network operational for as long as possible before it gets partitioned, by increasing the lifetime of the nodes. In any case, the routing algorithm must not only take into account the interests of a particular session, but the overall performance of the network and the quality of the communication, which are the actual goals. The routing protocols proposed to date are based on the single-cost (shortest path) idea, where a single metric is used to represent the cost of using a link. This link metric can be a function of several network parameters, but it is still a scalar. Routing algorithms of this kind calculate the path that has the minimum cost for each source-destination pair, but they cannot optimize performance with respect to general cost functions, and they do not easily support QoS differentiation. Also, they usually yield only one path per source-destination pair, leading to non-uniform traffic and energy distribution, and possible instability problems in the network.

To overcome these problems we introduce the multi-cost routing approach for wireless ad hoc networks. In the multi-cost routing approach proposed, a cost vector V_l consisting of several cost parameters is assigned to each link *l*. A cost vector can then be defined for a path by combining component-wise the cost vectors of its links, according to some associative operator. The multi-cost routing algorithms that we propose consist of two phases: they first compute a set of candidate non-dominated paths (to be defined later) for a given source-destination pair, and then they select the path that minimizes a certain optimization function. Different optimization functions give rise to different routing algorithms. The function to be optimized is chosen based on the interests of the network, but it may also depend on the user QoS requirements. So multi-cost routing supports service differentiation for sessions with varied QoS requirements (in the performance results presented, however, we assume a single QoS class). Finding the optimal path for a different QoS class does not require the recalculation of the set of non-dominated paths, but only the application of the new QoS metric on the non-dominated paths already found. The multi-cost algorithms are optimal, in the sense that they can optimize any desired function of the parameters of interest, provided that the optimization function is monotonic in each of these parameters. Multi-cost routing, can be considered as a generalization of the multi-constrained routing considered in earlier works mainly in the context of wired networks. In the multi-constrained routing problem, a constraint is specified for each of the cost parameters of a path and the target is to find paths that satisfy all the constraints. In the multi-cost routing problem there are no constraints and the target is to find paths that are better than all other paths for all or some of the cost parameters, based on the optimization function used. Multi-cost routing is therefore, a generalization of multi-constrained routing, in the sense that the latter case can be obtained from the former case, by choosing the function to be optimized so as to have infinite cost at the constraint points.

At the MAC layer, the strategies followed by many protocols can also affect the total energy consumed and cause many problems, possibly resulting in low network throughput. The usage of the RTS/CTS/ACK frames has solved many negative issues caused by the hidden terminal problem. However, this approach negatively impacts the channel utilization by completely disallowing concurrent transmissions over the reserved floor, resulting in a degraded overall QoS. Consider for example Figure 1, where node A wants to transmit some data packets to node B. Node A initially sends a RTS frame to node B and node B replies to node A with a CTS frame. As nodes C and D hear B's CTS frame, refrain for transmitting. Although, it is clear that both transmissions A→B and C→D could in principle take place simultaneously without causing interference to each other, these transmissions are not permitted due to the reception of the CTS frames by the nodes C and D. This reduces the maximum throughput that can be achieved, and also increases the energy consumption in the network. In particular, since the energy consumption is straightforwardly related to the transmission power, then the energy consumed by the nodes is usually more than the

Figure 1. The transmission floor of the RTS and CTS packets is illustrated with solid lines. Dashed lines illustrates that both transmissions A→B and C→D could in principle take place simultaneously without interference

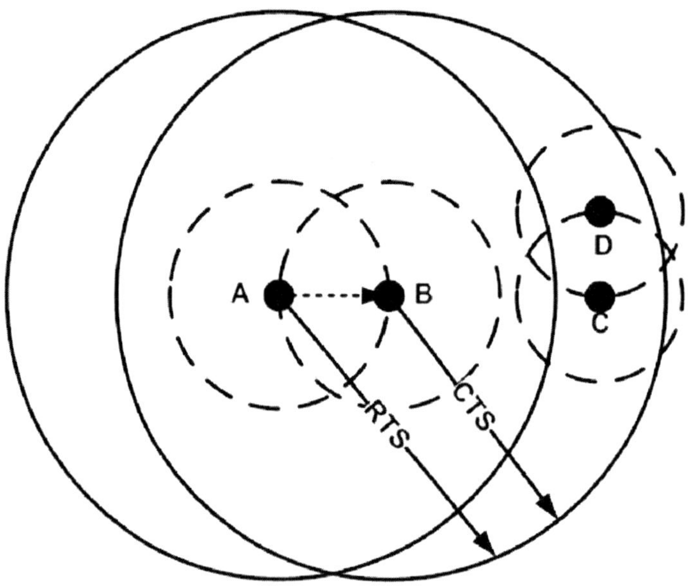

required, in the case where fixed transmission power is used and the distance of the intended receiver is smaller than the transmission range of the transmitter.

As a result, the necessary modifications are needed at the MAC-layer to overcome these problems. A solution would be to permit to the nodes receiving a RTS or CTS frame to make a packet transmission, if certain conditions are applied, if for example, the interference caused by these transmissions is not large enough to cause a packet collision. Also, the use of power control techniques can reduce the unnecessary energy consumption that can be observed in wireless ad hoc networks. In particular, using these techniques, the nodes can adjust their transmission powers to the minimum required, minimizing in this way the energy consumed in the network compared to the case where fixed transmission power is used. Power control techniques can also be combined with routing algorithms, reducing in this way the energy consumption and increasing the achieved throughout in the network.

The Multi-Cost Routing Approach

In multi-cost routing, each link of the network is assigned a cost vector consisting of several cost parameters. The cost vector of a path is obtained from the cost vectors of the links that comprise it by applying, component-wise, a monotonic associative operator to each cost vector parameter. The parameters that may be included in the path cost vector are categorized by the way they are obtained from the link cost vectors, that is, by the associative operator used for each cost vector component, and by the criterion applied to them (maximization or minimization) to select the optimal path. To be more specific, we denote by $v_l = (v_{1l}, v_{2l}, \ldots v_{kl})$ the link cost vector of link l, by $V(\mathrm{P}) = (V_1, V_2, \ldots, V_k)$ the cost vector of the path P that consists of links l=1,2,…,L and by f(V) the optimization function that has to be minimized in order to select the optimal path. The cost vector $V(\mathrm{P}) = (V_1, V_2, \ldots, V_k)$ of a path P consisting of links l=1,2,…,L is then obtained from the cost vectors of the links that comprise it by applying component-wise a monotonic associative operator $\otimes$ to each cost vector parameter:

$$V_m = \otimes_{l=1}^{L} v_{ml} \tag{1}$$

The associative operator may be different for different cost vector components. For example, the m-th parameter of the cost vector may be of one of the following types:

- additive cumulative, where

$$V_m = \sum_{l=1}^{L} v_{ml}, v_{ml} \geq 0 \tag{2}$$

and f is monotonically increasing in V_m (so our objective is to minimize V_m),

- restrictive, where

$$V_m = \min_{l=1,\ldots,L}\{v_{ml}\} \tag{3}$$

and f is monotonically decreasing in V_m (so our objective is to maximize V_m), and

- maximum representative, where

$$V_m = \max_{l=1,\ldots,L}\{v_{ml}\} \tag{4}$$

and f is monotonically increasing in V_m (so our objective is to minimize V_m).

Additive cumulative parameters include several important cost measures used in practice. For example, if v_{ml} is the delay on link l, then V_m represents the delay of the path, which in most practical situations has to be minimized. If $v_{ml} = 1$ for all links l, then V_m corresponds to the number of hops on the path. Since paths that use a small number of links are more economical in terms of resource utilization, it is natural to assume that the cost function f is an increasing function of V_m. If v_{ml} is the energy consumed on link l of a wireless network, then V_m represents the energy consumed for transmitting a packet on the path, which has to be minimized. Another interesting case arises when $v_{ml} \in [0,1]$ represents the probability that link l is operational, and $V_m = \prod_{l=1}^{L} v_{ml}$ is the probability that all links on a path are operational (assuming links fail independently of each other). For the routing algorithm to favor reliable paths, the cost function f should be a decreasing function in V_m. This problem can be reduced to a problem involving cumulative additive components by defining new cost components $v'_{m1} = -\ln v_{m1}, \ldots, v'_{ml} = -\ln v_{ml}$, where $v'_{ml} \geq 0$. Then maximizing the reliability V_m of a path is equivalent to minimizing $V'_m = \sum_{l=1}^{L} v'_{ml}$.

Restrictive cost parameters appear in routing problems when capacities or transmission rates are considered. In particular, if v_{ml} is the available capacity on link l, then $V_m = \min_{l=1,\ldots,L}\{v_{ml}\}$ represents the capacity of a path, defined as the minimum of the capacities available on the links of the path. For the routing algorithm to favor less congested paths, the cost function f should be a decreasing function of V_m. Another interesting case arises when v_{ml} represents the remaining energy at the transmitting node of link l, in which case V_m represents the minimum energy available over all nodes of a path, which in most practical cases we want to maximize.

An example of a maximum representative parameter is the case where v_{ml} is the energy consumed for transmitting a packet on link l, in which case $V_m = \max_{l=1,\ldots,L}\{v_{ml}\}$ represents the most energy-expensive transmission on the path. Another example is the case where v_{ml} is the Bit Error Rate (BER) on link l, in which case $V_m = \max_{l=1,\ldots,L}\{v_{ml}\}$ represents the link with the highest BER on the path, which is often a good approximation of (or at least of the same order of magnitude with) the path BER.

It is important to note that the path that optimizes $f(V_1, V_2, \ldots, V_k)$ is *not* in general the same with the path that optimizes $\sum_{l=1}^{L} f(v_{1l}, v_{2l}, \ldots, v_{kl})$ indicating that multi-cost routing is a generalization of single-cost (shortest path) routing. Also, in contrast to single-cost routing, multi-cost routing is not always compatible with distributed routing, since for some choices of the cost function f the optimal paths do not have the inclusion property that (found with single-cost routing) shortest paths have. These show that multi-cost routing is very different from single-cost routing both in terms of the decision it takes and the way it is implemented. A detailed comparison between single-cost and multi-cost routing is done in Section 3.2.3.

A multi-cost routing algorithm consists of two phases. In the first phase, an enumeration of an appropriate set of candidate paths for a given source-destination pair is performed. In the second phase, the optimal path is chosen from this set according to the optimization function $f(V)$ used. The set of candidate paths that a multi-cost routing algorithm produces at the end of the first phase consists of the so called non-dominated paths. These are paths for which it is impossible to find other paths that are better with respect to one cost parameter (of their cost vectors) without being worse with respect to some other cost parameter. This reduces to a large extent the algorithm's computational effort, since the optimization function does not need to be applied to every possible path between a certain source-destination pair. An example of the enumeration of the non-dominated paths is given in Figure 2, where an additive cumulative parameter h and a restrictive parameter R are assumed. The final cost of a path is given by a function $f(V)$ of its cost vector V, and the routing algorithm selects the path with the minimum cost from the set of non-dominated paths.

An algorithm that obtains the set of non-dominated paths is presented by Gutierrez et all (2000), which can be viewed as a generalization of Dijkstra's algorithm. In each round, the algorithm discovers new non-dominated paths between a source node and one of its neighbours or between a source node and one of the neighbours of a previously considered neighbour. The discovered paths optimize the first cost parameter, while in case of a tie, the second parameter is checked, and so on. These paths are added to the non-dominated set and the cost vectors for all the paths of the network that have common links with the discovered paths are updated accordingly. Next the paths in the set (either newly discovered, or found earlier in the process) that are found to be dominated are discarded and are not considered further. The algorithm is completed when there are no more paths to be checked. The basic difference of this algorithm with Dijkstra's algorithm is that a set of non-dominated paths between a source node and a destination node is obtained, instead of a single path. Also, a destination node for which a path has already been found may have to be considered again later. The result of this process is a set of non-dominated paths from a given origin node to each other node of the network. When the origin node has to transmit a packet to a destination node, the optimization function is applied to each path in the set of non-dominated paths, and the path with the minimum cost is chosen. The preceding algorithm results in a set of non-dominated paths from a given source node to any other node of the network. When the source node wants to transmit a packet to a destination node, then an optimization function is applied to every path in the non-dominated set, and the path with the minimum cost is chosen.

Energy and Joint Interference/Energy Aware Multi-Cost Routing Algorithms

At this section various energy and joint interference/energy-aware multi-cost routing algorithms, for wireless ad hoc networks are presented. First, the cost parameters used by the multi-cost routing algorithms are defined, and then the various optimization functions of every algorithm, are described.

Figure 2. Enumeration of the set of non-dominated paths, for the case of two cost parameters. Each dot represents the cost vector of a path. The parameters of this vector are an additive cumulative parameter h (representing, for example, the number of hops or the delay of the path) and a restrictive parameter R (representing, for example, the minimum residual energy on the nodes of the path or the available capacity on the path). (a) The set of all paths. (b) Obtaining the non-dominated paths. (c) The set of non-dominated paths; paths that have both larger R and smaller h than some of the other paths that have been discarded

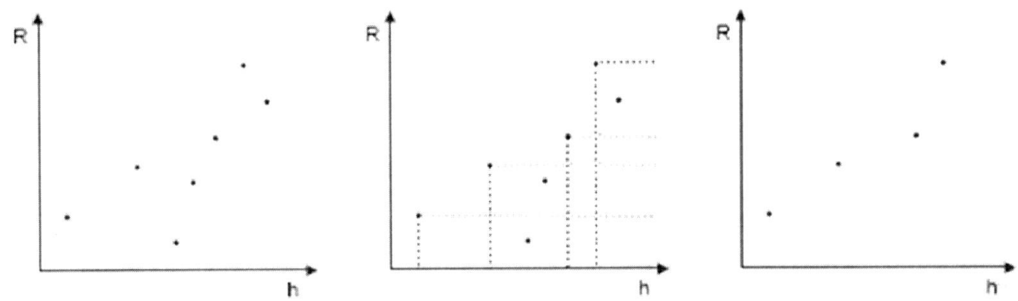

(a) (b) (c)

Cost Parameters

The cost parameters used by multi-cost routing algorithms can vary depending on the criteria optimized. In the case of energy-aware multi-cost routing, we propose the following cost parameters:

- The number of hops h of a path. The associative operator $\odot$ used in this case is the addition:

$$h = \sum_{l=1}^{j} h_l = j \tag{5}$$

where h_l=1 for all links l. The number of hops h of a path is obtained by counting the links that belong to it. Paths with a small number of hops are generally preferable to longer paths.

- The minimum residual energy R of a path. Here we use the residual energy R_l at the transmitting node of link l as the link cost metric. The minimum residual energy on the path is then obtained by applying the minimization operator to the link cost metrics to obtain:

$$R = \min_{l=1,\ldots,j} R_l \tag{6}$$

The minimum residual energy R indicates the degree to which a path is energy-critical. Paths with large minimum residual energy are generally preferable.

- The sum T_1 or the maximum T_∞ of the transmission powers used by the nodes on a path. If we denote by T_l the transmission power required for correct reception over link l, then T_1 is obtained by combining the link metrics using the additive operator, while T_∞ is obtained by combining the link metrics using the maximization operator:

$$T_1 = \sum_{l=1}^{j} T_l \tag{7}$$

or

$$T_{\infty} = \max_{l=1,\ldots,j} T_l \tag{8}$$

Paths with small values of T_1 consume little total energy, and are therefore preferable. Similarly, paths with small values of T_{∞} avoid energy-critical nodes and are also preferable.

For the case of the joint interference/energy-aware multi-cost routing we assume that the nodes can adjust their transmission power to the minimum required for the efficient communication with the target nodes. Moreover, except the previously described cost parameters, new cost parameters are defined which take into account the interference caused by the transmissions:

- The total interference I_1, or the maximum interference I_{∞} caused by using a path. Similar to the work by Burkhart et al (2004), we define the interference I_l caused by using link l as the number of nodes (other than the transmitter and the receiver) that are within the transmission range of the end nodes of link l. If we denote the distance between the transmitter a and the receiver b of link $l = (a,b)$ as $|a,b|$, then:

$$I_l = I_{(a,b)} = \left|\left\{c \in V, |b,c| \leq |a,b| \cup |a,c| \leq |a,b|\right\}\right| - 2 \tag{9}$$

Of course one could suggest that some of the neighbours of nodes a and b may be inactive; however the metric assumes of the worst case scenario where all nodes are active and can transmit at any time. Also in our metric we count the nodes that are within the transmission range of both the transmitter and the receiver and not only those of the transmitter. This is consistent with the RTS/CST mechanism of the 802.11 protocol we use, which informs and deters from transmission the neighbouring nodes of both ends of a link. The total interference I_1 or the maximum interference I_{∞} caused by using a path is obtained by employing the additive or the maximization operator, respectively, for combining the interference metrics of the links on the path:

$$I_1 = \sum_{l=1}^{j} I_l \tag{10}$$

or

$$I_{\infty} = \max_{l=1,\ldots,j} I_l \tag{11}$$

Paths that create little total interference I_1 or little maximum interference I_{∞} are generally preferable.

Optimization Functions - Algorithms

The aforementioned cost parameters can be combined in different ways to produce various multi-cost routing algorithms. All of them select the path P with the minimum cost returned from the corresponding function. In the case of the energy-aware multi-cost routing, the optimization functions examined, generally try to select paths that have a small number of hops, consume little energy and pass through nodes that have large residual energy. In the case of the joint interference/energy-aware multi-cost routing, the optimization functions examined generally try to select paths using the same criteria as before, plus one new, that is choosing paths which cause little interference. The optimization functions presented (that in fact correspond to different optimization algorithms), differ in the relative importance they give to these metrics. We examined a number of different optimization functions, the most interesting of which are the following:

- *Minimum-Hop* algorithm: The optimization criterion is the number of hops h on the path P.

$$\min_p h(p) \tag{12}$$

- *MAX/MIN Energy* algorithm: The optimization criterion is the ratio of the maximum of the transmission powers of the nodes on path P, over the minimum amount of residual energy at the nodes of the path,

$$\min_p \frac{T_\infty(P)}{R(P)} \tag{13}$$

- *SUM/MIN Energy* algorithm: The optimization criterion is the ratio of the sum of the transmission powers of the nodes on path P, over the minimum residual energy at the nodes of the path:

$$\min_p \frac{T_1(P)}{R(P)} \tag{14}$$

- *MAX/MIN Energy-Hop* algorithm: This algorithm is a hybrid between the *Minimum-Hop* and the *MAX/MIN Energy* algorithms. The cost criterion optimized is the product:

$$\min_p \frac{h(P) \cdot T_\infty(P)}{R(P)} \tag{15}$$

- *SUM/MIN Energy-Hop* algorithm: This algorithm is a hybrid between the *Minimum-Hop* and the *SUM/MIN Energy* algorithms. The cost function optimized is the product:

$$\min_p \frac{h(P) \cdot T_1(P)}{R(P)} \tag{16}$$

- *MAX/MIN Energy-Half-Hop* algorithm: The optimization function is similar to that in the *MAX/MIN Energy-Hop* algorithm, but uses $\sqrt{h(P)}$ instead of $h(P)$, in order to give a smaller dependence on the cost of the number of hops:

$$\min_p \cdot \frac{\sqrt{h(P)} \cdot T_\infty(P)}{R(P)} \tag{17}$$

- *SUM/MIN Energy-Half-Hop* algorithm: The optimization function is as in the *SUM/MIN Energy-Hop* algorithm, but uses $\sqrt{h(P)}$ instead of $h(P)$ in the product:

$$\min_p \cdot \frac{\sqrt{h(P)} \cdot T_1(P)}{R(P)} \tag{18}$$

  - *Minimum Interference* algorithm: The criterion optimized is the sum of the interference of all the links on the path:

$$\min_p I_1(P) \tag{19}$$

  - *Minimum Transmission Power* algorithm: The criterion optimized is the sum of the transmission powers of the nodes on the path:

$$\min_p T_1(P) \tag{20}$$

  - *SUM/MIN Energy-Interference* algorithm: The optimization that takes place is:

$$\min_p \frac{T_1(P) \cdot I_1(P)}{R(P)} \tag{21}$$

which tends to select paths that cause small total interference, use low total transmission power, and pass through nodes that have large energy reserves.

- *SUM/MIN Energy-Half-Interference* algorithm: The function optimized is similar to the one used in the *SUM/MIN Energy-Interference* algorithm, but has a smaller dependence on the interference metric:

$$\min_P \frac{T_1(P)\cdot\sqrt{I_1(P)}}{R(P)} \quad (22)$$

- *SUM/MIN Energy-Interference-Half Hop* algorithm: The optimization function is equal to the *SUM/MIN Energy-Interference* function, multiplied by $\sqrt{h(P)}$ so as to discourage, to a certain extent, the use of long paths:

$$\min_P \frac{\sqrt{h(P)}\cdot T_1(P)\cdot I_1(P)}{R(P)} \quad (23)$$

- *SUM/MIN Energy-Half-Interference-Half Hop* algorithm: The optimization function used is equal to that in the *SUM/MIN Energy-Half-Interference* algorithm, multiplied by $\sqrt{h(P)}$:

$$\min_P \frac{\sqrt{h(P)}\cdot T_1(P)\cdot\sqrt{I_1(P)}}{R(P)} \quad (24)$$

- *MAX Interference* algorithm: The function optimized is the maximum of the interferences of the links on the path:

$$\min_P I_\infty(P) \quad (25)$$

- *MAX/MIN Energy-Half-Interference* algorithm: The optimization function is similar to that in the *SUM/MIN Energy-Half-Interference* algorithm, except that the transmission power and the interference are used as *maximum representative* instead of *additive* cost metrics:

$$\min_P \frac{T_\infty(P)\cdot\sqrt{I_\infty(P)}}{R(P)} \quad (26)$$

- *MAX/MIN Energy-Half-Interference-Half Hop* algorithm: The optimization function is similar to that in the *SUM/MIN Energy-Half-Interference-Half Hop* algorithm, except that the transmission power and the interference are used as *maximum representative* instead of *additive* cost metrics:

$$\min_P \frac{\sqrt{h(P)\cdot T_\infty(P)}\cdot\sqrt{I_\infty(P)}}{R(P)} \quad (27)$$

Optimization functions (12)-(18) are used for the energy-aware multi-cost routing algorithms, while equations (19)-(27) are used for the joint interference/energy-aware multi-cost routing algorithms. In what follows, we will refer to functions *MAX/MIN Energy* and *SUM/MIN Energy* as *Energy functions* and to the corresponding routing algorithms as *Energy algorithms*, while we will refer to the *MAX/MIN Energy-Hop* and the *SUM/MIN Energy-Hop* functions as *Energy-Hop functions* and to the corresponding routing algorithms as *Energy-Hop algorithms*. Also, we will refer to the functions *SUM/MIN Energy-Interference* and *SUM/MIN Energy-Half-Interference* as *Energy-Interference functions*, and to the corresponding routing algorithms as *Energy-Interference algorithms*. For the sake of brevity, we will also refer to the functions *SUM/MIN Energy- Interference-Half- Hop* and *SUM/MIN Energy-Half-Interference- Half Hop* as *Mixed functions* and to the functions *MAX/MIN Energy- Half- Interference* and *MAX/MIN Energy- Half- Interference- Half Hop* as *MAX/MIN functions*.

Based on the definition of the various cost parameters at Section 3.2.1, it is obvious that the cost parameters *h*, T_1 and I_1 are additive metrics, while the *R*, T_∞ and I_∞ are concave. According to Wang & Crowcroft (1996), the complexity of any optimization function using at least two additive metrics is exponential, except from the case where one of the two metrics is the hop count. Also, when one additive and one concave metric are used, then the complexity of the algorithm using the corresponding optimization function is polynomial. As a result some of the *SUM/MIN* (*Energy-Interference* and *Mixed*) algorithms are exponential, while all the other algorithms (e.g., *Energy, Energy-Hop*) are polynomial. One of the main reasons we examine both additive and concave metrics is that additive metrics usually result in somewhat better performance but longer algorithmic complexity, while the opposite is true for concave metrics.

In all cases, the algorithms first find the set of non-dominated paths with cost parameters (h,T,R) (for the case of energy-aware multi-cost routing), or (h,T,R,I) (for the case of joint interference/energy-aware multi-cost routing) and then use the corresponding optimization function $f(h,T,R)$ (for the case of energy-aware multi-cost routing) or $f(h,T,R,I)$ (for the case of joint interference/energy-aware multi-cost routing) to select the optimal path. In other words the process of computing the set of non-dominated paths is common to all algorithms and the selection of the optimal path is done at the end in a way that is different for each of the algorithms proposed. The function to be optimized at the last step may depend on the QoS requirements of the user. The optimization functions considered penalize paths that use a large number of hops, or consume a large amount of energy, or pass through nodes that have small energy reserves, or pass through links that cause large interference, differentiating however from each other by giving different importance to each of these parameters.

Multi-cost vs. Single-cost Routing

Multi-cost routing should not be confused with and is more general and powerful than single-cost routing. In single-cost routing one or more parameters (link characteristics) are combined in a single-cost metric characterizing the link. Multi-cost routing optimizes $f(V_1,V_2,\ldots,V_k)$ where $V_m = \otimes_{l=1}^{L} v_{ml}$, which is different than optimizing $\sum_{l=1}^{L} f(v_{1l}, v_{2l},\ldots, v_{kl})$ as single-cost routing does. The paths discovered by multi-cost routing are optimal for the specific optimization function f, which is not always the case with single-cost routing, except from linear optimization functions. Multi-cost routing better captures the meaningfulness of each cost parameter by considering it for the whole path than for a single link. Also, in the multi-cost approach operators like minimization (restrictive parameters) and maximization (maximum representative parameters) can be used, something that is not possible in the single-cost approach. For example, consider the *SUM/MIN Energy-Hop* multi-cost algorithm that uses the optimization function

$$h(P)\cdot\frac{T_1(P)}{R(P)} \tag{28}$$

and compare it to the single-cost algorithm that uses the cost metric

$$h_l\cdot\frac{T_l}{R_l} \tag{29}$$

for link (i,j), where h_l equals 1. Using the network of Figure 3 it can be shown that these algorithms select different paths, for a given source-destination pair, even though they use the same parameters and in similar way.

The example of Figure 3 also shows that the inclusion property does not hold for multi-cost routing. This property states that every subpath of an optimal (shortest) path is an optimal (shortest) path. In the network of Figure 3 and the *SUM/MIN Energy-Hop* algorithm, the optimal path for the A, D source-destination pair is path (A, B, C, D), while for the B, D source-destination pair it is path (B, D) and not (B, C, D); therefore, the inclusion property, which is true for single-cost routing, does not hold in general for the optimal paths produced by multi-cost routing. This also shows that the distributed implementation of multi-cost routing is not possible for some optimization functions, since intermediate nodes

Figure 3. Example network for the use of the various cost optimization functions

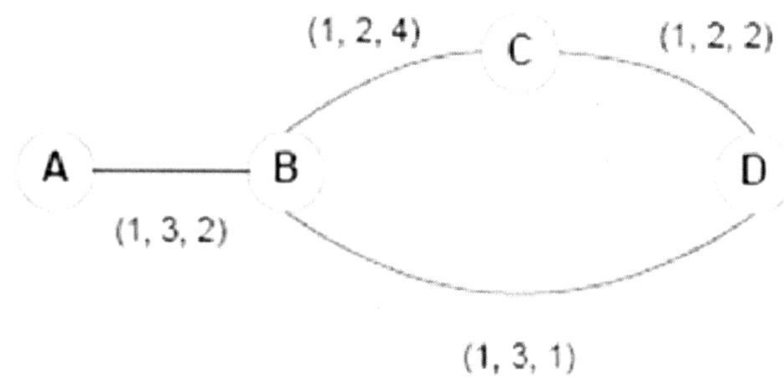

may not choose the paths originally intended by the source node.

Multi-cost routing supports service differentiation for sessions with different QoS requirements, where each optimization function can be thought of as representing a different QoS class. For example, if both delay and energy are important for the session, then we may use the *Energy-Hop* algorithm, while if only energy is important, we may choose the *Energy* or the *Energy-Half-Hop* algorithms. Finding the optimal path for a different QoS class does not require the recalculation of the sets of non-dominated paths, but only the application of the corresponding optimization function on the sets of non-dominated paths already found. In the single-cost approach each path is characterized by a single scalar, which is the sum of the scalar costs that characterize each link of the path. In order to find the optimal path for a different QoS class, a different cost metric has to be applied to every link of the network and the shortest paths need to be recalculated.

The Slow Start Power Control MAC protocol

In this section we turn our attention to the MAC layer, and present an energy-aware power control MAC protocol, called the Slow Start Power Controlled (SSPC) protocol. One of the main characteristics of the SSPC protocol is the usage of the slow start principle for the calculation of the minimum required power for the packets' transmission. The nodes in the network, by using the slow start technique do not have to know the topology of the network, nor the channel conditions, in order to compute the minimum required DATA frames transmission, while in parallel the power consumed in the network is kept to the minimum required. The motivations that lead us to design a power control MAC protocol with a slow start principle were the need to keep to the minimum possible degree (i) the power consumption in the network without degrading the network's Quality of Service and (ii) the overhead in order to incorporate the SSPC protocol in the IEEE 802.11 standard.

The SSPC protocol has two main advantages over the IEEE 802.11 protocol: (i) the energy consumed and the transmission floor reserved is close to the minimum required, (ii) the CTS frame silences only those nodes that are going to cause to the transmitter of the CTS frame interference greater than its interference tolerance, enlarging in this way the set of nodes that can communicate simultaneously. As, will be approved at Section 4.2, these two factors result in higher reuse factor, less power consumption, and larger network lifetime compared to IEEE 802.11 protocol. At the following sections, the operation of the SSPC protocol is described.

The Slow Start Feature of the SSPC Protocol

To illustrate the operation of the SSPC protocol, consider the situation depicted in Figure 5, where node A wants to transmit some DATA frames to node B. Node A senses the medium for a DIFS interval and if the medium is idle, A sends an RTS frame to B using transmission power that follows the slow start principle to be described shortly. The RTS frame informs its recipients that a DATA frames transmission will occur at node B, and is sensed by all the nodes in the coverage area of node A.

The format of the RTS frame is given in Figure 4. The common fields that are used at both the RTS frames of IEEE 802.11 (see ANSI/IEEE Std 802.11, 1999 Edition Part 11) and of SSPC are the following:

- *Frame Control*: It is comprised of various subfields, such as Protocol Version, Type, Subtype etc.
- *Duration*: time, in microseconds, required to transmit the pending data or management frame, plus one CTS frame, plus one ACK frame, plus three SIFS intervals.
- *RA*: Address of the receiver of the RTS frame.
- *TA*: Address of the transmitter of the RTS frame.
- *FCS*: Frame Check Sequence used for error control.

The new field SSPC adds in the RTS frame is the field P^{i}_{RTS}, which is the transmission power of the current *i*-th RTS transmission attempt.

At its first attempt node A sends the RTS frame with power P^{0}_{RTS}, hoping that node B is near enough to reach it and sets a timer equal to T_{RTS}. Typical values of the power P^{0}_{RTS} are, for example, 15dbm for the D-Link AirPlus™ G DWL-G630 Wireless Cardbus Adapter operating at 2.4GHz (http://www.dlink.com) and 14dbm for the IEEE 802.11b Wireless LAN PC Card operating at 2.4GHz (http://www.ansel.com.mx). The value of the timer T_{RTS} is $T_{RTS}=2T_{PROP}+T_{SIFS}+T_{CTS}$ which is the sum of the propagation delay required for the RTS frame to reach the destination (T_{PROP}), the time the receiver must wait before sending back the CTS frame (T_{SIFS}[INSERT FIGURE 004], the propagation delay it takes for the CTS frame to reach the sender (T_{PROP}), and the CTS transmission duration (T_{CTS}). We also define T_{DATA} as the time it takes the DATA frame to reach the destination and T_{ACK} as the time it takes the ACK frame to reach the transmitter of the DATA frame. If after period T_{RTS} node A has not received a correct CTS frame, it concludes that the transmission of the RTS frame has failed, invokes its back-off procedure, and retransmits the RTS frame, but now with transmission power P^{1}_{RTS} that has been increased by S dbm compared to the previous transmission power P^{0}_{RTS}.

The parameter S is referred to as the step of the slow start principle. Node A sets again the timer equal to T_{RTS} and waits for a CTS frame. Node A continues to send the RTS frame with increased transmission power P^{i}_{RTS} until it receives a CTS frame indicating that node B has successfully received the RTS frame, or until the transmission power reaches its maximum value. An unsuccessful RTS frame transmission can occur if the

Figure 4. The RTS frame format in SSPC

Octets:	2	2	6	6	1	4
	Frame Control	Duration	RA	TA	P_{RTS}^{i}	FCS

Figure 5. The Slow Start mechanism at SSPC protocol

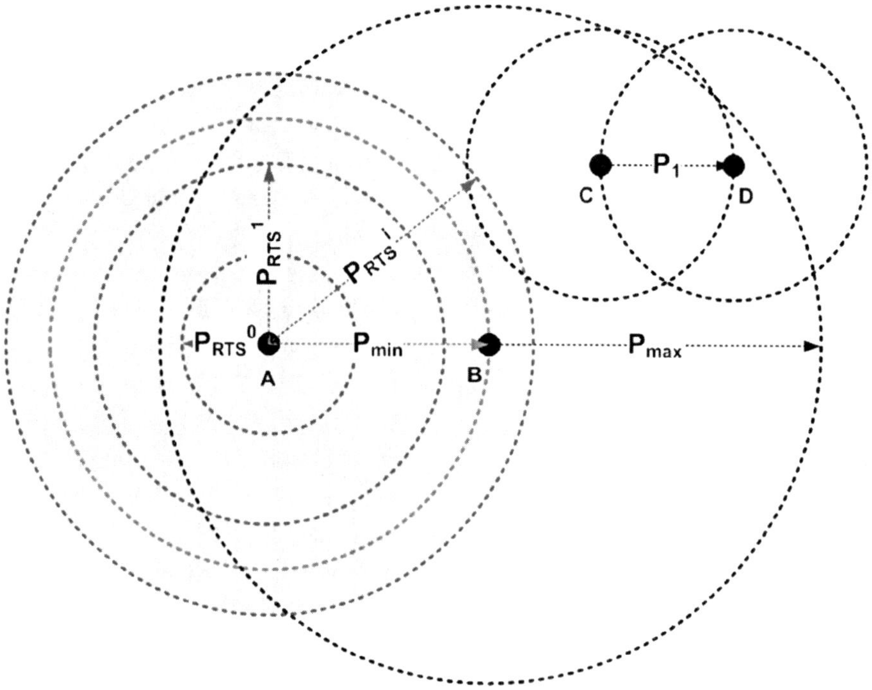

intended receiver is unreachable with the use of the current transmission power, or although the current transmission power is large enough to reach the intended receiver, the RTS frame has collided with a frame from another node. The latter means that is possible for a node to be considered as unreachable, though it is not. However, for something like that to occur, all the attempts of sending a RTS frame (using different levels of power; ranging from the minimum power to the maximum, with increment step *S*) should fail due to continuous collisions; something that has small probability. In general, a number of continuous collisions may lead to the selection of larger transmission power, than the minimum required, however this would again be smaller than the maximum one, used by the classic IEEE 802.11 protocol. The simulation experiments conducted (Section 4.2) show that even with the existence of such "misleading" collisions, the SSPC protocol performs quite well.

The minimum initial power P_{RTS}^{0} can be different for different nodes. The smaller P_{RTS}^{0} is, or the larger the distance between two nodes is, the more RTS retransmissions a node may have to undertake in order to reach the intended receiver. Also, the minimum initial power of a RTS frame transmission P_{RTS}^{0} does not depend on the utilized transmission power of the previous frame, however this could be an extension of the proposed protocol in order to reduce the number of the required RTS frames retransmissions.

Most commercial IEEE 802.11 wireless adapters and access points that support automatic or manual transmit power control mechanisms use, when controlling output power, a step size varying from 1 to 3 db; see, for example, the "Alvarion BreezeAccess VL" (www.alvarion.com) or the "Cisco Aironet 1240AG Series 802.11A/B/G Access Point" (www.cisco.com) datasheets. We propose the parameter *S* to take values between 1 to 3 db, depending on the accuracy we want to have in the estimation of P_{RTS}^{i}. The smaller the value of S, the more accurate is the computation of P_{RTS}^{i}, but also the larger is the number of RTS transmissions that may be needed before the intended receiver is reached.

All nodes that are in the transmission range of node A and correctly decode the RTS frame, set their Network Allocation Vectors (NAV) to the value $NAV_{RTS}^{TR} = T_{SIFS} + T_{CTS} + T_{SIFS} + T_{DATA} + T_{SIFS} + T_{ACK}$ as given in the ANSI/IEEE Std 802.11, 1999 Edition Part 11, and defer their transmissions for that period. We must mention that the value of the NAV_{RTS}^{TR} parameter is equal to the value included in the field Duration of the RTS frame. Instead, nodes that lie in the carrier sensing zone of node A, that is nodes that detect a frame but cannot decode it, set their NAVs to the value $NAV_{RTS}^{CS} = EIFS$. This is because nodes in the carrier sensing zone do not know the duration of the frame transmission. The main purpose of the EIFS is therefore to provide enough time for a source node to receive the ACK frame. As per IEEE 802.11, the EIFS is obtained using the SIFS, the DIFS, and the length of the time to transmit an ACK frame at the physical layer's lowest mandatory rate.

$$EIFS = SIFSTime + (8 \cdot ACKsize) + PreambleLength + PLCPHeaderLength + DIFS \quad (30)$$

where $ACKsize$ is the length in bytes of an ACK frame and $(8 \cdot ACKsize) + PreambleLength + PLCPHeaderLength$ is the transmission time of an ACK frame at the physical layer's lowest mandatory rate. See ANSI/IEEE Std 802.11, 1999 Edition Part 11 for the explanation of the other terms of equation (30).

Note that IEEE 802.11 does not completely prevent collisions due to a hidden terminal: nodes in the receiver's carrier sensing zone, but not in the sender's carrier sensing zone or transmission range, may cause a collision with the reception of a DATA frame at the receiver (Jung & Vaidya, 2005). This problem is also inherited by the SSPC protocol, as it also handles the hidden terminal problem using the RTS/CTS frames handshake. The use of the RTS/CTS frames in the IEEE 802.11 protocol can negatively impact channel utilization by completely disallowing concurrent transmissions over the reserved floor. In contrast, in the SSPC protocol, concurrent transmissions over the reserved floor can take place, provided that they do not use power more than a computed threshold. As a result, the negative performance impact of the hidden terminal problem is relatively limited in the SSPC protocol compared to the IEEE 802.11 protocol, as we will also see in Section 4.2.

In the example of Figure 5, Node B upon receiving the RTS frame checks the *RA* field to see if it is the intended receiver and the *TA* field to find the address of the transmitter. It also examines the *FCS* field to see if the received RTS frame contains errors. We must note that when we say that node B received the RTS frame we mean that it received and decoded correctly the RTS frame. The ability of node B to correctly decode the received RTS frame depends on its sensitivity, which is the minimum signal level required at the receiver for adequate reception. For example, if an SNR of 9dbm is required to achieve sufficient signal quality and the noise floor at the receiver is -111dbm, then the minimum signal or sensitivity for good reception is -102dbm. The sensitivity is typically supplied by manufacturers, and minimum acceptable levels can be found in the technical specifications of its device. For example the sensitivity of the D-Link AirPlus™ G DWL-G630 Wireless Cardbus Adapter is -84dbm operating at 11Mbps and at the band of 2.4GHz (http://www.dlink.com), while the sensitivity of the IEEE 802.11b Wireless LAN PC Card is -80dbm operating at 11Mbps and at the band of 2.4GHz (http://www.ansel.com.mx).

We denote by SNR_{RTS}^{i} the SNR at the receiver for the current (*i-th*) attempt of the RTS frame transmission when power equal to P^{i}_{RTS} is used at the transmitter and by $SNR^{\min}_{RTS}$ the SNR at the receiver when power equal to the minimum power $P^{\min}_{RTS}$ that guarantees the connectivity between node A and node B is used. In other words, $SNR^{\min}_{RTS}$ is the minimum received SNR that results in a desired frame error rate FER (it depends among other things on the error correction codes used), while $P^{\min}_{RTS}$ is the minimum power that should be used at the transmitter to result in SNR equal to $SNR^{\min}_{RTS}$ at the receiver. The transmitter would like to know the value of $P^{\min}_{RTS}$ so that it can use it in its transmission, but of course it cannot use the definition:

$$SNR = 10 \log P_{received} \,/\, N \tag{31}$$

where $P_{received}$ is the received signal power and N is the sum of the power of the thermal noise plus the interference noise caused at the receiver, to compute it, because it does not know the value of N or the channel characteristics. This minimum transmission power $P^{\min}_{RTS}$ is instead estimated at the receiver and is communicated back to the transmitter through the RTS/CTS frames exchanged during the slow start mechanism, as it will be explained below.

Consider now the current attempt (i-1) of node A to send the RTS frame with transmission power equal to P^{i-1}_{RTS} and let SNR^{i-1}_{RTS} be the corresponding SNR at the receiver. If the timer T_{RTS} that node A sets upon sending the RTS frame expires and node A does not receive a CTS frame, then node A invokes its back-off procedure and sends again the RTS frame with power P^{i}_{RTS}. This is node A's (i-th) current attempt to send the RTS frame, and let us assume that this time node B decodes correctly the RTS frame. This means that the power P^{i}_{RTS} used by node A at its current attempt to send the RTS frame is greater than or equal to the minimum power $P^{\min}_{RTS}$ that guarantees the connectivity between nodes A and B. At the same time we know that $P^{\min}_{RTS}$ is greater than P^{i-1}_{RTS} since the (i-1)-th attempt failed, assuming that no collisions occur. In the SSPC protocol the receiver uses the approximations $SNR^{\min}_{RTS} \approx SNR^{i}_{RTS}$ and $P^{\min}_{RTS} \approx P^{i}_{RTS}$, where i is the first successful RTS transmission attempt. We must underline that this is only an approximation of the minimum power that guarantees the connectivity between nodes A and B and not its accurate value. The accuracy of this approximation depends on the step size S used in the slow start mechanism. The smaller the value of S, the more accurate is the estimation of $P^{\min}_{RTS}$, but also, the larger is the number of the RTS retransmissions needed before the intended receiver is reached. We must mention that although several failed RTS transmissions may occur till the transmitter reaches the intended receiver, the overall network performance is better in the case that nodes adjust their transmission power to the minimum required through the SSPC protocol, compared to the case where the transmission power of the nodes is fixed, as it will be shown in the related performance evaluation section (Section 4.2). Finally, note that this estimated value of $P^{\min}_{RTS}$ takes into account all thermal and interference noise N present at B when it received the RTS frame.

The CTS Mechanism in the SSPC Protocol

When node B correctly decodes the RTS frame, it replies with a CTS frame that includes the transmission power P_{DATA} that node A must use to transmit DATA frames to node B. This power is given by the equation:

$$P_{DATA} = P^{i}_{RTS} + M \tag{32}$$

Note that in order to compute the DATA frames transmission power, a node does not need to know the exact location of all the nodes in the network or the channel conditions. The DATA frame transmission power is simply derived from the value of P^{i}_{RTS}, according to equation (32).

Figure 6. CTS frame format in the SSPC protocol

Octets:	2	2	6	6	1	1	1	4
	Frame Control	T_{CTS}	RA	TA	P_{DATA}	P_{INTERF}	P_{max}	FCS

The term M is used as a safety margin and also to allow for any future interference at node B (interference tolerance). In other words, to allow for a number of future interfering transmissions to take place in the vicinity of node B, node B requests node A to increase by M the transmission power of the DATA frames. M is a design parameter and determines the future interference margin that the receiver will be able to accept from its neighbouring nodes. The larger M is, the more nodes in the vicinity of a DATA frame receiver will be allowed to transmit, but also the larger will be the interference caused by the transmitting node to nodes other than the intended receiver.

Node B sends the CTS frame using the maximum transmission power. The CTS frame format is described in Figure 6. The fields that are common in the IEEE 802.11 and in the SSPC protocol are the following:

- *Frame Control*: It is comprised of several subfields, such as Protocol Version, Type, Subtype, etc.
- T_{CTS}: The CTS frame transmission duration.
- *RA*: Address of the receiver of the CTS frame.
- *FCS*: Frame check sequence.

The new fields SSPC adds in the CTS frames are:

- *TA*: Address of the transmitter of the CTS frame.
- P_{DATA}: The power that should be used by the transmitter to send the DATA frames, as computed by the receiver.
- P_{max}: The CTS frame transmission power.
- P_{INTERF}: The additional interference power that each neighbour/future interferer can add to the receiver.

The interference power P_{INTERF} that neighboring nodes are allowed to contribute to the receiver B will be computed shortly.

All nodes that are in the transmission range of node B set their NAVs to the value $NAV_{CTS}^{TR} = T_{SIFS} + T_{DATA} + T_{SIFS} + T_{ACK}$. Not all these nodes, however, but only the nodes that are neighbors to node B and are going to cause to it interference power greater than P_{INTERF}, to be specified later, defer their transmissions. To ensure node B that its CTS frame was received successfully by node A, node B sets a timer to a timeout value $T_{CTS} = 2T_{PROP} + T_{SIFS}$. If after this time node B has not started receiving a DATA frame it concludes that the transmission of the CTS frame has failed.

Nodes that are in the carrier sensing zone of node B set their NAVs to the value $NAV_{CTS}^{CS} = EIFS$ and defer their transmissions for this duration. As already mentioned, nodes in the carrier sensing zone cannot decode the received frame, so they do not know the duration of the frame transmission (a difficulty present in IEEE 802.11 as well), and also they cannot compute the maximum power they can use. These nodes set their NAV's for the EIFS duration, to prevent a collision with the DATA frame at the receiver. Note that we do not want all the nodes in the carrier sensing zone, but only those that are going to cause to node B excessive interference to defer their transmissions. We will present a solution to this problem later.

An important issue that must be addressed is the computation of the interference power P_{INTER} that each neighbour node can contribute to the receiver in the future, from the total interference margin M allowed at the receiver. Jung & Vaidya (2005) present a method to compute the interference power. The future interference M that is allowed must be equitably distributed among the future potentially interfering users in the vicinity of B. Let $N^R(t)$ be the number of nodes in the vicinity of node B at time t that are to share the interference M. Node B keeps track of the instantaneous number of simultaneously active transmissions in its neighbourhood at time t, which we denote by $N^R_{inst}(t)$. This can be easily computed by monitoring the RTS/CTS frames exchange. We denote by $N^R(RTS)$ the value of $N^R_{inst}(t)$ when the RTS packet is received at the receiver; the interference of these $N^R(RTS)$ nodes was present when the RTS packet was received and is already accounted for. In addition node B keeps track of a moving average of $N^R_{inst}(t)$, denoted by $N^R_{avg}(t)$. Then $N^R(t)$ is calculated as follows:

$$N^R(t) = \max\left\{N^R_{avg}(t), N^R_{inst}(t)\right\} - N^R\left(RTS\right) \tag{33}$$

The rationale behind the above equation is the following. When the CTS message was sent, there were $N^R(RTS)$ active transmissions in the neighbourhood of B. The future interference margin M is to be shared by future interferers, other than the $N^R(RTS)$ interferers already accounted for. The interference power that each neighbour can add to node B is finally given by the equation:

$$P_{INTERF} = \max\left\{\frac{M}{N^R(t)}, P^{\min}_{INTERF}\right\} \tag{34}$$

where $P^{\min}_{INTERF}$ is a lower bound we pose on P_{INTERF}. The rationale for posing this lower bound is the following. If the margin M is equitably distributed among a large number of neighbouring nodes then the interference power that every node will be allowed to cause to node B may be too small, and may be unusable. Also, if no lower bound on P_{INTERF} is given, then nodes in the carrier sensing zone of B, which do not correctly decode the CTS frame and therefore do not learn the accurate value of P_{INTERF}, would be prevented from transmitting since they would not be able to estimate a safe power level to use. These nodes are however at relatively large distance from B and should therefore be allowed to transmit up to a given power level, something that can be achieved by imposing the lower bound $P^{\min}_{INTERF}$, as we will see. Therefore, P_{INTERF} is the additional interference power that each future interferer can add to a receiver, and is chosen to be the same for all the neighbouring nodes, while M is the aggregate future interference that the receiver can tolerate, which is equitably distributed to its neighbouring nodes.

The value of P_{INTERF} depends on the values of $N^R(t)$, M and $P^{\min}_{INTERF}$ according to equation (34). The values of M and $P^{\min}_{INTERF}$ are design parameters, while $N^R(t)$ is the number of nodes at time t that are in vicinity of the node under consideration and which are to share the interference margin. In order to compute the value of $N^R(t)$, a node monitors the RTS/CTS frames exchange in its vicinity during an interval of a given duration T_{WAIT}, which is also a design parameter. The future interference allowed per node depends on the interval duration and the number of RTS/CTS frames sensed during that interval. The more RTS/CTS frames are sensed, the larger is the estimated number of neighbouring nodes $N^R(t)$ and, consequently, the smaller is the value of P_{INTERF}.

The duration for which a node must listen to collect information before making the access decision must be chosen by fine tuning the system parameters using simulation and experimental data. If nodes listen to the channel even when they do not have data to transmit, then they already have a good estimate of $N^R(t)$. If they do not listen when they are not active, then they may either wait for an interval T_{WAIT}, or use a less accurate estimate of $N^R(t)$. The parameter T_{WAIT} must be at least equal to the duration a node listens the channel before transmitting a RTS frame according to the IEEE 802.11 protocol, that is, a DIFS interval (T_{DIFS}). The value of T_{WAIT} could be chosen to be a small multiple of $T_{RTS}+2T_{PROP}+T_{SIFS}+T_{CTS}$ that is the time a node must wait in order to receive a CTS frame. Note that an inaccurate estimate of $N^R(t)$ does not affect the correctness of the protocol; it only results in less or more allowed future interferers, each being permitted a larger or smaller, respectively, interference margin. If however, the estimate of $N^R(t)$ is close to the actual number of nodes (and therefore, to the number of potential future interferers) in the receiver's vicinity, we expect better performance results. Alternative ways to determine the number $N^R(t)$ of future interferers to be allowed, other than the one proposed here could also have been used.

As previously mentioned, node B replies to the RTS frame with a CTS frame using power equal to $P_{max}(B)$ (node B does not use power P_{min} to transmit the CTS frame as it wants to inform as many stations as possible for the intended transmission that will occur). This frame informs all the nodes in the coverage area of node B that a DATA frame transmission will occur at node B. One important difference from the IEEE 802.11 protocol is that the CTS frame will not cause all the nodes that hear it to defer their transmissions, but only those nodes that are going to cause to node B interference greater than P_{INTERF}.

Another issue that has to be specified is the way a neighbour node, say node C in Figure 5, determines the maximum transmission power which can use without resulting in interference greater than P_{INTERF} at the receiver B. We assume that every node has the ability to compute the strength of the received signal. There are a lot of commercial chips that among others can compute the signal strengths. Example of those chip are the Atheros AR6001X Radio On Chip which integrates the RF transceiver, baseband, MAC, central process and peripheral control functions (http://www.atheros.com), and the POLARIS™ TOTAL RADIO™ solution from RF Micro Devices which is a highly integrated transceiver that performs all the functions of a handset radio section (http://www.rfmd.com/).

Every node, say node C, that hears the CTS frame sent by node B at power $P_{max}(B)$, has the ability to compute the received signal strength $P_{CTS\text{-}received}$, and consequently the channel gain between B and C by the equation:

$$G_{B,C} = \frac{P_{CTS-received}}{P_{\max}} \tag{35}$$

Based on that, and assuming that the channel gain is approximately the same in both channel directions (this is a reasonable assumption when node mobility is small during the duration of a RTS/CTS/DATA transmission, so that the distance and channel attenuation do not change significantly during the control and data packets exchange), node C can compute the maximum transmission power $P_{max}(C/B)$ that it can use without causing excessive interference at node B as:

$$P_{\max}(C / B) = \frac{P_{INTERF}}{G_{C,B}} \approx \frac{P_{INTERF}}{G_{B,C}} \tag{36}$$

Every node, like node C, maintains an interference table, where it records for each of its neighbours, say nodes $B_1, B_2, \ldots, B_N$, from which it has received a CTS frame, the maximum transmission power it can use without causing excessive interference to that neighbor (See Table 1).

The maximum power at which node C can transmit is then found as:

$$P_{\max}(C) = \min_i P_{\max}(C / B_i) \tag{37}$$

The interference table of node C together with the NAV data structure that it maintains (to record

Table 1. Description of the interference table that node C maintains

B_1 $P_{max}(C/B_1)$	B_2 $P_{max}(C/B_2)$	...	B_N $P_{max}(C/B_n)$	$P_{max}(C)$

the durations of the ongoing transmissions in its neighbourhood) is used to dynamically update the maximum transmission power P_{max}(C) that node C can use. When node C uses the slow start mechanism to transmit to another node D, it can increase its power up to P_{max}(C).

For example, let us consider the situation described at Figure 7. The dashed lines represent the receiving areas of the CTS frames when the maximum transmission power is used, the solid black lines represents the transmission range of the nodes A, C, E and F when the minimum required power for coherent reception of DATA frames is used, and the solid blue lines represent the receiving areas of the RTS frames from node C when the slow start principle is applied. Nodes A, F and E have already gone through the RTS/CTS procedure and have started transmitting some DATA frames to nodes B_1, B_2 and B_3, respectively. During the RTS/CTS exchange that preceded these DATA frames transmissions, nodes B_i, i = 1,2,3 have computed the future interference power $P_{INTERF,i}$ i = 1,2,3 that each of their neighbours is allowed to cause to them, and transmitted it in the CTS frames they have sent at power $P_{max,i}$ i = 1,2,3, respectively.

Node C wants to compute the maximum transmission power that it can use without causing excessive interference to its neighbours. Node C has received three CTS frames from nodes B_1, B_2, B_3. Using the received signal strength of each CTS frame, it computes the maximum transmission power it can use without causing excessive interference to its neighbours as:

$$P_{\max}(C) = \min\left\{P_{\max}\left(C/B_1\right), P_{\max}\left(C/B_2\right), P_{\max}\left(C/B_3\right)\right\} \quad (38)$$

where

$$P_{\max}\left(\frac{C}{B_i}\right) = \frac{P_{INTERF,i}}{G_{B_i,C}} \text{ for i=1,2,3} \quad (39)$$

P_{max}(C) is the maximum transmission power node C can use to transmit its DATA frames without causing excessive interference to its neighbours. So when it invokes the slow start mechanism to transmit to node D, it will increase its transmission power in steps of S dbm up to that power level.

Assuming that the power required for C to reach node D is less than the maximum power P_{max}(C) that C is allowed to use, node D replies with a CTS frame informing C about the power level P_{DATA} it should use. When node C receives D's CTS frame, it waits for SIFS duration and transmits the DATA frames with power level P_{DATA}. After transmitting the DATA frame, node C sets a timer to $2T_{PROP}+T_{ACK}$ seconds (see the ANSI/IEEE Std 802.11, 1999 Edition Part 11). If after this period node C has not received a correct ACK frame, it concludes that the transmission of the DATA frame has failed, and hence it invokes its back-off procedure. From the above it is obvious that several transmissions can take place simultaneously if every node that receives CTS frames adjusts its transmission power at the appropriate value using the information included in the CTS frames. At the previously described scenario, it is possible to have a collision between the CTS frame transmitted from node D and a frame from node F to node B_2, if these nodes transmit their frames

Figure 7. Usage of the CTS frame

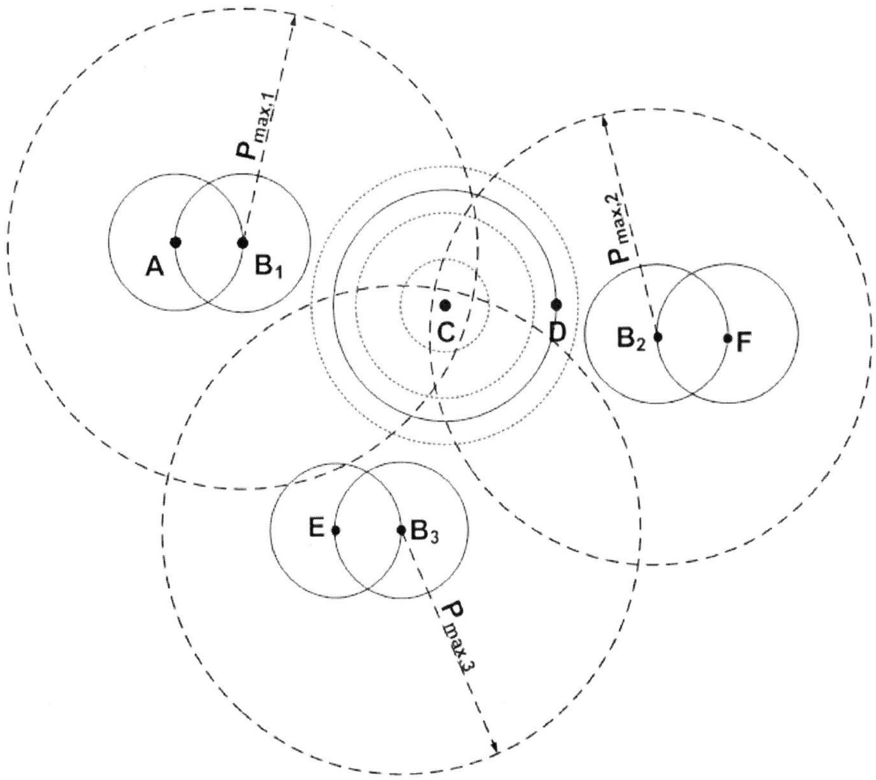

at the same slot, a situation which can also occur when using the classic IEEE 802.11 protocol. However, by using the proposed SSPC protocol the probability of a frame collision, in general, is smaller due to the usage of the minimum required DATA frames transmission power, compared to the case of the classic IEEE 802.11 protocol, where the transmission power of the frames may be much larger than the minimum required.

A node updates its interference table every time it senses and decodes a new CTS frame originated from one of its neighboring nodes or every time an interference source is gone. For example, in Figure 7, when the DATA frame transmission from node A to node B_1 is completed, node C removes from its interference table the record related to node B_1. Node C knows when the transmission of the DATA frame from node A to node B_1 is completed by using its network allocation vector which indicates the remaining time of an ongoing transmission. More specifically, node C knows that after time $NAV_{CTS}^{TR} = T_{SIFS} + T_{DATA} + T_{SIFS} + T_{ACK}$ from the time instance at which it sensed the CTS frame from node B_1, the DATA frame transmission from node A to node B_1 will have been completed.

Consider finally, a node G that is in the carrier sensing zone of node B (Figure 8) and receives its CTS frame. Since this node cannot correctly decode the received CTS frame, it does not know the value of the interference power P_{INTERF} it is allowed to cause to node B, or the transmission power of the CTS frame it received. We assume however, that node G can still compute the received signal strength $P_{CTS\text{-}received}$ of the CTS frame. Even though node G cannot compute the exact value of the maximum transmission power it is allowed to use, it can compute a looser bound on the maximum power allowed to it as follows:

$$P_{\max}\left(\frac{G}{B}\right)=\frac{P_{INTERF}}{G_{G,B}}=P_{INTERF}\cdot\frac{P_{\max}}{P_{CTS-received}}\geq P_{INTERF}^{\min}\cdot\frac{P_{RTS}^{0,\min}}{P_{CTS-received}} \tag{40}$$

where $P_{RTS}^{0,\min}$ is the minimum initial transmission power used in the slow start mechanism for the RTS transmission. In other words, node G uses the equation (35) together with the upper bound

$$G_{G,B}\leq\frac{P_{CTS-received}}{P_{RTS}^{0,\min}} \tag{41}$$

on the channel gain between G and B, for which we assume the approximation $G_{G,B}\approx G_{B,G}$. The upper bound on the channel gain is obtained from the received signal strength $P_{CTS\text{-}received}$, since we know that any transmission of an RTS frame must have used power at least equal to $P_{RTS}^{0,\min}$. Recall that $P_{INTERF}^{\min}$ is the lower bound on the interference power P_{INTERF} and is also a known design parameter. So although nodes in the carrier sensing zone cannot compute the exact value of their maximum allowable transmission power, they can still compute a lower bound [namely, the right hand side of inequality (40)] on the power they can safely use without causing excessive interference to node B. Note that in the classic IEEE 802.11 standard, nodes in the carrier sensing zone defer their transmissions for the EIFS duration. The EIFS is the largest of all the IFSs, and is used to reduce the probability of a collision with the ACK frame at the source. Instead, in the SSPC protocol, nodes in the carrier sensing zone can compute a threshold on the maximum allowable transmission power and defer their transmissions for an EIFS duration only if they intend to use power more than that threshold. Note that as the number of terminals that cannot correctly decode the CTS frame increases, the performance of the SSPC protocol will degrade, since such terminals will use a rather pessimistic upper bound on the maximum power they can use; however, performance will still be better than that of IEEE 802.11 protocol, where all nodes in the carrier sensing zone of a CTS frame are prevented from transmitting.

PERFORMANCE RESULTS

At this chapter, the performance of the previously described algorithms is evaluated. First, the performance evaluation of the proposed energy-aware and joint interference/energy-aware multi-cost routing algorithms is done. Then, the performance of the proposed power control MAC protocol is compared to the performance of the standard IEEE 802.11 protocol.

Multi-Cost Routing Algorithms Performance Evaluation

We evaluate the performance of the proposed energy-aware multi-cost routing algorithms under the network evacuation model. In this model, the network starts with a certain number of packets that have to be routed and a certain amount of energy per node, and the objective is to serve the packets in the smallest number of steps, or to serve as many packets as possible before the energy at the nodes is depleted. The proposed algorithms has been implemented and carried out corresponding experiments using the Network Simulator (http://www.isi.edu/nsnam/ns/). The routing agent running on each node calculates the set of non-dominated paths to all destinations at periodic time intervals. When a data packet is generated at a node, the node applies the optimization function to the cost vectors of the corresponding non-dominated paths to select the optimal path, and the packet is sent on that path. If no route to the destination can be found, the packet is discarded.

In general, the routing process involves two levels: the routing information exchange level and the routing algorithm level. Routing information protocols deal with collecting and disseminating

Figure 8. Transmission and carrier sensing zone of node B

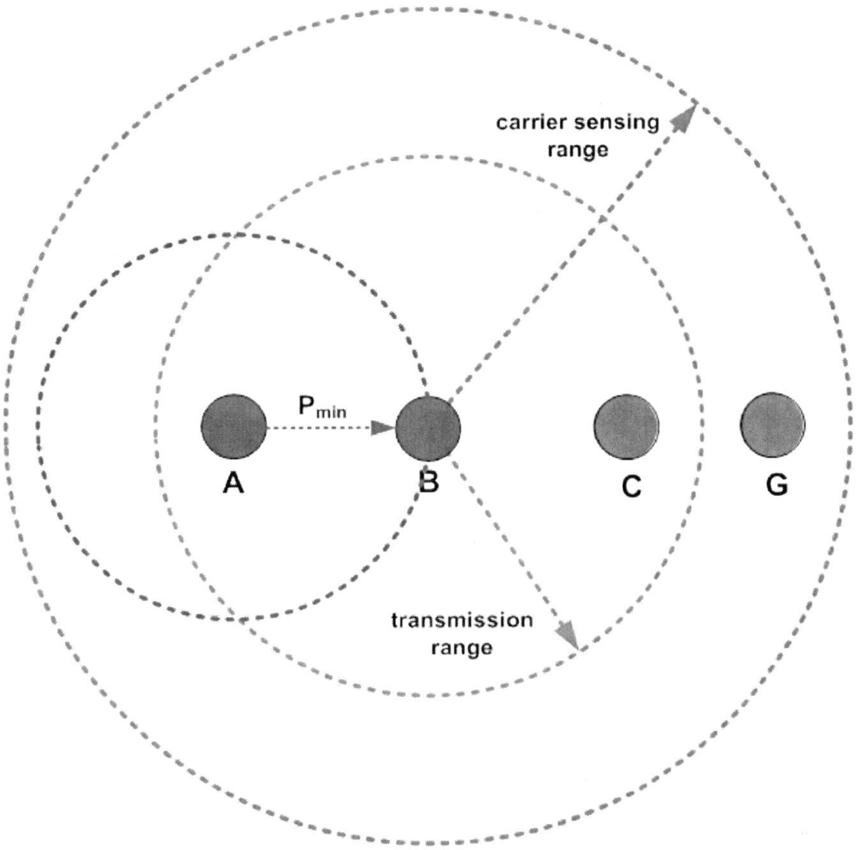

network state information, while routing algorithms compute the optimal-best path(s) using this information. Our focus is on the routing algorithm level and thus in our simulations it is assumed that each node has global knowledge of the network topology and all other information it needs for making routing decisions. Actually, the operation of an information exchange mechanism that gathers network-related information may introduce network and energy overheads, while the network information available at each node may not always be accurate or updated. Papageorgiou et al. (2008) assessed the performance of the multi-cost routing approach based on energy-related parameters for mobile ad hoc networks by embedding its logic at the Dynamic Source Routing (DSR) algorithm, which is a well-known fully distributed routing algorithm. In this way, the multicost-DSR algorithm relies on the DSR's mechanisms of route discovery and maintenance in order to collect the necessary information regarding the network topology and state, while for the path selection the multicost approach is used. The simulation results showed that though the information exchange mechanism increases the energy consumption in the network, the multi-cost routing with energy-aware metrics still prolongs its lifetime. It has been observed that the increased node mobility can improve the accuracy and the validity of the information (e.g., energy) located at each node and used for routing decisions. On the other hand in case the wireless nodes are static (e.g., in mesh networks) then a distributed information exchange mechanism may be not necessary. Instead, information dissemination can be achieved by a periodical beacon-like protocol according to which each node sends its local information to its direct neighbors. Eventually this information will be broadcasted to all the nodes in the network. Given the fixed topology of wireless mesh networks, the energy and communication overhead of this process can be minimized with a simple scheduling of the packet transmissions. Another choice could also be to incorporate the information about the cost parameters in the data exchanged anyway between the nodes of such networks. In both cases the only concern is the validity of the infromation. Kokkinos et al. (2005) studied the effect that the duration of the information update interval has, on the multi-cost algorithm's performance.

The energy-aware multi-cost routing algorithms were evaluated using an ad hoc network consisting of 49 stationary nodes placed along a 7x7 grid. The distance between neighbouring grid points is set at 50m. The topologies studied in the experiments are either a regular grid topology (where the transmission range of the nodes is fixed at 50m) or a random topology (where the transmission range varies from node to node, and is uniformly distributed between 50m-100m in one set of experiments, and between 50m-150m in another set of experiments). For the joint interference/energy-aware multi-cost routing and power control algorithms the nodes are capable of dynamically adjusting their transmission power. Specifically, we assume that all nodes can communicate directly with each other and hence the network is fully connected, but depending on the routing algorithm employed, a node may choose not to use the direct link to the destination, and use a multi-hop path instead. In a wireless network, the signal power at a receiver located at distance d from the sender is given by:

$$P_r(d) = \frac{P_t \cdot G_t \cdot G_r \cdot \lambda^2}{4 \cdot \pi^2 \cdot L \cdot d^a} \qquad (42)$$

where P_t is the power of the transmitted signal, G_t and G_r are the gains of the sender's and receiver's antennas, respectively, L is the system loss, and λ is the wavelength used. In our experiments we used $G_t=1$, $G_r=1$ and $L=1$. The parameter a is the path loss constant, and is typically between 2 and 4 depending on the wireless channel.

All packets have equal length (taken to be 500 bytes) and require one unit of time (slot) for transmission over a link. The amount of energy expended for a packet transmission is taken equal to the transmission power multiplied by the packet transmission duration. We assume that the energy consumed for the reception of a packet is constant and same for all the nodes. More specifically, the energy expenditure for a packet reception at a node is taken to be equal to 10% of the transmission power required for the minimum transmission range (50m) multiplied by the packet duration, which is fixed since the packet lengths are fixed at 500 bytes. We also assume that the processing energy cost is included in this energy expenditure and that an idle node's energy consumption in negligible. The initial energy of the nodes was taken to be 2 joules, which represents a scenario where some nodes run out of energy during the experiments (finite energy), depending on the amount of traffic they end up serving. The number of packets per node that have to be delivered to their destinations ("evacuated" from the network) in our experiments, varies from 100 to 1000 (at steps of 100) packets per source node. Packet destinations are uniformly distributed over all remaining network nodes and the packet generation rate at each node is equal to 0.1 packets/sec.

In the experiments conducted we measured the average residual energy E remaining at the nodes at the end of each experiment, the variance σ^2 of the node residual energies, the time when a node runs out of energy, referred to as the node depletion time DT, the network throughput T and the received-to-sent packets ratio RS.

Energy-Aware Multi-Cost Routing Algorithms

Figure 9 illustrates the average residual energy in the network at the end of an evacuation experiment, as a function of the number of packets evacuated per node. The *Minimum-Hop* algorithm results in a higher average residual energy E at the end of the evacuation experiments than the other routing algorithms examined. However, the *Minimum-Hop* algorithm also results in less uniform energy consumption in the network (Figure 10) and in smaller energy depletion times DT than the other algorithms, as is indicated by Figure 11 and Figure 12. As a result, the *Minimum-Hop* algorithm also achieves lower throughput and higher dropping ratio. Note in Figure 9, that when more than 400 packets are generated per source node, then the average residual energy stops decreasing for all algorithms examined. This happens because many nodes run out of energy and consequently the network becomes disconnected and no more packets are served.

The *Minimum-Hop* algorithm uses the same path for the entire duration of a session, or until the energy of a node on the path is depleted. As a result, a relatively small subset of nodes has more active participation, in the transmission of packets, than other nodes. The *Energy* and *Energy-Hop* algorithms, on the other hand, are based on parameters (namely R_l) that change over time, and the path selected may not remain the same for all the packets of a session. In this way the traffic forwarding and the resulting energy consumption are spread more uniformly, over a larger number of nodes, leading to smaller average residual energy E and smaller variance σ^2 of the residual energy than the *Minimum-Hop* algorithm. Regarding the time the energy of the nodes is depleted, the Energy-Hop algorithms exhibit the best performance in all the experiments, while the *Minimum-Hop* algorithm seems to result in the worst DT. Note that with the *Energy* and *Energy-Hop* algorithms, when nodes start running out of energy, this happens almost simultaneously for all the nodes. This is because these algorithms

Figure 9. Illustrates the average residual energy at the end of the evacuation experiment for the Minimum-Hop, MAX/MIN Energy and MAX/MIN Energy-Hop algorithms

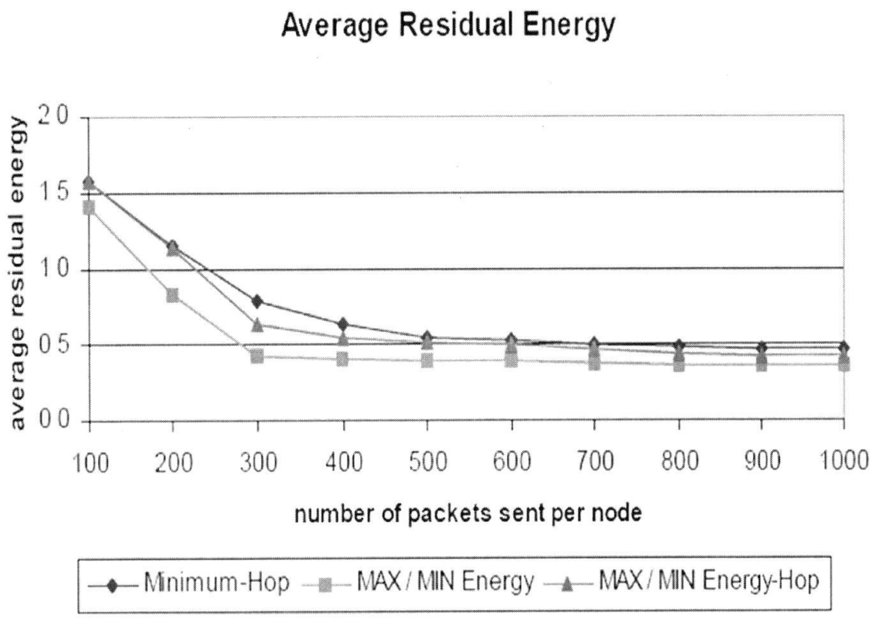

Figure 10. Illustrates the variance of the residual energy at the end of the evacuation experiment for the Minimum-Hop, SUM/MIN Energy and SUM/MIN Energy-Hop algorithms

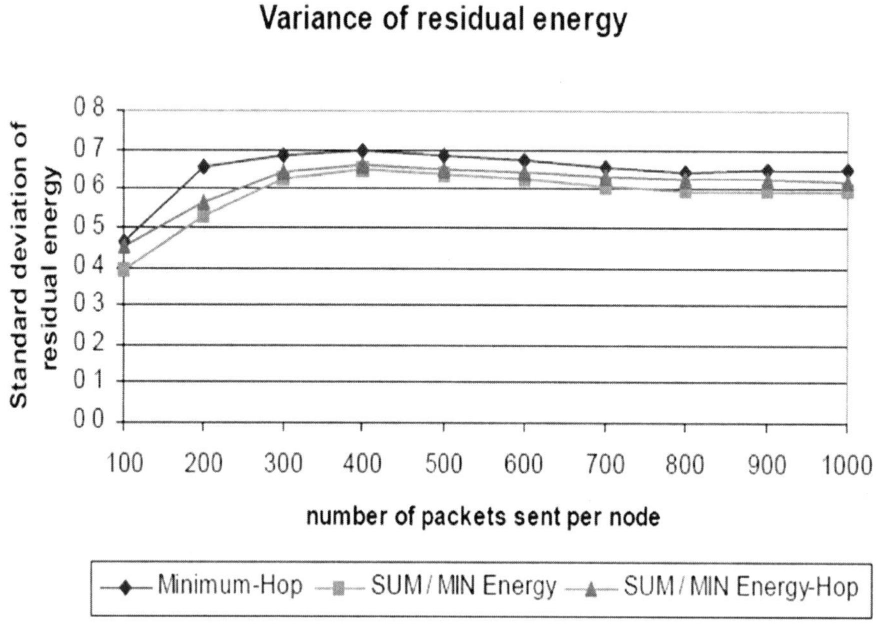

spread the energy consumption uniformly in the network, so that when one node is at the point of first running out of energy, most other nodes are at the same energy-critical situation.

In most of the experiments conducted, it was found that the performance of the *Energy-Hop* algorithms was between that of the *Minimum-Hop* algorithm and that of the Energy algorithms, and,

Figure 11. Illustrates the current number of nodes with depleted energy at the end of an evacuation experiment for the Minimum-Hop, MAX/MIN Energy, and MAX/MIN Energy-Hop

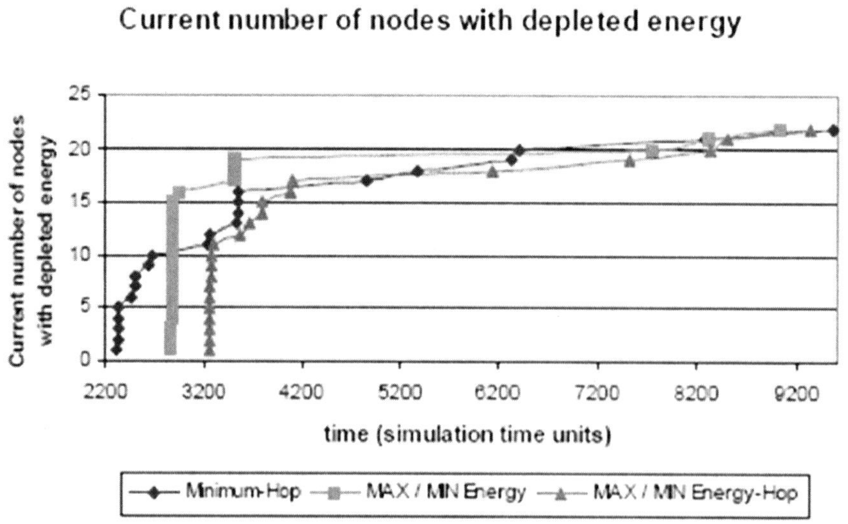

actually, in most cases it was closer to that of the *Minimum-Hop* algorithm. The *Energy-Half-Hop* algorithms were found to behave very similarly to the *Energy* algorithms in all cases considered. It seems that the 0.5 exponent on the number of hops in the former algorithms effectively eliminates the impact of the hop term on the cost function. This is the reason we chose not to present in great detail the results on the *Energy-Half-Hop* algorithms.

Figure 13 and Figure 14 shows the received-to-sent packets ratio *RS* for various algorithms examined. We observe that when the initial nodes' energy is finite, the fraction of packets delivered to their destinations decreases after a certain number of packets have been inserted in the network. The reason is that nodes run out of energy, limiting the ability of the network to route packets. We observed that the *Energy-Hop* algorithms achieve the best *RS* ratio in almost all the experiments, since with these algorithms the network nodes remain alive for longer periods of time. The *MAX/MIN Energy* algorithm seems to give the worst results among the algorithms in this class. The performance of the *SUM/MIN Energy* algorithm is considerably better and it surpasses, even marginally, that of the *SUM/MIN Energy-Hop* algorithm.

Another observation concerns the choice of the associative operator (sum or maximum) that should be used for combining the transmission power T_l parameters in the optimization functions. Figure 11, Figure 12, Figure 13 and Figure 14 show that the *SUM/MIN Energy* algorithm behaves considerably better than the *MAX/MIN Energy* algorithm. This indicates that summing up the transmission powers of the nodes on a path is a more representative measure of the path's energy cost, than taking the maximum value of them. For similar reasons, the *SUM/MIN Energy-Hop* algorithm was also found to behave better than the *MAX/MIN Energy-Hop* algorithm, even though the difference between these two was not so significant. This is because the hops parameter dominates the optimization function in the *SUM/MIN Energy-Hop* and the *MAX/MIN Energy-Hop* algorithms.

Figure 12. Illustrates the current number of nodes with depleted energy at the end of an evacuation experiment for the Minimum-Hop, SUM/MIN Energy and, SUM/MIN Energy-Hop algorithms

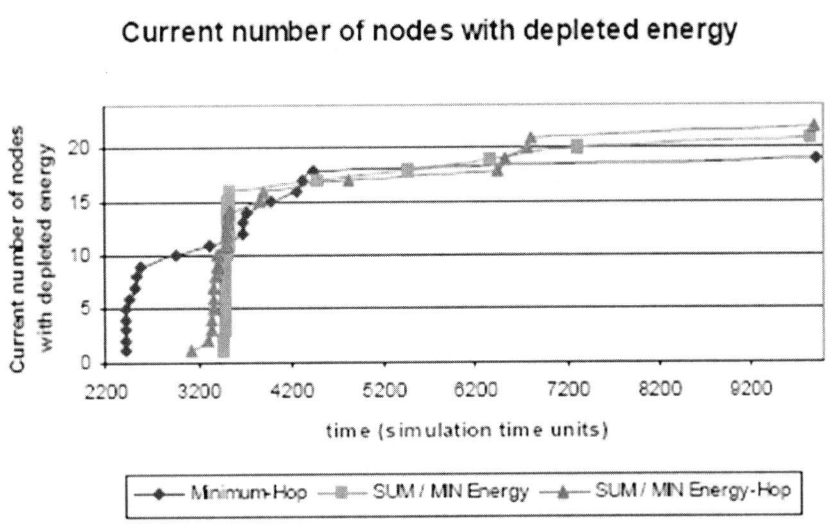

Figure 13. Illustrates the received-to-sent packets ratio for the Minimum-Hop, MAX/MIN Energy, and MAX/MIN Energy-Hop

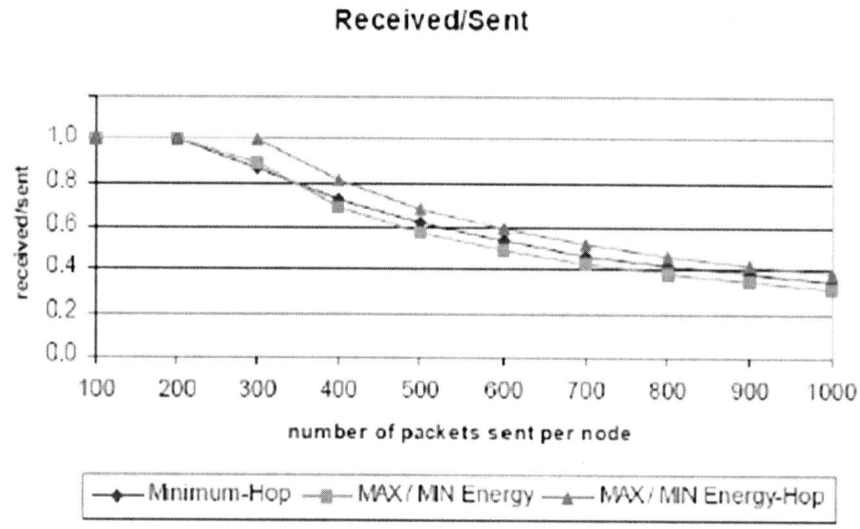

Joint Interference/Energy Aware Multi-Cost Routing Algorithms

Figure 15 and Figure 16 illustrate the average residual energy at the end of an evacuation period, and the variance of the residual energies, respectively, as a function of the number of packets that are evacuated per node. We observe that the *Minimum Transmission Power* algorithm outperforms the other algorithms examined with respect to the average residual energy, since it selects paths consisting of many short links

Figure 14. Illustrates the received-to-sent packets ratio for the Minimum-Hop, SUM/MIN Energy and SUM/MIN Energy-Hop algorithms

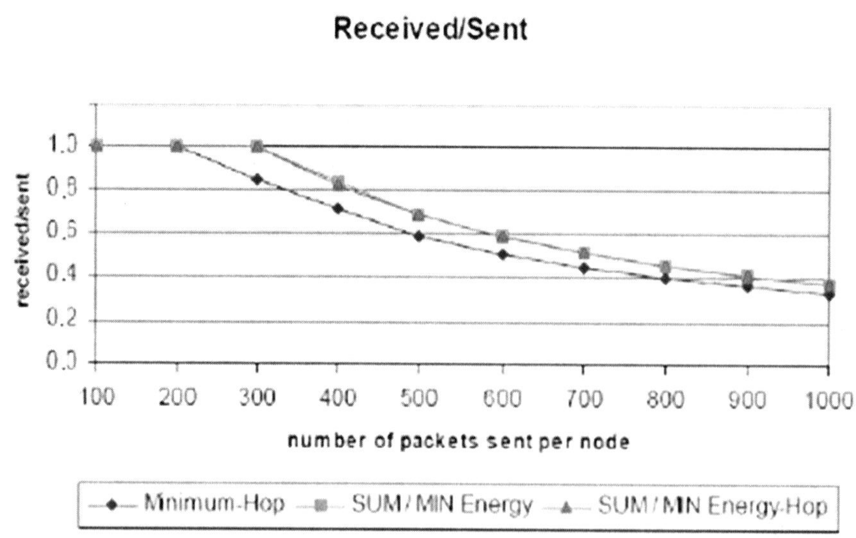

Figure 15. Illustrates the average residual energy at the end of the evacuation problem, as a function of the number of packets evacuated per node, and different choices of the routing algorithms

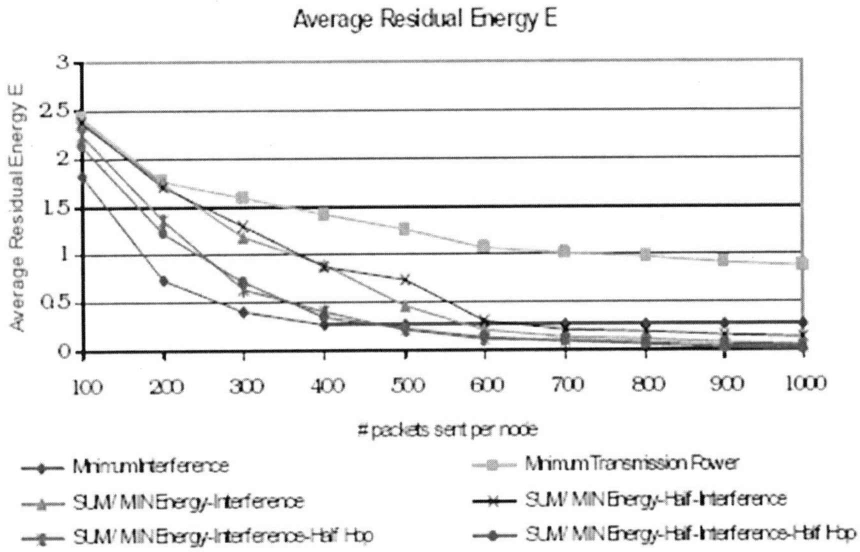

(chooses long paths) in order to minimize the total transmission power used. This way it also minimizes the energy consumption at the nodes. Similarly, the *Mixed* algorithms that consider the hop metric (choose short paths), behave worse than the *Energy-Interference* algorithms that do not consider this metric. The *Minimum Interference* algorithm on the other hand exhibits the worst performance, since it does not consider any energy related metric.

Regarding the variance of the residual energy, shown in Figure 16, for almost all the algorithms examined, the variance initially increases, but then starts decreasing rapidly as the number of packets that are evacuated increases. This happens because as more packets are evacuated, then more nodes' energy is depleted, resulting in small residual energy variance. Also, energy is a critical resource and the algorithms that incorporate the path residual energy *R* in their cost functions (that is, the *Mixed* and *Energy-Interference* algorithms) perform better, achieving small variance in the residual energy. This is because the paths these algorithms select are adjusted over time to reflect changes in the energy reserves of the nodes, resulting in a more even distribution of energy consumption across the network. In contrast, the *Minimum Interference* and the *Minimum Transmission Power* algorithms do not change their paths when nodes start running out of energy, resulting in earlier depletion of the energy at some nodes, while there are other nodes that still have significant energy reserves.

Also, from the experiments performed, it was observed that the *Mixed* and the *Energy-Interference* algorithms achieve better throughput than the *Minimum Interference* algorithm. This is because these algorithms use energy more efficiently, by taking into account the node residual energies, and thus manage to deliver more packets to their destinations before the energy is depleted. Moreover, the *Minimum Transmission Power* algorithm gives the worst results, in terms of the throughput achieved, since it does not consider the interference caused.

Figure 16. Illustrates the variance of the residual energy at the end of the evacuation problem, as a function of the number of packets evacuated per node and different choices of the routing algorithms

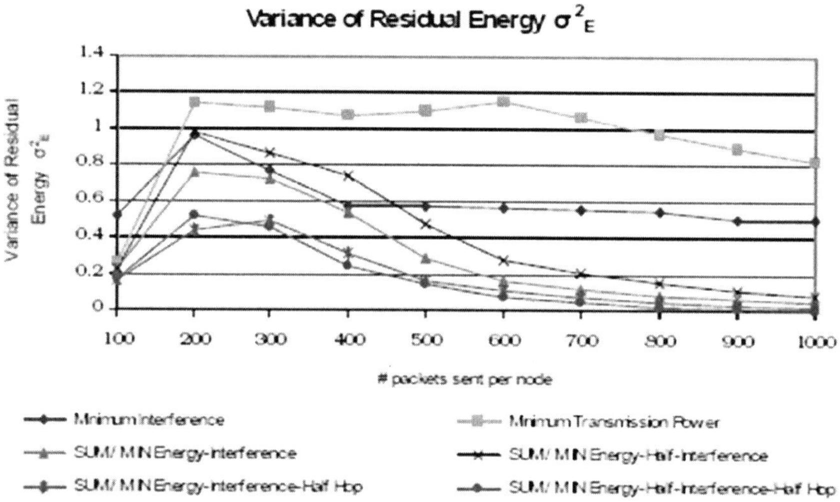

Power Control MAC Protocol Performance Evaluation

In the experiments done the Network Simulator was used to simulate a wireless multi-hop network of 36 nodes distributed according to a two dimensional uniform distribution in a 500x500m^2 area. To obtain results that are easier to interpret, we mainly focused on the performance benefits that can be obtained through the use of power control, and not on the protocol overhead involved. Therefore, we assumed in our simulations that nodes have global knowledge of the network topology and other information they need for adjusting their transmission power and no control packets and related overhead were included. Note that the SSPC protocol's overhead is somewhat larger than that of IEEE 802.11, due to the repetitive RTS frame transmissions of the slow start mechanism and the additional fields used in the RTS and CTS frames. The received signal power at distance *d* from the transmitter can be again calculated by using the equation (42).

The MAC protocol we used is a power controlled version of IEEE 802.11. The amount of energy expended for a packet transmission was taken to be equal to its transmission power multiplied by the duration of the packet transmission. Some constant (independent of the distance) energy was also consumed for packet reception and processing. When a node is idle we assume that it consumes no energy. We considered the following two cases:

- Static power case: Each node uses static transmission power for its transmissions (as in IEEE 802.11).
- Adjustable power case: Each node adjusts its transmission power to the minimum required value that guarantees the reliable reception of the frames at the sink nodes (by using the SSPC protocol).

The performance of these two approaches was evaluated in the setting of the evacuation problem. In the experiments done, the number of packets evacuated from the network varies from 100 to 1000 (at steps of 100) packets per node. Packet destinations are taken to be uniformly distributed over all remaining nodes of the network. The

Figure 17. Illustrates the average consumed energy per node at the end of an evacuation period for the static and the adjustable power case

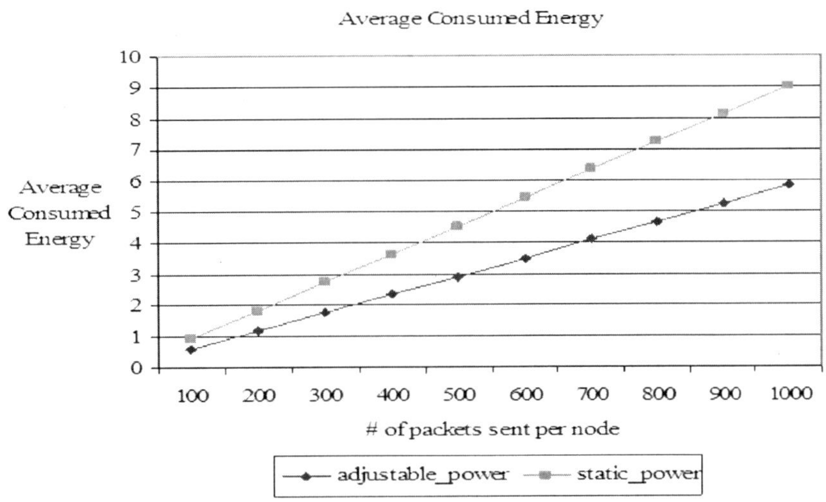

threshold for the received power required for correct reception (that is, the receiver sensitivity) is the same for all nodes. The routing protocol used, is the minimum hop algorithm. To compare the two approaches the same paths were chosen for the packets delivery from the source to the sink nodes. We must note here that if the SSPC protocol is adopted, the routing algorithm could be improved to better exploit the advantages of SSPC over the usual IEEE 802.11. We chose to use the same (minimum hop) routing algorithm in our performance results for both protocols so that the results can be directly comparable. Also, the energy consumption due to RTS/CTS frames exchange is not accounted for, since we focus on the performance benefits that can be obtained through the use of power control, and not on the protocol overhead involved.

In the experiments conducted for the static power and the adjustable power cases we measured: (i) the average energy E consumed by the nodes at the end of an evacuation period, (ii) the variance σ_E^2 of the energy consumed by the nodes, (iii) the average packet delay D, defined as the average time between the beginning of an evacuation instance and the time a packet reaches its destination, averaged over all packets delivered to their destinations (iv) the frequency of the RTS frames collisions and (v) the received-to-sent packets ratio RS.

The first two performance parameters are related to energy considerations, while the remaining four are directly related to network performance. We choose to compare the performance of the SSPC protocol to that of the usual IEEE 802.11 protocol, as is done in most other related works, which also use IEEE 802.11 as a reference point. We believe that the results would be harder to interpret if comparisons were made to other (not equally used in practice) protocols.

Figure 17 illustrates the average energy consumed per node (measured in Joules) after all packets have been evacuated from the network, for the adjustable and the static power case. It shows that using adjustable power results in considerably smaller energy consumption than using static power. The energy savings increase (linearly) with the number of transmitted packets.

Figure 18. Illustrates the variance of the energy consumed at the nodes for the adjustable power and the static power case

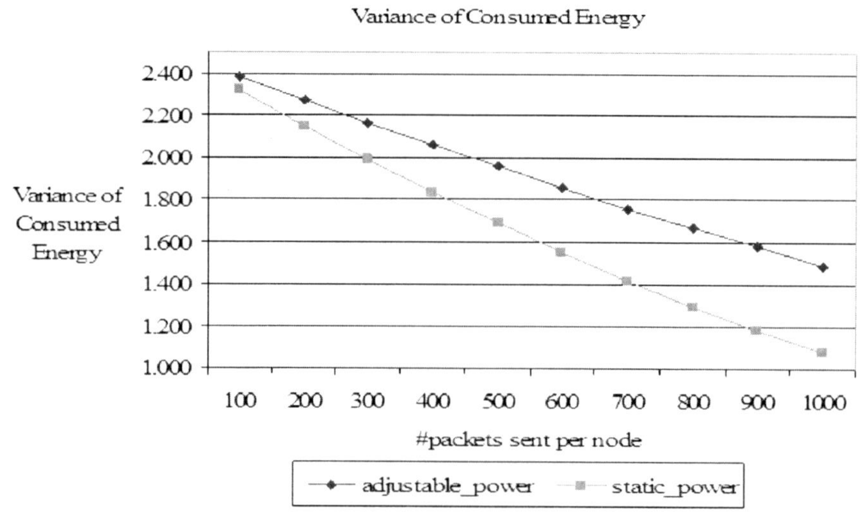

This is because in the adjustable power case every node adjusts its transmission power, via the slow start mechanism of the SSPC protocol, to the minimum required level for coherent reception at the receiving node, so that nodes consume only the necessary amount of energy. Instead, when static power is used, and the desired recipient is at a smaller distance than the static transmission range used, a node may expend an unnecessarily large amount of energy and cause unwarranted interference to other nodes.

The variance of the energy consumed by each node is shown in Figure 18. The variance of the consumed energy indicates how the power consumption in the network is distributed among the various network's nodes. This is an important parameter, since we want power consumption to be uniformly distributed among the nodes of the network. From Figure 18, it can be seen that static power approach results in smaller variance in the energy consumed per node than the adjustable power approach. This indicates that SSPC tends to spread energy consumption more uniformly among the nodes.

Figure 19 illustrates the average delay D of the packets delivered to their destinations for the adjustable power and the static power case, as a function of the number of packets evacuated per node. For both approaches, the average packet delay increases as the number of packets that are evacuated increases. It can be seen that the adjustable power approach used in SSPC outperforms the static power approach used in standard IEEE 802.11. This is because the SSPC protocol allows more concurrent packet transmissions as long as they do not cause excessive interference to ongoing transmissions.

The number of the RTS frames collisions due to the MAC protocol and the hidden terminal problem is illustrated in Figure 20, which shows that the frequency of RTS frame collisions is smaller when adjustable power is used. This is because, in the case of the static power approach every node sends its RTS frame with maximum transmission power, while in the case of the adjustable power, the transmission power of the RTS/DATA frames is adjusted at the minimum required, resulting in fewer RTS collisions. The performance difference

Figure 19. Illustrates the average packet delay for the adjustable and the static power cases

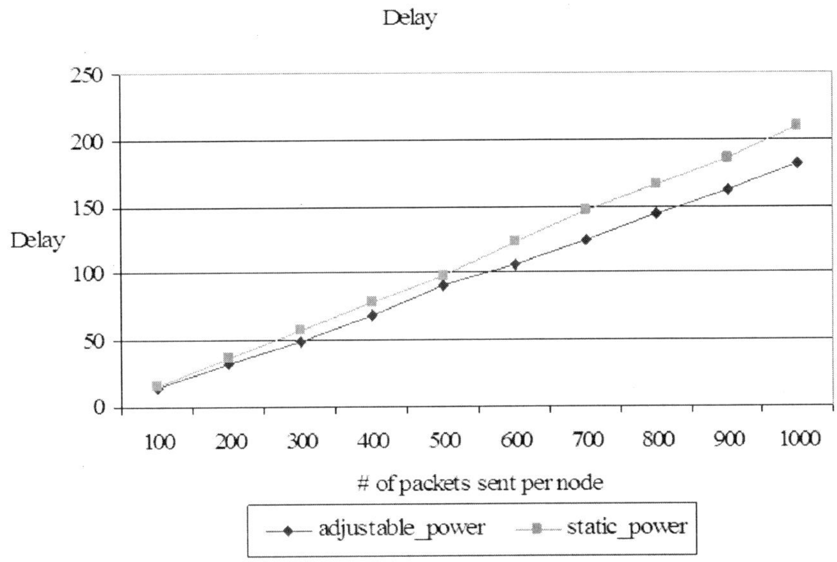

between the adjustable power and the static power approach increases (linearly) as the number of packets evacuated increases.

Finally, the received-to-sent packets ratio is shown in Figure 21. We observe that more packets are delivered to their destinations in the adjustable power case than in the static power case. This is because nodes spend less energy by using the minimum required power for their DATA frames transmissions, prolonging in this way the lifetime of the network. Since the nodes remain alive for longer time, the network capability of delivering packets to their destination is increased.

CONCLUSIONS AND FUTURE RESEARCH DIRECTIONS

This chapter discusses Quality of Service (QoS) and energy saving mechanisms for wireless networks. A cross-layer approach has been described in order to save energy at the network, while in parallel providing an acceptable QoS.

The multi-cost routing approach was described, where several multi-cost energy-aware routing algorithms for wireless ad hoc networks were proposed. As it has been proven at Section 4.1.1, the multi-cost energy-aware routing algorithms distribute the traffic more uniformly across the network, prolonging its lifetime, and improving its performance. More specifically, the *Energy-Hop* algorithms were found to have better performance than both the *Energy* and the *Minimum-Hop* algorithms, under the energy and network related performance measures used. Energy-aware routing and medium access control protocols have been investigated also, in the case where nodes have variable transmission power capabilities. The joint multi-cost routing and power control has been considered and a number of energy- and interference-aware multi-cost routing algorithms that use the power adjustment capability of the nodes has been presented. A number of cost parameters, including hop count, interference caused, residual energy of the nodes and variable transmission power were considered. The

Figure 20. Illustrates the number of the RTS frames collision for the adjustable and the static power cases

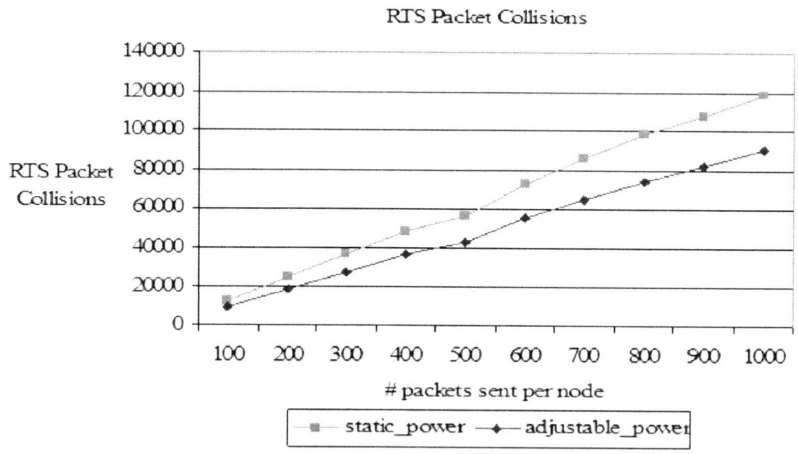

proposed algorithms were evaluated (see Section 4.1.2) using both performance and energy related measures, exhibiting large throughput, reduced interference and energy consumption, and increased network lifetime. Also, the *SUM/MIN Energy-Half-Interference-Half-Hop* multi-cost algorithm, which incorporates hop count, interference and

Figure 21. Illustrates the received-to-sent packets ratio for the adjustable power and the static power case

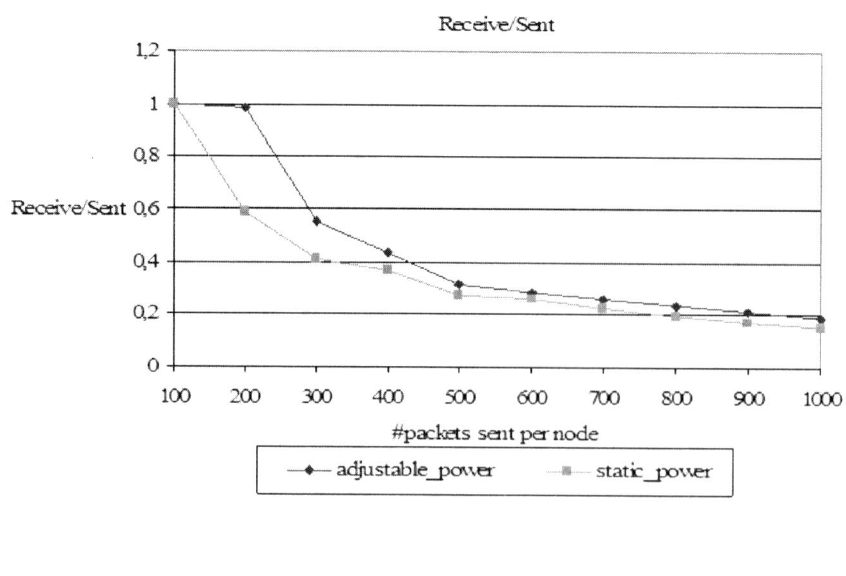

energy related metrics, outperforms, in most cases, all the other algorithms considered.

Also, the case of power control has been investigated for energy savings in wireless ad hoc networks. A Slow Start Power Controlled MAC protocol was described for energy consumption reduction, while in parallel providing QoS. Reduced energy consumption is achieved by adjusting the nodes transmission power to the minimum required value for reliable reception of the packets, while increase in network throughput is achieved by allowing more transmissions to take place simultaneously. It has been shown (see Section 4.2) that this protocol achieves a significant reduction in energy consumption and average packet delay and also a significant increase in received-to-sent packets ratio compared to IEEE 802.11 protocol.

Future research work can concentrate in the development of energy-aware broadcasting and multicasting algorithms, which are two of the most usual communications tasks that take place at a wireless network. Also, mobility is a parameter that must be taken into consideration and examine how it affects the design of various energy saving and QoS protocols. Future work can also include the investigation of the proposed energy saving mechanisms, in the field of the wireless sensor networks. A wireless sensor network is a wireless network consisting of spatially distributed autonomous devices using sensors to cooperatively monitor physical or environmental conditions. Although, wireless ad hoc networks and wireless sensor networks share many characteristics, there are some obvious differences most notably in the number of nodes, which may be thousands, and the very high deployment density of the wireless sensor networks. These characteristics, can affect the design of new energy-aware protocols and the applicability of the existed at the field of the wireless sensor networks.

REFERENCES

Agarwal, S., Krishnamurthy, S., Katz, R., & Dao, S. (2001). Distributed Power Control in Ad-Hoc Wireless Networks. In *12th IEEE International Symposium on Personal, Indoor and Mobile Radio Communications* (pp. 59-66), San Diego, CA. Alvarion BreezeAccess. (n.d.). *VL datasheet.* Retrieved from www.alvarion.com

ANSI/IEEE Std 802.11, 1999 Edition Part 11: Wireless LAN Medium Access Control and Physical Layer Specifications. (1999). *Atheros AR6001X Radio On Chip.* (n.d.). Retrieved from http://www.atheros.com

Badis, H. & Al Agha, K. (2006). *Quality of Service for Ad hoc Optimized Link State Routing Protocol (QOLSR).* IETF Internet Draft (draft-badis-manet-qolsr-03.txt).

Banerjee, S., & Misra, A. (2002). Minimum energy paths for reliable communication in multi-hop wireless networks. In *The 3rd ACM International Symposium on Mobile Ad Hoc Networking and Computing* (pp. 146-156), Lausanne, Switzerland.

Burkhart, M., Von Rickenbach, P., Wattenhofer, R., & Zollinger, A. (2004). Does topology control reduce interference? In *ACM International Symposium on Mobile Ad Hoc Networking and Computing* (pp. 9–19), Tokyo, Japan.

Chang, J.-H., & Tassiulas, L. (2000). Energy conserving routing in wireless ad-hoc networks. *IEEE Conference on Computer Communications,* (pp. 22-31), Tel-Aviv, Israel

Chang, J-H. & Tassiulas, L. (2004). Maximum lifetime routing in wireless sensor networks. *IEEE/ACM Transactions on Networking, 12,* 609-619.

Chen, B., Jamieson, K., Balakrishnan, H., & Morris, R. (2001). Span: An energy-efficient coordination algorithm for topology maintenance in Ad Hoc wireless networks. In Mobile Computing and Networking (pp. 85-96).

Chen, S., & Nahrstedt, K. (1998). On finding multi-constrained paths. *IEEE International Conference on Communications,* (pp. 874-879), Atlanta, USA. *Cisco Aironet 1240AG Series 802.11A/B/G Access Point.* (n.d). Retrieved from www.cisco.com.

D-Link AirPlusTM DWL-G630 Wireless Cardbus Adapter Manual. (n.d.). Retrieved from http://www.dlink.com

ElBatt, T. A., Krishnamurthy, S. V., Connors, D., & Dao, S. (2000). Power Management for Throughput Enhancement in Wireless Ad-Hoc Networks. In *IEEE International Conference on Communications* (pp. 1506 - 1513), New Orleans, LA.

Ephremides, A. (2002). Energy concerns in wireless networks. In IEEE Wireless Communications, (pp. 48 – 59)

Garey, M., & Johnson, D. (1979). Computers and Intractability: A Guide to the Theory of NP-Completeness. New York: W.H. Freeman and Company.

Gupta, N., & Das, S. (2002). Energy-aware on-demand routing for mobile ad hoc networks. In *4th International Workshop on Distributed Computing* (pp. 164-173), Calcutta, India.

Gutierrez, F., Varvarigos, E., & Vassiliadis, S. (2000). Multicost routing in max-min fair share networks. In *38th Annual. Allerton Conference on Communication, Control and Computing,* (pp. 1294—1304).

Huang, X., & Fang, Y. (2006). Multi-constrained soft-QoS provisioning in wireless sensor networks. In *3rd international conference on Quality of service in heterogeneous wired/wireless networks,* (pp. 465-478), Canada. *IEEE 802.11b Wireless LAN PC Card User's Guide Version 5.0.* (n.d.). Retrieved from http://www.ansel.com.mx

Jaffe, J. (1984). Algorithms for finding paths with multiple constraints. *Networks Journal, 14,* 95–116. doi:10.1002/net.3230140109

Jung, E.-S., & Vaidya, N. H. (2005). A Power Control MAC Protocol for Ad Hoc Networks. In *The 11th Annual International Conference on Mobile Computing and Networking* (pp. 55-66), Cologne, Germany.

Karagiorgas, N., Kokkinos, P., Papageorgiou, C., & Varvarigos, E. A. (2007). Routing in Wireless Ad-hoc Networks with Variable Transmission Power. *IEEE 18th International Symposium on Personal, Indoor and Mobile Radio Communications* (pp. 1-6). Athens, Greece

Kokkinos, P., Papageorgiou, C., & Varvarigos, E. (2005). Energy aware routing in wireless ad hoc networks. *IEEE 6th International Symposium on a World of Wireless, Mobile and Multimedia Networks,* (pp. 306- 311), Naxos, Greece.

Korkmaz, T., & Krunz, M. (2001). Multi-Constrained Optimal Path Selection. In *12th Annual Joint Conference of the IEEE Computer and Communications Societies,* (pp. 834 - 843), Anchorage, AK.

Li, Z., & Garcia-Luna-Aceves, J. (2007). Finding multi-constrained feasible paths by using depth-first search. *Wireless Networks, 13*(3), 323–334. doi:10.1007/s11276-006-7528-8

Liu, L., & Feng, G. (2007). Simulated Annealing Based Multi-constrained QoS Routing in Mobile ad hoc Networks. *International Journal of Wireless Personal Communications, 41*(3), 393–405. doi:10.1007/s11277-006-9149-z

Mieghem, P. & Kuipers, F. (2004). Concepts of Exact QoS Routing Algorithms. *IEEE/ACM Transactions on Networking, 12*(5), 851-864.

Muqattash, A., & Krunz, M. (2003, April). A Distributed Transmission Power Control Protocol for Mobile Ad Hoc Networks. *The 22nd Annual Joint Conference of the IEEE Computer and Communications Societies,* (pp. 113 - 128), San Fransisco.

Papageorgiou, C., Kokkinos, P., & Varvarigos, E. (2006). Multicost Routing over an Infinite Time Horizon in Energy and Capacity Constrained Wireless Ad-hoc Networks. In *European Conference on Parallel Computing* (pp. 931-940), Dresden, Germany.

Papageorgiou, C., Kokkinos, P., & Varvarigos, E. (2008). Implementing distributed multicost routing in mobile ad hoc networks using DSR. In 6th ACM international symposium on Mobility management and wireless access, (pp. 35-42), Vancouver, Canada.

Perkins, C., Royer, E., & Das, S. (2000). *Quality of Service in Ad-hoc On-demand Distance Vector Routing.* IETF Internet Draft, (draft-ietf-manet-qos-00.txt). *POLARIS™ TOTAL RADIOTM solution from RF Micro Devices*. (n.d.). Retrieved from http://www.rfmd.com/

Rahimi, M., Shah, H., Sukhatme, G., Heidemann, J., & Estrin, D. (2003). Studying the Feasibility of Energy Harvesting in a Mobile Sensor Network. *IEEE International Conference on Robotics and Automation* (pp. 19- 24). Taipei, Taiwan.

Ramanathan, R., & Rosales-Hain, R. (2000). Topology Control of Multihop Wireless Networks Using Transmit Power Adjustment. In *IEEE Conference on Computer Communications* (pp. 404 - 413), Tel-Aviv, Israel.

Rodoplu, V., & Meng, T. (1998). Minimum energy mobile wireless networks. *IEEE International Conference on Communications,* (pp. 1333 – 1344).

Shah, R., & Rabaey, J. (2002). Energy aware routing for low energy ad hoc sensor networks. In *IEEE Wireless Communications and Networking Conference* (pp. 350-355), Florida.

Singh, S., Woo, M., & Raghavendra, C. (1998). Power-Aware Routing in Mobile Ad Hoc Networks. In *The Fourth Annual ACM/IEEE International Conference on Mobile Computing and Networking* (pp. 181 - 190), Dallas, TX.

Stojmenovic, I., & Lin, X. (2001). Power-aware localized routing in wireless networks. *IEEE Transactions on Parallel and Distributed Systems, 12*(11), 1122–1133. doi:10.1109/71.969123

Toh, C.-K. (2001). Maximum Battery Life Routing to Support Ubiquitous Mobile Computing in Wireless Ad Hoc Networks. *IEEE Communications Magazine, 39*(6), 138–147. doi:10.1109/35.925682

Wang, Z., & Crowcroft, J. (1996). Quality-of-Service Routing for Supporting Multimedia Applications. *Journal of Selected Areas in Communications, 14*(7), 1228–1234. doi:10.1109/49.536364

Woo, K., Yu, C., Youn, H. Y., & Lee, B. (2001). Non-Blocking, Localized Routing Algorithm for Balanced Energy Consumption in Mobile Ad Hoc Network. In *9th IEEE International Symposium on Modeling, Analysis and Simulation of Computer and Telecommunication Systems* (pp. 117-124), Cincinnati, OH.

Ye, W., Heidemann, J. & Estrin, D. (2003). Medium Access Control with Coordinated, Adaptive Sleeping for Wireless Sensor Networks. *IEEE/ACM Transactions on Networking, 12*, 493-506.

Younis, M., Youssef, M., & Arisha, K. (2003). Energy-aware management for cluster-based sensor networks. *Computer Networks, 43*(5), 649–668. doi:10.1016/S1389-1286(03)00305-0

Yuan, X. & Liu, X. (2002). Heuristic algorithms for multi constrained quality-of service routing. *IEEE/ACM Transactions on Networking, 10*(2), 244-256.

ADDITIONAL READING

Bononi, L., Conti, M., & Donatiello, L. (1999), A distributed contention control mechanism for power saving in random-access ad-hoc wireless local area networks, IEEE International Workshop on Mobile Multimedia Communications.

Bononi, L., Conti, M., & Donatiello, L. (2001), A Distributed Mechanism for Power Saving in IEEE 802.11 Wireless LANs, ACM/Kluwer Mobile Networks and Applications.

Chen, S., & Nahrstedt, K. (1999). Distributed Quality-of-Service Routing in Ad-Hoc Networks. *IEEE Journal on Selected Areas in Communications*, *17*(8), 1488–1505. doi:10.1109/49.780354

Chiasserini, C. F., & Rao, R. R. (2000), A distributed power management policy for wireless *ad hoc networks, IEEE Wireless Communications and* Networking Conference, (pp. 1209 - 1213). Chicago, U.S.A.

Das, S., Perkins, C., & Royer, E. (2000), Performance Comparison of Two On demand Routing Protocols for Ad Hoc Networks, Nineteenth Annual Joint Conference of the IEEE Computer and Communications Societiesm (pp. 16 - 28), Tel Aviv, Israel.

De Couto, D., Aguayo, D., & Bicket, J. R., Morris (2003), A High-Throughput Path Metric for Multi-Hop Wireless Routing, The 9th Annual International Conference on Mobile Computing and Networking, (pp. 134 - 146). San Diego, California, U.S.A.

Elbatt, T., & Ephremides, A. (2002), Joint scheduling and power control for wireless ad-hoc networks, The 21st Annual Joint Conference of the IEEE Computer and Communications Societies, (pp. 976-984). New York, U.S.A.

Iannone, L., Khalili, R., Salamatian, K., & Fdida, S. (2004), Cross-layer routing in wireless mesh networks, 1st International Symposium on Wireless Communication Systems, (pp. 319- 323). Mauritius

Ilyas, M., & Dorf, R. (2003), The Handbook of Ad Hoc Wireless Networks.

Ingelrest, F., & Simplot-Ryl, D. (2008). Localized broadcast incremental power protocol for wireless ad hoc networks. *Wireless Networks*, (3): 309–319. doi:10.1007/s11276-006-9817-7

Johansson, T. CarrMotyckova L., (2005), Reducing Interference in Ad hoc Networks through Topology Control, Joint Workshop on Foundations of Mobile Computing, Portland, Oregon, USA

Li X.S., Demmel J.W., Bailey D.H., Henry G., Hida Iskandar Y., J., Kahan W., Kapur A., Martin M. C., Tung T., and Yoo D. J., (2002), Design, implementation and testing of extended and mixed precision BLAS, ACM Transactions on Mathematical Software

Li, Y., & Ephremides, A. (2007). Joint Scheduling, Power Control, and Routing Algorithm for Ad-Hoc Wireless Networks. *Ad Hoc Networks*, *5*(7), 959–973. doi:10.1016/j.adhoc.2006.04.005

Meyer auf der Heide, F., Schindelhauer, C., Volbert, K., & Grunewal, M. (2002), Energy, congestion and dilation in radio networks, Fourteenth ACM Symposium on Parallel Algorithms and Architectures, (pp. 230 - 237) Winnipeg, Manitoba, Canada.

Mieghem, P., & Kuipers, F. (2003). On the complexity of QoS routing. *Computer Communications*, *26*(4), 376–387. doi:10.1016/S0140-3664(02)00156-1

KEY TERMS AND DEFINITIONS

Wireless Ad Network: A decentralized wireless network, where every node forwards data for other nodes and so, the determination of the forwarding nodes is made dynamically based on the network connectivity.

Quality of Service: A technology which provides different priority to different applications, users, or data flows, or guarantees a certain level of performance to a data flow. QoS technologies give the ability to measure bandwidth, detect changing network conditions, and prioritize or throttle traffic.

Energy-Efficiency: The consideration of energy as a critical resource, when designing a protocol, mainly in wireless networks where energy is an important parameter for network lifetime.

Power Control: The ability of adjusting transmission power in a communication system, in order to achieve specific performance goals. The performance parameters can depend on context and may include optimizing metrics such as link data rate, network capacity, geographic coverage and range, energy and lifetime of the network and network devices.

Chapter 8
Mobility Management and QoS Support in Wireless Environments:
Trends and Open Issues

Alexandros Kaloxylos
University of Peloponnese, Greece

Sarantis Paskalis
University of Athens, Greece

Dimitra Vali
OTE Research, Hellenic Telecommunications Organization - OTE S.A, Greece

Anthony Boucouvalas
Science and Technology, University of Peloponnese, Greece

ABSTRACT

Researchers are trying to find viable means to support the quality of services for fixed IP users for quite some time. This is a difficult task and even up until today there is no universally accepted mechanism to assure the quality of an operational service from the one end to the other. As years were passing, the need of IP users to keep their network connectivity while on the move, introduced a family of mobility management protocols. However, it was soon noted that these mobility management protocols were inter-working rather inefficiently with the protocols for the quality of services' support. Thus, new protocols are under design to tackle this issue. However, even with these new protocols there are important issues left unchallenged. This chapter provides all the necessary information for this research area and its current status.

DOI: 10.4018/978-1-61520-771-8.ch008

INTRODUCTION

The wireless and mobile communication devices industry sector is experiencing an enormous growth. People are getting accustomed to be productive while on the move, utilizing the capabilities their wireless and mobile devices offer. The connectivity support, one of the most fundamental requirements, is certain to rely on the ubiquitous Internet Protocol (IP). There are, however, some fundamental challenges that need to be overcome in order to be able to use the same protocol architecture as the fixed users do.

Mobility support, the first of them, stems from the users' need to communicate in every imaginable way, while on the road, on the train, at home or in the office. The IP protocol suite needs to adapt to nowadays' era and start offering uninterruptible connectivity to devices and users, irrespective of their location and movement conditions. Several years of research work have been performed to accommodate mobility management in IP. The primary efforts focused on the ability of the mobile node to communicate with any other Internet connected node while being attached to a different network, which led to a procedure satisfying the target set, but not the actual mobility requirements of moving and maintaining an uninterrupted, smooth communication. After the baseline for mobility support was set, multiple optimization efforts began to achieve minimization of disruption time, optimization of resources used and generally satisfaction of the mobile users.

Quality support, the second of the essential requirements, has a more complicated history. The IP protocol stack follows the end-to-end principle, which dictates to keep functionality and complexity out of the core of the network and push it to the end-points. In other words, end-devices can and should bear the complexity of evolution and capability additions, whereas internal devices, i.e. the routers, should be kept as simple as possible. The simplicity requirements for the routers extend to simple processing for data packets, and memory-less operation. Following the guidelines strictly, every incoming packet would receive the same priority and would be forwarded to the same output queue waiting for transmission. In today's Internet, however, not all packets are created equal. Some packet flows can cope with long delays and/or packet loss, whereas other flows can only bear extremely small delays and jitter (real-time voice or video communication). The desire to offer prioritized treatment to certain packets, so as to offer either guarantees or just better service, led to Quality of Service (QoS) schemes, which add processing and possibly state-fullness requirements to the routers.

The aforementioned efforts to provide mobility support and QoS guarantees in the Internet began – and mostly continued independently, leading to inefficiencies and/or incompatibilities. The most obvious and cited example is the usage of the end-points' IP addresses to refer also to a QoS state along the data path (Balakrishnan, Lakshminarayanan, Ratnasamy, Shenker, Stoica & Walfish, 2004). This QoS state identifies the packets that will receive a certain priority treatment and needs to be modified in an end-to-end fashion when mobility causes the modification of an end-point's IP address.. Thus, the net result is the invalidation of the existing QoS setup along the data path, and, thus, the need to re-establish the QoS reservation according to the new IP address and the need to tear down the now invalidated QoS state throughout the data path containing the previous IP address.

The mobility management and QoS incompatibilities have been identified in the research literature practically from the beginning of the individual standardization efforts. However, the relative isolation between the QoS and mobility management specification groups, the experimental nature of the schemes, and the lack of operational use cases, prevented any harmonization attempt at least in the relevant standardization avenues. This finally has been changed with the creation of the IETF "Next Steps in Signaling" working group (Next Steps In Signaling IETF Working Group), which undertook the difficult task to propose a generic signaling architecture for the Internet, capable of dealing with the multiple and contradictory signaling needs for the Internet infrastructure. Among them is also the undisturbed interaction between mobility management and QoS signaling and state maintenance.

Another interesting issue that has to be dealt with stems from the fact that dynamic control of end-to-end QoS schemes requires that signaling has to travel from one end to the other each time a new session is to be established. However, if individual flow states are maintained at each router along the data path, scaling issues are raised, especially for the core network routers. To ameliorate this problem, aggregation of signaling state information is required. The scheme that one should use to minimize the processing load and the signaling information stored in the routers still remains an open issue, but is definitely something that needs to be specified if QoS provisioning is ever to be deployed in a real end-to-end fashion. Unfortunately, regarding their interaction with mobility management, the proposed aggregation schemes exhibit similar incompatibilities as the existing individual QoS proposals. This is a rather important drawback since it is expected that in the future, users will be able to vertically handoff their connections from one radio access technology to another and even from one Internet Service Provider to another, thus moving partly away from an aggregation path.

All the aforementioned issues are described in this chapter that is organized as follows. In the next section we provide a description of the mobility management protocols and what is expected to prevail in the near future. Then, a discussion follows about the advantages and inefficiencies of QoS mechanisms. We present those that have already been standardized or have been proposed to deal with specific issues (e.g., aggregation of signaling states). Next, we discuss on current trends to tackle the interworking issues between QoS protocols and mobility management mechanisms. Finally, we conclude this chapter.

PRESENTATION OF MOBILITY MANAGEMENT PROTOCOLS

As it is well known, the core protocols of the Internet (i.e., IP and TCP) were not designed to handle mobile terminals. In IP networks, the destination IP address has topological information and is used to determine the next hop to forward a packet. On the transport layer, the TCP protocol maintains information in the form <source IP address, source port number, destination IP address, destination port number>. Thus, without any additional provision, supporting mobility in TCP/IP networks faces a conflict. From the one hand, a terminal entering a new network has to receive a new IP address in order to be reachable. On the other hand, a TCP connection has to keep the source IP address constant because otherwise the connection will fail since the new information will not correspond to the old one. As we will present, the Internet community came up with a mobility management protocol that makes mobility transparent to the higher-level protocols and requires minimum changes to the existing infrastructure.

Mobility management consists of two distinct sets of operations. The first one has to do with the location management of the terminal. This set of operations includes all the procedures for updating the knowledge of the network about the current location of a terminal as well as the procedures for finding the current location of a terminal when it is required to deliver data to it. The second set has to do with the handover of an active flow from the old data path to the new one.

Mobile IP

All location management mechanisms, independently if they are deployed in cellular or IP networks, follow some common principles. The overall network is divided into different areas. For each terminal one of these areas is called the "home area". Inside the home area there is an entity that is aware at any time for the current location of a terminal. In terms of Mobile IPv4 (Perkins, 2002), this entity is called the home agent (HA) and maintains a mobility binding table with the following information: <permanent home address, temporary care-of address, association lifetime>. This information is kept for each terminal currently located outside its home area. To maintain the information of this table up to date, each time a terminal moves into a "foreign area" it will receive a temporary IP address that is called care of address (COA). As soon as the terminal receives this address it will inform the home agent. This way a terminal may have two IP addresses. The permanent home address is used to identify the terminal while the temporary care-of address represents the current location of the host and used mainly for routing purposes (i.e., reaching the terminal in its current location). This temporary address is usually assigned to a terminal by another specialized entity called "foreign agent" (FA).

In summary, when a terminal enters into a new area, it discovers the existence of a new FA. Then, it requests a registration with the FA and send its home address, its media address (e.g., Ethernet MAC address) and the address of its HA. The FA then sends a registration to the HA by sending a message that contains the home address of the terminal and its own address. This way, location update is performed in the HA. When a node, called correspondent node (CN) in the MIP terminology, wishes to communicate with the mobile terminal, it will send packets using the permanent home IP address of the terminal. The packets will end up as expected in the home area of the terminal. There, they will be intercepted by the HA. The HA, after consulting the mobility binding table, will construct a new IP packet in which the payload will be the original IP packet sent by the CN. The header of the new IP packet will have the temporary COA as its destination address. This process is called encapsulation (Perkins, 1996) or tunneling. The new packet will reach the FA that is responsible to de-capsulate the original packet and use the media address of the terminal to forward the packet to it. Note that if the mobile node wants to send a packet to the CN, this can be routed directly. This forms an asymmetry in the routing of packets between the two communicating end nodes and is called *triangular routing*. The problem of the triangular routing has been solved in the MIPV6 (Johnson, Perkins & Arkko, 2004) since the COA can be communicated to the CN directly. Another important difference from MIPv4, is that in MIPv6 there is no need to use any FA, since the required functionality is supported by the mobile terminals themselves.

As we already mentioned, apart from location management, mobility management has to do handing over an active flow from the old data path to the new path. This functionality by definition built in the MIP. If a terminal receives a new COA while having an active communication flow then the new packets will reach the terminal in its new location as soon as the bindings have been updated with the new COA. Of course, if no special provision is taken then, until this binding update takes place, some packets will be forwarded to the old location of the terminal, even though it has already change its location, and will be lost. To tackle this issue several alternatives have been proposed like packet forwarding from the old location to the new one, or prediction schemes that start sending copies of packets to possible future locations of a terminal.

Mobile IP Enhancements and Extensions

Because of the dominance of the IP protocol and the simplicity of MIP, it is expected that it will be the de-facto standard even for future mobile cellular networks. However, MIP faces some serious problems in terms of the flow disruption time and packet losses. This is why alternative solutions like Fast MIP (Koodli, 2008), Hierarchical MIP (Soliman, Castellucia, Malki & Bellier, 2005) and Proxy MIP (Gundavelli, Devarapalli, Chowdhury & Patil, 2008) have been introduced to deal with the minimization of the disruption time during a HO execution, the localization of signaling exchanges inside a pre-defined area and the transparency of mobility support for the end users.

More specifically, a handover involves not only layer 3 operations (e.g., acquire a new IP address, notify the HA etc), but also some layer 2 actions (e.g., scan for neighboring access points, associate and de-associate in link layer etc). FMIP has been specifically designed to take into consideration these operations and minimize as much as possible the disruption time by having the terminal collecting information and the network components to be configured appropriately before the execution of a handover. In FMIP, the terminal can request the serving access router (called Previous Access Router – PAR) to send it information about a target router for a handover (called New Access Router – NAR). With the received information the terminal can form a perspective new COA. Moreover, it can request the PAR to start transmitting any received packets for the terminal to the NAR, by having the PAR to establish an appropriate tunnel between the two routers. To do this, the PAR will communicate with the NAR to check several issues (e.g., the validity of the perspective COA). When everything is ready the terminal is notified by the PAR to hand-off to the NAR. To do this, the terminal will perform a layer 2 association to the new network (e.g., a WLAN association) and will send to the NAR a notification to inform it that it has been attached under its control area and that the NAR should start forwarding packets for it. Up to this point, the packets are arriving to the terminal following the tunnel between the PAR and the NAR. The tunnel between the two nodes will remain valid until the terminal completes its binding updates with its CNs. Obviously, the aforementioned description is valid only when there is enough time for the terminal to collect all the required information through the old path and the routers to be configured appropriately. If however there is not sufficient time then it is most likely that several operations will be made through the NAR, resulting in an increased disruption time.

In the case of Hierarchical MIP, the target is to achieve smaller disruption times by keeping any signaling exchange inside a pre-defined area. This is achieved with the use of two different COAs namely the Regional COA (RCOA) and the Link COA (LCOA). More specifically, inside a pre-defined area a new component is introduced, called Mobility Anchor Point – MAP. This component actually acts as a "local HA" for the terminals that remain inside the domain of the MAP. To reach any terminal, CNs are using a globally visible address (i.e., RCOA). Using this address, the packets will eventually arrive in the MAP entity. Note that as long as a terminal remains inside the same domain, it uses the same RCOA. Any packets targeted to the terminal are intercepted by the MAP entity and encapsulated using the LCOA. When a terminal moves inside a domain it may acquire a new LCOA. This information has to be transferred up to the MAP entity. It does not however needs to be transferred up the HA of the terminal. This is because the HA only knows about the RCOA that is unchanged as long as the terminal keeps moving inside the same domain. The only case the HA needs to be updated is when the terminal moves into the area of a new MAP entity and receives a new RCOA. HMIP presents several advantages. First of all, signaling does not usually require traveling long distances and remains local. This means that the network is burdened with less signaling exchanges and more importantly experiences a smaller handover execution time. Moreover, there is some location privacy for the users since their exact location inside a domain is not known to the outside world. However, this solution requires the introduction of

Figure 1. Basic architecture for FMIPv6 and HMIPv6

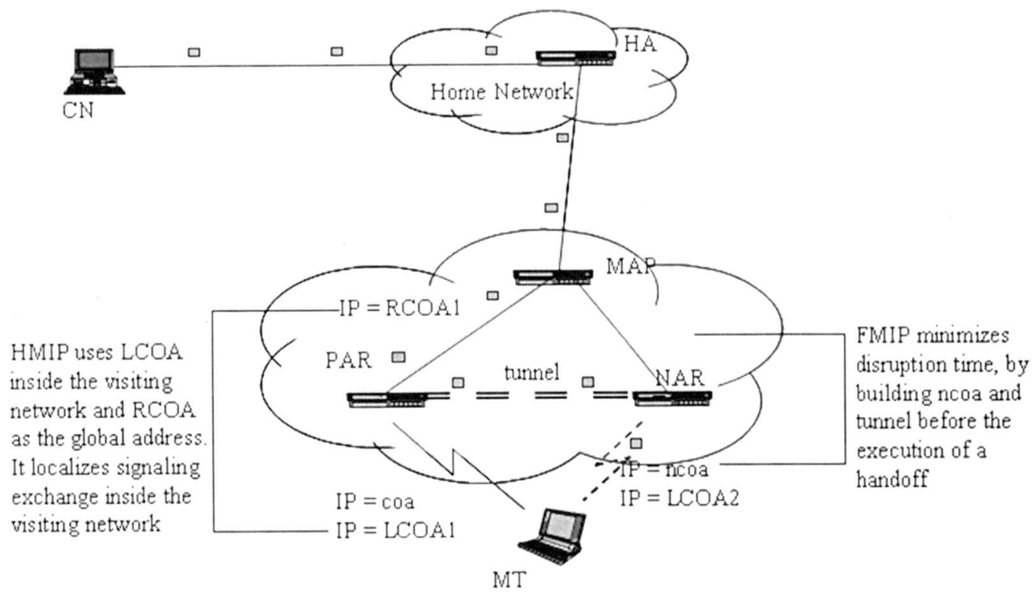

a new component (i.e., MAP). Figure 1 illustrates the operation of FMIP and HMIP.

Finally, we briefly present the case of Proxy MIP since it has been selected for the interworking of future mobile cellular networks (i.e., LTE/SAE). The main goal of this mechanism is to execute mobility management functions in a totally transparent way for the terminals. In other words additional functions are added inside the network in order to leave the terminals totally unaware of any mobility management procedures. For this, two new functions are introduced namely the Local Mobility Anchor (LMA) and the Mobile Access Gateway (MAG). The former acts as the "home agent" for a terminal located inside a specific domain. The latter is a function located on an access router and handles all mobility related signaling for a terminal that is attached to the access router. When a terminal enters a Proxy MIP domain, the terminal communicates with a MAG and after being authenticated, MAG is responsible to contact LMA and create a bi-directional tunnel among them for the specific terminal. The MAG function is responsible to keep track of any movement from the terminal and update accordingly the LMA.

Although there have been vast research efforts to develop mobility management protocols, until now there has been no real adoption of these protocols in every day life. There are many reasons for this. First of all, standard MIP cannot easily support real time services when the disruption time during a handover is measured to be in the order of some seconds. Moreover, the specifications are considered to be quite "heavy" to be executed on small mobile end devices. Finally, operators are in favor of network based MIP solutions such as PMIPV6. Obviously there are several open issues to be dealt like how to combine the aforementioned techniques to achieve better results and also how these protocols can be combined with the appropriate QoS support mechanisms. Some mobility protocol reviews are available at (Saha, Mukherjee, Misra, Chakraborty & Subhash, 2004).

QUALITY OF SERVICE MECHANISMS

QoS mechanisms rely on prioritizing some traffic over other less important/urgent data. The prioritization usually aims at providing guaranteed minimum bandwidth, and occasionally other parameters, such as guaranteed maximum delays and jitter. The specifics of QoS provided depend directly on the underlying link-layer technology and the provisions it is able to make. However, the common attribute in all QoS service provisioning schemes is that in order to provide different services to packets belonging to different service groups, the network, i.e. the routers that comprise it, must have a way to differentiate between the packets. In other words, the existence of some packet classification criteria at each router is the single common attribute in every QoS architecture solution in the Internet.

The existence of the packet classification mechanisms in every router along the data path implies:

- QoS state maintenance at each router, and
- QoS signaling, to establish, maintain and teardown the QoS state.

Integrated and Differentiated Services

The Internet community introduced the Integrated Services (IntServ) architecture (Braden, Clark & Shenker, 1994) to implement the vision of end-to-end QoS services into specifications. IntServ supports end-to-end signaling, QoS state establishment and management for per-flow differentiated treatment in intermediate routers along the data path. The signaling protocol that emerged to meet the Integrated Services requirements is RSVP (Resource reSerVation Protocol) (Braden, Zhang, Berson, Herzog & Jamin, 1997). The IntServ architecture was designed to facilitate every QoS element (router functionality, signaling, and accounting) in a fine-grained manner. To achieve this goal, IntServ was founded on the underlying assumption that a homogeneous Internet environment equipped with IntServ enabled routers and end hosts would be the common case.

The IntServ architecture in general and the RSVP protocol in particular received criticism, mainly due to the scalability issues raised by the state maintenance for every data flow in intermediate routers across the end-to-end path. The Internet community considered, therefore, other alternatives to the QoS provision problem. This time, the target was a lightweight QoS architecture putting as little burden in the routers as possible and providing coarse-grained traffic prioritization based on the statically contracted Service Level Agreements (SLAs) between users and the network. SLAs specify the amount and types of traffic each side has agreed to send and receive. The outcome was DiffServ (Differentiated Architecture) (Blake, Black, Carlson, Davies, Wang & Weiss, 1998).

DiffServ networks are statically configured to support a small set of QoS levels (PHBs – Per Hop Behaviors) and do not use any QoS signaling for state establishment and maintenance in routers. DiffServ routers prioritize the data packets according to a 6-bit field in the IP packet header (DSCP, DiffServ Code Point) that reflects the requested QoS level. This procedure results in aggregating reservations for different users sharing the same QoS level. Appropriate packet marking takes place either at end-hosts or at DiffServ edge routers before the traffic enters the DiffServ network. DiffServ edge routers perform, in addition, traffic classification and traffic conditioning procedures (including metering, marking, shaping and policing) based on the contracted SLAs. In other words DiffServ establishes a minimum, static QoS state at each router, eliminating any QoS signaling and QoS state differentiation, the opposite of the fine-grained IntServ approach.

Figure 2. Two-tier QoS signaling

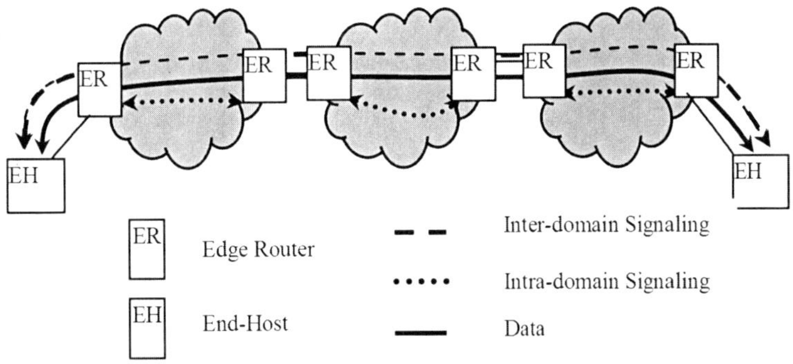

MPLS, a Traffic Engineering Approach

MPLS (Rosen, Viswanathan & Callon, 2001) employs a somewhat similar treatment to packets as does DiffServ. It assigns each packet an MPLS label (32-bit header) and forwards it in the interior of an MPLS network solely based on the value of that label, without any packet inspection. The difference to the DiffServ approach is that the MPLS network creates MPLS paths for each different label throughout the MPLS network, and forwards the packets through the established paths. Traditional IP routing is not employed inside the MPLS network, and the provider has ample opportunities to configure the service received as well as the path taken for each individual MPLS label path. Note that MPLS was not designed as a means to provide QoS but rather a faster way to route packets. However, it has been recognized that its built-in functionality can be used for QoS purposes if handled appropriately.

QoS Aggregation

The realization of the fact that the Internet is a concatenation of technologically and administratively different domains (Autonomous Systems - ASs) led to the identification of separate QoS techniques for the efficient support of intra- and inter-domain QoS. Thus, a two-tier resource management model was proposed in (Terzis, Wang, Ogawa & Zhang, 1999), with the intra-domain QoS signaling performing resource management inside a domain, and the inter-domain signaling managing resource allocation between domains. The two tiers must be closely coordinated to enable provision of the necessary end-to-end QoS support. The two-tier model increases the degrees of freedom regarding end-to-end QoS support, since each domain is free to choose any QoS support mechanism for allocating resources internally, as long as proper co-operation takes place with the respective inter-domain signaling protocol.

The two-tier signaling architecture, illustrated in Figure 2, implies that each domain is allowed to use its own QoS mechanism or protocol internally, allowing for concatenation of the various heterogeneous domains. The provision, however, of end-to-end QoS requires that appropriate interworking between the intra- and the inter-domain QoS protocols take place at the domain boundaries. Multiple configurations are possible with this approach, including both fine- and coarse-grained QoS schemes, in the intra- and the inter-domain signaling, depending on the flexibility and simplicity desired to achieve.

One of the important trends in multi-tier QoS schemes is the de-coupling of QoS state (as in packet classification and resource reservation) and QoS signaling state. An example of that scenario is the maintenance of individual flow signaling

Figure 3. Signaling state aggregation toward domain D1

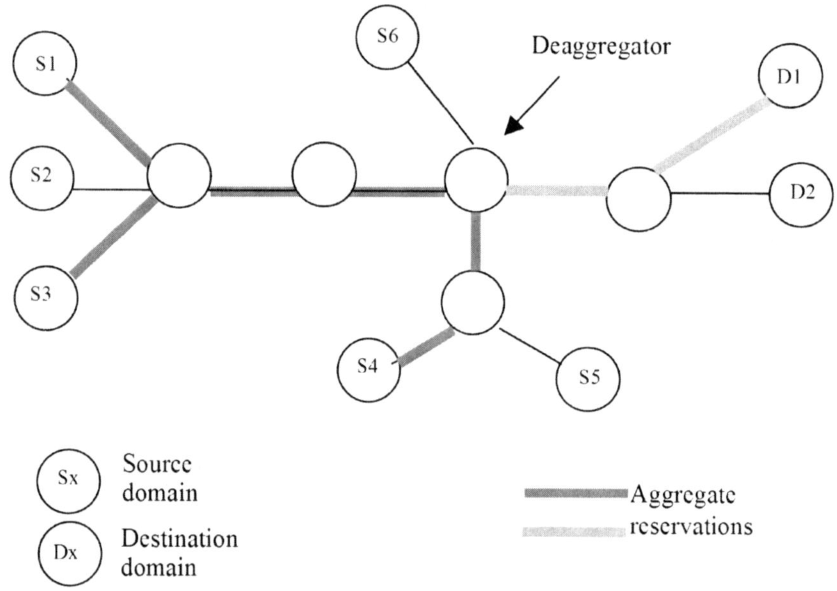

state at the QoS signaling capable border routers and packet marking to impose intra-domain specific QoS handling (i.e. DiffServ marking or MPLS labeling).

A side-effect of the two-tier architecture is the need to maintain both intra- and inter-domain QoS state at the border routers as well as to perform the necessary mapping and parameter translation between them. As such, the QoS state (and by extension the QoS signaling if needed) needs to be implemented for both intra- and inter-domain signaling at the border routers performing the necessary transitional steps. The gain from imposing such a complexity weight on the border routers is the bare simplicity of the internal routers and the minimal QoS state and signaling needed. An extensive review of the tradeoffs imposed on the specific paradigm can be found in (Vali, Paskalis, Kaloxylos & Merakos, 2004).

There are only a few alternatives for aggregating signaling information. Arguably, the most straightforward way to aggregate signaling is to re-use the existing signaling protocol (Baker, Iturralde, Faucheur & Davie, 2001). Using a single RSVP reservation to aggregate other RSVP reservations across a transit routing region consolidates the signaling state inside that region in just one entry. Other approaches completely dismiss the use of RSVP and propose their own mechanisms for aggregation purposes. The DARIS (Dynamic Aggregation of Reservations for Internet Services) architecture (Bless, 2002) assumes the existence of a central resource management entity inside each DiffServ domain that has a complete knowledge and control of the resources inside the domain. DARIS enables the creation of an aggregate between two arbitrary domains as soon as a threshold of active common reservations between the two domains is exceeded.

A couple of other aggregation techniques focus on the reservation at the AS (Autonomous System) level. A scheme designed for aggregate inter-domain usage between heterogeneous domains (Autonomous Systems) is the Border Gateway Reservation Protocol (BGRP) (Pan, Schulzrinne & Hahne, 2000). BGRP operates end-to-end only between domain border routers and aims at aggregating reservations between domains improving scalability. BGRP uses the sink-tree aggregation approach and performs reservation aggregation by building a sink tree for each destination domain (Figure 3). Reservations from different source domains that are destined towards the same destination domain are aggregated along the path, forming a sink-tree rooted at the destination domain edge router. The Shared-segment based Inter-domain Control Aggregation Protocol (SICAP) (Sofia, Guerin & Veiga, 2003) is another approach for supporting aggregate inter-domain reservations between Autonomous Systems (ASs).

The need for a generic Internet signaling framework led to the development of the NSIS framework (Hancock, Karagiannis, Joughney & den Bosch, 2005). NSIS consists of two layers, the underlying General Internet Signaling Transport (GIST) (Schulzrinne & Hancock, 2009 and the application specific NSIS Signaling Layer Protocol Protocol (NSLP), which in the QoS case is the QoS-NSLP (Manner, Karagiannis & McDonald, 2009). The QoS-specific signaling protocol of NSIS is similar in concept to RSVP, dealing with individual flows, and maintaining soft state for each of them. The important difference regarding mobility interaction is the choice of a Session ID irrelevant to the end-point location identifiers (IP addresses).

Each of the proposed QoS schemes usually provides a different packet classification option for the packet flows. Table 1 lists some of those.

Two- or multiple tier models usually employ a simple packet classification approach (DiffServ or MPLS), enjoying the fast and stateless operation for the intra-domain packet classification. The inter-domain packet classification, as well as the necessary signaling state required, is a much bigger issue. If end users are capable of QoS signaling, i.e. requesting and receiving specific time- and service-based QoS, then either RSVP or QoS-NSLP is used, the packet classification criteria are session-based and the signaling state contains information about each session. If only AS border routers handle QoS provisioning, then the packet classification may still be simple (i.e. DiffServ), but the signaling state contains information about session aggregations.

The most relevant to QoS and mobility management interaction attribute is the content of QoS signaling state, and especially the identity of each state. Table 2 presents the signaling state maintained for some of the presented QoS proposals.

MOBILITY AND QOS INTERACTION

The most important clash between IP mobility management assumptions and IP QoS assumptions is the consideration of the IP address as an immutable identifier for the end host. The historic dual consideration of the IP address as a geographic/topological qualification and a unique identifier provides many advantages when networked devices do not move. In our post-classic era, though, this is no longer the case, and mobility considerations dictate the use of several IP addresses from a node as it moves and changes points of attachment. The same logic applies to multi-homed devices, which connect to multiple networks simultaneously, seeking optimal connectivity.

Table 1. Packet classification arguments

RSVP	5-tuple <Protocol, Source Address, Source Port, Destination Address, Destination Port>
DiffServ	DiffServ code point (DSCP) – a 6-bit field in the IP TOS field
MPLS	MPLS Label stack (shim header)
QoS-NSLP	Packet classifier object (more flexible than RSVP's 5-tuple)

Table 2. Signaling state maintained at each intermediate router

RSVP	5-tuple <Protocol, Source Address, Source Port, Destination Address, Destination Port>
DiffServ	None
MPLS	None
BGRP	Autonomous System Number
SICAP	Intermediate border router (Intermediate De-aggregation Location – IDL)
DARIS	Intermediate router or AS
QoS-NSLP	Session ID (random 128 bit number)

The optimum utilization of access networks resources is a critical issue, especially in large domains, containing a significant number of wireless devices. In such networks, the administrator has to deal not only the scarce resources on the wireless links, but also with the efficient utilization of the expensive resources in the link(s) to the upstream ISP.

This task becomes more challenging when the wireless devices are also mobile ones.If no special care is provided mobile terminals face long service disruption times. Moreover, the delay to reorganize reserved resources in the end-to-end path (reserve resources in the new path and release the ones in the old path) results in a waste of network resources for this time period. This percentage is expected to be significant when a large number of mobile terminals resides in the access network or when the terminals demonstrate a high mobility rate.

Interaction with RSVP

The existing QoS protocols, such as RSVP, do not interact well with the IP address change, imposed by Mobile IP. If a mobile host (MH), with established RSVP data flows, performs a network layer handoff, it acquires a new IP address (Care of Address, COA) (Perkins, 2002; Johnson, Perkins & Arkko, 2004), and a new round of RSVP signaling exchanges must be triggered. RSVP creates soft session states in every intermediate router of the traffic flow. Each session is uniquely identified by the "Session" object, which is defined by the triplet <DestAddress, DestPort, [ProtocolId]>. Thus, the downlink reservation (the packet flow toward the mobile) becomes invalid, when the DestAddress parameter is modified due to mobility. The new uplink re-establishment is also affected, since the RSVP "Path" messages sent by the MH contain its new IP address in the "Sender Template" object. These messages sent and received by the MH at its new location are considered to correspond to a new session, and generate a new "Path" state (Braden & Zhang, 1997).

The major problems emerging from this mobility-QoS interaction are: (a) the sessions from and to the MH may not receive expected QoS treatment, and (b) the reserved resources that correspond to the old COA will not be available to other traffic until the RSVP soft states expire thus, becoming what we will refer to as *stale sessions* and *stale reservations*, respectively. As illustrated in Figure 4, when the MH switches to a different point of attachment and assigns itself a new IP address, end-to-end RSVP signaling (Path/Resv messages) must traverse the network in order to re-establish the QoS reservation pointing to the current COA of the MH. Moreover, explicit teardown of the stale sessions (through PathTear/ResvTear messages) may be initiated, either by the MH or the correspondent node, in order to avoid resource waste until the stale reservations expire.

The research community has identified the need for efficient interaction between mobility and QoS, and has proposed a number of different

Figure 4. Handoff and RSVP interaction

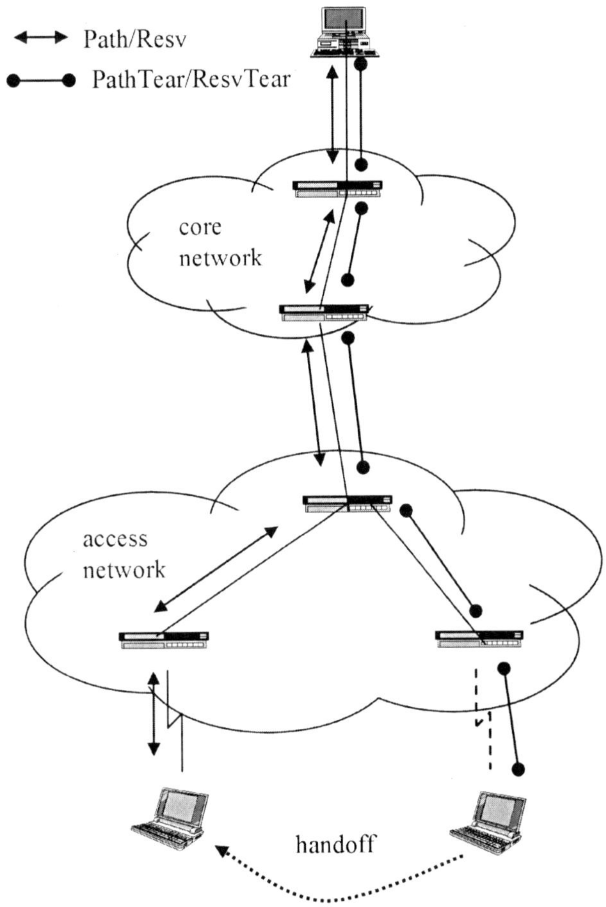

approaches to the problem. The various proposals try to meet the diverse requirements (Chaskar, 2003) from different angles.

To minimize the time needed for re-organization of the network resources, proposals included RSVP modifications and extensions that deal with context transfers between successive access routers (Kempf, 2002), proactive reservations to neighboring access routers with the use of mobile agents for reserving passive or pseudo-reserved resources, or multicasting in hierarchical or not domains (Talukdar, Badrinath & Acrarya, 2001; Huang & Chen, 2003; Lee, Kim, Chanson, Yu & Lee, 2003; Chang, Lee & Lee, 2005; Tseng, Lee, Liu & Wang, 2003; Chen & Huang, 2000). These solutions address the time minimization issue on the cost of complex procedures.

Other researchers suggested that one needs to find the cross-over router for a moving terminal and simply reconfigure the network in a way that the resources of the old branch are now reserved in the new branch (Moon & Aghvami, 2001, 2004). Some proposals limit their scope to an administrative domain (e.g. an access network) and propose either modifications to RSVP (Manner & Raatikainen, 2003) or the addition of functionality to cross-over routers (Paskalis, Kaloxylos, Zervas & Merakos, 2003).

All of the above proposals contain either qualitative or quantitative performance comparison results which show the required improvement regarding the proposed scheme and the original standard. More extensive reviews comparing the qualitative aspects of each scheme can be found at (Belhoul, Şekercioglu & Mani, 2008) and (Manner, Toledo, Mihailovic, Munoz, Hepworth & Khouaja, 2002).

The obvious solution to the state modification after a handoff is the de-coupling of state session identification from end point IP addresses. One of the proposals suggested modifying RSVP signaling, so that a unique session identity (possibly a random integer) was included in the Session and Sender Template fields (Thomas, 2002). The message processing rules should also be modified to deliver the same QoS to packets originating or destined to different IP addresses, but conforming to the same session ID. Similar approaches were suggested by (Shen, Lo, Seah & Ko, 2000), where the immutable Home Address assumes the role of session ID, and (Kuo & Ko, 2000), where the IPv6 Flow Label is the session reference identifier. The urge to create a signaling protocol with a mobility-immutable Session ID gave birth to the creation of the IETF NSIS Working Group and the proposal of a suite of signaling protocols, which dealt with RSVP shortcomings.

Interaction with QoS-NSLP

The development of QoS-NSLP had one particular feature regarding mobility considerations. QoS-NSLP decouples QoS state and flow identification. Session ID (SID), a cryptographically random number, which is probabilistically globally unique, is the state reference object. The NSIS Transport Layer Protocol (GIST) notices when a routing path associated with a SID changes, and provides a notification to the NSLP. It is then up to the NSLP to update the state information in the network (T. Sanda, Fu, Jeong, Manner & Tshofenig, 2009). Thus, the effect is an update to the states, not a full new request.

The actual signaling and state re-establishment/maintenance burden imposed on the network depends on the location of the cross-over router (the router where the old path and the new path converge), and whether the mobile end point changed IP addresses or not.

Figure 5 illustrates the typical message exchange for QoS-NSLP after a handoff, where the mobile host (MH) switched points of attachment and possibly changed its IP address. To re-establish a QoS reservation with its corresponding node (CN), it needs to re-initiate the reservation procedure sending RESERVE messages along the new data path toward the CN. The first QoS-NSLP aware router that receives a RESERVE message that contains the same Session ID, but different SII (Source Identification Information) or MRI (Message Routing Information), realizes that one of the end points performed a handoff and that itself is the crossover router (CRN) for the specific session involving a handoff.

For the part of the path that did not contain any session state, (i.e. the MH→CRN part), Connection Admission Control, and state establishment must be processed as with any new QoS request. For the part of the path, though, after the crossover router (i.e. the CRN→CN part), the implications of QoS-NSLP signaling are just state updates to reflect the new conditions regarding the IP address of the mobile host, and, consequently, performed much faster and easier.

The crossover router can also issue a NOTIFY toward the previous location of the MH, which travels hop-by-hop, until it reaches its old access router (R1). R1 figures out that MH has left its network service and sends a teardown request RESERVE(T) toward the other end of the session, i.e. the correspondent node, which destroys QoS states along the way. CRN intercepts this message and discards it, since its purpose was to free the reserved resources in the now stale part of the old data path (i.e. R1→CRN).

The use of QoS-NSLP solves one of the biggest incompatibility problems between mobility management and QoS setup. Despite these efforts, the need for end-to-end signaling has not disappeared, since state updates regarding the packet classification scheme employed, must be

Figure 5. Message exchange after a handoff in QoS-NSLP

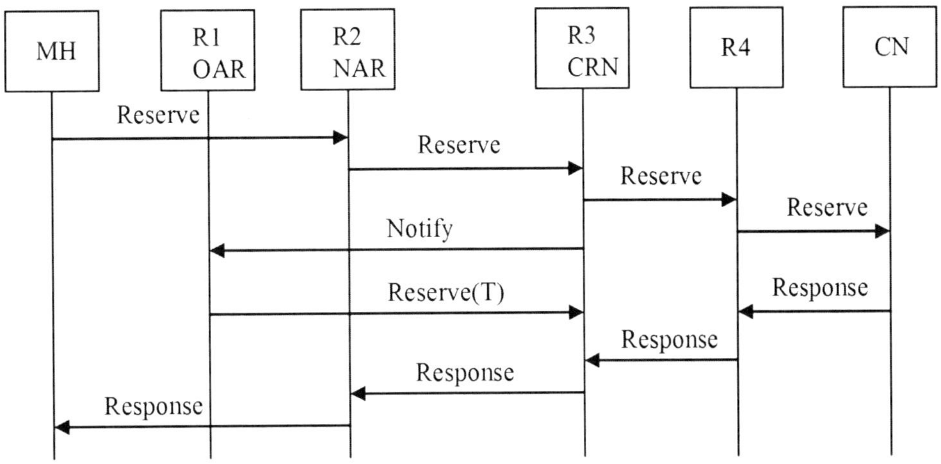

re-installed along the data path. Note also that if localized signaling is desired, one must deploy a mobility management protocol such as HMIP, that "hides" the movement of the mobile host inside an HMIP domain, presenting a stable IP address as the MH COA to the outer environment.

Aggregating Resources and Mobility

As already mentioned, there are only a few alternatives for aggregating signaling information. Their main operation is to aggregate a large number of signaling states into a single state through a specific path located between one aggregation and one de-aggregation point. As it is obvious, between these two points any signaling information for a specific flow is lost. This functionality is advantageous for fixed terminals since it eliminate a vast number of signaling states, especially from the core network routers.

Future scenarios expect users to be able to vertically handoff their connections from one radio access technology to another or even dynamically select to switch from one Operator, or Internet Service Provider, to another (e.g., handover from/to WLAN to/from WiMAX, UMTS etc). In these cases it is possible that the new end-to-end path, although sharing a large segment with the old one, will not include the previous aggregation or de-aggregation points. Since the intermediate routers do not have any means to recognize that some resources have already been reserved, under an aggregated signaling state, they will attempt to reserve from scratch new resources. Thus, resources need to be re-established in an end-to-end fashion, despite the fact that a portion of the previously established path could be re-used. This is actually a problem similar to the one arisen for the interworking between MIP and RSVP.

The existing standard and the few alternative proposals have not been designed to deal with this problem. Thus, additional functionality is needed. A possible solution to this problem has been proposed in (Kaloxylos, Vali, Paskalis, Panagiotou, Gonianakis & Zervas, 2006) where appropriate extensions to BGRP have been designed. However, since the signaling aggregation problem has not solved yet, we note that any future solution should take into consideration the inefficiencies that may be caused by the mobile terminals.

FUTURE RESEARCH DIRECTIONS

In the future mobility management and QoS provisions need to be integrally designed in the system in order to be fully and flexibly functional. Although the problem spaces (for mobility and quality of service) are distinct, their intersection is not trivial. Perhaps, some hard design choices need to me made breaking backwards compatibility with existing standards and functionality.

One such attempt is to distinguish the host identity from its topological association, making the IP address a routing only designator. Such an attempt is proposed by HIP (Moskowitz & Nikander, 2006), the Host Identity Protocol, which tries (among other things) to create a new identity layer, not connected to IP semantics. There is, however, little work done about any QoS needs regarding this proposal, and the security/privacy semantics it may need to offer.

Another aspect worth taking into account is the value of signaling aggregation (QoS signaling mostly) and its repercussion on mobility management. Since individual QoS state is removed and possibly re-instated at a later time, how will this affect any mobility state? Should mobility state maintain a much lighter QoS state? How will any movement affect the aggregated QoS state. All these questions are still open.

Most research has been done on achieving effective and fast handoff functionality, so that little or no content is dropped during movements. Whereas, this is the most user-visible metric, the smooth handoff challenge is still ongoing with the constant modification of the underlying parameters, rendering the problem a moving target.

In conclusion, most if not all, of the individual subjects in mobility/QoS interaction are still open and in need of integration in a grander view.

CONCLUSION

In this book chapter we present the main mobility management protocols that are present in the Internet and the QoS protocols that attempt to solve the end-to-end QoS support. In the last category we also present protocols that deal with the aggregation of QoS signaling states. Our aim is to identify the problems that arise from the interworking of these mobility management and QoS support protocols. The chapter identifies open issues, even in the latest standardization attempts and provides hints on how these can be tackled.

REFERENCES

Baker, F., Iturralde, C., Faucheur, F. L., & Davie, B. (2001). *RFC 3175: Aggregation of RSVP for IPv4 and IPv6 Reservations.*

Balakrishnan, H., Lakshminarayanan, K., Ratnasamy, S., Shenker, S., Stoica, I. & Walfish, M. (2004). A Layered Naming Architecture for the Internet. *SIGCOMM '04, 34*, 343–352.

Belhoul, A., Şekercioglu, Y. A., & Mani, N. (2008). Mobility protocols and RSVP performance in wireless IPv6 networks: Shortcomings and solutions. *Wireless Communications and Mobile Computing, 8*(9), 1183–1189. doi:10.1002/wcm.560

Blake, S., Black, D., Carlson, M., Davies, E., Wang, Z., & Weiss, W. (1998). *RFC 2475: An Architecture for Differentiated Services.*

Bless, R. (2002). Dynamic Aggregation of Reservations for Internet Services. In *International Conference on Telecommunication Systems – Modeling and Analysis,* (ICTSM), (pp. 26–38).

Braden, R. (Ed.). Zhang, L., Berson, S., Herzog, S. & Jamin, S. (1997). RFC 2205: Resource ReSerVation Protocol (RSVP) — Version 1 Functional Specification

Braden, R., Clark, D., & Shenker, S. (1994). *RFC 1633: Integrated Services in the Internet Architecture: an Overview.*

Braden, R., & Zhang, L. (1997). *RFC 2209: Resource ReSerVation Protocol (RSVP) — Version 1 Message Processing Rules.*

Chang, M., Lee, M., & Lee, H. (2005). An Efficient Resource Reservation Scheme Based on Gray-Cell in Wireless Mobile Networks. In *IEEE Wireless Communications and Networking Conference (WCNC), 3*, 1311–1316.

Chaskar, H. (Ed.). (2003). RFC 3583: Requirements of a Quality of Service (QoS) Solution for Mobile IP.

Chen, W.-T., & Huang, L.-C. (2000). RSVP Mobility Support: A Signalling Protocol for Integrated Services Internet with Mobile Hosts. In IEEE/ACM INFOCOM (pp. 1283–1292), Tel Aviv, Israel.

Gundavelli, S. (Ed.). Devarapalli, V., Chowdhury, K., & Patil, B. (2008). RFC 5213: Proxy Mobile IPv6.

Hancock, R., Karagiannis, G., Joughney, J., & Van den Bosch, S. (2005). RFC 4080: Next Steps in Signaling (NSIS), Framework.

Huang, N.-F., & Chen, W.-N. (2003). RSVP Extensions for Real-Time Services in Hierarchical Mobile IPv6. [MONET]. *Mobile Networking and Applications*, *8*, 625–634. doi:10.1023/A:1026022325876

Johnson, D., Perkins, C., & Arkko, J. (2004). *RFC 3775: Mobility Support in IPv6.*

Kaloxylos, A., Vali, D., Paskalis, S., Panagiotou, G., Gonianakis, G., & Zervas, E. (2006). Mobility Support for a QoS Aggregation Protocol. In *International Symposium on Communication Systems, Networks and Digital System Processing (CSNDSP),* Patra, Greece.

Kempf, J. (Ed.). (2002). RFC 3374: Problem Description: Reasons For Performing Context Transfers Between Nodes in an IP Access Network.

Koodli, R. (Ed.). (2008). RFC 5268: Mobile IPv6 Fast Handovers.

Kuo, G.-S., & Ko, P.-C. (2000). Dynamic RSVP for Mobile IPv6 in Wireless Networks. In *IEEE Vehicular Technology Conference (VTC) Spring*, Tokyo, Japan, *1*, 455–459.

Lee, K., Kim, M., Chanson, S., Yu, C., & Lee, J. (2003). CORP – A Method of Concatenation and Optimization for Resource Reservation in Mobile Internet. *IEICE Transactions on Communications, Special Issue on Internet Technology. E (Norwalk, Conn.)*, *86-B*(2), 479–489.

Manner, J., & Raatikainen, K. (2003). Localized QoS Management for Multimedia Applications in Wireless Access Networks. In IASTED Internet and Multimedia Systems and Applications (IMSA) (pp. 193–200), Honolulu, HI.

Manner, J., Toledo, A., Mihailovic, A., Munoz, H., Hepworth, E., & Khouaja, Y. (2002). Evaluation of mobility and quality of service interaction. *Computer Networks*, *38*(2), 137–163. doi:10.1016/S1389-1286(01)00255-9

Manner, ed., J., Karagiannis, G., & McDonald, A. (2009). *NSLP for Quality of Service signalling.* Internet Draft, work in progress. draft-ietf-nsis-qos-nslp-10.txt.

Moon, B., & Aghvami, H. (2001). RSVP Extensions for Real-Time Services in Wireless Mobile Networks. *IEEE Communications Magazine*, *39*(12), 52–59. doi:10.1109/35.968812

Moon, B., & Aghvami, H. (2004). Quality of Service Mechanisms in all-IP Wireless Access Networks. *IEEE Journal on Selected Areas in Communications*, *22*(5), 873–888. doi:10.1109/JSAC.2004.826924

Moskowitz, R., & Nikander, P. (2006). *RFC 4423: Host Identity Protocol (HIP) Architecture. Next Steps In Signaling IETF Working Group*. (n.d.). Retrieved from http://www.ietf.org/html.charters/nsis-charter.html

Pan, P., Schulzrinne, H., & Hahne, E. (2000). BGRP: A Tree-Based Aggregation Protocol for inter-Domain Reservations. *Journal of Communications and Networks*, *2*(2), 157–167.

Perkins, C. (1996). *RFC 2003: IP Encapsulation within IP*.

Perkins, C. (Ed.). (2002). RFC 3344: IP Mobility Support for IPv4.

Rosen, E., Viswanathan, A., & Callon, R. (2001). *RFC 3031: Multiprotocol Label Switching Architecture*

Saha, D. (2004, November-December). Mukherjee, A. Misra, I.S. Chakraborty, M. Subhash, N. (2004). *Mobility support in IP: a survey of related protocol. IEEE Network*, *18*(6), 34–40. doi:10.1109/MNET.2004.1355033

Sanda T., E., Fu, X., Jeong, S., Manner, J., & Tshofenig, H. (2009). *Applicability Statement of NSIS Protocols in Mobile Environments*. Internet Draft, work in progress. draft-ietf-nsis-applicability-mobility-signaling-12.txt.

Schulzrinne, H., & Hancock, R. (2009). *GIST: General Internet Signalling Transport*. Internet Draft, work in progress. draft-ietf-nsis-ntlp-19.txt.

Shen, Q., Lo, A., Seah, W., & Ko, C.-C. (2000). On Providing Flow Transparent Mobility Support for IPv6-based Wireless Real-Time Services. In *IEEE International Workshop on Mobile Multimedia Communications (MoMuC),* (pp. 2B41–2b46), Tokyo, Japan.

Sofia, R., Guerin, R., & Veiga, P. (2003). SICAP, A Shared-segment Inter-domain Control Aggregation Protocol. In *Workshop on High Performance Switching and Routing (HPSR),* (pp. 73–78).

Soliman, H., Castellucia, C., Malki, K. E., & Bellier, L. (2005). *RFC 4140: Hierarchical Mobile IPv6 Mobility Management (HMIPv6).*

Talukdar, A., Badrinath, B., & Acrarya, A. (2001). MRSVP: A Resource Reservation Protocol for an Integrated Services Network with Mobile Hosts. [WINET]. *Wireless Networks*, *7*(1), 5–19. doi:10.1023/A:1009035929952

Terzis, A., Wang, L., Ogawa, J., & Zhang, L. (1999). A Two-Tier Resource Management Model for the Internet. In Global Internet.

Thomas, M. (2002). *Analysis of Mobile IP and RSVP Interactions*. Internet Draft, work in progress. draft-thomas-nsis-rsvp-analysis-00.txt.

Tseng, C.-C., Lee, G.-C., Liu, R.-S., & Wang, T.-P. (2003). HMRSVP: A Hierarchical Mobile RSVP Protocol. [WINET]. *Kluwer Wireless Networks*, *9*(2), 95–102. doi:10.1023/A:1021833430898

Vali, D., Paskalis, S., Kaloxylos, A., & Merakos, L. (2004). A Survey on QoS Signaling. *IEEE Communications Surveys and Tutorials, 6*(4).

ADDITIONAL READING

Armitage, G. (2000). *Quality of Service in IP Networks: Foundations for a Multi-Service Internet.* MTP.

Bernet, Y. (2000). The Complementary Roles of RSVP and Di erentiated Services in the Full-Service QoS Network. *IEEE Communications Magazine*, *38*(2), 154–162. doi:10.1109/35.819910

Campbell, A., Gomez, J., Kim, S., Wand, C.-Y., Turanyi, Z., & Valko, A. (2002). Comparison of IP Micromobility Protocols. *IEEE Wireless Communications*, *9*(1), 72–82. doi:10.1109/MWC.2002.986462

Castellucia, C. (2000). *Extending Mobile IP with Adaptive Individual Paging: A Performance Analysis*. In IEEE International Symposium on Computers and Communications (ISCC). Antibes, France.

Chen, W.-T., & Huang, L.-C. (2000). RSVP Mobility Support: A Signalling Protocol for Integrated Services Internet with Mobile Hosts. In IEEE/ACM INFOCOM (pp. 1283-1292). Tel Aviv, Israel.

Chiruvolu, G., Agrawal, A., & Vanderhoute, M. (1999). *Mobility and QoS Support for IPv6-based Real-time Wireless Internet Traffic*. In IEEE International Conference on Communications (ICC). Vancouver, Canada.

Clark, D., Shenker, S., & Zhang, L. (1992). Supporting Real-Time Applications in an Integrated Services Packet Network: Architecture and Mechanism. In ACM SIGCOMM (pp. 14-26). Baltimore, MA, USA.

Csaszar, A., Takacs, A., Szabo, R., & Henk, T. (2005). Resilient Reduced-State Resource Reservation. *Journal of Communications and Networks*, *7*(4), 509–524.

Huang, C., & Lee, C. (2007). Layer 7 Multimedia Proxy Handoff Using Anycast/Multicast in Mobile Networks. *IEEE Transactions on Mobile Computing*, *6*(4), 411–422. doi:10.1109/TMC.2007.55

Huang, N.-F., & Chen, W.-N. (2003). RSVP Extensions for Real-Time Services in Hierarchical Mobile IPv6. [MONET]. *Kluwer Mobile Networking and Applications*, *8*, 625–634. doi:10.1023/A:1026022325876

Ji, P., Ge, Z., Kurose, J., & Towsley, D. (2003). A Comparison of Hard-state and Soft-state Signaling Protocols. In ACM SIGCOMM (pp. 251-262). Karlsruhe, Germany.

Karagiannis, G., & Rexhepi, V. (2000). *A Framework for QoS & Mobility in the Internet Next Generation*. Technical Report TR-CTIT-00-10, Centre for Telematics and Information Technology (CTIT), University of Twente, Enschede, The Netherlands. http://www.ub.utwente.nl/webdocs/ctit/1/00000024.pdf.

Kurose, J., & Ross, K. (2000). Computer Networking: A Top Down Approach Featuring the Internet (2nd Ed.). Addison-Wesley.

Leu, S.-J., & Chang, R.-S. (2003). Integrated Services Mobile Internet. [MONET]. *Kluwer Mobile Networking and Applications*, *8*(6), 635–642. doi:10.1023/A:1026074309946

Liu, C., Qian, D., Liu, Y., & Xiao, K. (2004). A Framework for End-to-End QoS Context Transfer in Mobile IPv6, Volume 3260 of Lecture Notes in Computer Science, PWC 2004, I. Niemegeers and S. Heemstra de Groot (eds.) (pp. 466-475). Springer Verlag.

Ma, W., & Fang, Y. (2004). Dynamic Hierarchical Mobility Management Strategy for Mobile IP Networks. *IEEE Journal on Selected Areas in Communications*, *22*(4), 664–676. doi:10.1109/JSAC.2004.825968

NSF-NetWorkshop 2003 (2003). *Report of the National Science Foundation Workshop on Fundamental Research in Networking*. http://www.cs.virginia.edu/~jorg/workshop1.

Park, S. (2006). *DiffServ Quality of Service Support forMultimedia Applications in Broadband Access Networks*. In International Conference on Hybrid Information Technology (ICHIT). Cheju, Korea.

Perkins, C. (1998). Mobile IP: Design Principles and Practices. Addison-Wesley.

Reinbold, P., & Bonaventure, O. (2001). *A Comparison of IP Mobility Protocols*. In IEEE Symposium on Communications and Vehicular Technology in the Benelux (SCVT).

Saltzer, J., Reed, D., & Clark, D. (1984). End-To-End Arguments in System Design. *ACM Transactions on Computer Systems*, *2*(4), 277–288. doi:10.1145/357401.357402

Solomon, J. (1997). Mobile IP the Internet Unplugged. Prentice Hall.

Stevens, W. R. (1994). TCP/IP Illustrated, Volume 1. Addison Wesley.

Taha, A., Hassanein, H., & Mouftah, H. (2005). Extensions for Internet QoS Paradigms to Mobile IP: A Survey. *IEEE Communications Magazine*, *43*(5), 132–139. doi:10.1109/MCOM.2005.1453434

Terzis, A., Wang, L., Ogawa, J., & Zhang, L. (1999). A Two-Tier Resource Management Model for the Internet. In Global Internet.

Thomas, M. (2001). *Analysis of Mobile IP and RSVP Interactions*. Internet Draft, work in progress. draft-thomas-seamoby-rsvp-analysis-00.txt.

Tseng, C.-C., Lee, G.-C., Liu, R.-S., & Wang, T.-P. (2003). HMRSVP: A Hierarchical Mobile RSVP Protocol. [WINET]. *Wireless Networks*, *9*(2), 95–102. doi:10.1023/A:1021833430898

Chapter 9
Management, Monitoring and QoS in Multi-Cell Centralized WLANs

Filippo Cacace
Università Campus Bio-Medico di Roma, Italy

Giulio Iannello
Università Campus Bio-Medico di Roma, Italy

Luca Vollero
Università Campus Bio-Medico di Roma, Italy

ABSTRACT

Large deployments of access points in wireless local area networks (WLANs) based on the IEEE 802.11 standard require management, configuration and control mechanisms. Centralized WLANs are defined as multi-cell wireless access networks that implement some of these functions in a centralized manner. In this chapter the authors illustrate how the mechanisms designed for the management of centralized WLANs can also be used for monitoring parameters related to QoS support and for pursuing QoS goals. They describe the Control and Provisioning Wireless Access Protocol (CAPWAP), a recent IETF standard for the management of centralized WLANs which is currently in the final stages of the definition process, its implementation for the existing types of centralized WLANs, and its use for monitoring and QoS management. The authors discuss the QoS goals that can be pursued in this framework, such as access control, load balancing, cell resizing, and Medium Access Control parameters adaptation, as well as the algorithms and strategies that can be used to fulfill them.

INTRODUCTION

The notion of Mobile Internet relies on scenarios where mobile users may have access anytime and anywhere to conventional and emerging Web applications requiring multimedia and interactive communications. Wireless access networks are the key element to implement these scenarios and much work has been done in recent years to develop and improve wireless technologies, including the introduction of support for mobility and Quality of Service. In this respect, special interest has been paid

DOI: 10.4018/978-1-61520-771-8.ch009

to wireless local area networks (WLANs) based on the IEEE 802.11 standard (I.S. 802.11, 2007), that exhibits characteristics useful to pursue the goals of the Mobile Internet, including a large number of installations in a wide range of contexts.

Three main trends have emerged in the last years in the evolution of wireless access networks. The first one is the deployment of large access networks based on wireless technologies, with the aim of providing uniform coverage and connectivity to wider areas. This aim has been pursued either by the installation of new large access networks, or by integrating existing networks in a single administrative federation. A major problem in this respect is the lack of standard configuration tools to manage the access network, and the consequent need for inter-operable solutions. A second trend is the integration of networks based on heterogeneous technologies to enhance connection opportunities. The most cited case is the integration of cellular data networks with IEEE 802.11 WLANs. This trend has triggered much research effort in the area of mobility management, specifically vertical hand-off among heterogeneous networks, and in the area of adaptive behavior to variations of network conditions. The third trend is the enhancement of the service provided by wireless access networks with the introduction of features like security, fast and transparent mobility support, quality of service, access control, accounting, etc.

In this chapter we discuss how the introduction of a network management protocol specifically designed for wireless networks can be of help in pursuing the above goals. To this end, in the rest of the chapter we describe the current proposal for the CAPWAP protocol and its current implementations, together with its utilization as a tool to monitor and manage wireless access networks. An attractive feature of this approach is that the resulting schema can be considered as a framework in which specific solutions to individual problems can be integrated to provide a global enhancement in the functionalities of wireless access networks.

BACKGROUND

The Rationale Behind Centralized WLAN Architecture

Born as a wireless extension of small office or home networks, WLANs have experimented a growing success as access networks that are efficient and easy to deploy Their popularity has triggered the deployment of larger wireless networks in areas such as university campus, corporate buildings and also metropolitan areas. The early implementations of a WLANs followed a pattern known as *Autonomous WLAN architecture*: a single Access Point (AP), or a small collection of Access Points, related to some geographically restricted area such as an office or a building. An autonomous network is manually configured, it defines its own access rules and it is substantially unaware of other nearby WLANs. However, the management of a large multi-cell WLAN as a collection of separate autonomous networks is both complex and inefficient. On one side, for lack of coordinated management functionalities, there is the need to manually configure each cell, which may prove unfeasible or too expensive. On the other hand, the problems or opportunities due to the interaction of the cells cannot be managed. A classical example is the problem of planning the frequencies used by the Access Points in a large WLANs: it is intuitive that the problem of reducing the interference among overlapping channels cannot be solved at each Access Point since it requires some kind of global coordination. Another typical issue is related to the load balancing among adjacent cells. In situations when many users are concentrated in a single cell (think of a classroom that is using wireless connectivity for some task), it makes sense to balance the number of associated users by moving some of them to adjacent cells. This in turn may require to increase the transmission power of nearby Access Points while reducing the power of the Access Point of the original cell to facilitate the migration. A positive outcome of this strategy is not only to improve the service offered in presence of unexpected concentrations of users, but also to decrease the need of the over-provisioning of wireless resources, thus leading to less expensive deployments.

The introduction of WLANs belonging to the *Centralized WLAN architecture* family stems from this need to enable network-wide monitoring, improve management scalability, and facilitate dynamic configurability. In order to clarify the issues related to centralized networks and the differences with respect to autonomous networks we briefly summarize the network functions and services of the original IEEE 802.11 standard. A WLAN architecture can be considered as a type of cellular architecture, where each cell is called Basic Service Set (BSS) and is controlled by a base station called Access Point. Each network has a name called Service Set Identifier (SSID) which is advertised by the AP in special control messages called beacons. When two or more APs using the same SSID are connected via a broadcast layer 2 network, called Distribution System (DS), an Extended Service Set (ESS) is created. The standard also defines a set of networking services, which are categorized into station services and distribution services. Specifically, station services are the Authentication, Deauthentication, Confidentiality, and MSDU Delivery services, while the distribution services include Association to the access point, Disassociation, Reassociation, Distribution in the whole ESS, Integration towards non-802.11 networks. Additional Medium Access Control (MAC) services are defined in the standard for optimizing and protecting the use of the wireless resources by means of rate adaptation, quality of service differentiation, encryption/decryption, and so on.

Contrary to the Autonomous WLAN Architecture, where the IEEE 802.11 functions and network control functions are all implemented within a single physical device, the Centralized WLAN Architecture employs one or more centralized controllers, called *Access Controllers* (Yang, 2005), referred to in the following as ACs. In short, the Centralized Architecture distributes between two categories of devices the logical functions that in the IEEE 802.11 standard are broadly described under the term "Access Point". For this reason, in the context of the Centralized architecture, the base station that provides to the mobile stations the wireless connectivity is referred to as *Wireless Termination Point* (WTP) instead of Access Point. The Access Controller and the WTPs are connected either trough a direct connection, or a layer 2 switched, or a layer 3 routed network. The Access Controller exchanges configuration and control information with the WTP devices, allowing the management of the network from a centralized point. The forth coming protocol used for the communication among the Access Controller and the WTPs is called *and Provisioning Wireless Access Protocol* (CAPWAP) and it is currently undergoing the final stages of definition (Calhoun, 2008). Designs of the Centralized WLAN Architecture family do not presume that the Access Controller necessarily intercedes in the data plane to/from the WTP, even though this is not excluded. Even if the Access Controller is a single logical entity, its functionalities may of course be distributed across multiple physical devices, due to their different nature in terms of resource requirement (such as CPU, storage, etc.). The use of multiple Access Controllers can also increase the resilience of the network, avoiding the presence of a single point of failure.

Since the introduction of the Centralized Architecture was originally motivated by the practical need of providing configuration tools and other value added services to the network administrators, its definition has come after the diffusion of proprietary solutions developed by vendors of network services. The increased demand for monitoring and consistent configuration of large wireless networks has resulted in a set of 'value-added' services provided by the various vendors, most of which share common design properties and service goals. Nonetheless, these proprietary solutions differed for the level of centralization they offer, which is a direct consequence of the flexibility characterizing in the IEEE 802.11standard the implementation of the logical functions attributed to the AP. Needless to say, the protocols developed by vendors for the communication among the Access Controller and the WTPs were also not inter-operable, and this led to the standardization effort carried on by a IETF Working Group that focused on analyzing the existing solutions and resulted in the ongoing definition of the CAPWAP protocol.

In (Yang, 2005) the solutions belonging to the Centralized Architecture family are categorized in three distinct groups according to their level of centralization, namely the *Local MAC*, *Split MAC*, and *Remote MAC* approaches. The naming of Local MAC, Split MAC, and Remote MAC reflects how the functions, and especially the 802.11 MAC functions, are mapped onto the network entities. Local MAC indicates that the MAC functions stay intact and local to WTPs, while Remote MAC denotes that all the MAC has moved away from the WTP to a Access Controller (This solution is restricted to very few industrial applications and it will be not considered further). Split MAC shows the MAC being split between the WTPs and Access Controller. Typically, Split MAC designs choose to put real-time functions on the WTPs while leaving non real-time functions to the Access Controllers. IEEE 802.11 does not clearly specify what constitutes real-time functions versus non real-time functions, and so a clear and definitive separation line does not exist. However, all Split MAC designs agree on the characterization of the majority of MAC functions. For example, wireless channel access mechanisms are always regarded as real-time functions.

- The main motivation of the Local MAC architecture model is to offload network access policies and management functions to the Access Controller without splitting the IEEE 802.11 MAC functionality between WTPs and Access Controller. The whole IEEE 802.11 MAC resides on the WTPs locally, including all the IEEE 802.11 management and control frame processing for the stations. On the other hand, information related to management and configuration of the WTP devices is communicated with a centralized Access Controller to facilitate management of the network and maintain a consistent network-wide configuration for the WTP devices.
- The main idea behind the Split MAC architecture is to implement part of the IEEE 802.11 MAC functionality on a centralized Access Controller instead of the WTPs, in addition to providing the required services for managing and monitoring the WTP devices. In the Split MAC architecture, the WTP terminates the infrastructure side of the wireless physical link, provides radio-related management, and also implements time-critical functionality of the IEEE802.11 MAC. In addition, the non real-time management functions are handled by a centralized Access Controller, along with higher level services, such as configuration, QoS, policies for load balancing, and access control lists.

As shown in Table 1, the key distinction between Local MAC and Split MAC relates to non real-time functions: in Split MAC architecture, the Access Controller terminates IEEE 802.11 non real-time functions, whereas in Local MAC architecture, the WTP terminates the IEEE 802.11 non real-time functions and consequently sends appropriate messages to the Access Controller. There are several motivations for taking the Split MAC approach. The first is to offload functionality

Table 1. Split MAC and local MAC comparison

Function	Location	
	Split MAC	Local MAC
Distribution Service	AC	WTP/AC
Integration Service	AC	WTP
Beacon Generation	WTP	WTP
Power Management	WTP	WTP
Association/disassociation	AC	WTP
IEEE 802.11 QoS	AC	WTP/AC
Scheduling	WTP/AC	WTP
Queuing	WTP	WTP
IEEE 802.11 RSN	AC	AC
IEEE 802.11X/EAP	AC	AC
RSNA key management	AC	AC
IEEE 802.11 Encryption/decryption	WTP/AC	WTP

that is specific and relevant only to the locality of each BSS to the WTP, in order to allow the Access Controller to scale to a large number of 'light weight' WTP devices. Moreover, real-time functionality is subject to latency constraints and cannot tolerate delays due to transmission of IEEE 802.11 control frames (or other real-time information) over multiple-hops. Another consideration is cost reduction of the WTP to make it as cheap and simple as possible. Finally, moving functions like encryption and decryption to the Access Controller reduces vulnerabilities from a compromised WTP, since user encryption keys no longer reside on the WTP. As a result, any advancement in security protocol and algorithm designs do not necessarily obsolete the WTPs; the Access Controller implement the new security schemes instead, which simplifies the management and update task. Additionally, the network is protected against LAN-side eavesdropping.

The CAPWAP Protocol: Features, Design and Implementation Issues

The Control And Provisioning of Wireless Access Points working group (Calhoun, 2008) is a recent effort of IETF aiming at defining an inter-operable protocol, enabling an Access Controller to manage and control a collection of possibly heterogeneous WTPs. Although originating from IEEE 802.11 architectures, the CAPWAP specifications aim at being independent of a specific WTP radio technology. The goals explicitly stated in current specifications of CAPWAP are the following:

- centralize authentication and policy enforcement functions for a wireless network;
- move processing away from the WTPs, leaving there only time critical functions;
- provide a generic encapsulation and transport mechanism.

Currently the CAPWAP protocol defines the communication and general management functions among the Access Controller and WTPs. CAPWAP control and data messages are sent using UDP, over separate UDP ports, and secured using DTLS, the Datagram Transport Layer Security (Dierks, 2006).

CAPWAP defines a discovery protocol for the automatic association of WTPs to the Access

Controller. As soon as the WTP is turned on, it sends a Discovery Request message (*Discovery* phase). Any Access Controller receiving this request responds with a Discovery Response message. Hence, the WTP selects the Access Controller, if any responded, with which it wants to interact and it establishes a DTLS session with it. Once the DTLS session has been established, both devices exchange their configurations and capabilities. The WTP is then ready to send and receive CAPWAP messages to/from the Access Controller (*Run* phase).

The CAPWAP protocol transport layer introduces resiliency using a request/response paradigm, where timeouts schedule retransmissions when a response does not follow a certain request. Two types of payload may be managed by this transport protocol: CAPWAP data messages and CAPWAP control messages. CAPWAP data messages encapsulate wireless frames forwarded by the WTP to the Access Controller or by the Access Controller to the WTP. CAPWAP control messages are CAPWAP management, control or monitoring messages exchanged among Access Controller and WTPs.

The CAPWAP protocol has been designed to accommodate the needs of any wireless technology. Indeed, the specifications in (Calhoun, 2008) do not assume any specific message related to any particular wireless technology. The implementation of CAPWAP for a specific wireless technology is called a "binding". For instance, (Calhoun, 2008b) defines the binding for IEEE 802.11 WLANs. Specifically, (Calhoun, 2008b) considers two of the Centralized Architectures described in the previous section: Split MAC and Local MAC. In a CAPWAP Split MAC architecture, the distribution and the integration services reside on the Access Controller, whereas in a Local MAC architecture, the integration service exists on the WTP, while the distribution service may reside on either the WTP or the Access Controller. The distribution service enables the MAC layer to transport data units between stations, when those stations cannot communicate directly over a single instance of the wireless medium. The integration service enables the delivery of data between IEEE 802.11 and non-IEEE 802.11 devices.

The Local MAC mode of operation allows for the data frames to be either locally bridged, or tunneled as 802.3 frames. When the distribution service resides on the Access Controller, frames generated by end-user wireless devices are not tunneled to the Access Controller in their native format, but encapsulated as 802.3 frames. When the distribution service resides on the WTP, frames generated by end-user wireless devices are locally bridged. In both cases, the layer 2 wireless management frames are processed locally by the WTP, and then forwarded to the Access Controller.

In Split MAC mode the distribution and integration services reside on the Access Controller and therefore all layer 2 wireless data and management frames are encapsulated by the WTP via the CAPWAP protocol and directly forwarded to the Access Controller. Real-time IEEE 802.11 services, including the Beacon and Probe Response frames, are handled on the WTP. All remaining IEEE 802.11 MAC management frames are supported on the Access Controller, including the Association Request frame which allows the Access Controller to be involved in the access policy enforcement portion of the IEEE 802.11 protocol.

With respect to the IEEE 802.11 QoS functionalities, the Local MAC architecture requires the WTP to support them. In Split MAC mode, the queuing function is also in charge of the WTP, whereas the real time scheduling can be managed either by the WTP or by the Access Controller. For both architectures, the IEEE 802.1X and RSNA Key Management functions reside in the Access Controller. Therefore, the WTP needs to forward all IEEE 802.1X-RSNA Key Management frames to the Access Controller and forward the corresponding responses to the end user device.

When needed, the transport of IEEE 802.11 frames is realized using the CAPWAP data messages encapsulation rules. IEEE 802.11 header and payload are encapsulated, while the FCS checksum is excluded. An optional CAPWAP header may be added. When the frame is encapsulated by the WTP, the optional header contains information on the Received Signal Strength Information, the Signal to Noise Ratio and the data rate used by the sending station. When the frame is encapsulated by the Access Controller, the optional header informs the WTP about the WLAN identifier to be used when sending the frame. As explained above, encapsulation is not used for data messages in a Local MAC architecture, since in this case the WTP acts a traditional AP and forwards the data in a frame with a new MAC layer header (usually, a IEEE 802.3 header).

CAPWAP control messages for IEEE 802.11 are enriched by the introduction of specific information for radio control, management and monitoring. Moreover, special control frames for the Quality of Service management are defined in (Calhoun, 2008b). The Quality of Service management is based on the Enhanced Distributed Channel Access mechanism (EDCA), originally defined in the IEEE 802.11e standard and recently included in the IEEE 802.11 standard (I.S. 802.11, 2007).

EDCA prioritizes traffic according to four Access Categories: *Voice*, *Video*, *Best Effort* and *Background*. The priority of each Access Category is controlled by five MAC layer parameters: *AIFS*, *CWmin*, *CWmax*, *TXOPlimit* and *q_depth*. To be more specific, each Access Category within each station implements a slotted CSMA/CA channel access protocol. The *AIFS* defines the fixed waiting time for carrier sensing that a specific Access Category has to use before any attempt to transmit. The *CWmin* and *CWmax* control the exponential backoff algorithm implemented by each Access Category. They aim at mitigating network congestion. The *TXOPlimit* is a maximum channel holding time. When an Access Category transmits successfully, that Access Category can transmit other frames without contending with other stations until this time has expired. Eventually, the *q_depth* parameter is not directly specified by EDCA, but it represents the maximum number of packet that may be buffered in each Access Category queue. It can be used to control the trade-off between delay and loss in the sending queue.

To the best of our knowledge, the only complete open source CAPWAP implementation is *OpenCAPWAP*, reported in (Bernaschi et el., 2009), that we briefly describe.

The OpenCAPWAP implementation consists of two Linux applications, one running on the Access Controller and one running on the WTPs. Each WTP acts as a client of an Access Controller. Each Access Controller manages both the communication with WTPs already registered and new requests sent by WTPs not-yet registered. The life-cycle of a WTP-Access Controller session is implemented with a Finite State Machine (FSM), as defined in the protocol specification (Calhoun, 2008, section 2.3). Use of DTLS by the CAPWAP protocol results in the juxtaposition of two nominally separate yet tightly bound state machines. The DTLS and CAPWAP state machines are coupled through an API consisting of commands and notifications. Certain transitions in the DTLS state machine are triggered by commands from the CAPWAP state machine, while certain transitions in the CAPWAP state machine are triggered by notifications from the DTLS state machine. The same Finite State Machine is defined for both the WTP and the Access Controller, although some states and transitions are implemented only on either the WTP or the Access Controller.

The applications that implement the WTP and the Access Controller use a multi-threaded programming model (See Figure 1) to achieve a reasonable trade-off between modularity and efficiency. The Access Controller application uses a *receiver thread* that is in charge of receiving any packet arriving from the WTPs, and one *session manager thread* for each of the WTPs with an ongoing session. Upon receiving a message, the receiver thread passes it to the corresponding session manager, or, if it is the first message sent by a WTP when it want to establish a CAPWAP session with the Access Controller, a new session manager thread is created. The number of threads used by the WTP application is always limited to three. The *principal thread* is in charge of establishing a CAPWAP session with an available Access Controller. After the start of the session, a *receiver thread* is created, whose task is to manage the requests sent by the Access Controller. The principal thread, to the contrary, is in charge to send to the Access Controller the requests of the WTP and to receive the corresponding answers. A third thread, the *receiver-from-STA thread* is required to intercept IEEE 802.11 management frames sent by the wireless stations to the WTP, that need to be forwarded to the Access Controller, like, for example, the *Associations Requests* frames.

The CAPWAP protocol, like other existing control and management protocols (e.g., the SNMP), relies on UDP as transport protocol. Since UDP does not guarantee reliable communications, the implementation described in (Bernaschi et al., 2009) uses, as required by the CAPWAP protocol, a retransmission mechanism for requests whose response does not arrive within a specified timeout. As to the security requirements imposed by the CAPWAP protocol, in case of timeouts and consequent re-transmission of packets it is necessary to cipher them again to avoid that the receiver could assume that a *replay attack* is in progress.

Two components of the WTP application are, to some extent, platform dependent. The first is the module in charge of "capturing" the IEEE 802.11 management frames. The second is the module in charge of modifying one or more parameters of the WTP when the Access Controller sends a *Configuration Update Request*. Both the access to received management frames and the enforcing of the configuration update require low level operations that depend on the specific driver used. In absence of standard interfaces to perform these operations, it can be expected that some specific modules of the CAPWAP implementation must be rewritten for use on new wireless platforms. The CAPWAP implementation has been reported to work with both wireless cards with well documented open source drivers like the MADWiFi diver as well as with cards with small proprietary drivers, and this supports the claim that the operations needed by the CAPWAP implementation are available on any wireless platform used as WTP.

A final remark is deserved by the problem of the question of Split MAC vs Local MAC. Presently, the implementation of (Bernaschi et al., 2006) is Local MAC, and it is reasonable to forecast that open source implementations of Split MAC will

Figure 1. Software architecture of AC (left) and WTP (right) applications

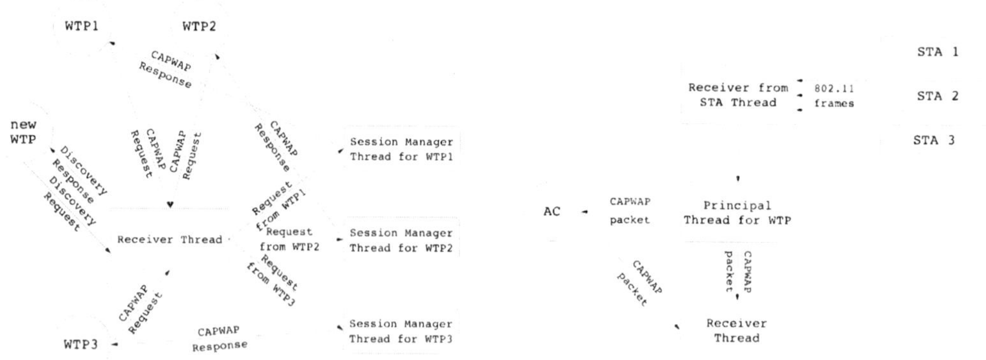

not exist in the short term. As detailed previously, in Local MAC mode the whole 802.11 MAC resides on the WTPs, including all the 802.11 management and control frame processing for the stations whereas in Split MAC mode only real-time MAC functions should be managed by WTPs. Such distinction appears quite sharp but actually in the IEEE 802.11 specifications there is no a clear definition of which 802.11 MAC functions are considered real time, so that each vendor has decided to use his own interpretation. The major problem, as already mentioned above, is the lack of standard interfaces that makes some operations on a WTP platform-dependent. However the difficulties in the Split MAC case are more serious since on many wireless devices there is no way at all (not even proprietary) to prevent them from carrying out tasks that in a Split MAC architecture should be in charge of the Access Controller. As a consequence, we expect that full support of the Split MAC mode for heterogeneous environments will be possible, probably, only when vendors understand the issue and address it.

QoS Issues in WLANs and Multi-Cell WLANs

Quality of Service support is increasingly important in wireless networks of any nature, due to the growing demand of services like VoIP, streaming, gaming and videoconferencing and to the limited capacity of wireless networks. They are especially challenging for WLANs of the IEEE 802.11 family, that use distributed and contention based algorithms like DCF and EDCA as channel access mechanisms.

Since the focus of our work is on multi-cell centralized networks, we review the QoS problems and solutions for the basic case of single-cell WLAN, and then move to the more general situation of a multi-cell WLAN belonging to the Centralized Architecture family, with the aim of explaining the additional QoS issues that are introduced in this case.

As it is well known, the distributed algorithm for the channel access control of the original IEEE 802.11 standard, DCF, has no traffic differentiation mechanism. The answer to QoS issues was delegated to the polling channel access mechanism, named PFC (Point Coordination Function), that however received scarce attention in real implementations. The recently introduced EDCA provides the mechanism for traffic differentiation, while retaining the distributed and contention-based features of DCF. As detailed in the previous section, EDCA introduces the concept of access category, and it allows a flexible setting of the channel access parameters that differentiates the behavior of access categories. However, the statistical mechanism for traffic differentiation of EDCA provides less predictable performance than a reservation-based mechanism and also suffers from network congestion. When the traffic load increases, EDCA cannot provide strict QoS guarantees. In addition, it is neither trivial to predict the performance of the traffic on some access category given an assignment of the EDCA parameters, nor it is easy to specify the adjustment to these parameters needed to solve a given QoS problem.

A number of QoS issues have been considered in the recent literature. They can be schematically outlined as follows:

- **Admission control for real-time traffic.** Real-time traffic is inelastic and without admission control an excess of offered load would downgrade the quality of the communications for all the stations in a wireless cell.
- **Protection of high priority traffic against low priority traffic.** QoS guarantees for high priority traffic such as real-time flows should not be jeopardized by sudden surges of best-effort traffic.
- **Fairness between high priority and low priority traffic.** Differentiation mechanisms can provide QoS guarantees to high

priority traffic but they should not starve low priority users.

- **Fairness between up-link and down-link traffic**. Contention-based mechanisms provide each station with the same channel access opportunities, but the down-link traffic is transmitted only by the base station. Hence, fairness between up-link and down-link traffic is an issue.

Specific solutions have been proposed for all the above issues, even though there is not at present a convincing unified framework for pursuing all these QoS goals, that are moreover conflicting with each other. The solutions proposed for QoS support in a single-cell WLAN resort to a variety of methods, that we group for convenience in four categories:

- **Analytical models**. Analytical models have been proposed for EDCA (Banchs & Vollero, 2006; Bellalta et al., 2006; Engelstad & Østerbø, 2005; Harsha, Kumar & Sharma, 2006; Kong et al., 2004; Robinson & Randhawa, 2004; Xiong & Mao, 2007), but the complexity of the protocol, the difficulty of modeling all the relevant issues and the often unpredictable behavior of a real wireless channel make these efforts more interesting from a theoretical point of view rather than for practical usage in real scenarios. Theoretical models are however valuable to provide insights on the impact of each MAC parameter on the traffic differentiation. For example (Bellalta et al., 2006) derives from the analytical model an optimization algorithm for the MAC parameter setting. Some of these analysis (Bellalta et al., 2006; Clifford et al., 2006; Engelstad & Østerbø, 2005) are also applicable to the case of non-saturated sources.
- **Measurement based methods**. Measurement-based methods were originally proposed in the context of wired IP networks to perform admission control of real-time flows with the goal of providing bounded delay and loss rate with high channel utilization. More recently, several probing-based admission techniques for EDCA wireless networks have been proposed (Gao & Ngan, 2005; Gu & Zhang, 2003; Pong & Moors, 2003; Xiao & Li, 2004; Xiao, Li & Li, 2007; Zhai, Chen & Fang, 2006). Some of these methods do not use probing, but rather define indicators related to the channel occupancy to perform predictions and take decisions about the impact of new flows. One limit of these approaches is that the additional overhead due to increased contention is difficult to estimate *a priori*: in general the amount of time wasted in collisions does not grow linearly with the load offered to the channel, thus it is difficult to estimate with precision the impact of a new flow. An integrated measurement-based scheme for admission of real-time traffic and bandwidth sharing control for voice/video/data traffic is presented in (Xiao, Li & Li, 2007). The approach requires that QoS-enabled AP and stations measure the time spent in transmitting and colliding. A *guard* period is introduced to prevent bandwidth allocation from over-provisioning and for best-effort traffic protection.
- **Scheduling algorithms**. Scheduling algorithms approaches propose modifications to the EDCA protocol (Banchs & Perez, 2002; Fan, Huang & Tseng, 2006) or implement custom service differentiators above the MAC layer to pursue QoS goals. The method to enforce fair channel sharing among stations presented in (Park, Kim, Choi & So, 2007) defines a service level manager (SLM) that is implemented on top of a scheduling based differentiation algorithm. SLM monitors the channel access

time of each station and adjusts the service level of packets dropping those exceeding the assigned fair share. The approach cannot directly control up-link flows, but it is proposed to use packet dropping at layer 3 as a control mechanism for TCP best effort traffic. These approaches are efficient and flexible, but they require modifications to the standard protocol and must be deployed on both the AP and the stations, and this hinders their practical deployment.

- **Dynamic setting of MAC parameters**. Dynamic tuning of EDCA parameters (Banchs & Perez, 2002; Dangerfield, Malone & Leith, 2006; El Housseini & Alnuweiri, 2005; Freitag, da Fonseca & de Rezende, 2006; Malli, et al., 2004; Ramos, Panigrahi & Dey, 2003; Romdhani, Ni & Turletti, 2003; Tanigawa, Kim, Tode & Murakami, 2006; Zhang & Zeadally, 2004) is a powerful technique to achieve QoS guarantees, but there is no general method for determining the optimal configuration of parameters in a given scenario. All the cited approaches solve some QoS problems but are not guaranteed to work in the general case.

The QoS issues and solutions discussed above are relevant for the multi-cell case in a Centralized Architecture. As a matter of fact the algorithms proposed for an Autonomous WLAN can be inherited by Centralized architectures, with the important different that now part of the algorithm is executed by the Access Controller. Of course, some of the cited approaches are more suited to a distributed architecture with a centralized manager than others. Moreover, in a Centralized Architecture there is more flexibility in deciding where to allocate the tasks related to the QoS management, with respect to an Autonomous network. These issues are discussed in the next section.

In a Centralized Architecture there are several additional issues related to QoS management that are less relevant in an Autonomous network:

- **Load balancing**. Several studies showed that in multi-cell scenarios users are often unevenly distributed in space and, hence, the number of associated users may vary widely from WTP to WTP. This may translate in an uneven load distribution which can severely degrade the quality of services that users experience, especially if some WTPs result highly congested. Load Balancing aims at mitigating this problem by forcing some users to roam toward less loaded neighbor WTPs, avoiding congestion.
- **Frequency planning**. Since the number of transmission channels that can be used simultaneously without cross-interference is very limited (for example in the case of IEEE 802.11b there are only 3 such channels) frequency planning is in itself a relevant configuration problem, with important consequences on the QoS that the network can guarantee. When configuring or upgrading large deployments of WTPs, the attribution of frequencies to each WTP may be a major problem in optimizing overall network performance. Optimal or sub-optimal frequency reuse is desirable to reduce interference among adjacent cells, but it is not always possible. An additional difficulty in this context is due to the possible presence of "external" WLANs that are not under the administrative control of the Access Control and that use some of the available channels.
- **Mobility management**. In many scenarios, user mobility is an important requirement: for example, in a corporate WLAN that supports VoIP on small portable devices, the users would expect a quality of service similar to the one experienced on mobile

phones when moving in the area covered by the WLAN. Due to hand-off latencies, both at layer 2 (a change of the WTP that is transparent at the network layer) and at layer 3 (a change of the IP address), this expectation would be disappointed in absence of mechanisms to support cell migration in presence of QoS requirements.

STRATEGIES AND MECHANISMS FOR QOS SUPPORT IN MULTI-CELL CENTRALIZED WLANS

In this section we present a framework for the QoS management in a Centralized WLAN architecture. As a first step we introduce a conceptual schema to identify the functionalities that are needed in this respect and their location in the network. We assume that one or more Access Controller and their respective WTPs are present in the network, and that they communicate through the CAPWAP protocol, as outlined above. The schema presented here can be broadly defined as measurement-based, in the sense that it assumes that some measurement on the performance of the network is available to take decisions about QoS issues. Even though analytical models, scheduling algorithms or other techniques can be used to implement the decisional part of the schema, the overall framework differs from a a purely analytical one because it uses a *reactive* approach similar to the feedback-driven stabilization of a dynamic controller. This approach has several attractive features. First, it can leverage results from previous work on analytical models, but, unlike approaches relying on models only, it does not require precise knowledge of the parameters characterizing the wireless link at both PHY and MAC layer. Second, it is robust to both unexpected or difficult to model workload and to variations of channel conditions. Third, it does not require modifications to the existing protocols and it is easy to implement using available hardware.

We do not present any specific model for the QoS specifications or guarantees, rather we generically require that each traffic flow can specify some requirements in terms of network resources (*traffic specifications*) as well as bounds on a set of traffic parameters (*QoS requirements*), like minimum throughput, loss ratio, average delay or jitter. For example, in the IEEE 802.11 standard (I.S. 802.11, 2007) a station can generate a ADDTS Request Frame that contains a TSPEC information element. A TSPEC information element may contain both traffic specifications, for instance "peak data rate", and QoS requirements, like "delay bound". The TSPEC parameters are compatible with the IP RSVP protocol for resource reservation in IP networks (Braden, R., ed., *et al.*, 1997).

Of course, traffic flows have different nature, and in some case it is easy to identify their resource requirements, whereas in others this can be rather difficult. For example, real-time flows are *inelastic*, in the sense that their throughput needs are approximately constant throughout flow existence, and their needs are therefore easy to identity. To the contrary, other kinds of traffic are *elastic*, in the sense that they are managed by means of adaptive algorithms, and this makes less meaningful the notion of guaranteed throughput: for example, the throughput of a TCP connection can be limited by factors that are not related to the availability of the wireless link, like, for instance, the rate at which the data are generated at the source. The same problem exists with best-effort traffic, that by definition has no QoS requirements, but for which it is usually necessary to reserve a certain amount of available resources. It is possible, also in this case, to extend the notion of resource requirements to include one or more classes of elastic/ best-effort traffic, and associate to each class a threshold of guaranteed throughput. In this way, a set of elastic flows are transformed into a unique inelastic class with a default QoS requirement of the kind "*best-effort traffic should be granted at least 800 kbps of traffic*". At the same time, it is necessary to use some heuristic or statistical inference to assess when the throughput of an elastic class is under the threshold due to the scarcity of network resources rather than to the lack of offered load.

In the three parts of this section, we first introduce the general schema and then present the issues, methods and references related to, respectively, the monitoring and the control functionalities.

A Framework for Multi-Cell QoS Monitoring and Management

In this section we describe an architecture to integrate QoS support, in the meaning outlined above, in a multi-cell WLAN belonging to the Centralized Architecture family. A conceptual overview of the reference schema is illustrated in Figure 2. Basically, the schema includes the following entities:

- a *Monitoring Module*, on the Access Controller, in charge of receiving information about relevant QoS parameters from the controlled WTPs. The information may be related to many parameters measured at the WTP, such as the transmitted and received throughput, average queue lengths at the WTP, number of packets dropped due to queue overflow or collisions, frequency of collisions, physical errors, interference on the channel used by the WTP, etc. The information provided by the WTP may be related to distinct entities (access categories, stations, wireless channels). The periodicity of the statistics can be set by the Monitoring Module, according to the entity to which they refer and to the situation on the channel state.
- A *Controller Module*, is the active entity that provides QoS support to the centralized WLAN. Conceptually it may be split in two kinds of controllers, one in charge of providing intra-cell QoS support, by means of admission control and MAC parameters setting, and the other in charge of inter-cell QoS support, which is concerned with problems related to frequency planning, load balancing and mobility. The Controller Module operates by issuing configuration commands to the WTP. These may refer to the MAC parameters to use for the Access Categories of the WTP or of the stations, to admission decisions concerning new flows, to the channel to be used in the cell, etc., and they are communicated through CAPWAP messages.

Figure 2. QoS monitoring and management in the centralized architecture

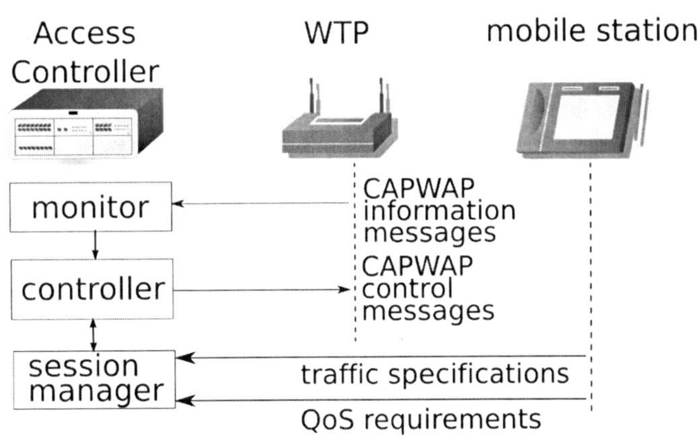

- A *Session Manager*, that interacts with the Controller Module and it is in charge of implementing the QoS policies of the WLAN. The Session Manager may require, for example, that all flows needing QoS support must previously declare their needs in terms of QoS requirements, and/or specify the features of the traffic flows they will generate (for example in terms of average of maximum throughput, packets per second, etc.). The Session Manager might also enforce the *Service Level Agreements* (SLAs) of a given category of users, by allowing or not them to send traffic on a high-priority Access Category, etc. The Session Manager does not interact with the WTPs, but only with the Controller Module. In essence, the Session Manager is in charge of fixing the goals and constraints of the WLAN management, whereas the task of the Controller Module is to choose the optimal configuration in the context fixed by the Session Manager.

This schema reflects a classical feedback approach, with the WTPs in the dual role of sensors-actuators and the Access Controller acting as the controller. It should be remarked that the advantage of the Centralized Architecture with respect to an Autonomous architecture is not only the possibility to perform intra-cell QoS optimization, but also that the control algorithms may be more complex without requiring expensive WTPs with large computational capacity.

Obviously, the conceptual schema outlined has many variants. In the first place, since the monitoring function is conceptually distributed between the Access Controller and the WTPs, the choice between Local MAC and Split MAC solution influences how monitoring is performed. In a Split MAC WLAN, many monitoring functions are performed directly at the Access Controller (for example, traffic monitoring), whereas the WTP collects only real-time information, for example about the length of the sending queues, or about the state of the wireless channel. Moreover, it is more coherent in this setting to let the WTP send only raw information, performing all the necessary elaboration at the Access Controller. Conversely, in a Local MAC WLAN, it is possible to decide to locate more advanced monitoring functions on the WTP. This means that WTPs execute a local monitoring application, with the aim of aggregating or filtering the raw data, and send to the Access Controller a more high-level description of the status of the wireless cell. At one end of the solution spectrum, the Control Module of the cell could even be co-located with the WTP, with the Access Controller in charge of executing modules for inter-cell QoS support only.

In the second place, the functionalities of the Control Module and its interactions with the Session Manager may rely on different solutions. As already remarked, a Control Module must exist in the Access Controller for each WTP, in order to perform local optimization tasks. Further Control Modules can be in charge of specific tasks related to the whole WLAN. For example it may be convenient to deploy, as separate modules, a Mobility Manager, to manage mobility among adjacent cells, a Frequency Planning module and a Load Balancing Module. The actions undertaken by these modules to perform inter-cell modifications could interfere with the activity of the local Control Module of the WTPs. For example, a station migration decided by the Load Balancing Module can make useless a MAC parameter setting decided by the Control Module of the overloaded cell to cope with channel saturation. This interference may cause the instability of the QoS control or lead to frequent and unnecessary changes of configuration. In order to solve this problem, it is necessary to design some form of coordination and hierarchy among the modules, or to resort to a central decision function, that has the ultimate responsibility of deciding which module will cope with each specific problem. To the best of our knowledge, there are no general solutions to this problem in the literature, and more research in this area is still needed. Whereas much work has been devoted to the solution of QoS issues in specific situations or with a given technique, the decision problem of which tool is the most effective in a given scenario has not been fully solved, with the result that commercial products usually recur to heuristic rules that in practice lead to dubious and often counterproductive effects.

Issues and Experiences in Multi-Cell QoS Monitoring

In this section we review approaches that allow to monitor at real-time the quality of service that the traffic in a wireless cell is experiencing. It should be remarked that the QoS monitoring that we consider is limited to the wireless portion of the WLAN. It is usually a reasonable assumption to suppose that the wireless link is the bottleneck and that the wired portion of the WLAN does not contribute to the QoS issues.

A first set of parameters that are useful to estimate is related to individual traffic flows[1].

Throughput. The amount of successfully data sent is in principle an easy to measure yet obviously fundamental parameter to monitor the QoS in a WLAN. Up-link throughput can be measured either at the Access Controller or at the WTPs, whereas down-link throughput is only available at the WTPs, since losses due to queue overflow or retransmissions must be considered. The information about aggregate data transmitted and received per Access Category is readily available by means of driver functions, but this is only sufficient for best-effort traffic monitoring, where the parameter of interest is usually the aggregate throughput. Throughput of individual flows to the contrary, it is necessary for flows with QoS requirements on throughput, and this requires more processing and memory and it is usually performed at the Access Controller. For elastic flows the instantaneous throughput is not necessarily meaningful, thus some form of filtering should be performed, for example by means of autoregressive filters.

Down-link delay. The down-link delay includes queuing delay and service delay. It may be accurately measured at the WTP for each Access Category, and two approaches are possible. The first one (Dangerfield, Malone & Leith, 2006) is to associate a time-stamp to individual packets that arrive at the queue, and use this value to measure the delay of each individual packet. The second one (Cacace, Iannello, Vellucci & Vollero, 2008) monitors the queue length at fixed intervals and it approximates the average delay in the interval with the expression

$$d = (1 + q_{AC})\, n_{AC} / \Delta t,$$

where q_{AC} is the queue length of the Access Category, n_{AC} is the number of packets sent by the access category in the interval, and Δt is the interval duration. The approximation consists in using a constant service time of the queue during the interval. This approach is less accurate, and it does not provide useful results when the queue is empty and the delay is low, but it is more simple and adequate for all practical purposes.

An important property of the down-link delay is that in principle it does not depend on the mobile station to which the packet is addressed, but only on the Access Category, since the predominant component of the delay is the queuing delay, which does not depend on the specific flow or packet size. This simplifies the monitoring of this delay, since the estimate is the same for all the flows.

The down-link delay is a crucial parameter for monitoring the QoS for at least three reasons. The first one is that, in most traffic scenarios, the WTP is the first to become saturated. The reason lies in the fact that down-link traffic is usually predominant over up-link one, and even when the two are comparable it is sent by one node only, namely the WTP. Conversely, up-link traffic is divided more or less equally among all the mobile stations. In real scenarios it is thus reasonable to assume that, using the same channel access parameters, the WTP is the first to experiment congestion. Consequently, down-link delay is a reliable indicator of the congestion of the wireless cell. The second motivation for measuring down-link delay is that other QoS parameters are related to it: for example, packet loss due to queue overflow is can only happen in situations with very large delays. A final motivation is that delay is the most common QoS requirement for real-time traffic.

Up-link delay. For the reasons outlined above, up-link delay is less crucial to assess the QoS experienced by traffic in a wireless cell. Nonetheless, strict QoS guarantees require up-link delay to be monitored. Since up-link traffic is sent by the mobile stations, there is no direct and general way to estimate its value without modifying the MAC protocol or performing active probes in the wireless cells. Since in practice both options are not feasible, indirect approaches can be used. In (Cacace & Vollero, 2009) the number of packets sent in each transmission opportunity is used to estimate the queue length and thus the average delay for each Access Category of the mobile station. The method is not computationally complex, but it requires to keep trace of flows for every station, thus it is better suited for execution on the Access Controller.

Loss rate. Losses on the wireless links have two causes: queue overflow and collisions. The first situation is more common for low-priority Access Categories that cannot have access to the channel, whereas the second situation happens when there are many stations competing for the channel access and the contention windows are set to a small value, causing many repeated collisions. For down-link traffic, both rates are easy to obtain from the driver of the wireless device. Conversely, it is very hard to have reliable estimates of loss rates on the mobile stations, for the same reasons for which it is difficult to estimate up-link delay. The only available method for estimating the loss rate due to queue overflow resorts to the estimate of the queue length on the stations following the approach of (Cacace & Vollero, 2009) and derive the probability that the queue is completely full. For losses due to collisions, it is possible to rely on the *retransmission bit* contained in the MAC header of IEEE 802.11 to estimate of the collision probability (see below), which can be easily related to the loss rate.

Jitter. Jitter is defined as the difference of the one-way trip time between a pair of consecutive packets of the same flow. The average jitter is the average of the jitter of all the pairs of consecutive packet across the measurement window. For traffic flows at constant packet rate, the jitter may be evaluated from the arrival time only. Obviously, here we are mainly concerned to the contribution to the jitter of the wireless portion of the access network. Thus, whereas for down-link traffic the delay estimation methods presented above may be sufficient to estimate the jitter, for up-link traffic the average of the variations of the arrival times can provide a reliable estimate of the jitter on a per-flow basis in the case of constant packet rate. For up-link flows at variable packet rates a direct estimate is not possible, and it may be necessary to resort to indirect approaches, for example using the time-stamp contained in higher layer headers (such as the RTP header).

Besides the parameters that are directly related to traffic flows, it is possible to monitor the parameters that are related to the state of the wireless channel.

Collision rate and physical errors. The rate of packets incorrectly received or sent it is again easy to monitor for down-link traffic, even though it is not possible to distinguish between collisions and physical errors (due to interference or other causes). Usually, the wireless card provide statistics about the number of sent packets and the number of retransmission, from which an estimate of the combined collision-physical error rate can be derived. The collision rate is a useful parameter to monitor because it can be related to the rate of packets discarded due to retransmissions. It must be noticed that the WTP can measure its own collision rate per Access Category, but this variable depends on its traffic load. In general, given a certain channel occupancy and in absence of physical errors, a station will see a larger collision rate when it access the channel with less frequency, and vice-versa. Since the WTP usually has a larger traffic load than the mobile stations, the collision rate measured at the WTP underestimates the collision rates at the stations. An indirect way of estimating the collision rate at the specific station is that of relying on the retransmission bit contained in the MAC header of the packets received from it. A global channel parameter that can be efficiently measured by the WTP, or, in fact, by any station on the channel, is the *conditional collision probability*, defined as the probability of experiencing a collision when transmitting on any available time slot. The available time slots are those following a transmission after an *AIFS* waiting time that depends on the Access Category, thus the conditional collision probability depends on the Access Category. This probability p can be estimated as proposed in (Bianchi, & Tinnirello, 2003) through the number of experienced collisions, C_c, and the number of busy slots C_b, in which an eventual packet transmission would have failed, divided by the number N of observed slots, $p = (C_c + C_b) / N$.

A practical issue to be consider in practice is that when the *TXOPlimit* of an Access Category allows to send more than one packet per channel access, it becomes important to distinguish between packets and channel access, because the collision rate must be referred to channel accesses, but the statistics provided by the wireless card driver usually are on a per packet basis.

Wireless interference. Information related to the physical state of the wireless channels can be measured by means of periodic scanning and reported to the Access Controller for frequency planning purposes. Relevant information include the signal level from nearby WTPs (either belonging to th same WLAN or not), and the noise level on the channel.

Signal-to-Noise ratio. Information about the quality level of frames received by the mobile stations can be collected at the WTP. This information is related to the PHY layer, and therefore it is in principle independent from the channel occupancy. The analysis of this information on a per station basis can be useful in two regards. The first one is related to mobility: when the Access Controller detects that signal level of a station becomes weaker, this information can be used by the Mobility Manager to raise a hand-off warning. In the second place, if the Access Controller has some knowledge about the topology of the network, in the sense of the physical position of the WTPs, it can use this information to decide which stations to migrate to a nearby WTP for load balancing purposes.

An important design parameter for the QoS monitoring is a set of values for the measurement windows. Since the measurement procedure implies some overhead it is important to choose measurement intervals that do not impair the efficiency of the WTP and that, at the same time, allow for a timely intervention of the Access Controller. The rate of variation in time of the parameters described so far is different, thus it is important to choose the appropriate value for each one. Moreover, as already remarked, the instantaneous values of many parameters may display wide variations, thus also some form of filtering is often needed. In general, run-time estimation is provided by simple mechanisms, such as AR (Auto Regressive) or ARMA (Auto Regressive Moving Average) filters. In practice, it is often used the estimator

$$p(t+1) = \alpha\, p(t) + (1 - \alpha)\, m(t+1),$$

where $p(t)$ is the smoothed estimate at the discrete time t, $m(t)$ is the measured value, and α is a memory parameter. More sophisticated approaches can be used to achieve a better accuracy/tracking ability trade-off. For example, (Bianchi, & Tinnirello, 2003) use an Extended Kalman Filter to estimate the number of competing stations in a wireless cell using a measurement of the conditional collision probability. This method estimates the number of stations that are actually competing for channel access, a figure that is in general less or equal to the number of associated stations.

The estimate of a suitable set of parameters related to the traffic and the state of the wireless channel can be used by the Control Module to deduce a global higher-level picture of the state and evolution of the quality of service in the wireless cell that it controls. This global picture must be expressed as a state of some formal QoS model, and the task of the Control Module is that of react to state changes in the appropriate way. This *QoS state* can in general be expressed with several service goals:

- **Fair allocation of network resources**. This goal is related to the fairness in allocating network resources expressed with respect to traffic categories (up-link vs down-link, high priority vs low-priority, etc.) and a given set of criteria on network parameters (for example, throughput).
- **Network utilization**. This goal concerns the network utilization, that should obviously tend towards the maximum possible fraction of the offered load. In particular, when admission control is active, a low value for the *blocking probability*, that is, the probability of rejecting a flow with QoS requirements, is an important goal for QoS management. In a multi-cell WLAN it is relevant not only the network utilization of the individual cells, but also the variance of this parameter across the cells, since load balancing should tend to reduce the differences across the network.
- **QoS guarantees for high-priority traffic**. This goal is related to satisfying the QoS guarantees of all the traffic flows with QoS requirements.

Issues and Experiences in Multi-Cell QoS Management

In this section, we consider the problem of implementing typical techniques for QoS management using the monitoring framework described so far. It is quite obvious that a QoS monitoring approach is useful to control the performance of a QoS management system of any nature. We now consider more specifically a QoS management system that is based on a monitoring part, in the sense that takes as input to pursue its goals the estimation of the QoS state provided by the monitoring system. In the remaining part of the section we therefore present measurement-based approaches to specific QoS management tasks, in the context of multi-cell WLANs.

Admission Control

Admission control is an important feature in a WLAN with QoS support, since the number of traffic flows with QoS guarantees that can admitted is limited either by network resources, or by constraints on the resource allocation, or by access policies. For instance, a flow of a given category may be rejected in order to maintain a fair proportion of traffic with other categories. Conversely, rejection due to access policies happens when a station starts an exceeding number of high-priority flows, or when the traffic specifications of the new flow do not match those agreed for the service level of the station.

The set-up of a new flow with QoS guarantees can follow different patterns. The station may be required or not to notify to some authorization entity its will to start the flow. In the first case, the station may be required or not to provide the traffic specifications for the new flow. It also possible that the flow is initiated from a correspondent node outside the WLAN. For this reason it is desirable for a mechanism of admission control to be sufficiently flexible to handle the different cases, even if the efficiency of the mechanism may vary in the different cases.

In a Centralized WLAN it is reasonable to assume that the Access Controller, and specifically the Session Manager module, takes care of performing admission control on the basis of the access policies, that consist of static information about the rights of the individual stations. Conversely, the Controller Module will be in charge of deciding if the new flow can be admitted for a given state of the wireless channel. Consequently, the decision taken by the Controller Module is more dynamic in nature since it reflects the current situation in the wireless cell.

In the context of the QoS monitoring and management architecture described previously, the admission control functionalities in the Controller Module take advantage of the information about the occupancy of the wireless cell reported by the Monitoring Module. The admission functionalities can be designed according to several approaches, that we briefly outline. It is worth remarking that the following methods are not mutually exclusive, and may be combined in a mixed approach.

- The information about the state of the wireless channel, possibly integrated with the traffic specifications of the new flow, can be used as the input of an analytical model to decide whether the new flow can be admitted. This approach has the drawback that in realistic scenarios the number of parameters that should be included into the model is very large and some approximations are needed. Moreover, it is likely that a precise analytical model will require information that cannot be provided by the Monitoring Module. For these reasons, an approach based only on analytical models is complex and, in some situations, probably also not sufficiently accurate.
- The information provided by the Monitoring Module can be used to derive one or more indicators of the occupancy of the channel that is used to perform the admission decision using some heuristic. The latter can be based on analytical models, table-based algorithms, or other categories of decision algorithms. The effectiveness of the approach may be jeopardized by the nonlinear behavior of the wireless channel in proximity of the saturation zone.
- The Control Module may rely on a *probe phase*, and use the Monitoring Module to detect the impact of the new flow on the state of the wireless channel. To this aim is possible to inject in the channel a probe flow with the same features contained in the flow specification. Another approach is to start the real flow and to monitor the channel behavior for a short period. The advantage of this method is to measure the actual impact of the new flow, at the expenses of an additional overhead (the network resources wasted in the probe phase) and of the delay due to the probe phase.

In all cases, the Control Module must notify its decision. When the station has issued an explicit request, the Control Module can send its answer using the same signaling protocol. Conversely, when the station just starts transmitting (or receiving) a high-priority flow, the Control Module must enforce its decision in some way. In the case of a negative decision, down-link traffic may be discarded at the WTP or at the Access Controller, whereas the case of up-link traffic this is not possible. If no other means are available, the Control Module can instruct the WTP to disassociate the station.

In order to provide a quantitative evaluation of the QoS monitoring approach to admission control, we present performance figures about the admission of VoIP flows in a IEEE 802.11 EDCA

WLAN. These results extend those presented in (Cacace, Iannello, Vellucci & Vollero, 2008), where more details about the experimental set-up can be found.

The experiments focus on admission of VoIP flows with or without interfering best-effort traffic. The Control Module executes the following admission algorithm, which is based on a probe phase:

1. Each new VoIP session undergoes a probe phase whose maximum duration is 2 seconds.
2. A Monitoring Module measures the QoS parameters, and specifically bidirectional throughput and down-link loss and delay, every 0.1 s.
3. The probe phase fails if either the down-link delay exceeds the threshold of 50 ms in a single measurement, or the aggregate best-effort throughput falls under a fixed threshold (in the experiments this threshold was 4 Mbps and the throughput was averaged over 5 measurement samples).

This algorithm uses the actual traffic flow to perform the probe phase, thus no traffic specification is required. This approach has the advantage of a faster setup for a new VoIP session which is eventually admitted, since the probe phase and the session setup are concurrent. When the flow is rejected, the VoIP session will be dropped after 2 seconds or less of conversation.

EDCA is used in the test-bed as channel access mechanism, and IEEE 802.11b at the PHY layer. The VoIP and best-effort traffic are mapped, respectively, onto the VO and BE Access Category with standard values for the MAC parameters. Elastic best-effort traffic is generated by TCP connections, either up-link or down-link.

The performance figures that have been investigated in the experiments are related to (i) the number of admitted VoIP flows; (ii) the duration of the probe phase in case of rejection; (iii) the impact of the probe phase on the ongoing traffic. To test the precision of the algorithm we compare the admission figures with the limit obtained in simulations performed by the *ns-2* network simulator. The results of the experiments are reported in Table 2. The VoIP flows were bi-directional at constant packet rate in the range 40-100 packets per second (pps) and were generated using G.729 codec (8 kbps), G.711 codec (64 kbps) and a videoconferencing codec at 200 kbps. In Table 2, the ns-2 admission limit is computed in the conditions and a flow is considered not admissible when the resulting packet loss rate in either direction exceeded 1%, or down-link delay exceeded 50 ms. As it can be seen, the admission limit reached by the Control Module matches very well the capacity estimated by the network simulator. The probe duration (averaged across 5 attempts to add an extra station) is reported with its 90% confidence interval. It turns out that the flow rejection can be decided in approximately 1 second. This delay corresponds to the time needed to detect that the sending queue at the WTP is building up in a congested situation. The impact of the probe phase on the ongoing traffic, reported in the *Loss rate* column of Table 2, refers to the probe phases. It can be appreciated that the probe phase has no perceivable effect on the ongoing VoIP conversations, since the worst case corresponds to a loss rate of less than 0.4% for less than a second. Finally, when there is best-effort traffic the Control Module correctly protects its guaranteed throughput limit of 4 Mbps.

Dynamic MAC Parameter Setting

The dynamic setting of MAC parameters is a powerful tool that can be used by the Control Module to protect QoS guarantees. The introduction of EDCA a channel access mechanism has triggered many research efforts aimed at the defining algorithms for the dynamic setting of the parameters, see for instance the proposals of (Banchs & Perez, 2002; Banchs, Iannello, Serrano

Table 2. Admission control limits of VoIP flows with best-effort traffic through QoS monitoring

VoIP flows	Best-effort traffic	Admission limit	ns-2 admission limit	Probe duration (s)	Loss rate (%)	Best-effort throughput (Mbps)
G.729 @ 60 pps	-	16	16	1.11±0.02	0.01±0.01	-
G.729 @ 80 pps	-	14	14	1.31±0.07	0.02±0.02	-
G.729 @100 pps	-	12	12	1.41±0.01	0.07±0.03	-
G.711 @ 60 pps	-	15	15	0.73±0.01	0.02±0.02	-
G.711 @ 80 pps	-	12	12	0.86±0.31	0.03±0.02	-
G.711 @100 pps	-	10	10	1.43±0.50	0.05±0.04	-
200 kbps @ 40 pps	-	12	12	0.98±0.07	0.11±0.05	-
200 kbps @ 60 pps	-	10	10	1.30±0.18	0.10±0.04	-
200 kbps @ 80 pps	-	9	9	0.61±0.04	0.22±0.06	-
200 kbps @100 pps	-	8	8	0.58±0.02	0.38±0.18	-
G.711 @ 50 pps	3 TCP up	4	4	1.65±0.12	0.01±0.01	4.71
G.711 @ 50 pps	3 TCP down	4	4	1.14±0.03	0.01±0.01	4.30
G.711 @ 100 pps	1 TCP up	2	2	1.66±0.07	0.02±0.00	4.39
G.711 @ 100 pps	5 TCP up	2	2	0.13±0.01	0.00±0.00	4.83
G.711 @ 100 pps	1 TCP down	2	2	0.73±0.04	0.03±0.03	4.30
G.711 @ 100 pps	5 TCP down	2	2	1.28±0.09	0.00±0.00	4.37
G.729 @ 100 pps	1 TCP up	4	4	1.49±0.12	0.03±0.04	4.05
G.729 @ 100 pps	5 TCP up	4	4	0.58±0.03	0.01±0.02	4.37
G.729 @ 100 pps	1 TCP down	4	4	0.90±0.03	0.04±0.06	4.07
G.729 @ 100 pps	5 TCP down	4	4	0.59±0.04	0.00±0.01	4.01

& Vollero, 2006; Dangerfield, Malone & Leith, 2006; El Housseini & Alnuweiri, 2005; Freitag, da Fonseca & de Rezende, 2006; Tanigawa, Kim, Tode & Murakami, 2006). Specifically, the approach proposed in (Banchs, Iannello, Serrano & Vollero,2006) concerns explicitly the situation of Centralized WLANs managed through the CAPWAP protocol. A discussion about the respective merits of these algorithms and the best techniques to use to pursue specific QoS goals is outside the scope of this chapter. In our context, based on QoS monitoring, it is sufficient to assume that the Control Module will rely on some information about the wireless state to take decisions about the best MAC parameters setting. This information can be provided by static traffic specifications only, or it may also include the state of the wireless channel measured by the Monitoring Module. In the latter case, the design of an algorithm for the dynamic setting of MAC parameters includes the following steps.

1. Definition of the information reported by the Monitoring Module and of its frequency.
2. Classification of the state of the wireless channel in a limited number of categories, according to the QoS goals and to the values of the channel parameters reported by the Monitoring Module. These states correspond to categories of broad macro-situations such as "QoS ok", or "up-link/down-link unfairness in best-effort traffic", etc.
3. Definition of the algorithm for the parameters setting as a set of Event-Condition-Action

rules, that can be specified as follows:

3.1. Event: the transitions between pair of QoS states;
3.2. Condition: the parameter whose variation caused the transition;
3.3. Action: a change of the value of some EDCA parameters.

The Condition part is important to specify the rules: for instance, a sudden unbalance between up-link and down-link best-effort traffic can be detected either when there is a surge in up-link traffic or when there is a decrease in down-link traffic, and the reaction to these two situations will in general be different. Even though the actual algorithm for the dynamic setting of MAC parameters can be specified with formalisms that are not rule-based, for example by means of analytical functions of the measured traffic and channel parameters, the above schema is sufficiently general to compare distinct algorithms.

The action part specifies a new set of MAC parameters for the Access Categories of either the WTP or the stations or both. The differentiation of the MAC parameters of the WTP and the stations is used to cope with issues related to the amount of up-link and down-link traffic. The *AIFS* and *CWmin*/*CWmax* pair of parameters allow the prioritization of the Access Categories, whereas a larger value of the *TXOPlimit* parameter can be used to reduce the collision rate or to decrease the delay. Three desirable properties for a dynamic adaptation strategy are:

- **Stability**. The MAC parameter setting algorithm is not influenced by transient fluctuations but only by permanent variations of the traffic and of the physical state of the channel. An unstable algorithm may cause a periodic and unnecessary change of MAC settings.
- **Optimality**. The MAC setting chosen by the algorithm is the best for the present traffic situation and wireless state. Thus an optimal algorithm is also invariant with respect to the past history.
- **Fast convergence**. Ideally, the Control Module should be able to immediately determine the new set of MAC parameters. However, some approaches include a *tuning phase*, to determine the best configuration. Since this requires a sequence of parameter changes and the evaluation of their impact on the channel state, this approach can have cause serious delays in reaching the optimal configuration.

These properties are not verified trivially. The reason is that, on the one side, a wireless channel exhibits the typical unpredictable fluctuations in presence of high traffic load, and on the other, the range of the MAC parameters is reduced and their discretization is coarse. Consequently, there is a trade-off between stability and optimality, due to the need of introducing some hysteresis threshold to prevent frequent parameter changes. An algorithm that is stable but not invariant will find a sub-optimal set of MAC parameters: optimality is traded for stability.

In (Cacace, Iannello, Vellucci & Vollero, 2008) an approach for the dynamic setting of MAC parameters is presented. The method is based on QoS monitoring and aims at the fair allocation of network resources between real-time and best-effort traffic, as well as between up-link and down-link traffic. This reference also provides performance figures obtained in a test-bed and details about the synchronization between admission control and dynamic setting within the Control Module. The proposed algorithm is stable but not invariant (and thus, not optimal). The sub-optimal solution is reached with an iterative procedure, that however aims at restoring in the first place the QoS guarantees for high priority traffic.

The Control Module can implement the parameter adaptation strategy by issuing CAPWAP control messages:

- the Control Module can change the EDCA parameters on the WTP by sending a Configuration Update Request message. This command contains a WTP Quality of Service message element with a new set of EDCA parameters for the WTP.
- The Control Module can change the EDCA parameters on the stations by sending a WLAN Configuration Request message which contains an Update WLAN message element with the new set of EDCA parameters. These parameters will be communicated to the stations with the IEEE 802.11 beacon.

Frequency Planning

When managing large Centralized WLANs, the configuration of frequencies used by the WTPs may be a major problem in optimizing overall network performance. For example, IEEE 802.11b define eleven (or thirteen, depending on the geographic area) transmission channels, but at most three of them can be used simultaneously without cross interference. Optimal or suboptimal frequencies reuse is desirable to reduce interference among adjacent cells but it is not always possible. In this context, a Centralized approach is desirable, since a frequency optimization algorithm needs a global knowledge of the channel allocation in the whole network that is not available at individual WTPs.

Two cases are possible, requesting slightly different approaches:

- the Access Controller manages all the WTPs that operate in the area covered by the WLAN;
- other WTPs/APs not under the control of the Access Controller are operating in the area covered by the WLAN.

The second situation adds additional constraints and it is not uncommon due to the widespread diffusion of wireless home or personal networks. Most existing proposals, however, focus on the first case (Arunesh, Shrivastava, Banerjee & Arbaugh, 2006; Chandra, Bahl & Bahl, 2004; Levanti, Giordano & Tinnirello, 2007; Liu, Yu, Liu, Chuah & Mohapatra, 2007; Youngseok, Kyoungae & Choi, 2002).

Frequency planning can be easily introduced in the schema of multi-cell QoS Management outlined above. Each WTP periodically sends information about the signal strength measured on the available channels, by means of CAPWAP information messages. This information is elaborated by the intra-cell Controller Module that can decide to issue requests to change the channel in use by means of CAPWAP control messages. The feasibility of the approach is investigated in (Levanti, Giordano & Tinnirello, 2007) that implements a variant of the algorithm proposed in (Arunesh, Shrivastava, Banerjee & Arbaugh, 2006). They conclude that the exploitation of a Centralized WLAN Architecture can avoid not only the transient channel adjustments due to subsequent and independent WTP decisions, but also the so-called *blocked cell* problem, thanks to the complete knowledge of the network topology. The blocked-cell phenomenon happens when a WTP is located in proximity of a set of other WTPs that use the three orthogonal channels, a situation not unrealistic in practice. In this conditions, if the adjacent WTPs do not hear each other, it is very likely that their transmissions are operated asynchronously, and in presence of saturated traffic, the considered WTP would sense the channel busy during all the time. The blocked-cell effect can reduce the throughput of the affected cell to zero, with the WTP waiting indefinitely for channel release. In the schema proposed by

(Levanti, Giordano & Tinnirello, 2007), each WTP periodically scans the channels in order to know all the potential interferences. Once the scansion is done, the WTP sends a report to the Access Controller indicating the interfering WTP identifiers and the reception power levels. Two different scanning report solutions are proposed. According to the first one, each WTP forwards to the Access Controller the beacon frames herd by overlapping WTPs. Through a specific field in the CAPWAP header, in the forwarding it is possible to include PHY layer information about the frame reception (Received Signal Strength Indicator, *RSSI*, and Signal to Noise Ratio, *SNR*, at the receiver), thus specifying the interfering power from the neighbor cells. According to the second solution, some message elements are encapsulated in a standard CAPWAP control frame, in order to simultaneously report all the interferences heard from the WTPs in range. Each message element indicates the cell identifier (SSID), the MAC address, the operating channel and an SNR measure. Collecting the scans from all the WTPs, the Access Controller can be aware of the network configuration. On the basis of this information, the intra-cell Controller Module assigns the channel to the WTPs, while minimizing the number of total reciprocal interference in the network. Since the number of orthogonal channel may not be sufficient for covering the interfering cells, the planning algorithm should try to assign orthogonal channels to the subset of interfering cells which result most loaded, while trying to avoid the blocked-cell phenomenon.

Load Balancing

Several studies have shown that in large WLAN scenarios users are often unevenly distributed in space and, hence, the number of associated users may vary widely from WTP to WTP. This may translate in an uneven load distribution which can severely degrade the QoS. Load balancing aims at mitigating this problem by forcing some users to roam toward less loaded neighbor WTPs, avoiding congestion.

IEEE 802.11 does not provide any explicit mechanism to control the association of users to a specific WTP. For this reason, Cell Breathing (Bejerano & Han, 2006) gained consensus due to its simple and standard implementation. Cell Breathing is based on the assumption that stations choose to associate with the WTP from which they receive beacons with the highest power. Hence, Cell Breathing tries to redistribute users association varying the power that each WTP uses for transmitting beacons, by both increasing the power of beacons sent by low loaded WTPs, and reducing the one of beacons generated by highly loaded WTPs.

The intra-cell Controller Module can also be used to implement Load Balancing strategies based on Cell Breathing, by using monitoring functionalities and commands for setting transmission power provided by CAPWAP. Applying the Load Balancing strategy, we assume that the Controller Module knows the position of each station. This assumption is based on the observation that WTPs in highly populated networks allow for a quite precise localization of users. Moreover, the localization of stations can be implemented using monitoring functionalities based on the IEEE 802.11 Information Element message of CAPWAP. The data frames forwarded from the WTP to the AC include the optional IEEE 802.1 Frame Info CAPWAP header field. This field is used to include radio and PHY specific information associated with the frame. This information is used as an input for the Load Balancing algorithm, that uses the Cell Breathing approach to control the number of stations associated to each WTP. For instance, the algorithm may perform an exhaustive search among all possible configurations, choosing the one with the lowest maximum number of stations associated to a single WTP. The corresponding configuration is then enforced by using the IEEE 802.11 Tx Power message element within a CAPWAP Configuration Update

Request from the AC to the WTP. A comparison among the two power allocation strategies is presented in (Bernaschi et al., 2009) by means of simulations with a number of stations ranging from 20 to 200 located randomly in a WLAN. The performance metric concerns the maximum number of stations associated to each WTP. The Load Balancing strategy outperforms the fixed distribution, in which each WTP sends a beacon at its maximum power level, by reducing the maximum number of stations associated to WTP by an average 20-25%, depending on the number of active WTP in the WLAN.

Mobility Manager

Wireless local area networks are typical access networks for mobile users. When a user moves across wireless cells belonging to the same IP network, the hand-off procedure is managed by the MAC layer. Conversely, when the roaming involves an IP network change, a mobility protocol is needed and the hand-off latency can be substantially higher. In recent years much research effort has been devoted to the support of network mobility, leading to the definition of a protocol for general mobility at the network level, called Mobile IP (Johnson, Perkins & Arkko, 2004; Perkins, 2002). Mobility at the network level is transparent to higher layers of the protocol stack, for instance to the transport and application layers. At the same time, for situations in which mobility is limited to local movements in administrative domains, for instance a corporate network, a number of micro-mobility protocols have also been proposed to enhance the hand-off performance (Campbell et al., 2002). Mobility across networks based on heterogeneous MAC layers, called vertical hand-off, has also been extensively studied (Bernaschi et al., 2005).

In this context, the choice of a Centralized Architecture and the presence in the WLAN of the Access Controller is relevant in two respects:

1. The Access Controller can play a role in improving the hand-off performance given its privileged position, that allows to maintain information related to the position and traffic generated by the stations.
2. From a QoS point of view, once the users has obtained certain QoS guarantees with the access networks these should be maintained also when he/she moves around.

The first point has been extensively studied in the context of hand-off optimization, and many available results can be readily applied to the Centralized WLAN case. Conversely, the integration of mobility and MAC parameter setting for the QoS support is still an open research issue and only a few works propose solutions that can be useful in the context of Centralized WLANs. Most of the work in this area has focused on the integration of local mobility with reservation protocols commonly used in wired networks. In (Paskalis, Kaloxylos, Zervas & Merakos, 2003) the efficient integration of the hand-off protocol with the RSVP signaling process in the wired portion of the network is considered. Performance evaluation is focused on the reservation phase, since it is assumed that traffic flows obtain the QoS requested after that reservation is successfully completed. In (Sroka & Karl, 2003) a mobility management protocol is extended to deal with hand-off requests towards WTPs that have not enough resources to provide the desired QoS. Again, the proposal considers QoS mechanisms available at routing, providing traffic differentiation only in the wired portion of the WLAN infrastructure. In (Garcia-Macias, Rousseau, Berger-Sabbatel, Toumi & Duda, 2003) a hierarchical QoS architecture that extends the DiffServ QoS model to mobile hosts in a IEEE 802.11 WLAN is presented. The approach provides a quite comprehensive solution to the integration of mobility and QoS. However, the problem of managing QoS requirements inside WLANs is solved through mechanisms working at the network layer or above.

In the specific context of Centralized WLANs managed through the CAPWAP protocols an approach is presented in (Acanfora, Cacace, Iannello & Vollero, 2006). The general schema involves the following steps:

1. The Session Manager on the Access Controller maintains an information repository about ongoing traffic flows. This global *session directory* is built either from explicit traffic specifications sent by the stations, or through statistics provided by the Monitoring Module.
2. The Access Controller becomes aware of the beginning of the hand-off procedure of a given station from the control frames that the WTP forwards to it.
3. At this point, the information about the ongoing traffic flows of the mobile station is retrieved from the Session Manager and used by the Control Module of the WTP to which the station is moving to determine if it is necessary to change the MAC parameter settings and to compute the new configuration.

Even though this schema is straightforward, in practice there are some crucial issues. In the first place, it may be difficult for the Control Module to determine the new setting of MAC parameters, due to the cited limitations of analytical and heuristic methods. One solution to this problem is to delay the setting decision after that the hand-off is completed. In this case, for the Control Module the hand-off procedure is indistinguishable from the start of a new flow. Consequently, if admission control is in effect and the arrival cell is overloaded the moving station may see its traffic flows blocked by the admission control mechanism, in contrast with the principle that ongoing traffic should be granted a higher priority than new flows, and that blocking should only happen before the session starts. Notice that in general it is not convenient to prevent the station from performing the hand-off, for in this case the station would see a traffic disruption after losing the connection with the original WTP.

Another approach is to protect the existing flows of the moving station. In this case, the Control Module does not perform admission control on the ongoing traffic of the moving station. However, if the latter would jeopardize the traffic in the arrival cell, the station may be temporarily *downgraded*, in the sense that it is granted a reduced quality of service as long as the load of the destination WTP does not allow to provide the level of QoS requested.

As it can be seen, the integration of QoS and mobility introduces some complications in the admission control and dynamic MAC management of the Control Modules. Actually, the simulation results reported in (Acanfora, Cacace, Iannello & Vollero, 2006) show that in many common situations a table-based approach can improve the overall performance. However, additional research and experiments are needed in order to devise a reliable and comprehensive solution.

FUTURE RESEARCH DIRECTIONS

The area of QoS monitoring and management in multi-cell wireless networks is highly active and we forecast it will be of high interest for the research community also in the next future. Two major reasons explain this forecasting: the growing diffusion of such networks and their increasing complexity.

Probably, one of the major trends in the management of wireless infrastructures is represented by the ability of controlling a high number of co-located systems while applying cooperative strategies for the distribution and access of network resources. In this context, several activities can be devised involving CAPWAP and CAPWAP bindings. The first problem to be addressed is the joint management of such infrastructures. CAPWAP extensions, with properly designed bindings, may represent a good solution in those scenarios. The problem of CAPWAP extensions has been partially considered for the management of Wireless Mesh Networks (WMNs) and WiMAX infrastructures, while it is still missing an analysis of CAPWAP functionalities employed in classical and next generation cellular networks. It can be expected that this will be an active research field in the next future in order to provide solutions in realistic scenarios (for instance, in urban areas and university campus). Strictly related to this research field is the field of algorithms design for the management and configuration of heterogeneous and cooperative wireless infrastructures. While several works have analyzed specific optimization problems present in such scenarios, the analysis and design of systems and algorithms solving problems in a wide variety of working conditions is still missing. The high impact of such research fields on services let forecast a growing research activity on them in the near future.

Another area of interest is the monitoring and management of community networks. A community network is a virtual network where involved users share some common interest: from the simple Internet access, to common resources for storage and media distribution. Community networks are usually built on top of physical networks by mean of virtualization protocols (for instance, Virtual Private Networking). All the community networks using the same physical infrastructure share its resources, and require the definition of proper monitoring and controlling strategies both for the traffic internal to the community and for the traffic interactions among communities. The definition and design of such monitoring and controlling strategies is currently under study, but several issues require to be addressed in order to find suitable solutions in realistic scenarios, where the community can span over several communication technologies and a high number of communities interact and compete for the available resources. Also in this field of research CAPWAP can be used profitably: indeed several elements, from encryption/encapsulation to virtual access points management, can be adapted to community networks scenarios. Moreover most of CAPWAP monitoring strategies can be extended in order to provide suitable tools to be used in the definition of control and management algorithms.

Eventually, another scenario of great interest is represented by the joint management of network infrastructures and users. This scenario requires that all the functionalities provided by CAPWAP are integrated with protocols controlling and monitoring users devices (for instance protocols based on the IEEE 802.21 standard). Although very few works analyzed such network scenarios, we can expect a growing interest on it, basing this expectation on the performance levels that network operators can provide to users in this context.

CONCLUSION

Large deployments of access points based on the IEEE 802.11 standard require management, configuration and control mechanisms. In this chapter we discussed the using of the CAPWAP standard as a viable protocol providing solutions to traditional problems of such scenarios in the context of centralized management architectures. In particular we analyzed several issues related to the management of the networks, and we showed the CAPWAP suitability under three different time scale: packet, for the management of QoS in every wireless cell, session, for the load balancing of users, and network deployment, for the interferences reduction when planning frequencies distribution. In all cases CAPWAP proves to be a flexible tool allowing the monitoring of networks (interferences and experienced quality levels) and the enforcing of QoS policies. As every protocols providing functionalities, CAPWAP relies on intelligent management and control systems for a particular strategy to be enforced. In this context we also discussed its integration in an architecture for the management of users mobility, showing the flexibility and suitability of the protocol also in such a scenario.

The concrete applicability of analyzed solutions has been discussed referring also to an open-source implementation of the protocol. Based on this implementation analysis, issues have been identified and related, as constraints, to all the scenarios analyzed in the chapter. In the chapter we showed that those constraints have a very low impact on the flexibility of the protocol and its practical employment in realistic scenarios.

Future trends on access networks infrastructures let to forecast an increasing demand on monitoring and policy enforcing strategies with flexible characteristics. In this context, centralized strategies, relying on protocols like CAPWAP, will represent good solutions suitable for most scenarios of interests.

ACKNOWLEDGMENT

This article is the outcome of many years of work on Quality of Service in wireless networks. Several people helped us in this activity and we would like to thank them all. In particular, we would like to thank Simona Acanfora, whose pioneering work on QoS in multi-cell wireless access networks represented the starting point of our research in this area, Massimo Bernaschi, who has been working with us for many years on mechanisms supporting QoS in different wireless technologies, and Massimo Vellucci, whose contributions to the experimental activity we carried out made it possible to greatly enrich our results.

REFERENCES

Acanfora, S., Cacace, F., Iannello, G., & Vollero, L. (2006). Dynamic Configuration of MAC QoS Mechanisms in 802.11 Access Networks. In Next Generation Teletraffic and Wired/Wireless Advanced Networking, (LCNS 4003, pp. 542-553). Berlin: Springer-Verlag.

Arunesh, M., Shrivastava, V., Banerjee, S., & Arbaugh, W. (2006). Partially overlapped channels not considered harmful. In *Proceedings of the joint international conference on Measurement and modeling of computer systems* (pp. 63-74). New York: ACM Press.

Banchs, A., Iannello, G., Serrano, P., & Vollero, L. (2006). A CAPWAP Architecture for Dynamic Configuration of QoS MAC Parameters. In *Proceedings 1st Workshop on Multimedia Applications over Wireless Networks (MediaWiN).*

Banchs, A., & Perez, X. (2002). Providing Throughput Guarantees in IEEE 802.11 Wireless LAN. In [New York: IEEE Press.]. *Proceedings of IEEE Wireless Conference on Networking Communication WCNC, 1*, 130–138.

Banchs, A., & Vollero, L. (2006). Throughput analysis and optimal configuration of 802.11e EDCA. *Computer Networks, 50*(11),1749–1768. doi:10.1016/j.comnet.2005.07.008

Bejerano, Y., & Han, S.-J. (2006). Cell Breathing Techniques for Balancing the Access Points Load in Wireless LANs. In *Proceedings IEEE INFOCOM2006, Twenty-fifth Annual Joint Conference of IEEE Computer and Communications Societies,* (pp. 1-13). New York: IEEE Press.

Bellalta, B., Cano, C., Oliver, M., & Meo, M. (2006). Modeling the IEEE 802.11e EDCA for MAC parameter optimization. In *Proceedings 4th International Working Conference on performance modeling and evaluation of heterogeneous networks*, Ilkley, UK.

Bernaschi, M., Cacace, F., Iannello, G., Vellucci, M., & Vollero, L. (2009). OpenCAPWAP: An open source CAPWAP implementation for the management and configuration of WiFi hot-spots. *Computer Networks, 53*(2),217–230. doi:10.1016/j.comnet.2008.09.016

Bernaschi, M., Cacace, F., Iannello, G., Za, S., & Pescapé, A. (2005). Seamless internetworking performance of WLANs and cellular networks: architecture and performance issues in a Mobile Ipv6 scenario. *IEEE Wireless Communications, 12*(3),73–80. doi:10.1109/MWC.2005.1452857

Bianchi, G., & Tinnirello, I. (2003). Kalman Filter Estimation of the Number of Competing Terminals in an IEEE 802.11 network. In *IEEE INFOCOM 2003, Twenty-Second Annual Joint Conference of the IEEE Computer and Communications Societies: Vol. 2* (pp. 844-852). New York: IEEE Press.

Braden, R. (Ed.). Zhang, L., Berson, S., Herzog, S., & Jamin, S. (1997). *Resource ReSerVation Protocol (RSVP) -- Version 1 Functional Specification* (RFC 2205). IETF Network Working Group.

Cacace, F., Iannello, G., Vellucci, M., & Vollero, L. (2008). A Reactive Approach to QoS Provisioning in IEEE 802.11e WLANs. In *Proceedings of the 2008 4th international telecommunication networking workshop on QoS in multiservice IP networks, 4th EURO-NGI Conference on Next Generation Internet Networks* (pp. 253-260). New York: IEEE Press.

Cacace, F., & Vollero, L. (2009). A Delay Monitoring Method for Up-link Flows in IEEE 802.11 Networks. In *Proceedings of the 2nd International Conference on Simulation Tools and Techniques 2009* (pp. 1-8), ICST Brussels.

Calhoun, P., Montemurro, & M., Stanley, D. (2008). *CAPWAP Protocol Specification, version 15* (Internet draft). IETF Network Working Group.

Calhoun, P., Montemurro, & M., Stanley, D. (2008b). *CAPWAP Protocol Binding for IEEE 802.11 version 12* (Internet draft). IETF Network Working Group.

Campbell, A.T., Gomez, J., Kim, S., & Turanyi, Z.L., Valko', A.G., & Wan, C.-Y. (2002). Internet micromobility. *Journal of High Speed Networks, 11*, 177–198.

Chandra, R., Bahl, P., & Bahl, P. (2004). Multinet: Connecting to multiple IEEE 802.11 networks using a single wireless card". In *Proceedings IEEE INFOCOM2004, Twenty-third Annual Joint Conference of IEEE Computer and Communications Societies: Vol. 2* (pp. 882-893). New York: IEEE Press.

Dangerfield, I., Malone, D., & Leith, D. (2006). Experimental Evaluation of 802.11e EDCA for Enhanced Voice over WLAN Performance. In *Proceedings of the 4th International Symposium on Modeling and Optimization in Mobile, Ad Hoc and Wireless Networks,* (pp. 1-7).

Dierks, T., & Rescorla, E. (2006). *The Transport Layer Security (TLS) Protocol Version 1.1* (RFC 4346), IETF Network Working Group.

El Housseini, S., & Alnuweiri, H. (2005). Adaptive Contention-Window MAC Algorithms for QoS-Enabled Wireless LANs. In *Proceedings of 2005 International Conference on Wireless Networks, Communications and Mobile Computing: Vol.1* (pp. 368-374).

Engelstad, P. E., & Østerbø, O. N. (2005). Non-Saturation and Saturation Analysis of IEEE 802.11e EDCA with Starvation Prediction. In *Proceedings of the 8th ACM Symposium on Modeling, Analysis and Simulation of Wireless and Mobile Systems MSWIM 05* (pp. 224-233). New York: ACM Press.

Fan, Y., Huang, C., & Tseng, Y. (2006). Multimedia Services in IEEE 802.11e WLAN Systems. In *Proceedings of the ACM International Wireless Communications and Mobile Computing Conference IWCMC 06* (pp. 401-405). New York: ACM Press.

Freitag, J., da Fonseca, N. L. S., & de Rezende, J. F. (2006). Tuning of 802.11e network parameters. *IEEE Communications Letters, 10*(8). doi:10.1109/LCOMM.2006.1665127

Gao, D., Cai, J., & Ngan, K. N. (2005). Admission Control in IEEE 802.11e Wireless LANs. *IEEE Network, 19*(4), 6–13. doi:10.1109/MNET.2005.1470677

Garcia-Macias, J. A., Rousseau, F., Berger-Sabbatel, G., Toumi, L., & Duda, A. (2003). Quality of Service and Mobility for the Wireless Internet. *Wireless Networks, 9*, 341–352. doi:10.1023/A:1023647311052

Gu, D., & Zhang, J. (2003). A New Measurement-Based Admission Control Method for IEEE 802.11 Wireless Local Area Networks. In *Proceedings IEEE International Symposium on Personal, Indoor, and Mobile Radio Communications: Vol. 3* (pp. 2009-2013). New York: IEEE Press.

Harsha, S., Kumar, A., & Sharma, V. (2006). An Analytical Model for the Capacity Estimation of Combined VoIP and TCP File Transfers over EDCA in an IEEE 802.11e WLAN. In *Proceedings 14th IEEE International Workshop on Quality of Service* (pp. 178-187). New York: IEEE Press.

I.S. 802.11 (2007). *IEEE Standard for Information Technology - Telecommunications and Information Exchange between Systems - Local and Metropolitan Area Networks - Specific Requirements Part 11: Wireless LAN Medium Access Control (MAC) and Physical Layer (PHY) Specifications*, IEEE802.11-REVma/D9.0, Revision of Std. 802.11-1999, August 2007.

Johnson, D., Perkins, C., & Arkko, J. (2004). *Mobility Support in IPv6* (RFC 3775), IETF Network Working Group.

Kong, Z., Tsang, D. H. K., Bensaou, B., & Gao, D. (2004). Performance analysis of IEEE 802.11e contention-based channel access. *IEEE Journal on Selected Areas in Communications, 22*(10), 2095–2106. doi:10.1109/JSAC.2004.836019

Levanti, A. Giordano, F.; & Tinnirello (2007), I., A CAPWAP-Compliant Solution for Radio Resource Management in Large-Scale 802.11 WLAN. *In Proceedings of IEEE Global Telecommunications Conference, 2007.* (pp.3645-3650), 26-30 Nov. 2007.

Levanti, A., Giordano, F., & Tinnirello, I. (2007) A CAPWAP Architecture for Automatic Frequency Planning in WLAN. In *Proceedings 12th IEEE Symposium on Computers and Communications, ISCC 2007* (pp 51-56). New York: IEEE Press.

Liu, H., Yu, H., Liu, X., Chuah, C., & Mohapatra, P. (2007). Scheduling Multiple Partially Overlapped Channels in Wireless Mesh Networks. In *Proceedings IEEE International Conference on Communications, ICC '07*, (pp. 3817-3822). New York: IEEE Press.

Malli, M., et al. (2004). Adaptive Fair Channel Allocation for QoS Enhancement in IEEE 802.11 Wireless LANs. In *Proceedings 2004 IEEE International Conference on Communications* (pp. 3470-3475). New York: IEEE Press.

Park, E.-C., Kim, D.-Y., Choi, C.-H., & So, J. (2007). Improving Quality of Service and Assuring Fairness in WLAN Access Networks. *IEEE Transactions on Mobile Computing, 6*(4), 337–350. doi:10.1109/TMC.2007.53

Paskalis, S., Kaloxylos, A., Zervas, E., & Merakos, L. (2003). An Efficient RSVP-Mobile Internetworking Scheme. *Mobile Networks and Applications, 8*, 197–207. doi:10.1023/A:1023385429587

Perkins, C. (2002). *IP Mobility Support in IPv4* (RFC 3344). IETF Network Working Group.

Pong, D., & Moors, T. (2003). Call Admission Control for IEEE 802.11 Contention Access Mechanism. In *Proceedings IEEE Global Telecommunications Conference, Globecom 2003: Vol. 1* (pp. 174-178). New York: IEEE Press.

Ramos, N., Panigrahi, D., & Dey, S. (2003). ChaP-LeT: Channel-dependent Packet Level Tuning for Service Differentiation in IEEE 802.11e. In *Proceedings International Symposium on Wireless Personal Multimedia Communications: Vol. 3* (pp. 89-90).

Robinson, J. W., & Randhawa, T. S. (2004). Saturation Throughput Analysis of IEEE 802.11e Enhanced Distributed Coordination Function. *IEEE Journal on Selected Areas in Communications*, *22*(5),917–928. doi:10.1109/JSAC.2004.826929

Romdhani, L., Ni, Q., & Turletti, T. (2003). Adaptive EDCF: Enhanced Service Differentiation for IEEE 802.11 Wireless Ad Hoc Networks. In *Proceedings 2003 IEEE Wireless Communications and Network Conference: Vol. 3* (pp. 1373-1378). New York: IEEE Press.

Sroka, S., & Karl, H. (2003). Performance Evaluation of a QoS-Aware Handover Mechanism. *Proceedings 8th IEEE International Symposium on Computers and Communications: Vol. 1* (pp. 117- 124). New York: IEEE Press.

Tanigawa, Y., Kim, J., Tode, H., & Murakami, K. (2006). Proportional Control and Deterministic Protection of QoS in IEEE 802.11e Wireless LAN. In *Proceedings International Conference on Wireless Communications and Mobile Computing, IWCMC 06* (pp. 1147 – 1152). New York: ACM Press.

Xiao, Y., & Li, H. (2004). Voice and Video Transmissions with Global Data Parameter Control for the IEEE 802.11e Enhanced Distributed Channel Access. *IEEE Transactions on Parallel and Distributed Systems*, *15*(11),1041–1053. doi:10.1109/TPDS.2004.72

Xiao, Y., Li, H., & Li, B. (2007). Bandwidth Sharing Schemes for Multimedia Traffic in the IEEE 802.11e Contention-Based WLANs. *IEEE Transactions on Mobile Computing*, *6*(7),815–831. doi:10.1109/TMC.2007.1054

Xiong, L., & Mao, G. (2007). Saturated Throughput Analysis of IEEE 802.11e EDCA. *Computer Networks*, *51*(1),3047–3068. doi:10.1016/j.comnet.2007.01.002

Yang, L., Zerfos, P., & Sadot, E. (2005). *Architecture Taxonomy for Control and Provisioning of Wireless Access Points (CAPWAP),* (RFC 4118). IETF Network Working Group.

Youngseok, L., Kyoungae, K., & Choi, Y. (2002). Optimization of AP placement and channel assignment in wireless lans. In *Proceedings of the 27th Annual IEEE Conference on Local Computer Networks, LCN 02* (pp. 831-836). Los Alamitos: IEEE Computer Society Press.

Zhai, H., Chen, X., & Fang, Y. (2006). A call admission and rate control scheme for multimedia support over IEEE 802.11e wireless LANs. *ACM Wireless Networks*, *12*(4),451–463. doi:10.1007/s11276-006-6545-y

Zhang, L., & Zeadally, S. (2004). HARMONICA: Enhanced QoS Support with Admission Control for IEEE 802.11 Contention-based Access. In *Proceedings 10th IEEE Real-Time and Embedded Technology and Applications Symposium* (pp. 64–71). Los Alamitos: IEEE Computer Society Press.

ADDITIONAL READING

Byun, S. Seungbeom Lee; & Sin-Chong Park (2007), Performance Compensation of IEEE 802.11e QoS stations with Coexistence of Legacy 802.11 stations, *In Proceedings of 9th International Conference on Advanced Communication Technology*. (pp.18-22), 12-14 Feb. 2007.

Cano, C., Bellalta, B., & Oliver, M. (2007), Adaptive Admission Control Mechanism for IEEE 802.11e WLANs, *In Proceedings of IEEE 18th International Symposium on Personal, Indoor and Mobile Radio Communications*. (pp.1-5), 3-7 Sept. 2007.

Chung, A. Y. T., & Hassan, M. (2005), Traffic distribution schemes for multi-homed mobile hotspots, *In Proceedings of IEEE 61st Vehicular Technology Conference*. vol. 4, (pp. 2127-2131), 30 May-1 June 2005.

Clancy, T. (2008), Secure handover in enterprise WLANs: capwap, hokey, and IEEE 802.11R, *IEEE Wireless Communications, 15* (5), (pp.80-85), October 2008.

Clifford, P., Duffy, K., Foy, J., Leith, D., & Malone, D. (2006). Modeling 802.11e for data traffic parameter design. In *Proceedings 4th International Symposium on Modeling and Optimization in Mobile, Ad Hoc and Wireless Networks WiOpt 06,* (pp. 1-10).

Der-Jiunn Deng; & Hsu-Chun Yen (2005), Quality-of-service provisioning system for multimedia transmission in IEEE 802.11 wireless LANs, *IEEE Journal on Selected Areas in Communications, 23* (6), (pp. 1240-1252), June 2005.

Diwakar, K., & Iyer, S. (2004), Supporting real-time applications with better QoS guarantees in 802.11, *In Proceedings of 1st International Symposium on Wireless Communication Systems*. (pp. 373-377), 20-22 Sept. 2004.

Doufexi, A.; Tameh, E.; Nix, A.; Armour, S.; & Molina, A. (2003), Hotspot wireless LANs to enhance the performance of 3G and beyond cellular networks. *IEEE Communications Magazine, 41* (7), pp. 58-65, July 2003.

Elaoud, M., Famolari, D., & Ghosh, A. (2005), Experimental VoIP capacity measurements for 802.11b WLANs, *In Proceedings of 2nd IEEE Consumer Communications and Networking Conference*. (pp. 272-277), 3-6 Jan. 2005.

Eng Hwee Ong; & Khan, J.Y. (2009), An Integrated Load Balancing Scheme for Future Wireless Networks, *In Proceedings of 4th International Symposium on Wireless Pervasive Computing*. (pp.1-6), 11-13 Feb. 2009.

Eng Hwee Ong; & Khan, J.Y. (2009), A Unified QoS-Inspired Load Optimization Framework for Multiple Access Points Based Wireless LANs, *In Proceedings of IEEE Wireless Communications and Networking Conference*. (pp.1-6), 5-8 April 2009.

Halpert, B. J. (2005), Interference effects on wireless link availability, *In Proceedings of IEEE SoutheastCon, 2005*. (pp. 537-541), 8-10 April 2005.

Huang, C.-M., & Li, J.-W. (2008), A Context Transfer Mechanism for IEEE 802.11r in the Centralized Wireless LAN Architecture. *In Proceedings of 22nd International Conference on Advanced Information Networking and Applications, 2008*. pp.257-263, 25-28 March 2008.

Koukoutsidis, I., & Siris, V. A. (2008), Access point assignment algorithms in WLANs based on throughput objectives, *In Proceedings of 6th International Symposium on Modeling and Optimization in Mobile, Ad Hoc, and Wireless Networks and Workshops*. (pp.375-383), 1-3 April 2008.

Le, L. Albayrak, S.; Elkotob, M.; & Toker, A.C. (2007), Improving TCP Goodput in 802.11 Access Networks, *In Proceedings of IEEE International Conference on Communications*, (pp.4494-4499), 24-28 June 2007.

Me, G., & Verdone, D. (2006), An overview of some techniques to exploit VoIP over WLAN, *In Proceedings of International Conference on Digital Telecommunications*, (pp.67-67), 29-31 Aug. 2006.

Mi Kyung Han. Overstreet, B.; & Lili Qiu (2007), Greedy Receivers in IEEE 802.11 Hotspots, *In Proceedings of 37th Annual IEEE/IFIP International Conference on Dependable Systems and Networks*. (pp.471-480), 25-28 June 2007.

Na, C. Chen, J.K.; & Rappaport, T.S. (2004), Hotspot traffic statistics and throughput models for several applications, *In Proceedings of IEEE Global Telecommunications Conference.* (pp. 3257-3263) Vol.5, 29 Nov.-3 Dec. 2004.

Pattara-Atikom, W., Krishnamurthy, P., & Banerjee, S. (2003), Distributed mechanisms for quality of service in wireless LANs, IEEE Wireless Communications, 10 (3), (pp. 26-34), June 2003

Ping-Chi Wang. Kuochen Wang; & Lung-Sheng Lee (2005), A QoS scheme for digital home applications in IEEE 802.11e wireless LANs, *In Proceedings of 16th IEEE International Symposium on Personal, Indoor and Mobile Radio Communications*. vol. 3, (pp.1845-1849), 11-14 Sept. 2005.

Ranjan, A. (2005), MAC issues in 4G, *In Proceedings of IEEE International Conference on Personal Wireless Communications*. (pp. 487-490), 23-25 Jan. 2005.

Sarikaya, B. & Xiao Zheng (2006). CAPWAP Handover Protocol, *In Proceedings of ICC '06. IEEE International Conference on Communications, vol.4*, (pp.1933-1938), June 2006.

Sung Won Kim; Byung-Seo Kim; & Yuguang Fang (2005), Downlink and uplink resource allocation in IEEE 802.11 wireless LANs, *IEEE Transactions on Vehicular Technology, 54* (1), (pp. 320-327), Jan. 2005.

Tartarelli, S., & Nunzi, G. (2006), QoS Management and Congestion Control in Wireless Hotspots. *In Proceedings of 10th IEEE/IFIP Network Operations and Management Symposium*. (pp.95-105), 3-7 April 2006.

Varone, S. C. (2005, Aug. 22-24). Collaborative channel affectation vs greedy method for 802.11 b/g networks, *In Proceedings of IEEE International Conference on Wireless And Mobile Computing, Networking. Communications*, *1*, 55–57.

Villalon, J., Cuenca, P., & Orozco-Barbosa, L. (2005), Limitations and capabilities of QoS support in IEEE 802.11 WLANs, *In Proceedings of IEEE Pacific Rim Conference on Communications, Computers and signal Processing.* (pp. 633-636), 24-26 Aug. 2005.

KEY TERMS AND DEFINITIONS

CAPWAP: The IETF "Control And Provisioning of Wireless Access Points" protocol

Centralized WLAN: A centralized architecture for wireless local area networks

Wireless Networks: A computer network based on a wireless communication technology

Network Management: Activities devoted to configuration and control of a computer network

QoS Monitoring: Activities devoted to to gather quantitative data about network performance

QoS Support: Mechanisms useful to enforce specific performance levels of the network

Performance Evaluation: Quantitative characterization of network performance

ENDNOTE

1 We consider the principal metrics used to characterize QoS performances.

Chapter 10
QoS Support in Multi-hop Ad-hoc Networks

Marek Natkaniec
AGH University of Science and Technology, Poland

Katarzyna Kosek-Szott
AGH University of Science and Technology, Poland

Szymon Szott
AGH University of Science and Technology, Poland

ABSTRACT

The chapter contains an overview of existing QoS solutions for multi-hop ad-hoc networks. Firstly, an introduction and short motivation are presented. The authors present an analysis of the QoS aspects of the physical layer because the wireless communication channel is constantly changing and inherently prone to errors. QoS provisioning at the data link layer is studied next. The authors focus on protocols which enable traffic differentiation, solve the hidden node problem and provide fair medium access. The chapter also deals with QoS issues at the network layer, where the authors mostly discuss QoS routing protocols. Additionally, cross-layer solutions for QoS support in multi-hop ad-hoc networks are analyzed. Finally, the expected direction of future work and a brief summary are presented.

INTRODUCTION

Multi-hop ad-hoc networks are distributed, wireless networks without infrastructure in which every node acts as both terminal and router. They are a rapidly evolving telecommunications technology which will assure connectivity for popular mobile devices (laptops, PDAs, cell phones, etc.). Ad-hoc networks can provide spontaneous communications for users which are out of reach of infrastructure networks. They can also be used as extensions to existing networks. For example, community networks can be used to offer Internet access in a neighborhood. Finally, multi-hop ad-hoc networks can provide communications in emergency situations, in which the infrastructure networks have failed or are unavailable.

Currently existing wireless networks have demonstrated that it is possible to efficiently deal with data services (e.g., Internet connectivity). Therefore, there is a growing expectation that future wireless networks will efficiently deal with multimedia

DOI: 10.4018/978-1-61520-771-8.ch010

services as well. This is caused by the growing popularity of such applications as VoIP, multimedia streaming, peer-to-peer file sharing, etc. However, the nature of ad-hoc networks makes the task of serving delay sensitive or bandwidth consuming traffic with a proper QoS very complex. In comparison to wired networks, ad-hoc networks offer much smaller bandwidth and, therefore, their design requires much more attention. Additionally, such factors as mobility of devices, unpredictable channel conditions, the hidden and exposed node problems, limited battery power, and heterogeneity of devices make QoS provisioning in ad-hoc networks a very complicated challenge.

We begin with background information regarding the challenges of QoS provisioning in multi-hop ad-hoc networks. Then, we describe QoS solutions proposed for the physical, data link and network layers. Additionally, we discuss cross-layer solutions, which combine features of the previously presented protocols. Finally, we sketch future research directions and present the most important conclusions.

BACKGROUND

QoS is a term which has been widely used in modern telecommunications. QoS is the ability to provide different priorities to different applications or flows to guarantee a certain level of performance. QoS guarantees are especially important when the network capacity is insufficient or the network is exposed to congestion. QoS is most commonly measured by the following metrics: bit rate, delay, variation of delay (jitter), packet dropping probability and bit error rate (BER). In multi-hop ad-hoc networks providing QoS is particularly difficult because of the challenges at the following layers:

- **Physical layer.** The use of wireless technologies makes links susceptible to fluctuations in the radio channel. As a result such factors as fading or interferences may lead to low bit rates and high BERs. The physical layer should quickly respond in such situations to prevent high frame error rate (FER) at the data link layer. Furthermore, random movement of mobile nodes introduces unpredictable link failures which lead to network reconfiguration. Additionally, mobile nodes are usually limited by their battery power. Power consumption can be one of the QoS attributes, because it has a strong influence on all QoS metrics.
- **Data link layer.** With the help of adequate MAC protocols, nodes need to support service guarantees for multiple traffic classes and efficiently share a common radio channel with their neighbors. Additionally, traffic scheduling schemes for real-time traffic should be used to avoid starvation of best effort traffic. The protocol should also promptly react to transmission errors and collisions. The automatic repeat request (ARQ) or adaptive error correction methods should also be used when transmission quality degrades on the data link layer.
- **Network layer.** Nodes can move in a random way. Therefore, the network topology changes unpredictably and routing protocols need to quickly adjust. Additionally, there should be a signaling protocol responsible for admission control, resource reservation, reaction to congestion and negotiation of QoS parameters.

All the mentioned features make assuring QoS in multi-hop ad-hoc networks both a challenging task and an interesting research problem. Providing a complete QoS solution for the ad-hoc networking environment requires the interaction and cooperation between three OSI/ISO layers, i.e., the physical, data link and network layers. The first two layers allow for QoS support in a single-hop connection, the third layer is responsible for end-to-end QoS. Therefore, in a multi-hop environment a cross-layer approach seems to be mandatory. However, before describing several cross-layer approaches, we first look at the QoS solutions for each layer separately.

PHYSICAL LAYER PROTOCOLS

Wireless communication is very unpredictable, because the physical properties of the wireless channel change continuously. A signal transmitted over the wireless channel is vulnerable to interferences, fading and background noise. As a result, the quality of a wireless link is considerably lower and less stable than of a comparable wired link. In addition, the operation of neighboring nodes in multi-hop ad-hoc networks, where communication usually occurs in a common channel, decreases the available capacity of wireless links. Furthermore, any link changes in a multi-hop path can easily affect the quality of an end-to-end connection. To assure a proper QoS level at the physical layer (PHY), more sophisticated control of wireless links is needed. There are a number of PHY parameters that can directly influence the considered QoS metrics, namely: transmission power, receiver sensitivity, signal-to-noise ratio, and transmission rate. They are presented next.

Transmission Power

A transmitter is an electronic device which generates electromagnetic waves, usually with the aid of an antenna. The transmitter emits these signals with a certain power level, referred to as the transmission power. The strength of these signals decreases with distance. In theory (for short, line-of-sight, LOS, distances), the receiving power is proportional to $1/d^2$, where d is the distance between the transmitter and the receiver. In practice (for long, non line-of-sight, NLOS, distances), the receiving power is proportional to $1/d^{\alpha}$, where α is the path loss exponent and $\alpha \in [4, 6]$. When a receiver moves from a LOS to a NLOS condition the received power drops (typically by 15-25 dB).

National regulation agencies set an upper limit on the transmission power to protect human health and to avoid interferences. These limits usually depend on the type of devices and frequency bands used.

Ad-hoc transmitters are usually equipped with a limited power source such as batteries or accumulators. Their transmission power should be small enough to extend their operation time and high enough to achieve acceptable signal quality at the receiver.

Receiver Sensitivity

Receiver sensitivity determines its ability to discern low-level signals. It is a measurement of the weakest signal that can be received and correctly recognized by the receiver. Therefore, this parameter is one of the key specifications of any radio device. The larger the absolute value of the negative number, the better the sensitivity. For example, a sensitivity of -95 dBm is better than a sensitivity of -92 dBm by 3 dB, or a factor of two. This also means that at a specified data rate, a receiver with a -95 dBm sensitivity can hear signals that are half the strength of those heard by a receiver with a -92 dBm sensitivity. The impact of this parameter on network performance in multi-hop ad-hoc wireless networks is studied by Ferrari, Tonguz, and Bhatt (2004).

Signal-to-Noise Ratio

The signal-to-noise ratio (SNR) is defined as the power ratio between a signal and the background noise. The higher the ratio, the less obtrusive the noise is. Due to the nature of signals, mostly characterized by a very wide dynamic range, SNR is usually expressed in the logarithmic decibel scale. The noise is calculated as the sum of the background noise and the level of interferences at the receiver. The SNR at the receiver can be improved by reducing noise at the receiver or increasing the transmission power.

Transmission Rate

Transmission rate or bit rate is defined as the number of bits that are transmitted over a wireless link within a unit of time. It is usually quantified in bits per second. There are a number of bit rate definitions depending on the protocol layer. They are listed next. The physical layer gross bit rate (also known as the raw bit rate or data signaling rate) is the total number of physically transferred bits per second over a wireless link, including both useful data as well as protocol overhead. The net bit rate (also known as the useful bit rate or information rate) is the link capacity excluding the PHY layer protocol overhead, for example framing bits, equalizer training symbols, forward error correction (FEC) codes and other channel coding. The relationship between the gross bit rate and net bit rate can be expressed using the following formula: gross bit rate $\times$ FEC code $\geq$ rate. For example, in one of the PHY layers defined by IEEE 802.11, for a net bit rate of 6 Mbps the gross bit rate is 12 Mbps.

The PHY layer of modern wireless standards (e.g. IEEE 802.11, IEEE 802.16) can support multiple transmission rates. In order for a wireless device to utilize high transmission rates, the received signal needs to be greater than a given threshold, which is highly dependent on receiver sensitivity. It is up to the transmission rate selection algorithm to decide which rate to choose given the current channel conditions. This has a direct influence on the QoS metrics of the channel, especially throughput. Unfortunately, modern wireless standards do not specify the method of automatic rate selection in the presence of multi-rate capable devices. As a consequence, there are several existing methods of choosing the appropriate transmission rate and vendors of wireless devices are free to choose one of them or design their own. Several of the most popular approaches are presented next.

Statistics-Based Rate Selection Algorithms

Auto Rate Fallback (ARF) is an example of a rate selection protocol based on channel statistics. According to Awerbuch, Holmer, and Rubens (2003), it is one of the most common rate selection protocols. It was developed for Lucent's WaveLAN II devices (Kamerman & Monteban, 1997) and it uses the FER to determine the quality of the channel. After successful reception of a given number of consecutive ACKs from a neighboring node, the transmission rate is increased. Similarly, after a consecutive number of ACKs have been lost, the rate is decreased. This protocol requires no changes in the IEEE 802.11 standard because the sender imposes the transmission rate. However, ARF is not the optimal strategy because it is very slow to adapt to the channel conditions. Additionally, even if the channel conditions are stable, it will unnecessarily try to change the rate. Furthermore, it can mistake collisions for channel losses. A slight improvement over ARF is a retry-based approach (Van der Vegt, 2002). In comparison to ARF, it differs in that down-scaling is performed after a number of unsuccessful retransmissions. This results in a very short response time to deteriorating links. However, the protocol behavior is pessimistic. The rate will increase only after a FER threshold has been reached. This takes longer than the down-scaling procedure. There are also other well known statistics-based algorithms such as Onoe (http://madwifi-project.org/), Adaptive Multi Rate Retry (AMRR) (Lacage, Manshaei, & Turletti, 2004), and SampleRate (Bicket, 2005). Onoe is similar to ARF but not as sensitive to individual packet loss. It looks for the highest bit rate that has a loss rate less than 50%. AMRR uses binary exponential backoff and works well for high latency systems. SampleRate uses aggressive probe packets to estimate the optimum transmission rate.

SNR-based Rate Selection Algorithms

Numerous SNR-based alternatives to the statistics-based approach have been proposed. One of them is Receiver Based Auto Rate (RBAR) proposed by Holland, Vaidya, and Bahl (2001). In this protocol, the receiver can determine the transmission rate on the basis of the SNR of each received RTS frame. Then it informs the sender about the desired rate in the CTS frame. This estimation is precise because it is done just before the transmission of a data frame. RBAR requires both changes to the IEEE 802.11 standard and the use of RTS/CTS even when there are no hidden nodes. On the other hand, it allows faster adaptability then ARF.

The Opportunistic Auto Rate (OAR) protocol (Sadeghi et al., 2002) uses a different, more efficient approach. It utilizes the coherence times of good channel conditions to send high-rate multi-frame bursts. This is similar to the TXOP feature of IEEE 802.11 (IEEE, 2007). The overhead in OAR is low because there is no contention period or sending of RTS/CTS frames in these bursts. Changing the burst size can also increase fairness (in terms of bandwidth allocation time) within the network. However, the downside to these advantages is that OAR requires modifications to the IEEE 802.11 standard. Additionally, both RBAR and OAR suffer from using pre-selected SNR thresholds, therefore they may not perform well under different channel conditions.

DATA LINK LAYER PROTOCOLS

The data link layer is responsible for establishing the physical and logical communication between network nodes. This layer consists of two sub-layers: the Medium Access Control (MAC) sub-layer and the Logical Link Control (LLC) sub-layer. Most of the issues related to QoS provisioning occur at the MAC sub-layer. This includes the aspects of efficient and fair channel access, the problem of hidden and exposed nodes, traffic differentiation, resource reservation, and traffic scheduling. The perfect MAC protocol should provide suitable mechanisms to efficiently share the available bandwidth among nodes, achieve high system throughput, support different traffic classes with the required QoS metrics, and perform well in a multi-hop environment affected by hidden and exposed nodes. The QoS solutions at the upper layers (discussed in the next subchapters) usually assume the existence of a QoS-aware MAC protocol which supports reliable unicast transmission and scheduling of real-time traffic.

The problem of QoS support at the MAC layer has been a broad topic of research in recent years. The IEEE 802.11 EDCA (IEEE, 2007) protocol has been widely studied in the literature and is the only commercially available QoS MAC protocol. It is described elsewhere in this book. Many other MAC protocols supporting QoS have also been proposed and the most interesting solutions are shortly presented next.

MACA/PR

Multiple Access Collision Avoidance with Piggyback Reservation (MACA/PR) (Lin & Gerla, 1997) is a MAC protocol which provides guaranteed bandwidth to real-time traffic in a single hop network. Additionally, with the cooperation of QoS routing and fast connection setup mechanisms it can be used to support end-to-end multimedia delivery in a multi-hop network.

To transmit real-time frames, the sender initiates an RTS/CTS handshake and then, after receiving the CTS frame, proceeds with the DATA/ACK frames. The RTS/CTS frame transmission is used only to set up reservation for the first data exchange. If the ACK frame is not received, e.g. due to a collision, the DATA frame is not retransmitted. Moreover, if the sender fails to receive a number of consecutive ACKs, which

Figure 1. The MACA/PR protocol operation (Adapted from (Lin & Gerla, 1997))

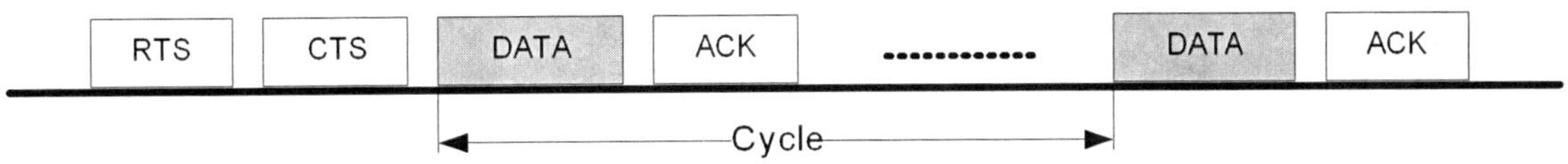

is a configurable protocol parameter, it restarts the connection with the RTS/CTS exchange. The reservation scheme requires that each node has a Reservation Table (RT) which keeps track of transmit and receive reserved windows (for any node within transmission range). The real-time scheduling information is carried in the headers of DATA and ACK frames. A node recognizes the next transmit time from the DATA or ACK headers and records it in its RT. This allows to avoid conflicts with ongoing reservations. The sender should piggyback the reservation information for the next DATA frame transmission on the current DATA frame. The receiver reads this information, puts it in its RT and confirms it with the ACK frame. The information transmitted in the ACK frame also prevents other nodes from transmitting at the time when the receiver is scheduled to receive the next DATA frame from the sender. The propagation and maintenance of RTs among neighbors overcomes the hidden node problem for real-time traffic. A typical frame transmission cycle is presented in Figure 1.

For the transmission of best effort traffic the operation of MACA/PR is similar to IEEE 802.11 DCF. The sender must first wait for a free window in the RT. Additionally, it waits a random time in the order of a single-hop round-trip delay. Then it starts sensing the channel. If the channel is free, it initiates the transmission of RTS/CTS/DATA/ACK frames. If the channel is busy, the whole procedure is delayed until the channel becomes idle.

To summarize, in MACA/PR best effort and real-time frame transmissions can be mixed at each node, with priority given to real-time traffic. For this traffic, the protocol behaves like a Time Division Multiplexing (TDM) system. Best effort frames can easily fill all empty windows in the cycle to achieve high overall protocol efficiency.

IEEE 802.11 DCF with a Multi-Priority Scheme

A variation of the IEEE 802.11 DCF protocol to support different traffic classes is proposed by Deng and Chang (1999). There are four traffic classes differentiated by their inter-frame space (IFS) and backoff periods. For higher priority traffic a node waits for the channel to be idle for PIFS, for lower priority traffic it waits for DIFS. Even if a node is waiting for PIFS it can still lose the contention if it chooses a backoff larger than other nodes (in particular, nodes which waited for DIFS).

The proposed scheme is simple and can be easily implemented in IEEE 802.11 devices. Simulation results (Deng & Chang, 1999) show that the proposed protocol has better performance than DCF in terms of throughput, access delay, and loss probability for higher priority traffic. Unfortunately, the considered scheme cannot provide deterministic delay bounds for higher priority traffic. Moreover, the lowest priority traffic suffers from much higher delay compared to DCF because longer backoff periods are selected even when no higher priority traffic is being transmitted.

Black Burst Contention Scheme

The Black Burst protocol (Sobrinho & Krishnakumar, 1999) provides a bounded time delay for real-time traffic in ad-hoc networks. This protocol is distributed and based on the CSMA access method. It ensures collision-free transmission of real-time frames. Nodes sending real-time traffic use pulses of energy, which are called Black Bursts (BB), to contend for medium access. The length of these pulses is proportional to the time the nodes had to wait for the channel to become idle. This delay is measured from the first attempt to access the channel by a node until its transmission starts. After transmitting its BB, the node waits for a specified time interval to see if any other node is transmitting a longer BB. If the channel is perceived idle after this interval, then the node can immediately transmit its real-time frame. Otherwise, it waits for the next channel access cycle and repeats the algorithm. A round-robin discipline among nodes transmitting real-time frames is enforced, which results in bounded access delays. The BB protocol can also support asynchronous data transmission. Nodes transmitting asynchronous data frames use a longer IFS than nodes sending real-time traffic. The BB contention scheme guarantees that real-time frames are always favored over asynchronous data frames. The BB protocol can be easily combined with DCF and implemented, with minor modifications, in WLAN cards. Unfortunately, the protocol does not consider the exposed node problem.

PUMA

The Priority Unavoidable Multiple Access (PUMA) protocol (Natkaniec & Pach, 2002) enhances DCF to support strict priority isochronous traffic transmission in ad-hoc mode. Three different time intervals are defined: SIFS, PIFS, and DIFS, where DIFS>PIFS>SIFS. This is similar to the IEEE 802.11 standard. Every active station measures these intervals after the end of each frame to determine the moment it can start its own transmission. The station proceeds with its isochronous transmission if the medium is determined to be idle for an interval that exceeds PIFS. All stations sending isochronous frames should start its transmission simultaneously and send the JAM signal. The JAM signal consists of pulses of energy (similarly to the BB protocol) and has the length of one slot. This signal informs all other stations (especially stations sending asynchronous frames) that in their neighborhood an isochronous transmission will begin. This means that all other stations have to defer their transmission until the reception of an RTS or CTS frame to update their network allocation vector (NAV). A random backoff interval is then selected and used to initialize the backoff timer. The backoff timer is decremented only when the medium is idle. It is frozen when the medium is busy until the next PIFS period. A station initiates an RTS frame transmission when the backoff timer reaches zero. To increase the efficiency of PUMA in a scenario with high load and a large number of contending stations, a backoff scheme called DIDD was used as the default backoff mechanism (Natkaniec & Pach, 2000). A typical isochronous frame transmission scenario is presented in Figure 2.

On reception of an RTS frame the receiver responds with a CTS frame, which can be transmitted after the channel has been idle for a time interval exceeding SIFS. After a successful exchange of RTS and CTS frames the transmitter sends its DATA frame in a collision free manner. In the case when a CTS frame is not received within the predetermined time interval, the RTS is retransmitted according to the backoff rules. Additionally, a multiple frame transmission mechanism is implemented in PUMA to increase the protocol performance measures. Data frames are transmitted in sequence without the risk of collision after a successful medium reservation through the RTS/CTS exchange. The number of data frames transmitted in sequence is configurable.

PUMA has the following additional features. The life-time of each isochronous frame is measured. If it reaches its limit and the frame cannot be sent to its destination it is treated as useless and removed from the station buffer. Furthermore, PUMA allows controlling the minimal amount

Figure 2. The operation of PUMA for isochronous traffic (Adapted from (Natkaniec & Pach, 2002))

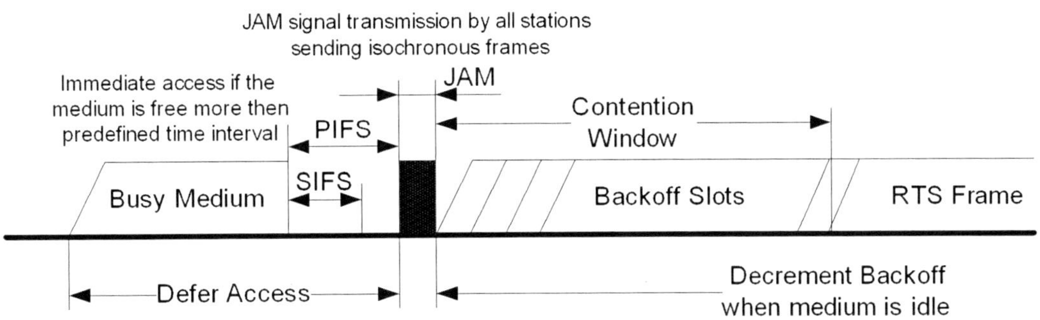

of asynchronous traffic by introducing an additional timer. It is used to measure the life-time of asynchronous frames located in the source station buffer. An asynchronous frame located in the head of the queue gets a higher priority if its life-time is reached. Its priority becomes equal to the priority of isochronous frames.

ES-DCF and DB-DCF

Two variants of DCF that incorporate explicit support of real-time traffic are proposed by Pal, Dogan, and Ozguner (2002). Both protocols use deterministic collision resolution algorithms in order to provide QoS guarantees for different traffic classes. The interesting fact is that both schemes do not apply any backoff mechanism.

The Elimination by Sieving DCF (ES-DCF) protocol defines three phases of operation: elimination, channel acquisition and collision resolution. In the elimination phase, every node receives a grade which depends on the deadlines and priorities of its real-time frames. A lower numerical grade means that a frame has a closer deadline, which also means that it has waited in the queue for a longer time, and its *channel-free-wait-time* parameter is smaller. This parameter also depends on real-time frame priority because ES-DCF defines two classes of real-time traffic. Additionally, finer sub-grades are assumed by choosing random numbers from a specified interval to avoid the existence of two or more real-time frames from different nodes with the same grade (similar frame deadlines usually mean the same grade). After a node has waited for the *channel-free-wait-time* the channel acquisition phase begins. If the channel is idle, the node transmits an RTS frame. After receiving the RTS frame all other nodes defer their accesses until the next channel acquisition phase. If the node receives a CTS frame, it can begin the transmission phase, in which its real-time frame can be sent. Otherwise, the collision resolution phase is initiated by transmitting BBs of lengths equal to the unique node ID numbers. The node that sends the longest BB wins the contention and accesses the channel at the subsequent attempt. It should be pointed out that the collision resolution phase introduces the blocked-access feature, where all nodes that have experienced a collision in the channel acquisition phase use the smallest *channel-free-wait-times*. This means that they can pre-empt all other nodes during the collision resolution phase. The operation of ES-DCF is presented in Figure 3.

The Deadline Bursting DCF (DB-DCF) protocol is similar in operation to ES-DCF. In the first phase, called the BB contention phase, a real-time node starts the transmission of a BB proportional to the urgency of its real-time frames (which corresponds to their deadlines). The BB lengths are defined as multiples of a BB slot time. After sending its BB transmission, the node checks the

Figure 3. Phases of ES-DCF operation (Adapted from (Pal, Dogan, & Ozguner, 2002))

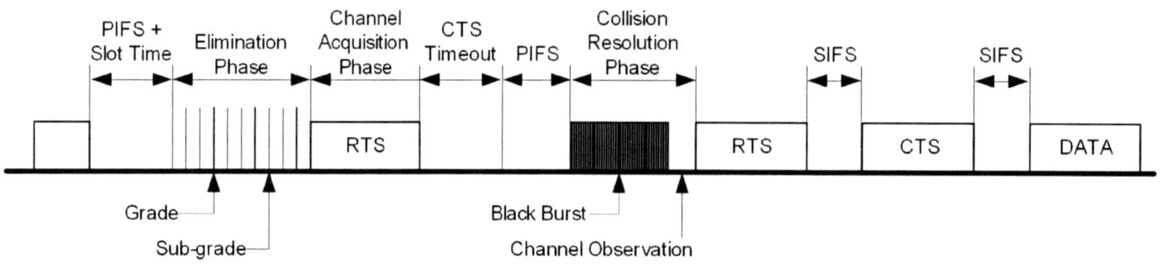

channel to determine any longer BB duration. If it senses any other BB transmission, it defers its channel access until the next channel access cycle, where a new BB length is calculated and transmitted. The channel acquisition and collision resolution phases of DB-DCF are exactly the same as ES-DCF. The operation of the DB-DCF protocol is presented in Figure 4.

Both protocols assume asynchronous frame transmission after the DIFS period. However, the ES-DCF protocol cannot be directly combined with an IEEE 802.11 DCF implementation because the *channel-free-wait-time* intervals for asynchronous data frames are longer than DIFS. Moreover, a high volume of real-time traffic can completely suppress asynchronous data transmission. Simulation results show that ES-DCF is more efficient for hard real-time traffic (i.e., real-time frames are dropped when expired), while DB-DCF behaves better for soft real-time traffic (Dogan & Ozguner, 2002).

QoS Enabled MAC for Multi-Hop Ad-Hoc

The MAC protocol proposed by Ying, Anand, and Jacob (2003) provides service differentiation for real-time constant bit rate traffic, real-time variable bit rate traffic and asynchronous non-real-time traffic. For real-time traffic it uses a distributed mechanism for scheduling and reserving the radio channel. According to the proposed scheme every non-real-time frame and the first frame from a real-time session (or burst) begins a typical RTS/CTS/DATA/ACK sequence. The subsequent frames in a real-time burst are transmitted using the DATA/ACK exchange. The protocol differentiates between ACK and DATA frames of real-time (R-ACK, R-DATA) and non-real-time traffic (D--ACK, D-DATA). R-ACK performs a reservation for the next R-DATA frame because RTS/CTS frames are transmitted only at the beginning of the transmission of real-time traffic.

Figure 4. Phases of DB-DCF operation (Adapted from (Pal, Dogan, & Ozguner, 2002))

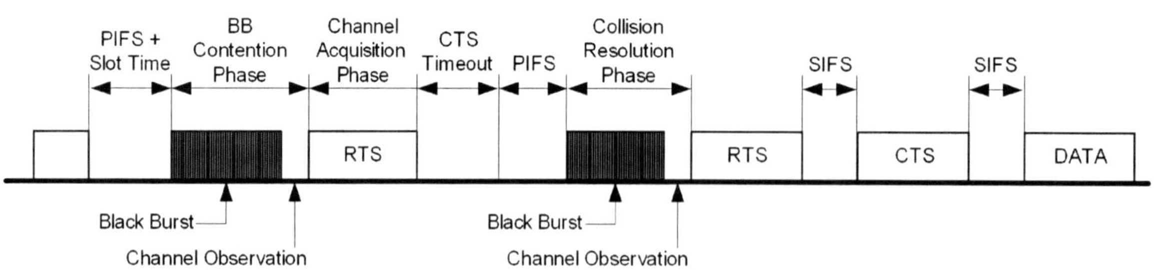

Figure 5. QMA protocol operation (Adapted from (Wang & Liu, 2007))

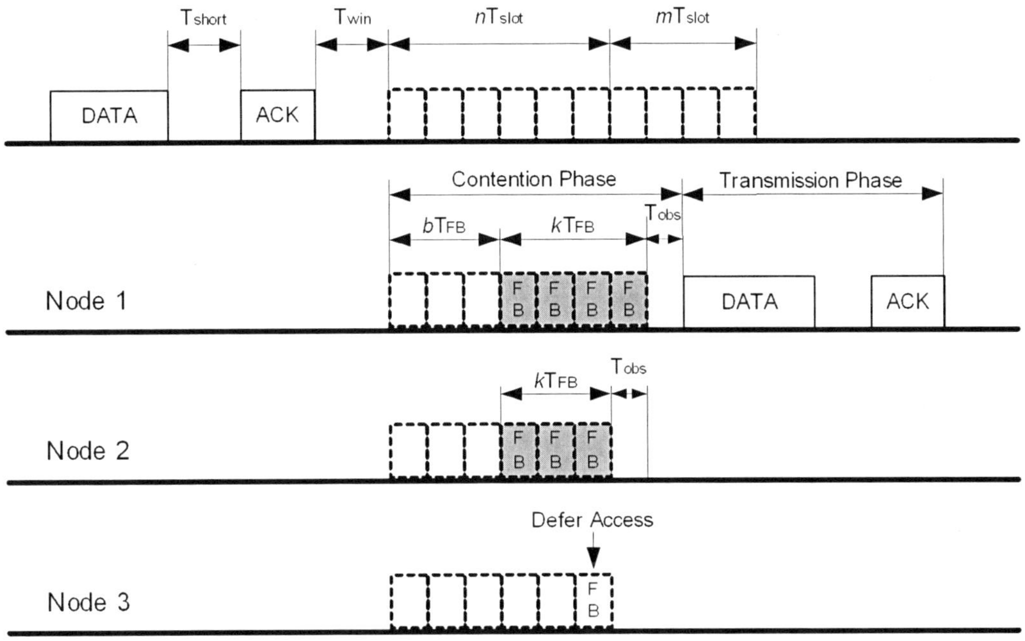

Each node maintains a receive and transmit reservation table (Rx RT and Tx RT, respectively). They contain reservation windows in which the neighboring nodes have scheduled to receive and transmit real-time frames. When a node receives R-DATA, it estimates the next Tx time and writes it into its Tx RT. After receiving R-ACK, it writes the reservation into its Rx RT. If a node wants to initiate a new non-real-time or real-time session, it has to check its Rx RT and Tx RT to find a free slot in which no neighbor is scheduled to receive or transmit during the time needed for the RTS/CTS/DATA/ACK transmission. In the case of receiving an RTS frame, it checks its Rx RT and Tx RT tables before responding with a CTS frame. After a successful DATA transmission, an ACK frame is expected. If no ACK is received, the sender node assumes that a collision has occurred and enters the backoff stage. The protocol uses the same binary exponential backoff mechanism as defined in the IEEE 802.11 standard.

The proposed protocol guarantees bounded delays for real-time traffic, however, its effectiveness highly depends on overhearing R-DATA and R-ACK frames. The authors have proven that their protocol achieves lower maximum and average delays for real-time traffic than EDCA, BB, and MACA/PR. The reservation tables help avoid collisions in hidden node scenarios, which results in small packet loss rates.

QMA

Wang and Liu (2007) have proposed a QoS-based Multiple Access (QMA) protocol for ad-hoc networks. This protocol supports two types of traffic: real-time and best effort. In QMA, the channel access cycle is divided into a contention and transmission phase (Figure 5). Each node is obliged to sense the medium for a time interval T_{win} before accessing it. If the medium is idle, the node can start the contention phase. The contention phase is composed of $n + m$ slots which are assigned to real-time (n slots) and best effort traffic (m slots). The nodes sending best effort traffic are allowed to broadcast a forecast burst (FB) in m slots only when all n slots are idle. This assures priority of real-time over best effort traffic.

A node chooses a number b, which is a random variable with a truncated geometric distribution. If a node senses the first b slots idle, it immediately starts transmission of $k \cdot$ FB slots. Otherwise, it stops its backoff. The k parameter depends on the frame lifetime. The frame with the earliest deadline has the largest k value. After transmission of FBs the node senses the medium. If the channel is busy, it means that there must be at least one contending node with a higher priority frame, and the node with a lower priority frame has to backoff. Otherwise, if the channel is idle, the node can start its DATA transmission. Successful reception of DATA is confirmed with an ACK frame. An example of the operation of QMA is presented in Figure 5.

The QMA protocol guarantees that only the nodes that start transmission of FBs in the same slot can successfully survive the contention phase. The node which sends the largest number of FBs wins the overall contention. The simulation results show that the QMA protocol with its well-organized collision resolution mechanism obtains a higher efficiency than IEEE 802.11 EDCA (Wang & Liu, 2007). Unfortunately, the protocol supports only two types of traffic. Furthermore, high real-time traffic can completely starve best effort traffic.

NETWORK LAYER PROTOCOLS

The network layer is mostly responsible for ensuring QoS routing, admission control and signaling. In this subchapter we mostly discuss QoS routing protocols, because they are the main focus of research in this layer. The main goals of such protocols are the following:

- Estimate the available network capacity. This information is often used to perform admission control.
- Find loop free routes which satisfy QoS requirements of flows. QoS constraints typically taken into account are jitter, delay, bandwidth and power consumption.
- Reserve the required resources.
- Maintain routes by utilizing redundant routes, predicting route breaks, and using a route recovery mechanism.

In multi-hop ad-hoc networks routing is very challenging mostly because of two reasons. Firstly, bandwidth is very limited and, therefore, a QoS routing protocol must have small overhead. Secondly, the topology is constantly changing and, therefore, the reserved resources cannot be hard guaranteed. The most important QoS routing protocols are described next in chronological order.

CEDAR

Core-Extraction Distributed Ad-hoc Routing (CEDAR) (Sinha, Sivakumar, & Bharghavan, 1999) is a hierarchical QoS routing algorithm for MANETs. It consists of three key components: core extraction, link state propagation, and route computation.

The main goals of CEDAR are, firstly, to compute routes quickly and, secondly, to react to the network changes without large amounts of state propagation. Therefore, the protocol focuses on rapid reaction to network changes rather than on the optimality of routes. Furthermore, there are several basic assumptions made in CEDAR. Firstly, nodes communicate on the same channel. Secondly, transmitters have a fixed transmission range. Thirdly, networks are small or of a medium size (tens to hundreds of nodes). Finally, the MAC-link layer can be used to estimate available link bandwidth.

Figure 6. Network with core nodes (a) and corresponding core graph (b). (Adapted from (Sinha, Sivakumar, & Bharghavan, 1999))

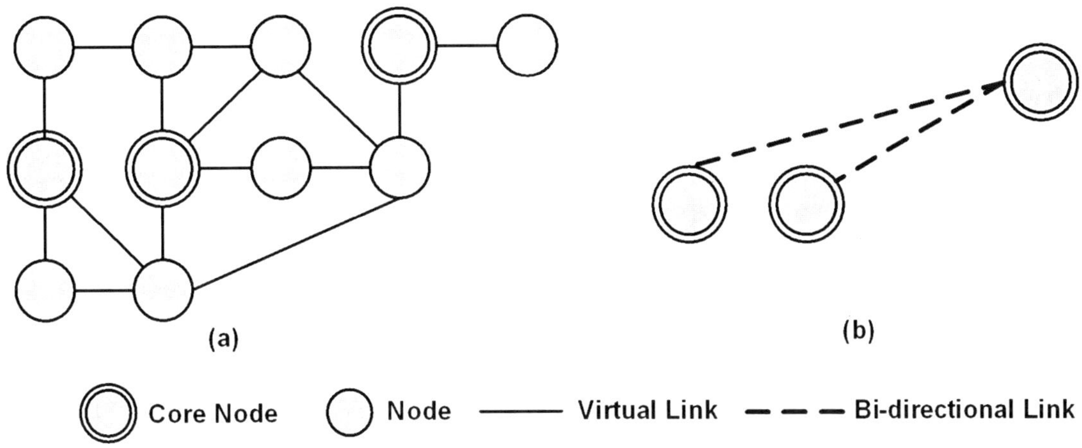

Route Discovery and Maintenance

CEDAR uses a greedy algorithm to create an approximate minimum dominating set (DS) of the core nodes (Figure 6). This set is chosen in a distributed manner. Each core node has enough local topology information to reach the domain of its nearby core nodes and set up paths (virtual links) to them. Other MANET members need to choose a *dominator* from the DS because only the core nodes maintain local topology information, participate in the exchange of network state information, discover and maintain routes, and perform admission control. When a host loses connectivity with its *dominator* (due to mobility) it either finds a new *dominator* from the neighboring core nodes, nominates one of its neighbors to join the core, or itself joins the core.

CEDAR assumes that each core node not only has up-to-date information about its local topology but also about stable high bandwidth distant links. To achieve this goal, it adopts *increase* and *decrease waves*. The former provide information about an increase of the available bandwidth. They are propagated locally and they are periodical. The latter provide information about a decrease of the available bandwidth. They are propagated distantly and they are sent immediately after each bandwidth change. These waves are generated every time when an estimate of the available bandwidth changes by some threshold value. CEDAR propagates information about the state of stable high bandwidth links throughout the core, and keeps information about the sate of low bandwidth or unstable links locally. This is possible because for an unstable link the *decrease wave* stops the *increase wave* from propagating.

The QoS route computation scheme in CEDAR involves the following three phases:

- **Establishment of the core path:** Firstly, a source node sends a request to its *dominator*. Then the *dominator* forwards this message to each of its nearby core nodes using the core broadcast algorithm. After the *dominator* of the destination node receives this message it responds with a source routed unicast *core_path_ack* message. When the *core_path_ack* message is received by the source *dominator*, the core path establishment phase is finished and the QoS route computation phase can be started.

Figure 7. Reestablishment of a route (Adapted from (Sinha, Sivakumar, & Bharghavan, 1999))

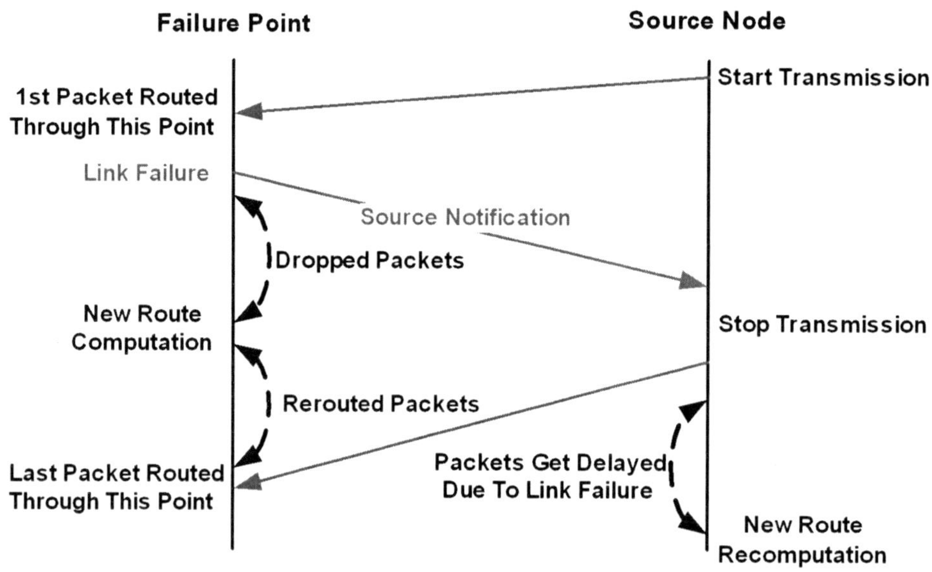

- **QoS route computation:** The *dominators* of the source and the destination nodes have partial knowledge about the local topology and the remote stable high bandwidth links. On this basis the *dominator* of the source node is able to compute a path to the furthest, intermediate core node in the core path which can guarantee the requested bandwidth. The route selection is based on a two phase Dijkstra's single source shortest path algorithm in order to find the shortest-widest path. Then, the intermediate core node starts a QoS route computation using its local state. As a result, the concatenation of the partial paths computed by the core nodes provides a QoS core route from the *dominator* of the source node to the *dominator* of the destination node.
- **Dynamical re-establishment of routes for ongoing connections** (Figure 7)**:** In the case of link failure or a topology change two mechanisms can be used: QoS route re-computation at the failure point and QoS route re-computation at the source. The former is suitable for failures occurring near the destination. The latter is effective when the failures occur near the source.

TBP

Ticket-Based Probing (TBP) (Shigang & Nahrstedt, 1999) is a multipath distributed routing scheme which uses tickets to limit the number of candidate paths. It can handle different QoS constrains (such as bandwidth and delay). The basic scheme of TBP is as follows. When a source node wants to find QoS paths to a destination node, based on the available network state information, it issues routing messages (probes, P) with a given number of tickets (T). When a probe message with more than one ticket is received by an intermediate node, based on its local state information, the node may decide to split the probe and forward different tickets on different downstream sub-paths. When the destination node receives the probe, a possible path from the source to the destination is found. TPB assumes that one ticket is a permission to search a single path and that one probe should

Figure 8. TBP operation (Adapted from (Shigang & Nahrstedt, 1999))

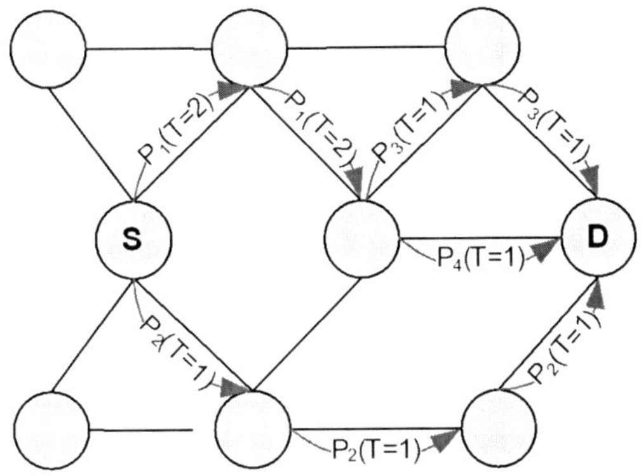

carry at least one ticket. Therefore, the maximum number of probes and the maximum number of searched paths are dependent on the number of tickets issued from the source node.

An example of TBP operation is presented in Figure 8 where S is the source node, and D is the destination node. Two probes are initiated at S: P1 with two tickets, and P2, with one ticket. P1 is split into two probes (P3 and P4) at one of the intermediate nodes. As a result three paths are found.

BR

The Bandwidth Routing (BR) protocol (Lin & Liu, 1999) uses bandwidth QoS constraints to establish paths between pairs of nodes. The protocol is designed for TDMA networks. Therefore, bandwidth is measured in terms of free timeslots available. BR works in conjunction with the Destination-Sequenced Distance Vector (DSDV) routing scheme.

The performance of BR is based on the source node's knowledge of end-to-end bandwidth available to any possible destination, which enables efficient support of real-time applications and helps establish QoS routes. Additionally, the protocol supports admission control. The assumptions made by Lin and Liu are the following. Transmissions are half-duplex, i.e., each node can either transmit or receive data. The channel is time slotted and either a time synchronization mechanism or a global clock is provided. In each data slot one data packet can be transmitted.

Bandwidth Reservation and Slot Assignment

In this protocol each node has its own set of free slots. A common set of free slots between two adjacent nodes denotes the *link bandwidth* between these two nodes. The *path bandwidth* is calculated on a hop-by-hop basis along the whole path from a source node to a destination node and it is the set of available slots between the two nodes.

The protocol assumes that each frame is divided into two phases: the control phase and the data phase. The control phase is used to perform the control functions (e.g., slot and frame synchronization, power measurement, setup of virtual connections, building of routing tables). The amount of slots/frames assigned to a path is determined by a QoS requirement. The control phase uses pure TDMA with full power transmission in a common code, i.e., each node broadcasts its routing information (obtained by DSDV) and its QoS requirements to its neighboring nodes in predefined timeslots. In noisy environments an additional ACK mechanism is employed to assure correct data exchange. At the end of the control phase each node can schedule free slots, verify the failures of reserved slots and drop expired packets. This is possible because nodes have information about channel reservations made by their neighbors. In the data phase, the required bandwidth resources are first pre-allocated and then traffic exchange may take place. Therefore, the protocol assumes that the amount of bandwidth along the path is computed and known to all nodes. This information becomes useful when a new request enters the network because it can immediately determine if a new flow can be accepted or not.

The free slots are assigned during session setup by the slot assignment algorithm. Every intermediate node and the destination node, after receiving a reservation request from the source node, checks whether it has enough free slots to receive and forward the data packets. If the required number of slots is available, they become reserved, the routing table gets updated and the session setup is forwarded to the next neighbor. Otherwise, the reservation fails and all current reservations on the path back to the source node are cancelled with the use of a RESET packet. When the end-to-end path reservation is successful, the destination node sends a REPLY packet to the source node to acknowledge a positive connection setup.

Route Maintenance

Each node holds two secondary paths in its routing table which can be used when the primary route fails. The secondary path is chosen for a new primary path if it satisfies the QoS requirements of a particular flow. The primary route does not have to be the highest bandwidth path; it must be the shortest one meeting the QoS requirements.

BRuIT

Bandwidth Reservation under InTerferences influence (BRuIT) (Chaudet & Guérin Lassous, 2001) is a distributed signaling protocol for bandwidth reservation which takes into account the existence of interferences between nodes. The authors concentrate on the bandwidth metric because it may affect such parameters as delay or jitter. The performance of BRuIT is based on periodically determining which nodes interfere with other nodes and what are their bandwidth reservations. BRuIT is implemented over a reactive routing protocol.

Neighborhood Discovery

In order to obtain knowledge about its neighborhood, each node periodically broadcasts a *Hello* packet. This packet contains the sender's address, the total bandwidth which it will use for already accepted flows and the information learnt from its neighbors which are k hops away. Propagation of *Hello* packets within two hops is presented in Figure 9. The reception of these packets helps each node compute the remaining bandwidth and allows more precise admission control.

Route Discovery

Every source node, to reserve bandwidth for a single flow, broadcasts a route request message including information about the destination node's address and the amount of requested bandwidth. Then, the admission control procedure is performed at each intermediate node. If it fails, the request is dropped. Otherwise, the request is further broadcasted until it reaches the destination node. Upon reception of more than one request for a particular flow, only the first one is accepted. After the destination node receives the request packet, it does admission control and, if the amount of free bandwidth is sufficient it replies with a route reply message. When intermediate nodes receive the route reply packet, they check if they still have enough bandwidth to satisfy the QoS requirements of the flow. If the amount of available bandwidth is not satisfactory, the reply message is dropped. Otherwise, the requested amount of bandwidth is reserved. Finally, after the source node receives

Figure 9. Broadcasting of Hello packets (Adapted from (Chaudet & Guérin Lassous, 2001))

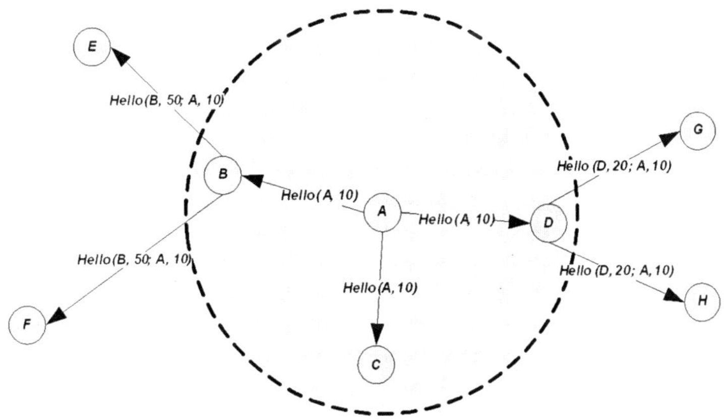

the reply from the destination node, it starts its data transmission.

Route Maintenance

To deal with mobility each node periodically sends *Hello* packets. Lack of *Hello* packets from a given neighbor informs about its movement or failure. Additionally, to deal with route breaks, reservations made by intermediate nodes have a soft-state. Therefore, if there is no data exchange on a given path for a certain time, the reserved bandwidth is released. Finally, to deal with misbehaving applications (i.e., applications using more bandwidth than requested) and to avoid frequent interferences caused by the exchange of false information, each sender node shapes its traffic using token bucket filters.

TDR

Trigger-based Distributed Routing (TDR) (De et al., 2002) is an on-demand routing protocol designed to support QoS-aware real time applications. It aims to effectively deal with link failures. Additionally, to reduce control traffic, TDR utilizes GPS-based location information and on-demand route discovery.

Database Management

In TDR all hosts must keep local neighborhood and routing information. Therefore, each host maintains two databases:

- **Local Neighbor Database**. It stores information about the location and the mobility of neighboring nodes (carried in periodically sent beacons) and the power level of these beacons.
- **Activity-Based Database**. It stores routing information valid for each session. Depending on the role of the node which stores the routing table it is called the source, destination or intermediate node database. Obviously, nodes may require maintaining some or all types of databases for different ongoing sessions.

Route Discovery

In TDR the source node floods route discovery packets to its neighbors. In order to reduce the signaling overhead, only some of the neighboring nodes are considered as possible next hops in the route. They are selected on the basis of the received power level which has to be greater than a predefined threshold.

At the beginning of the route discovery process, the source node must check if it has enough available bandwidth to satisfy the bandwidth requirement in the request. If enough bandwidth is available it is temporarily reserved by the source node. The reservation time is equal to the time in which it is expected that an acknowledgement packet from the destination node will be received. A valid route to the destination node is found with the help of a modified breadth first search algorithm. All intermediate nodes, upon receiving the first discovery packet, perform admission control based on available bandwidth. When the destination node receives the discovery packet and meets the QoS requirements of the session, it accepts the discovered route and sends an acknowledgement to the source node. If the location of the destination node changes it sends an appropriate location update message. Additionally, in order to avoid routing loops, TDR requires intermediate nodes to accept only one route discovery packet per session.

Route Maintenance

Route maintenance in TDR is based on three received power levels: P_{th1}, P_{th2}, P_{cr}, where $P_{th1} > P_{th2} > P_{cr}$. When the downstream received power level is lower than the critical limit P_{cr}, the source-destination route gets disrupted until an alternate route is set up by the source node. If the power level is between P_{th2} and P_{cr}, the source node is notified by an intermediate node with a rerouting request. Finally, if the power level is between P_{th1} and P_{th2}, the intermediate node initiates the rerouting process.

QoS-AODV and QoS-TORA

Gerasimov and Simon (2002) propose QoS extensions to the AODV and TORA (Temporally Ordered Routing Algorithm) routing protocols. These extensions add scheduling and resource reservation for a TDMA-like data link mechanism. The modified protocols are called QoS-AODV and QoS-TORA, respectively. They combine information from the data link layer and the network layer.

QoS-AODV

At the beginning of the path discovery procedure a source node checks if it has enough residual bandwidth to any of its neighbors to meet the requirements of an application. If there is enough bandwidth available the source node floods a modified route request (RREQ) packet to its neighboring nodes. The modifications include application ID and number of slots required for successful reservation. Upon receiving RREQ, each intermediate node performs admission control based on available bandwidth. Additionally, each intermediate node checks if it has an entry in its routing table corresponding to the received application ID. If the entry does not exist, it is created. Otherwise, the node checks if the RREQ received is newer than the one it has and, if necessary, updates its routing table. Each entry in a routing table contains addresses of three downstream nodes (in the direction to the source) as well as bandwidth schedules between those nodes. The bandwidth information is included in order to inform of the calculated bandwidth, prevent direct collisions and avoid the hidden node problem. Finally, upon receiving the RREQ packet, the destination node checks its residual bandwidth. If the admission control succeeds, it starts the reservation protocol. Additionally, if more than one RREQ is received by the destination node, it chooses the one with enough bandwidth, not the one with the fewer number of hops as in the original AODV.

QoS-TORA

There are two possible means of route discovery in QoS-TORA. Firstly, if a best-effort path from a source to a destination node does not exist a TORA *Query* packet is sent. Secondly, if any path exists, the source node sends a *Bandwidth Query* packet, which contains the number of slots needed and the application ID, on a known path. Upon receiving this packet, the destination node does admission control based on its residual bandwidth and broadcasts an *Update Bandwidth* packet, which contains the application ID, number of slots required and the source node ID. After an intermediate node receives this packet it calculates a new path bandwidth, checks if it has an entry for the received application ID, and updates its routing table with the best QoS path. The source node has to wait for several *Update Bandwidth* packets from its neighboring nodes before it can start the reservation protocol. This is done in order to make the selection of a QoS path possible and, upon path break, skip the route discovery procedure and immediately use an alternative path.

QoS-AODV

QoS-AODV (Chenxi & Corson, 2002) is a routing protocol using TDMA. Its operation is limited to small networks. The basic idea is that QoS routes are built only if necessary. The authors assume that applications are session-oriented and have a constant bandwidth requirement. The QoS requirement of a session is specified by the number of time slots needed on a route from a source to a destination node. Therefore, QoS-AODV finds both the route between the two nodes and the slots for each link on a path.

Bandwidth Calculation

The source node specifies the required number of slots along a QoS path. Each node along this path must find at least the required number of free slots for a transmission to its downstream neighbor. The algorithm looks for non-conflicting slots only on three adjacent links. Therefore, QoS-AODV aims for the local rather than the global maximum bandwidth. After the local maximum bandwidth is found, the calculation is propagated along the path to the destination node.

Route Discovery

Network bandwidth is calculated in conjunction with route discovery, i.e., to find a QoS path a source node floods a route request (RREQ) packet and, simultaneously, bandwidth is calculated on a hop-by-hop basis. If the requested bandwidth is not available at any intermediate node, RREQ is dropped. Otherwise, upon receiving the request, a destination node sends a route reply (RREP) packet to the source node and reserves necessary transmission slots. Additionally, if the destination node receives multiple RREQs, the first request satisfying the bandwidth requirement is accepted and the others are ignored. This is done in order to reduce the delay of the route discovery procedure.

Bandwidth Reservation

QoS-AODV proposes to use soft-sate bandwidth reservations for QoS paths. If a particular path is not used for some time, its entry is dropped from the routing tables. The authors define several states of a QoS route, e.g., indicating that the route is established or broken. Transitions between states are triggered by receiving or transmitting a packet or by the expiration of a timer associated with a particular state. In addition, QoS-AODV uses two timers, *Route setup time* and *Route life time*. The former is used during route discovery and route repair. It is equal to the round-trip time from the source to the destination node. The second timer is used as a maximum interval for data arrival. It helps detect broken paths.

PLBQR

Predictive Location-Based QoS Routing (PLBQR) (Shah & Nahrstedt, 2002) consists of an update

Figure 10. Location prediction using last two updates (Adapted from (Shah & Nahrstedt, 2002))

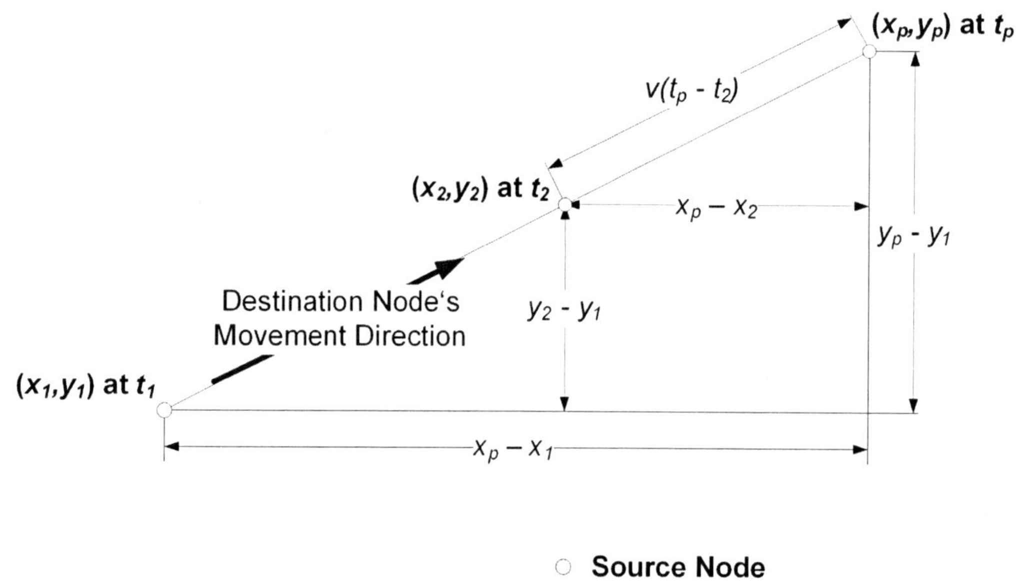

protocol, a location-delay prediction scheme and a QoS routing protocol.

Update Protocol

Distribution of geographical location (obtained through GPS) and resource information is done by the update protocol. PLBQR considers two types of updates: a *Type 1 update* (generated periodically by each node) and a *Type 2 update* (generated when there is a considerable change in a node's speed or direction). All update packets contain timestamps, current geometric coordinates, movement direction, velocity, resource information (for QoS routing) and a single-bit *motion stability parameter*. This parameter informs the QoS routing protocol about the type of an update, i.e., it helps assign dynamic nodes as intermediate nodes only for connections without strict delay or jitter constraints.

Location-Delay Prediction Scheme

Before any source node can establish a connection to a particular destination node, it has to predict the geographical location of all intermediate nodes and the destination node at time t_p when a packet from the source node will reach them. Additionally, the propagation delay must be known in order to estimate t_p. Location predictions are performed based on updates received from other nodes.

An exemplary location prediction is presented in Figure 10. In the figure, $(x_1; y_1)$ *at* t_1 and $(x_2; y_2)$ *at* t_2, where $t_2 > t_1$, are the latest two updates from a destination node received by the source node and v is the velocity of the destination node. The value of t_p is set to a sum of the current time and the time to reach the destination node from the correspondent node. Using simple calculations (based on triangle similarity and the Pythagoras theorem), the location $(x_p; y_p)$ of the destination node at time t_p can be predicted.

The destination node uses the same calculations to judge if there is a considerable change in its location and decide if a *Type 2 update* message must be sent to its neighbors. Additionally, PLBQR assumes that the end-to-end delay of a data packet transmission from the source to the destination node will be the same as the delay experienced by the latest update from the destination to the source node.

QoS Routing Protocol

Based on the information received from the update messages, each node has up-to-date information about the whole network. On this basis it can compute a route to any destination. Each node maintains two tables. The update table contains information learnt from the update messages and, for each node, a *proximity list* with a list of nodes lying within a distance of one and a half transmission range. The route table contains information about active connections set up by the source node. Thanks to the *proximity list* a given source node, during its QoS routing process, may take into account nodes which moved into the transmission ranges of other nodes as possible intermediate nodes.

When an update message is received at the source node, it checks if any of the known routes is damaged or is about to be broken by either node movements or by being unable to satisfy the QoS requirements of a connection. In both situations a route re-computation must be initiated. Because such information as remaining battery power, transmission range, and CPU utilization are exchanged in the update packets, the re-computation of a route may begin before it is really broken.

The routing algorithm itself works as follows. At first the source node runs the location and delay predictions for each node in its *proximity list*. On this basis it determines which nodes have enough resources available to satisfy the QoS requirements of a request. Then, it finds all possible routes towards the destination node by simultaneously performing route discovery and admission control on a hop-by-hop basis. If more than one route satisfies the QoS requirements, the geographically shortest one is selected for data transmission.

QoS-OLSR

Ying, Kunz, and Lamont (2003) propose several algorithms which allow OLSR (Optimized Link State Routing) to support QoS routing by selecting the highest-bandwidth paths between any two nodes. Their basic ideas are based on changing the way of selecting multipoint relays, depending on the bandwidth QoS constraints. Additionally, instead of using the shortest path algorithm, the maximum bandwidth spanning tree method is used. To achieve this goal, the bandwidth of each link is considered as its weight and all nodes compute trees in which the total weight of links are maximal among all possible trees.

AQOR

Ad Hoc QoS On-demand Routing (AQOR) (Xue & Ganz, 2003) is a reservation-based routing and signaling scheme. It provides end-to-end QoS support and admission control. Additionally it supports two QoS maintenance mechanisms: temporary reservation and destination-initiated recovery.

Neighborhood Maintenance

AQOR is based on the exchange of neighborhood information consisting of local topology, traffic and mobility information. This is crucial for traffic measurement, QoS violation detection and route recovery. Therefore, each node within a network periodically sends *Hello* packets with information about its traffic. Other nodes, upon receiving these packets, maintain lists of their neighbors with their corresponding traffic. If the *Hello* packets from a particular neighbor are not received during a predefined period, the connection to this neighbor is broken.

Route Discovery

Route discovery is done on-demand with the use of route request (RREQ) and route reply (RREP) packets exchanged between the source and the destination nodes. If the destination node is not within the neighborhood of the source node it broadcasts a RREQ packet. RREQ contains information about the minimum requested bandwidth and the maximum end-to-end delay. Every intermediate node does admission control on a hop-by-hop basis. If multiple routes are found during the exploration process the source node chooses the best path for its data transmission. Additionally, in order to avoid routing loops, all control packets are sequenced.

Admission Control and Temporary Reservation

During its operation AQOR takes two QoS constraints into account, namely bandwidth and end-to-end delay. In order to determine the available bandwidth the total traffic load is calculated for each node. The end-to-end delay is measured as the round-trip delay. After the route discovery phase is completed the path with the lowest end-to-end delay is selected for data transmission. If the source node, however, does not receive any answer to its RREQ during a predefined time interval it has to either backoff and initiate the route re-discovery procedure or resign from sending its flow. Finally, if the path is found, intermediate nodes make temporary reservations for the source node's data flow in order to guarantee the availability of the resources.

Route Maintenance

Route maintenance in AQOR includes the detection of the following end-to-end QoS violations:

- **Channel deterioration/congestion**. Detection is possible through one-way delay measurements. If the destination node receives a number of consecutive data packets with delay exceeding the maximum delay requirement, it triggers the QoS route recovery procedure.
- **Route breaks**. For best effort traffic detection is possible thanks to *Hello* messages. When a route break is detected, the source node is notified with an error message. Upon receiving the notification the source node can start the rerouting process. This approach is not appropriate for real-time traffic because of large delays. Additionally, in AQOR the bandwidth reservation timeout at the destination node is utilized. After the timeout is exceeded the source node starts the QoS route recovery procedure again.

In both cases, the destination node initiates the QoS route recovery procedure by sending an update message, which is treated in the same manner as a typical RREQ. When the message reaches the source node it can either immediately switch its flow to the new route or suspend it.

ADQR

Adaptive Dispersity QoS Routing (ADQR) (Youngki & Varshney, 2003) is a source initiated on-demand routing protocol. It uses signal strength information to predict link breaks and initiate fast data rerouting. ADQR assumes that lower layers provide information about the estimated bandwidth.

Sets and Classes

ADQR defines three signal strength levels: Th_1, Th_2, and S_R (Figure 11), where S_R is the minimum signal strength required to successfully receive a

Figure 11. ADQR thresholds (Adapted from (Youngki & Varshney, 2003))

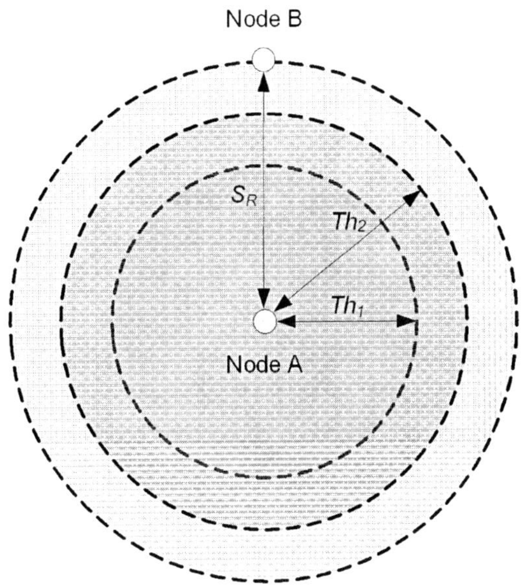

data packet from a neighbor and R is the transmission range to this neighbor.

Furthermore, ADQR defines three sets: *node*, *link* and *route*. Each set is divided into three classes. If for a given node the received signal strength from its neighbor is higher than Th_1, the node belongs to the *first node class* of this neighbor. Correspondingly, all connections between *first node class* nodes are in the *first link class*. If the received signal strength from a neighbor is between Th_1 and Th_2, the node belongs to the *second node class* of the neighbor. The connections between these nodes are in the *second link class*. Finally, if the received signal strength is between Th_2 and S_R, the node is in the *third node class* of the neighbor and the connections between such nodes are in the *third link class*. The route classes are determined by their weakest links.

Neighbor and Routing Table

In ADQR each node keeps two tables. The neighbor table contains an updated list of the node's neighbors and their corresponding received signal strength and the minimum signal strength S_R. The routing table contains a list of all possible routes to destination nodes. Each entry includes information about the available and reserved bandwidth along the path and the list of links in the link classes.

Route Discovery

Route discovery begins when a source node broadcasts a *Route_Request* packet to its neighbors in order to find multiple disjoint paths to a destination node. *Route_Request* is then appended with the addresses of all intermediate nodes. In addition, it is updated with available bandwidth information. Upon receiving this message, the destination node updates its routing table. The destination node replies with a *Route_Reply* packet. All intermediate nodes check *Route_Reply* and update their routing tables. If the source node receives multiple routes it selects the one with the best signal strength of the links (i.e., with the best *route class*). Finally, when an appropriate route is selected, a *QoS_Reserve* packet is sent along this path (from the source to the destination node) in order to reserve the required bandwidth. Additional *QoS_ACK* packets, sent to the neighbor from which the *QoS_Reserve* packet was received, guarantee correct reservation.

Route Maintenance

In AQDR route maintenance is based on the measurement of the received signal strength. When the received signal becomes too weak, *Route_Update* packets are generated for inactive paths. They are sent to the source node by nodes which detected the problem in order to update the *link class* information. For active paths AQDR proposes a fast route maintenance scheme called *two-phase monitored rerouting*. When the signal strength of a particular link on the active route becomes lower than Th_1 the *Pre_Routing* phase is started. If the signal strength does not stop to deteriorate (i.e., it drops below Th_2) the *Rerouting* phase is started. The *Pre_Routing* phase is introduced to prepare rerouting. It is invoked in order to find an alternative route before the currently used route becomes unavailable. In addition to these procedures, the source node caches all possible routes to the destination node and, therefore, alternative routes are practically always known. Furthermore, when a particular link breaks, a *Route_Error* packet is sent to the source node, which triggers all intermediate nodes to release the reserved bandwidth. Each intermediate node replies with a *Route_Ack* packet to its previous neighbor and forwards the packet to the next neighbor. Additionally, when a node reserves network resources for its currently active paths it sends a *QoS_Update* packet to the source node of each non-active path to which it belongs. This helps intermediate nodes and the source node to keep up-to-date network resource information in their routing tables.

QS-AODV

QS-AODV (Yihai & Gulliver, 2005) is a QoS routing protocol based on AODV which creates routes according to the QoS requirements of an application. In order to improve the packet delivery ratio QS-AODV employs a local repair mechanism. Its performance is comparable to AODV under light traffic conditions and considerably better (in terms of packet delivery ratio and signaling overhead) under heavy traffic conditions. On the other hand, under heavy network load, QS-AODV has longer end-to-end delays than AODV.

QS-AODV adds additional QoS information into route request (RREQ), route reply (RREP) and route error (RERR) packets of AODV in order to create and discover routes. Information about application bandwidth requirements and session ID are added.

Route Discovery

To begin route discovery, a source node sends a RREQ packet with the QoS extension to its neighbors which perform admission control based on available bandwidth. When a destination node receives the request, after updating its routing table and reserving the required bandwidth, it can send a RREP packet to the source node. However, if the destination node has already received a similar RREQ packet, the request is buffered in case of a route reply failure. Each intermediate node, upon receiving the RREP packet, checks if it still has the required bandwidth. If the admission control fails, a RERR packet is sent to the downstream node, otherwise, the RREP packet is forwarded to the upstream node. Upon receiving RERR, each node invalidates the route entry associated with the session ID carried in this RERR, releases the reserved bandwidth and forwards the RERR packet to its next hop neighbor. Finally, when the RERR packet reaches the destination node, it can use another available route to send a new RREP packet.

Route Maintenance

To provide fast rebuilding of routes, QS-AODV introduces the idea of local repair requests. Each repair request packet includes the session ID and the required amount of bandwidth of the considered flow and has its TTL (Time To Live) value set to 3 in order to limit the broadcast area. Upon detecting a link failure to the next hop neighbor, a repair request packet is sent to a node located after this neighbor. This is possible because QS-AODV assumes that routes to nodes which are two hops away are stored in the routing table of each node. If the local repair request expires, an error procedure is invoked.

QoS-aware AODV

QoS-aware AODV (also known as Bandwidth Estimation QoS-aware Routing, BEQR) (Lei & Heinzelman, 2005) is a QoS routing protocol based on the traditional AODV routing scheme, which utilizes a cross-layer design. It can either provide feedback about the available bandwidth to the application (feedback scheme) or admit a flow with the requested bandwidth (admission scheme). The former is suitable for applications which can adjust their coding rate on the basis of the received feedback. The latter is suitable for applications which have a predefined minimal required bandwidth. Both schemes require knowledge of the available end-to-end bandwidth along the path from a source to a destination node. Bandwidth estimation is therefore a key concern of this protocol.

Bandwidth Estimation

The goal of bandwidth estimation is to find the minimal available bandwidth along the path. Two methods of bandwidth estimation are possible:

- **"Listen" bandwidth estimation**. Each node estimates the residual bandwidth by listening to the channel using physical and virtual carrier sense. This method is inaccurate when a route is broken because nodes do not know how much bandwidth is consumed by other nodes in the broken path and, therefore, do not know the amount of bandwidth released.
- **"Hello" bandwidth estimation**. Information about the sender's current bandwidth usage and its one-hop neighbors' current bandwidth usage is piggybacked onto a modified *Hello* message. The residual bandwidth is based on the information from nodes within two-hops because, typically, the interference range is twice the transmission range.

Route Discovery

QoS-aware route discovery is initiated when the source node sends a route request (RREQ) packet. After an intermediate node receives this packet, it performs admission control based on available bandwidth. For the adaptive scheme, the intermediate node compares its residual bandwidth with the minimum bandwidth specified by the RREQ packet. The packet is forwarded if the residual bandwidth is greater than the minimum bandwidth. Otherwise, the RREQ is updated with the residual bandwidth. Later, after receiving the RREQ packet, the destination node does the same checking procedure. After the procedure is completed an additional checking procedure is invoked in which the upper bound of the minimum available bandwidth is re-estimated. In the end, the destination node sends the route reply (RREP) packet with an updated minimum bandwidth value to the source node. Upon receiving RREP, all intermediate nodes enable the path and update their routing tables with the new value of the minimum available bandwidth.

Route Maintenance

AODV detects a broken path by monitoring *Hello* packets. If a node stops receiving *Hello* packets from its neighbor, it sends an error message to its upstream neighbors. Upon receiving the error message, only the source node reinitiates a routing discovery procedure. This scheme works correctly for the *Listen* bandwidth estimation method. However, it has to be modified for the *Hello* bandwidth estimation method. In this scheme the neighboring nodes' caches are not updated in a timely fashion and, therefore, there is a high probability that the bandwidth used by the broken path will not be released before a new RREQ will arrive. As a remedy, QoS-aware AODV introduces *Immediate Hello* messages which have the same function as the modified *Hello* packets but they are sent immediately after detecting each broken link to allow faster cache updates.

QMRPCAH

Layuan and Chunlin (2007) propose a QoS multicast routing protocol for clustering mobile ad-hoc networks (QMRPCAH). The protocol establishes paths based on QoS constraints in a scalable way and reduces the signaling overhead. In QMRPCAH each node maintains local multicast routing information and/or summary information of other clusters. Additionally, the protocol supports mobility of nodes, i.e., each node being a member of a multicast group can join and leave it dynamically. Finally, the protocol supports several QoS metrics although mainly delay and bandwidth are considered. The protocol does not provide hard QoS guarantees.

Clustering

Figure 12 shows an exemplary clustering of an ad-hoc network. All nodes within the network are divided into clusters (domains) of different levels. A 1st-level cluster consists of nodes with similar mobility characteristics. When several 1st-level clusters are combined a 2nd-level cluster is created etc. Clusters of different levels do not overlap with each other. Nodes within the 1st-level clusters are called local nodes. Nodes lying within the transmission range of one or more clusters are called bridge nodes. Each cluster contains also a single cluster head. The cluster head is a coordinator which decides on channel assignments, performs power control, maintains time division frame synchronization, and deals with the spatial reuse of bandwidth.

Mobility Support

QMRPCAH assumes that each local node periodically measures the delay of its outgoing links and broadcasts the gathered information to all nodes within its cluster in the form of update messages. Upon receiving an update message the local nodes update their intra-cluster routing information. Similarly, each bridge node performs the same procedure and periodically sends update messages to other bridge nodes. On this basis the inter-cluster information becomes updated. Additionally, in order to handle mobility, each node joining a new cluster must subscribe to its multicast tree to become a local multicast node of the new domain.

QMRPCAH uses a receiver-initiated flooding algorithm. The algorithm assures that for all joining nodes only the paths which satisfy bandwidth requirements will be included in the multicast tree. Additionally, QMRPCAH can guarantee that all messages which have a larger path delay will arrive after the messages which have a smaller path delay. Flooding messages are forwarded only if all QoS constraints are met.

TORA-SHORT

Asokan, Natarajan, and Venkatesh (2008) integrate TORA with SHORT (Self-Healing and Optimized Routing Techniques) in order to improve QoS routing. This is done by monitoring routing paths and, if a shortcut route is available, redirecting the path. The performance of TORA-SHORT is based on the exchange of query and update packets. The

Figure 12. Exemplary clustering of an ad-hoc network

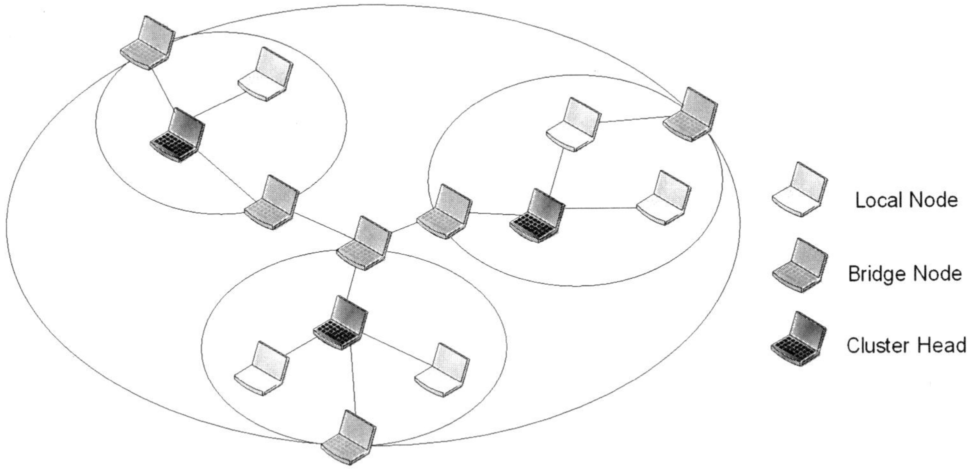

query packets are generated by source nodes in order to begin the route discovery process. The update packets are generated by destination nodes in order to terminate the query packets and by intermediate nodes to inform about route breaks.

Two classes of SHORT are proposed: path aware (PA)-SHORT and energy aware (EA)-SHORT. The former optimizes hop count, whereas, the latter conserves power. By incorporating (PA)-SHORT and (EA)-SHORT in TORA, QoS routing can be optimized by means of path lengths and the residual energy level of nodes. It assures that both the shortest paths and the nodes with sufficient available power are chosen during the route selection process. In order to meet this goal, TORA-SHORT makes use of two tables: the hop comparison table and the overhear table. Additionally, it is defined that each new entry has three obligatory fields, i.e., hop counter, residual energy level of a transmitting node, and the sender's address.

CROSS LAYER QOS ARCHITECTURES

Providing QoS in mobile ad-hoc networks is a challenging task because of the many factors involved. The previous parts have shown solutions to problems apparent at separate layers of the OSI/ISO model. However, these solutions usually aim to solve only a single problem. In order to provide end-to-end QoS in multi-hop ad-hoc networks a cross-layer approach is needed. Such a framework needs to integrate the singular solutions into a complete approach for QoS provisioning. The advantage of using this cooperative approach is being able to share relevant information between layers and provide feedback between components of the architecture. This leads to a more responsive, scalable, and flexible system and adaptability is required because MANETs are dynamic environments. There is one disadvantage of using a cross-layer approach. Namely, unintended interactions between components can occur (e.g., in the form of feedback loops) which makes locating problems troublesome.

The cross-layer approach offers a wide range of design possibilities. However, most of the solutions that are presented in this part have been influenced by each other and there are many similarities among them. Especially the first two approaches, INSIGNIA and SWAN, have been very influential. Table 1 presents the important building blocks of a cross-layer architecture.

Table 1 Building blocks of a cross-layer architecture

ISO/OSI Layer	Building blocks
PHY	channel monitoring (rate, SNR, BER), dynamic rate control
MAC	bandwidth estimation, priority queuing, traffic differentiation
Network	QoS routing, QoS signaling (resource reservation), traffic classification, traffic shaping, admission control

Most of them are present in all the mentioned solutions.

FQMM

Flexible Quality of service Model for Mobile ad-hoc networks (FQMM) (Xiao et al., 2000) is a QoS model developed for MANETs. It represents an interesting and unique approach to the problem of QoS provisioning. It is not associated with QoS negotiation procedures, however, it can be adopted to enhance existing solutions.

FQMM is a hybrid solution combining per-flow and per-class service provisioning. Therefore, it can be treated as a combination of the IntServ and DiffServ models. High priority traffic (which is a small percentage of overall traffic) is given per-flow provisioning. Traffic of other priorities is given per-class provisioning. FQMM defines three types of nodes (based on DiffServ): ingress, interior, and egress. The sender is an ingress node while the destination is an egress node. Ingress nodes perform classification, marking, policing and shaping. Interior nodes forward traffic of others and perform traffic shaping according to traffic profiles. The goal of the traffic profiles is to keep consistent differentiation between sessions. A profile is defined as the relative percentage of the effective link capacity, in order to keep the differentiation between classes predictable and consistent under different network dynamics.

INSIGNIA

INSIGNIA (Lee et al., 2000) is a QoS framework which operates mostly at the IP layer. Its main design consideration is the support of adaptive multimedia services. INSIGNIA aims to provide *base* QoS or *enhanced* QoS depending on available resources. It is based on in-band signaling and soft-state reservations. Its features include resource reservation, restoration control, and session adaptation between communicating nodes. Figure 13 presents an overview of the INSIGNIA architecture.

Admission control in INSIGNIA allocates bandwidth to flows based on information provided by the service (i.e., minimum and maximum bandwidth). This allows providing either *base* or *enhanced* QoS, respectively. The decision of admitting a flow to the network is based on the requested bandwidth, the measured channel capacity and current channel utilization. This ensures that new reservation requests do not impact existing reservations. Admission control is done on a hop-by-hop basis. Each intermediate node, upon receiving a reservation packet, accepts or denies the request. After a positive decision, the node maintains per-flow soft-state reservations and subsequent packets are scheduled accordingly. Reservations are maintained for the duration of the packet flow. If packets do not arrive before a certain timeout, the reserved resources are released. Packets of denied reservations are treated as best-effort. After receiving a reservation packet, the destination node sends a QoS report to the source to complete the reservation phase. Such reports are also sent at time intervals (specified by applications) and whenever requested. Furthermore, packets are scheduled to be sent to the network using a weighted round-robin scheduling discipline.

Figure 13. The INSIGNIA architecture (Adapted from (Lee et al., 2000))

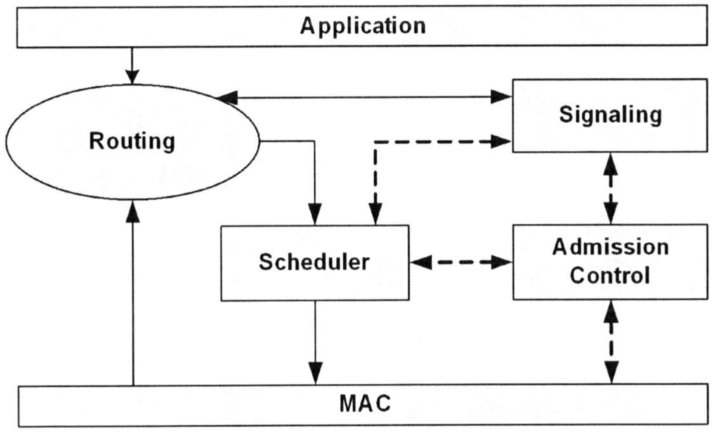

INSIGNIA uses in-band signaling for establishing, adapting, restoring, and tearing down end-to-end QoS sessions. The IP option field of the IP packet header is used to deliver the signaling protocol commands. Packets are sent either in reservation mode or as best-effort.

INSIGNIA aims to be very flexible to varying network conditions. Whenever the available bandwidth changes adaptation algorithms are invoked. Flow restoration algorithms are used to respond to dynamic route changes. Once the routing protocol updates the routing table, admission control and resource reservation is performed on the new paths. Flows may have immediate restoration, or partial/permanent degradation of QoS depending on available resources.

In this QoS framework neither the routing nor the MAC protocol are defined. However, INSIGNIA can be used with already existing protocols. There exists a combination of INSIGNIA and TORA known as INORA (Dharmaraju & Roy-Chowdhury, 2002). The TORA routing protocol provides multiple paths (between sender and destination) to the signaling protocol, and the latter checks if they meet the necessary QoS requirements.

SWAN

SWAN (Stateless Wireless Ad-hoc Networks) (Ahn et al., 2002) is a distributed, cross-layer QoS framework for MANETs. It provides service differentiation, QoS negotiation, admission control, and dynamic regulation in case of congestion. Two traffic classes are considered: real-time and best-effort traffic. Figure 14 presents an overview of the SWAN architecture.

The SWAN model consists of a number of mechanisms present in every node. The classifier differentiates traffic between the two classes and marks packets accordingly. Differentiation is achieved because real-time packets go directly to the MAC layer, while best-effort packets are sent to the shaper. The shaper is a leaky bucket which delays packets accordingly to the rate calculated by the rate controller. The operation of the rate controller is based on delay measurements in the MAC layer.

SWAN is stateless because intermediate nodes do not keep any per-flow or aggregate state information. It is only the source node which performs admission control and, therefore, efficient estimation of bandwidth availability is required. This is provided by the admission controller which sends request/response probes to the destination to determine the bottleneck bandwidth along the

Figure 14. The SWAN architecture (Adapted from (Ahn et al., 2002))

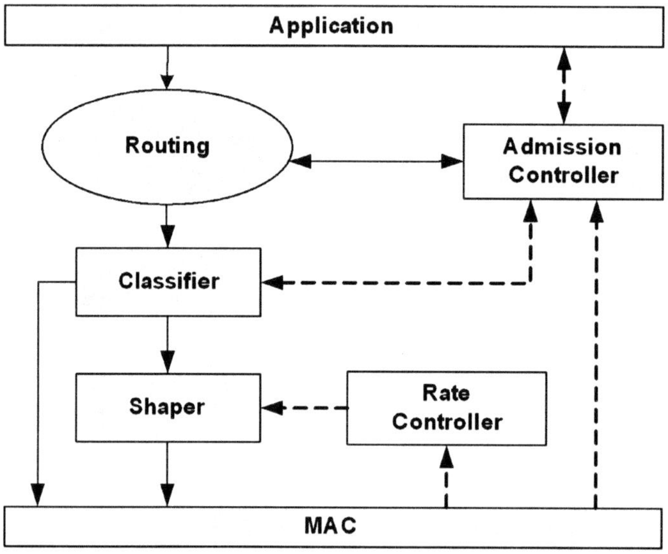

path. Based on the received information an admission decision is made. If a session is admitted, its packets are marked by the classifier as real-time packets; otherwise they are treated as best-effort packets.

Dynamic regulation of real-time sessions is needed when congestion conditions occur. This can be caused by node mobility and dynamic rerouting. When a mobile node detects a violation of the real-time traffic utilization limits it begins marking ECN (Explicit Congestion Notification) bits in the IP header of real-time packets. As a result, the destination node, which monitors incoming packets, can send a regulate message to inform the source node about the congestion. The source node should then re-establish the real-time session by sending a new probing request to the destination. The result of this re-establishment is either achieving the previous QoS or the session being dropped. Bandwidth adaptation of real-time sessions is not considered.

2LQoS

2LQoS (Two-Layered Quality of Service) (Nikaen et al., 2002) is a cross-layer QoS routing scheme with traffic differentiation and shaping. The network and application layers cooperate to determine the most suitable path through the network. Metrics from both layers are used. Path discovery is based on network layer metrics, such as: hop count, power level, buffer level, and stability level. The first metric is related to resource consumption, the second and third to load balancing, while the final one is a measure of mobility. Path selection is done according to application layer metrics: delay, throughput and cost (which is a function of power and buffer level of a node). Traffic is categorized into three classes. The first provides low delay (for voice applications), the second – high throughput (for video applications) and the last one has no constraints (best effort). Traffic can be shaped to meet QoS conditions in the network. This scheme does not perform any resource reservation. Service differentiation is done at each ad-hoc node through scheduling. Packets from each class are assigned to their appropriate queues, with each queue having a different, user defined weight. Packet classification is done at the source node.

DS-SWAN

DS-SWAN (Differentiated Services-Stateless Wireless Ad Hoc Networks) (Domingo & Remondo, 2004) provides end-to-end QoS in ad-hoc networks which are connected to a fixed IP network. DS-SWAN utilizes SWAN in the ad-hoc network and DiffServ in the infrastructure network. Additionally, it allows both these mechanisms to cooperate. The parameters of SWAN change dynamically according to the conditions in both the ad-hoc network and the infrastructure network. When packet delays exceed a given threshold special messages (*QoS Lost*) are sent. These messages cause more aggressive shaping of best effort traffic. The authors also propose a new routing protocol, SD-AODV (Service Differentiation AODV). It is aware of the *QoS Lost* messages and is able to re-route new flows away from congested zones.

The DAIDALOS Approach

Another QoS architecture was developed within the DAIDALOS I research project (Crisostomo et al., 2005). It was further developed in the DAIDALOS II project (Natkaniec, Gozdecki, & Sargento, 2007). The goal of this architecture is the integration of ad-hoc networks with infrastructure networks, a very useful scenario in hotspot environments. It supports the following set of features. It is a cross-layer solution providing stateless QoS to MANETs integrated with infrastructure. The architecture supports MAC layer traffic differentiation in the four IEEE 802.11 EDCA access categories. The implemented end-to-end signaling is not only simple but also allows resource reservations to originate from either the ad-hoc or the infrastructure part. The ad-hoc signaling mechanism is integrated with the Next Steps in Signaling (NSIS) protocol suite (Hancock et al., 2005) in the infrastructure network. NSIS provides flexibility in end-to-end signaling, supporting different resource management models and bidirectional reservations. Additionally, the architecture provides MAC layer measurements, traffic shaping, dynamic regulation, and admission control. Finally, it features the first real-life implementation of EDCA in a MANET (Natkaniec et al., 2009).

The ad-hoc part of the proposed QoS architecture consists of two main logical units, the mobile node (MN) and the gateway (GW) (Figure 15). MN has a double role acting both as a user terminal which provides QoS support to the end user, and as a router which forwards the traffic of neighboring nodes. The physical interconnection between the MANET and the infrastructure planes occurs in GW. The GW provides connectivity with infrastructure, participates in admission control and dynamic regulation. It is also responsible for the interaction of QoS signaling. Separate models exist for MN and for GW. End-to-end QoS resource management is supported through the interoperation between the MN and the GW (Figure 15). More specifically, this occurs through the interaction between the QoS Client (QoSC) in the MN and QoS Manager (QoSM) in the GW. QoSC retrieves the necessary QoS parameters from applications (through an API) and maps them to network QoS parameters. QoSC also performs per-flow end-to-end QoS signaling, controls the Classification and Marking module (which marks the Traffic Class and Flow Label fields in IPv6 headers), notifies the applications of the state of network interfaces, and synchronizes network resource reservation. QoSM provides the same functionalities as QoSC but without the interface to the application layer.

The other modules have the following functionalities. The Ad-hoc QoS Controller (AHQoSC) coordinates the work of the other modules, collects information about available resources in the wireless medium, and performs admission control in

Figure 15. The DAIDALOS model for the mobile node (a) and gateway (b) (Adapted from (Natkaniec, Gozdecki, & Sargento, 2007))

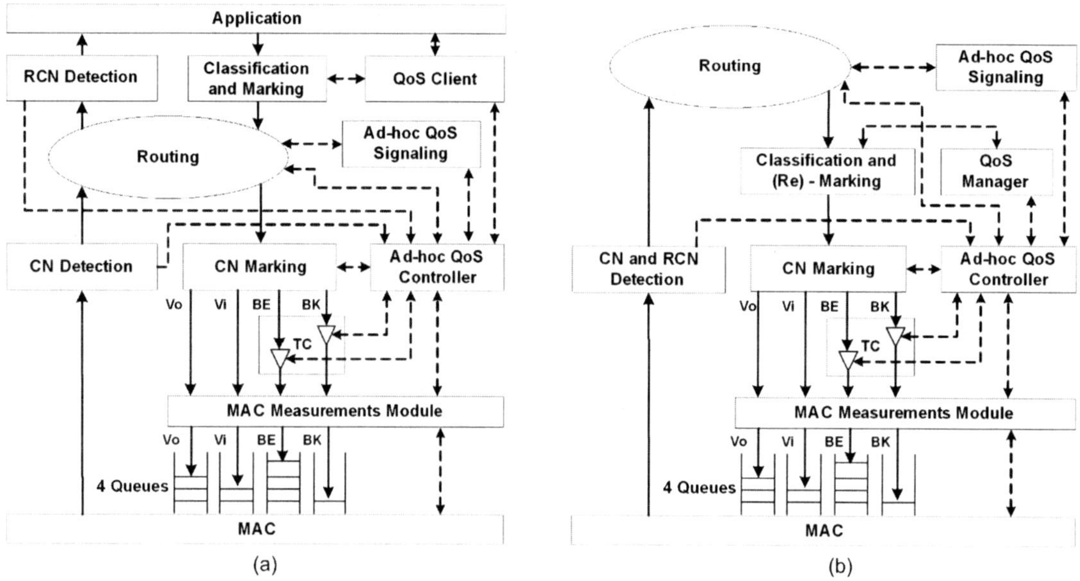

the ad-hoc path. It is also responsible for traffic control, reaction to congestion, and participating in resource management. The MAC Measurements Module (MMM) provides AHQoSC with information regarding bandwidth utilization, transmission delay, current transmission rate, frame statistics and idle intervals. The Ad-hoc QoS Signaling (AHQoSSig) module is responsible for QoS negotiations, probing for available bandwidth (like in SWAN), and session setup between infrastructure and ad-hoc. Traffic differentiation is performed by the IEEE 802.11 EDCA function and the Traffic Controller (TC) module, which is used to shape lower-priority flows. The architecture is aware of overload situations through the Congestion Notification (CN) signaling mechanism, which is implemented in three modules: CN Marking (CNM), CN Detection (CND), and Receiver CN Detection (RCND).

The GW is able to support the same functionalities as the MN, but does not interact with applications (since it works only at the IP layer and below) (Figure 15b). Additionally, it collects, generates and processes QoS signaling messages in the ad-hoc and infrastructure part. It is also responsible for the enforcement of QoS in the infrastructure network.

HQMM

HQMM (Hybrid QoS Model for Mobile Ad-hoc Networks) (He & Abdel Wahab, 2006) combines the IntServ and DiffServ models for ad-hoc networks. In this approach, per-flow provisioning (based on INSIGNIA) is given to traffic of the highest priority. Other priorities receive per-class provisioning (based on DiffServ). Differentiation is done based on bandwidth. HQMM defines three types of nodes (as DiffServ): ingress, interior and egress. Since each node can perform any of these functions, even simultaneously, they all have the same architecture (Figure 16). Traffic is DSCP marked in the conditioner, according to a pre-established agreement (which is out of scope of HQMM). The choice of routing protocol is also beyond the scope of this scheme. The signaling module provides the capabilities of INSIGNIA signaling. The link management module monitors the

Figure 16. The HQMM architecture (Adapted from (He & Abdel Wahab, 2006))

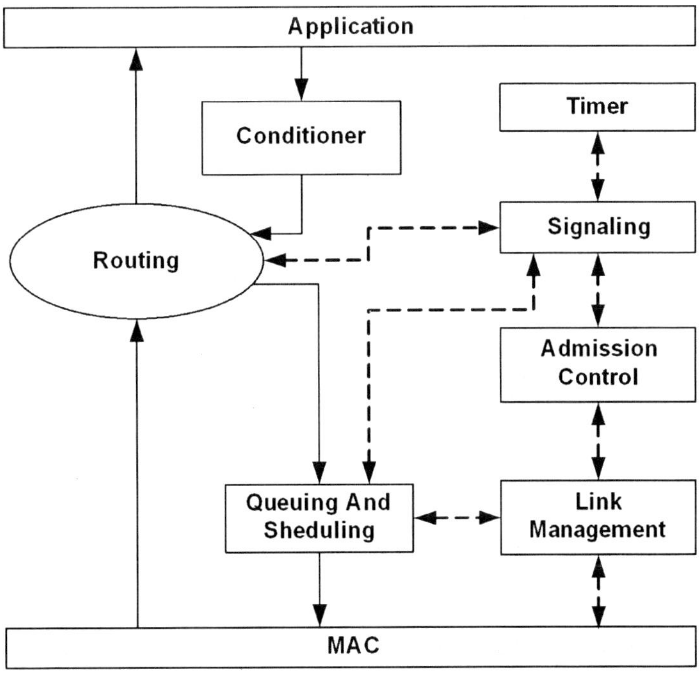

wireless channel and reports current and average bandwidth availability. Service differentiation is achieved through scheduling, queue management algorithms, and packet classification. In particular, INSIGNIA reservation packets are treated with the highest priority, while other packets are put into DiffServ classes.

FQM

FQM (Framework for QoS Multicast) (Saghir, Wan, & Budiarto, 2006) aims to support QoS for multicast applications in MANETs. Multiple paths meeting QoS requirements are found through a new on-demand routing protocol. FQM implements features from both the IntServ and the DiffServ models. For every accepted QoS route request, IntServ is used. Data packets from other sources are given DiffServ. Bandwidth is therefore divided between fixed-bandwidth for flows that have been admitted, and shared-bandwidth for other data. Depending on resource availability, either redundant or single paths are used. Route requests are intelligently flooded through the network. Cross-layer bandwidth estimation is used for admission control. This estimation is performed at each node by passively listening to the channel. The route request reaches the destination if there is enough bandwidth on the path. In such a case, a route reply message originates at the destination and traverses intermediate nodes on the path. This constructs the multicast tree. Service differentiation in FQM is achieved with the use of a classifier, shaper, dynamic rate control, and priority queues. Packets are mapped into two traffic classes: real-time and best effort. The mapping is based on the Type of Service (ToS) field in the IP header. Shaping is applied only to best effort traffic. The rate of best effort traffic is changed dynamically with an additive increase multiplicative decrease algorithm and, additionally, a drop tail algorithm is used in the queues.

CLQM

Sarma and Nandi (2008) propose a Cross-layer QoS Mapping (CLQM) framework for MANETs. It works on three layers: application, network and data link. Four service classes are considered. At the application layer the QoS classes are characterized by maximum end-to-end delay, minimum throughput or as best-effort. These metrics are mapped to network layer metrics which are in turn mapped to data link layer metrics. The QoS metrics at the network layer are path bandwidth, path delay, path stability and hop count. The data link layer metrics are MAC delay, link bandwidth and link stability. They are used in the process of admission control.

Admission control is performed during route discovery, based on available throughput. Modified AODV packets (RREQ and RREP) carry network layer QoS metrics. RREQ packets are dropped if their QoS requirements are not met. Network monitoring is periodically done at both the data link and network layers. At the data link layer, in order to adapt to network conditions, the framework dynamically adjusts the values of CW_{min} and CW_{max} for each class with the use of the *CW* adaptor module. Bandwidth estimation is performed at each node by measuring network utilization, i.e., the fraction of the time the channel is busy to the total measured time. MAC delay is calculated as the time from when a packet arrives at the MAC layer to the time an acknowledgement is received. An exponentially weighted moving average is used to estimate the average delay. At the network layer, the RSQR (Route Stability based QoS Routing) protocol is used to choose the best path for each class. Service differentiation among the QoS classes is achieved by a class-based scheduler. Congestions in the network detected using the ECN mechanism. They are resolved through flow rerouting or termination. The architecture of CLQM is presented in Figure 17.

Summary

In this part we have presented multiple cross-layer solutions to the problem of QoS provisioning in ad-hoc networks. Even though the problem is complex, most solutions are based on a mixture of similar features, such as bandwidth estimation, traffic differentiation, QoS routing, resource reservation, traffic shaping and admission control. Table 2 gives a comparison of the presented solutions and the features they support.

FUTURE RESEARCH DIRECTIONS

Even though there are a number of protocols and solutions for QoS-aware ad-hoc networks, further exploration in this field is required. There are several unsolved challenges that need to be addressed when providing QoS support in future multi-hop ad-hoc networks. Some of the possible research directions are shortly discussed next. We look at the three lower OSI/ISO layers and at cross-layer challenges.

The physical layer is responsible for effective bit transmission. Even when the BER of the radio channel is high the physical layer parameters, e.g., modulation and channel coding, need to be correctly selected. Therefore, efficient multi-rate algorithms are required. They should interact with the MAC sub-layer and schedule the transmitted traffic classes to avoid bad channel conditions.

A number of QoS-aware MAC schemes have been proposed, however, most of them have drawbacks. They focus on specific QoS features and assume a simple network topology, one traffic priority per node, limited node mobility, ideal channel conditions, etc. In a multi-hop ad-hoc network, nodes usually forward traffic which belongs to different flows, and support many traffic classes with different delay bounds and bandwidth requirements. A MAC protocol for multi-hop ad-hoc networks should therefore optimize the trade-off between fairness, efficient

Figure 17. The CLQM framework architecture (Adapted from (Sarma & Nandi, 2008))

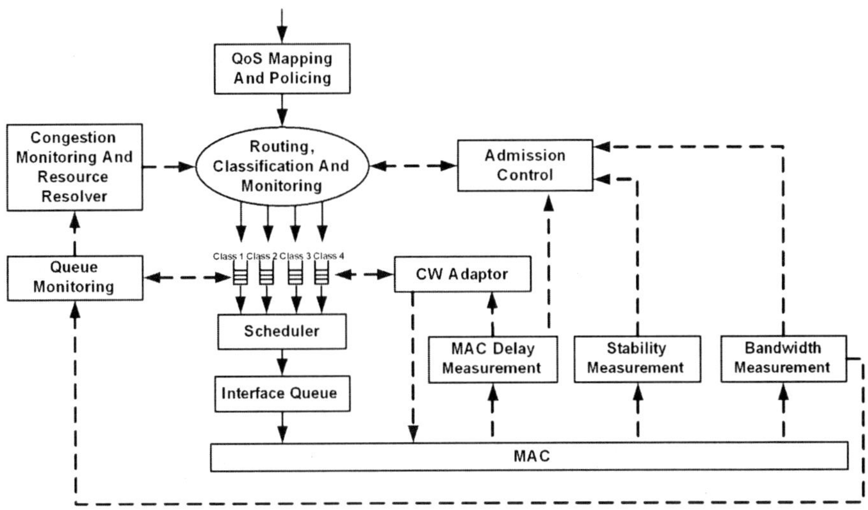

resource utilization, support of multiple traffic classes, strict priority guarantees, bandwidth on demand allocation, traffic scheduling, and fast reaction to transmission errors. It seems that close interactions between the MAC protocol, and the physical and network layers are needed to achieve all these requirements.

The designers of new network layer QoS mechanisms must take into account several requirements. Firstly, the accuracy of QoS routing protocols must be maximized. Secondly, control overhead should be kept as small as possible. Thirdly, because intermediate nodes are forwarders of traffic they may limit available resources and, furthermore, their unreliability may lead to route breaks. Another concern is that currently not all QoS routing protocols estimate the available network resources. Some of them assume that the network capacity is known *apriori*. As a result, all schemes based on bandwidth and delay estimation (e.g., admission control, location prediction) are not done appropriately. Route maintenance is also a challenging task due to the frequently changing topology of MANETs. The best way to deal with this problem would be to have mechanisms which could predict possible route breaks, find redundant routes, and perform re-computation of broken routes. However, each of these techniques increases overhead and, therefore, it is crucial to find a trade-off between full mobility support and required routing overhead. Additionally, it has to be decided if global or local state information has to be obtained to perform QoS routing. The former estimates network resources more accurately, but the latter requires lower control overhead. Furthermore, it has not been decided yet if reactive, proactive or hybrid routing protocols are the best way to find feasible paths. Finally, an appropriate resource reservation scheme must be decided on as well.

Because a cross-layer solution seems to be the most promising approach to provide end-to-end QoS provisioning in multi-hop ad-hoc networks, the interaction between protocols at different layers is crucial. Even though QoS frameworks have been proposed, more advanced solutions should be investigated. Cross layer design should obviously integrate QoS solutions from the physical, data link and network layers. In particular, most current QoS schemes lack dynamic PHY rate control, accurate channel measurements, a QoS-aware MAC, restoration control, session adaptation

Table 2. Comparison of the presented cross-layer solutions

	Channel Monitoring (rate, SNR, BER)	MAC Measurements	QoS MAC (e.g., EDCA)	Scheduler (Priority Queuing)	Number of Traffic Classes	QoS Routing	QoS Signaling (Resource Reservation)	Traffic Shaping	Admission Control	Restoration Control	Session Adaptation / Dynamic Regulation (Like ECN)	Soft/Hard QoS Guarantees	Complies with 802.11	Per-session or Per-class	Verification (Implementation or Simulation)	Integration with Infrastructure
2LQoS	-	-	-	X	3	X	-	X	-	-	-	S	X	C	-	-
CLQM	-	X	-	X	4	X	-	-	X	-	X	S	X	C	-	-
Daidalos	X	X	X	-	4	-	X	X	X	-	X	S	X	C	I	X
FQM	-	X	-	X	2	X	-	X	X	-	-	S	X	C	S	-
FQMM	-	X	-	X	2	-	-	X	X	-	-	B	X	Both	S	-
HQMM	-	X	-	X	5	-	X	X	X	X	X	S	X	Both	S	-
INSIGNIA	X	X	-	X	-	-	X	-	X	X	X	S	X	S	S	-
SWAN (DS-SWAN)	-	X	-	-	2	-	-	X	X	-	X	S	X	C	S/I	X

and integration with infrastructure. Additionally, QoS frameworks are difficult to implement and verify because of their complex architectures. Many architectures can be proposed, but until they are tested by users they are only theoretical accomplishments. Furthermore, it is up to standard organizations and equipment vendors to attempt to deploy reliable QoS solutions in multi-hop ad-hoc networks.

CONCLUSION

In this chapter we have discussed QoS issues for multi-hop ad-hoc networks. Supporting appropriate QoS for these networks is a very complex problem and has become an active area of research in recent years. Several QoS protocols designed for the physical, data link and network layers have been shortly presented. The analysis of single layer QoS solutions shows that they are unsuitable for multi-hop ad-hoc networks, where complex QoS mechanisms are needed. We have also described and compared the most interesting cross-layer QoS solutions. It is clear that these approaches should provide a general model, which can be dynamically tuned to support applications with different QoS requirements. We found that there are a number of unsolved challenges that need to be addressed to design complete QoS-aware solutions for multi-hop ad-hoc networks. Therefore, there are many research possibilities in this field of engineering. Solving the existing

QoS issues will allow future ad-hoc networks to meet user expectations.

REFERENCES

Ahn, G. S., Campbell, A., Veres, A., & Sun, L. H. (2002). Supporting Service Differentiation for Real-Time and Best-Effort Traffic in Stateless Wireless Ad Hoc Networks (SWAN). *IEEE Transactions on Mobile Computing*, *1*(3), 192–207. doi:10.1109/TMC.2002.1081755

Asokan, R., Natarajan, A. M., & Venkatesh, C. (2008). Quality-of-service routing using path and power aware techniques in mobile ad hoc networks. *Journal of Computer Systems, Networks, and Communications*, *2008*(1).

Awerbuch, B., Holmer, D., & Rubens, H. (2003). High throughput route selection in multi-rate ad hoc wireless networks. Tech. Rep., Johns Hopkins University, Computer Science Dept., Baltimore, MD.

Bicket, J. (2005). *Bit-rate Selection in Wireless Networks*. Unpublished master's thesis, Massachusetts Institute of Technology, Cambridge, MA.

Chaudet, C., & Guérin Lassous, I. (2001). BRuIT: Bandwidth Reservation under Interferences Influence. (Research Rep. No. 4308), INRIA.

Chenxi, Z., & Corson, M. S. (2002). QoS routing for mobile ad hoc networks. In *Proceedings of Twenty-First Annual Joint Conference of the IEEE Computer and Communications Societies, IEEE INFOCOM 2002* (pp. 958-967). New York: IEEE Inc.

Crisostomo, S., Sargento, S., Natkaniec, M., & Vicari, N. (2005). A QoS architecture integrating mobile Ad-Hoc and infrastructure networks. In *Proceedings of the ACS/IEEE 2005 international Conference on Computer Systems and Applications* (pp. 897-903). Washington, DC: IEEE Computer Society.

De, S., Das, S. K., Wu, H., & Qiao, C. (2002). A resource efficient RT-QoS routing protocol for mobile ad hoc networks. In *The 5th International Symposium on Wireless Personal Multimedia Communications,* (pp. 257-261).

Deng, D. J., & Chang, R. S. (1999). A priority scheme for IEEE 802.11 DCF access method. *IEICE Transactions on Communications. E (Norwalk, Conn.)*, *82-B*(1), 96–102.

Dharmaraju, D., & Roy-Chowdhury, A. (2002). INORA - A Unified Signaling and Routing Mechanism for QoS Support in Mobile Ad hoc Networks. In *Proceedings of the 2002 International Conference on Parallel Processing Workshops* (pp. 86-93). Washington, DC: IEEE Computer Society.

Domingo, M. C., & Remondo, D. (2004). A cooperation model between ad hoc networks and fixed networks for service differentiation. In *29th Annual IEEE International Conference on Local Computer Networks,* (pp. 692-693). New York: IEEE Inc.

Ferrari, G., Tonguz, O. K., & Bhatt, M. (2004). Impact of receiver sensitivity on the performance of sensor networks. In IEEE Vehicular Tech. Conf. VTC Spring, Milan, Italy.

Gerasimov, I., & Simon, R. (2002). Performance analysis for ad hoc QoS routing protocols. In International Mobility and Wireless Access Workshop, MobiWac'02, (pp. 87-94).

Głowacz, A., Natkaniec, M., Kosek, K., & Szott, S. (2008). Guarantying QoS in MANETs using MAC Layer Measurements. In 17th ICT Mobile & Wireless Communications Summit. Dublin: IIMC International Information Management Corporation.

Hancock, R., Karagiannis, G., Loughney, J., & Van den Bosch, S. (2005). *Next Steps in Signaling (NSIS): Framework*. RFC 4080.

He, Y., & Abdel Wahab, H. (2006). HQMM: A Hybrid QoS Model for Mobile Ad-hoc Networks. *11th IEEE Symposium on Computers and Communications, ISCC '06*, (pp. 194-200). New York: IEEE Inc.

Holland, G., Vaidya, N. H., & Bahl, P. (2001). A rate-adaptive MAC protocol for multi-hop wireless networks. *International Conference on Mobile Computing and Networking* (pp. 236-251). New York: ACM.

IEEE. 802.11 Standard for Wireless LAN: Medium Access Control (MAC) and Physical Layer (PHY) Specification. (2007). New York: IEEE Inc.

Kamerman, A., & Monteban, L. (1997). WaveLAN-II: A high-performance wireless LAN for the unlicensed band. *Bell Labs Technical Journal*, *2*(3), 118–133. doi:10.1002/bltj.2069

Lacage, M., Manshaei, M. H., & Turletti, T. (2004). IEEE 802.11 Rate Adaptation: A Practical Approach (Research Rep. No. 5208), INRIA.

Layuan, L., & Chunlin, L. (2007). A QoS multicast routing protocol for clustering mobile ad hoc networks. *Computer Communications*, *30*(7), 1641–1654. doi:10.1016/j.comcom.2007.01.017

Lee, S., Ahn, G. S., Zhang, X., & Campbell, A. T. (2000). INSIGNIA: An IP-Based Quality of Service Framework for Mobile Ad Hoc Networks. *Journal of Parallel and Distributed Computing, Special issue on Wireless and Mobile Computing and Communications, 60*, 374-406.

Lei, C., & Heinzelman, W. B. (2005). QoS-aware routing based on bandwidth estimation for mobile ad hoc networks. *IEEE Journal on Selected Areas in Communications*, *23*(3), 561–572. doi:10.1109/JSAC.2004.842560

Lin, C. R., & Gerla, M. (1997). Asynchronous multimedia multihop wireless network. In *Sixteenth Annual Joint Conference of the IEEE Computer and Communications Societies*, INFOCOM '97 (pp. 118-125). New York: IEEE Inc.

Lin, C. R., & Liu, J. S. (1999). QoS routing in ad hoc wireless networks. *Journal on Selected Areas in Communications*, *17*(8), 1426–1438. doi:10.1109/49.779924

Natkaniec, M., Gozdecki, J., & Sargento, S. (2007). A QoS Architecture Integrating Ad-Hoc and Infrastructure in Next Generation Networks. In B. Simak, R. Bestak, & E. Kozowska (Eds.), IFIP International Federation for Information Processing, Personal Wireless Communications (pp. 25–37). New York: Springer.

Natkaniec, M., Kosek-Szott, K., Szott, S., Gozdecki, J., Głowacz, A., & Sargento, S. (2009). Supporting QoS in Integrated Ad-Hoc Networks. *Wireless Personal Communications*, Springer, ISSN 1572-834X (Online).

Natkaniec, M., & Pach, A. R. (2000). An Analysis of Modified Backoff Mechanism in IEEE 802.11 Networks. *Polish-German Teletraffic Symposium* (pp. 89-96).

Natkaniec, M., & Pach, A. R. (2002). PUMA - a new channel access protocol for wireless LANs. *International Symposium on Wireless Personal Multimedia Communications*, (vol. 3, pp. 1351-1355).

Nikaen, N., Bonnet, C., Moret, Y., & Rai, I. A. (2002). 2LQoS - Two-layered QoS model for reactive routing protocols for mobile ad-hoc networks. In *Proceedings of SCI - 6th World Multiconference on Systemics, Cybernetics and Informatics*.

Pal, A., Dogan, A., & Ozguner, F. (2002). MAC Layer Protocols for Real-time Traffic in Ad hoc Wireless Networks. *IEEE International Conference on Parallel Processing* (pp. 539-546). New York: IEEE Inc.

Sadeghi, B., Kanodia, V., Sabharwal, V., & Knightly, E. (2002). Opportunistic media access for multirate ad hoc networks. *International Conference on Mobile Computing and Networking* (pp. 24-35). New York: ACM

Saghir, M., Wan, T. C., & Budiarto, R. (2006). A new cross-layer framework for QoS multicast applications in mobile ad hoc networks. *International Journal of Computer Science and Network Security*, *6*(10), 142–151.

Sarma, N., & Nandi, S. (2008). A priority based QoS-Aware MAC protocol (PQAMP) in mobile ad hoc networks. 4th ACM symposium on QoS and security for wireless and mobile networks, Q2SWinet (pp. 79-82), New York: ACM.

Shah, S. H., & Nahrstedt, K. (2002). Predictive location-based QoS routing in mobile ad hoc networks. *IEEE International Conference on Communications, ICC'02* (pp. 1022-1027). New York: IEEE Inc.

Shigang, C., & Nahrstedt, K. (1999). Distributed quality-of-service routing in ad hoc networks. *IEEE Journal on Selected Areas in Communications*, *17*(8), 1488–1505. doi:10.1109/49.780354

Sinha, P., Sivakumar, R., & Bharghavan, V. (1999). CEDAR: a core-extraction distributed ad hoc routing algorithm. In *Proceedings of Eighteenth Annual Joint Conference of the IEEE Computer and Communications Societies, INFOCOM '99*, (pp. 202-209). New York: IEEE Inc.

Sobrinho, J. L., & Krishnakumar, A. S. (1999). Quality-of-service in ad hoc carrier sense multiple access wireless networks. *IEEE Journal on Selected Areas in Communications*, *17*(8), 1353–1368. doi:10.1109/49.779919

Van der Vegt, A. J. (2002). Auto Rate Fallback Algorithm for the IEEE 802.11a Standard, (Tech. Rep.), Utrecht University.

Wang, D., & Liu, K. (2007). A Novel QoS MAC Protocol for Multi-hop Ad Hoc Networks. *International Conference on Wireless Communications, Networking and Mobile Computing, WiCom* (pp. 1484-1487).

Xiao, H., Seah, W. K. G., Lo, A., & Chua, K. C. (2000). A Flexible Quality of Service Model for Mobile Ad hoc Networks. In *IEEE Vehicular Technology Conference Proceedings,* (pp. 445-449). New York: IEEE Inc.

Xue, Q., & Ganz, A. (2003). Ad hoc QoS on-demand routing (AQOR) in mobile ad hoc networks. *Special Issue on Routing in Mobile and Wireless Ad Hoc Networks*, *63*(2), 154–165.

Yihai, Z., & Gulliver, T. A. (2005). Quality of service for ad hoc on-demand distance vector routing. *IEEE International Conference on Wireless And Mobile Computing, Networking And Communications, WiMob'05* (pp. 192-196).

Ying, G., Kunz, T., & Lamont, L. (2003). Quality of service routing in ad-hoc networks using OLSR. In *Proceedings of the 36th Annual Hawaii International Conference on System Sciences.*

Ying, Z., Anand, A. L., & Jacob, L. (2003). A QoS enabled MAC protocol for multi-hop ad hoc wireless networks. *IEEE International Performance, Computing, and Communications Conference, IPCCC'03* (pp. 149- 156). New York: IEEE Inc.

Youngki, H., & Varshney, P. (2003). An adaptive QoS routing protocol with dispersity for ad-hoc networks. In *Proceedings of the 36th Annual Hawaii International Conference on System Sciences.*

ADDITIONAL READING

Aad, I., & Castelluccia, C. (2001). Differentiation mechanisms for IEEE 802.11. In *Proceedings of Twentieth Annual Joint Conference of the IEEE Computer and Communications Societies,* IEEE INFOCOM'01, vol. 1, (pp. 209-218). New York: IEEE Inc.

Ahn, C. W., Kang, C. G., & Cho, Y. Z. (2000). Soft reservation multiple access with priority assignment (SRMA/PA): a novel MAC protocol for QoS-guaranteed integrated services in mobile ad hoc networks. *IEEE Vehicular Technology Conference*, vol. 2, (pp. 942–947).

An, H.-Y., Lu, X., Gong, Z., & Peng, W. (2005). A Cluster-Based QoS Multipath Routing Protocol for Large-Scale MANET. High Performance Computing and Communications (pp. 321-330).

An, H.-Y., Zhong, L., Lu, X.-C., & Peng, W. (2005). A cluster-based multipath dynamic source routing in MANET. *IEEE International Conference on Wireless And Mobile Computing, Networking And Communications* (pp. 369-376).

Badis, H., & Khaldoun, K. A. (2007). *Quality of Service for Ad hoc Optimized Link State Routing Protocol (QOLSR)*. (Internet draft, expired August 9, 2007). March 2007. The URL was http://ietfreport.isoc.org/all-ids/draft-badis-manet-qolsr-05.txt.

Baldwin, R. O., Davis, N. J., & Midkiff, S. F. (1999). A real-time Medium Access Control protocol for ad hoc wireless local area networks. *Mobile Computing and Communications Review*, *3*(2), 20–27. doi:10.1145/584027.584028

Barua, G., & Chakraborty, I. (2002). Adaptive routing for ad hoc wireless networks providing QoS guarantees. *IEEE International Conference on Personal Wireless Communications* (pp. 196-200).

Bernhardt, R. C., Cain, J. B., & Windham, W. A. (2007). QoS Architecture for a Mobile Ad Hoc Network. In [New York: IEEE Inc.]. *Proceedings of Military Communications Conference*, *2007*, 1–6. doi:10.1109/MILCOM.2007.4455142

Canales Compes, M., Gallego Martinez, J. R., Hernández Solana, A., & Valdovinos Bardaji, A. (2008). QoS provision in mobile ad hoc networks with an adaptive cross-layer architecture. Wireless Networks.

Cardei, I., Varadarajan, S., Pavan, A., Graba, L., Cardei, M., & Min, M. (2004). Resource Management for Ad-Hoc Wireless Networks with Cluster Organization. *Cluster Computing*, *7*(1), 91–103. doi:10.1023/B:CLUS.0000003946.65880.25

Cerda, L., Voorhaen, M., Guimaraes, R., Barcelo, J. M., Garcia, J., & Blondia, C. (2005). A Reservation Scheme Satisfying Bandwidth QoS Constraints for Ad-Hoc Networks. EURO-NGI network of excellence, International Workshop, (pp. 176-188).

Chen, K., Shah, S. H., & Nahrstedt, K. (2002). Cross-Layer Design for Data Accessibility in Mobile Ad Hoc Networks. *Journal of Wireless Personal Communications. Special Issue on Multimedia Network Protocols and Enabling Radio Technologies*, *21*, 49–75.

Chen, L., & Heinzelman, W. B. (2004). Network architecture to support QoS in mobile ad hoc networks. In *Proceedings of IEEE International Conference on Multimedia and Expo* (pp. 1715-1718). New York: IEEE Inc.

Chen, L., & Heinzelman, W. B. (2007). A Survey of Routing Protocols that Support QoS in Mobile Ad Hoc Networks. *IEEE Network Magazine*, *21*(6), 30–38. doi:10.1109/MNET.2007.4395108

Chen, N., Li, L., & Dong, W. (2006). Multicast routing algorithm of multiple QoS based on widest-bandwidth. *Journal of Systems Engineering and Electronics*, *17*, 642–647. doi:10.1016/S1004-4132(06)60110-2

Choi, S., del Pedro, J., Shankar, N. S., & Mangold, S. (2003). IEEE 802.11e contention-based channel access (EDCF) performance evaluation. *IEEE International Conference on Communications* (pp. 1151-1156). New York: IEEE Inc.

Hanzo-II, L., & Tafazolli, R. (2007). A Survey of QoS Routing Solutions for Mobile Ad Hoc Networks. *IEEE Communications Surveys & Tutorials*, 9(2), 50-70.

Jawhar, I., & Wu, J. (2005). QoS support in TDMA-based mobile ad hoc networks. *Journal of Computer Science and Technology*, *20*(6), 797–810. doi:10.1007/s11390-005-0797-3

Jurdak, R. (2007). Wireless Ad Hoc and Sensor Networks: A Cross-Layer Design Perspective. New York, NY: Springer Science+Business Media.

Kosek, K., Szott, S., Natkaniec, M., & Pach, A. R. (2008). Practical Analysis of IEEE 802.11b/g Cards in Multirate Ad-hoc Mode. *Journal of Telecommunications and Information Technology. JTIT*, *08*, 1.

Kumar, S., Raghavan, V. S., & Deng, J. (2004). Medium Access Control protocols for ad hoc wireless networks: A survey. [Elsevier.]. *Ad Hoc Networks*, *4*(3), 326–358. doi:10.1016/j.adhoc.2004.10.001

Mangold, S., Choi, S., May, P., Klein, O., Hiertz, G., & Stibor, L. (2002). IEEE 802.11e wireless LAN for quality of service. *In Proceedings of the European Wireless* (pp. 32-39).

Marwaha, S., Indulska, J., & Portmann, M. (2008). Challenges and Recent Advances in QoS Provisioning, Signaling, Routing and MAC protocols for MANETs. *Australian Telecommunication Networks and Applications Conference,* ATNAC'08 (pp. 97-102).

Masoudifar, M. (2009). A review and performance comparison of QoS multicast routing protocols for MANETs. *Ad Hoc Networks*, *7*, 1150–1155. doi:10.1016/j.adhoc.2008.10.004

Mishra, A. (2008). Security and Quality of Service in Ad Hoc Wireless Networks. New York, NY: Cambridge University Press.

Mohapatra, P., & Krishnamurthy, S. (Eds.). (2005). Ad Hoc Networks: Technologies and Protocols. Boston, MA: Springer Science+Business Media.

Mohapatra, P., & Li, J., & Gui, C. (2003). QoS in Mobile Ad hoc Networks. Special Issue on QoS in Next-Generation Wireless Multimedia Communications Systems in IEEE Wireless Communications Magazine, 10(3).

Perkins, C. E. (2003). *Quality of Service for Ad hoc On-Demand Distance Vector Routing*. (Internet draft, expired April 14, 2004). October 14, 2003. The URL was http://www.psg.com/~charliep/txt/aodvid/qos.txt.

Reddy, T. B., Karthigeyan, I., Manoj, B. S., & Murthy, C. S. R. (2006). Quality of service provisioning in ad hoc wireless networks: a survey of issues and solutions. *Ad Hoc Networks*, *4*(1), 83–124. doi:10.1016/j.adhoc.2004.04.008

Sarkar, S. K., Basavaraju, T. G., & Puttamadappa, C. (2008). Ad Hoc Mobile Wireless Networks: Principles, Protocols, and Applications. Boca Raton, FL: Auerbach Publications.

Sarma, N., & Nandi, S. (2006). QoS support in mobile ad hoc networks. IFIP International Conference on Wireless and Optical Communications Networks.

Sarma, N., & Nandi, S. (2008). A Cross-layer QoS Mapping Framework for Mobile Ad Hoc Networks. [Mumbai, India: Medknow Publications.]. *IETE Technical Review*, *25*(6), 346–358. doi:10.4103/0256-4602.45427

Sethi, P., & Barua, G. (2003). CRESQ: providing QoS and security in ad hoc networks. In *Proceedings Eleventh Euromicro Conference on Parallel, Distributed and Network-Based Processing* (pp. 544-550).

Sheu, S. T., & Sheu, T. F. (2001). A bandwidth allocation/sharing/extension protocol for multimedia over IEEE 802.11 ad hoc wireless LANs. *IEEE Journal on Selected Areas in Communications*, *19*(10), 2065–2080. doi:10.1109/49.957320

Sun, B., & Li, L. (2006). QoS-aware multicast routing protocol for Ad hoc networks. *Journal of Systems Engineering and Electronics*, *17*, 417–422. doi:10.1016/S1004-4132(06)60071-6

Sun, B., Pi, S., Gui, C., Zeng, Y., Yan, B., & Wang, W. (2008). Multiple constraints QoS multicast routing optimization algorithm in MANET based on GA. *Progress in Natural Science*, *18*, 331–336. doi:10.1016/j.pnsc.2007.11.006

Vaidya, N. H., Bahl, S., & Gupta, S. (2000). Distributed fair scheduling in a wireless LAN. *6th Annual International Conference on Mobile Computing and Networking* (pp. 167-178).

Vishnumurthy, V., Sandeep, T., Manoj, B. S., & Murthy, C. S. R. (2004). A Novel Out-of-band Signaling Mechanism for Enhanced Real-time Support in Tactical Ad hoc Wireless Networks. In *Proceedings of the 10th IEEE Real-Time and Embedded Technology and Applications Symposium* (pp. 56-63). Washington: IEEE Computer Society.

Yeh, C. H., & You, T. (2003). A QoS MAC Protocol for Differentiated Service in Mobile Ad Hoc Networks. *International Conference on Parallel Processing*, ICPP'03 (pp. 349-356).

You, T., Yeh, C. H., & Hassanein, H. (2005). DRCE: A High Throughput QoS MAC Protocol for Wireless Ad Hoc Networks. *10th IEEE Symposium on Computers and Communications*, ISCC'05 (pp. 671- 676). New York: IEEE Inc.

Zhang, L., Cai, L., Li, M., & Wang, F. (2009). A method for least-cost QoS multicast routing based on genetic simulated annealing algorithm. *Computer Communications*, *32*, 105–110. doi:10.1016/j.comcom.2008.09.023

Chapter 11
Efficient Obstacle Avoidance for Sensory Data Propagation in Wireless Systems

Pierre Leone
University of Geneva, Switzerland

Luminita Moraru
University of Geneva, Switzerland

Sotiris Nikoletseas
University of Patras and CTI, Greece

Jose D. P. Rolim
University of Geneva, Switzerland

ABSTRACT

The authors study the problem of fast and robust data propagation in wireless sensor networks in the presence of obstacles obstructing communication. They survey representative state of the art techniques, such as variations of geographic routing, which is known to scale well, mainly due to its greedy nature and low memory requirements. Still, most of these algorithms are concerned with finding some path, while the optimality of the path is difficult to achieve. Towards improving QoS, and especially latency for time-critical applications, in this chapter, the authors are presenting a novel geographic routing algorithm with obstacle avoidance properties. It aims at finding the optimal path from a source to a destination when some areas of the network are unavailable for routing due to low local density or obstacle presence. It locally and gradually with time (but, as we show, quite fast) evaluates and updates the suitability of the previously used paths and ignores non optimal paths for further routing. The performance comparison to existing state of the art protocols shows that this approach performs much better in terms of path length, thus minimizing latency and space, while introducing low overhead and being energy efficient.

DOI: 10.4018/978-1-61520-771-8.ch011

INTRODUCTION

Recent advances in micro-electromechanical systems (MEMS) and wireless networking technologies have enabled the development of very small sensing devices called sensor nodes. Unlike traditional sensors that operate in a passive mode, sensor nodes are smart devices with sensing, data-processing and wireless transmission capabilities. Data can thus be collected, processed and shared with neighbors. Sensor nodes are meant to be deployed in large wireless sensor networks instrumenting the physical world. Originally developed for tactical surveillance in military applications, they are expected to find growing importance in civil applications such as building, industrial and home automation, infrastructure monitoring, warehousing and data mining, agriculture and security. Their range of applications makes sensor networks heterogeneous in terms of size, density and hardware capacity. Their great attractiveness follows from the availability of low cost, low power and miniaturized sensor nodes that pervade, instrument and augment the physical world without requiring human intervention: sensor networks are self-organizing, self repairing and highly autonomous. Those properties are implemented by designing protocols addressing the specific requirements of sensor networks.

Common to all sensor net architecture and applications is the need to propagate data inside the network. Almost every scenario implies multi-hop data transmission because of the low power used for radio transmission mostly to limit depletion of the scarce energy resources of battery powered sensor nodes. Needless to say, sensory data exchange is usually done in an environment in which wireless communication and data routing is obstructed, e.g. by walls, narrow corridors, voids due to physical failures of some devices etc. The theme of this chapter is exactly how to efficiently route sensory data in the presence of obstacles.

Computing the shortest path between two nodes in a graph is a problem intensively studied in graph theory. For the geometric domain, where only local information is available and the path is built based on the geographical position of the destination and of the one hop neighbors, a plethora of solutions and applications exists as well. We will consider the behavior of path finding algorithms in the geometric domain, in the case when there are obstacles and local irregularities in the initial network configuration. Because of the severe limitations of sensor network devices and the lack of global network knowledge, our purpose is to find in a distributed manner the shortest path that avoids the obstacles between two points using only local geographical information.

In the context of a global computing approach, our aim is to include wireless nodes in an Internet-based overlay computer in a transparent and efficient way.. We envision diverse interacting wireless elements (mobile users/phones, sensors, robots, actuators), hybrid/hierarchical compositions and heterogeneous wired interconnections. The context is highly dynamic (due to mobility, failures, obstacles), heterogeneous, and the wireless devices are resource-constrained and low-cost. Yet, we come up with algorithmic solutions that are efficient (QoS), robust, secure and easy to manage.

QoS basically refers to fast propagation of sensory data (the low latency challenge), since most applications are time-critical. However, we notice the existence of inherent trade-offs between latency and other crucial performance metrics, mostly energy dissipation. Hence, satisfactory latency vs. energy tradeoffs is crucial, and can be achieved by several methods (and their hybrid combination). Such methods relate to several network layers, including power saving schemes at the individual device level, communication optimization at the routing layer, as well as low additional communication overhead[1]. In this chapter, we focus on the latter, i.e. our algorithms aim to minimize the number of hops needed to propagate data around obstacles, to both improve latency and energy efficiency, while incurring a small communication overhead.

In our perspective, the Global Computer is in fact the Internet/WEB enhanced by mobile users and embedded sensors. The global network computer is structured in three interacting layers: the P2P (Overlay) Network (where applications are executed on powerful devices that communicate via fixed networks), the Wireless Peers (mobile/resource-limited devices that communicate over wireless mediums) and the Gateway Peers (an intermediate layer for wireless access).

BACKGROUND

Computing the shortest path between two nodes in a graph is a problem intensively studied in graph theory (Ahuja et al., 1993). For the geometric domain, where only local information is available and the path is built based on the geographical position of the destination and of the one hop neighbors, a plethora of solutions and applications exists as well (Sack and Urrutia, 2000). We will consider the behavior of path finding algorithms in the geometric domain, in the case where there are obstacles and local irregularities in the initial network configuration. Because of the severe limitations of sensor network devices and the lack of global network knowledge, our purpose is to find in a distributed manner the shortest path that avoids the obstacles between two points using only local geographical information.

In sensor networks there is no fixed predeployed infrastructure, therefore each node relays on his peers for sending a message to the destination, strategy called hop by hop routing. At each step, the node currently holding the message will choose one of its neighbors to relay the message, based on a predefined routing strategy. Due to limited transmission range of the sensor nodes, each node can communicate directly only with a limited number of peers - the one hop neighbors.

Geographic routing is considered as one of the best solutions for routing in sensor networks. It needs only local information about the one hop neighbors. This makes it a scalable and low complexity[2] solution.

The simplest form of geographic routing, greedy, chooses for forwarding the neighbor closest to the destination. This routing strategy ensures that the routing path is unique and that the propagation of data progress toward the destination. However, the routing strategy does not provide the existence of a neighboring node closer to the destination. This situation appears often when there is an obstacle or a void in the network, in low density network areas and even in the case of medium densities. In that case, a solution consists in changing the routing strategy and use perimeter routing which get round the obstacle by using the right hand rule until greedy routing can be used.

But these paths are not optimal in terms of length, and in fact can be quite long and thus inefficient. Additionally, it cannot guarantee delivery for any shape of the obstacle, i.e. it fails in the presence of some hard obstacles. Although several solutions have been proposed for routing around the voids, they represent a tradeoff between optimal path routing and network topology maintenance traffic.

This chapter describes the state of the art in geographic routing and outlines the importance of efficient obstacle avoidance techniques. Additional qualitative comparisons are presented by Chen and Varshney (2007a) and Stojmenovic (2005).

Greedy Routing

Greedy is the simplest form of geographic routing. For each forwarding decision, a node will need information only about the position of one

Figure 1. Greedy metrics

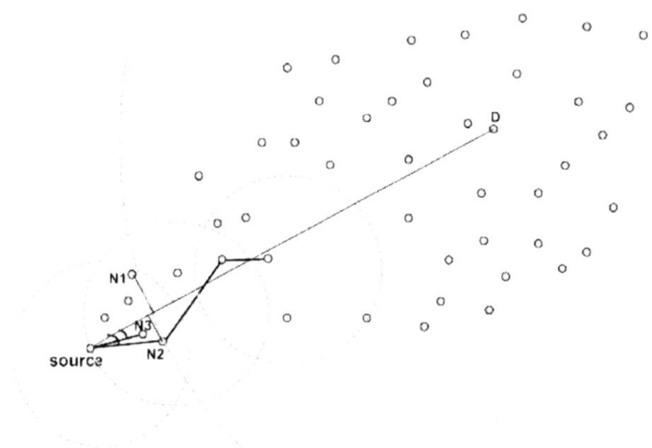

hop neighbors and the destination. The decision is made based on the optimization of a specific metric.

The first metric, called progress and was proposed by Mathar and Mattfeldt (1996), is defined as the distance between the current node and the projection of the neighbors on the line connecting the current node and the destination.

The second criterion, used by Finn (1987) for selecting a neighbor is the Euclidean distance. The node will choose for forwarding the neighbor closest to the destination. Kranakis et al. (1999) consider the angle between the next hop, current node and destination. The method selects the node that minimizes this angle. It is proved to reach destination on Delaunay triangulations (described below), when a path exists between the source and the destination.

Figure 1 shows an example of greedy routing. Source represents the current node, D represents the destination. N_1 is the neighbor with optimal progress, the projection of N_1 is the closest to the destination. N_2 is the neighbor closest to the destination; the circle centered in D with the radius DN_2 does not contain any direct neighbor of the source. N_3 has the smallest angle with the line CD.

Planar Graph Transversal Techniques

Planar graph[3] transversal techniques are geographic routing strategies alternative to greedy. They use two routing modes: greedy routing and perimeter (face) routing. Initially they use greedy routing. When greedy fails, they will switch to perimeter.

We consider that the current relay, P_1 knows its geographical coordinates, the coordinates of the previous node, of the destination D and of all its one hop neighbors N_i. As mentioned before, during greedy forwarding the choice of the next relay is a local optimum decision: the selected neighbor is the one closer to the destination. In Figure 2, the first three transmissions are in greedy mode.

Faced with an obstacle, a node will switch to perimeter routing and the next hop will be the first node located counterclockwise with respect to the P_1D line. This heuristic will continue until the message reaches an edge crossing the P_1D line. At this point it will switch to another face. The purpose is to route the message around faces progressively closer to the destination. The node will return to greedy when the current node is

Figure 2. Planar graph based routing

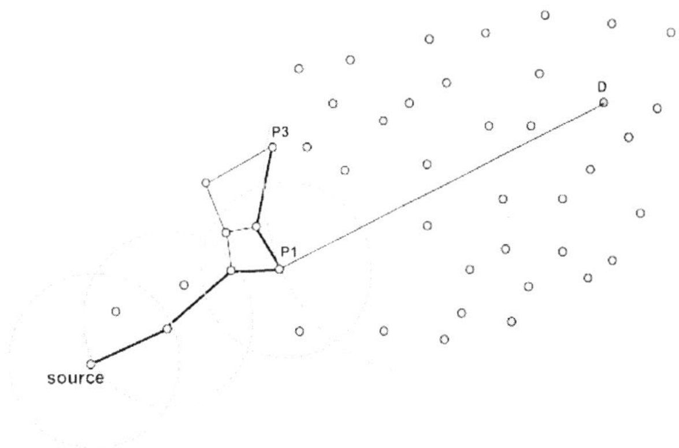

closer to the destination than the node where perimeter routing started. In Figure 2, P_3 is closer to D than P1; therefore the algorithm will switch to greedy.

Perimeter routing has a constraint - it works only on planar graphs with no crossing edges. Indeed, the fact that the graph is planar allows proving that perimeter routing cannot cycle indefinitely and that the messages are always delivered. Since most of the networks are initially not planar, the algorithm can only be applied after an initial planarization phase.

In this section we will discuss several aspects of planar graph traversal techniques: different network models proposed for sensor networks, planarization techniques and face (perimeter) routing strategies.

Network Model

We consider the case of a static network, where the entire network traffic is oriented to and from the base station. Each node can transmit at the maximum distance R. The network model is a graph G = (V, E), where V represents the set of nodes and $E \subseteq VxV$ represents the set of edges. The distance between nodes is the Euclidean distance, and the path length is the sum of the distances between the intermediary hops.

The information about the presence of the direct (one hop) neighbors is obtained by the means of short beacon messages, called hello messages (Stojmenovic, 2005). Sent with maximum signal strength, hello messages are used to detect the neighbors of a node in the network. They advertise both the presence of a node and its physical location.

The classical representation for network is the unit disk graph model Clark et al. (1990) - it makes the assumption that an edge $e \in E$ exists only if the distance between two nodes u and v is below a certain value R. The model presented by Kuhn et al. (2002) makes the additional assumption that the distance between two connected nodes has a lower bound d0. The effect of radio fluctuations on topological change rate and link stability was studied by Schmitz and Effelsberg (2004). The results are showing that they may have a significant impact on the network performance.

A different approach was proposed by Kuhn and Zollinger (2003) and further discussed by Kuhn et al. (2004). A link between two nodes in Quasi Unit Disk Graph exists only if the distance between the nodes is lower than a threshold d, between 0 and 1. If the distance is between d and 1 the connection may exist or not with a given probability and f the distance is larger than 1 no link exists.

Schmid andWattenhofer (2006) proposed ten different models for sensor network communication. They take into account different parameters: directional antenna, interferences, adaptive transmission power and link failures.

Another representation is the realistic physical layer model (Stojmenovic et al., 2005, Nadeem and Agrawala, 2004) - a message is received with a probability depending on the distance between nodes. This model uses the log normal shadow fading model to represent the probability p(x) of receiving a packet successfully. In this new context the transmission radius R is defined as the distance for which the probability of receiving a message is 0.5.

Graph Planarization

The representation of the network is not always a planar graph. Therefore planar graph transversal algorithms use different distributed graph planarization techniques. This can be done at the network level, in the network setup phase, or it can be done on demand, only for the set of nodes where greedy forwarding cannot be used.

The requirements for a planarisation algorithm are the low computational complexity and the possibility of being computed in a distributed and localized way. Santi (2005) analyse a series of algorithms that compute the proximity graphs used in geographic routing: Relative Neighborhood Graph (Gabriel and Sokal, 1969), Gabriel Graph (Toussaint, 1991), Delaunay Triangulation).

The proximity graphs are graphs where the neighbors of the nodes can be computed based on their geographical coordinates. They are subgraphs of the initial graph representing the network.

In a Gabriel Graph, a link between two nodes u and v exists only if there is no other node in the circle of diameter uv. In a Relative Neighborhood Graph, a link between u and v exists only if no other node exists at the intersection of the circles centered at u and v, with radius uv. RNG is a subset of GG.

A Triangulation of a set P of nodes in the plane is a planar graph where no edge connecting two nodes in P can be added without compromising planarity.

A Delaunay Triangulation for a set P of points in the plane is a triangulation DT(P) such that no point in P is inside the circumcircle of any triangle in DT(P).

Delaunay Triangulation is also defined as the dual graph of a Voronoi diagram (Sack and Urrutia, 2000). A Voronoi diagram is a partition of the plan in cells, such that each cell contains all the points closer to a specific node $p \in P$.

Gabriel Graph is a subset of Delaunay Triangulation. A DT has an edge between two nodes if their Voronoi cells are adjacent. Kranakis et al.(1999) define a routing strategy called compass routing which uses the Delaunay Triangulation and prove that it finds a path between any source and destination.

In Karp and Kung (2000a) both algorithms are evaluated as planarization method. They argue in favor of RNG mainly due to the impact of lower graph density on MAC layer performances - the efficiency of the MAC layer is reduced with the increase of the number of links in the network. Cal et al., Kim et al. (2005b) analyze the factors that influence the failure of graph planarization techniques and identify those robust to the presence of these factors, when the network model is unit disk graph.

The alternative we use for constructing the planar graph the crossed link detection protocol algorithm, CLDP, proposed by Kim et al. (2005a). The algorithm is proved to incur moderate communication overhead, to converge quickly and to choose low loss paths. Another advantage of this protocol is the robustness to arbitrary localization error and radio irregularities. In CLDP each node will send messages along the faces containing its links. If a crossing link is detected and its removal will not disconnect the graph, the link is removed.

In Govindan et al. (2006), the authors remove the link crossings only when they interfere with geographic routing. The algorithm improves the performance of CLDP since it is observed that only in particular situations crossing edges are harmful for geographic routing. However, the algorithm needs to detect possible loops in order to remove the harmful crossing edges.

Perimeter Routing

Planar graph traversal techniques (Karp and Kung (2000b), Heissenbuttel et al. (2003), Bose et al. (2001), Datta et al. (2002)) are used since they were proved to guarantee delivery if a path exists. Planar graph based obstacle avoidance strategies use greedy as long as a node has a neighbor closer to the destination.

Otherwise, one of existing planar graph traversal algorithms (Urrutia, 2002, Kuhn et al., 2003c, Kuhn et al., 2003a) is used. These algorithms are described by Stojmenovic (2005) as memoryless guaranteed delivery algorithms.

There are different approaches to perimeter routing, the difference between them consisting mainly in the condition used to switch from face to face. They all work in the unit disk model. FACE routing was first introduced by Kranakis et al. (1999), (further analyzed by Feng Zhao, 2004). A message is routed along a face until it finds an edge that intersects the line between the source and destination. At this point it switches faces and continues routing. The message will route on faces progressively closer to the destination. The algorithm works under the assumption that the faces are convex, and is proved to reach the destination in linear time if a path exists.

The drawback of this algorithm is the increase of path length with the network density, since the links in the planar graph will be shorter. Face routing performances were improved by Bose et al. (2001) who propose GFG (Greedy-Face-Greedy). GFG uses greedy until the message gets stuck at a node, when it switches to FACE routing. Karp and Kung (2000a) introduce Greedy Perimeter Stateless Routing (GPSR) that implements GFG with IEEE 802.11.

An algorithm that works under the more general assumption of arbitrary planar graphs, Original Face Routing, is also presented by Kranakis et al. (1999) - each message traverses all the edges of a face and chooses from the edges that intersect the line between the source and the destination the one that minimizes the distance toward the destination.

With the Bounded Face Routing (BFR) algorithm, (Kuhn et al. (2002)), the quality of the path of face routing is improved by limiting how far away a path is allowed to deviate from the line between the source and the destination. The path has to be inside an ellipse with the source and the destination as foci. If the path intersects the ellipse, the message will switch direction between clockwise and counterclockwise routing.

BFR tries to build the ellipse around an optimal path. However it does not always find a solution in the space limited by the ellipse. In this case, the message is routed back to the source. Adaptive Face Routing (AFR) improves this method, by restarting the same routing algorithm when a message returns to the source, with a double sized ellipse.

Kuhn et al. (2003c) present a less complex alternative, Other Face Routing (OFR): each message traverses all the edges of a face and switches face at the vertex closest to the destination. Other Adaptive Face Routing (OAFR) works similarly with AFR. The message will use OFR for routing inside an ellipse, increasing the size of the ellipse if the path is not found.

Greedy Other Adaptive Face Routing (GOA-FR) alternates greedy and OFR, while resizes the initial ellipse when needed. The algorithm starts in greedy mode. When the next hope is outside the ellipse, the ellipse is doubled. If a node is a communication void, then OFR is used only for one face. When the closest node to the destination is detected on the current face, the algorithm changes to greedy - the algorithm falls back to greedy as soon as possible. Kuhn et al. (2003b) limit the searchable area by replacing the ellipse with a circle centered at destination and with the radius proportional with the distance between the source and the destination. Whenever possible, the radius of the circle is decreased. The algorithm alternates greedy and face routing. Once in the face routing, it keeps two counters: one of the nodes of the current face closer to the destination than the face entry point and one for the nodes further to the destination than the same reference. If the ratio of closer nodes is above a specified threshold, the algorithm switches back to greedy at the closest node to the destination visited so far on the current face.

Kuhn et al. (2002), Kuhn et al. (2003c) discus the optimality of these algorithms. The AFR algorithm is proved to have a path cost bounded by a function dependent of the optimal route. The authors prove a routing path with quadratic cost compared with the shortest path. Additionally, Kuhn et al. (2003c) proved that GFG is not asymptotically optimal, while GOAFR is.

The performances of these strategies depend on two factors: the performances of graph traversal algorithm and of the distributed planarization algorithm.

Thus most of the algorithms are concerned with improving the planar graph traversal algorithms. The optimality of the path is not considered. The gain in path length (compared with the path chosen when the complete network topology is known) becomes significant when obstacles are present and it is proportional with the size of the obstacle. Our approach is different: while keeping both the greedy and planar graph based strategies, our algorithm progressively marks and avoids the non-optimal paths.

Geographic Routing Without Planarization

Geometric obstacle detection is proposed by Fang et al. (2006). It uses the geometric properties of a node to determine if a message can be stuck at that node. This happens if no neighboring nodes are closer to the destination and greedy routing fails. The nodes at which this happen are called weakly stuck[4]. They define the strongly stuck nodes as the nodes for which there exist a location in the plane such that greedy fails to route data to this location (even if there are no nodes in the location). Weakly stuck nodes are strongly stuck nodes and strongly stuck nodes do not necessarily cause problems since the location which is unreachable with greedy routing can contain no nodes. However strongly stuck nodes are easier to detect. A local rule, TENT, is developed to find voids in the network, defined as areas of the network bounded by the nodes at which greedy routing fails. At the same time a distributed algorithm, BOUNDHOLE, is proposed to route around the voids.

A weak stuck node n is a node for which there exists a node n1 outside its transmission range such that none of the one hop neighbors of n is closer to n1 than n. The authors prove that a node is not a weakly stuck node if its adjacent edges in a Delaunay Triangulation (DT) of the network graph are no longer than 1. The voids are represented by faces in a Restricted Delaunay Graph (a DT where links longer than 1 are removed) with at list 4 vertices. TENT is used for detection of strongly stuck nodes. It first orders counterclockwise the neighbors' n_1, n_2 of a node n and verifies each adjacent pair. It checks the position of the intersection point of the bisectors of the links n_1n and nn_2 of the current node with the two selected neighbors. If the intersection point is not in the transmission range of the current node, then there is a direction for which a packet can get stuck at the node. Starting with a stuck node, the entire border of the void is identified using only local information. The routing algorithm starts in greedy mode. If it reaches a stuck node, it uses the information about the void to select the adjacent neighbor on the void border. The message switches back to greedy when the current node is closer to the destination than the initial stuck node.

The disadvantage of this technique is the high complexity of the void detection and the high amount of required memory. Additionally, it does not guarantee delivery when the destination is inside the void. The advantage is that it maintains local the complexity generated by the presence of the void, while some planar graph traversal algorithms planarize the entire network. Additionally the routing algorithm will choose the direction with the shortest path.

Cost based approach Ye et al. (2001) consists in assigning a cost to each node, equivalent with the minimum cost to forward a message to the destination on the optimal path. Each message stores the minimum cost between the source and the destination and the cost of routing between the source and the current node. In order to relay a message, a node will check if the sum of the cost of routing to the destination and the cost of routing to the current node is equivalent with the initial cost.

In L. Zou and Lu (2004), a mechanism (called shadow spread phase) is proposed to detect the nodes close to a void, called shadow nodes. A neighbor of a stuck node is considered shadow if all its greedy neighbors are shadow or stuck nodes. The cost is assigned through a cost spread phase. It is proportional to the distance to the destination for normal nodes and higher than Euclidean cost for shadow nodes. When greedy forwarding fails, a node will forward a packet to a neighbor with a lower cost than itself. Although the message complexity of the algorithm is rather medium, it does not choose optimal paths.

S. Chen and Cui (2006) introduce a distance downgrading algorithm, to detect the optimal routing path. Additionally, the algorithm is robust to network changes.

Flooding based techniques (Stojmenovic and Lin, 2001, Stojmenovic et al., 2000, Jain et al., 1999) are using broadcast to forward the message, once a packet is stuck. Although the algorithm is simple, the communication overhead is high. They guarantee delivery, but path optimality is not a concern. Stojmenovic and Lin (2001) use flooding to forward the message to one hop neighbors when a message is stuck. Each neighbor will further use greedy forwarding to handle the message. Flooding is based on memorization: the packet ID is stored at each node that used flooding and discarded if another copy of the same message arrives.

Jain et al. (1999) use route discovery, by sending a broadcast message when a message is stuck at a node. Each node that receives a route discovery request adds its ID on the path discovered and rebroadcasts the message. Duplicate messages are discarded. This method is called breadth first search.

A depth first search algorithm was proposed simultaneous by Stojmenovic et al. (2000) and Jain et al. (1999). The difference from the previous method is that a node will forward the request to a single node, the one that minimizes the sum of distances between the current node and source and the current node and destination.

Chen and Varshney (2007b) propose a partial path discovery mechanism that aims at discovering a node closer to the destination than the current node. A path discovery message is sent to one hop neighbors. If the query is not successful, another message is send, targeting two hops neighbors. The node extends the flooding range progressively, until a path is found.

Multipath techniques (Lin et al. (2001), Chatzigiannakis et al. (2006)), explore several paths toward the destination, to trade-off efficiency with fault tolerance. Similar with the case of flooding techniques, the overhead may be high.

Heuristic techniques use additional resources or information to route the message. ? propose a second communication device at each node. Chen et al. (2005) increase the transmission power of a stuck node until a greedy neighbor is detected.

Another alternative (Zorzi and Rao, 2003) is for a void node to stop forwarding messages. At the same time it announces its neighbors about its impossibility of forwarding messages, technique called reverse propagation. The technique was proved to be ineffective for low density randomly deployed messages.

An alternative was proposed by He et al. (2003). A void node will advertise its status backwards. The propagation is finished at the first node that has a valid alternative for routing.

A probabilistic solution, Intermediate Node Forwarding, is proposed by Couto and Morris (2001). A void node will send a message to the source when it discards a message. The source will select a random node in a circle centered at the middle of the distance between the source and the destination. If the message is discarded again, the radius of the circle is increased.

Blazevic et al. (2002) detect a set of nodes called anchors, such that the probability of reaching the destination is high if the messages are forwarded between the anchors.

Hybrid techniques use at least a combination of two obstacle avoidance methods. The motivation is the improved efficiency of the path and the guaranteed delivery of the message. They are used when only one of the two techniques is not enough to achieve these requirements. The disadvantage is the increased overall complexity due to the supplementary messages necessary for coordinating the nodes.

Multihop data knowledge Spanning tree based techniques build a spanning tree when a message is stuck at a node. Radhakrishnan et al. (1999) forward messages using flooding, while Leong et al. (2006) aggregate the locations covered by subtrees using convex hulls to decide which direction in the tree is closer to the destination.

The methods described above are mainly concerned with guaranteeing delivery. In contrast, we aim at providing high quality paths, by keeping track of previous evaluations in a distributed manner. Additionally, our technique preserves the properties of the network, like scalability and low complexity since it works only with local information about the direct neighbors of the node currently propagating data. Previous related work is published in Moraru et al. (2008).

Obstacle Avoidance

GRIC (Georouting around Obstacles) algorithm was proposed by Nikoletseas and Powell (2007) to handle obstacles. It uses a combination of inertia and direct movement towards the destination in order to bypass the obstacles. The algorithm fails in the presence of concave obstacles; therefore a rescue mode was introduced to handle these cases, based mainly on the right hand rule.

GRIC requires only one hop information and does not use planarization. The authors claim that it uses a path close to the optimal one in the absence of global knowledge. Additionally the algorithm performs well in the case of highly dynamic networks, due to the lack of the planarization phase.

BEHAVIOR BASED OBSTACLE AVOIDANCE

The purpose of the algorithm is to obtain a gradual convergence to the optimal path. We intend to obtain minimal path lengths by dynamically evaluating the frequency of every routing method used by a node and assigning further routing tasks based on this evaluation.

As mentioned before, geographical routing uses greedy method to choose the next neighbor. When greedy fails, another strategy - rescue mode – is enforced. We are choosing as the rescue mode strategy a planar graph based traversal algorithm to route around obstacles. We are making this decision considering the guaranteed delivery and the stateless nature of this class of protocols. Greedy routing will be considered as good quality routing method, while face routing will be evaluated as a poor quality one.

The protocol gradually evaluates the performance of a path, detecting dynamically the nodes around holes, and progressively redefining the routing paths. Each node is evaluating itself and it is spreading locally information about its performance. Once the non optimal nodes are detected and advertised, each node in greedy routing mode will avoid choosing non optimal neighbors for forwarding, thus redirecting the message outside the non optimal nodes area. The shape of the non optimal area is presented in Figure 3(a), for convex obstacles, and Figure 3(b), for concave ones. Both figures represent intermediary stages during the convergence process. The black areas represent nodes marked as optimal, while the gray area contains non optimal nodes. The greedy nodes located at the border of the non-optimal area represent the path that the messages originated at the left side of the obstacle will follow. The upper area of the void for the concave shape is marked faster, since the messages are routed only counterclockwise around the obstacle.

The evaluation of a node as non optimal depends on the relative position between the source, object and destination. The point of incidence with the object influences the number of hops a message is routed with perimeter, since the message will exit perimeter only when it finds a node closer to the destination then the perimeter entry point. This is also the reason why a small area of nodes above the concave object is marked.

Further, we will present a new routing protocol which will take into account the optimality of a node for choosing the next relay and several behavior based non optimality detection methods, outlining their advantages and disadvantages.

Reputation - Path Ranking Analogy

The usage of previous behavior for the prediction of the future behavior of a node is a technique used in reputation mechanisms. The current reputation mechanisms are used to evaluate the willingness of a node to execute collaborative tasks in the network. We aim at using previous behavior to evaluate the ability of a node to forward messages on an efficient path in terms of specific cost metrics important for the sensor networks. Therefore we are building our evaluation system by borrowing the tools from reputation building mechanisms. We will further give an overview of reputation systems and we will describe the analogy between reputation and path optimality evaluation.

Reputation systems are enforced in collaborative environments and are used to help predicting the behavior of an entity based on the previous experience with that entity. Each interaction is evaluated in terms of only two possible results: positive or a negative. Therefore, their behavior can be modeled as a statistical process with binary events. Before interacting with an entity, a node will infer all the information obtained from the previous interactions and it will predict the outcome of future interactions. Based on this prediction, a decision is made about interacting with this entity or not.

Beta reputation system (Ismail and Josang, 2002) is a reputation representation system, which uses beta probability density function to build entities' ratings. Reputation, defined as the opinion of an entity about another, is represented as a probabilistic distribution.

Trust is the expectation of an entity about the actions of the other. It is obtained by taking the statistical expectation of the probability distribution representing the reputation. If α is the number of positive outcomes and β is the number of negative outcomes, then the expected value of beta distribution is:

$$E(p) = \alpha/(\alpha + \beta),$$

with p a random variable which has a beta distribution with parameters α and β. Based on the trust value, an entity will decide if it will cooperate

Figure 3. Non optimal regions: (a) Convex void ; (b) Concave void

(a)

(b)

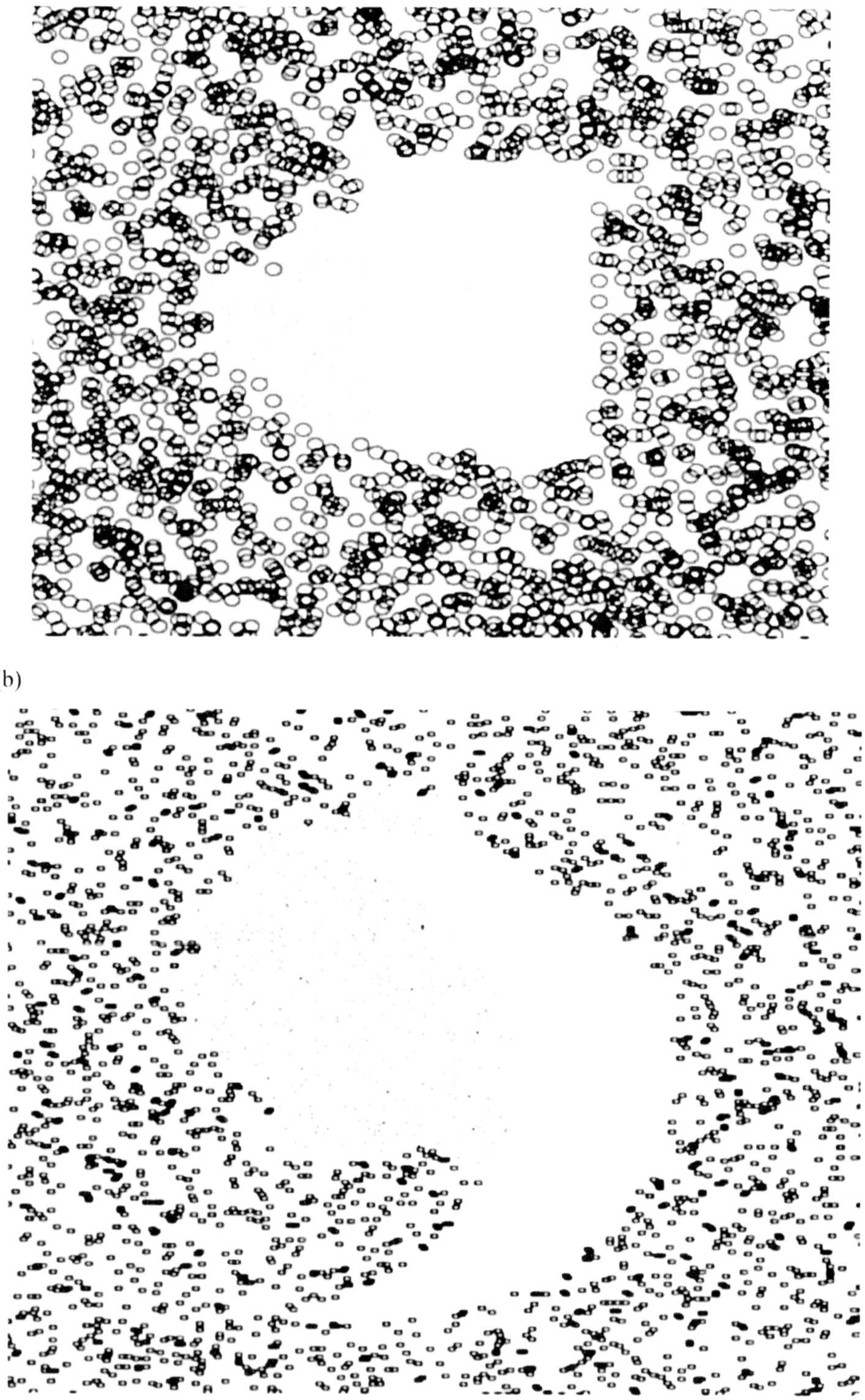

or not. A threshold is set and compared with the expected value. If the expected value is under the threshold, then the node will not cooperate.

We take the mathematical tools used by beta reputation system, but we give a different meaning. We are interested in evaluating optimality / non optimality of a path passing through a certain node.

We are using the same model of binary events. Each node can choose from two routing decisions; one of them is evaluated as optimal, thus it will represent a positive outcome; the other is evaluated as non-optimal, and it will represent a negative outcome. Each node will choose the next forwarding neighbor based both on the suitability with the selection method and on its expected behavior.

The difference between the two applications of beta probability representation consists in the location where the two systems are gathering, storing and inferring the information.

In a system where the willingness to cooperate is evaluated, it is reasonable that the trust is computed by the peers. But when the frequency of the used routing methods has to be evaluated, the node can make the decision by itself. Thus, the estimation of the routing method optimality can be made at the node level. Furthermore, the evaluation made at the neighbors' level will increase the network traffic. The advantages of this method are the simplicity of implementation in practical applications and the strong statistical background. Simplicity of implementation is important in sensor networks which are scarce in computational and energy related resources. The statistical background will provide robustness to the evaluation mechanism.

Routing

This section describes the modifications we brought to the state of the art geographic routing protocol in order to consider the quality of the path. Algorithm 1 describes our routing protocol. It starts by checking what the current routing mode is. If the message is in the perimeter mode, there are two possible situations:

a. The current node is closer to the destination than the perimeter entry point and the routing mode is switched to greedy,
b. Otherwise the current node will forward the message counterclockwise to the next neighbor on the face of the planar graph.

If the message is in the greedy mode, it will select next relay between the optimal neighbors. If greedy fails in finding the next relay, the perimeter mode is enabled.

Based on the routing mode used, the optimality of the current node is updated.

Algorithm 1 Trust Based Routing Strategy

Variables:

- *this:* The identifier of the current node executing the routing strategy.
- *entry_point:* The identifier of the node which first switch from greedy to perimeter mode.
- *neighs:* The set of neighboring nodes of the current node *this*.
- *sel_neighs:* store the identifier of the next node selected for transmitting the message to.
- *routing_mode:* store the value of the selected routing mode, can be *perimeter* or *greedy*.

Procedure:

- *is_closer(x,y):* Boolean valued procedure, returns true if the node with identifier x is closer to the destination than the node with identifier y.

- *get_next_hop(a,b):* Returns the next node to transmit the message among the set of node *b* using the routing mode *a*.
- *filter(a):* Return the set of nodes belonging to *a* which are optimal nodes.
- *update_optimality:* Switch the mode *routing_mode* if necessary.

Code:

```
if routing _mode is "perimeter"
then
   if is _closer(this, entry_
point) then
      routing _mode ← greedy
   end if
end if
if routing _mode is "perimeter"
then
   next ← get _next_
hop("perimeter", neighs)
else
   sel neighs ← filter(neighs)
   next ← get _next _
hop("greedy", sel_neighs)
   if ! ∃ next then
      next ←  get _next_
hop("perimeter", neighs)
   end if
end if
update _optimality(routing _
mode)
```

The modifications to the behavior of the routing protocol are as follows.

1. A node in greedy mode will choose a successor only between optimal nodes. First a selection of candidates for greedy routing is made: only nodes that are closer to the destination than the current node and that are marked as optimal are included. The reason is that, even if they could represent a local optimum, since afterward they use perimeter routing, they are nodes on non-optimal paths. The condition to enter perimeter routing remains unchanged. When a message enters the perimeter mode, it will choose the first neighbor counterclockwise, with the line between the source and the destination as a reference. Perimeter routing selection of the next hop remains unchanged as well.

With this configuration in mind, the behavior of the algorithm is similar to the classical perimeter routing algorithms for both the marked and unmarked area. We define the border as the area composed of nodes with neighbors in both marked and unmarked areas. The only difference is represented by the behavior at the border: by giving priority first to the optimal nodes, the messages will be routed around the marked area. But this behavior is possible only when the nodes at the border have greedy optimal neighbors toward the destination.

Otherwise, the node will enter into the marked area, without any significant improvement on the routing path.

2. When the perimeter entry point chooses the next relay, it will check first the status of its neighbors. If it has both optimal and non optimal neighbors, it will select an optimal neighbor to continue routing.

The second change in the routing algorithm aims at improving the path, even when the border nodes have no greedy optimal neighbors toward the destination. Since the probability that a node on a current face is closer to the destination than the current node is high, the node will switch to greedy with high probability. Therefore the message is kept on the border, advancing toward the destination in greedy or perimeter mode. Further we will address another important issue of the algorithm. We will sketch three different classes of optimality evaluation methods, that achieve different compromises between the efficiency and the cost of the solution achieved.

Behavior Model - Beta Probabilistic Function

First method (Entire History Evaluation), as described in Algorithm 2, is to use two counters to evaluate the interactions. Each time the greedy routing is used, the counter for positive interactions, is increased. Each time perimeter routing is used, the counter for negative interactions is increased.

We will use Bayesian probability for interference. If g represents the number of messages sent by greedy and p the number of messages sent by perimeter, then the performance is calculated by the formula $p/(p + g)$ and represents the expected value, a number between 0 and 1. The optimality is determined by comparison between the expected value and a specific THRESHOLD. Initially all nodes are assigned the rating of 1.

Algorithm 2 Bayesian Computation Trust Update

```
if routing_ mode is "perimeter"
then
   p ← p + 1
else
   g← g + 1
end if
rank=p/(p+g)
if rank ≥ THRESHOLD then
   optimality ← 0
else
   optimality ← 1
end if
```

In the context of stationary traffic, this method offers a robust evaluation and faster convergence to a stable state than the evaluation in one step. Additionally, the current state reflects the behavior of the network, the traffic density distribution. Another factor that influences both the convergence time and the number of nodes marked as non optimal is the threshold value. By decreasing the THRESHOLD, we decrease the convergence time and mark more nodes as non optimal.

Threshold Selection

The evaluation of a node as optimal or non optimal depends on the rank value, compared with the THRESHOLD. Therefore the choice of the THRESHOLD will influence the performances of the algorithm.

There are two factors that are directly influenced: the convergence time and the number of non optimal nodes. The convergence time should decrease with a lower THRESHOLD. This should happen especially after the first messages routed and the first layer of non optimal nodes detected. The layers of nodes further away from the object already used greedy, therefore it will take more perimeter decisions to reach the ratio represented by the THRESHOLD. In this situation, a lower threshold will allow faster convergence.

The marking of the nodes at the border in the optimal or non optimal area depends mainly on the traffic patterns. If more messages are routed from an area with little incidence on the object, then the rank is lower.

Example

Figure 4(a) represents the path of the first message. When the message arrives at node n1[5], it switches to perimeter, since node n1 does not have any greedy neighbors toward the destination. The message is forwarded in perimeter mode until n6 which is closer to the destination than the perimeter entry point, therefore it switches to greedy. The negative counter is increased to one for the nodes n1 – n5, which are marked as non optimal.

Figure 4. Message path: (a) First scenario; (b) Second scenario; (c) Third scenario

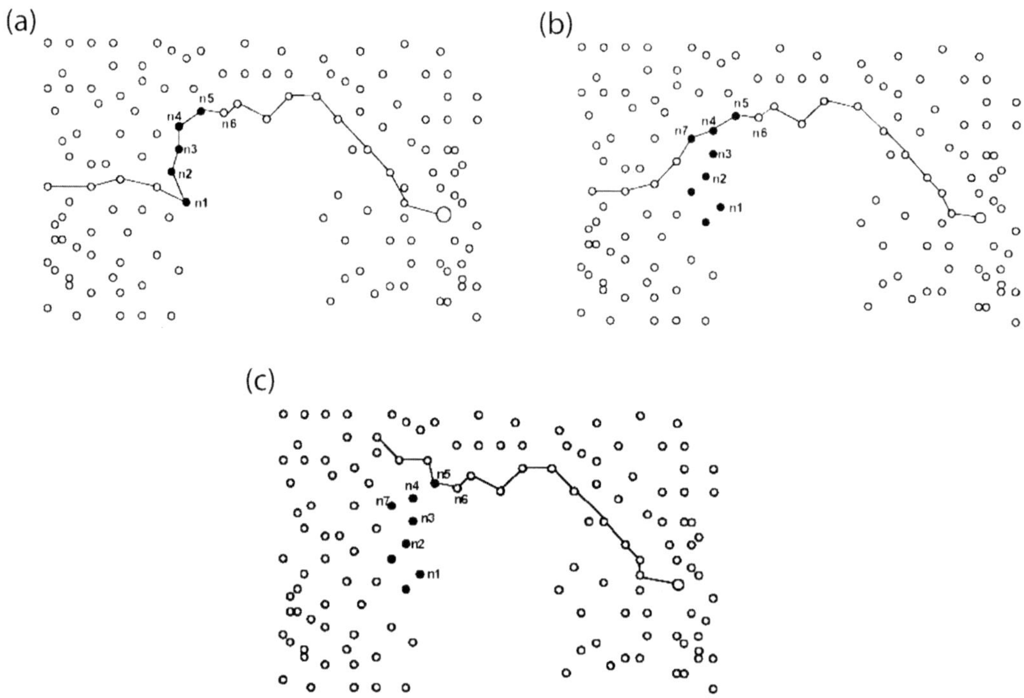

We will further present the behavior of the network and the detection of the non optimal nodes. We consider few other messages sent from the same source in Figure 4(a), having as a result the marked area in Figure 4(b). Figure 4(b) represents the path of a message that avoids the marked area, choosing a path close to optimal.

Node n7 has no optimal greedy neighbors toward the destination; therefore its negative counter is increased. Similar for n4, n5 is closer to the destination than n4, therefore it will switch to greedy, and its positive counter is increased.

The last step, Figure 4(c), shows the path of a message with a different source than the previous ones. The node n8 has no optimal neighbors closer to the destination than itself, therefore its negative counter will increase and it will start routing in perimeter. n5 is closer to the destination than n8, and it has optimal greedy neighbors, therefore it will increase the positive counter and route in greedy mode.

Second method (Last Step Trust), described in Algorithm 3, is to build a reputation based only on evaluation of the last interaction. Each time a node is used as a relay, it will evaluate the quality of the path and it will set the optimality value to 1 if greedy is used and 0 otherwise.

The advantage of this method is its simplicity, no computing effort is necessary.

Algorithm 3 Last Step Trust Update

```
if routing_mode is "perimeter"
then
   optimality ← 0
else
   optimality ← 1
end if
```

The disadvantage is the increased number of transmissions needed to notify the neighbors about the state changes. Additionally, the convergence time can be quite long, especially for the border nodes. The state of a node can change often at the border between the nodes for which the obstacle intersects the direct line between the source and the destination and the other nodes.

The third algorithm (Bayesian with Propagation) uses neighbors' optimality in building the optimality of each node. We are using the same method as the previously described reputation system uses to combine feedback from multiple sources. The method is presented in Algorithm 4. R is the optimality of a node, Rd is the optimality value obtained as a result of the own decisions and R_i is the optimality inferred from that of the neighbors. The method *getrank(x)* returns the rank of node *x*.

Algorithm 4 Bayesian with Inference Trust Update

```
Rd ← get rank(this)
for each neighbor i do
   Ri ← get rank(i)
end for
R ← (1 -WEIGHT) * Rd + Σi WEIGHT
* Ri
if R ≥ THRESHOLD then
   optimality ← 0
else
   optimality ← 1
end if
```

Weight Selection

The weighting factor, WEIGHT, is a measure of the influence of the neighbors. It improves the convergence time of the algorithm, since the non optimal nodes in the vicinity of an already detected area will be marked faster. The value of the weighting factor is a trade off between the convergence time and the number of nodes detected. It has to be small enough, such that the influence of the node owns behavior is more significant than the influence of the neighbors.

State Dissemination

By exploiting the existence of hello messages, each node maintains an accurate image of the path rating of each neighbor. Nodes can locally send information about the optimality status, by adding this information to the content of the hello messages. If the messages are sent periodically, this additional information is transmitted with only one additional bit, added at the end of the hello message. A second alternative for optimality status dissemination is to send an update message each time the state is changed. The trade off in choosing one of these methods is between convergence time and the overhead. If the algorithm waits until the beacon is send, then the convergence time is delayed, otherwise, more overhead is added into the network. Other factors to consider when choosing the dissemination method are the number of nodes marked as non optimal and the number of state switches of each node, since both influence directly the overhead added in the network when reactive updates are used.

Performance Evaluation

Network Configuration

We evaluate experimentally the performances of the algorithms mentioned above. The network configuration used is as follows. The size of the network is 50x50 units; with the number of nodes varying between 2800 and 6700. The nodes are placed randomly in the 50x50 square with uniform distribution. We choose rather high densities, in order to minimize the number and the impact of the communication voids. This way, the alteration of routing performances is mainly caused by the behavior of the protocol around the object. Two types of obstacles in the network topology are tested. The first one is a convex shape, a rectangle with the size of 10x20 units. The second type is a concave object, with the shape of a half moon, with the radius of the generator of 10.

The transmission radius of a node has 2 units, thus the ratio between the obstacle size and the transmission range is 5, and therefore the obstacles are quite large. Moreover, the density measured as the average number of neighboring nodes ranges between 15 and 35. The traffic in the network is generated in an area of size 3x3 units. The destinations are located in a 3x3 units square area on the other side of the obstacle reported to the source. The reason for this configuration is to have an evaluation of the worse case influence of the obstacles on the traffic.

The results are presented as the median of 50 experiments, after the convergence of the network. For each experiment, a message is sent into the network at each step. If the number of non optimal nodes remains constant after 300 steps, the algorithm is considered convergent, and the performance of the first message routed in the network after convergence is measured. For each experiment we are changing the network configuration: the position of the nodes and the links between the nodes.

The performances of our protocol in terms of path gain are independent of the model used. The physical layer model has a direct impact on the routing protocol performance, but not on the optimization method.

Experimental Results

We are evaluating the performances of our three protocols: Last Step Trust, Entire History Evaluation, and Bayesian with Propagation. The comparison is made based on criteria like the path quality and the time to convergence.

Our goal is to devise algorithms that produce paths close to optimal. We compare the quality of a discovered path with the one generated by a topology aware strategy which selects the shortest path. In the case of obstacles, the optimal path will have as intermediary destinations the extremities of the obstacle. Each intermediary reference is chosen as the closest extremity to the line between the current node and the destination. The optimal hop count is measured as the ratio between the Euclidean distance and the communication range.

The second reference for comparison is the greedy perimeter stateless routing, as the state of the art in planar graph traversal based routing protocols.

The planar graph traversal algorithm is convex face routing: a node walks on faces progressively closer to the destination.

Hop Count

First simulation results, Figure 5 present the average hop distance in the vicinity of the obstacle, as an estimator of the latency of the message. The first image represents the measurements for the convex obstacle, while the second represents the measurements for the concave one. The x axis represents the network density, measured as the average number of neighboring nodes. The y axis represents the average number of hops traversed by the message.

Greedy perimeter stateless routing introduces a latency of more than 200% in the vicinity of the obstacle. For convex obstacles, all the trust based algorithms are introducing at most 50% of latency for the highest densities. For lower densities, the One Step Evaluation has the best performance, while Bayesian with Propagation algorithm is closed to Greedy Perimeter Stateless Routing.

For concave obstacle, the gain in hop distance compared to GPSR is even higher. This is due to the effort of attaining the node closest to the destination inside the convex shape, added to the perimeter routing along the obstacle. For highest densities, the increase in hop count compared with the ideal case is 25% for ranking mechanisms, while for

Figure 5. Hop distance after convergence

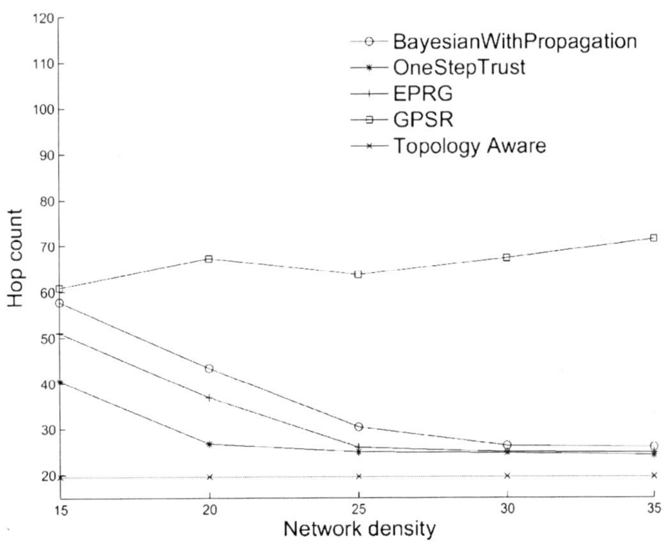

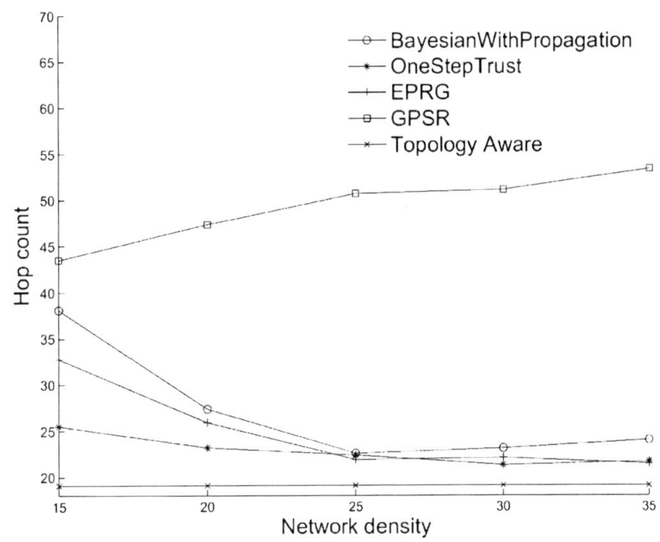

Greedy perimeter stateless routing is 200%. For lower densities, One Step evaluation mechanism has the closest to optimal performances.

Another observation is that GPSR does not have a linear dependence on the network density. This can also be observed in Leong et al. (2005) for GPSR and other suggested algorithms, the performance of the algorithms are sensitive to the density due the impact of this parameter on the planarization process.

Thus, the explanation resides in the number of perimeter routing decisions. Actually, one can claim that as the network density is not high enough, voids cause difficulties to GPSR. As the network density increases, voids disappear and the total hop count decreases. Indeed, numerical experiment tends to confirm this since trust based algorithm avoid using perimeter routing and do not show similar behavior.

Figure 6. Path length after convergence

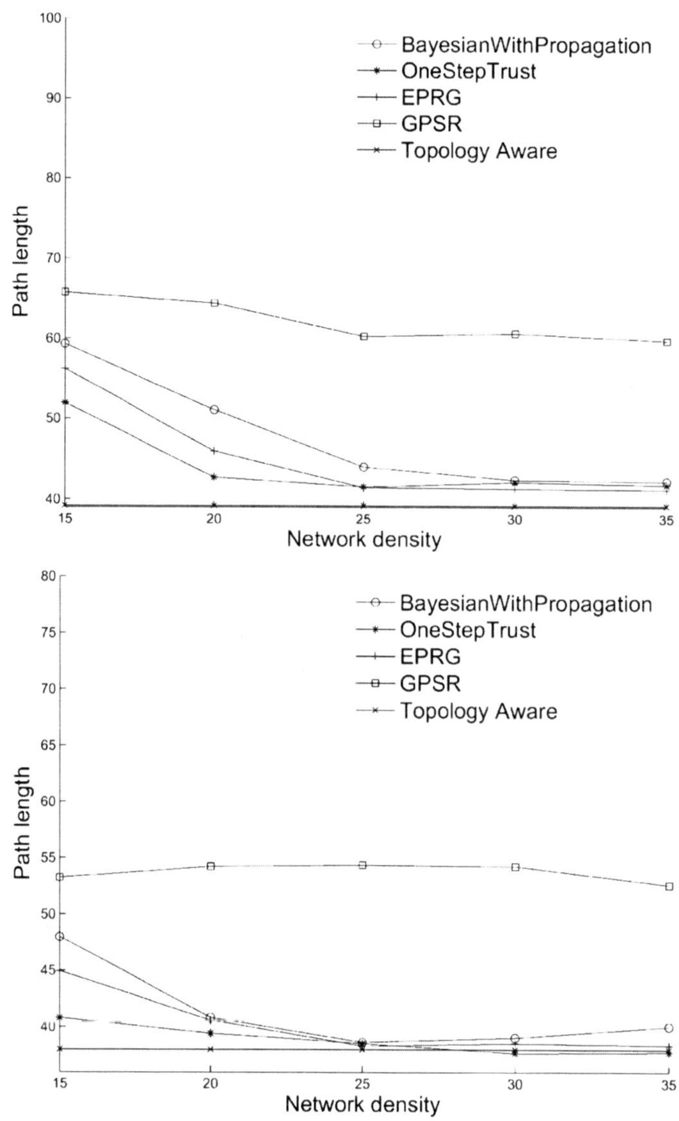

The final remark is that our class of algorithms has a behavior closer to the optimum when the network density increases.

Path Length

The second criterion for path quality is the path length, evaluated as a sum of all intermediary Euclidean distances. This parameter is important for energy consumption evaluation, since energy is more or less proportional to the sum of squares of distances. Of course, to take advantage of the performance of trust based algorithms one still have to tune the energy of transmission according to the physical conditions. Figure 6 shows the median of the path length obtained by each of the algorithms. The x axis represents again the density of the network.

One Step has better results than all the other algorithms in the same class. For high densities, all the ranking based algorithm have smaller average

Figure 7. Convergence time for ranking algorithms

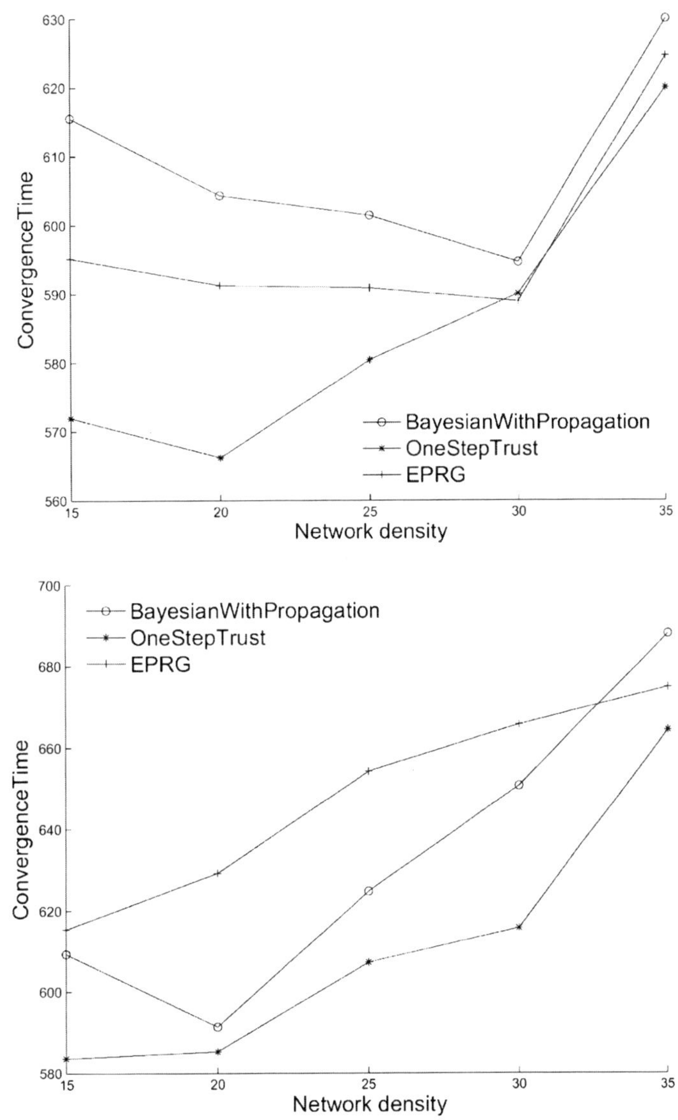

path length and their behaviors are similar. The path length decreases as the density of the network is increasing. Notably, for densities higher than 20, they are less sensitive to the particularities of the network's topology and the average path length is almost constant, as the density increases. However, the density of the network has a smaller impact on the ranking based algorithms than on GPSR.

Convergence Time

The convergence time, Figure 7 is also evaluated for different trust-based algorithms. The convergence condition is that all untrustworthy nodes are marked as non optimal. If the network configuration is static then our algorithm is convergent.

While for lower densities, the performance of the three algorithms is different, they start to have the similar performances for high densities.

Figure 8. Untrustworthy nodes before convergence

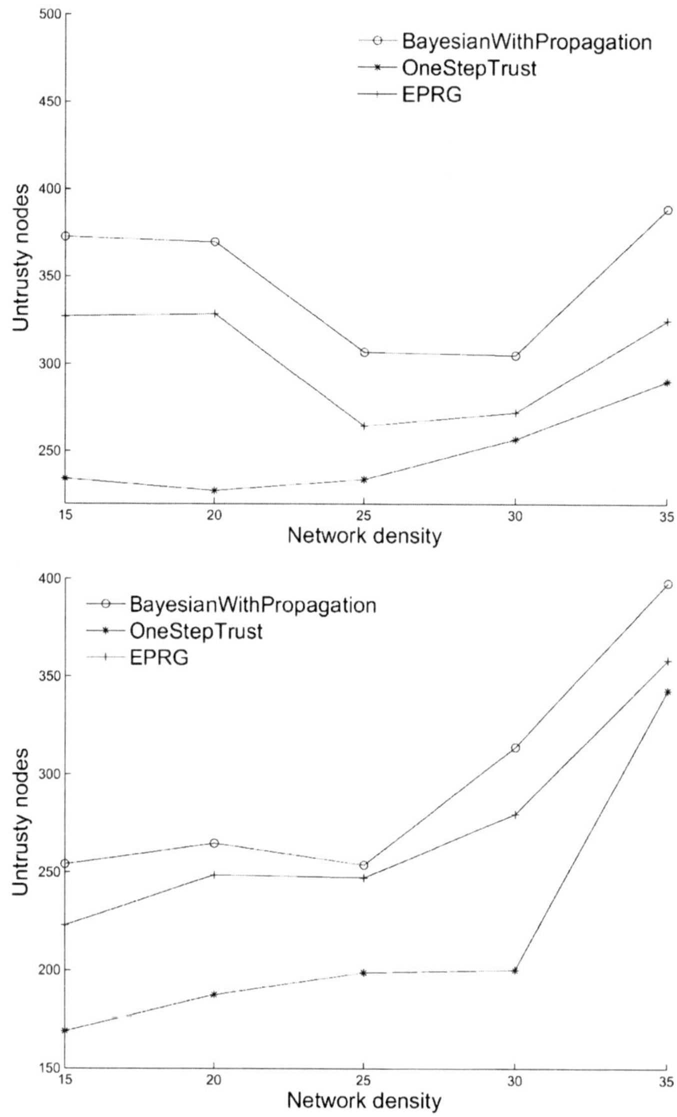

Additionally, for high densities, the convergence time increases significantly.

The Bayesian with Interference algorithm has the highest convergence time. This was expected, since the state of the non optimal nodes influences the detection time for their neighbors.

Influence of the WEIGHT Parameter on the Routing Performance

The next set of simulations is provided to show the impact on the WEIGHT parameter on the performance of Algorithm 4 (Baysesian with inference trust update). The set of experiments are done with the parameter WEIGHT having the different values 0.1, 0.3, 0.5. The results are plotted with the labels propag1, propag2 and propag3 on Figure 9, 10, 11 and 12. Each set of experiments is conducted with the convex and the concave obstacles.

The influence of neighbors' reputation will increase the number of untrustworthy nodes. This is explained by the fact that some of the nodes

Figure 9. Hop distance after convergence

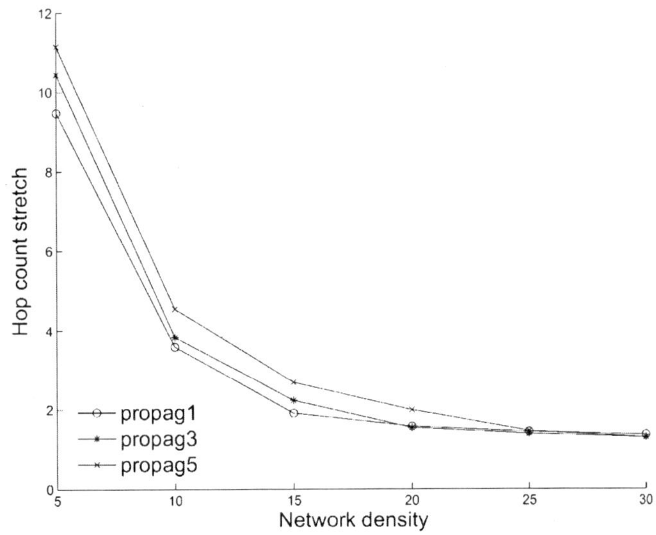

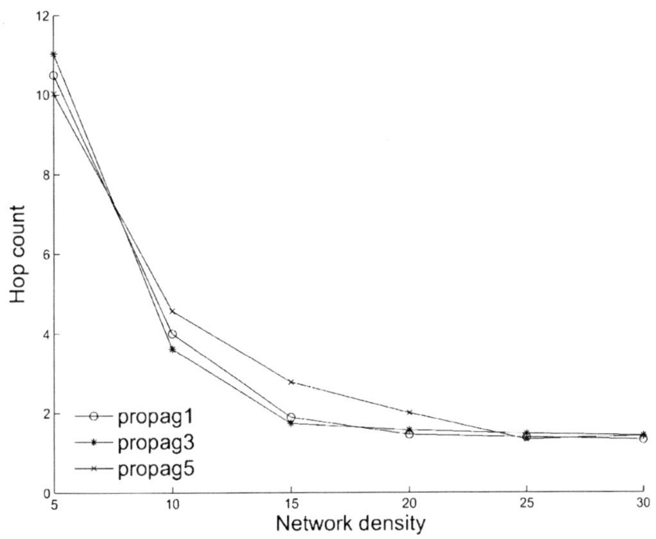

at the edge of the untrustworthy area are having a small difference between the optimal and non optimal routing decisions. If the number of non optimal neighbors is significantly higher than of optimal ones, than they can determine a change in the state of the node.

Hop Count

First simulation results, Figure 9 present the average hop distance in the vicinity of the obstacle. The first image represents the measurements for the convex obstacle, while the second represents the measurements for the concave one. The x axis represents the network density. The y axis represents the average number of hops traversed by the message. The results show little impact of the neighbors' state on the hop count.

Path Length

Figure 10 shows the median of the path length obtained by each of the algorithms. The x axis represents again the density of the network. It

Figure 10. Path length after convergence

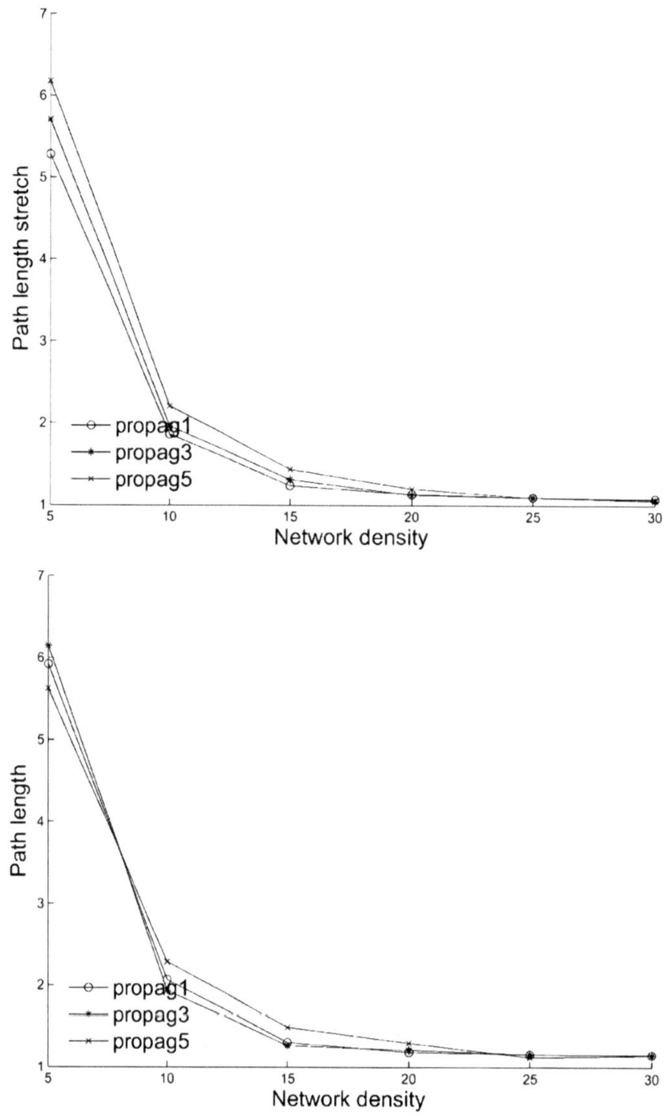

shows little impact of the weighting factor on the path length.

Convergence Time

The convergence time, Figure 11 is also evaluated for different weight factors for the influence of the neighboring nodes. It is presented as the ratio between the real convergence time and the number of nodes in the network. It decreases with network density and with the value of the weight factor.

Marked Area Size

In Figure 12 we notice a decrease of number of non optimal nodes with the density. At the same time the percent of marked nodes decreases with the value of weight factor.

Figure 11. Convergence time for ranking algorithms

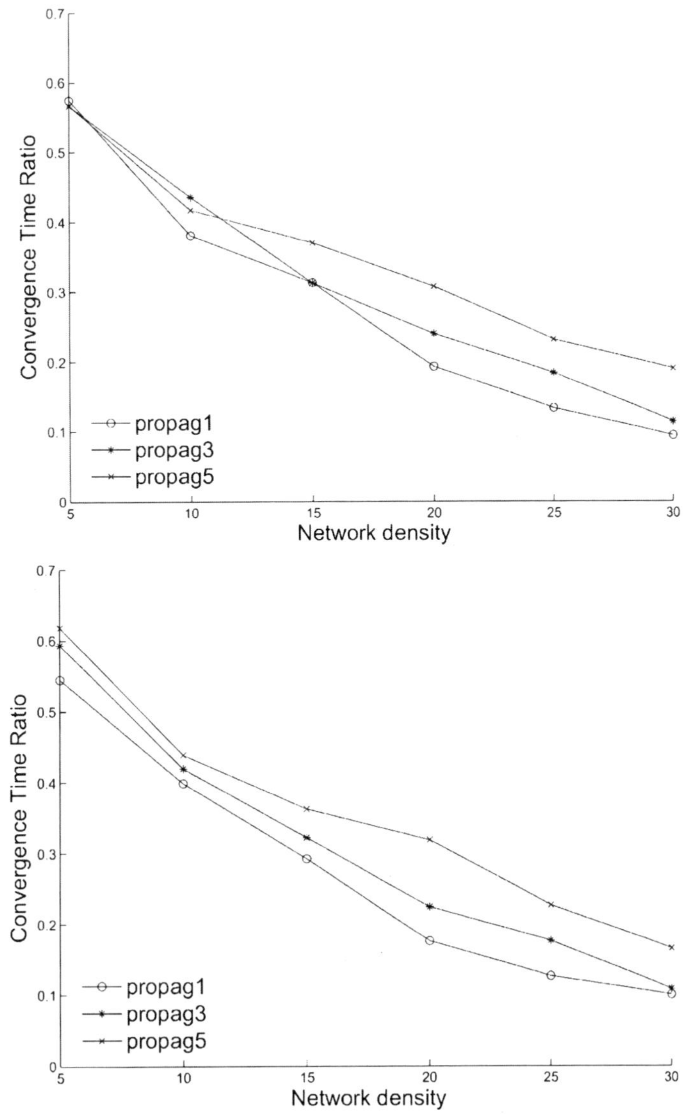

FUTURE RESEARCH DIRECTIONS

The trust-based strategies presented in this chapter are used to find optimal path from many nodes to a single fixed destination. In the future, we are interested to investigate how such methods can adapt to the case where there are multiple destinations. In this case one trust value is not enough. Indeed, the optimality or non optimality of the nodes depends on the relative position of the nodes and the destination. The situation where the nodes can choose between many destinations is relevant from various points of view. Indeed, a classical scenario is that the destination is a particular node, called base station, collecting data from the various nodes composing the network. Having many such base stations helps in managing the network traffic and avoids overusing some nodes which are prone to act as relay because of their proximity to the base station. This further line of research links

Figure 12. Untrustworthy nodes before convergence

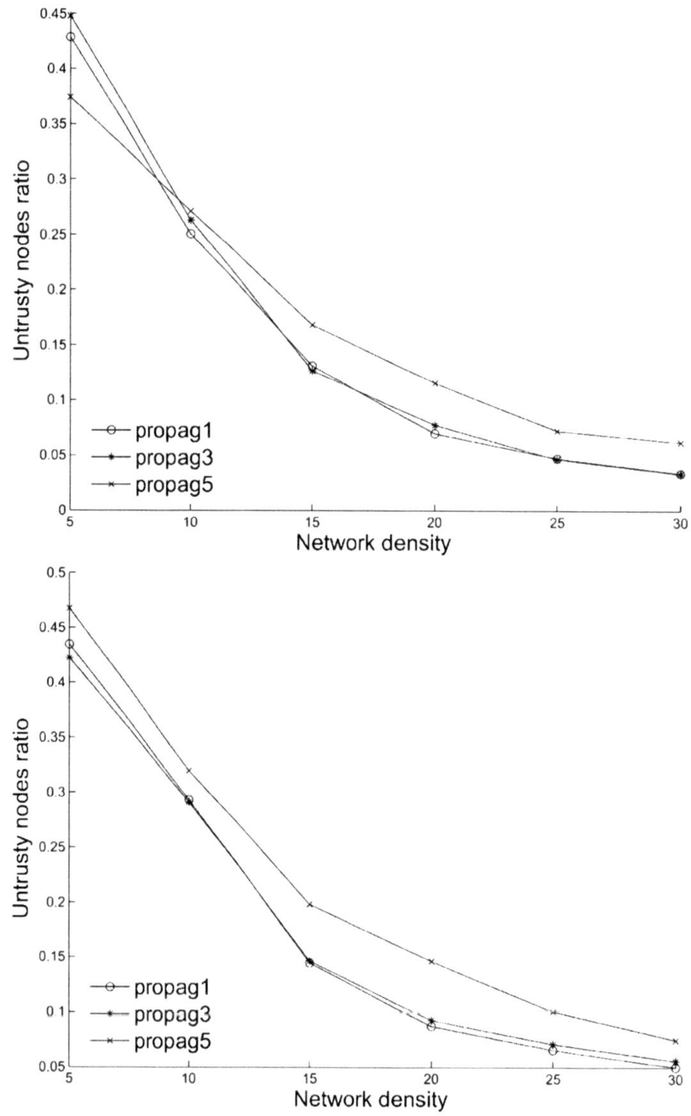

with energy balanced algorithms. Basically, such algorithms try to balance the energy consumption due to routing data among the nodes.

More generally, we feel that considering the all-to-all communication pattern is relevant. Even in the case where we have a fixed number of base stations, the all-to-all communication pattern allows for the mobility of the base stations.

Another assumption was that the nodes are cooperative. In computing the trust value of the nodes we assume that the information provided by neighboring nodes is correct. Further work could deal with the security/fault tolerance of the algorithm and the impact of non collaborative entities in the network.

CONCLUSION

In this chapter, we present a class of algorithms, heuristically based, which significantly improve the performance of the geographic routing with obstacle avoidance protocols. We introduce a new measure of the performance by considering the length of the routing paths. Generally, geographic routing strategies focus on the guarantee of delivering the data.

As the simulations show, the algorithms achieve better performances than the greedy perimeter stateless routing (GPSR) the reference protocol. The good performances can be observed after the paths converge to an optimum. The convergence time is an important factor, and again, our algorithms converge quite quickly. The quality of the path is not the only comparison criterion for the geographic routing algorithms. The overhead and the complexity of the algorithm are also important factors that we consider and the proposed mechanisms comply with both requirements.

REFERENCES

Ahuja, R. K., Magnanti, T. L., & Orlin, J. B. (1993). Network Flows: Theory, Algorithms, and Applications. Englewood Cliffs, NJ: Prentice Hall.

Blazevic, L., Giordano, S., & Boudec, J.-Y. L. (2002). Self organized terminode routing. *Cluster Computing*, *5*(2), 205–218. doi:10.1023/A:1013998030317

Bose, P., Morin, P., Stojmenovic, I., & Urrutia, J. (2001). Routing with guaranteed delivery in ad hoc wireless networks. *Wireless Networks*, *7*(6), 609–616. doi:10.1023/A:1012319418150

Cal, F., Conti, M., & Gregori, E. (1998) Ieee 802.11 wireless lan: Capacity analysis and protocol enhancement. In *Fourth Annual ACM/IEEE International Conference on Mobile Computing and Networking (MobiCom '98)*.

Chatzigiannakis, I., Dimitriou, T., Nikoletseas, S., & Spirakis, P. (2006). A probabilistic algorithm for efficient and robust data propagation in smart dust networks. *Ad-Hoc Networks Journal, 4*(5).

Chen, D., Deng, J., & Varshney, P. (2005). On the forwarding area of contention-based geographic forwarding for ad hoc and sensor networks. Sensor and Ad Hoc Communications and Networks, (pp. 130 – 141).

Chen, D., & Varshney, P. (2007a). A survey of void handling techniques for geographic routing in wireless networks. In Communications Surveys and Tutorials, (pp. 50–67). Washington, DC: IEEE.

Chen, D., & Varshney, P. K. (2007b). On-demand geographic forwarding for data delivery in wireless sensor networks. *Computer Communications*, *30*(14-15), 2954–2967. doi:10.1016/j.comcom.2007.05.022

Chen, S. G. F., & Cui, J. (2006). Avoid void in geographic routing for data aggregation in sensor networks. International Journal of Ad Hoc and Ubiquitous Computing (IJAHUC), Special Issue on Wireless Sensor Networks.

Clark, B. N., Colbourn, C. J., & Johnson, D. S. (1990). Unit disk graphs. *Discrete Mathematics*, *86*(1-3), 165–177. doi:10.1016/0012-365-X(90)90358-O

Couto, D. D., & Morris, R. (2001). Location proxies and intermediate node forwarding for practical geographic forwarding. Technical report, MIT Laboratory for Computer Science.

Datta, S., Stojmenovic, I., & Wu, J. (2002). Internal node and shortcut based routing with guaranteed delivery in wireless networks. *Cluster Computing*, *5*(2), 169–178. doi:10.1023/A:1013985610753

Fang, Q., Gao, J., & Guibas, L. J. (2006). Locating and bypassing holes in sensor networks. *Mobile Networks and Applications*, *11*(2), 187–200. doi:10.1007/s11036-006-4471-y

Feng Zhao, L. G. (July 2004). *Wireless Sensor Networks: An Information Processing Approach*. San Francisco: Morgan Kaufmann.

Finn, G. G. (1987). Routing and addressing problems in large metropolitan-scale internetworks. Technical report, University of Southern California, Marina del Rey, Information Sciences Institut.

Gabriel, K., & Sokal, R. R. (1969). *A new statistical approach to geographic variation analysis*.

Govindan, Y.-J. K. R., Karp, B., & Shenker, S. (2006). Lazy cross-link removal for geographic routing. In *SenSys '06: Proceedings of the 4th international conference on Embedded networked sensor systems*, (pp. 112–124). New York: ACM.

He, T., Stankovic, J., Lu, C., & Abdelzaher, T. (2003). Speed: A stateless protocol for real-time communication in sensor networks. In *International Conference on Distributed Computing Systems* (ICDCS).

Heissenbuttel, M., Braun, T., Bernoulli, T., & Walchli, M. (2003). *BLR: Beacon-less routing algorithm for mobile ad-hoc networks.*

Ismail, R., & Josang, A. (2002). The beta reputation system. In *Proceedings of the 15th Bled Conference on Electronic Commerce.*

Jain, R., Puri, A., & Sengupta, R. (1999). *Geographical routing using partial information for wireless ad hoc networks.*

Karp, B., & Kung, H. T. (2000a). Gpsr: greedy perimeter stateless routing for wireless networks. In *MobiCom '00: Proceedings of the 6th annual international conference on Mobile computing and networking*, (pp. 243–254). New York: ACM Press.

Karp, B., & Kung, H. T. (2000b). GPSR: greedy perimeter stateless routing for wireless networks. In Mobile Computing and Networking, (pp. 243–254).

Kim, Y.-J., Govindan, R., Karp, B., & Shenker, S. (2005a). Geographic routing made practical. In *NSDI'05: Proceedings of the 2nd conference on Symposium on Networked Systems Design & Implementation*, (pp. 217–230). Berkeley, CA: USENIX Association.

Kim, Y.-J., Govindan, R., Karp, B., & Shenker, S. (2005b). On the pitfalls of geographic face routing. In *DIALM-POMC '05: Proceedings of the 2005 joint workshop on Foundations of mobile computing*, (pp. 34–43). New York: ACM.

Kranakis, E., Singh, H., & Urrutia, J. (1999). Compass routing on geometric networks. In *Proc. 11 th Canadian Conference on Computational Geometry*, (pp. 51–54), Vancouver.

Kuhn, F., Moscibroda, T., & Wattenhofer, R. (2004). Unit disk graph approximation. In *DIALM-POMC '04: Proceedings of the 2004 joint workshop on Foundations of mobile computing*, (pp. 17–23). New York: ACM.

Kuhn, F., Wattenhofer, R., Zhang, Y., & Zollinger, A. (2003a). *Geometric ad-hoc routing: Of theory and practice.*

Kuhn, F., Wattenhofer, R., Zhang, Y., & Zollinger, A. (2003b). Geometric adhoc routing: of theory and practice. In *PODC '03: Proceedings of the twenty second annual symposium on Principles of distributed computing*, (pp. 63–72). New York: ACM Press.

Kuhn, F., Wattenhofer, R., & Zollinger, A. (2002). Asymptotically optimal geometric mobile ad-hoc routing. In *DIALM '02: Proceedings of the 6th international workshop on Discrete algorithms and methods for mobile computing and communications*, (pp. 24–33). New York: ACM.

Kuhn, F., Wattenhofer, R., & Zollinger, A. (2003c). Worst-Case Optimal and Average-Case Efficient Geometric Ad-Hoc Routing. In *Proc. 4 th ACM Int. Symposium on Mobile Ad-Hoc Networking and Computing (MobiHoc).*

Kuhn, F., & Zollinger, A. (2003). Ad-hoc networks beyond unit disk graphs. In *DIALM-POMC '03: Proceedings of the 2003 joint workshop on Foundations of mobile computing*, (pp. 69–78). New York: ACM.

Leong, B., Liskov, B., & Morris, R. (2006). Geographic routing without planarization. In *NSDI'06: Proceedings of the 3rd conference on 3rd Symposium on Networked Systems Design & Implementation*, (pp. 25–25). Berkeley, CA: USENIX Association.

Leong, B., Mitra, S., & Liskov, B. (2005). Path vector face routing: Geographic routing with local face information. In *ICNP '05: Proceedings of the 13TH IEEE International Conference on Network Protocols (ICNP'05)*, (pp. 147–158). Washington, DC: IEEE Computer Society.

Lin, X., Lakshdisi, M., & Stojmenovic, I. (2001). Location based localized alternate, disjoint, multi-path and component routing schemes for wireless networks. In *MobiHoc '01: Proceedings of the 2nd ACM international symposium on Mobile ad hoc networking & computing*, (pp. 287–290). New York: ACM.

Mathar, R., & Mattfeldt, J. (1996). Optimal transmission ranges for mobile communication in linear multihop packet radio networks. *Wireless Networks*, *2*(4), 329–342. doi:10.1007/BF01262051

Moraru, L., Leone, P., Nikoletseas, S., & Rolim, J. (2008). Path Quality Detection Algorithms for Near Optimal Geographic Routing in Sensor Networks with Obstacles. *Wireless Communications and Mobile Computing Journal,* 1–13.

Nadeem, T., & Agrawala, A. (2004). Ieee 802.11 fragmentation-aware energy efficient ad-hoc routing protocols. In *2004 IEEE International Conference on Mobile Ad-hoc and Sensor Systems*, (pp. 90–103).

Nikoletseas, S., & Powell, O. (2007). Simple and efficient geographic routing around obstacles for wireless sensor networks. In *Proceedings of the 6th International Workshop on Efficient and Experimental Algorithms (WEA)*, (LNCS pp. 161–174). Berlin: Springer-Verlag.

Radhakrishnan, S., Rao, N., Racherla, G., Sekharan, C., & Batsell, S. (1999). Dst - a routing protocol for ad hoc networks using distributed spanning trees. In *IEEE Wireless Communications and Networking Conference,* (pp. 100– 104).

Sack, J.-R., & Urrutia, J. (2000). Handbook of computational geometry. Amsterdam: North- Holland Publishing Co.

Santi, P. (2005). Topology Control in Wireless Ad Hoc and Sensor Networks. New York: Wiley.

Schmid, S., & Wattenhofer, R. (2006). Algorithmic models for sensor networks. In *20th International Parallel and Distributed Processing Symposium, 2006. IPDPS 2006.*

Schmitz, R. T.-M. M. H. H., & Effelsberg, W. (2004). The impact of wireless radio fluctuations on ad hoc network performance. In *Proceedings of the 29th Annual IEEE international Conference on Local Computer Networks*, (pp. 594–601), Washington, DC.

Stojmenovic, I. (2005). Handbook of Sensor Networks: Algorithms and Architectures. Chichester, UK: John Wiley and Sons, Inc.

Stojmenovic, I., & Lin, X. (2001). Loop-free hybrid single-path/flooding routing algorithms with guaranteed delivery for wireless networks. *IEEE Transactions on Parallel and Distributed Systems*, *12*(10), 1023–1032. doi:10.1109/71.963415

Stojmenovic, I., Nayak, A., Kuruvila, J., Ovalle-Martinez, F., & Villanueva- Pena, E. (2005). Physical layer impact on the design and performance of routing and broadcasting protocols in ad hoc and sensor networks. In *Computer Communications,* (Vol. 28).

Stojmenovic, I., Russell, M., & Vukojevic, B. (2000). Depth first search and location based localized routing and qos routing in wireless networks. In *ICPP '00: Proceedings of the 2000 International Conference on Parallel Processing*, (pp. 173), Washington, DC: IEEE Computer Society.

Toussaint, G. (1991). *Some unsolved problems on proximity graphs.*

Urrutia, J. (2002). Routing with guaranteed delivery in geometric and wireless networks, (pp. 393–406).

Ye, F., Chen, A., Liu, S., & Zhang, L. (2001). A scalable solution to minimum cost forwarding in large sensor networks. In *Proceedings of Tenth International Conference on Computer Communications and Networks*, (pp. 304 –309).

Zorzi, M., & Rao, R. R. (2003). Geographic random forwarding (geraf) for ad hoc and sensor networks: multihop performance. *IEEE Transactions on Mobile Computing*, *2*(4), 337–348. doi:10.1109/TMC.2003.1255648

Zou, L. Z. X., & Lu, M. (2004). Pager: A distributed algorithm for the deadend problem of location-based routing in sensor networks. In *International Conference on Computer Communications and Networks*, Chicago, IL.

ADDITIONAL READING

Carlsson, N., & Eager, D. L. (2007, Sep.). Non-Euclidian geographic routing in wireless networks. *Ad Hoc Networks*, *5*(7), 1173–1193. doi:10.1016/j.adhoc.2006.07.001

Chatzigiannakis, I., Mylonas, G., & Nikoletseas, S. (2007, Aug.). A Model for Obstacles to be used in Simulations of Wireless Sensor Networks and its Application in studying Routing Protocol Performance. *Simulation*, *83*(8), 587–608. doi:10.1177/0037549707082330

Hornsberger, J., & Shoja, G. C. 2006. Geographic grid routing: designing for reliability in wireless sensor networks. In *Proceedings of the 2006 international Conference on Wireless Communications and Mobile Computing* (Vancouver, British Columbia, Canada, July 03 - 06, 2006). IWCMC '06. ACM, New York, NY, 281-286

Huang, X., & Ma, J. (2006, Jul.). Optimal distance geographic routing for energy efficient wireless sensor networks. *Int. J. Ad Hoc Ubiquitous Comput.*, *1*(4), 203–209.

Kim, Y., Lee, J., & Helmy, A. 2003. Impact of location inconsistencies on geographic routing in wireless networks. In *Proceedings of the 6th ACM international Workshop on Modeling Analysis and Simulation of Wireless and Mobile Systems* (San Diego, CA, USA, September 19 - 19, 2003). MSWIM '03. ACM, New York, NY, 124-127

Kim, Y., Lee, J., & Helmy, A. (2004, Jan.). Modeling and analyzing the impact of location inconsistencies on geographic routing in wireless networks. *SIGMOBILE Mob. Comput. Commun. Rev.*, *8*(1), 48–60. doi:10.1145/980159.980168

Lee, S., Bhattacharjee, B., & Banerjee, S. 2005. Efficient geographic routing in multihop wireless networks. In *Proceedings of the 6th ACM international Symposium on Mobile Ad Hoc Networking and Computing* (Urbana-Champaign, IL, USA, May 25 - 27, 2005). MobiHoc '05. ACM, New York, NY, 230-241

Leong, B. W. 2006 *New Techniques for Geographic Routing*. Doctoral Thesis. UMI Order Number: AAI0810094., Massachusetts Institute of Technology.

Li, Q., & Rus, D. (2005, Aug.). Navigation protocols in sensor networks. *ACM Trans. Sen. Netw.*, *1*(1), 3–35. doi:10.1145/1077391.1077393

Liu, K., Abu-Ghazaleh, N., & Kang, K. (2007, Feb.). Location verification and trust management for resilient geographic routing. *Journal of Parallel and Distributed Computing*, *67*(2), 215–228. doi:10.1016/j.jpdc.2006.08.001

Nath, S., & Gibbons, P. B. 2007. Communicating via fireflies: geographic routing on duty-cycled sensors. In *Proceedings of the 6th international Conference on information Processing in Sensor Networks* (Cambridge, Massachusetts, USA, April 25 - 27, 2007). IPSN '07. ACM, New York, NY, 440-449

Philip, S. J. 2005 *Scalable Location Management for Geographic Routing in Mobile Ad Hoc Networks*. Doctoral Thesis. UMI Order Number: AAI3185289., State University of New York at Buffalo.

Qu, H. 2008 *Co-Designed Localization and Geographic Routing Algorithm for Sensor Networks*. Doctoral Thesis. UMI Order Number: AAI3317468., Cornell University.

Rao, A., Ratnasamy, S., Papadimitriou, C., Shenker, S., & Stoica, I. 2003. Geographic routing without location information. In *Proceedings of the 9th Annual international Conference on Mobile Computing and Networking* (San Diego, CA, USA, September 14 - 19, 2003). MobiCom '03. ACM, New York, NY, 96-108

Seada, K., Helmy, A., & Govindan, R. (2007, Aug.). Modeling and analyzing the correctness of geographic face routing under realistic conditions. *Ad Hoc Networks*, *5*(6), 855–871. doi:10.1016/j.adhoc.2007.02.008

Wu, S., & Candan, K. S. (2007, Sep.). Power-aware single- and multipath geographic routing in sensor networks. *Ad Hoc Networks*, *5*(7), 974–997. doi:10.1016/j.adhoc.2006.04.007

Xi, F., & Liu, Z. 2009. Small World Topology-Aware Geographic Routing in Wireless Sensor Networks. In *Proceedings of the 2009 WRI international Conference on Communications and Mobile Computing - Volume 02* (January 06 - 08, 2009). CMC. IEEE Computer Society, Washington, DC, 116-120

Younis, M., & Akkaya, K. (2008, Jun.). Strategies and techniques for node placement in wireless sensor networks: A survey. *Ad Hoc Networks*, *6*(4), 621–655. doi:10.1016/j.adhoc.2007.05.003

Zamalloa, M. Z., Seada, K., Krishnamachari, B., & Helmy, A. 2008. Efficient geographic routing over lossy links in wireless sensor networks. ACM Trans. Sen. Netw. 4, 3 (May. 2008), 1-33

Zeng, K., Ren, K., Lou, W., & Moran, P. J. (2009, Jan.). Energy aware efficient geographic routing in lossy wireless sensor networks with environmental energy supply. *Wireless Networks*, *15*(1), 39–51. doi:10.1007/s11276-007-0022-0

ENDNOTES

1. In this chapter, we use the term overhead, or communication overhead, to refer to the number of messages needed for a node to implement an algorithm.
2. In this chapter, we use complexity to refer to the computational resources that a node needs to implement the algorithm. Sometimes, we distinguish between space and time complexity.
3. Basically, the techniques which provide guarantee delivery turn round the obstacle by using the right (or left) hand rule. However, the heuristic does not prevent the messages to cycle in some pathological situations. For this reason, more sophisticated strategies are used which, applied to planar graphs, are proved to guarantee the delivery of the data.

[4] Weakly stuck nodes are also simply called stuck nodes when the distinction between weakly and strongly stuck nodes is not relevant.

[5] We denote by n1, n2, … the first, the second, … nodes marked as non optimal.

Section 2

QoS Provision in Wireless Wide Area Networks

Chapter 12
TPC and Non-TPC Based Topology Control Approaches for QoS Improvement in MR-WMN

Vinod Mirchandani
University of Technology, Australia

Ante Prodan
University of Technology, Australia

Olivier Marcé
Alcatel-Lucent, Centre de Villarceaux, France

ABSTRACT

A grand challenge in Multi-Radio Wireless Mesh Networks (MR-WMN) is to limit the interference such that its net capacity increases without compromising scalability and stability. In this chapter, the authors first provide an insight into the implications of transmit power control (TPC) on the MR-WMN topology and QoS. In this regard, a review of some of the key work is carried out they then explore the approach of non-TPC based topology control schemes for limiting the interference in a static nodes based MR-WMN system that uses a distributed, light-weight, cooperative multiagents. A path reduction (PR) algorithm is the principle behind our topology control scheme and its viability is explained through NetLogo tool based simulation results. The effectiveness of the PR algorithm is shown in terms of improved interference cost reduction and decrease in path length. The focus of this chapter is mainly on non-TPC approach rather than the TPC approach.

INTRODUCTION

A wireless mesh network (WMN) is a networking paradigm, which enables inexpensive peer wireless network nodes to relay frames from one node to another by leveraging the broadcast nature of wireless medium. Essentially, WMN facilitates to cost-effectively and quickly extend a wired network by creating multiple topologically diversified wireless connections between peer nodes. This feature

DOI: 10.4018/978-1-61520-771-8.ch012

coupled with the dynamic reconfiguration of links, increases robustness. Mesh networks are very reliable and self-healing (Hossain & Leung, 2007), i.e., the network can still operate even when a node breaks down or a specific link is not usable any more. A mesh network whose nodes are all connected to each other is known as a *fully connected* WMN.

The drawback of a *single radio* mesh network is that the throughput of the link between each hop progressively decreases due to the co-channel interference (BelAir networks, 2006). As such, the core network considered in this chapter and shown in Figure 1 is a multi-radio wireless mesh network (MR-WMN) in which multi-radio nodes, which are also known as wireless routers, are used to result in a decrease of the interference between the channels of two adjacent routers (Mobileman, 2005). In Figure 1 the multi-radio routers with an *exclusive* wireless connectivity are termed as *client nodes*. The root node in the mesh networking terminology is known as the *mesh portal node.* MR-WMN is used to facilitate broadband wireless connectivity to the heterogeneous access networks such as 2G/3G, PSTN, Cdma etc. In the MR-WMN node, we are considering multiple 802.11 based interfaces corresponding to the 802.11 radios in a node. The 802.11 radios are: 802.11a, 802.11b, 802.11g. Each of the radio interfaces can communicate with an interface of the same radio type by using a channel within the frequency range of the radio type. The mesh nodes i.e. mesh routers which also have an access point functionality are termed as the *mesh access point* (MAP). The MAPs in Figure 1 essentially multi-hop the traffic to and from the access networks and the wired Internet.

A *self-organization* (defined later in section on multi-agent systems) process that methodically allocates the channels to the MR-WMN nodes in the system facilitates to make the throughput of the links, as much as possible, less susceptible to the channel interference.

Transmit power control (TPC) has a bearing on the *transmission range* of the node's radio. As such, it influences the *physical distance of the link* between the radios of two nodes. This leads to a topology control (TC) of the wireless mesh networks. The topology of a MR-WMN has a bearing on the overall system capacity and other Quality of Service (QoS) metrics viz. end-to-end packet transfer latency, jitter and throughput. TC is an extensively studied topic in the area of wireless telecommunication networks, particularly in the mobile ad hoc and sensor networks wherein *energy* is considered as a critical resource. Reference (Mobileman, 2005) informally defines TC as:

The art of coordinating nodes decision regarding their transmitting ranges, in order to generate a network with the desired properties (e.g. connectivity) while reducing node energy consumption and/or increasing network capacity.

There are two key reasons that motivate to have a proper TC in a MR-WMN. These are enumerated below:

- The power used to transmit messages has a bearing on the amount of interference caused between two simultaneous transmissions on the same channel within a common interference range. Limiting the amount of interference between channels in the MR-WMN system is a major factor that contributes towards the gain in overall *network capacity.* Ref (Santi, 2005) has shown that a decrease in interference hence a gain in network capacity is possible by communicating over short multi-hops between the sender and receiver nodes. The purpose of topology control in this regard is to prune the MR-WMN so as to have only "capacity-efficient" links.
- Generally, in an ad hoc and sensor networks energy is a critical resource. Santi (2005) has shown that transmitting

Figure 1. Multi-Radio wireless mesh network (MR-WMN) infrastructure

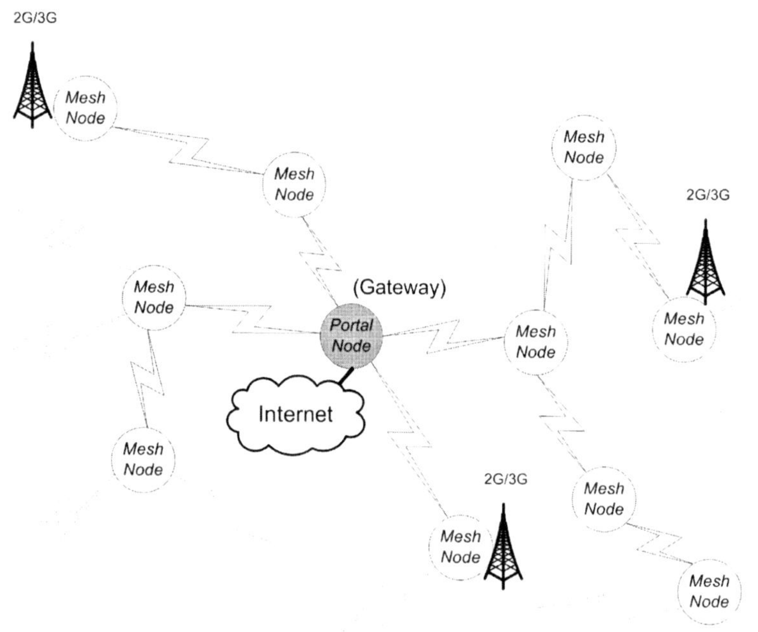

messages across *short multi-hops* between the source and destination nodes is more energy efficient than a direct transmission between the source and destination nodes. The objective of TC here is to prune "energy inefficient" links in a MR-WMN and thereby help to conserve energy resource of the nodes.

The TC approaches of TPC and non-TPC assume significance in the area of energy conservation, interference reduction (capacity increase) and hence QoS improvement for MR-WMN. However, our work pertains to the non-TPC approach, as such the focus of this chapter is more on non-TPC approach rather than the TPC approach.

This chapter is organized as follows: In the section on background, first a review of key work on TPC done in the literature is provided along with their shortcomings. The section also discusses the implications of TPC on topology control (TC) and hence QoS by explaining the advantages and disadvantages of TPC. The dependence of QoS on TC in the MR-WMN is highlighted in a separate sub-section of the background section. A sub-section on the influence of network management on TC is also included, which is followed by a sub-section on Multi-agent systems (MAS). We have discussed the concept of MAS because our MR-WMN infrastructure network is based on it. A sub-section on network initialization gives an overview of the schemes that we have used for a non-TPC based topology control. We have devoted an entire section on our proposed non-TPC based approach that includes: description of the algorithms used, its pseudo code, implementation in NetLogo simulation tool and presentation and discussion of performance evaluation results to show the effectiveness of our proposed non-TPC approach.

BACKGROUND

Transmit power control (TPC) is a technique at the *physical layer* that regulates the signal transmission power with the objective that the signal of the desired quality reaches the destination while causing minimal interference to the other nodes. Minimizing interference level leads to a better QoS performance of the wireless networks. The measures for managing interference can be broadly classified as *passive and active*. The most common *active measures* directed towards reducing interference include: TPC, efficient channel assignment, i.e. allocation of the frequencies in the available spectrum, and re-transmission of a message signal when significant bit errors occur in the signal that can't be corrected at the receiver.

On the other hand, the *passive measures* such as signal encoding schemes used for error correction do not have a bearing on the interference directly; instead they decrease sensitivity to interference by increasing the transmission signal's robustness.

Literature Review

In the Introduction section the influence of TPC on topology control was brought into perspective. To accomplish TC by means of TPC the most common approach is to reduce the transmission power. In the simplest theoretical approach for TPC in ad hoc and sensor networks all the nodes are assumed to have the same transmitting range r and the problem is to find the minimum value of r - the critical transmitting range (CTR) that assures selected network property/properties. The property that is most often selected is the connectivity and as such, the problem of CTR for connectivity is to find the minimum value of r such that resulting network is fully connected (Santi, 2005).

COMPOW protocol developed in (Narayanaswamy et al., 2002) represents an example of a CTR based protocol. It is implemented through a centralized algorithm in which main deficiencies are messaging overheads and a slow reaction on topology changes. Furthermore, the constraint of using common transmit power may force most of the network nodes to use unnecessary high transmit power. The analysis presented in (Santi, 2005) indicates the characterization of the CTR, based on the *idealized point graph* where the radio coverage area is the same. The assumption that all the nodes have the same r is not acceptable for many types of networks deployed in heterogeneous radio in particular for MR-WMNs that we are concerned with. To address this issue, an approach of finding a separate transmitting range for each node has been developed. This method is known as the range assignment (RA) problem that was first time studied by Kirousis et al. (2000). It is important to note, in (Kirousis et. al., 2000) the path loss model is identical for all the network nodes, and the shadowing and fading effects are not considered (Santi, 2005).

Informally, RA problem can be stated as the problem of finding a minimal nodes' range assignment that generates a fully connected network. Where minimal RA indicates the lowest energy cost. The optimal solution for the RA problem for one-dimensional networks can be found in polynomial time (Kirousis et al., 2000). The same reference provides the proof of NP-hardness for the RA problem in three-dimensional networks. Furthermore, Clementi et al. (Clementi et al., 1999) present the proof of NP-hardness of the same problem for two-dimensional networks. CLUSTERPOW proposed by Kawadia and Kumar (Kawadia and Kumar, 2003) is the protocol developed along the lines of the RA problem.

Neighbor based topology control protocols rely on the nodes' ability to classify its neighbors according to predefined or dynamically defined criteria such as link quality or distance. These protocols are always localized, distributed and asynchronous; thus have practical relevance in ad hoc and mesh networks. In addition, localized communication in such protocols enables their deployment on very large networks. KNeigh protocol introduced in (Blough et al., 2003) is a protocol based on distance to neighbors estimation. This protocol is simple, does not require high quality information and it is lightweight. However, in

Figure 2. Taxonomy of topology control techniques (Adapted from (Santi, 2005))

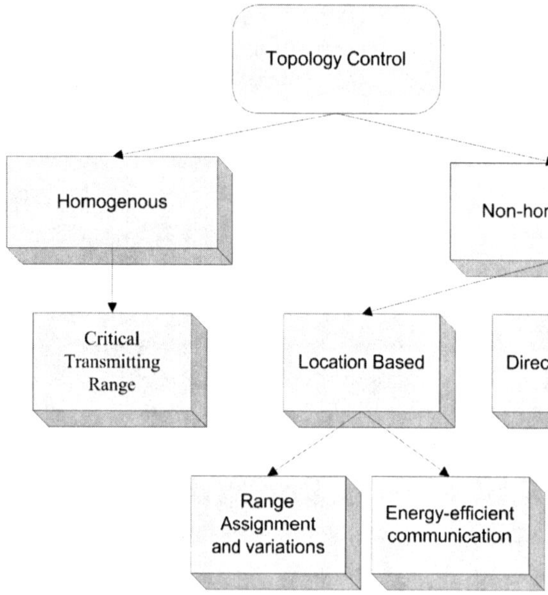

worst case KNeigh does not preserve network connectivity (Santi, 2005).

XTC algorithm developed in (Wattenhofer & Zollinger, 2004) is similar to Kneigh protocol i.e. it is simple and local. It does not assume the network graph to be a unit disk graph and proves correct on general weighted network graphs. Furthermore, it does not require availability of node position information. Instead, XTC operates with a general notion of order over the neighbors' link qualities. We consider this as an important feature; therefore have implemented it in our own approach.

The taxonomy of TC solutions discussed above is summarized by means of Figure 2 (Santi, 2005).

A brief description is given below for each of the boxes represented in Figure 2 .(Santi, 2005).

- Homogenous topology Control - All the network nodes must use the same transmitting range r and then the TC problem is to determine the Critical transmitting range (CTR). CTR has already been defined above in this section.
- Non-Homogeneous Control – The nodes can choose different transmitting ranges up to the maximum range of transmission.
- Location Based – In location based TC approaches the location of the nodes is determined, which is then used either in a centralized or a distributed manner to evaluate the optimal topology.
- Direction based approach – In this approach the nodes do not know their location but can estimate the relative direction of their neighbors.
- Neighbor based approach – The nodes at least have minimum information about their neighbors such as: ID, link quality, and distance. This approach is generally useful for mobile ad hoc networks.

We state below the benefits and deficiencies of TPC with regards to TC. Specifically, we look into two issues that arise: increase in transmission path length and decrease in network throughput. Some of the positive attributes of a reduction in transmission power are:

- The reduction of transmission power confines interference and consequently reduces collisions and retransmissions that occur for a media access layer (Burkhart et.al, 2004). This is also true for MR-WMN, as MR-WMN are similar to the ad hoc networks considered by (Burkhart et.al, 2004).
- As the energy required to transmit a message increases quadratically with distance i.e. link distance to the power of 4. This is valid for a realistic *urban environment* as it depends on the propagation model used. So from an energy conservation perspective it seems preferable to replace longer links with multiple shorter links (Burkhart et.al., 2004).

Some of the drawbacks of TPC are:

- As noted in (Kawadia and Kumar, 2003) a reduction in transmission power leads to longer paths with a larger number of hops, which may linearly increase end-to-end latency due to the packetization delay that occurs at each hop.
- TPC affects routing as the transmitter communication range depends on a transmit power used. Thus it is imperative to combine it with an appropriate reactive routing protocol (Kawadia and Kumar, 2003).
- An often neglected fact in the reviewed TPC literature is that the reduction of transmission power on a pair of nodes may still not result in an elimination of contention interference. This is due to a significant difference between *communication range* and *carrier sensing range.*
- Another shortcoming of a TPC based solution is an overall increase in complexity of network management due to an additional parameter (transmission power) that has to be assigned to each radio transmitter.

In the literature that we have reviewed, a relationship between the power mechanism contributed by TPC and a QoS metric i.e. network throughput is often neglected or just partially analyzed. A notable example of a more comprehensive approach that studies the effect of TPC on a network throughput is provided in (Park & Sivakumar, 2002) In this reference, the authors analyze the implicit assumption that the optimal throughput performance in ad-hoc networks can be achieved by using the minimum transmit power required to keep the network connected.

Authors in (Park & Sivakumar, 2002) conclude that such an assumption remains valid only under certain circumstances, specifically in networks with very high node densities. This is the case in a small number of real-world wireless network deployments. Furthermore, in this case the reduction of transmission power leads to a drop in signal to noise plus interference ratio (SNIR), which as a consequence can often lead to the selection of more robust and slower modulation scheme. *Essentially, the use of a TPC control will frequently lead to a reduction in network throughput.*

The advantages and disadvantages of transmission power reduction are summarized by means of Figure 3.

A co-relationship between the TPC and TC is provided in (Marina & Das, 2005). The authors extend the definition of TC on all configurable parameters of a link. They consider transmission power, bit rate, frequency band/channel and beam direction (if beam forming or directional antennas are used) as being all parts of a TC problem. Although, this formulation at first seems to be too broad but after a careful analysis, we were able to verify its correctness.

An appropriate example can be provided by analyzing the influence of a link data rate, which is considered as an unrelated parameter to a network topology. It is easy to imagine a simple scenario where, because of the increased interference, the data rate on specific link is reduced. This leads

Figure 3. Effects of transmit power reduction, + indicates advantages and – the disadvantages (Adapted from (Santi, 2005))

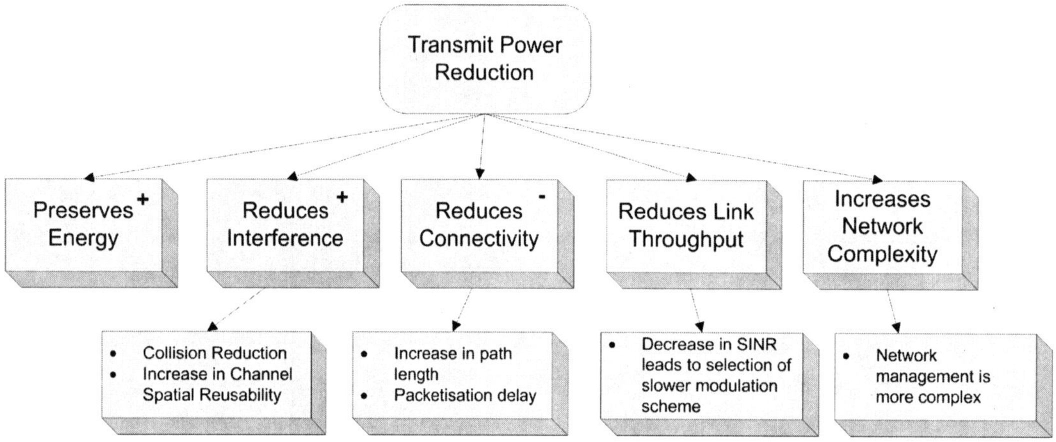

to congestion and packet loss. To address this issue an additional link or links have to be created to satisfy throughput requirements between two segments of the network. In other words, the network topology has to be changed. Although, the relationship is not explicit however, it is evident from this example that parameters other than the transmission power may influence topology and thus have to be considered as parameters relevant for the TC.

QoS in a MR-WMN – Motivation and Dependence

The importance of having Quality of Service (QoS) in MR-WMN is made clear from the following example (Santi, 2005) Consider a MR-WMN scenario in which the fast notification of an abnormal detected event to the external user is vital such as in the case of intrusion detection, or to monitor nuclear plants. The QoS requirement of the network in this example is a guarantee on the delivery time of the alarm message to the gateway node. What is the optimal network topology and/or routing strategy in this context? This depends on features such as expected node density, expected traffic pattern etc.

A great deal of attention has been devoted to the identification of efficient network topologies, with a particular emphasis on energy efficiency. However, a common approach in the literature (Santi, 2005) is to consider the TC problem as a stand-alone problem, which is analyzed and solved from a graph-theoretic perspective. This type of approach has resulted in the message conveyed by current TC literature: *the sparser the network topology, the better* (provided certain spanning properties of the graph are satisfied).

QoS requirements in a MR-WMN is dependent upon *routing and resource management control*. Specifically, the QoS measures, such as delay bounds, throughput, jitter depends on the:

- Quality of the chosen route.
- The use of a MAC that can support QoS for different types of traffic.
- Robustness of the MAC to support successful transmission of packets under high mobility and/or heavy load cases.

The aim of a routing protocol in a MR-WMN is to transfer the packets between the source and the destinations while ensuring low latency and a high network capacity. This aim of the routing protocol can be met if it can (Aggelou, 2009):

- Estimate the network flows.
- Determine the network topology.
- Determine the information about the residual capacity of links.

The routing protocol can be made capable of knowing the above listed factors if some degree of QoS support is incorporated in the routing protocol. In addition, the routing protocol should be coupled with mechanisms that can predict the future state of the network such as links going down. This would enable the network to seamlessly provide the servers and/or clients with the agreed on services. The efficiency as well as the performance of the application can be further improved if the *timing* of the links going down can be predicted as well.

In this regard, (Aggelou, 2009) has proposed a methodology to determine the network disconnection and to estimate the future connectivity states of a time-varying topology. The methodology is based on self-learning techniques in pattern recognition practices and the theory of time-series. To explain this further the mechanism first identifies the location of the mobile nodes and then uses this information to quantitatively model the topology changes. Following this, a time series-based classifier derives statistical data on future network connectivity states (Aggelou, 2009).

Network Management and Topology Control

The telecommunication management networks (TMN) standards have been successfully used in the management of circuit-switched and IP based wireline networks. While the TMN models provide valuable abstractions for managing any type of network, some aspects important for adhoc as well as wireless mesh networks pertaining to *network management* have not been addressed by these models or by other network management systems in use. The critical deficiency of these models is the *lack of automation* in network management that dynamically address specific issues in MR-WMN. This is important because the characteristics of MR-WMN, such as fluctuations in throughput requirements and the interference on a radio channel, demand frequent network reconfiguration and rigorous monitoring.

Considering the network management paradigm, we can distinguish two functions of the proposed system:

1. Flexible autonomous monitoring that affects all aspects of network management
2. Adaptation that affects operation and maintenance.

The autonomous monitoring represents a base of the feedback loop that is created to enable reactive and proactive network management activities. The activities of the monitoring system in a MR-WMN can also be classified as *passive* and *active*; where passive activities are not consuming the network capacity, i.e., do not involve transmission, and active activities require transmission, thus consume network capacity.

Due to limited resources in MR-WMNs its monitoring system should be highly flexible. In other words, it should be able to reduce the frequency of active monitoring in the case of the adverse radio channel and other similar conditions. The capacity preserved by such flexibility should result in an extended operating region for MR-WMN. Similarly, since wireless channel conditions cannot be predicted adequately, management functionality must be able to swiftly react to information provided by monitoring. For example, if the amount of available network capacity drops suddenly, e.g., due to node failure, increased demand from users or weather deterioration, the management system needs to adapt the network topology to new conditions by creating new *virtual links*, changing channels, and accordingly optimizing network paths. By virtual link, we mean that a certain fraction of bandwidth is reserved on each channel for network management purposes rather than reserving one exclusive channel.

This adaptation process, from the network management perspective, represents the operations management because it is keeping the network and the services that the network provides available. In comparison to different routing schemes the distinctive feature of this approach is the channel assignment and the TC that can be achieved through it. Other implemented techniques like link quality assessment or interference awareness are present in some wireless routing schemes, although in somewhat limited extent.

Conversely, provisioning, another class of network management concerned with configuring resources in the network to support a given service, is of a secondary interest in this chapter. Herein, we limit its scope to address the provisioning only in the form of a load balancing. This comprises of assigning specific radio channels to node's interfaces, thus adapting topology so required amount of data throughput is available on certain links. The actions that facilitate this are focused on the reduction of interference and/or path lengths, and the circumvention of failed links.

The objective of network monitoring combined with corrective and preventive measures, such as adjusting configuration parameters of wireless interfaces, is maintenance. The network maintenance is also concerned with performing repairs and upgrades. In our context this means discovering situations when equipment must be replaced, and when a network is expanded with additional routers.

Multiagent Systems (MAS)

The heterogeneous field of multiagent systems (MAS) has its roots in distributed problem solving in Artificial Intelligence (AI). In the mid-eighties it emerged as a distinctive discipline (Kirn et al., 2006) and since then has been under prolific development. During this period a number of theoretical frameworks based on different perspectives such as sociological, AI, and economic, have been developed.

A. What is an Intelligent Agent?

In this chapter, we will adhere to the following definition provided in (Weiss, 1999, Wooldridge, 2002).

An agent is a computer system that is situated in some environment, and that is capable of autonomous action in this environment in order to meet its design objectives.

Agents embody a strong sense of autonomy; i.e., agents are able to act without the intervention of humans or other systems; they have control over their own internal state as well as over their behavior. The intelligent behavior of an agent emerges from interactions of various simple behaviors and results with agent's overall flexibility (Kirn et al., 2006). Now that we know what an *intelligent agent* is, we give below an overview of a Multiagent system (MAS) as our MR-WMN architecture of Figure 1 is based on a *lightweight MAS*.

B. What is a Multiagent System?

Generally, we can define a MAS as a system that consists of multiple interacting intelligent agents. MAS exhibits high flexibility since agents can join and leave the system at any stage. However the flexibility is not free, its cost might be in additional resources or increase in the overall complexity and reduced maintainability of the system as well as the induction of periods of instability.

In this chapter, we are particularly interested in *lightweight multiagent systems*, also known as *collectives*. Some authors (e.g. Larman and Galstyan in (Tumer and Wolpert, 2004)) differentiate collectives and MAS by its constituents' ability to deliberate; where agents in a MAS are able to deliberate and agents in a collective are not. It is generally considered that a collective consist of less sophisticated agents or agents exhibiting lesser degree of complexity.

Two central terms to the paradigm of collectives are *emergence* and *self-organization*. In particular, we are interested in the term *self-organization* because our MR-WMN architecture makes use of a self-organization algorithm proposed in (Mirchandani et. al., Oct 2007) for Intereference cost reduction (ICR). This algorithm is briefly explained as part of the section on Network Initialization.

Both the terms i.e. *emergence* and *self organization* are used often, sometimes as synonyms, but rarely clearly defined. Thus, we have adopted definitions from (De Wolf and Holvoet, 2005):

The essence of **emergence** is the existence of a global behavior that is novel with respect to the constituent parts of the system. In other words, the global behavior cannot simply be traced back to individual parts (Crutchfield, 1993; Crutchfield, 1994; Holland, 1998).

Self-organization *is a dynamical and adaptive process where systems acquire and maintain structure themselves, without external control.* Where 'structure' can be a spatial, temporal or functional structure, and 'no external control' refers to the absence of direction, manipulation, interference, pressures or involvement from outside the system (Haken, 2000, Shalizi, 2001).

The desired features for the MAS in order of priority are:

- Stable- Stability is a particularly important feature of any distributed system. Specifically, interactions between different feedback loops in a MAS are known to cause stability issues.
- Scalable- Its functioning should be as independent of the number of nodes involved.
- Lightweight- It is not demanding on computational and memory resources, so it can function on low cost devices.
- Simple- It should be simple to incorporate and test new algorithms for the architecture.

Our MR-WMN architecture incorporates the above listed attributes of a MAS.

PROPOSED NON-TPC APPROACH

In the following sections, we systematically discuss our Non-TPC based TC approach. We start off with Literature review and an explanation of Network initialization, which also forms an important precursor to the proposed PR algorithms for TC in our non-TPC based approach.

Literature Review

We focus in this section on the non-TPC based approach for TC, which is also based on the channel assignment process in a MR-WMN. Specifically, we examine the literature in terms of performance, complexity, scalability and stability of WMN.

The work of (Ko and Rubenstein, 2003) specifically targets the channel assignment problem on WMN. Authors have adopted their theoretical work in (Ko and Rubenstein, 2003) and created a self-stabilizing distributed protocol and an algorithm for channel assignment.

The method of (Ko and Rubenstein, 2003) assumes that the interference is symmetric and is based on an interference range of three hops. Their method results in improvements of only 20% compared to random channel assignment. In reality, most of the times interference will be asymmetric because neighboring node interface may transmit on the same channel at different powers. In contrast, a better proposal would not assume symmetric interference and would not require a dedicated channel for frequency co-ordination, which is a significant advantage. Further the interference cost function in (Ko and Rubenstein, 2003) has not been justified i.e. the cost function has not been based on an interference model. The other main limitation of their proposal, as well as the one by (Raniwala and Chiueh, 2004) is the usage of a common channel on each node for the management of channel assignment. We believe that his approach should be avoided because it can be wasteful of bandwidth and imposes severe limitations on network capacity especially when nodes have only two interfaces. Furthermore, a strong source of interference on the frequency that is used for the coordination of channels can render the throughput of parts or the whole network unsatisfactory.

In (Jain et.al., 2003), contrary to previous findings, authors state that the addition of new nodes can actually improve a per-node throughput because the richer connectivity provides increased opportunities for routing around interference "hotspots" in the network that offsets the increase in traffic load caused by the new nodes. They explain this by the fact that previous research has been done under the assumption that nodes always have data to send and are ready to transmit as fast as their wireless connection will allow. However, in realistic settings, sources tend to be bursty, so nodes on an average transmit at a slower rate than the speed of their wireless link.

In (Jain et.al., 2003) authors use liner programming to model maximum achievable flow between the source and destination in absence of wireless interference and heuristics to obtain lower and upper bounds on the throughput. Also, they avoid making assumptions about the homogeneity of nodes with regards to radio range or other characteristics as well as regularity in communication patterns. The conclusion from the work in (Jain et.al., 2003), was that neither multi-path routing nor doubling the range of the radio increases cumulative throughput. On the other hand, by using two channels instead of one, the network may achieve the maximum possible throughput. Scenarios provided in (Jain et.al., 2003) illustrate that the model they have developed could be a useful tool for analysis and capacity planning in wireless multi-hop networks. However, we believe that their model suffers from oversimplification. Although valid theoretical capacity bounds can be produced most of real world deployments are more complex – that involve neighboring or co-located networks with unknown interference characteristic. Consequently in realistic circumstances we can not simply obtain information that has to be fed into the proposed model and the capacity predicted by the model and in reality may vary significantly.

In (Leung and Kim, 2003) the authors have deployed heuristics that is based on interference measurements. Still, they do not define a threshold value range and a mechanism to keep a channel change under control. This may result in an infinite loop of channel changes that is caused by a slight variation in noise or by cyclical interference. Furthermore, it is not clear how much time (steps) the algorithm needs to achieve acceptable results as well if it can cope with dynamic environment.

Raniwala and Chiueh, (2004) considers a combined solution for channel assignment and routing issues in (Raniwala and Chiueh, 2005) and extend their previous proposal with the usage of a virtual control network instead of dedicated interface-channel on each router. In other words this means that certain fraction of bandwidth is used on each channel for channel assignment and other management purposes rather than dedicating one exclusive channel. Their work contains careful analysis of all aspects of resource allocation problems relevant for 802.11 based WMN.

In (Subramanian, 2005) authors propose the usage of partially overlapping channels. Their interference model is theoretically based on a conflict graph and the interference data is acquired through the measurement of link pair interference. Subramanian, (2005) uses integer linear programming to obtain bounds of optimal solution and evaluate the proposed algorithm.

Kyasanur and Vaidya, (2005) approach is based on the assumption that an interface can dynamically switch over from one channel to another.

They present a distributed interface hybrid assignment strategy and their routing strategy selects routes that have low switching and diversity cost. However, a coordination protocol is required to assign the channels to fixed channels in the hybrid nodes i.e. nodes having an interface assigned with fixed channel and the other interfaces with switchable channels.

Raniwala and Chiueh, (2004) also propose a load aware based channel assignment. Although the work presented in their paper is of a good value but it assumes a centralized method for channel assignment, which also needs to keep track of the load in different parts of the WMN. Their work is thus not scalable. However, they have shown a circular dependency between channel assignment, load on each link and routing.

In (Ramachandran et. al., 2006) the authors base interference estimate on the number of *interfering radios* on each channel supported by each router. An interfering radio is defined as a simultaneously operating radio that is *visible* to a router but external to the mesh. A visible radio is one whose packet(s) pass frame check sequence (FCS) checks and are therefore correctly received. However, this method is incomplete since it neglects interference caused by transmissions that are too weak for the signal to be decoded but still result with degradation of SNR on particular link. The other main drawback of last two proposals is the scalability since centralized algorithms are used. However, both proposals motivate further investigation since they indicate a 40% performance gains in comparison to static assignment.

Network Initialization

The objective of the network initialization is to connect all wireless nodes to portal nodes. The priority of this goal is always 100% connectivity. This is because we expect a node to be connected to a *portal node* to be able to attempt achieving any other goal, e.g., a load optimization or an interference reduction. In the mesh networking terminology, *portal node is the root node*, which is illustrated in Figure 1.

In addition to basic methods and functions, we use an interference cost reduction (ICR) algorithm (Self-organization algorithm) outlined later that ensures that no two interfaces belonging to a single node are assigned with the same channel. This procedure is carried out to avoid an increase in interference and consequent reduction in network capacity. Thus, whenever an interface changes channel this algorithm is invoked.

Initialisation in A MR-WMN Infrastructure Network

Initialization process is an important element of a *non-TPC based* power saving approach. It results in a *topology control* (TC) of the MR-WMN by way of spatial distribution of connectivity between the mesh nodes. The term *initialization process* herein refers to the way by which the multi-radio router nodes are selected to establish wireless connectivity at start up of the MR-WMN system. In (Prodan et. al., 2008), we had introduced the concept of a non-TPC based TC achieved by means of *sequential and random initialization* based schemes.

The sequential initialization process involves the construction of spanning trees in the MR-WMN from each of the mesh portal node. The nodes of spanning tree, we termed them as *seed nodes*, which create a cluster of connected nodes around itself. These seed nodes are selected so as to cover wide spread regions in the MR-WMN and hence facilitate maximum connectivity. The network operator could also use the spanning trees in the MR-WMN for network management purposes.

The drawbacks of sequential algorithm are that:

- It results in a higher number of links between adjacent nodes. Consequently, low spatial diversity of the links between the neighboring nodes causes a higher level of channel interference amongst the node clusters.

- It does not ensure a higher number of links between the mesh portal nodes and the neighboring nodes. As mesh portal nodes carry the overall aggregate traffic of the WMN to the wired Internet therefore a high degree of connectivity to the mesh portal nodes is vital.

After realizing the drawbacks of the sequential algorithm, we proposed an improved initialization process that we termed as *random initialization* process. Each of the nodes in the WMN performs the random initialization process asynchronously and autonomously depending on its internal state such as busy, idle, etc. Note: Even if all the nodes are in the same internal state it will still be limited by the inherent half duplex nature of the 802.11 protocol. The random initialization algorithm operates along the following steps:

The node, which wants to establish link connectivity with the neighboring nodes, we term it as the link creator (LC) node.

- The LC node randomly selects one of the neighboring node's interfaces.
- The LC node then creates a connectivity with the selected interface of the neighboring node.
- A blocking process described (Prodan et. al., 2008) is used by the selected node to prevent it from participating in forming a connectivity with some other LC node(s); it is important to bear in mind that the connectivity formation in random initialization occurs *autonomously* and *asynchronously*.

A comparative study of the impact of random and sequential initialization algorithms on the performance of our self-organization process is given in (Mirchandani et. al., Oct 2007).

ICR Algorithm (Self-Organization Algorithm)

We give here an overview of our ICR algorithm (Mirchandani et. al., Oct 2007) that is used in the Results section.

The ICR algorithm has 3 steps (1) Initialisation (3) Proactive logic and (3) Reactive logic. During the initialisation phase, connectivity is formed between two node interfaces by selectively considering the frequency at which the highest value of SNIR is obtained.

The proactive logic starts operating in response to trigger criteria (Mirchandani et. al., Oct 2007) that indicates to the node that the measured link performance for one of its node interface is lower than the expected performance. During this phase the node interface measures the sum of non-symmetrical interference cost function for a frequency say f_1. If this is below a threshold range then f_1 is assigned to the node interface for which the proactive logic was initiated otherwise the process is repeated until the sum of interference cost function is below a threshold range.

The reactive logic is concerned dealing with unexpected changes in the agent's environment. The aim of our reactive module is simply to restore communication to a workable level that may be substantially sub-optimal.

Impact on Network Management

In the sub-section on Network management earlier in the chapter, we had outlined some of the techniques by which network management could be carried out in a MR-WMN. In this sub-section, we briefly state the ways in which the performance of network management process in a MR-WMN can be affected due to the initialization and/or the ICR algorithm. The impact will be in terms of:

- The topology created, which will influence the speed of dissemination of network management information

- Reliability of reaching all the concerned nodes. This will also depend on if a dedicated channel is used between the nodes for network management purposes or a virtual link is created. Virtual link was defined earlier in the chapter in the section on Network management.
- Efficiency of the ICR algorithm in alleviating interference and the reliability of the connection formed during initialization process will have an impact on the frequency of repeating network management information. A high repetition frequency of transferring network management information will lead to the consumption of additional capacity, which will essentially affect the performance of non-management services.
- Flexibility in accommodating new nodes.
- Availability of the system in case of outages of a link or node.
- Frequency of network monitoring.

The above listed factors will also be influenced by the implementation technique of network management in the MR-WMN system, which could be either *centralized* or decentralized (distributed).

The benefits of centralization depend on the:

- Character of the network,
- The types of services being carried,
- The penalty for outages.
- The requirement for, and availability of trained people to administer the system at remote locations
- Nature of the organization that owns the network.

When diagnosing problem requires information to be gathered and analyzed on a central point, the centralization can be justified as usually the most effective method of managing a network. Furthermore, communication overhead associated with the centralized management, and a delayed reaction caused by it, is not significant in simple wireline networks since they have high data throughput, low loss, and do not require frequent changes of network parameters. In contrast, wireless media has limited bandwidth and is much less reliable, so we think that the management of a MR-WMN should be distributed.

Proposed Non-TPC Approach - Issues, Solutions & Results

MR-WMN that is considered in our work (refer Figure 1) is an *infrastructure network i.e. the nodes* are **static** and have access to a permanent power source. The current work in *sensor networks* focuses on power consumption because sensors are standalone devices that run on batteries. Whereas, in our work, *we are not using sensor networks*, thus energy conservation is not of concern but the effect of interference on the network throughput is the most important performance factor. Other important factors that we focus on in our work are quality and length of the paths that packets traverse by moving between portal and client nodes. The length of the path is measured in hops and the probability of a packet reaching destination is dependant on quality of each link on the path. Furthermore, it has to be considered that TPC affects packet end-to-end latency. In view of all this the TC in MR-WMN can be defined as *a trade-off between network connectivity (in particular multi-hop path length and link quality) and interference reduction.*

As part of our further study, we have identified the need to improve on the concept of TC in addition to that already discussed in (Mirchandani et. al., Aug 2007). This could be possible by means of suitable algorithms to increase the number of shortest paths between the client nodes and the portal nodes.

Furthermore in (Prodan et. al., 2008), we have investigated through our research that the distribution of shortest paths from the client to the portal nodes is very uneven, which motivates the need for

Figure 4. Frequency distribution of the path length (in hops) without and with link substitution algorithm plus 10 additional links for 100 node network density

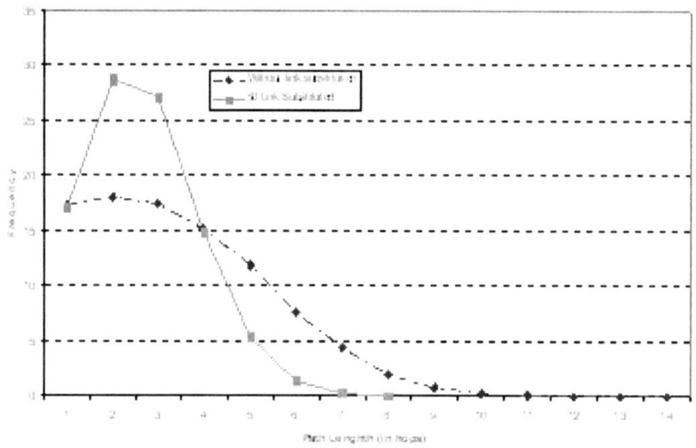

the above stated algorithms. Such a TC coupled with an algorithm that evenly distributes the client nodes amongst the available portal nodes would result in a noticeable increase of the overall WMN capacity. We base this anticipated outcome on the simplistic preliminary premise that the client nodes generate the traffic load evenly.

Link Substitution and Addition Algorithms

Our preliminary algorithm for path reduction (PR) consists of two separate mechanisms: (i) *link substitution* and (ii) *link addition*. It is important to note that both of these mechanisms can be triggered only when the path length to portal nodes is known. Furthermore, these mechanisms rely on node blocking, self-blocking and interference cost measurement techniques, which are discussed in our earlier work (Mirchandani et. al., Aug 2007). The first mechanism enables a node to initiate a link substitution whenever it discovers a free radio interface in the transmission range *that can provide a shorter path link.* This is a preferred mechanism for PR as it can be deployed in a way that does not result in an increase of the overall interference.

The second mechanism used for preliminary PR algorithm is link addition Similar to link substitution mechanism, a node looks for available radio interfaces only this time additional links are created so that the resulting *additional paths are shorter*. However, this approach almost always results in an increase of the overall interference and thus care has to be taken that the benefit of PR is not compromised by an increase of interference. For this purpose, we use it only in conjunction with interference cost measurements.

In Figure 4, the comparison of path lengths with and without link substitution is shown. From Figure 4, we can observe that our method significantly reduces path length by eliminating longer paths - maximum path length is 8 with the link substitution and 14 without it. This method also increases the number of shortest paths such as 2 and 3 hops long. It is emphasized here that these results are independent of a mesh topology of 100 node density in which 50 links are substituted and 10 links are additional. Our experiments also found that the link substitution process has no effect on interference cost (IC) or ICR algorithm discussed in (Prodan et. al., 2008) as well as briefly above in this section.

Enhanced Topology Control

In this section, we explain our non-TPC based *enhanced algorithm* for TC in wireless mesh networks through its implementation in a multi-agent simulation tool called NetLogo. Results are also provided to show that the *enhanced algorithm* provides significant path and interference cost reductions in a MR-WMN system. As the enhanced algorithm for TC is implemented in NetLogo, we first provide a concise overview of NetLogo.

Netlogo

NetLogo (Netlogo website, 2007) is a cross-platform, multi-agent programmable modelling environment used for simulating natural and social phenomena. The primary purpose of NetLogo tool has been to provide a higher-level platform that allows modellers to build and learn from simple agent-based models (ABM).

Motivation for using NetLogo

To evaluate the TC approaches discussed above, we had created and used a Java framework. This was also used to carry out a feasibility study of the self-organization algorithm that we proposed in (Prodan et. al., 2008). However, it soon became apparent that building and modifying algorithms within the existing framework was getting tedious due to the complexity of the tasks related to the integration of additional distributed algorithms (e.g. topology control algorithms) into an essentially sequential framework.

The approach of a multi-agent, multi-threaded simulation environment offered by NetLogo is very significant as it corresponds more realistically to the natural representation of a self-organization system. Though it was complex to create an implementation of our self-organization scheme for MR-WMN (Prodan et. al., 2008), its main benefit has been that it now enables us to improve, extend and test the algorithms interactively with considerably less programming. We have built a model for wireless channel assignment scheme for multi-radio mesh networks (MR-WMN) as proposed in (Prodan et. al., 2008) by using NetLogo tool, which is essentially consistent with the model built in Java framework earlier.

Topology Control Algorithms implementation in NetLogo

The NetLogo implementation of our *enhanced topology control algorithm* for MR-WMN system that results in a non-TPC based topology control leverages the functionalities offered by NetLogo tool. Its logic however is independent of any simulations tool.

Our *enhanced topology control* process is composed of two independent algorithms- (i) Portal first and (ii) new PR algorithm which is based on link substitution. We have only given a wider scope of our algorithm rather than describing the details of its operation.

Portal First

As a matter of fundamental principle in a MR-WMN system the portal nodes are not linked with other portal nodes however each portal node is linked to its neighbouring client node until all its radio interfaces are exhausted. In our algorithm implementation in NetLogo, we achieve this by making use of *blockFree* and *blockBusy* node attributes. In this algorithm the initialisation process ensures that the portal nodes *first* form connectivity with their neighbouring client nodes subsequent to which the connectivity is formed in the rest of the MR-WMN. As such, this algorithm is termed *portal first* (PF).

New Path Reduction Algorithm

Earlier in this section, we have explained our PR algorithm, which essentially consists of two separate mechanisms: (i) *link substitution* and (ii) *link addition*. Our *new PR algorithm* differs from our previously explained *PR algorithm* in the following ways:

- In our *new PR algorithm* only link substitution method is used which makes it different from our previous PR algorithm, discussed earlier in this section in which *link substitution is used along with link addition*.
- Another distinction is that previously, we selected only those substituted links that would not increase the interference cost (IC), whereas in the current algorithm, we include substituted links irrespective of their effect on the IC.

The general steps involved in the operation of the *new PR algorithm* are:

- Selection by the creator (initiator) node of one of its available radio interfaces on the basis of strongest transmission power
- The selected interface creates a list of available interfaces in its communication range;
- From amongst these interfaces the interfaces that have a path length to the portal node longer than the current shortest path are filtered out.
- The remaining ones are short listed and the available interface from amongst these that offers the best SNIR is selected, the new link between the two interfaces is created and the previous shortest path link is switched off and their interface's operational attributes are reset to their default values.
- This process occurs simultaneously across the MR-WMN system.

We reiterate here that the process described above just gives a gist of operation and does not explain the various decisions that may need to be taken based on specific situations in the detailed algorithm.

Link Substitution Algorithm- Pseudo Code

In the presented pseudo-code (refer Figure 5) for link substitution, which is part of our *new PR algorithm* in *enhanced TC,* use is made of two node attributes - *blockBusy* and *blockFree*. The *blockBusy* parameter value is set to be non-zero when a node is engaged in an activity, such as creating a link or taking part in a measurement process. This signifies that the node can neither

Figure 5. Pseudo code for link substitution in new PR algorithm

```
For node n_x ∈ N select one of its interfaces i_x
that is free and has the strongest signal of all
its free interfaces;
if i_x = ∅
        for n_x set blockFree ← bc;
        end;
else
        set blockBusy ← bc;
set I_free ← communicationRange(i_x , I) ;
if I_free = ∅
        endAction.
set i_y ← bestSNR(I_free)
if i_y = ∅
        endAction.
l_{i_x i_y} ← createLink(i_x , i_y)
for n_x linkCounter  linkSubCounter + 1;
remove (l_{i_x i_y});
reset (i_u , i_v);
set p ← newShortestPath(n_x)
end.
```

N	a set of all nodes
I	a set of all interfaces
L	a set of all links
I_{free}	a subset of interfaces
i_x	an available interface.
n_x	a node which contains i_x.
i_y	an available interface.
n_y	a node which contains i_y.
i_u	an interface with shortest path on n_x
n_v	a node that contains i_v.
i_u	an interface on n_z that is linked to i_u
p	a shortest path for a n_x
$l_{i_v u_v}$	a link between i_u *and* i_v
bc	a blocking constant

```
endAction
        for n_x
                set blockBusy ← 0;
                set blockFree ← bc;
end.
```

initiate an action nor participate in an action. Whereas, if the *blockFree* parameter has a non-zero value it means that the node can not initiate an action but is free to participate in actions initiated by other nodes.

The routines in the above pseudo-code are briefly explained below:

- shortestPath(p, interface, set of interfaces)

This function selects a *set of interfaces* that has a shortest path from the *interface* to the *set of interfaces*. If the path between the *interface with a shortest path* and the *interface* is not shorter than path *p* the function returns ø. Otherwise it returns a *set of interfaces* with a shortest path.

- bestSNR(set of interfaces)

This function selects an interface with a best SNIR from the set of interfaces. If there is more than one such interface this function randomly selects one.

- createLink(interface A, interface B)

This function creates a link between *interface A* and *interface B*. Returns a newly created link

- remove(link)

This function removes a *link* from the set *L*.

- reset(interface A, interface B)

This function resets attributes of *interface A* and *interface B* to default attributes.

- newShortestPath(node)

This function returns a shortest path value for the *node*.

- cummuncationRange(interface, set of interfaces)

This function selects all interfaces from the set of interfaces that are free and within the range of interface.

Results & Disscussion

First, we enumerate the attributes and the conditions under which the experiments were conducted. These are:

- The self-organizing channel assignment process was limited to a single channel change per link.
- All radio interfaces were static, deployed with omni-directional antennas, based on 802.11g standard, and transmits power for each interface was generated randomly with a 50% variation.
- Calculation of interference cost was based on the following parameters:
- Distance between interfaces.
- Signal strength of transmitting interface (consequently it is not symmetrical).
- Interference factor between partially overlapping channels as provided in (Subramanian, 2005).
- All networks generated occupied an equal size area of 750 X 500 meters. Three different densities of routers per sq. unit of area were deployed in each topology: 35, 70 and 100.

Three different topologies were generated:

- The simple grid - the routers were positioned from each other in a uniform grid with their in between distances randomly varying 5%. An example of simple grid is the cellular network.

Figure 6. Result showing the increase in IC reduction by using ICR+PR in comparison to the use of ICR algorithm only

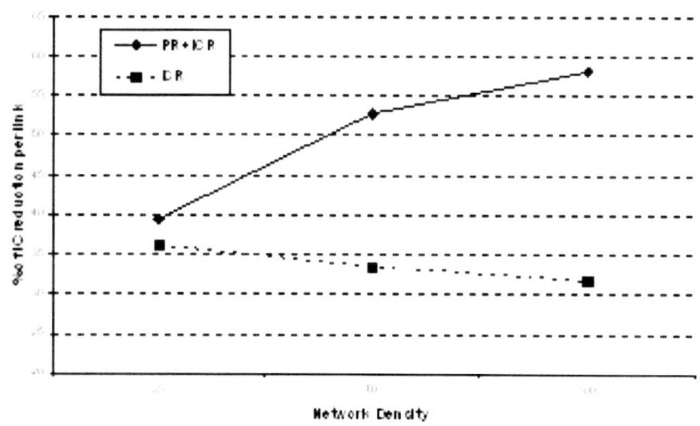

- The random grid – the same as previous only with 50% of random variation.
- The completely random – in this topology the arrangement of the routers was generated completely randomly. An example of completely random topology is the ad hoc network.

It is emphasized again that in our work, all the mesh nodes (routers) of the MR-WMN were considered to be static. We present below some of the key results obtained to illustrate the usefulness of our *new PR algorithm*.

The performance is measured in terms of *interference cost reduction* (ICR) that means higher the ICR better is the performance. In this regard, Figure 6 shows a graph of interference cost reduction (ICR) vs. network density for the case involving just *ICR algorithm* (Prodan et. al., 2008) and *ICR algorithm* when it operates after *new PR algorithm* has been invoked.

This comparative study clearly shows that without using the *new PR algorithm* and by just using the *ICR algorithm* results in much higher interference cost (IC). The reason *new PR algorithm* + *ICR algorithm* has a better performance could be because the *new PR algorithm* as discussed earlier inherently: (i) has an increased choice of interfaces that the creator (initiator) node can use (2) can select the interface based on the highest SNIR value.

Furthermore, as the network density increases the performance of the *new PR algorithm* followed by *ICR algorithm* significantly increases, whereas if just the *ICR algorithm* is used then the performance is relatively lower. This is due to the increase in proximity of the nodes with an increase in node density that creates a higher likelihood for having more shortest paths for the *new PR algorithm* to select from.

Figure 7 essentially compares the path reduction that is achieved by using just the *PF algorithm* and the *new PR algorithm*. It can be seen that the *new PR algorithm* is much more effective at reducing the path length relative to that achieved by *PF algorithm*. Although, *PF algorithm* results in a decrease in path length, *new PR algorithm* provides 2-5 times more decrease.

Figure 8 shows a comparison between the degree of *IC reduction* which is achieved by using (i) *new PR algorithm* along with *ICR algorithm* and (ii) *PF algorithm* along with new *PR algorithm* and *ICR algorithm*. It can been seen that the reduction in IC is marginally better in case of (i). A possible main reason for this is that in case of *PF algorithm* the availability of interfaces to

Figure 7. A comparative performance study of portal first (PF) algorithm versus new algorithm for PR through link substitution in terms of percentage of PR per link

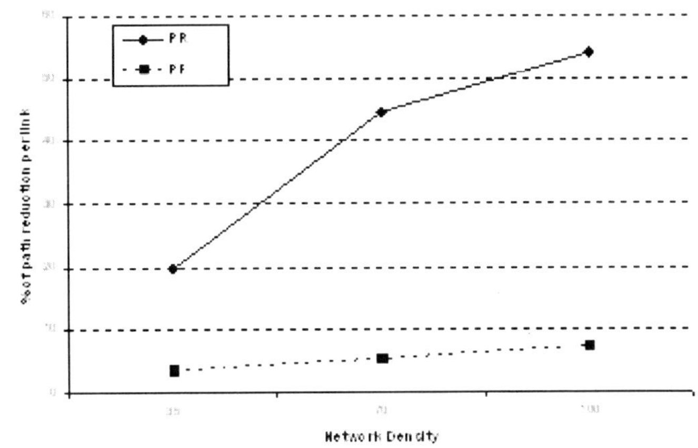

make a selection is zero because portal nodes need to provide maximum connectivity to the neighbouring client nodes. Whereas, in case of *new PR algorithm + ICR algorithm* there is more availability of interfaces to make a selection from. As such, this manifests itself as a slight increase in the IC reduction. This means that in a MR-WMN incorporation of a *new PR algorithm* will be more useful rather than *PF algorithm*.

Impact on QoS

The results obtained above show that the *new PR algorithm*, which is part of our *enhanced TC algorithm* is effective in the reduction of interference cost i.e. ICR. The reduction in interference cost translates to higher capacity in the system. The higher capacity will facilitate to support QoS dependent real-time services such as multimedia due to an increased resource i.e. bandwidth. As the *new PR algorithm* also reduces the path length

Figure 8. A comparative study of PR + ICR algorithm performance vs. PF+ PR+ ICR algorithm in terms of IC reduction

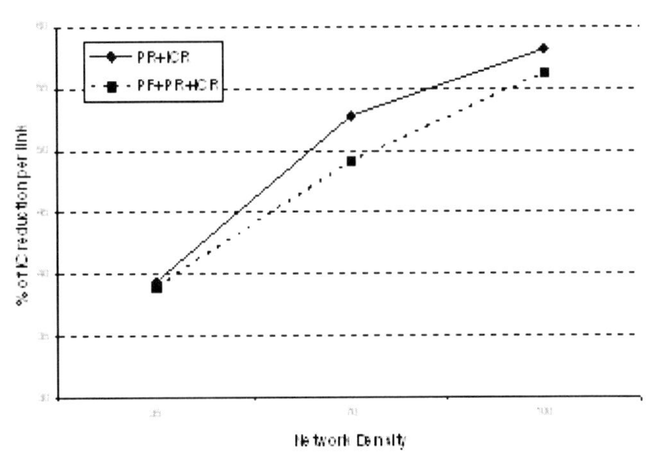

this means few number of hops will be required for the packets to flow from the access networks to the backbone Internet via the backhaul MR-WMN. Few number of hops will facilitate to decrease the end-to-end delay of packet transfer and hence lead to better QoS for the delay sensitive services.

FUTURE RESEARCH DIRECTIONS

The issues that remain open, pertaining to the material presented and discussed in this chapter, on which future research could be carried out are categorically listed below (Santi, 2005):

1. Realistic but simple modelling of radio link & energy consumption

The representative models used for radio link and node energy consumption are either too complex to obtain meaningful results or are too simple and are not accurate enough for practical environments. Thus a good balance between *simplicity* and *representativeness* of the model should be kept. Models that strike this balance are challenging to build for MR-WMN and as such are a fascinating area of open research.

2. Node Mobility Implications

The overall effect of node mobility on the network capacity needs to be studied thoroughly. Note: In our work, presented in this chapter, we had considered only static MR-WMN nodes.

3. Determination of the optimal frequency for reconfiguration

When the nodes are mobile, the more frequently the TC protocol is re-executed, the higher the quality of the constructed topology thus there is a trade-off between the message overhead generated by the repeated execution of a TC protocol and the quality of the constructed topology. This trade-off needs to be investigated further.

4. Optimal topology

What determines an *optimal topology* depends on several factors, such as: the expected network traffic, the desired level of QoS that the network should provide etc., which are disregarded by the current approaches to the TC problem. The problem of identifying energy optimal and/or interference-optimal network topologies is challenging which can perhaps be addressed by using a general approach that optimizes the network topology *and* the routing strategy as a function of certain target properties such as, extending network lifetime, or minimizing packet delivery time. We believe much work along this line of research is still to be done.

5. Symmetric Link Issue

The implicit assumption of symmetric links presented in literature does not hold in practice. In particular, we cite two shortcomings of the above cited approaches to interference-optimal TC: (i) reliance on a specific radio channel model and (ii) disregarding the effect of multi-hop communications. The issues associated with the determination of interference-optimal topologies and investigation of the differences/similarities between energy-optimal and interference-optimal topologies have not been addressed in the current literature (Santi, 2005).

6. Experimental study through testbeds

A lot of research that has been carried out for TC, which includes simulation study, shows the potential for reducing node energy consumption and radio interference. Unfortunately, as yet there is no *experimental* evidence that TC techniques can be actually used for these purposes in prcatice.

The future research must aim to experimentally demonstrate the usefulness of TC mechanisms.

Also, the creation of test beds so as to conduct experiments with TC will require the setting up of a network topology that is composed of an adequate number of nodes that communicate along *multihop* paths. The setting up of a wireless network with medium to a large scale might be very difficult in practice because of *cost* as well as *logistic reasons*.

CONCLUSION

In the chapter, the significance of topology control (TC) from the perspective of network capacity hence QoS was clearly highlighted. Some of the key work conducted in the literature was reviewed and their limitations were explained. In particular, it was explained that the TPC mechanism generally cannot achieve the goal of a minimum energy use and minimum interference. This is due to the dependence of an optimal wireless network topology on several factors, such as: the expected network traffic, the desired level and type of QoS that the network should provide and the routing strategy. The existing TC solutions suffer from the following drawbacks:

- Assumption of symmetric links and a point graph model, does not hold in practice.
- The proposed solutions often (i) rely on a specific radio channel model and (ii) neglect the complex relationships between signal strength, modulation scheme, link data throughput and bit error rate.
- Neglect the effect of multi-hop communications.

Interference cost reduction (ICR) in the multi-radio wireless mesh network (MR-WMN) is a major task, which a non-transmit power control (TPC) based topology control can manage by scheme(s) that create connectivity between the mesh nodes of MR-WMN. A salient aspect of this chapter is that it demonstrates a non-transmit power control (TPC) strategy based on our *new PR algorithm* (part of our enhanced TC algorithm), which operates in a distributed multi-agent system based MR-WMN to be an effective TC approach. The effectiveness is demonstrated through several results that point to an increase of system capacity. In particular, a comparative performance study of the algorithm by simulations was conducted in terms of path reduction (PR) and ICR. The results of this study showed that a significant ICR and PR is achieved, which in turn increases QoS.

ACKNOWLEDGMENT

This research has been performed as part of an Australian Research Council Linkage Grant between Alcatel-Lucent (Bell labs, Paris) and the University of Technology, Sydney, Australia. We also thank the referees of this chapter for their constructive comments and suggestions.

REFERENCES

Aggelou, G. (2009). *Wireless Mesh Networking*. New York: McGraw Hill Publishers. *Capacity of Wireless Mesh Networks*. (n.d.). White paper, BelAir Networks. Retrieved Aug 2006, from http://www.belairnetworks.com/resources/pdfs/Mesh_Capacity_BDMC00040- C02.pd

Blough, D. M., Leoncini, M., Resta, G., & Santi, P. (2003). The k-Neighbors Approach to Interference Bounded and Symmetric Topology Control in Ad Hoc Networks. In ACM MobiHoc, Annapolis, MD.

Burkhart, M., Rickenbach, P. V., Wattenhofer, R., & Zollinger, A. (2004). Does Topology Control Reduce Interference? *International Symposium on Mobile Ad Hoc Networking and Computing (MOBIHOC),* Roppongi Hills, Tokyo, Japan.

Clementi, A. E. F., Penna, P., & Silvestri, R. (1999). Hardness Results for the Power Range Assignment Problem in Packet Radio Networks. In G. Goos, J. Hartmanis, & J. V. Leeuwen, (Eds.), Randomization, Approximation, and Combinatorial Optimization. Algorithms and Techniques. Berlin: Springer.

Crutchfield, J. P. (1993). The calculi of emergence: Computation, dynamics, and induction. Oji International Seminar Complex Systems — from Complex Dynamics to Artificial Reality. Numazu, Japan.

Crutchfield, J. P. (1994). Is anything ever new? Considering emergence. Santa Fe Institute.

De Wolf, T., & Holvoet, T. (2005). Emergence Versus Self-Organization: Different Concepts but Promising When Combined. In S. G. D. M. S. Brueckner, A. Karageorgos & R. Nagpal, (Eds.), Engineering Self Organizing Systems: Methodologies and Applications

Haken, H. (2000). Information and Self-Organization: A Macroscopic Approach to Complex Systems. Berlin: Springer Verlag.

Holland, J. (1998). Emergence: from Chaos to Order. Reading, MA: Addison-Wesley.

Hossain, E., & Leung, K. (Eds.). (2007). Wireless Mesh Networks: Architectures and Protocols. Berlin: Springer.

Jain, K., Padhye, J., Padmanabhan, V. N., & Qiu, L. (2003). Impact of interference on multi-hop wireless network performance. MobiCom.

Kawadia, V., & Kumar, P. R. (2003) Power Control and Clustering in Ad Hoc Networks. IEEE INFOCOM 2003, San Francisco, CA.

Kirn, S., Herzog, O., Lockermann, P., & Spaniol, O. (Eds.). (2006). Multiagent Engineering. Heidelberg: Springer.

Kirousis, L. M., Kranakis, E., Krizanc, D., & Pelc, A.(n.d.). Power consumption in packet radio networks. *Theoretical Computer Science, 243*, 41–64.

Ko, B.-J., & Rubenstein, D. (2003). A Distributed, Self-stabilizing Protocol for Placement of Replicated Resources in Emerging Networks. In *11th IEEE International Conference on Network Protocols (ICNP),* Atlanta, GA.

Kyasanur, P., & Vaidya, N. H. (2005). Routing and Interface Assignment in multi-channel multi-interface Wireless networks. In *Proc. WCNC, 4,* (pp. 2051-2056), New Orleans, LA.

Leung, B. J., & Kim, K. K. (2003). *Frequency Assignment for IEEE 802.11 Wireless Networks.* Presented at 58th IEEE Vehicular Technology Conference (VTC 2003 - Fall), (pp. 1422-1426), Orlando, FL.

Marina, M. K., & Das, S. R. (2005). A Topology Control Approach for Utilizing Multiple Channels in Multi-Radio Wireless Mesh Networks. In IEEE Broadnets 2005, Boston.

Mirchandani, V., Prodan, A., & Debenham, J. (Aug 2007) A Method and Study of Topology Control based Self-Organization in Mesh Networks. In *Proc. IEEE AccessNets (BWIA Workshop)*, Ottawa, Canada.

Mirchandani, V., Prodan, A., & Debenham, J. (Oct 2007). Impact of Topology Control on the Performance of a Self-Organization Scheme for Wireless Mesh Networks. In *Proc. ACM WICON,* Austin, TX.

MOBILEMAN. (2005). *Mobile Metropolitan Ad hoc Networks*, IST-2001-38113, Oct 2005.

Narayanaswami, S., Kawadia, V., Sreenivas, R. S., & Kumar, P. R. (2002). Power Control in Ad-Hoc Networks: Theory, Architecture, Algorithm and Implementation of the COMPOW Protocol. In *European Wireless,* Florence, Italy. *NetLogo*. (n.d.). Retrieved March 2007 from http://ccl.northwestern.edu/netlogo/

Park, S.-J., & Sivakumar, R. (2002). Load-sensitive transmission power control in wireless ad-hoc networks. IEEE Globecom 2002, Taipei, Taiwan.

Prodan, A. (2009). *Topology Control on Multi-radio Infrastructure Wireless Mesh Networks,* Unpublished doctoral dissertation, University of Technology, Sydney (UTS), Australia.

Prodan, A., Mirchandani, V., & Debenham, J. (2008). A Topology Control based Self-Organization in Wireless Mesh Networks. International Journal of Communication Networks and Distributed Systems (IJCNDS).

Ramachandran, K. N., Belding, E. M., Almeroth, K. C., & Buddhikot, M. M. (2006). Interference-Aware Channel Assignment in Multi-Radio Wireless Mesh Networks. Infocom, (pp. 1-12), Barcelona, Spain.

Raniwala, A., & Chiueh, T.-c. (2004). Architecting a High-Capacity Last-Mile Wireless Mesh Network. MobiCom, Philadelphia.

Raniwala, A., & Chiueh, T.-c. (2005). Architecture and Algorithms for an IEEE 802.11- based Multi-channel Wireless Mesh Network. In *Proceedings IEEE Infocom,* (vol. 3, pp. 2223-2234), Miami, FL.

Raniwala, A., Gopalan K., & Chiueh, T-c. (n.d.). Centralized Channel Assignment and Routing Algorithms for Multi-channel wireless mesh networks. *Mobile Computing and Communications Review, 8*(2).

Santi, P. (2005). Topology Control in Wireless Ad Hoc and Sensor Networks. Chichester, UK: Wiley.

Shalizi, C. R. (2001). Causal Architecture, Complexity and Self-Organization in Time Series and Cellular Automata. University of Wisconsin at Madison.

Subramanian, A. P., Krishnan, R., Das, S. R., & Gupta, H. (2005). Minimum Interference Channel Assignment in Multi-Radio Wireless Mesh Networks. In *13th International Conference on Network Protocols (ICNP 2005),* Boston, MA.

Tumer, K., & Wolpert, D. H. (2004). Collectives and the design of complex systems. New York: Springer.

Wattenhoffer, R., & Zollinger, A. (2004). XTC: A Practical Topology Control Algorithm for Ad-Hoc Networks. In *IEEE Parallel and Distributed Processing Symposium,* Santa Fe, NM.

Weiss, G. (Ed.). (1999). Multiagent Systems A Modern Approach to Distributed Artificial Intelligence.Cambridge, MA: Massachusetts Institute of Technology.

Wooldridge, M. (2002). An Introduction to Multiagent Systems. Chichester,UK: John Wiley & Sons.

ADDITIONAL READING

Akyildiz, I. F., Wang, X., & Wang, W. (2005) Wireless mesh networks: a survey. *Computer Networks*, pp. 445-487.

Bollobas, Á. S. B. (2001) Random graphs, Cambridge; New York, Cambridge University Press.

Dousse, O., Mannersalo, P., & Thiran, P. (2004). Latency of wireless sensor networks with uncoordinated power saving mechanisms. *Proc. ACM Mobihoc*, pp. 109–120, Tokyo, Japan.

Dousse, O., Thiran, P., & Hasler, M. (2002). Connectivity in ad hoc and hybrid networks. *Proc. IEEE Infocom 02*, pp. 1079–1088., New York. USA.

Farag'o, A. (2002). Scalable analysis and design of ad hoc networks via random graph theory. *Proc. ACM DIAL-M 02*, pp. 43–50, Atlanta, GA, USA.

Gast, M. (2005) *802.11 Wireless Networks: The Definitive Guide*, O'Reilly.

Goldsmith, A. (2005) Wireless communications, New York, Cambridge University Press.

Goodman, J., & O'Rourke, J. (1997). Handbook of Discrete and Computational Geometry. CRC Press, New York. USA.

Gupta, P., & Kumar, P. (1998). Critical power for asymptotic connectivity in wireless networks. Stochastic Analysis, Control, Optimization and Applications, Birkhauser, pp. 547–566, Boston, MA. USA.

Hajek, B. (1983). Adaptive transmission strategies and routing in mobile radio networks. *Proc. Conference on Information Sciences and Systems*, pp. 373–378, Baltimore, MD. USA.

Hajiaghayi, M., Immorlica, N., & Mirrokni, V. (2003). Power optimization in fault-tolerant topology control algorithms for wireless multi-hop networks. *Proc. ACM Mobicom 03*, pp. 300–312, San Diego, CA, USA.

Held, G. (2005) Wireless Mesh Networks, Boca Raton, Auerbach Publications.

Horak, R. (2007) Telecommunications and data communications handbook, Hoboken, N.J., Wiley-Interscience.

Hou, T., & Li, V. (1986). Transmission range control in multihop packet radio networks. *IEEE Transactions on Communications, COM-34*, 38–44.

Judd, G., & Steenkiste, P. (2007) Understanding Link-level 802.11 Behavior: Replacing Convention With Measurement. WICON Austin, Texas, USA.

Jung, A., & Vaidya, N. (2002). A power control mac protocol for ad hoc networks. *Proc. ACM Mobicom*, pp. 36–47, Atlanta, GA, USA.

Karl, H., & Willig, A. (2005) Protocols and Architectures for Wireless Sensor Networks, John Wiley and Sons.

Katzela, & Naghshineh, M. (1996). Channel Assigment Schemas for Cellular Mobile Telecommunication Systems- a Comprehensive Survey, *IEEE Personal Communications*, vol. 3, pp. 10-31.

Kaufmann, B., Baccelli, F., Chaintreau, A., Papagiannaki, K., & Diot, C. (2005). Self Organization of Interfering 802.11 Wireless Access Networks, INRIA Technical Report.

Kawadia, V., & Kumar, P. R. (2005) A cautionary perspective on cross-layer design. IEEE Wireless Communications.

Mishra, A., Banerjee, S., & Arbaugh, W. (2005). "Weighted Coloring based Channel Assignment for WLANs," ACM SIGMOBILE Mobile Computing and Communications Review.

Mishra, A., Brik, V., Banerjee, S., Srinivasan, A., & Arbaugh, W. (2006). A Client-driven Approach for Channel Management in Wireless LANs, *IEEE Conference on Computer Communications (INFOCOM)*, pp. 1-12, Barcelona, Spain.

Mishra, E., & Rozner, S. Banerjee, & Arbaugh, W. (2005). Exploiting partially overlapping channels in wireless networks: Turning a peril into an advantage, *ACM/USENIX Internet Measurement Conference*.

Palmer, E. M. (1985) Graphical evolution: an introduction to the theory of random graphs, New York, Wiley.

Proakis, J. G. (2002) Wiley Encyclopedia of Telecommunications.

Radha Krishna Rao, G. S. V., & Radhamani, G. (2008) WiMAX: a wireless technology revolution, Boca Raton, Auerbach Publication

Russell, S. J., & Norvig, P. (2003) Artificial intelligence: a modern approach, Upper Saddle River, N.J., Prentice Hall/Pearson Education.

Sklar, B. (2001) Digital Communications, Englewood Cliffs, NJ, Prentice-Hall.

Smith, C., & Meyer, J. (2004) 3G Wireless with WiMAX and Wi-Fi, McGraw-Hill.

Tse, D., & Viswanath, P. (2005) Fundamentals of Wireless Communication, New York, Cambridge University Press.

Walke, B. H., Mangold, S., & Berlemann, L. (2006) IEEE 802 Wireless Systems: Protocols, Multi-Hop Mesh/Relaying, Performance and Spectrum Coexistence, John Wiley & Sons, Ltd.

Wattenhofer, R., & Zollinger, A. (2004) XTC: A Practical Topology Control Algorithm for Ad-Hoc Networks. *Parallel and Distributed Processing Symposium.* Santa Fe, New Mexico, USA, IEEE.

Chapter 13
QoS Support Mechanisms in WiMAX

Maode Ma
Nanyang Technological University, Singapore

Jinchang Lu
Nanyang Technological University, Singapore

ABSTRACT

Quality of service (QoS) provisioning is an important issue in the deployment of broadband wireless access networks e.g. WiMAX (IEEE Std 802.16-2004, 2004) networks with real-time and non-real-time traffic integrated. To design a QoS support framework tailored for WiMAX networks is more challenge as wireless channel has unique characteristics such as time-varying channel and limited channel capacity. This chapter presents various QoS support mechanisms in WiMAX networks. Existing proposals with the state-of-the-art technology have been classified into three main categories: QoS support architecture, bandwidth management mechanism, and packet scheduling schemes. Representative schemes from each of the categories have been evaluated with respect to major distinguishing characteristics of the WiMAX MAC layer and PHY layer as specified in the IEEE 802.16d standard. Suggestions and research trends on QoS support in WiMAX networks are highlighted.

INTRODUCTION

The Worldwide Interoperability for Microwave Access (WiMAX) system has been specified by the IEEE 802.16d and its amendment IEEE 802.16e standard. The IEEE 802.16d and IEEE 802.16e standard defines the physical (PHY) and medium access control (MAC) layer of the fixed and mobile broadband wireless access systems, respectively.

DOI: 10.4018/978-1-61520-771-8.ch013

Broadband wireless access (BWA) systems like WiMAX networks have been deployed not only to be complement and extension of existing last mile wired networks such as cable modem and xDSL but also to be competitor to wired broadband access networks. Due to the upcoming air interface technologies, which promise to deliver high transmission data rates, BWA systems become an attractive alternative.

IEEE 802.16d supports both frequency division duplex (FDD) and time division duplex (TDD)

Table 1. AMC vs. receiver SINR min sensitivity requirement

AMC Index AMC(q)	Receiver SINR(dB) γ(q)	Receiver Min Sensitivity RSS(q) (dBm)	Modulation/ Coding Scheme	E(q) (bit/symbol)	Bit Rate R(q) (Mbps)
AMC(1)	8.3-11.6	- 80	BPSK1/2	0.5	20
AMC(2)	11.7-13.2	- 80	QPSK1/2	1	40
AMC(3)	13.3-18.9	- 78	QPSK3/4	1.5	60
AMC(4)	19.0-21.9	- 73	16-QAM1/2	2	80
AMC(5)	21.0-28.0	-71	16-QAM3/4	3	120
AMC(6)	28.1-29.1	- 66	64-QAM2/3	4	160
AMC(7)	>=29.2	-65	64-QAM3/4	4.5	180

PHYs. IEEE 802.16d specifies 4 different PHY specifications, namely, WirelessMAN-SC PHY specification, WirelessMAN-SCa PHY specification, WirelessMAN-OFDM PHY specification, WirelessMAN-OFDMA PHY specification. At the PHY layer, a WiMAX system can take an adaptive modulation policy selecting 1 from 3 different modulation schemes. On the uplink (UL), QPSK is mandatory, while 16-QAM and 64-QAM are optional. The downlink (DL) can support QPSK and 16-QAM, while 64-QAM is optional. To fully utilize the flexible and robust PHY layer, a WiMAX system equips a flexible radio link control (RLC) scheme, which is responsible for transition from one PHY scheme to another. The system also uses the receiver sensitivity (RS) as a parameter together with the signal to interfernece and noise ratio (SINR) thresholds of receivers used by an adaptive modulation and coding (AMC) scheme to select the different burst profiles in order to maximize the network throughput and maintain the bit error rate (BER) under a preset level e.g. BER$<=10^{-5}$.

WiMAX transceivers support different transmission modes with different modulation and coding schemes corresponding to different data transmission rates. For each modulation scheme, there is one relationship between the theoretical bit error rate and the ratio of energy per bit (E_b) to the spectral noise density (N_o) (E_b/N_o). E_b/N_o can be expressed in terms of *SINR* as:

$$\frac{E_b}{N_o} = SINR * \frac{W}{R_b} \tag{1}$$

where W is the channel bandwidth, R_b is the transmission bit rate of a transmission mode corresponding to a modulation and coding scheme. Different modulation schemes have different curves of theoretical bit error rates versus E_b/N_0.

The range of the received SINR values will be classified into seven non-overlapping scales corresponding to an adaptive modulation and coding index $AMC(q)$ where $q=1, 2,\ldots,7$ on a target prescribed BER e.g. BER$<=10^{-5}$ in a WiMAX PMP network. Further compiling with the RS requirement specified by IEEE 802.16d standard, a lookup table shows the relationship among AMC, SINR and RS requirement with its corresponding data rate as shown in Table 1. It is obvious that the data transmission rate can be dynamically changed from 20 Mbps to 180 Mbps if the channel capacity of WiMAX network is set to 40Mbps.

The MAC Layer of WiMAX networks can be designed to meet the requirements of very-high-data-rate applications with a variety of quality of service (QoS) requirements. The MAC layer is composed of three sub-layers. From bottom to top, they are the privacy sub-layer (PS), the MAC common part sub-layer (CPS), and the service specific convergence sub-layer (CS). The PS deals with security and network access authentication procedures. CPS carries out the major MAC functions. The CS sub-layer provides the interface to the upper layer to decide the MAC service class for the specific connection and to initialize the resource allocation requests of the CPS.

The MAC frame is divided into UL and DL sub-frames. Service flows which are uniquely identified by a 32-bit service flow identifications (SFID) may be transmitted in either the UL or DL sub-frames. Service flows can be created, changed, or deleted, which can be accomplished by a series of MAC management messages referred to as dynamic service addition (DSA), dynamic service change (DSC), and dynamic service deletion (DSD). There are three types of service flows namely, provisioned service flows, admitted service flows, and active service flows. They are associated with a set of QoS requirement parameters, namely, provisioned QoS parameter set, admitted QoS parameter set and active QoS parameter set, respectively. Admitted and active service flows are mapped to a 16-bit connection identification (CID). They are controlled and maintained by a connection admission control (CAC) scheme in an active connection list (ACL) at the base station (BS). The connections are classified into four types of MAC layer services, namely, Unsolicited Grant Service (UGS), Real-time Polling Service (rtPS), Non-Real-time Polling Service (nrtPS) and Best Effort (BE) service. The mandatory QoS parameters of the service flows with UGS are:

- Maximum sustained traffic rate (MSTR)
- Maximum latency (ML)
- Tolerated jitter
- Request/transmission policy

The mandatory QoS parameters of the service flows with rtPS are:

- Minimum reserved traffic rate (MRTR)
- MSTR
- ML
- Request/transmission policy

The mandatory QoS parameters of the nrtPS service flows are:

- MRTR
- MSTR
- Traffic priority
- Request/transmission policy

The mandatory QoS parameters of the BE service flows are:

- MSTR
- Traffic priority
- Request/transmission policy

The bandwidth is always requested on a CID basis and bandwidth is allocated on a subscriber station (SS) basis. A connection represents either an individual application or a group of applications at one SS sending data with the same CID. All packets from the application layer at the SS are classified by the connection classifier based on their CIDs and are forwarded to an appropriate queue. At the SS, the scheduler will retrieve the packets from the queues and transmit them to the network as defined by the UL-MAP sent by the BS. The UL-MAP is determined by the uplink bandwidth allocation scheduling module based on the BW-request messages that report the current queue size of each connection in the SS.

In summary, in WiMAX networks, bandwidth is always requested on a CID basis. The bandwidth is aggregated into a single grant to an SS. There is an active connection list controlled and maintained by a connection admission control (CAC) scheme at the BS. The CAC policy is to decide whether the QoS requirements of a connection can be satisfied. The scheduler at the BS or SS schedules the bandwidth allocation to the applications with the connections in the ACL and the packets transmitted according DL/UP map via PHY at different data transmission rates which dynamically change according to different AMC schemes based on the real-time channel SINR condition.

IEEE 802.16d defines the connection signaling including connection request and response between SS and BS but it has not defined the admission control process. IEEE 802.16d medium access control, which is based on the concepts of connections and service flows, specifies QoS signaling mechanisms (per connection or per station) such as bandwidth requests and bandwidth allocation. However, IEEE 802.16d standard left the details of the QoS based packet scheduling algorithms and reservation management, which determines the UL and DL bandwidth allocation, undefined. IEEE 802.16d PHY provides AMC and the conceptually power control. It also left details of AMC and adaptive power control algorithm undefined.

This chapter presents various QoS support mechanisms in WiMAX networks. Existing proposals with the state-of-the-art technology have been classified into three major categories. They are:

- QoS support architecture
- Bandwidth management mechanism
- Packet scheduling schemes

Representative schemes from each of the categories have been evaluated with respect to major distinguishing characteristics of the WiMAX MAC layer and PHY layer as specified in the IEEE 802.16d standard.

BACKGROUND

The QoS term can be interpreted in different ways. In general, QoS can be described from two perspectives: user perspective and network perspective. In user perspective, QoS refers to the application quality as perspective by the user. In network perspective, QoS refers the service quality that the network offers to applications or users in term of network QoS parameters that include: latency or delay of packets traveling across the network, reliability of packet transmission and throughput.

From the network perspective, the networks' goal is to provide the QoS services that adequately to meet the users' needs while maximizing the network resources' utilization. To achieve this goal, the networks analyze the application requirements, manage the network resources and deploy various network QoS mechanisms.

QoS parameters quantitatively represent the applications' QoS requirements .They are:

- Throughput
- Delay
- Delay jitter
- Error rate
- Packet loss rate

Networks may use a combination of QoS services, i.e., per-flow and quantitative, per-class and quantitative. Some networks may include multiple types of QoS services in order to support a wide range of applications.

Providing quality of service (QoS) simultaneously to services with different requirements is a much more difficult task in wireless mediums as compared to wired networks as wireless access networks have unique characteristics which are time-varying channel conditions and multi-user diversity. To cope with such issues, QoS in wireless networks is handled at the medium access control (MAC) layer as well as at the PHY layer.

In recently years, QoS support architecture, bandwidth management mechanism and QoS support packet scheduling algorithms in WiMAX system have been proposed.

QOS SUPPORT ARCHITECTURE IN WIMAX

Alavi, Mojdeh & Yazdani (2005) have proposed an inclusive architecture to provide QoS support in WiMAX. Some compatible methods for specific modules such as Scheduler, Traffic Shaper, and Request and Grant Manager to optimize Delay, Throughput and Bandwidth Utilization metrics have been developed.

Protocols can be designed by violating the reference architecture, for example, by allowing direct communication between protocols at non-adjacent layers or sharing variables between layers. Such violation of a layered architecture is a cross-layer design approach with respect to the reference architecture. In order to map QoS parameters from layer 3 (L3) to layer 2 (L2), L3 and L2 in the reference architecture can be integrated (Mai, Y., Yang, C. & Lin, Y., 2007). The proposed MAC layer cross to Network layer QoS framework can enhance QoS supports in WiMAX networks. Major functional blocks in the QoS framework include QoS mapping from L3 to L2, Admission Control, Fragment Control, and Remapping. Fragment Control handles the data frames that belong to the same IP datagram in an atomic manner to reduce useless transmission. Remapping is concerned with the mapping rules from IP QoS to 802.16 QoS and is designed to reduce the impact of traffic burstiness on buffer management.

Kwon, T. et al. (2005) have presented a design example of primitives for cross-layer operation between its MAC and PHY layers. The proposed cross-layer design framework for IEEE802.16e OFDMA systems that are compatible with WiBro based on various kinds of cross-layer protocols for performance improvement. In the proposed model, the MAC layer contains a user grouper, scheduler, and resource controller. Each functional entity exploits physical layer information to increase system throughput. The physical layer consists of a diversity channel physical layer protocol data unit (PPDU) controller, AMC channel PPDU controller, control information controller, and hybrid automatic repeat request (HARQ) functional blocks. AMC sub-channel users and diversity sub-channel users are classified by the user grouper. Since the properties of AMC sub-channels and diversity sub-channels are quite different, the grouping of users into two channel types is essential if system throughput is to be increased. The scheduler determines the scheduling of users and the quantity of packets that should be scheduled in the current frame. For cross-layer optimization, the scheduler should be designed to exploit not only PHY information but also application layer information.

BANDWIDTH MANAGEMENT QOS SUPPORT MECHANISMS

Bandwidth management mechanisms are approaches that manage the network resources by coordinating and configuring network devices. The main mechanisms are:

- Resource reservation
- Connection admission control
- Cross-layer approach bandwidth management mechanisms

Resource Reservation Mechanisms

Resource reservation mechanisms inform the network entities on the QoS requirements on the network resources for various applications. The network devices will use this information to manage the network resources in order to meet such requirements. The resource reservation mechanisms include the following functions:

- Provisioning of resource reservation signaling that notifies all devices along the communication path on the QoS requirements of multimedia applications.

- Delivery of QoS requirements to the connection admission control mechanism that decides if there are available resources to meet the QoS requirements of a new request without violating the QoS requirement of existing connections.
- Notification of the application regarding the admission result.

The representative proposal tailored for WiMAX is Dynamic Resource Reservation (DRR) scheme by (Kamal, G., Mounir, A., & Annie, G., 2006). The basic principle of DRR is that the reserved bandwidth will vary between a minimum C_m and a maximum C_M value as per the bandwidth utilized by the clients. The proposal is able to optimize reservation and utilization of bandwidth for Committed Bandwidth (CB) type traffic. However, it is very difficult to select the parameters of C_m and C_M. And the fluctuations of the reserved bandwidth from C_M to C_m could increase signaling costs.

Connection Admission Control Mechanisms

An admission control scheme determines how bandwidth is allocated to control the traffic entering the network. The role of connection admission control (CAC) is to control the number of connection flows into the network. A new connection request is progressed only when sufficient resources are available at each successive network element to establish the connection through the entire network based on its service category, traffic contract, and QoS requirements, while the agreed QoS of all existing connections are still maintained. Admission control is useful in the situations where a certain number of connections may all share a link, while an even greater number of connections can cause significant degradation to all connections to the point of making them all useless such as in congestive collapse.

A lot of CAC schemes tailored for WiMAX system have been proposed. They can be classified into three categories.

The first category is the CAC schemes with degradation strategy based on service degradation (Ge, Y. & Kuo,G. 2006), bandwidth borrowing (Niyato, D. & Hossain, E., 2007; Wang, W., Liu, F., Ji,Y. & Nararat, R., 2007) or bandwidth stealing (Jiang, C. & Tsai, T., 2006) strategies. The main idea of these policies is to decrease the resources already allocated to ongoing connections in order to be able to accept a new service flow. This strategy could be combined by a threshold-based capacity sharing approach in order to avoid starvation or by a guard channel scheme that sets a small portion of total bandwidth aside for more bandwidth-sensitive flows. The drawback of those CAC schemes is the difficulty to set the suitable threshold to tradeoff the bandwidth efficiency and the connection blocking probability. If a large percentage of total bandwidth is set as guard channel, the system loses its efficiency. In contrast, if a small guard channel is set, the connection blocking probability increases.

The second strategy is more conservative. The algorithm maintains the QoS provisioning for ongoing connections and simply rejects a new service flow if the remainder bandwidth is less than the bandwidth requirement of the new service flow (Wongthavarawat, K. & Aura, G. 2003). This strategy is very simple, but it is very rigid without much flexibility.

The third strategy is more advanced which has adopted a cross-layer approach. The representative proposal is the AMC-induced CAC that incorporates the modulation type into the CAC process (Kwon, E., Jung, K., Lee, J. & Ryu, S., 2005). This proposal, however, supports only two types of modulations. And it has been proposed based on the assumption that all the connections have the fixed and equal bandwidth requirements which limits its applicability. Another CAC with cross-layer approach is TCP-Aware CAC which considers the effect of TCP applications on the packet-level performance (Wang, X., Eun, D. & Wang, W., 2007).One more example of this strategy is a CAC scheme for opportunistic scheduling with cross-layer approach. Although a CAC scheme has been always considered as an upper layer scheme to optimize the system performance, its application to the opportunistic scheduling in heterogeneous systems requires a cross-layer admission control process, which takes into consideration the physical layer and traffic pattern characteristics, together with the number of users tradeoff. Perez-Neira & Zorba (2007) proposed a CAC scheme for an opportunistic scheme based on minimum rate outage and delay outage to regulate the number of users or connections in a WiMAX system to obtain multi-users diversity.

Bandwidth Management by a Cross-Layer Approach

There are a few prior works that deal with the MAC layer cross to the higher layers for bandwidth management, i.e. the MAC to application and transport layers. Chen, J., Jiao, W., & Guo, Q. (2005) have used IntServ and DiffServ in connection oriented WiMAX PMP and mesh networks with the aim to provide end-to-end QoS guarantee services. This work maps the RSVP at the IP layer to DSA/DSC/DSD at the MAC layer. The authors have suggested that messages exchange for DSA and DSC can be deployed to carry QoS parameters of IntServ services for end-to-end resource (bandwidth/buffer) reservation. For DiffServ services, on the other hand, a number of per-hop behaviors (PHBs) for different classes of aggregated traffic can be mapped into different connections directly.

When a new application flow arrives at the IP layer, it will be firstly parsed according to the definition in PATH message (for InteServ) or differentiated services code point (DSCP) for DiffServ. Then it will be classified and mapped into one of the four types of services (UGS, rtPS, nrtPS or BE). By the proposal, the dynamic service module at an SS sends a request message to the BS, where the admission control determines whether this request will be approved or not. If it is not approved, the service module informs the upper layer to deny this service. Otherwise, the admission control notifies the scheduling module to make a provision based on the parameters in the request message. At the same time, the accepted service is transferred into a traffic grooming module. According to the grooming result, the SS will send a bandwidth request message to the BS. The centralized scheduling module in the BS will retrieve the requests and generate an UL-MAP message carrying the bandwidth allocation results. Finally, the SS will package service data units (SDU) from the IP layer into protocol data units (PDU) and upload them in its allocated UL slots to the BS.

Packet Scheduling QoS Support Mechanisms

Packet scheduling refers to the decision process used to choose which packets should be transmitted or dropped. Packet scheduling is the process of resolving contention for bandwidth. A scheduling algorithm has to determine the allocation of bandwidth among the traffic flows and the transmission order of the packets in the each flow. One of the most important tasks of a scheduling scheme is to satisfy the QoS requirements of each traffic flows while efficiently utilizing the available bandwidth.

Many legacy scheduling algorithms, capable of providing certain guaranteed QoS, have been developed for wired networks. However, these existing scheduling algorithms, such as fair queueing scheduling, virtual clock, and earliest-due-date (EDD), are not directly applicable in wireless networks because they have not considered the varying wireless link capacity and the location-dependent channel state. The characteristics of wireless communication pose special difficulties that do not exist in wired networks.

They include high error rate and bursty errors, location-dependent and time-varying wireless link capacity, scarce bandwidth, user mobility, and power constraint of the mobile hosts. All of the above characteristics make the development of efficient and effective scheduling algorithms in wireless networks very challenging.

WiMAX networks provide QoS services for heterogeneous classes of traffic with different QoS requirements. Currently, there is an urgent need to develop new technologies for providing QoS differentiation and guarantees in WiMAX networks. Among all the technical issues that need to be resolved, packet scheduling in WiMAX networks is one of the most important. Lin, Chou & Liu (2008) have conducted a comparison among a few proposed WiMAX scheduling algorithms. Belghith & Nuaymi (2008) evaluated the performance of some legacy scheduling algorithms for WiMAX systems.

The current proposed scheduling algorithms for QoS support in WiMAX networks can be classified into three categories with respect to the nature of scheduling as following:

- Holonomic approach
- Hierarchical approach
- Opportunistic packet scheduling with a cross-layer approach

The holonomic approaches use single layer scheduling scheme. In contrast, hierarchical approaches use several layers or stages for the scheduling. Opportunistic packet scheduling with a cross-layer approach takes advantage of instantaneous channel variations by giving priority to the users with favorable channel conditions.

Holonomic Approach for Packet Scheduling

In order to maximize throughput of non-real-time traffic flows with satisfying QoS requirements of real-time traffic flows, Ryu, B., Seo, H., Shi, M & Ryu, S. (2005) have applied holonomic approach and proposed one layer hybrid scheduling scheme. The proposed urgency and efficiency based packet scheduling (UEPS) algorithm has been designed not only to support multiple users simultaneously but also to offer real-time and non-real-time services to a user at the same time. The UEPS algorithm uses the time-utility function as a scheduling urgency factor and the relative status of the current channel to the average one as an efficiency indicator of radio resource usage. The proposed packet scheduler assigns priorities to the packets to be transmitted, based on the channel status reported by the user equipments as well as the QoS statistics maintained by the BS. Since the scheduler works in a global timeline, a time utility function (TUF) is used for the scheduling. Two scheduling factors, the urgency of scheduling and the efficiency of radio resource usage, are used to schedule real-time and non-real-time traffic packets at the same time. The TUF is used to represent the urgency of scheduling while the channel state is used to indicate the efficiency of radio resource usage.

Xergias, S.A., Passas, N. & Marekos, L. (2005) proposed Frame Registry Tree Scheduler (FRTS). The scheduler contains three operations:

- Packet/request arrival
- Frame creation
- Subscriber's modulation type change or connection QoS service change

The basic idea of the packet/request operation is to distribute packet transmissions in time frames, based on their deadlines. For UGS and rtPS services, the packet deadline is equal to the arrival time plus the latency of this packet. The sub-tree of the last time frame where this packet can be transmitted is updated, if it exists. Otherwise, it is created. For nrtPS and BE services, the packet deadline does not need to be calculated. Then, the sub-tree of the last existing time frame is updated.

Figure 1. Proposed 3-tier scheduling framework

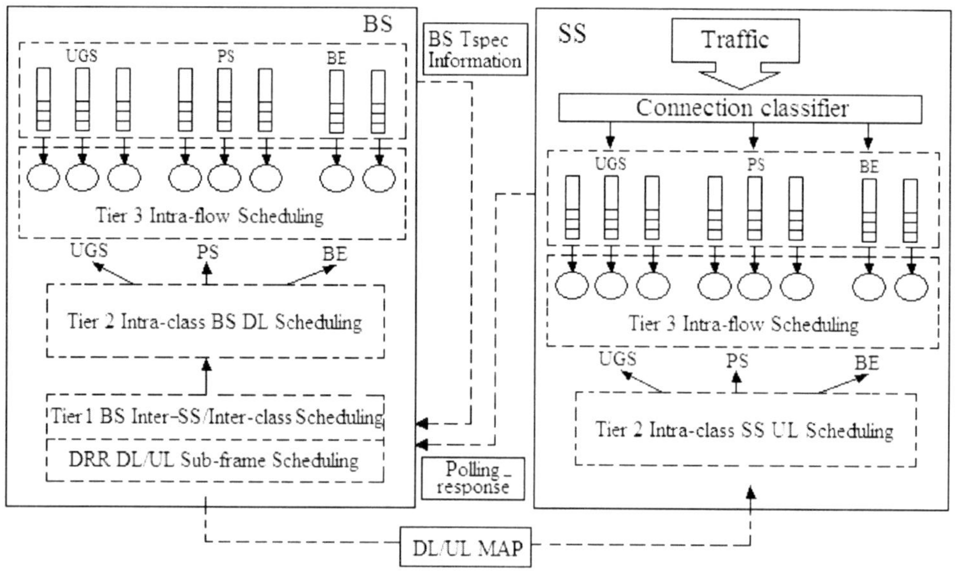

The function of the frame creation procedure is to decide on the frame contents. There are three cases:

- If the sub-tree of the first time frame contains a number of packets equal to one time frame, all these packets fill up the frame content.
- If the sub-tree of the first time frame contains a number of packets less than one time frame, the empty slots are occupied by packets from the next time frame sub-trees and/or will be left for contention.
- If the sub-tree of the first time frame contains a number of packets more than one time frame, packets for BE service are moved to the next time frame sub-tree. If there are still excess packets to transmit, first nrtPS packets, then rtPS packets and finally UGS packets are deleted until the number of packets fit exactly into one time frame. A change in a subscriber's modulation type or connection QoS service causes a moving of the corresponding sub-tree to the right modulation substructure or service substructure.

Hierarchical Approach for Packet Scheduling

A scheduling algorithm should posses the following features: efficient link utilization, bounded delay, enough fairness, high throughput, low implementation complexity, graceful performance degradation, strong isolation, more delay/bandwidth decoupling, and flexible scalability. With the objective to provide service differentiation between the real-time and non-real-time classes of traffic, Lu, J. C., Ma, M., Ng, B.C. & Sanjay (2007) have proposed a three-tier hierarchical framework and a scheduling scheme to support QoS in WiMAX as shown in Figure 1. The solution has completed the missing part of the framework specified in the IEEE 802.16d standard. This efficient QoS scheduling structure and corresponding scheduling schemes for WiMAX networks has taken the following principles into account:

- The existing WiMAX QoS signaling mechanisms should be complied with. DL/UL map mechanism should be fully applied.
- Per-connection scheduling overhead should be minimized.
- Service parameters for each connection should be guaranteed.

The novel solution has the following unique features. (1) The Pre-scale Dynamic Resource Reservation (PDRR) scheme has been proposed to allocate bandwidth to UL sub-frame and DL sub-frame dynamically. (2) The Priority-based Queue Length Weighted (PQLW) scheduling algorithm has been proposed for inter-class scheduling and the Max-Min Fair Sharing (MMFS) scheduling has been applied for inter-SS scheduling within each class of service at the BS as Tier 1 scheduling. (3) The Self-Clocked Fair Queuing (SCFQ) and Weighted Round Robin (WRR) scheduling schemes have been applied to inter-connection scheduling within each class of service at each SS as Tier 2 scheduling. (4) Earliest Deadline First (EDF) and Shortest Packet Length First (SPLF) scheduling have been applied to the packet scheduling within each of the connections carrying burst traffic as Tier 3 scheduling.

Opportunistic Packet Scheduling Algorithms with Cross-layer Approach

Wireless communication systems have unique characteristics – namely, time-varying channel conditions and multi-user diversity. Efficient MAC design and new scheduling solutions need to be developed that are specifically tailored for the wireless communication environment (Liu, X., 2003). Opportunistic MAC (OMAC) is the modern view of communicating over the spatiotemporally varying wireless links. The cross-layer nature embeds OMAC with the potential to revolutionize the design of wireless networks from physical to data link layers.

The wireless resources including bandwidth and power are scarce and more expensive than their wired counterparts because the overall system performance degrades dramatically due to multi-path fading, Doppler, and time-dispersive effects caused by the wireless air interface. Unlike wired networks, even if large bandwidth/power is allocated to a certain wireless connection, the loss and delay requirements may not be satisfied when the channel experiences deep fades. In this scenario, the scheduler plays an extremely important role. Among the competing users, an OMAC seeks to pick the one who is currently experiencing the relatively best channel condition at each scheduling instant. The judicious schemes should be developed to support prioritization and resource reservation in wireless networks in order to enable guaranteed QoS with efficient resource utilization.

Recently, various opportunistic scheduling schemes have been proposed for wireless communication systems. They can exploit the time-varying nature of the radio environment to improve the spectrum efficiency while maintaining a certain level of QoS satisfaction for each connection or user in various wireless networks. The proposed opportunistic scheduling schemes can be classified into two categories based on the functionality of scheduling algorithm (Vegard, H., 2006) as:

- Channel-aware only
- Channel-aware & queue-aware algorithm

Max Carrier-to-Noise Ratio Scheduling (MCS) (Bonald, T., 2005) and Proportional Fair Scheduling (PFS) (Jalali, A., Padovani, R. & Pankaj, R., 2000; Huang, J. & Niu, Z., 2007; Huang, J. & Niu, Z. 2007) are two typical channel-aware opportunistic scheduling schemes. The MCS scheme is to allocate resources to users with the best channel condition to achieve high system throughput and the PFS attempts to trade-off among the throughput, efficiency and fairness among users by taking packet length into account, or estimating future channel quality. On top of wireless channel characteristics, traffic characteristics also play an important role in the design of opportunistic scheduling algorithm. Andrews, M. et al.(2000) ; Hu, M., Zhang, J. & Sadowsky, J. (2004); Shakkottai, S. & Stolyar, A.(2001) proposed channel-aware & queue-aware opportunistic scheduling schemes. Modified Largest Weighted Delay First (M-LWDF) (Andrews, M. et al., 2000) is a modified version of PFS algorithm that tries to meet QoS requirement in terms of head-of-line packet delay. Traffic-Aided Opportunistic Scheduling (TAOS-1) (Hu, M., Zhang, J. & Sadowsky, J., 2004) is a heuristic opportunistic scheduling scheme that unifies file size information and wireless channel variations in order to reduce the completion time of file transmissions. Exponential Rule Scheduler (EXPRule) (Shakkottai, S. & Stolyar, A., 2001) attempts to equalize the weighted delays of all buffers when their differences become larger in a wireless system.

Cross-layer MAC and opportunistic scheduling designs tailored for WiMAX networks have been proposed (Lera, A., Molinaro, A. & Pizzi, S., 2007; Liu, Q., Wang, X. & Giannakis, G. B., 2006; Wan, L., Ma, W. & Guo, Z. 2007). Lera, A., Molinaro, A. & Pizzi, S. (2007) have addressed a channel-aware scheduling algorithm with a per-flow channel error compensation technique tailored for WiMAX system based on its typical features: class-based QoS guarantees and per-flow resource assignment. The Worst-Case Fair Weighted Fair Queuing + (WF2Q+) algorithm has been suggested to manage the flow-level and class-level granularity per-class queues to provide the target QoS to each WiMAX service class while achieving fairness of traffic flows belonging to the same class. It can achieve as high system throughput as about 95% of ideal system throughput. A priority-based scheduler at the MAC layer has been proposed (Liu, Q., Wang, X. & Giannakis, G. B. 2006) where each connection employs adaptive modulation and coding (AMC) scheme at the physical (PHY) layer. Authors have defined a priority function (PRF) for each connection admitted in the system. And the PRF is updated dynamically depending on the wireless channel quality, QoS satisfaction and service priority through MAC-PHY cross-layer manner. Wan, L., Ma, W. & Guo, Z. (2007) have proposed a joint packet scheduling and sub-channel allocation scheme. With the help of several additional parameters for the preference metrics, which are the scheduling priority of connections or users determined by the opportunistic scheduler for its transmission service, those proposed schemes have shown to be able to achieve satisfied performance in the given network conditions.

Rath, H. K., Bhorkar, A. & Sharma, V. (2006) have proposed an Opportunistic Deficit Round Robin scheduler (O-DRR). The scheduler is used in the uplink direction. The BS polls subscribers periodically. After each period, the BS determines the set of subscribers that are eligible to transmit and their bandwidth requirements. This set is defined as the eligible set. A number of conditions must be verified by an SS to be in this set:

- The queue is not empty
- The received SIR is above a minimum threshold, denoted *SIRth*

Once these conditions are satisfied, the subscriber is eligible to transmit during a given frame of the current scheduling epoch. The scheduled set changes dynamically depending on the wireless link state of subscribers. At the beginning of each scheduling epoch, the BS resets the eligible and scheduled sets and repeats the above mentioned process.

Some researchers have considered cross-layer scheduling using the MAC scheduler and the PHY resource allocator. Sang, J., Jeong, D. & Jeon, W. (2006) have proposed a cross-layer design of packet scheduling and resource allocation in OFDMA wireless networks which concentrates on DL scheduling in BS. In the OFDMA systems, each carrier is subdivided into a number of sub-carriers which can be controlled or allocated separately making the PHY very robust. In the proposed system, the sub-carriers are grouped into allocation units (AU) for allocation so that overheads are minimized. The base station (BS) estimates the sub-channel condition of each user and allocated resources to the users on a frame-by-frame basis. The authors have considered a system where some SSs have real-time traffic and others have non real-time traffic, i.e. real time and non real-time traffic do not co-exist at the same SS. The BS can estimate the channel gain of each user on a sub-channel and an AU can be independently allocated to a particular user. The aim is to maximize the overall utilization while satisfying the rate requirements of individual user. To do this, authors have formulated a linear programming method and subsequently obtained a sub-optimal solution which reduces the computational time. A packet scheduler has also been proposed at the BS MAC layer which provides equal priority to both real-time and non real-time traffic except if the ratio of the waiting time of a real-time packet in the queue and the delay constraint of the packet exceeds a certain threshold. Associated with the cross-layer packet scheduler, a 3-stage resource allocation algorithm has been proposed. At the first stage, the urgent packets will be scheduled. At the second stage, the non real-time packets and the real-time packets which are not urgent will be scheduled. The highest priority is given to the users whose channel quality is better regardless of the traffic type. At the third stage, the unallocated AU's if any will be allocated. After the packet scheduler decides the rate requirements of each user, the actual resources are allocated at the PHY layer. It is carried out by using a sub-optimal heuristic algorithm which first allocates the sub-channels constraints in order to maximize the utilization or minimize the transmission power. And then, the algorithm relocates (or swaps) sub-channels so as to satisfy the rate requirements of each user. Though this cross-layer scheme can be followed in the DL at the BS, it cannot be adopted for UL scheduling since individual node cannot decide the channel condition and cannot place request for those particular channels.

FUTURE RESEARCH DIRECTIONS

Different mechanisms can address different issues in QoS support in WiMAX networks. The scheduling algorithms provide mechanisms for bandwidth allocation and multiplexing at the packet level. Admission control policy is dependent on the specific scheduling disciplines used. For the UL traffic, the scheduling algorithm has to work in tandem with CAC to satisfy the QoS requirements. The issue to combine QoS support architecture, bandwidth management mechanisms, traffic scheduling mechanisms in a cross-layer structure to provide a new QoS support scheme in WiMAX is still an open issue.

The computational complexity of a proposed QoS support algorithm strongly influences its scalability. We suggest that the QoS support framework in WiMAX networks could be in a hierarchical structure to minimize per-connection scheduling overhead and to improve the scalability of a proposed QoS support framework.

The wireless communication systems have some unique characteristics such as time-varying channel conditions and multi-user diversity. The new scheduling solutions need to be developed that are specifically tailored for this environment. The new scheduling scheme must consider the impact of evolving traffic characteristics on scheduling and the impact of air interface on scheduling.

Opportunistic MAC (OMAC) is the modern view of communicating over spatiotemporally varying wireless link whereby the multi-user diversity is exploited rather than combated to maximize bandwidth efficiency or system throughput.

The cross-layer nature embeds OMAC with the potential to revolutionize the design of wireless data networks from physical to data link layers. We suggest that cross-layer opportunistic scheduling can combine with adaptive power control scheme and CAC scheme to provide QoS support in WiMAX networks.

Orthogonal frequency-division multiple access (OFDMA) is a multi-user version of the orthogonal frequency-division multiplexing (OFDM) digital modulation scheme. OFDM can combat multipath with more robustness and less complexity. OFDMA technique offers not only the frequency diversity by spreading the carriers all over the used spectrum but also the time diversity by optional interleaving of carrier groups in time. OFDMA technique can use the cell capacity to the utmost by adaptively using the highest modulation a user can use. To design a scheduling algorithm with MAC-PHY cross-layer approach to effectively fill time-spectrum block in WiMAX networks is a new emerging trend.

With the advance of wireless region area network (WRAN) driven by IEEE 802.22 standard, WiMAX networks can coexist with a WRAN. The end-to-end QoS support mechanisms by cognitive spectrum sensing and intelligent opportunistic spectrum or sub-carrier allocation is another new emerging trend.

CONCLUSION

QoS support in WiMAX is a fundamental design requirement, and is considerably more difficult than in wired networks, mainly because of the variable and unpredictable characteristics of wireless links.

In this chapter, we discuss various QoS support techniques, e.g. QoS support architecture, bandwidth management mechanism and packet scheduling schemes proposed in WiMAX networks. We evaluate representative schemes from each with respect to major distinguishing characteristics of the WiMAX MAC layer and PHY layer as specified in the IEEE 802.16d standard. We also discuss and highlight research trends on QoS support issues in WiMAX networks.

REFERENCES

Alavi, H. S., Mojdeh, M., & Yazdani, N. (2005). A Quality of Service Architecture for IEEE 802.16 Standards. In *Asia-Pacific Conference on Communications,* (pp. 249 – 253).

Andrews, M., Kumaran, K., Ramanan, K., Stolyar, A., Vijayakumar, R., & Whiting, P. (2000). CDMA data QoS scheduling on the forward link with variable channel conditions. Tech. Rep., Bell Labs.

Bang, H. J., Ekman, T., & Gesbert, D. (2008). Channel predictive proportional fair scheduling. *IEEE Transactions on Wireless Communications*, 7(2), 482–487. doi:10.1109/TWC.2008.060729

Belghith, A., & Nuaymi, L. (2008). Performance evaluation for scheduling algorithms in WiMAX networks. In The 14th European Wireless conference 2008, (pp. 1-6).

Bonald, T. (2005). Flow-level performance analysis of some opportunistic scheduling algorithms. *European Transactions on Telecommunications 2005,* (Vol.16, pp. 65- 75).

Chen, J., Jiao, W., & Guo, Q. (2005). An integrated QoS control architecture for IEEE 802.16 broadband wireless access systems. *Global Telecommunications Conference, (Globecom) 2005:* (Vol. 6, pp. 3330-3335).

Chen, J., Jiao, W., & Wang, H. (2005). A Service Flow Management Strategy for IEEE 802.16d Broadband Wireless Access Systems in TDD Mode. *Proceedings of IEEE International Conference on Communications, ICC 2005*, (Vol. 5, pp.3422 – 3426).

Ge, Y., & Kuo, G. (2006). An Efficient Admission Control Scheme for Adaptive Multimedia Services in IEEE 802.16e Networks. *IEEE 64th Vehicular Technology Conference, VTC-2006,* (pp.1–5).

Hu, M., Zhang, J., & Sadowsky, J. (2004). Traffic aided opportunistic scheduling for wireless networks: algorithms and performance bounds. *Computer Networks: The International Journal of Computer and Telecommunications Networking, 46*(4), 505–518.

Huang, J., & Niu, Z. (2007). A Cross-Layer Proportional Fair Scheduling Algorithm with Packet Length Constraint in Multiuser OFDM Networks. In. *Proceedings of IEEE International Conference on Global Telecommunications Conference, GLOBECOM, 07,* 3489–3493. doi:10.1109/GLOCOM.2007.662

IEEE Standard for Local and Metropolitan Area Networks Part 16: Air Interface for Fixed Broadband Wireless Access Systems. (2004). *IEEE Std 802.16-2004 (Revision of IEEE Std 802.16-2001-),*(pp.1-857).

Jalali, A., Padovani, R., & Pankaj, R. (2000). Data throughput of CDMAHDR: a high efficiency-high data rate personal communication wireless system. In. *Proceedings of IEEE VTC, 2000,* 1854–1858.

Jiang, C., & Tsai, T. (2006). Token bucket based CAC and packet scheduling for IEEE 802.16 broadband wireless access networks. *Consumer Communications and Networking Conference, CCNC 2006.* (Vol.1, pp.183 – 187).

Kamal, G., Mounir, A., & Annie, G. (2006). Dynamic Resource Reservation in IEEE 802.16 Broadband Wireless Networks. In IWQoS 2006, (pp.140 – 148).

Kwon, E., Jung, K., Lee, J., & Ryu, S. (2005). A Performance Model for Admission Control in IEEE 802.16. *Proceedings of WWIC 2005,* (Vol. 3510, pp.159–168).

Kwon, T., Lee, H., Choi, S., Kim, J., & Cho, D. (2005). Design and Implementation of a Simulator Based on a Cross-Layer Protocol between MAC and PHY layers in a WiBro Compatible IEEE802.16e OFDMA System. *IEEE Communications Magazine, 43*(12), 136–146. doi:10.1109/MCOM.2005.1561931

Lera, A., Molinaro, A., & Pizzi, S. (2007). Channel-Aware Scheduling for QoS and Fairness Provisioning in IEEE 802.16/WiMAX Broadband Wireless Access Systems. *IEEE Network, 21*(5), 34–41. doi:10.1109/MNET.2007.4305171

Lin, J., Chou, C., & Liu, C. (2008). Comparison of WiMAX scheduling algorithms and proposals for the rtPS QoS class. In *The 22nd international Conference on advanced Information Networking and Applications- Workshops,* (pp. 68-74).

Liu, Q., Wang, X., & Giannakis, G. B. (2006). A cross-layer scheduling algorithm with QoS support in wireless networks. *IEEE Transactions on Vehicular Technology, 55*(3), 839–847. doi:10.1109/TVT.2006.873832

Liu, X. (2003). Optimal Opportunistic Scheduling in Wireless Networks. In *IEEE 58th Vehicular Technology Conference,* (Vol.3, pp. 1417 -1421).

Ma, M. Lu, J.C., Sanjay, K., & Ng, B.C. (2007). A Three-Tier Framework and scheduling to Support QoS Service in WiMAX. In *IEEE 6th International Conference on Information, Communications & Signal Processing,* (pp.1-5).

Ma, M., & Ng, B. C. (2006). Supporting Differentiated Services in Wireless Access Networks. *Proceedings of IEEE International Conference on Communication Systems, ICCS, 2006,* 1–5.

Mai, Y., Yang, C., & Lin, Y. (2007). Cross-Layer QoS Framework in the IEEE 802.16 Network. *The 9th International Conference on Advanced Communication Technology,* (Vol. 3, pp. 2090 – 2095).

Niyato, D., & Hossain, E. (2007). Radio Resource Management Games in Wireless Networks: An Approach to Bandwidth Allocation and Admission Control for Polling Service in IEEE 802.16. *IEEE Wireless Communications*, *14*(1), 27–35. doi:10.1109/MWC.2007.314548

Rath, H. K., Bhorkar, A., & Sharma, V. (2006). An Opportunistic DRR (O-DRR) Uplink Scheduling Scheme for IEEE 802.16-based Broadband Wireless Networks. In *IETE, International Conference on Next Generation Networks (ICNGN)*.

Rong, B., Qian, Y., & Chen, Y. (2007). Adaptive power allocation and call admission control in multiservice WiMAX access networks. *IEEE Wireless Communications*, *14*(1), 14–19. doi:10.1109/MWC.2007.314546

Ryu, S., Ryu, B., Seo, B., & Shi, M. (2005). Urgency and efficiency based wireless downlink packet scheduling algorithm in OFDMA system. In *IEEE 61st Vehicular Technology Conference*, (Vol. 3 pp. 1456 – 1462).

Sang, J., Jeong, D., & Jeon, W. (2006). Cross-layer Design of Packet Scheduling and Resource Allocation in OFDMA Wireless Multimedia Networks. In *IEEE 63rd Vehicular Technology Conference, (VTC) 2006,* (pp. 309-313).

Shakkottai, S., & Stolyar, A. (2001). Scheduling algorithms for a mixture of real-time and non-real-time data in HDR. In 17th Int. Teletraffic Congress (ITC-17).

Vegard, H. (2006). *Tutorial: Opportunistic Multiuser Scheduling for Wireless Networks. Department of Electronics and Telecommunications, Norwegian University of Science and Technology (NTNU), Jun. 2006.* Retrieved May, 10, 2009, from www.unik.no/personer/porten/MoPSAR/MOPSARTutorial_Hassel.pdf

Wan, L., Ma, W., & Guo, Z. (2007). A cross-layer packet scheduling and subchannel allocation scheme in 802.16e OFDMA System. In *Wireless Communications and Networking Conference,* (pp. 1865 - 1870).

Wang, W., Liu, F., Ji, Y., & Nararat, R. (2007). Admission Control for Non-preprovisioned Service Flow in Wireless Metropolitan Area Networks. In *Fourth European Conference on Universal Multiservice Networks, ECUMN '07,* (pp. 243–249).

Wang, X., Eun, D., & Wang, W. (2007). A Dynamic TCP-Aware Call Admission Control Scheme for Generic Next Generation Packet-Switched Wireless Network. *IEEE Transactions on Wireless Communications*, *6*(9), 3344–3352. doi:10.1109/TWC.2007.06064

Wongthavarawat, K., & Aura, G. (2003). Packet scheduling for QoS support in IEEE 802.16 broadband wireless access systems. *International Journal of Communication Systems*, *16*(1), 81–96. doi:10.1002/dac.581

Xergias, S. A., Passas, N., & Marekos, L. (2005). Flexible Resource Allocation in IEEE 802.16 Wireless Metropolitan Area Networks. In *The 14th IEEE Workshop on Local and Metropolitan Area Networks, LANMAN 2005*, (pp.-6).

Zorba, N., & Perez-Neira, A. I. (2007). CAC for Multibeam Opportunistic Schemes in Heterogeneous WiMax Systems under QoS Constraints. In *Global Telecommunications Conference, GLOBECOM apos07,* (pp. 4296 – 4300).

ADDITIONAL READING

Bai, X., Shami, A., Meerja, K. A., & Assi, C. (2006). New Distributed QoS Control Scheme for IEEE 802.16d Wireless Access Networks. *IEEE Global Telecommunications Conference, GLOBECOM '06,* (pp.1 -5).

Chang, Y., Chien, F., & Jay Kuo, C. C. (2007). Cross-layer QoS Analysis of Opportunistic OFDM-TDMA and OFDMA Networks. *IEEE Journal on Selected Areas in Communications*, *25*(4), 657–666. doi:10.1109/JSAC.2007.070503

Chen, J., Jiao, W., & Wang, H. (2005). A service flow management strategy for IEEE 802.16 broadband wireless access systems in TDD mode. *IEEE International Conference, ICC 2005:* Vol. 5 (pp. 3422- 3426).

Chen, Y., & Li, L. (2005). A Random Early Expiration Detection Based Buffer Management Algorithm for Real-time Traffic over Wireless Networks. *Computer and Information Technology, CIT, 2005*, 507–511.

Cho, D., Song, J., Kim, M., & Han, K. (2005). Performance Analysis of the IEEE 802.16 Wireless Metropolitan Area Network. *First International Conference on Distributed Frameworks for Multimedia Applications, DFMA '05,*(pp.130 – 136).

Chu, G., Wang, D., & Mei, S. (2002). A QoS Architecture for the MAC Protocol of IEEE 802.16d BWA System. *Proceedings of IEEE International Conference on Communications, Circuits and Systems*: Vol. 1(pp. 435-439).

Fattah, H., & Leung, C. (2002). An Overview of Scheduling Algorithms in Wireless Multimedia Networks. *IEEE Wireless Communications*, *9*(5), 76–83. doi:10.1109/MWC.2002.1043857

Ghosh, A., David, R., Wolter, J., Andrews, G., & Chen, R. (2005). Broadband Wireless Access with WiMax/802.16: Current Performance Benchmarks and Future Potential. *Communications Magazine, IEEE*, *43*(2), 129–136. doi:10.1109/MCOM.2005.1391513

Hassanein, H., & Taha, A. (2009). Cognitive Radio Networks Technologies Enabling Dynamic Spectrum Allocations. Retrieved Jul, 10, 2009, from

Hou, F., Ho, P., & Shen, X. (2006). WLC17-1: Performance Analysis of a Reservation Based Connection Admission Scheme in 802.16 Networks. *IEEE GLOBECOM, 06*, 1–5.

Lakkakorpi, J., Sayenko, A., Karhula, J., Alanen, O., & Moilanen, J. (2007). Active Queue Management for Reducing Downlink Delays in WiMAX. *IEEE Vehicular Technology Conference, VTC-2007: Vol. 66* (pp.326 – 330).

Liu, N., & Li, X. Pei, C., & Yang, B. (2005). Delay Character of a Novel Architecture for IEEE 802.16 Systems. *Proceedings of the Sixth International Conference on Parallel and Distributed Computing, Applications and Technologies (PDCAT'05),* (pp. 293-296).

Liu, Q., Zhou, S., & Giannakis, G. B. (2004). Cross-layer modeling of adaptive wireless links for QoS support in multimedia networks. *Quality of Service in Heterogeneous Wired. Wireless Networks*, 68–75.

Niyato, D., & Hossain, E. (2006). A Queueing-Theoretic and Optimization-Based Model for Radio Resource Management in IEEE 802.16 Broadband Wireless Networks. *IEEE Transactions on Computers*, *55*(11), 1473–1488. doi:10.1109/TC.2006.172

Safa, H., Artail, H., Karam, M., Soudan, R., & Khayat, S. (2007). New Scheduling Architecture for IEEE 802.16 Wireless Metropolitan Area Network. *IEEE/ACS International Conference on Computer Systems and Applications, (*pp.203-210).

Seo, J., Kim, S., Lee, H., & Cho, C. (2006). An Efficient Capacity Allocation Scheme of Periodic Polling Services for a Multimedia Traffic in an IEEE802.16 System. [MASS]. *Mobile Adhoc and Sensor Systems, 2006*, 11–20.

Seo, J., Lee, H., & Cho, C. (2006). Performance of IEEE802.16 Random Access Protocol - Transient Queueing Analysis. *IEEE Global Telecommunications Conference, 2006, GLOBECOM'06*, (pp.1 – 6).

Shejwal, A., & Parhar, A. (2007). Service Criticality Based Scheduling for IEEE 802.16 WirelessMAN. *The 2nd International Conference on Wireless Broadband and Ultra Wideband Communications*,(pp. 12-12).

Song, G., & Li, Y. (2005). Cross-layer optimization for OFDM wireless networks-part II: algorithm development. *IEEE Transactions on Wireless Communications*, *4*(2), 625–634. doi:10.1109/TWC.2004.843067

Sun, J., Yao, Y., & Zhu, H. (2006). Quality of Service Scheduling for 802.16d Broadband Wireless Access Systems. *Proceedings of IEEE 63rd International Conference of Vehicular Technology 2006*:Vol. 3 (pp.1221-1225).

Tang, J., & Zhang, X. (2005). Cross-layer design of dynamic resource allocation with diverse QoS guarantees for MIMO-OFDM wireless networks. *6th IEEE International Symposium on World of Wireless Mobile and Multimedia Networks, WoWMoM 2005*, (pp.205 -212).

Wang, H., He, B., & Agrawal, D. P.(2006). Admission control and bandwidth allocation above packet level for IEEE 802.16 wireless MAN. *Parallel and Distributed Systems, ICPADS 2006.*

Wang, H., Li, W., & Agrawal, D. P. (2005). Dynamic admission control and QoS for 802.16 wireless MAN. *Wireless Telecommunications Symposium*, (pp.60 – 66).

Wong, W. K., Tang, H., & Guo, S. (2003). Scheduling algorithm in a point-to-multipoint broadband wireless access network. *Vehicular Technology Conference, VTC 2003: Vol.3* (pp. 1593 – 1597).

www.ieeegcc.org/Program/Tutorials/T1-COGNITIVE_RADIO_NETWORKS_2009.pdf.

Chapter 14
4G Wireless Networks:
Architectures, QoS Support and Dynamic Resource Management

Dimitrios G. Stratogiannis
National Technical University of Athens, Greece

Georgios I. Tsiropoulos
National Technical University of Athens, Greece

John D. Kanellopoulos
National Technical University of Athens, Greece

Panayotis G. Cottis
National Technical University of Athens, Greece

ABSTRACT

Fourth generation (4G) wireless networks aim at supporting various multiservice applications over IP architectures which satisfy enhanced users demands through innovative services of increased Quality of Service (QoS). QoS can be assured through independent optimal design of network components or by optimizing interoperability. The supported services impose also their classification into IP network service models and their specifications description. The integration of different wireless access technologies into the 4G network architecture leads to a heterogeneous network environment that raises several issues. An overview of various approaches employed to provide QoS in 4G networks concerning their architectures, different access technologies interoperability and resource management techniques are investigated in this chapter. Dynamic resource allocation, admission control, QoS provision using mobile management and pricing policies are presented. Concluding, in the demanding 4G environment under variable network conditions, appropriate schemes and architectures may provide a robust network management tool for QoS provision and efficient resource utilization.

DOI: 10.4018/978-1-61520-771-8.ch014

INTRODUCTION

The new social and economic trends having evolved worldwide such as the increasing demand for productivity and effectiveness, business travelling all over the world along with the pressure to respond immediately to customer needs have created an increased need for access to information any where in the world at anytime. The convergence of information communications technology and computing is creating demand and opportunities for ubiquitous computing via wireless and mobile equipment without technological restrictions. Next Generation Networks (NGNs) such as forth generation (4G) networks, also known as beyond third generation (Beyond 3G), aim at providing 'optimal connection anywhere, any time'. 4G networks support large scale global roaming across multiple wireless and mobile networks providing diverse applications over packet-switched networks.

One issue of utmost importance in 4G networks is the expected integration of all heterogeneous wired and wireless access technologies (e.g. cellular 3G/UMTS, IEEE 802.11 WLAN, Bluetooth, IEEE 802.16 WIMAX, BWA) into a common scalable network infrastructure. The terminals of 4G technology will incorporate almost all existing wireless options implemented in a unified environment assuring end – to – end Quality of Service (QoS). The rise of service-enabling technologies and platforms will help. These services include voice, data, message, video and world-wide web with high data rates and QoS, security measures, location awareness and energy efficiency ensuring better adaptation to users requirements and traffic conditions compared to existing 3G technologies (Varshney & Jain, 2001). 4G networks will be based on a common, flexible and seamless all-IP protocol (Wisely et al, 2003), where mobile terminals will need to be highly integrated multimode, multiband, and able to utilize a wide range of applications, incorporating better scheduling and Call Admission Control (CAC) techniques constituting a robust communication model and architecture. The use of the IP protocol will assure interoperability with existing widely accepted communication structures. These features offered by 4G networks will encourage new demand and create new technological and business opportunities not only for manufacturers and operators but, also, for service and content providers and above all for the end users.

In this book chapter a comprehensive survey on 4G networks regarding their characteristics, network architectures and dynamic resource management for QoS support is performed. The chapter is intended to provide a review into the most critical issues emerging in 4G networks. The chapter is composed of two main sections, organised as follows. In the first section, entitled "QoS provision in 4G networks" the key aspects of QoS provision techniques are presented, taking into account 4G networks specifications. In this section challenges related to the heterogeneity of 4G networks such as the need for seamless handoff among different access networks or between two access points of the same network and the best connectivity problem are described. Another important issue presented in the first section is the different access technologies interoperability, where the two most prevailing architectures are presented, focusing on the integration of Wireless Local Area Networks (WLANs) and cellular networks towards a unified heterogeneous network. The first section is concluded with a subsection concerning the Service Classes (SCs) supported by the network, which are mapped into the corresponding IP network service models. In the second section entitled "Dynamic resource allocation and call admission control", resource management issues related to 4G are surveyed. The diverse QoS requirements of the SCs supported and the dynamic nature of the wireless channel impose the use of efficient resource allocation techniques such as bandwidth reservation, bandwidth degradation and resource allocation with QoS renegotiation, which are presented in this section. Moreover, the integration of game theory in resource allocation and CAC schemes is demonstrated, since it constitutes a recent trend in dynamic resource management. Also, dynamic pricing and CAC schemes are demonstrated, denoting the role of pricing as an additional dimension of CAC for efficient use of resources in heterogeneous 4G networks. Finally, future research directions are presented regarding the design of cross-layer architectures for QoS provision in 4G networks.

BACKGROUND

In general, QoS provision in wireless networks is established based on two basic frameworks, namely Radio Resource Management (RRM) and mobility management. The resource management is the part of the network responsible for the efficient distribution of the available resources. It also guarantees QoS with regard to parameters like bandwidth, delay, reliability, whereas the mobility management addresses issues like handoff, location registration, roaming, etc. The above parameters and issues are critical for the demanding environment of future 4G networks. Radio resource management is again implemented in two levels: a) macro level involving CAC, gross traffic classification and bandwidth allocation that manage resources at session level and b) micro level including mechanisms dealing with the MAC layer and packet scheduling at timeslot level used to control QoS parameters such as delay and jitter (Wu, 2005) Of course, the physical layer mechanisms including channel coding and power control are also important as parts of micro level resource management. This chapter addresses macro level resource management and the problem of providing QoS for multiple SC applications in the unified communication environment of 4G networks.

Two are the dominant factors that influence QoS over wireless access networks. Related to the coupling of QoS assurance with mobility management, the first target is to preserve the same QoS level during and after a handover. As mobile channel is time-variant, the second target is to dynamically manage the wireless resources. Adaptive modulation techniques can adapt the terminal characteristics considering the physical layer to the varying channels and adjust the resources used to particular requirements of every SC. Adaptation strategies may also be implemented in other network elements. To adapt the network operation to the varying channel behaviour, cross-layer algorithms are employed to adjust configurable components to particular user applications, or system-performance requirements. Such algorithms are already present in legacy networks based either on the adaptation of the transmission codec to channel quality or on the selection of the proper transmission channel according to the bandwidth required by upper layers.

QoS in 4G networks can be defined and applied based on two different approaches concerning QoS management. The first is based on the relative treatment of the supported traffic flows. Each traffic flow corresponds to various SCs with different requirements as to system resources and QoS. Therefore, the various SCs must be independently treated according to their qualitative characteristics. For example, different priority levels may be assigned to each SC supported through an appropriate admission control mechanism. Appropriate prioritization criteria incorporated into the admission control mechanisms can give higher priority e.g. to Web traffic compared to an e-mail. This technique may be used by the network to provide a relatively faster forwarding of Web traffic, although no absolutes guarantees are given.

The second approach is quantitative and aims at assuring the transmission capabilities required based on the appropriate metrics such as throughput, delay or PLR. For example, video-streaming may require a guaranteed bit rate of 32 kbps so that it may be reproduced at the terminal without any interruption at an acceptable quality level.

This approach is merely based on QoS negotiation and through Service Level Agreement (SLA). The rationale behind this approach is that the service is provided based upon a set of metrics agreed by both the customers and the service providers. An SLA is a service contract between the customer and the service provider that specifies the forwarding service received (Blake, Black, Carlson, Davies, Wang & Weiss, 1998); it may include traffic rules and agreement in whole or in part. An SLA may contain both technical and non-technical terms and conditions. The technical specification concerning the IP connectivity service is given in Service Level Specifications (SLSs). An SLS contains a set of technical parameters and their agreed values, which define the QoS service level offered by network service models like the DiffServ model (Mykoniati, Charalampous, Georgatsos, Damilatis, Goderis, Trimintzios, Pavlou, Griffin, 2003; Moon & Aghvami, 2003). The SLS is associated with several attributes, for example, ingress and egress interfaces, flow identification, traffic envelop and traffic conformance parameters. The traffic conformance parameters include data flow specifications described by the FlowSpec that constitutes the basis for SLA negotiations.

QOS PROVISION IN 4G NETWORKS

QoS Provision & Specifications Related to 4G

The converged broadband wireless environment of 4G networks must take into account QoS specifications. QoS customization with regard to the variety of services supported has been one of the main goals in the development of 4G networks. QoS can be defined as the ability of a network to provide a satisfactory service with regard to specific criteria. Among these criteria are voice quality, signal strength, low call blocking and dropping probabilities and high data rates for multimedia and data applications. An efficient QoS design must treat different services or users in a specialized and cost-efficient way aiming at a good end-user experience when he uses a particular service. Considering that the end-users perception of the QoS varies according to the application, it is essential to define a set of common parameters which will translate the service requirements to the underlying protocols. These parameters incorporate the system compliance to users requirements. For example the reliability of 4G networks is expected to attain 99.999%, referred to as five nine reliability. The major challenges when considering QoS in 4G networks includes the efficient allocation and use of radio resources such as power and spectrum, the operation adaptive to channel variations characteristics, tolerance to systems faults and handoff support among heterogeneous wireless networks. QoS in heterogeneous 4G networks should comply with certain requirements imposed when network architecture incorporates different technologies:

- Independence of the specific radio access technologies.
- Inter-working with different mobility concepts for seamless handover.
- Independence of particular QoS provision techniques.

In wireless mobile networks, QoS is interpreted by good transmission quality, service availability and minimum delay. In packet-oriented wireless services, network overloading can cause excessive packet delay and/or jitter. The throughput level determined either at network or at user level may also drop to unacceptable levels. Hence, QoS must define the minimum performance levels with regard to latency, jitter and Packet Loss Rate (PLR) required by a certain service. These QoS specifications should be used in the CAC phase as the respective admission criteria. Assuring QoS for various applications with diverse characteristics is a primary objective in 4G networks. Diverse QoS requirements are usually expressed in terms of minimum data rates, delay bounds and respective probabilities. For network based services QoS depends on the following factors:

- *Throughput*: The packet rate through the network.
- *Delay*: Time required by a packet to travel from one end to the other.
- *Jitter*: The variability over time of packet latency through the network.
- *Packet Loss Rate*: The rate at which packets are lost.
- *Reliability*: The availability of a connection.

Apart from the factors affecting the perceived QoS, appropriate guarantees should establish the actual QoS provided to users. QoS guarantees are essential in defining upgraded system reliability. In applications where no packet must be lost or delayed during transmission, the strict guarantees usually imposed are called deterministic or hard guarantees. On the other hand, applications that do not require stringent QoS levels and can tolerate a certain degree of QoS relaxation have different types of QoS guarantees, commonly referred to as statistical or soft guarantees.

Challenges Related to Heterogeneous Networks

The interaction of different mobile access networks gives rise to new challenges (Buddhikot Chandranmenon, Han, Lee, Miller & Salgareli, 2003; Salkintzis, Fors & Pazhyannur 2002; Pahlavan, Krishnamurthy, Hatami, Ylianttila, Mareka, Pichna, 2000; Floroiou, Ruppelt, Sisalem, & Stephanopoli 2003; Zhuang, Gan, Loh & Chua 2003; Gao, Wu & Miki, 2004). One challenge is to implement seamless mobility management schemes (Zhang, Guo, Guo & Zhu, 2003; Pang, Chen, Chen & Agrawal, 2004; Akyildiz, Xie & Mohanty, 2004). The current dominant aspect is that Mobile IP will be used to provide global roaming within heterogeneous mobile networks (Perkins, 2002). However, even if the terminals can adapt to the different radio interfaces, maintaining the service continuity of mobile terminals through diverse environments is a complex issue that current relatively simplistic handover algorithms cannot support. Handoff methods shall enable mobile terminals to maintain connectivity when moving between cells, between systems, between frequencies and at the upper layers, between IP subnets. Moreover, mobile nodes must be able to interoperate within different networks regardless of the different signalling protocols, routing techniques and mobility management standards. The enhancement of basic Mobile IPv6 (MIPv6) as IP based mobility management with respect to handover latency it is been studied within the framework of the EU IST project Moby Dick (Gozdecki, et al., 2003; Jähnert, et al., 2005). Sophisticated handover policies and strategies are able to exploit the available network resources; optimal decisions can be made by taking into account factors such as system performance, network conditions, service type and user status and preferences.

Vertical and Horizontal Handoffs

Developing a seamless handoff procedure is an important challenge (Pahlavan, Krishnamurthy, Hatami, Ylianttila, Mareka, Pichna, 2000; Floroiou, Ruppelt, Sisalem, & Stephanopoli, 2003; Manzoni, Ghosal & Serazzi, 1995; Gupta, 2004). The problem of vertical handoff is defined in the context of heterogeneous network architectures as a handoff of a user that moves across different types of networks employing different access technologies (global handoff) (Akyildiz, Xie & Mohanty 2004). On the other hand, a horizontal handoff is a handoff between two access points or neighbouring cells using the same access network technology (local mobility). The two types of handoffs are depicted in Figure 1 as an example of their operation.

Figure 1. Vertical and horizontal handoff

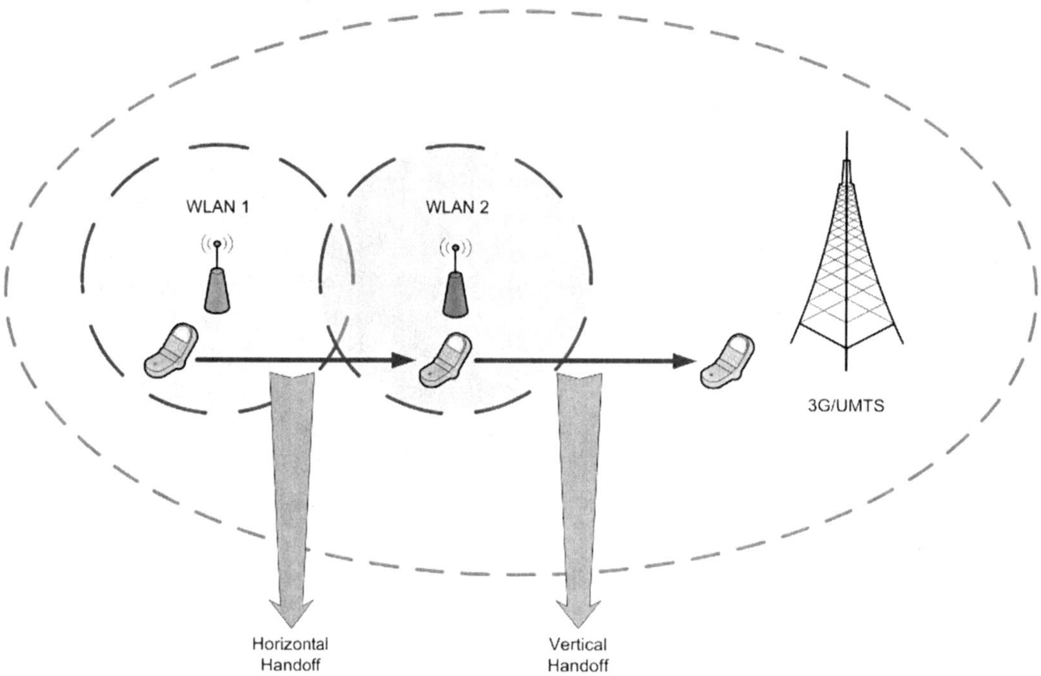

To make handoff decisions, the quality of the radio channel is estimated to detect any signal degradation and select the new cell to handoff. In heterogeneous networks, this means that the mobile terminal should be able to measure the quality over technologically different systems while keeping the original link. Handoff decisions depend on the signal strength received at the mobile terminal from both networks, taking also into account its velocity and its location. If a vertical handoff is triggered only by received signal strength, it may cause unnecessarily frequent handoffs, as in interoperation of 3G and WLAN networks, where frequent handoffs may degrade the performance of a user moving at a high velocity due to the small coverage of WLAN.

In this case, 3G networks may support a fast moving user more efficiently as they cover wider areas. Also, "ping-pong" effects, taking place near cell borders when mobile terminals are continually handed off between two cells should be prevented. Therefore, network selection criteria in heterogeneous networks must be defined taking into account the users location.

In heterogeneous wireless environments, a vertical handoff mechanism needs to consider not only the parameters related to the physical layer but also those related to the network and the transport layer. A framework for vertical handoffs was presented where the handoff decision criteria take into account service type, data rate requirement, network condition, and cost of handoff (McNair & Zhu, 2004). A dynamic optimization scheme was proposed to guarantee QoS to mobile users while maximizing the network utilization. A mobility management solution was proposed to handle vertical handoff and network roaming in heterogeneous wireless networks (Vidales, Baliosian, Serrat, Mapp, Stajano & Hopper, 2005). However, maximization of the network utility from the service providers point of view was not considered in these works. A very interesting concept for vertical handoff has been developed under the Simultaneous Multiple Access (SIMA) for Vertical HandOver (VHO) project (Ylitalo, Jokikyyny, Kauppinen, Tuominen, & Laine, 2003; Nikkilä, Langutin, Asplund & Ranta 2004). The SIMA describes the ability to use multiple network interfaces in a terminal simultaneously by routing different data flows through different interfaces depending on the users desire. In SIMA access and network selection could be based on user profiles or policies. The SIMA concept places the individual connections in the best available networks at any location, based on user/application/operator needs. The SIMA project constitutes a reliable solution for providing maximization in network capabilities utilization, providing QoS to end users.

The problem of integrating WLANs into cellular wireless networks was investigated in the literature, a hierarchical and a distributed RRM framework were designed to support seamless handoff and roaming across cellular networks and WLANs (Karetsos, Kyriazakos, Groustiotis, Giandomenico & Mura, 2005; Shenoy & Montalvo, 2005). The QoS mapping and Internet work message translation mechanisms were designed to support seamless handoff among multiple WLANs and cellular networks. In 2004, the development of the emerging IEEE 802.21 standard started, to enable handover and interoperability between heterogeneous networks including both 802 and non-802 networks (Gupta, 2004; Mussabbir, Yao, Niu, & Fu, 2007). This standard provides a reliable framework enabling interoperability and seamless handovers between networks of the same type as well as handoff between different network types, also known as Media Independent Handover (MIH). The scope of the IEEE 802.21 MIH is to develop a standard that could provide link layer intelligence and other network related information to upper layers to optimize handoffs between heterogeneous media. The main novelty, that the IEEE 802.21 standard proposes, is related with the MIH function that is defined as a shim layer between the L2 data link layer and the L3 network layer. Three types of MIH services are defined (event, command and information) that facilitate the mobility management and handover process in heterogeneous networks (Lim, Kim, Suh, & Won, 2009). Finally, the IEEE 802.21 standard deployment will enable innovative services in broadcasting and telecommunications convergence.

Another issue is how to manage the handoff triggering time. The delay of vertical handoffs is usually longer than that of horizontal handoffs, since an authentication procedure may be necessary when a terminal crosses different networks. The handoff latency is longer when loosely coupled integration is employed, though this is a more feasible solution. When an upward handoff from a smaller cell to a larger cell occurs, the host loses more data packets as the handoff delay increases while moving out of the coverage of the previous cell. When a downward handoff occurs, a host can still receive data packets since it is still in the larger cell. The Internet Engineering Task Force (IETF) Network Working Group proposed some recent and appropriate mobility management protocols such as the Hierarchical Mobile IPv6 Mobility Management (HMIPv6) and Fast Handovers for Mobile IPv6 (FMIPv6) protocols. The HMIPv6 protocol (Soliman, Castellucia, Malki & Bellier, 2005) is the enhancement of Mobile Internet Protocol version 6 (MIPv6) designed to reduce the amount of signaling required and to improve the handoff speed of mobile connections. The HMIPv6 concept is simply an extension to the MIPv6 protocol to handle local and global mobility separately. The MIPv6 handles global mobility and local mobility with the same mechanisms, causing inefficient use of resources in the case of global mobility. In HMIPv6, global mobility is managed by the MIPv6 protocols, while local handoffs are managed separately and locally. In HMIPv6, a new function called the Mobility Anchor Point (MAP) is introduced serving as local entity to aid in mobile handoffs. Just like the MIPv6 this solution allows mobility within or between different access technologies. MIPv6 involves a lot of signaling and processing for handoffs, requiring a lot of resources. The MAP in HMIPv6 helps to decrease handoff-related latency, providing seamless handoffs as local MAP can be updated more quickly than a remote agent. The FMIPv6 protocol aims at reducing the handover latency (Koodli, 2005, Koodli, 2008). This improved performance is beneficial to every supported type of applications. Through this protocol the problem of handoff latency is addressed adequately.

Best Connectivity Problem

Another major challenge is to obtain the best connectivity. It may be formulated as a complex optimization problem where several issues must be dealt with (Bari & Leung, 2007). The analysis is done from the point of view of every participant examining the cost and benefits of each access solution. The following normalized cost function is used and should be minimized.

$$f(n) = w_b \cdot \ln \frac{1}{B_n} + w_e \cdot \ln E_n + w_c \cdot \ln C_n, \quad (1)$$

where B_n is the bandwidth of the n network type, E_n is the network power consumption, C_n is the cost of network and w_b, w_e, w_c are weights assigned to each parameter $\left(\sum w_i = 1\right)$ (Wang, Katz & Giese, 1999). The terms of the cost function are independent of each other. Considering, an IP-network the network cost C_n will be a strong function of the requested and offered quality to the users and the available bandwidth. The bandwidth parameter estimates the current network condition. Power consumption and cost are parameters with fixed values; namely, the users terminal battery life and the maximum amount of money the user is willing to spend for a period of time, respectively. Following this approach the goals are: a) to maximize the level of perceived QoS and b) to minimize the cost provided that certain QoS constraints are satisfied. The cost may be evaluated by using the cost function defined above always comparing competitive networks (Wang, Katz & Giese, 1999). Then, an appropriate utility function is used to determine the best offer according to price and quality:

$$U = \left[a\left(q - \bar{q}\right) + \left(1 - a\right)\left(\bar{p} - p\right)\right]\Theta\left(\bar{p} - p\right)\Theta\left(q - \bar{q}\right) \quad (2)$$

$$\text{with } \begin{cases} \Theta(x) = 1 \;\; if \;\; x > 0 \\ \Theta(x) = 0 \;\; else \end{cases}, \quad (3)$$

where q is the evaluated quality value, $\bar{q}$ is the minimum quality that a user is willing to accept, p is the price of the network connection, $\bar{p}$ is the maximum price that a user is willing to accept and a is a parameter between 0 and 1 which indicates the users sensitivity between quality and cost. For each access network, through the application of the utility function, the network providing the maximum utility value u is selected. Otherwise, another network selection mechanism for 4G wireless networks is proposed using gray relation analysis to decide which network should be used by every terminal based on users preferences, application requirements, and network conditions without considering the bandwidth allocation problem (Song & Jamalipour, 2005).

Different Access Technologies Inter-Operability

Driven by the "service anywhere and anytime" concept, it is well accepted that 4G wireless networks will be heterogeneous, integrating different networks to provide seamless access for mobile users with multimode access capability. The cellular networks and WLANs deployed will be both included along with other access networks. One major challenge is to achieve inter-working of the cellular and WLAN in order to exploit their advantages and their unique features (Shi, Shen, & Mark, 2004; Song, Jiang, Zhuang & Shen, 2005;

Figure 2. Tight coupling architecture

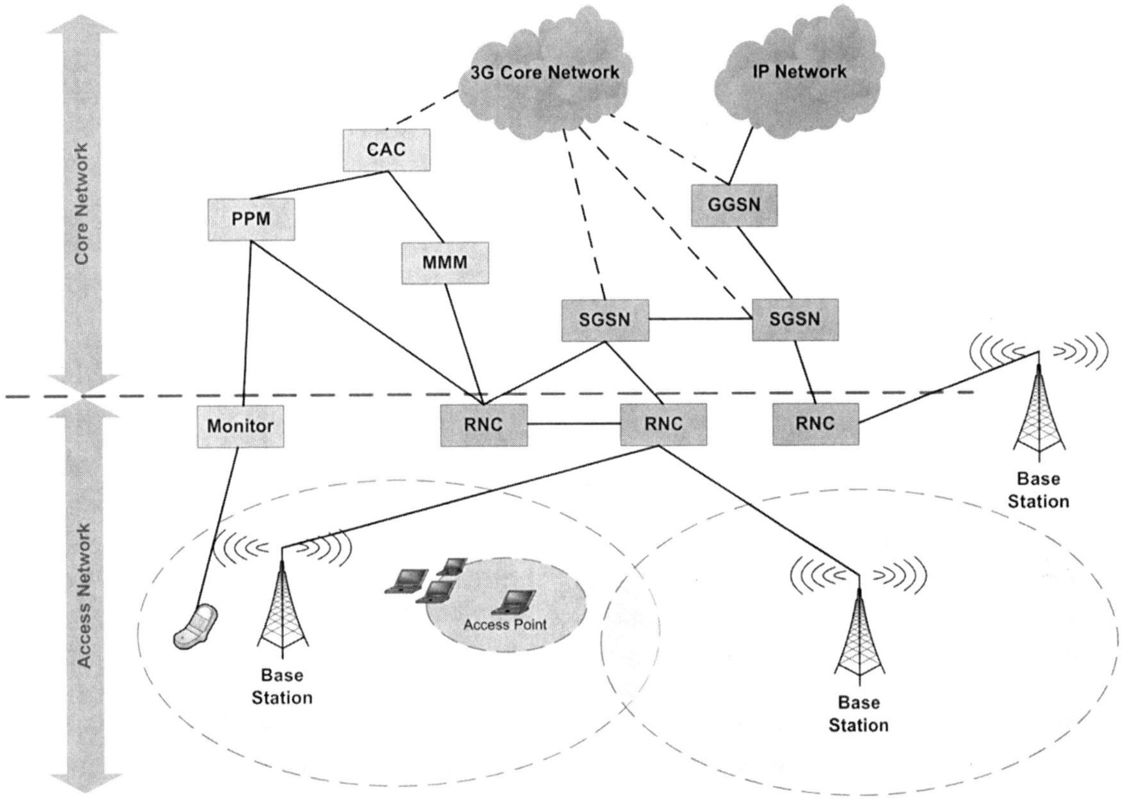

Song & Zhuang, 2005). Specifically, WLANs systems provide very high data rates but less mobility while on the other hand; cellular networks provide wide mobile coverage and universal roaming support. Their combination will result in a network providing very high data rates and mobility support. Based on the correlation between the two access networks, the main inter-working architectures can be classified into two categories: tight and loose coupling architectures.

Tight Coupling Architecture

In tight coupling architecture the WLAN is connected to the cellular core network as a cellular radio access network. In other words the WLAN is embedded into the cellular core network as shown in Figure 2. For example, the integration point of WLANs to a Universal Mobile Telecommunications System (UMTS) core network can be the serving GPRS support node (SGSN) (Salkintzis, Fors & Pazhyannur, 2002) or the gateway GPRS support node (GGSN) (Buddhikot, Chandranmenon, Han, Lee, Miller & Salgareli, 2003). Otherwise, the integration can be done by connecting the access point to the Radio Network Controller (RNC). A user roaming across the two domains is based on the mobility management protocols of the cellular networks, thus enhancing the inter-domain mobility management capability.

The main drawbacks of the tight coupling approach are (Song, Jiang, Zhuang & Shen 2005):

- An interface in the cellular core network exposed to WLANs is required, which is a challenge by itself since it is likely that the two domains are developed and deployed independently.

- A large volume of WLAN traffic will go through the cellular core network, rendering it to a network bottleneck.
- WLANs must have a protocol stack compatible with that of cellular networks. The induced complexity and cost may hamper the deployment of a tight coupling architecture.

A QoS model to support tight coupling architecture was proposed consisting of the following components: the Policy Provisioning Module (PPM) responsible for user mapping by deciding which traffic class a user belongs to corresponding to different priority level and then it handles the users request to the connection admission control module for further processing of the request as to allow the traffic or not, the connection admission control module that manages the traffic flows admitting the number of flows that can be served and allocates bandwidth to them maintaining QoS requirements of existing connections, the QoS Mobility Management Module (MMM) that inspects the terminals condition (connected, idle, disconnected) and those active nodes moving in high speed and the QoS monitoring module that monitors the satisfaction of QoS metrics providing the appropriate feedback (Wang, Mellor & Al-Begain, 2004). The connection admission control module first receives a connection request from the PPM and then it consults the MMM to know about the mobility status and the information about nodes. Then, it uses reservation protocols such as Resource Reservation Protocol (RSVP) to reserve connections.

The RSVP, described by Braden, Zhang, Berson, Herzog & Jamin, (1994), is a transport layer protocol designed to reserve resources across a network. RSVP provides a general facility for creating and maintaining distributed reservation state across a mesh of multicast and unicast delivery paths. RSVP can be used by either hosts or routers to request or provide specific levels of QoS for application data streams. It defines how applications place reservations and how they can relinquish the reserved resources once the need for them has ended. Nowadays, RSVP by itself is rarely deployed in telecommunications networks as RSVP Traffic Engineering (RSVP-TE), is becoming more widely accepted in many QoS-oriented networks (Awduche, Berger, Gan, Li, Srinivasan, & Swallow, 2001). RSVP-TE protocol is an addition to the RSVP protocol for establishing Label Switched Paths (LSPs) in Multi Protocol Label Switching (MPLS) networks taking into account network constraints such as the available bandwidth, signaling and the processing overhead (Lee, Kim, Park, & Kim, 2007). The RSVP-TE supports the instantiation of explicitly routed LSPs, with or without resource reservations. It is used as a general facility for creating and maintaining distributed forwarding and reservation state across a mesh of delivery paths. RSVP-TE transfer and manipulate traffic engineering control parameters as opaque data passing them to the appropriate module for interpretation. It also supports smooth rerouting of LSPs, preemption, and loop detection. The IETF working group, after 2003, concentrated purely on RSVP-TE abandoning other inefficient protocols.

Loose Coupling Architecture

In loose coupling architecture, the gateway directly connects the WLANs to the Internet backbone. There is no direct link between the WLANs and the cellular core network (Buddhikot, Chandranmenon, Han, Lee, Miller & Salgareli 2003). The loose coupling architecture is presented in Figure 3. The main advantage of this approach is the independent deployment of the two access networks and that each one will not be a bottleneck to the other. The most serious disadvantages is that roaming between networks is difficult; as the two domains are distinct, the handoff signal-

Figure 3. Loose coupling architecture

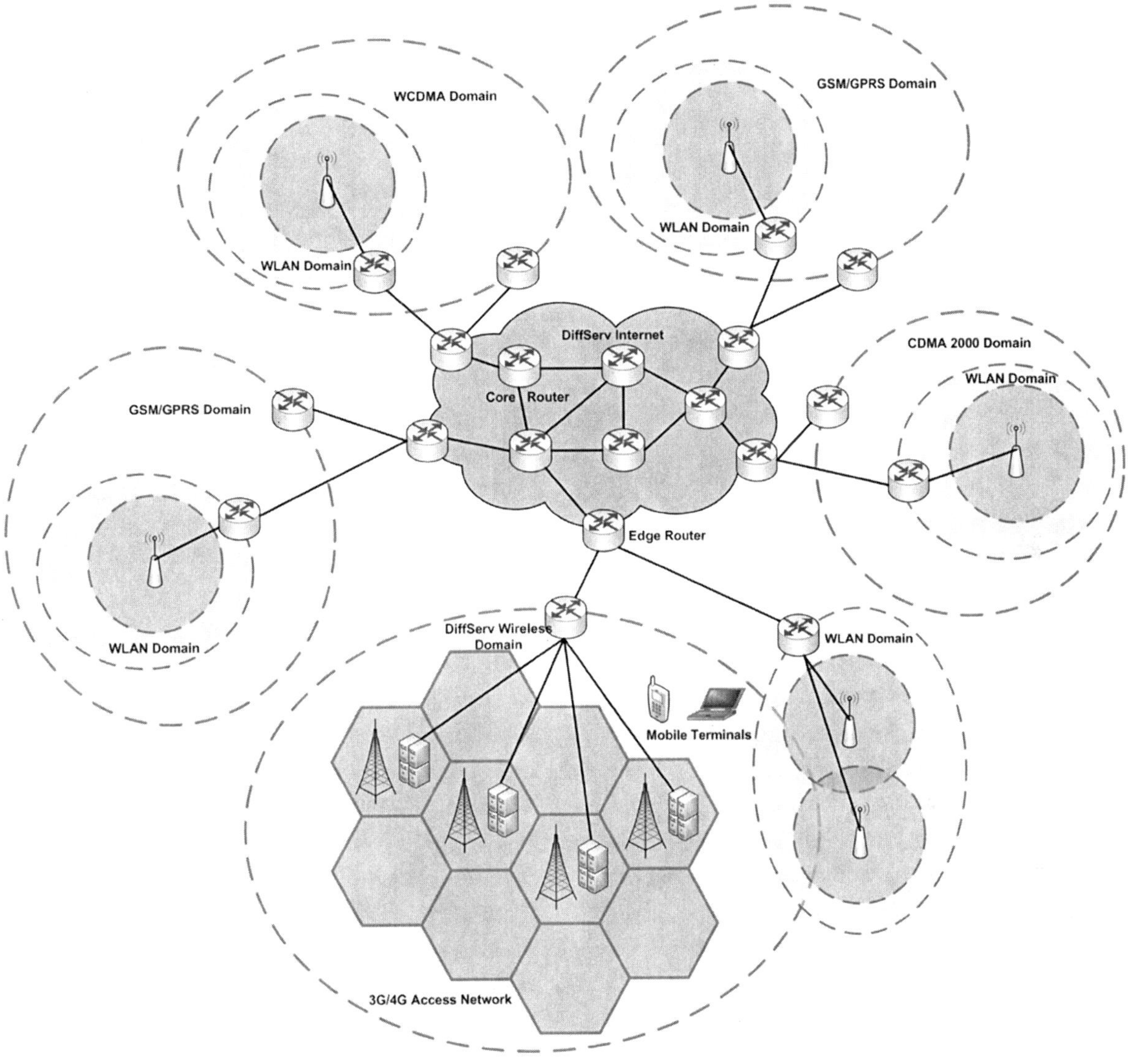

ling related to mobility may traverse a long path, causing high handoff delay.

To interconnect heterogeneous IP-based wireless access networks with the Internet backbone in 4G networks, it is well recognized that an all-IP DiffServ platform is most suitable (Moon & Aghvami, 2003). Loose coupling architecture for heterogeneous inter-working can cooperate perfectly with the DiffServ platform and the service models constituting a robust architecture for seamless broadband access (Cheng, Jiang, Zhuang, Niu & Lin 2005). The main reasons are:

- DiffServ is a scalable mechanism based on a limited number of SCs, as no per-flow processing is needed in core networks.
- Service mapping between the inter-working wireless access networks is reliable and the service models of the same categories have similar QoS requirements.

- The DiffServ platform adopts a domain-based architecture where each domain independently choose its own system mechanisms as long as its SLAs with neighboring domains are satisfied (Moon & Aghvami, 2003). Such a domain-based architecture allows flexibility and convenience in deploying each domain independently, and developing, modifying, or exchanging the techniques in a domain without a significant effect on the overall system. It fits very well with the loose coupling architecture where the inter-working networks are considered independent.
- A fast handoff procedure is required for seamless roaming in and among wireless access networks. The popular solution for fast handoff is to use Mobile IP for inter-domain (macro-) mobility and to use micro-mobility protocols for intra-domain mobility. Micro-mobility protocols can be seamlessly incorporated into the domain-based DiffServ platform.

Therefore, the loosely coupled inter-working of cellular networks and WLANs, implemented in a domain-based DiffServ platform, matches well with the evolution toward an all-IP 4G infrastructure.

Network Services Models

The introduction of new services with different QoS requirements requires the description of their specifications and their incorporation into network service models. The QoS requirements concerning the bandwidth and the tolerable delay of various types of multimedia traffic vary significantly over a wide range of values. For example, voice traffic requires a bandwidth of only 10-20 kb/s whereas high quality video may require 500-1000 kb/s. On the other hand, voice and video conference traffic are delay intolerable, whereas internet applications like file transfer can tolerate delay. As previously mentioned it is essential that service interoperability among different access technologies must be provided in 4G networks. This motivates to efficiently map the SCs onto the main wireless access technologies and the IP network core to achieve interoperability and efficient network operation.

Among the various service models and mechanisms proposed by the IETF concerning the IP part of a network, two approaches, namely Integrated Services (IntServ) and Differentiated Services (DiffServ), are the prevailing solutions (Niyato & Hossain, 2005). On the other hand, 3GPP described analytically the traffic classes for the existing 3G wireless mobile network. The use of these models proposed for network services can establish interoperability between 4G networks and the IP architecture.

Integrated Services (IntServ)

Aiming at providing end-to-end QoS, the IntServ architecture is characteristic of resource reservation. The IntServ framework for QoS was introduced in 1994 as an attempt to provide different types of traffic service in the same network (Braden, Clark & Shenker, 1994). After its introduction, several upgrades have been made. IntServ uses admission control, rate control and basic resource reservation mechanisms to deliver QoS. The framework takes into consideration the different QoS requirements employing service differentiation and resource reservation. The application must set up paths and reserve resources before data transmission. The IntServ approach is based on per-flow service provision, which informs the routers along an end-to-end route about the resources required by each flow. The architecture uses an explicit setup mechanism, such as the RSVP, to convey information to the routers so that they can provide requested resources to the flows (Braden, Zhang, Berson, Herzog & Jamin, 1994). The RSVP by itself is rarely deployed, since the RSVP-TE is becoming more widely accepted nowadays (Awduche, Berger, Gan, Li, Srinivasan, & Swallow, 2001). The original IntServ framework did not specify the SCs supported. In 1997, the Integrated Services (IntServ) Working Group of the IETF specified three types of service, namely, the Guaranteed Service (GS), the Controlled-Load service (CL) and the Best-Effort (BE) service, with distinct QoS requirements and specifications (Wroclawski, 1997; Shenker, Patridge & Guerin, 1997). These three SCs have different characteristics, concerning mainly the tolerance to delay and the various applications supported must be mapped onto these three SCs defined by IntServ.

The GS guarantees that the packets will arrive within the guaranteed delivery time and that they will not be discarded due to buffer overflows, provided that the flow traffic conforms to its specified traffic parameters. This SC is used for Delay Non-Tolerant (DNT) applications which require a hard guarantee that a packet will arrive not later than a certain time period after its transmission. That is, GS does not control the minimum or average delay of a packet but it merely limits the maximum queuing delay; hence, jitter is not controlled. Examples that have hard real-time requirements and require guaranteed service include certain multimedia broadband applications. The delay consists of two components, namely fixed delay and queuing delay. Fixed delay is dependent on the transmission path which is not determined by the guaranteed service which determines only the queuing delay.

CL service is intended to support numerous applications which have been developed for the Internet and are sensitive to heavy load conditions. To this class belong adaptive real-time applications offered by a number of vendors. These applications have been shown to work well in a lightly-loaded Internet environment but to degrade under heavy load conditions. The CL service does not specify any target QoS parameters. Instead, acceptance of a request for CL service implies a commitment by the network to provide a service closely approximating the QoS, which the same requested flow would achieve under light load conditions. Both GS and CL service are designed to support real-time applications which require different QoS guarantees by the network.

The BE SC offers the same type of service under the current Internet architecture. BE SC, do not include any kind of quality control notification or negotiation. That is, the network makes an effort to deliver data packets without guarantees. A service will be provided as long as the supported load and congestion is kept under certain limits. This type of service is sufficient for Non-Real-Time (NRT) applications which may use an end – to – end retransmission strategy (i.e., TCP) to ensure that all packets are delivered correctly. Among these NRT applications most popular are FTP, e-mail, internet browsing. All of these applications can work without guarantees concerning the timely delivery of data packets. Another term for such NRT applications is elastic, since they are able to stretch gracefully in case of increased delay. It must be noted that these applications may benefit from shorter routes but that they do not become useless as delay increases.

The implementation of the IntServ network models and its integration to 4G networks exhibits scalability problems. To overcome such problems, IETF proposed the DiffServ framework.

Differentiated Services (DiffServ)

The SCs supported by 4G networks can be mapped to those used in Differentiated Services (DiffServ), enabling a seamless integration of wireless networks with the IP-based architecture. DiffServ is one of the key network service models for QoS provision in the Internet, also employed in NGNs. DiffServ framework classifies the various network services differentiating them on the basis of users performance. The DiffServ architecture is based on a relative-priority scheme, which maps packets to predefined SCs. This is accomplished via DiffServ Code Point (DSCP) which is incorporated into the header of each IP packet. Packets belonging to different SCs are assigned different levels of QoS. If properly designed, the DiffServ architecture can offer great flexibility and scalability along with satisfaction of service requirements for multimedia streaming applications. Instead of providing QoS on a per-flow basis as in the IntServ model, DiffServ deals with group of flows by aggregating several IP-flows of the same QoS into the same group. The DiffServ routers identify the group an IP packet belongs to through the service type field in the IP packet header. Every group of flows with the same QoS requirements is characterized by the same DSCP in the IP header informing the routers about the resources required. The application of this traffic management scheme accomplishes the satisfaction of QoS requirements.

In the DiffServ model three groups of traffic services are provided: premium or expedited forwarding (EF), assured forwarding (AF) and best effort or default forwarding (DF). Through appropriate packet aggregation, the flow receives a particular forwarding treatment, Per-Hop-Behaviour (PHB), at each network node. To assure QoS, privileged treatment in buffer management and scheduling is reserved for flows belonging to the EF or AF classes (Jacobson, Nichols & Poduri 1999; Heinanen, Baker, Weiss & Wroclawski, 1999). Particularly, the IETF DiffServ working group has specified the AF PHB which intends to provide different levels of forwarding for IP packets at a node; hence AF PHB can be used to treat multiple priority SCs.

As mentioned above, the deployment of 4G heterogeneous networks is based on the interoperability between mobile 3G networks and IP networks either wired or wireless. The QoS classes of 3G networks, defined in the technical specifications of 3GPP, are four, namely the conversational, streaming, interactive and background classes (3GPP 23.107, 2003). The distinguishing characteristic of these SCs is the sensitivity of the traffic with respect to delay. For example, the conversational SC includes delay-sensitive traffic, whereas the background SC is the less delay-sensitive class. Each of the above SCs can be mapped into the respective SCs of the DiffServ platform, unifying service classification in 4G networks.

Conversational and streaming classes are intended for real-time traffic. The main difference between them is the degree of sensitivity with regard to the traffic delay. Conversational real-time traffic is the most delay-sensitive traffic. Voice telephony, voice over IP, and video conferencing belong to this traffic class. The maximum tolerable transfer delay is determined by the human perception of video and audio conversation and is very low and strict. Streaming class traffic is related to a user watching a real-time video or listening to real-time audio. This traffic is asymmetric and is arranged so that the intermediate time between information entities such as samples or packets is preserved (3GPP 23.107, 2003). Thus, conversational and streaming calls can be mapped to the expedited and assured forwarding SCs, respectively.

EF is a SC mainly used to describe low packet loss ratio, low-delay and low-jitter services. This SC is quite similar to the Constant Bit Rate (CBR) service in an Asynchronous Transfer Mode (ATM) network and can be used to build an enhanced best-effort service: traffic is subjected to losses due to transmission errors and to reordering caused by routing changes. However, through adequate queuing techniques, delay or delay variation is minimized. Therefore, the EF SC is generally used to carry, through the IP network, voice and data requiring high QoS standards similar to that of wired networks. AF is a SC that provides less strict guarantees as to PLR, delay and jitter. It is intended for networks that offer average SLAs, where traffic is expected to be elastic. In AF SC the receiver will detect losses and delay variations, providing proper feedback to the sender. Based on this feedback the sender adjusts its transmission rate to the available capacity.

Table 1. Service class mapping for interoperability in 4G networks

3GPP Classes	IntServ SCs	DiffServ SCs	Characteristics	Applications
Conversational	Guaranteed Service	Expedited (Premium Forwarding)	Intolerant Real Time Preserve time relation between entities Stringent low delay Minimize delay variation	Voice Calls Video Conference
Streaming	Controlled-Load	Assured Forwarding	Tolerant Real Time Minimize delay variation Preserve time relation between entities	Audio Streaming Video Streaming
Interactive	Best-Effort	Best Effort (Default Forwarding)	Non Real Time Minimize bit error rate Request response pattern Preserve payload content	Web browsing WAP browsing Server Access Data Bases
Background	Best-Effort	Best Effort (Default Forwarding)	Non Real Time Preserve payload content Minimize bit error rate Less delay sensitive than the interactive class	E-mail File Download

The interactive and background classes include traditional Internet applications such the World Wide Web (WWW) browsing, e-mail, telnet, File Transfer Protocol (FTP). These classes are less delay-sensitive compared to the conversational and streaming classes, but their content should be delivered with low error rate by means of channel coding and retransmission. The main difference between the interactive and the background class is that the first is intended for interactive applications such as web browsing, data base retrieval, or server access, whereas the second is intended for background traffic such as e-mail or file downloading. The round trip time is one of the key attributes of the interactive class. Therefore, traffic belonging to this class has higher scheduling priority compared to background class traffic; hence, background class traffic occupying system resources only when interactive applications do not need them. In background applications the destination does not expect data within a certain time period. Thus, interactive and background classes can both be mapped to best effort SCs with distinct priorities. Table 1 summarizes the above SC mapping with the corresponding SCs defined by 3GPP providing interoperability of IP-architecture in beyond 3G wireless mobile networks and the wired network part of a 4G network, satisfying the specific QoS demands of every supported SC.

The main difference between IntServ and DiffServ is their different treatment of packet streams. IntServ emphasizes on guaranteeing QoS on a per-flow basis requiring explicit signalling to reserve network resources along the end-to-end transmission. Not being able to monitor and process on a per-flow basis possibly millions of flow states, IntServ exhibits scalability problems. On the other hand, instead of focusing on per-flow treatment, DiffServ prioritizes the flows on an aggregate basis, i.e., a set of micro-flows with similar service requirements are treated equally based on the codepoint. Through appropriate distinct treatment of a limited number of SCs, the performance of the aggregated flow is guaranteed based on its PHB. However, DiffServ applies only to large scale networks and consequently, it cannot provide sufficient end-to-end QoS.

As a compromise between IntServ and DiffServ, a framework combining them has been proposed by IETF, where the DiffServ domains are viewed as a network element in the total IntServ end-to-end path (Bernet, et al, 2000). The idea behind this approach is to combine the distinct benefits of each framework, namely employ IntServ when possible (providing hard guarantees) and DiffServ where it is not possible to reserve resources, as for example in the backbone area. Based on this combined approach, a promising framework for end-to-end QoS could be created by embodying IntServ in the edge clouds and DiffServ in the core network.

The above basic service classification, provided by IETF for IP networks incorporated in 4G networks, is just the starting point for different research areas and for the development of a wide set of mechanisms and protocols oriented to classify, differentiate or provide a special treatment for each data flow.

DYNAMIC RESOURCE ALLOCATION AND CALL ADMISSION CONTROL

Dynamic Resource Allocation Schemes

Bandwidth Reservation Techniques

Despite the high data rate provided by the beyond 3G wireless networks, bandwidth allocation is still a major issue for most real-time multimedia services. Due to diverse QoS required by mobile users and to the dynamic nature of the wireless channel, adaptive bandwidth allocation is necessary to improve resource utilization. CAC schemes should be designed aiming at maximizing resource utilization while guaranteeing QoS. In most cases, bandwidth is reserved for the exclusive use of handoff calls while the rest is shared by new and handoff calls. The key factor that influences the performance of such schemes is how to determine the amount of bandwidth to reserve. If the reserved bandwidth is not sufficient, the QoS requirement concerning handoff CDP cannot be met. On the other hand, if the reserved bandwidth is unnecessarily high, a large number of new call requests will be rejected deteriorating the QoS concerning the CBP of new calls. Furthermore, if the reserved bandwidth remains out of use for long waste of the limited radio resources is observed. Fixed reservation schemes, set the reservation level during system design according to estimates of the area traffic and to the distribution of channel occupancy time (Hong & Rappaport, 1986; Kulavaratharasah & Aghvami, 1999). The reservation level remains fixed for each cell. However, since users are mobile, the configuration of the wireless network changes with time, resulting in over-reservation and/or under-reservation of bandwidth when the actual cell traffic diverges from its a priori estimation. Therefore, these schemes are not suitable for the demanding environment of NGNs such as 4G. Dynamic reservation schemes that make reservation adaptive to the changes of traffic conditions are designed to deal with the appropriate selection of the reservation level (Oliveira, Kim & Suda, 1998; Naghshineh & Schwartz, 1996) Employing adaptive bandwidth allocation, the quality of a call can be upgraded by assigning more resources under low traffic conditions. However, under heavy traffic, the additional bandwidth allocated to some ongoing calls will be taken back to accommodate more incoming calls, so that CDP and CBP are kept within the QoS requirements.

Since a dominant feature of 4G networks is mobility, the respective resource management process is much more complicated than that of fixed wireless networks. None of the previous schemes fully exploited the position/mobility information about mobile stations. Since handoffs are also due to users mobility, the bandwidth reservation strategy must take mobility into consideration. One of the major advantages provided by mobility information over traditional CAC schemes is "channel-reservation". The concept is simple: if a cell is informed that a channel will be required at a precise time instant, the cell can do its best to ensure that this channel will be available. Levine, Akyldiz, & Naghshineh (1997) suggested the incorporation of users mobility into the reservation making process. In this approach, it is the statistical information concerning mobility (such as the probability density function of the call residence time) instead of the real-time parameters (such as speed and position) of each user that is used to implement the reservation procedure. Reservation is made in both the immediate neighbouring cells and the next to them cells. Frequent changes of the users mobility pattern will result in inactive reservations in many cells. However, because channel-reservation is a waste of resource (a reserved-channel remains idle until the arrival of a call), an efficient reservation-based scheme should make the reservation at the latest possible moment. Therefore, it is necessary to have the most accurate possible knowledge about the call arrival time, and consequently, when the reservation must be made.

The study of the power received from different base stations and the analysis of the user population mobility pattern are many different ways of obtaining user behaviour information studied in the literature. In fact, when a mobile user moves, the transmission path between the source and the destination changes. If the new cell of a mobile is overloaded, the resources available may not be sufficient to guarantee its QoS. Consequently, the service might be interrupted. In certain cases when services with strict QoS specifications are required that should not be affected by users mobility, resource reservation must be performed at each cell the mobile user moves into. In some other cases, the mobile user may accept service with flexible QoS requirements specifying that the acceptable QoS ranges from a minimum to a maximum level. Then, the system may try to reallocate the resources so that the QoS requirements of all the mobile users are satisfied in the new cell. Ye, Hou & Papavassiliou (2002) proposed an integrated framework for bandwidth management based on mobile agents The Predictive Mobility - Based Bandwidth Reservation scheme (PMBBR) is introduced to overcome the problems of existing reservation schemes. The scheme under consideration supports multiple SCs that require various QoS levels including strict, flexible and soft QoS requirements. Adaptive bandwidth reservation is performed in the cell a mobile user goes into in order to support seamless handoff. User mobility is taken into account by the reservation scheme so that the efficiency of handoff mechanisms is optimized and the unnecessary reservation of resources is kept to a minimum. Hence, a considerable improvement in system capacity and throughput is achieved.

Bandwidth Degradation Techniques

The bandwidth reconfiguration processes intend to allow the efficient resource redistribution in a cell so that the QoS requirements of all the mobile users are satisfied, especially when users with flexible QoS requirements are supported. Degrading ongoing calls (or sessions) to yield bandwidth for incoming calls has been proposed in the literature to achieve better performance in saturated networks with regard to the call dropping and the call blocking rates. Thus, a system may release a certain amount of bandwidth for new users by lowering the QoS levels of ongoing users if possible. The bandwidth utilization or the service provider's revenues can be significantly improved by allowing QoS degradation. However, the impact of QoS degradation to individual users, which is a crucial characteristic of QoS was overlooked in most of the literature. For example, though users can tolerate quality degradation to a certain extent, it is still preferable to give them higher QoS when more resources become available. However, an important consequence of service degradation, namely the users behaviour with regard to the degraded service, has received relatively little attention. Though, the controlled degradation of ongoing calls enables the system

Figure 4. Trade-off between QoS degradation and user satisfaction

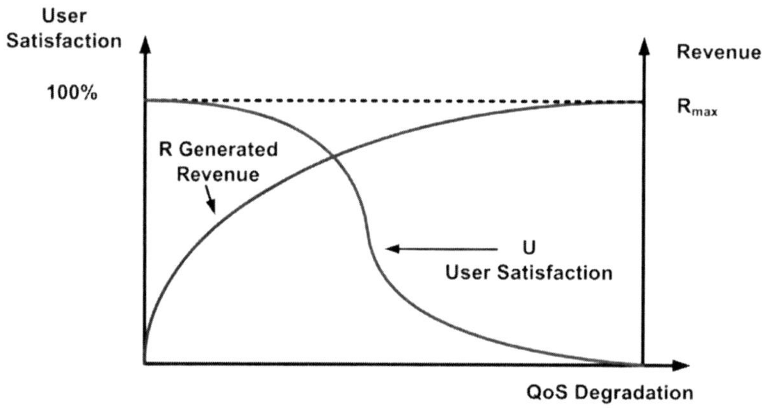

to carry through more traffic and, hence, increase its revenues, reducing the users satisfaction on an individual basis, as depicted in Figure 4. This might result in eventual revenue losses for the service provider.

Figure 4 illustrates this trade-off relationship, where the system revenue *R* increases as QoS decreases whereas the users satisfaction decreases. Thus, performance metrics reflecting the average QoS level should be considered. Of interest are quality-degradable connections as long as their resultant quality remains within the QoS range specified by the user. Assuming *k* QoS levels, the bandwidth requirement of the *ith* quality level is denoted as BW_i with $BW_{max}=BW_l>BW_i>BW_k=BW_{min}$. A base station may try to degrade the QoS of some ongoing users in order to admit more users and improve the overall system performance. For example, it may be possible to achieve high bandwidth utilization and, at the same time, maintain a small CBP. In systems where QoS may significantly degrade, a user may receive different QoS levels during the connection duration, depending on the cell load. Even if a user receives the highest QoS level when admitted to a cell, his QoS may still be degraded when some base stations encountered across his course decide to degrade his QoS in order to admit more users.

To this end, two widely accepted performance metrics are introduced (Chou & Shin, 2004):

- *The Degradation Ratio (DR):* defined as the fraction of time during which a user receives degraded QoS. For a multilevel QoS network DR is given by

$$DR = \frac{\sum_i \frac{\left(BW_{\max} - BW\right)_i}{BW_{\max}} \cdot T_i}{\sum_i T_i}, \tag{4}$$

if a user receives QoS level i for T_i.

- *The Upgrade/Degrade Frequency (UDF):* The frequency an ongoing user changes his QoS level

A bandwidth degradation framework attempts to estimate the impact of users behaviour on system revenues, formulating an effective revenue function to determine the optimal combined bandwidth degradation and admission policy that maximizes networks revenues (Das, Sen, Basu & Lin, 2003). The proposed framework aims at making an appropriate policy for CAC and resource sharing through combined modelling of high traffic and guaranteed QoS. These are two opposite goals since an improvement of either of them leads to the degradation of the other. A cost function is formulated to describe the total system revenue offered by a particular bandwidth degradation policy. A key point is how to determine an optimal policy maximizing the net revenue under various scenarios. The relevant simulations demonstrate that the system performance is highly sensitive to the proper choice of the degradation cost and revenue earned by every call admitted. Therefore, the system should be properly adjusted for maximum benefit. The framework is extended to a CAC scheme and since real-time traffic has a pre-emptive priority over non-real-time, a channel sharing scheme employing a Markov Modulated Poisson Process (MMPP)-based queuing model is proposed where average queue length is derived as a QoS metric (Das, Sen, Basu & Lin, 2003).

Dynamic Allocation Schemes Using QoS Renegotiation

QoS renegotiation upgrades the operation of dynamic allocation schemes. In the previous subsection only adaptive techniques concerning mainly bandwidth degradation of existing calls were described, aiming at temporarily increasing the network capacity. The main difference of QoS renegotiation compared to the previous techniques is that a continuous adaptive mechanism is adopted performing bandwidth adaptation either to degrade or to upgrade it depending on network conditions. Bandwidth degradation applies when a new or a handoff call arrives at an overloaded cell. According to the protocol, the bandwidth adaptation algorithm reallocates the bandwidth of existing calls to allocate the necessary bandwidth to the arriving call. The algorithm may extend the bandwidth of calls with lower bandwidth when an outgoing handoff call or a call completion takes place.

QoS renegotiation can be embedded as an independent module in various types of QoS provision and radio resource management schemes. Different CAC strategies have been introduced in the literature employing QoS renegotiation providing traffic regulation and QoS. QoS renegotiation has emerged as a promising solution for reliable CAC in heterogeneous networks supporting multiple SCs. In such networks, QoS renegotiation combined with prioritization schemes is employed for dynamic CAC. Therefore, in a system supporting multiple SCs, some calls are more likely to be upgraded or degraded. Both alternatives make certain assumptions concerning the hierarchy of upgrade or degrade assuring fairness and maximum resource utilization:

- Whenever an opportunity for upgrade exists, calls of higher priority are first upgraded.
- Among calls of the same SC, the call with the lowest QoS level will first be upgraded to the next higher one.
- Among calls of the same SC with the same QoS level, one will be randomly chosen to upgrade to a higher QoS level.
- In case of degradation, calls of lower priority will be degraded first.
- Among calls of the same SC, the one with the highest QoS level will be degraded to a lower QoS level first.
- Among calls of the same SC with the same QoS level, one will be randomly chosen to degrade to a lower QoS level if applicable (Li & Chao, 2007).

One of the most popular call characteristics that are renegotiated is the bandwidth necessary for each application support. Different call types have different bandwidth requirements; call admission strategies employing bandwidth renegotiation can meet the resource demands more efficiently, achieving high bandwidth utilization. QoS characteristics and several requirements like resource allocation and bandwidth utilization constitute the inputs for the renegotiation module. The renegotiation mechanism also prioritizes delay sensitive real time traffic over non delay sensitive. The renegotiation scheme utilizes unused resources reserved for high priority SCs and reallocates them to flows of lower priority that, at the time of admission, were assigned less than actually requested. In other words, bandwidth renegotiation is applied when available resources exist, aiming at increasing the overall bandwidth oriented for lower priority SCs while preserving the bandwidth oriented for higher priority SCs. Through the renegotiation technique, fairness among users of different SCs can be improved, assuring also high resource utilization and throughput. Simple CAC assigns different priorities to different QoS classes allocating bandwidth according to the class of the requesting application. CAC schemes employed, aim mainly at maximizing the number of admitted calls while satisfying the resource requirements. The renegotiation module introduces a dynamic character to CAC while it guarantees QoS. The various CAC schemes can also be extended to dynamically reallocate the bandwidth released after call terminations. The reallocation of resources allows the traffic accepted at lower bandwidth levels than requested to upgrade their transmission rate. The renegotiation can be based on various metrics concerning bandwidth, such as the minimum and maximum requirements or the average used bandwidth (Tragos, Tsiropoulos, Karetsos & Kyriazakos, 2008; Monego, Bodanese, Nacamura & Souza, 2005).

In dynamic CAC, real time DNT applications corresponding to the conversational class are assigned maximum priority. Real time Delay-Tolerant (DT) calls like streaming applications are assigned intermediate priority. These classes of maximum and intermediate priority will be admitted only if there is sufficient bandwidth to satisfy their requirements. If not, their requests will not be admitted. The lowest priority is assigned to non real time applications corresponding to the 3GPP interactive and background classes. Requests belonging to these classes are admitted even when the network can afford less bandwidth than requested. In simple CAC schemes that do not employ QoS renegotiation, when bandwidth has been allocated to calls of maximum and intermediate priority it cannot be transferred to a class of lower priority even when its freed after the termination of higher priority class call, resulting in a significant waste of resource. In a CAC scheme employing QoS renegotiation, a call of low priority can use more bandwidth that what was originally allocated. This is made possible because either unused resources destined to classes of higher priority or resources from terminated calls can be transferred to this class. When a new call of high priority arrives, the renegotiation mechanism can reduce the extra bandwidth allocated to low priority applications to the original bandwidth allocated to them by the admission control module. By taking back the extra bandwidth from low priority SCs and allocating it back to high priority SCs these SCs are not harmed. The renegotiation module just described employs efficiently the network resources to guarantee QoS, improve fairness and provide enhanced users satisfaction when possible.

Game Theory in Resource Allocation

One of the prevailing solutions for resource management and admission control with dynamic resource allocation in NGN networks is based on employing game theory. A cooperative game framework was proposed for bandwidth allocation in 4G heterogeneous wireless access networks formulating the bandwidth allocation problem as a cooperative game and the solution, namely, the amount of bandwidth offered to a new connection, obtained from the Shapley value (Niyato & Hossain, 2006). This formulation is different from a non-cooperative approach where each network is a rational and selfish player aiming at maximizing its own profit. In a non-cooperative environment, the players (users or networks) compete to achieve their objectives. In a cooperative approach, groups of players (users or networks) seek fair resource allocation whereas in a non-cooperative approach, allocation is done based on the individual users payoffs (bandwidth). A game-theoretic RRM framework following a non-cooperative approach for wireless access in heterogeneous networks like 4G is presented by Niyato &Hossain (2008).

The problem of resource management following a dynamic resource allocation strategy should be separated into two sub-problems, one considering resource management, at the network level and the other at the connection level. The objectives of dynamic resource allocation and reservation are to maximize network utilization through efficient resource allocation, achieve prioritization among different SCs supported and assure the QoS of ongoing connections when admitting new connections in a service area.

At the network level, the available network bandwidth must be allocated to each service area. Many service providers offer wireless access services in a non cooperative way trying to maximize their own revenues when allocating bandwidth to the connections. The bandwidth allocation problem is modelled as a non-cooperative game with the different access providers as its players. The solution is obtained from the Nash equilibrium if it exists. For seamless mobility across the service areas, a portion of the radio resources must be reserved for handoffs. Since new and handoff connections have to share a part of the bandwidth available in a service area, an arrangement of the bandwidth reserved should be made so that the required QoS performance with regard to the handoff CDP and the CBP is assured. Thus, the problem of bandwidth allocation among new, horizontal handoff, and vertical handoff connections can be formulated as a resource-sharing problem, where a negotiation among the players can be performed for efficient and fair bandwidth sharing. This negotiation is meaningful when new, horizontal, and vertical handoff connections belong to the same service area. Therefore, a bargaining game formulation may be employed to obtain the fair bandwidth allocation/ reservation thresholds for various types of connections. In this bargaining game, new, vertical and horizontal handoff connections negotiate with each other to determine the reservation thresholds so that QoS requirements at the connection level (namely CBP and handoff CDP) are satisfied. In this case, the equilibrium is considered as the solution of the game. Bandwidth reservation is used for service differentiation among new connections and vertical and horizontal handoff connections. Both bandwidth allocation and reservation at the network level can be done on a long-term basis based on the average (steady state) statistics at the connection level of various service areas.

On the other hand, the problem of bandwidth allocation at the connection level, which is allocating bandwidth to a new connection requesting admission to a certain service area, is modelled as a trading market. In a service area, each network and access provider offering bandwidth to a new connection is assumed to be rational and selfishly aiming at utility and profit maximization through an optimal strategy. At steady state (equilibrium), the various strategies must be stable in the sense that every participant in the game is satisfied with the solution given by the other participants strategies. A revenue function for every network offering bandwidth to an incoming connection should be established by taking into account the network gain when it allocates bandwidth to a new connection and its cost function, which accounts for the loss of utility due to revocation of some bandwidth from ongoing connections. Therefore, a non-cooperative game is formulated at connection level and the Nash equilibrium, if it exists, is its solution determining the bandwidth offered by each network to a new connection. In connection-level allocation, the required bandwidth is allocated to a connection arriving to a service area from different available access networks. To obtain the solution two algorithms are presented (Niyato & Hossain, 2008). The first is based on an optimization problem formulation and the second is an iterative method. Finally, the CAC module utilizes the results of bandwidth reservation together with the connection-level bandwidth allocation to decide whether an incoming connection should be admitted or not. Connection-level bandwidth allocation and CAC must be accomplished on a short-term basis and should be updated upon the arrival and departure of every connection in a service area.

Dynamic Pricing and Call Admission Control

Traditional Resource Allocation and CAC schemes are not sufficient, as they can not assure QoS when the network is congested. Pricing based models have also been considered for traffic management and congestion control. Regardless of the CAC scheme used, when the number of ongoing calls becomes very high, the blocking probability of new calls or incoming handoff calls becomes very high, too. Introducing pricing based schemes in admission control, the network may changes the users behavior.

The heterogeneous environment of 4G wireless networks raises certain issues to be considered when employing a pricing model (Niyato & Hossain, 2008). Capacity, coverage, data rate, mobility support and QoS provision are different for different access technologies causing inequality in service offering. Various access networks are operated by different service providers, who aim at maximizing their revenue and compete with each other to attract users. Wireless services from different access networks may not be completely interchangeable, since some mobile terminals may not capable of supporting all the radio interfaces incorporated into the network or because some users may prefer short range wireless access to long range ones. Finally, in dynamic pricing schemes for such networks vertical handoff among different access must be taken into account.

The following three approaches, namely the auction, optimization and demand/supply – based schemes have been used to develop pricing models for heterogeneous wireless networks. In Sallent et al, 2006, the auction-based approach was followed, where a user periodically bids for radio resources by informing the service provider of the price he is willing to pay and his QoS requirement. Then, the service provider decides which allocation of resources maximizes his revenue. In this multi-unit sealed-bid auction, a manager agent facilitates the negotiation between a mobile user and a service provider. The optimization-based scheme for service allocation and pricing in heterogeneous wireless network incorporating different access technologies was proposed by Zhang (2005). The capacity of the system was optimally allocated to different SCs. The objective was to maximize the revenue of a service provider and the network utilization under the capacity constraints of each of the access networks constituting the unified 4G network environment. The demand/supply-based scheme for resource allocation was proposed in (Chan, Fan & Cao, 2005) based on the well known theory of supply and demand of microeconomics. The supply function was obtained through maximizing the revenue of a service provider, whereas the demand function was obtained by solving a utility maximization problem for a certain number of active users. The equilibrium price was determined equating demand and supply. Based on this equilibrium price, network selection and admission control methods were developed.

Dynamic Pricing for Heterogeneous 4G Networks

A pricing based approach for dynamic admission control aims at maximizing the users utility of wireless network resources (Hou, Yang & Papavassiliou, 2002). It has been shown that for a certain wireless network a new call arrival rate exists that maximizes the users total utility. The admission control can dynamically adjust the price according to the actual network conditions, trying to avoid congestion. However, maximizing the users utility of the network might not maximize the revenue of the service provider. In particular, more resources should be allocated to the users to enhance their satisfaction. On the other hand, to maximize revenue under flat pricing, the resource allocation must be modified in order to accommodate more users.

The above framework is mainly destined for wireless cellular networks without considering vertical handoffs and competition among multiple service providers owning different radio access networks. Anyway, the above framework provides a well structured approach to dynamic pricing and CAC problems that can easily be incorporated in 4G networks. Due to the competition existing among providers in heterogeneous environments, a price offered by a service provider will affect the prices offered by the others. This situation can be modelled as an oligopoly market and, consequently, non-cooperative game models can be used for the relevant analysis aiming at obtaining the optimal pricing. Two cases of price competition among different service providers are considered. First, the case is the one where service providers offer their prices simultaneously namely the simultaneous-play game. The solution of this competition is given by Nash equilibrium where all service providers considered are satisfied by the solution with regard to prices. Second, the case where a service provider of one access network offers its price for a certain call before another service provider, namely the leader-follower game that is, when the first service provider has priority in offering its price. In particular, consider that WiMAX and Wi-Fi networks co-exist in an interworking 4G environment. The WiMAX service provider has absolute priority in offering price due to the larger service area covered by WiMAX networks (Niyato & Hossain, 2008). The Wi-Fi service provider may observe the price offered by the WiMAX service provider and then decide about its own price. The solution of this competition problem is the Stackelberg equilibrium that maximizes the profit of the first service provider (Niyato & Hossain, 2008).

The proposed pricing models are applicable in general scenarios involving many access providers of different types (e.g. WiMAX, 3G cellular, and Wi-Fi). Assume a heterogeneous wireless access environment where two different access networks compete in two service areas. The first service provider, assuming a WiMAX network provider, offers a premium real time SC to n_p users where the transmission delay is the major QoS performance metric. This provider charges these users at a flat rate of p_p (per unit of allocated bandwidth). In the second service area, the two providers (the WiMAX and a Wi-Fi provider) compete in offering wireless access services in the best-effort SC where a user through a dual-radio interface has the option to switch between the WiMAX and Wi-Fi networks.

Bandwidth demand is quantified through a quadratic utility function introduced for best-effort users by Singh & Vives (1984). The bandwidth demand function is obtained by maximizing utility. That is, the demand obtained as the optimal bandwidth that maximizes the utility of a best-effort user, given the qualities and the prices of the wireless access services and the degree of interchangeability between services. Let p_i denote the price per bandwidth unit offered by service provider *i*, where $i \in \{m, f\}$ corresponds to WiMAX and Wi-Fi networks, respectively. To capture the effect of QoS provision and service interchangeability, the bandwidth demand function is defined as follows (Niyato & Hossain, 2008):

$$D_i\left(p_i, p_j\right) = \frac{u_i\gamma_i - p_i - v\left(u_j\gamma_j - p_j\right)}{1 - v^2}, \quad (5)$$

where γ_i, $i \in \{m, f\}$ is the transmission quality for wireless access through the network of service provider *i*, u_i is a weighting factor and *v* denotes the degree of service substitutability – for best -effort services between WiMAX and Wi-Fi networks $v \approx 1$ –. The bandwidth demand from a best-effort user for a particular wireless access service increases as the transmission quality in the respective network improves. On the other hand, the bandwidth demand decreases as the price becomes higher. The bandwidth demand is also affected by the quality and the price offered by the other wireless access service networks. If for a certain access network, the transmission quality becomes better (e.g., a lower transmission error rate is achieved) or the prices becomes lower, some of the best-effort users will prefer this access network to the others. The proportion of the churning best-effort users is controlled through the substitutability factor.

A simultaneous-play and a leader-follower game is considered for two different access networks service (e.g. WiMAX and Wi-Fi) providers as players. The strategy of a player is to determine the optimal price per bandwidth unit denoted by p_i, $i \in \{m, f\}$. The respective payoff of each player is its profit denoted by π_i. The total profit of the first service provider is evaluated as based on the revenue obtained from the premium users minus the cost due to the transmission delay of the premium users plus the revenue obtained from the best-effort users (i.e., price multiplied by the bandwidth demand). Hence,

$$\pi_m\left(p_m, p_f\right) = n_p P_p - c_d \sum_{k=1}^{n_p} W_k\left(\left(B - D_m\left(p_m, p_f\right)\right) / n_p\right) + p_m D_m\left(p_m, p_f\right),\quad (6)$$

where B is the total bandwidth, $W_k(b)$ is the transmission delay of a premium class user k when the allocated bandwidth is b and c_d is the weighting factor related to the transmission delay performance. Similarly, the payoff of the second access network provider (e.g. Wi-Fi) is given by

$$\pi_f\left(p_m^*, p_f\right) = p_f D_f\left(p_m, p_f\right). \quad (7)$$

The best response of the first service provider to the price offered by the second one is the optimal price p_m^* maximizing its profit $\pi_m(p_m^*, p_f)$ given the price p_f offered by the second provider. Similarly, the best response of the second service provider is the optimal price p_f^* maximizing its payoff $\pi_f(p_m, p_f^*)$ given the price p_m offered by the first service provider. This best response is denoted by

$$B_i\left(p_j\right) = \arg\max_{p_i}\left\{\pi_i\left(p_i, p_j\right)\right\}. \quad (8)$$

When the two competing service providers offer their prices simultaneously, the Nash equilibrium determines the pair of prices so that a service provider can not increase his profit independently by choosing a different price, given that the price offered by the other service provider remains fixed. Hence, the Nash equilibrium (p_m^*, p_f^*) is determined from

$$B_m\left(p_f^*\right) = p_m^* \quad (9)$$

and $B_f\left(p_m^*\right) = p_f^*$. $\quad (10)$

When the first service provider offers its price before the other, the former will choose an appropriate strategy to maximize its profit based on the assumption that the latter will set the price based on its best response given the price offered by the first player. The price offered by the first service provider along with the best response of the second service provider constitutes the Stackelberg equilibrium that is:

$$p_m^* = \arg\max_{p_m} \pi_m\left(p_m, B_f\left(p_m\right)\right) \text{ and } p_f^* = B_f\left(p_m^*\right). \quad (11)$$

The resource allocation and dynamic pricing models employing game theory analyzed above, deals with the network challenges and demands for reliable QoS provision encountered in heterogeneous 4G networks. Through competitive dynamic pricing the network service providers can achieve high network utilization and efficient resource management, while guaranteeing QoS to network users.

FUTURE RESEARCH DIRECTIONS

QoS in 4G networks, where multiple access techniques and various network technologies are involved, should be handled as an end-to-end issue. End-to-end QoS provision constitutes a critical issue for the demanding environment of 4G networks. Especially, the design and implementation of suitable network architectures for end-to-end QoS provision in 4G networks is a promising and challenging research issue. Traditional schemes and QoS provision architectures are not able to provide for QoS in 4G networks due to the complicated network environment. The underlying platform of 4G networks is heterogeneous with varying topologies and standards. Therefore, QoS must be treated by all the communication layers (physical, MAC, IP, TCP and application), since each layer is required to provide a set of service guarantees at layer level so that the network becomes more flexible and tolerant to QoS issues. Moreover, cross-layer architecture for QoS provision in 4G networks must take into account the heterogeneous network environment and the relevant traffic complexity.

The cross-layer framework for QoS provision should be composed of a QoS manager and attain scheduling subsystems, namely the admission controller, the scheduler, the predictor and the feedback mechanism which regulate the MAC layer of heterogeneous networks. The QoS manager, that is the key component for cross-layer architecture, acts as middleware between applications and lower network layers and is able to dynamically manage the available resources under different load conditions transparently to the application level. In cross-layer architecture for QoS provision, the supported applications dynamically specify their own set of complex and flexible requirements, expressed through the SLAs negotiated with the underlying implementation, through the QoS manager. The subsystems may vary depending on the particular MAC(s) used (e.g. IEEE 802.11, IEEE802.16, MIMO, mobile public networks, wired networks, etc.).

The cross-layer framework operation should be based on the interaction between the application and the network layers. This architecture must incorporate these features to support the use of the same end equipment that should flexibly function in wireless access networks as well as in mobile cellular networks with optimal spectrum efficiency and resource management. The QoS support needs to be functional and offer differentiated service support according to strict latency/throughput requirements though the wireless medium is subjected to time and space variations. Due to the variability of radio resources hard guarantees for QoS can not be provided. Instead, soft guarantees regarding delay, jitter, PLR, throughput and bandwidth are provided. From the user perspective, the framework allows easy access to multimedia services, hiding the complexity of the lower MAC levels of the different networks. The design of a cross-layer integrated QoS provision architecture provides an interface to QoS support for any applications requiring tight guarantees. Cross-layer architectures for QoS provision constitute a promising research issue and an adequate solution to the need of ubiquity and diversity in 4G systems.

CONCLUSION

Wireless and Internet technologies continue to evolve at a fast pace adopting ubiquitous computing strategies. In this chapter, we look ahead, particularly at what might be the next generation of wireless networks not only in terms of QoS provision and the mechanisms related to, but also with regard to network architectures and system design. Common elements of 4G architecture are the order-of-magnitude bandwidth increase, new types of terminals, access-independent converging IP networks, new services and service-enabling platforms accompanied with reliable, robust and effective dynamic RRM schemes.

QoS requirements related to 4G networks have been examined. The coexistence of various wireless access technologies with IP networks imposes an appropriate SC mapping to establish interoperability between the well known service models of DiffServ and the widely used 3GPP SC classification. Heterogeneous networking challenges are examined focusing on the interoperability between different access technologies (e.g. 3G/UMTS-WLAN). To satisfy the requirements for multiservice support with diverse QoS characteristics, dynamic resource allocation and admission control is required. In this course, various techniques are presented including resource reservation, QoS renegotiation, game theoretic resource allocation and dynamic pricing.

Regardless how the next generation of wireless systems will be formed, a broad consensus exists that 4G heterogeneous networks based on mobile IP-structure will be the leading communication platform.

REFERENCES

Akyildiz, I., Xie, J., & Mohanty, S. (2004). A survey of mobility management in nextgeneration all-IP-based wireless systems. *IEEE Wireless Communications*, *11*(4), 16–28. doi:10.1109/MWC.2004.1325888

Awduche, D., Berger, L., Gan, D., Li, T., Srinivasan, V., & Swallow, G. (2001). RSVP-TE: Extensions to RSVP for LSP Tunnels. RFC 3209, IETF.

Bari, F., & Leung, V. C. (2007). Automated Network Selection in a Heterogeneous Wireless Network Environment. *IEEE Network*, *21*(1), 34–40. doi:10.1109/MNET.2007.314536

Bernet, Y., Ford, P., Yavatkar, R., Baker, F., Zhang, F., Speer, M., et al. (2000). A Framework for Integrated Services Operation over Diffserv Networks. RFC 2998, IETF.

Blake, S., Black, D., Carlson, M., Davies, E., Wang, Z., & Weiss, W. (1998). An Architecture for Differentiated Services. RFC 2475, IETF.

Braden, R., Clark, D., & Shenker, S. (1994). Integrated Services in the Internet Architecture: an Overview. RFC 1633, IETF.

Braden, R., Zhang, L., Berson, S., Herzog, S., & Jamin, S. (1997). Resource ReSerVation Protocol (RSVP) – Version 1 Functional Specification. RFC 2205, IETF.

Buddhikot, M., Chandranmenon, G., Han, S., Lee, Y. W., Miller, S., & Salgareli, L. (2003). Integration of 802.11 and third-generation wireless data networks. In IEEE INFOCOM 2003 (pp. 503-512). San Francisco: IEEE.

Chan, H., Fan, P., & Cao, Z. (2005). A Utility-Based Network Selection Scheme for Multiple Services in Heterogeneous Networks. In *International Conference on Wireless Networks, Communications and Mobile Computing, 2005 (IWCMC'05)* (pp. 1175-1180). Maui, Hawaii: IEEE.

Cheng, Y., Jiang, H., Zhuang, W., Niu, Z., & Lin, C. (2005). Efficient Resource Allocation for China's 3G/4G Wireless Networks. *IEEE Communications Magazine*, *43*(1), 76–83. doi:10.1109/MCOM.2005.1381878

Chou, C. T., & Shin, K. S. (2004). Analysis of Adaptive Bandwidth Allocation in Wireless Networks with Multilevel Degradable Quality of Service. *IEEE Transactions on Mobile Computing*, *3*(1), 5–17. doi:10.1109/TMC.2004.1261813

Das, S., & Sen, S., Basu, k., & Lin, H. (2003). A Framework for bandwidth degradation and call admission control schemes for multiclass traffic in next-generation wireless networks. *IEEE Journal on Selected Areas in Communications*, *21*(10), 1790–1802. doi:10.1109/JSAC.2003.815928

Floroiu, J. W., Ruppelt, R., Sisalem, D., & Stephanopoli, J. V. (2003). Seamless handover in terrestrial radio access networks: a case study. *IEEE Communications Magazine*, *41*(11), 110–116. doi:10.1109/MCOM.2003.1244930

Gao, X., Wu, G., & Miki, T. (2004). End-to-end QoS provisioning in mobile heterogeneous networks. *IEEE Wireless Communications*, *11*(3), 24–34. doi:10.1109/MWC.2004.1308940

Gozdecki, J., Pacyna, P., Marques, V., Aguiar, R. L., Garcia, C., Moreno, J. I., et al. (2003). An IP QoS Architecture for 4G Networks. In International workshop on architectures for quality of service in the Internet 2003 (Art-QoS 2003). 2698, (pp. 18-28). Warsaw, POLOGNE: Springer.

3*GPP23.107*. (2003). Quality of Service (QoS) Concept and Architecture (Release 5). TS 23.107 V5.11.0.

Gupta, V. (2004). *Media independent handover service*. Draft technical requirements 21-04-0087-08-0000, IEEE 802.21.

Heinanen, J., Baker, F., Weiss, W., & Wroclawski, J. (1999). Assured Forwarding PHB Group. RFC 2597, IETF.

Hong, D., & Rappaport, S. S. (1986). Traffic Model and Performance Analysis for Cellular Mobile Radio Telephone Systems with Prioritized and Nonprioritized Handoff Procedures. *IEEE Transactions on Vehicular Technology*, *35*(3), 77–92. doi:10.1109/T-VT.1986.24076

Hou, J., Yang, J., & Papavassiliou, S. (2002). Intergration of Pricing with Call Admission Control to Meet QoS Requirements in Cellular Networks. *IEEE Transactions on Parallel and Distributed Systems*, *13*(9), 898–910. doi:10.1109/TPDS.2002.1036064

Jacobson, V., Nichols, K., & Poduri, K. (1999). An Expedited Forwarding PHB. RFC 2598, IETF.

Jähnert, J., Zhou, J., Aguiar, R. L., Marques, V., Wetterwald, M., & Melin, E. (2005). The 'pure-IP' Moby Dick 4G architecture. *Computer Communications*, *28*(9), 1014–1027. doi:10.1016/j.comcom.2004.11.013

Karetsos, G. T., Kyriazakos, S. A., Groustiotis, E., Giandomenico, F. D., & Mura, I. (2005). A Hierarchical Radio Resource Management Framework for Integrating WLANs in Cellular Networking Environments. *IEEE Wireless Communications*, *12*(6), 11–17. doi:10.1109/MWC.2005.1561940

Koodli, R. (2005). Fast Handovers for Mobile IPv6, RFC 4068, IETF.

Koodli, R. (2008). Mobile IPv6 Fast Handovers, RFC 5268, IETF.

Kulavaratharasah, M. D., & Aghvami, A. H. (1999). Teletraffic Performance Evaluation of Microcellular Personal Communication Networks (PCN's) with Prioritized Handoff Procedures. *IEEE Trasactions on Vehicular Technology, 48* (1), 137-152.

Lee, Y. W., Kim, S., Park, J., & Kim, S. H. (2007). A Lightweight Implementation of RSVP-TE protocol for MPLS-TE Signaling. *Computer Communications*, *30*(6), 1199–1204. doi:10.1016/j.comcom.2006.12.003

Levine, D. A., Akyldiz, I. F., & Naghshineh, M. (1997). A Resource Estimation and Call Admission Algorithm for Wireless Multimedia Networks Using the Shadow Cluster Concept. *IEEE/ACM Transactions on Networking, 5* (1), 1-12.

Li, W., & Chao, X. (2007). Call admission control for an adaptive heterogeneous multimedia mobile network. *IEEE Transactions on Wireless Communications*, *6*(2), 515–524. doi:10.1109/TWC.2006.05192

Lim, W.-S., Kim, D.-K., Suh, Y. J., & Won, J.-J. (2009). Implementation and Performance study of IEEE 802.21 in Intergrated IEEE 802.11/802.16e Networks. *Computer Communications*, *32*(1), 134–143. doi:10.1016/j.comcom.2008.09.034

Manzoni, P., Ghosal, D., & Serazzi, G. (1995). Impact on mobility on TCP/IP: An integrated performance study. *IEEE Journal on Selected Areas in Communications*, *13*(5), 858–867. doi:10.1109/49.391748

McNair, J., & Zhu, F. (2004). Vertical Handoffs in Fourth-Generation Multinetwork Environments. *IEEE Wireless Communications*, *11*(3), 8–15. doi:10.1109/MWC.2004.1308935

Monego, H. I., Bodanese, E. L., Nacamura, L., & Souza, R. D. (2005). A dynamic Resource Allocation Scheme for Providing QoS in Packet-Switched Cellular Networks. *Second International Workshop Mobility Aware Technologies and Applications 2005 (MATA'05)* (pp. 117-126). Montreal, Canada: Springer.

Moon, B., & Aghvami, H. (2003). DiffServ extensions for QoS provisioning in IP Mobiltiy environments. *IEEE Wireless Communications*, *10*(5), 38–44. doi:10.1109/MWC.2003.1241096

Mussabir, Q., Yao, W., Niu, Z., & Fu, X. (2007). Optimized FMIPv6 Using IEEE 802.21 MIH Services in Vehicular Networks. *IEEE Transactions on Vehicular Technology*, *56*(6), 3397–3407. doi:10.1109/TVT.2007.906987

Mykoniati, E., Charalampous, C., Georgatsos, P., Damilatis, T., Goderis, D., & Trimintzios, G. (2003). Admission Control for Providing QoS in IP DiffServ Networks: the TEQUILA Approach. *IEEE Communications, special issue on QoS Advances: the European IST Premium IP Projects*, *41*(1), 38-44.

Naghshineh, M., & Schwartz, M. (1966). Distributed Call Admission Control in Mobile/Wireless Networks. *IEEE Journal on Selected Areas in Communications*, *14*(4), 711–717. doi:10.1109/49.490422

Nikkilä, R., Langutin, D., Asplund, H., & Ranta, M. (2004). Prototyping of Vertical Handover Enabled Services. Wireless World Research, (WWWRF11), Oslo, Norway.

Niyato, D., & Hossain, E. (2005). Call Admission Control for QoS Provisioning in 4G Wireless Networks: Issues and Approaches. *IEEE Network*, *19*(5), 5–11. doi:10.1109/MNET.2005.1509946

Niyato, D., & Hossain, E. (2006). A Cooperative Game Framework for Bandwidth Allocation in 4G Heterogeneous Wireless Networks. *International Conference on Communications 2006 (ICC'06)*. *2*, 1418-1422. Instabul: IEEE.

Niyato, D., & Hossain, E. (2008). Competitive Pricing in Heterogeneous Wireless Access Networks: Issues and Approaches. *IEEE Network*, *22*(6), 4–11. doi:10.1109/MNET.2008.4694168

Oliveira, C., Kim, J. B., & Suda, T. (1998). An Adaptive Bandwidth Reservation Scheme for High-Speed Multimedia Wireless Network. *IEEE Journal on Selected Areas in Communications*, *16*(6), 858–874. doi:10.1109/49.709449

Pahlavan, K., Krishnamurthy, P., Hatami, A., Ylianttila, M., Mareka, J., & Pichna, R. (2000). Handoff in hybrid mobile data networks. *IEEE Personal Communications*, *7*(2), 34–47. doi:10.1109/98.839330

Pang, A., Chen, J., Chen, Y., & Agrawal, P. (2004). Mobility and session management: UMTS vs. CDMA 2000. *IEEE Wireless Communications*, *11*(4), 30–43. doi:10.1109/MWC.2004.1325889

Perkins, C. (2002). Mobile IP. *IEEE Communications Magazine*, *40*(5), 66–82. doi:10.1109/MCOM.2002.1006976

Salkintzis, A. K., Fors, C., & Pazhyannur, R. (2002). WLAN-GPRS integration for next generation mobile data networks. *IEEE Transactions on Wireless Communications*, *9*(5), 112–124. doi:10.1109/MWC.2002.1043861

Sallent, O., Perez-Romero, J., Agusti, R., Giupponi, L., Kloeck, C., Martoyo, I., et al. (2006). Resource Auctioning Mechanisms in Heterogeneous Wireless Access Networks. *Vehicular Technology Conference 2006 (VTC '06), 1*, 52-56. Melbourne: IEEE.

Shenker, S., Patridge, C., & Guerin, R. (1997). Specification of Guaranteed Quality of Service. RFC 2212, IETF.

Shenoy, N., & Montalvo, R. (2005). A Framework for Seamless Roaming across Cellular and Wireless Local Area Networks. *IEEE Wireless Communications*, *12*(3), 50–57. doi:10.1109/MWC.2005.1452854

Shi, M., Shen, X., & Mark, J. (2004). IEEE 802.11 roaming and authentication in wireless LAN/cellular mobile networks. *IEEE Wireless Communications*, *11*(4), 66–75. doi:10.1109/MWC.2004.1325893

Singh, N., & Vives, X. (1984). Price and Quantity Competition in a Differentiated Duopoly. *The Rand Journal of Economics*, ▪▪▪, 546–554. doi:10.2307/2555525

Soliman, H., Castelluccia, C., El Malki, K., & Bellier, L. (2005). Hierarchical Mobile IPv6 Mobility Management (HMIPv6), RFC 4140, IETF.

Song, Q., & Jamalipour, A. (2005). A Network Selection Mechanism for Next Generation Network. *IEEE International Conference on Communications 2005 (ICC'05),* (pp. 1418-1422). Seoul: IEEE.

Song, W., Jiang, H., Zhuang, W., & Shen, X. (2005). Resource Management for QoS support in Cellular/WLAN Interworking. *IEEE Network*, *19*(5), 12–18. doi:10.1109/MNET.2005.1509947

Song, W., & Zhuang, W. (2005). QoS provisioning via Admision Control in Cellular/Wireless LAN interworking. *2nd International Conference on Broadband Networks 2005 (BroadNets 2005). 1*, (pp. 543-550). Boston: IEEE.

Tragos, E., Tsiropoulos, G., Karetsos, G., & Kyriazakos, S. (2008). Admission Control for QoS support in Heterogeneous 4G Wireless Networks. *IEEE Network Magazine*, *22*(3), 30–37. doi:10.1109/MNET.2008.4519962

Varshney, U., & Jain, R. (2001). Issues in Emerging 4G Wireless Networks. *IEEE Computer*, *34*(6), 94–96.

Vidales, P., Baliosian, J., Serrat, J., Mapp, G., Stajano, F., & Hopper, A. (2005). Autonomic System for Mobility Support in 4G Networks. *Journal on Selected Areas in Communications*, *23*(12), 2288–2304. doi:10.1109/JSAC.2005.857198

Wang, H. J., Katz, R. H., & Giese, J. (1999). Policy-Enabled Handoffs Across Heterogeneous Wireless Networks. In *2nd IEEE Workshops on Mobile Computing and Applications (WMCSA '99)* (pp. 51-60). New Orleans: IEEE.

Wang, X., Mellor, J., & Al-Begain, K. (2003). Towards Providing QoS for Intergrated Cellular and WLAN Networks. In *4th Annual Postgraduate Symposium on the Convergence of Telecommunications (PGNET'03),* (pp. 207-211). Liverpool, UK: JMU.

Wisely, D., Aghvami, H., Gwyn, S. L., Zahariadis, T., Manner, J., & Gazi, V. (2003). Transparent IP Radio Access for Next-Generation Mobile Networks. *IEEE Wireless Communication*, *10*(4), 26–35. doi:10.1109/MWC.2003.1224976

Wroclawski, J. (1997). Specification of the Controlled-Load Network Element Service. RFC 2211, IETF.

Wu, D. (2005). QoS provisioning in wireless networks. *Wireless Communications and Mobile Computing*, *5*(8), 957–969. doi:10.1002/wcm.359

Ye, J., Hou, J., & Papavassiliou, S. (2002). A Comprehensive Resource Management Framework for Next Generation Wireless networks. *IEEE Transactions on Mobile Computing*, *1*(4), 249–264. doi:10.1109/TMC.2002.1175539

Ylitalo, J., Jokikyyny, T., Kauppinen, T., Tuominen, A. J., & Laine, J. (2003). Dynamic Network Interface Selection in Multihomed Mobile Hosts. Thirty-Sixth Hawai'i International conference on System Sciences (HICSS-36). 9, (pp. 315). Big Island, Hawai'i: IEEE Computer Society.

Zhang, Q., Guo, C., Guo, Z., & Zhu, W. (2003). Efficient mobility management for efficient handoff between WWAN and WLAN. *IEEE Communications Magazine*, *41*(11), 102–108. doi:10.1109/MCOM.2003.1244929

Zhang, W. (2005). Bearer Service Allocation and Pricing in Heterogeneous Wireless Networks. *International Conference on Communications 2005 (ICC 2005)* (pp. 1367-1371). Seoul: IEEE.

Zhuang, W., Gan, Y., Loh, K., & Chua, K. (2003). Policy-based QoS management architecture in an integrated UMTS and WLAN environment. *IEEE Communications Magazine*, *41*(11), 118–125. doi:10.1109/MCOM.2003.1244931

ADDITIONAL READING

Anastasopoulos, M. P., Arapoglou, P.-D. M., Kannan, R., & Cottis, P. G. (2008). Adaptive Routing Strategies in IEEE 802.16 Multi-Hop Wireless Backhaul Networks: An Evolutionary Game Theory Approach, *IEEE JSAC. Special Issue on Game Theory in Communication Systems*, *26*(7), 1218–1225.

Anastasopoulos, M. P., Panagopoulos, A. D., & Cottis, P. G. (2008). A Distributed Routing Protocol for QoS provisioning in Wireless Mesh Networks. *Wireless Communications and Mobile Computing*, *8*(10), 1233–1245. doi:10.1002/wcm.562

Bharghavan, V., Lee, K., Lu, S., Ha, S., Li, J., & Dwyer, D. (1998). The TIMELY adaptive resource management architecture. *IEEE Personal Communications*, *5*(4), 20–31. doi:10.1109/98.709366

Chen, H., Cheng, C.-C., & Yeh, H.-H. (2008). Guard-Channel-Based Incremental and Dynamic Optimization on Call Admission Control for Next-Generation QoS-Aware Heterogeneous Systems. *IEEE Transactions on Vehicular Technology*, *57*(5), 3064–3082. doi:10.1109/TVT.2008.915521

Fitkov-Norris, E. D., & Khanifar, A. (2000). Dynamic pricing in mobile communication systems. *First International Conference on 3G Mobile Communication Technology* (pp. 416-420). London: IEEE.

Fitkov-Norris, E. D., & Khanifar, A. (2001). Dynamic pricing in cellular networks, a mobility model with a provider-oriented approach. *Second International Conference on 3G Mobile Communication Technology* (pp. 63-67). London: IEEE.

Gomez, G., & Sanchez, R. (Eds.). (2005). End-to-End Quality of Service over Cellular Networks. England, P.A.: John Wiley & Sons Ltd.

Guo, Y., & Chaskar, H. (2002). Class-based quality of service over air interfaces in 4G mobile networks. *IEEE Communications Magazine*, *40*(3), 132–137. doi:10.1109/35.989773

Ho, C. J., Copeland, J. A., Lea, C. T., & Stuber, G. L. (2001). On call admission control in DS/CDMA cellular networks. IEEE Trasactions on Vehicular Technology, 50 (6), 1328-1343.ference on 3G Mobile Communication Technology (pp. 416-420). London: IEEE.

Ibrahim, W., Chinneck, J. W., & Periyalwar, S. (2003). A QoS-based charging and resource allocation framework for next generation wireless networks. *Wireless Communications and Mobile Computing*, *3*(7), 895–906. doi:10.1002/wcm.179

Imre, S. (2007). Dynamic Call Admission Control for Uplink in 3G/4G CDMA-Based Systems. *IEEE Transactions on Vehicular Technology*, *56*(5), 2617–2629. doi:10.1109/TVT.2007.899932

Ishizaki, F., & Hwang, G. U. (2007). Cross-Layer Design and Analysis of Wireless Networks using the Effective Bandwidth Function. *IEEE Transactions on Wireless Communications*, *6*(9), 3214–3219. doi:10.1109/TWC.2007.06030108

Kim, I. M., Shin, B. C., & Lee, D. J. (2000). SIR-based call admission control by intercell interference prediction for DS-CDMA systems. *IEEE Communications Letters*, *4*(1), 29–31. doi:10.1109/4234.823540

Koskie, S., & Gajic, Z. (2005). A Nash Game Algorithm for SIR-Based Power Control in 3G Wireless CDMA Networks. *Trasactions on Networking*, *13*(5), 1017–1026. doi:10.1109/TNET.2005.857068

Liang, S., & Hatzinakos, D. (2007). A Cross-Layer Architecture of Wireless Sensor Networks for Target Tracking. *IEEE/ACM Transactions on Networking, 15* (1), 145-158.

Lina, A., Cecchetti, R., & Cecchetti, G. (2007). Toward the QoS Support in 4G Wireless Systems. *18th Tyrrhenian International Workshop on Digital Communications, TIWDC 2007* (pp. 245-252). Naples, Italy: Springer US.

Liu, Q., Zhou, S., & Giannakis, G. B. (2005). Cross-Layer Scheduling with Prescribed QoS Guarantees in Adaptive Wireless Networks. *IEEE Journal on Selected Areas in Communications*, *23*(5), 1056–1066. doi:10.1109/JSAC.2005.845430

Liu, Z., & Zarki, M. E. (1994). SIR-based call admission control for DS-CDMA cellular systems. *IEEE Journal on Selected Areas in Communications*, *12*(4), 638–644. doi:10.1109/49.286670

Pagani, M. (Ed.). (2005). Mobile and Wireless Systems Beyond 3G: Managing New Business Oppurtinities. Hershey, PA: IRM Press.

Paschos, G. S., Politis, I. D., & Kotsopoulos, S. A. (2005). A Quality of Service Negotiation-Based Admission Control Scheme for WCDMA Mobile Wireless Multiclass Services. *IEEE Transactions on Vehicular Technology*, *54*(5), 1875–1886. doi:10.1109/TVT.2005.853455

Ramakrishna, D., Mandayam, N., & Yates, R. (2000). Subspace based estimation of the signal-to-interference ratio for CDMA cellular systems. *IEEE Transactions on Vehicular Technology*, *49*(5), 1732–1742. doi:10.1109/25.892578

Shakkottai, S., Rappaport, T. S., & Karlsson, P. C. (2003). Cross-layer design for wireless network. *IEEE Communications Magazine*, *41*(10), 74–80. doi:10.1109/MCOM.2003.1235598

Stratogiannis, D. G., Tsiropoulos, G. I., & Cottis, P. G. (2008). Call admission control in Wireless networks: Probabilistic approach and efficiency evaluation. *International Conference on Wireless Communications and Mobile Computing Conference 2008 (IWCMC '08)* (pp. 712 - 717). Crete Island: IEEE.

Tang, J., & Zhang, X. (2007). Cross-Layer resource allocation over wireless relay networks for quality of service provisioning. *IEEE Journal on Selected Areas in Communications*, *25*(4), 645–656. doi:10.1109/JSAC.2007.070502

Tsiropoulos, G. I., Stratogiannis, D. G., Cottis, P. G., & Kanellopoulos, J. D. (2008). Efficiency evaluation of class-based call admission control schemes for wireless communications. *IEEE International Symposium on Wireless Communication Systems (ISWCS '08)* (pp. 69 - 73). Reykjavik: IEEE.

Zhang, Q., & Zhang, Y. Q. (2008). Cross-Layer design for QoS support in multihop wireless networks. *Proceedings of the IEEE*, *96*(1), 64–76. doi:10.1109/JPROC.2007.909930

Chapter 15

QoS-Predictions Service:
QoS Support for Proactive Mobile Services

Katarzyna Wac
University of Geneva, Switzerland & University of Twente, The Netherlands

Melanie Hilario
University of Geneva, Switzerland

Bert-Jan van Beijnum
University of Twente, The Netherlands

Richard Bults
University of Twente, The Netherlands & MobiHealth B.V., The Netherlands

Dimitri Konstantas
University of Geneva, Switzerland

ABSTRACT

The success of emerging mobile services depends on the serviceability of the underlying wireless networks, expressed in terms of Quality of Service (QoS) provided by a network available to service user at a given geographical location and time. In general, this serviceability is a priori unknown. As a solution the authors propose a QoS-predictions service, providing predictions of QoS of networks available at a given user's geographical location and time. In a case study they prove the feasibility of deriving predictions from historical data collected by a mobile service user. They have chosen this mobile user to be a patient, who uses a personalized health telemonitoring service in his daily environments for a period of one month. They consider the QoS-predictions service as a novel support for mobile services operational in 4G heterogeneous network environments.

INTRODUCTION

The emergence of new wireless broadband networks and diverse miniaturized and personalized networked multimedia devices has given rise to a variety of new interactive mobile applications and mobile services (Hansmann et al., 2003). These new applications allow the mobile users to not only access information anywhere-anytime-anyhow, but are also able to adapt their functionality based on different context information. This information can range

DOI: 10.4018/978-1-61520-771-8.ch015

from user-related context, like user profile, role and activity and geographical location and time, to device-related context, like available network interfaces or screen size and resolution, and up to the network-related context, like network signal strength and its current load. By having access to rich context information, mobile applications are able to adapt themselves and increase the quality of the user's experience (defined as QoE by (ITU-T, 2007b). Although user and device related context information are to a great extend available to the applications, network related context information (*i.e.* Quality of Service (QoS by (ITU-T, 1993, 2008)) provisions) is only roughly and approximately available (Chalmers & Sloman, 1999a).

Current trends are that mobile applications become highly interactive, and users become highly mobile, using different underlying networks as available in his environment (Ortiz, 2007). Hence, the success of application service delivery depends upon QoS provided by these underlying networks, which tend to be highly heterogeneous (Chalmers & Sloman, 1999b). In this situation, we consider that one of the most critical elements contributing to an improvement of mobile applications and services functionality is precise information about the QoS provided by these underlying networks. This information can be used for the application's QoS management and assurance of user satisfaction.

Today's wireless network providers (for example Mobile Network Operators (MNOs)), provide coarse-grained, marketing-based, static QoS information about their networks (*e.g.* network nominal capacity), refusing to give any detailed information to mobile users, mainly due to marketing reasons (Gomez & Sanchez, 2005). As a result, mobile applications base their QoS management and adaptation on best case scenarios for networks' QoS, as derived from the network's nominal capacity. Moreover, some mobile applications are using the signal strength to estimate the networks QoS; still this does not provide any information on, *e.g.* actual network congestion. Taking into consideration that the mobile users are by definition moving in space and time, the task of managing QoS by figuring out and trying to anticipate QoS of the underlying wireless network(s) on a continuous basis becomes impossible, leading to worst case scenario assumptions and lowering the QoE of the user provided applications and services.

On the other hand, mobile users are no longer passive information and content consumers but are now able (and willing) to create themselves new geo-referenced information and content and make it available to other (mobile) users. Many mobile collaborative applications (*i.e.* so-called Mobile Web 2.0 applications (Lin, 2007; O'Reilly, 2005)) are available on the market today allowing mobile users to exchange geo-information like points of interest, virtual geo-tags or geographical locations of traffic radars and so on (Want et al., 1999). Based on the need of the emerging mobile applications to have access to high quality fine-grained QoS information, and the ability of users to create and share geo-referenced information, we conclude that nowadays it is possible for mobile users to create and share any QoS information in a fully collaborative way. In particular, sharing information regarding fine-grained QoS provided by wireless networks at given geographical locations and times, as observed when using a particular application. This information can be further used for predicting the QoS, which in turn could be used by demanding mobile applications to choose the most suitable network and adapt themselves to provide services to the users on a higher QoE level. As for precise geographical location-determination techniques, many of them are currently available, and even more are emerging (Hightower & Borriello, 2001).

Towards this direction, we propose a platform for a QoS-predictions service. It is based on the Mobile Web 2.0 paradigm and supports collection of reliable, user- and application-transparent information about network provided QoS. Based on this information set, the service provides

QoS-predictions back to mobile applications. Particularly, this information is collected from and provided to a mobile user's device and stored in the form of a geo-spatial information grid; for different geographical locations, times and for different networks. Based on these collected historical data and with the use of data mining methods, we are able to predict the anticipated QoS at a certain geographical location and time, for a given wireless access network provider and access technology.

In this chapter we present a case study of feasibility assessment for QoS-predictions derived from historical data collected from a MobiHealth system (van Halteren et al., 2004b) service user; *i.e.*, a patient using a personalized health telemonitoring service in his daily environments, along one month. We consider the QoS-predictions service as a novel support for mobile service users in wireless heterogeneous networks and in particular in 4G environments.

BACKGROUND

Quality of Service Management

Internet was since its beginning, providing a 'best-effort' QoS level. In the late 90's, first solutions for QoS provisions of Internet-based service providers have been proposed, particularly for providers of real-time multimedia services (Hutchison et al., 1997; Shepherd et al., 1996). From the technical perspective, these solutions were based on use of rigorous and complex QoS management frameworks, including functions like QoS negotiation and resource reservation (Andersen et al., 2000; Xiao & Ni, 1999). In this situation, many service providers ignored the proposed solutions and learned to manage 'best-effort' QoS level provided by Internet to assure meeting the QoS requirements (and the expected QoE) of their service users. They particularly relied on estimations of QoS provided by the Internet. This approach was feasible due to at least two factors: firstly, QoS provided by Internet exhibits regularities and long-term estimations for *e.g.* months can be derived relatively accurately (claffy et al., 1998). Secondly, if necessary, service providers could easily acquire information on provided QoS via dedicated QoS measurements and all that without any degradation in the quality of their provided services (Michaut & Lepage, 2005).

With the dawn of the mobile era, history is repeating itself. Provisions of QoS by mobile service providers have been identified as critical to the business viability of mobile service providers already in the late 90's (Chalmers & Sloman, 1999b), when only basic voice and data services existed. The most critical factor in QoS provision is related to the user's mobility: a mobile user relies on the availability of different wireless network providers and wireless technologies at different geographical locations and times along his trajectory (Dekleva et al., 2007). To deal with this, providers of mobile services are (again) advised to employ complex QoS management frameworks or other solutions. For example (Q. Han & Venkatasubramanian, 2006; Soh & Kim, 2003) propose to employ in a mobile service delivery predictions of user mobility path acquired from a wireless network provider, or to employ a QoS broker for the reservation of network resources on behalf of mobile users (Nahrstedt et al., 2001). Simultaneously, Mobile Network Operators (MNOs) propose new concepts like Universal Mobile Access, Generic Access Network or IP-Multimedia-System (Cuevas et al., 2006), aiming at a technical as well as a business solutions for mobile service providers. Such solutions contradict the 'mobile' nature of services, because they limit mobile service providers' customer base and service-usage area to the MNOs customer base and its coverage area.

The mobile service providers not following the network provider-centric business models emerge on a growing scale (Tan, 2004). Naming few, Skype is a VoIP provider (Osterwalder et al.,

2005), MobiHealth.com is a mobile healthcare service provider (MobiHealth, 2007), while Digital Chocolate is a mobile gaming service provider (Digital Chocolate, 2008). They all struggle to assure meeting their mobile users' QoS requirements over unpredictable 'best-effort' QoS provided by wireless technologies (Bults et al., 2005b). Dedicated QoS measurements are not feasible due to the dynamic nature of QoS provided by wireless network technologies (Dood, 2005) and the limited resources (*e.g.* battery) of mobile devices.

At the same time, new wireless network providers appear and new long-range wireless technologies (*e.g.* High Speed Protocol Access) are deployed by the existing providers. These are steps towards the vision of 4G, where wireless communication infrastructures are going to be plentifully available for mobile service users (De Vriendt et al., 2002; Dekleva et al., 2007; Ortiz, 2007; Tachikawa, 2003). In 4G environment a mobile service user can access any network anytime-anywhere. Yet, any new network provider or wireless technology when launched provides only a 'best-effort' QoS level (Gomez & Sanchez, 2005) and uses 'drive-tests' as performance tests. Therefore, despite the 4G vision, still the question remains: how a mobile service user chooses a network best matching his user's QoS requirements and QoE-expectations at a given geographical location and time. Our solution aims at prediction the QoS provided by networks to a mobile service user. To the best of our knowledge, the kind of solution we propose has not yet been proposed in the literature. To justify this, in the following section we present state of the art approaches for QoS-predictions.

Quality of Service Predictions

There exists limited research on predictions of QoS, usually done for a given mobile service being used in a fixed lab environment in a limited interval. It is important to notice that much of research, in which authors claim they aim to provide predictions, we categorize as research on inference. After data mining literature (Alpaydin, 2004) we define *inference* as prediction provided only for a given moment of time (*i.e.* now), and not for the future.

For example (Bremler-Barr et al., 2003) aim to infer when a significant deviation of RTT between client and web-server will occur. They use Hidden Markov Models (HMM), using predictors that have exponential or polynomial decay of history. They prove that for some servers, the recent history (*i.e.* 2 minutes) is enough to predict such RTT degradation. (Gao & Wu, 2005) propose to select web-services based on artificial neural networks (ANN), which, based on service availability, reliability, data rate and a time of a service request (*i.e.*, a time of the day), would predict its availability and delay (if the service is available). Similarly, (Yang et al., 2008) aim to predict web-services delay, based on a time of a service request, data rate and a state of service (*e.g.* number of other users). They predict this delay based on recent history (*i.e.* hours) modelled via a semi-Markov model. Their experimentation shows that predictions can be done with up to 90% of accuracy. (Salvo Rossi et al., 2003a, 2003b), using only the recent history (*i.e.* minutes) model delay, loss function for Internet paths using the Hidden Markov Model. Moreover, (Iannello et al., 2005) model UDP application-traffic (that can represent *e.g.* a video stream) using Input-Output HMM – nodes. They particularly model delay or loss QoS measures. In their model they assume a Gamma distribution for delay; however they admit that this assumption has not been proven for data exchange with use of wireless access networks.

(Amirijoo et al., 2006) aim to infer a) data rate available for user; and b) utilization of a channel, for the purpose of dynamic reconfiguration of real-time systems. As an input they assume a reconfiguration specification and data rate specification for a given service. They do not use any historical data. The simulations of scenarios show possible advantage of their approach. (Mody et al.,

2007) research towards dynamic use of frequency spectrum; they aim to learn which frequencies are free and possible to be used without interference to others. So far, they have conducted only limited simulations, without real-user scenarios.

(Hegge, 2007), based on simple statistical methods (means and standard deviation) aim to predict FTP downloads for mobile users. Namely they aim to predict RTT, data loss and data rates in different geographical locations, times and using different access networks. The prediction accuracy was low due to a major flaw in the experiments; it could not be distinguished when and where the mobile device was using which technology: in particular UMTS or GPRS. Similarly, in (Mirza et al., 2007), the authors aim to predict the (TCP-level) data rates for arbitrary file sizes exchanged between two nodes on the Internet. They aim to predict this data rate from the measured delay and loss. They use Support Vector Machines (SVM) techniques. (Y. Sun et al., 2008), based on measured data rate, aim to predict HSPA-based access network's channel utility and accuracy and dependability using NN.

FEASIBILITY ASSESSMENT OF QOS-PREDICTIONS

Motivation, Context and Approach

The goal of our explorative case study is to asses a feasibility of deriving QoS-predictions for an operational mobile service being used by its representative user in his daily environments. As a mobile service, we have chosen the MobiHealth health telemonitoring service that enables ambulatory patient's vital signs monitoring (van Halteren et al., 2004b). In the study we focus on a typical MobiHealth service user. Namely, this service is used by a Chronic Obstructive Pulmonary Disease (COPD) patient living in Geneva city (Switzerland). The mobile patient follows his daily routine of work and home activities, while having his health state being continuously monitored. His service is operational in mobile environments with 802.11-WLAN and 2.5G-GPRS wireless network providers available throughout the city.

The approach towards the QoS-predictions feasibility study is as follows. Having the MobiHealth system at hand, we focus on collection of QoS data by means of system measurements, rather than by means simulation or modelling. Particularly, we focus on (end-to-end) delay as a QoS measure of importance for the operational MobiHealth services. Hence, after instrumentation of the MobiHealth system for delay measurements, we collect delay measurement data for a patient continuously using two systems (thus emulating two different users being in same geographical location and time) along one month interval. As there are two systems, two different wireless networks could be used at a geographical location. Moreover, having two systems, we attempt to test the hypothesis if based on data collected by one of them; we can obtain accurate predictions for the other one.

Based on the collected data, we analyze feasibility of delay predictions for a set of prediction algorithms. Namely we use 48 different data mining methods (assuming different set of parameters). Particularly, we assess the predictions feasibility in terms of prediction accuracy, its learning and prediction times and model's complexity. Based on these results, we derive conclusions upon delay-predictions feasibility for an operational health telemonitoring service.

Background Information and Methodology for Predictions Study

Classification and prediction are two forms of data mining that can be used to extract models describing important data classes or to predict future data trends (Alpaydin, 2004; J. Han & Kamber, 2006). Effective and scalable methods have been developed for decision trees induction, Naive Bayesian classification, Bayesian belief network, rule-based classifier, back-propagation, Support Vector Machine (SVM), associative classification, nearest neighbour classifiers, case-based reasoning, and other classification methods such as genetic algorithms, rough set and fuzzy set approaches (Alpaydin, 2004). Statistical significance tests are necessary for model selection. Issues such as model accuracy, training time, robustness, interpretability, and scalability must be considered in a model selection and these can involve trade-offs, which needs to be considered given a classification task at hand (Witten & Frank, 2005).

Classification predicts categorical class labels (discrete or nominal). It classifies data (constructs a model) based on the training set and the values (class labels) in a classifying attribute and uses it in classifying new data. Prediction models continuous-valued functions, *i.e.*, predicts model construction describing a set of predetermined classes. For prediction, each tuple/sample is assumed to belong to a predefined class, as determined by the class label attribute. The set of tuples used for model construction is training set. The model is represented as classification rules, decision trees, or mathematical formulae. Model is used for classifying future or unknown objects. To estimate accuracy of the model we compare the known label of test sample with the classified result from the model. The *accuracy rate* is the percentage of test set samples that are correctly classified by the model. Test set needs to be independent of training set, otherwise overfitting will occur. If the accuracy of the model is acceptable, we can use the model to classify data tuples whose class labels are not known.

Classification is a *supervised learning*. Supervision means that the training data (*i.e.* samples, measurements) contains labels indicating the class of the observations. The new data is classified based on the training set. In contrary, *clustering* is an *unsupervised learning*. The class labels of training data are unknown. Given a set of measurements, the aim is to establish the existence of classes or clusters in the data.

The QoS predictions problem addressed in this chapter is based on classification of delay (*i.e.* a QoS measure) observed in health telemonitoring service provided by the MobiHealth system. The classification of delay is done along one of nine schemas defined along the delay requirements posed on the MobiHealth system by its end user *i.e.* healthcare practitioner.

System Overview

The MobiHealth system is a distributed system for telemonitoring of a patient's health condition (MobiHealth, 2007; van Halteren et al., 2004b). A patient is wearing a Body Area Network (BAN) (Bults et al., 2004; van Halteren et al., 2004a) consisting of a sensor-set and a Mobile Base Unit (MBU, *i.e.*, central controlling unit of a BAN worn by a patient). The sensor-set usually consists of specialized sensors monitoring the patient's vital signs, a geographical location sensor (*e.g.* a GPS receiver (Hightower & Borriello, 2001)) for his location determination and an event-notification sensor. The sensor-set is specific for a patient's health condition, *e.g.* Chronic Obstructive Pulmonary Disease (COPD, Figure 1), cardiac condition, epilepsy or chronic pain.

The MBU is the central unit of a BAN, usually in the form of a mobile phone or PDA. The MBU has three functionalities: collecting sensor-set data, processing it (*e.g.* filtering, shaping, correlating) and sending (processed) data to a remote application server - backend-system (BEsys, *i.e.*, security, control, management, and data storage sub-system) located in a healthcare centre. It is specific for the MobiHealth system is that all these tasks are performed in real time. Once the sensor-set data has been send to the BEsys, it is made available (in a near real-time) to other applications, *e.g.* for storage, display for a medical decision support systems.

Figure 1. MobiHealth system overview: COPD care example

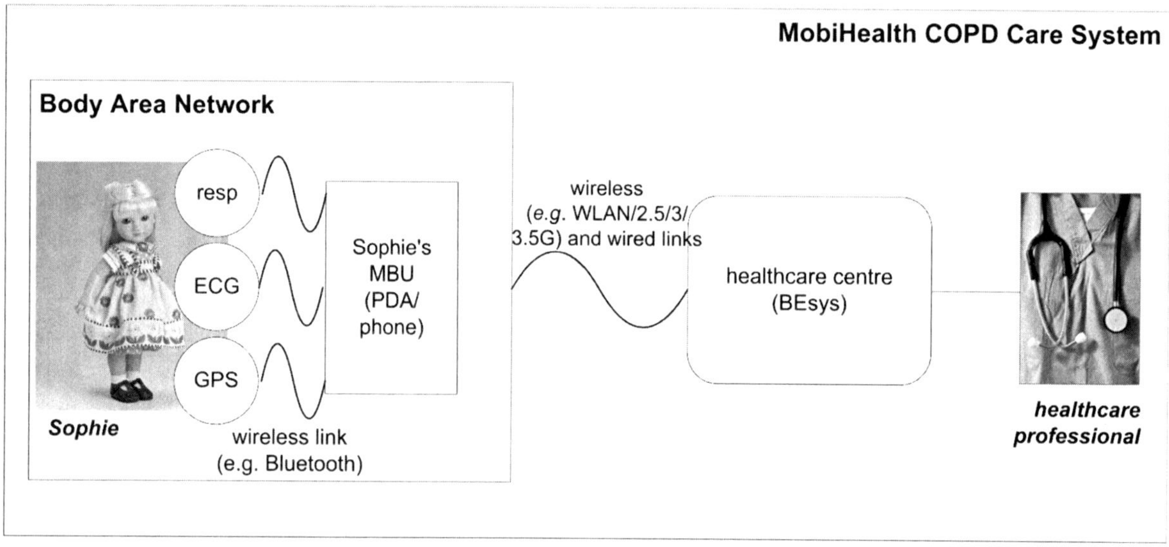

A patient's health hazardous event can be defined differently for each patient, based on his health condition. The event notification can be determined by a) patient's activating the event-notification sensor or b) based on the patient's vital signs analysis done i) at the patient's BAN or ii) at the BEsys.

Compliance with JINI Surrogate Architecture

A MobiHealth health telemonitoring service is implemented in Java and conforms to the JINI Surrogate Architecture specifications as extensively presented in (Dokovsky et al.; van Halteren & Pawar, 2006).

Patient's mobile device, *i.e.* MBU is a service provider. It continuously transmits BAN data from the MBU to the BEsys in a healthcare centre. According to the JINI Surrogate Architecture, a service provided by a device with limited resources (*e.g.* like MBU), called Device Service (DS), is represented on the Internet by its Surrogate Object (SO) deployed on a Surrogate Host (SH). The co-called JINI network consists of a SH (hosting SOs), service registry via which service users can find the required services, and service users themselves. Therefore the SO acts as service provider on behalf of the DS and shields service users (in the MobiHealth case: healthcare practitioners' viewer application) from the specific means to communicate with the device. At the health telemonitoring application level, this yields that the DS, representing health telemonitoring service provider is deployed at the MBU. Moreover, a healthcare practitioners' viewer application at a healthcare centre transparently retrieves BAN data from the SO at the BEsys (hosting a SH). It is important to notice that there can be multiple DSs deployed on one mobile device and made available to service users.

The SO rely on the SH for a service life cycle management. Device specific communication between the SH and the DS is implemented by the MobiHealth Interconnect Protocol (MHIP). It is an application protocol and its previous version has been presented in (van Halteren & Pawar). The protocol component at MBU is called MHIP-IO. This component is responsible for a DS-SO communication, even in case if there is a Network Address Translator (NAT) between the mobile device and the fixed server (which is very

Figure 2. MobiHealth health telemonitoring service components

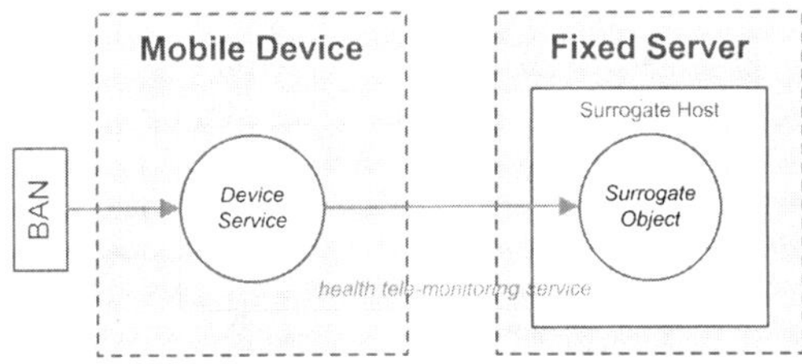

often a case especially in GPRS wireless access networks).

Summarising, the MobiHealth health telemonitoring service consists of two components (Figure 2):

- health telemonitoring Device Service on the mobile device
- health telemonitoring Surrogate Object hosted by the Surrogate Host in the fixed network, *i.e.*, on the BEsys in a healthcare centre.

The surrogate architecture specification requires the MHIP to support at least three mechanisms: device services discovery, surrogate object upload, and keep-alive. The purpose of the discovery mechanism is to make a SH aware of the DS existence and vice versa. Particularly, it is MHIP-IO component that discovers SH. After the SH discovery, DS provides SH with the SO that will act in the JINI network on the behalf of this device. After a SO has been instantiated and activated by the SH, the device must maintain its availability to provide its DS.

Consequently, the MHIP must implement a Keep-Alive mechanism, which implies that the DS needs to send at the fixed frequency so-called Keep-Alive messages to inform the SH that the service is (still) available. As soon as the device cannot confirm its availability, *i.e.* a Keep-Alive message is not received by SH within and expected time interval, the SH can deactivate the corresponding SO. The Keep-Alive message size is negligible comparing to the total volume of application-data being sent (Pawar et al., 2008; Wac et al., 2009a).

The MobiHealth system adapts its volume of data being sent by DS to its SO based on the QoS provided by underlying end-to-end communication path. Namely, there exists an application-level buffer at the MBU, via which data is being sent by the MHIP-IO to the SO. The state of the buffer is measured by the DS. When the QoS provided by the path, and especially its speed is not sufficient to support volume of data being sent, the buffer fills. Based on the buffer fill level, the DS adapts application flow by excluding some sensor-set data and hence reducing volume of data being sent (Bults et al., 2005a).

End-to-End Communication Path

In our case study we focus on the MobiHealth service for telemonitoring of a patient's health condition. Particularly, the BAN uses the extra-BAN communication network, like WLAN or 2.5/3/3.5G (*i.e.* GPRS/UMTS/HSxPA) for exchange of the application and control data between the DS at the MBU and the SO at the BEsys. Patient's vital signs data, *i.e.* telemonitoring application-data is sent continuously from the DS at the MBU to the SO at the BEsys (*i.e.* in an 'uplink' direction). The extra-BAN end-to-end data communication path is heterogeneous as it consists of wireless (*e.g.* 2.5/3/3.5G) and wired network technologies (Figure 3).

Figure 3. MobiHealth extra-BAN end-to-end communication path

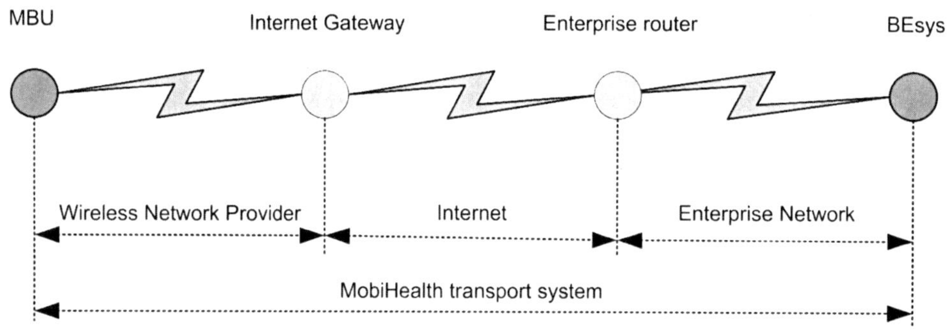

The QoS provided by this path influences the QoS provided by the MobiHealth application to the patient. Particularly, the QoS is influenced by the choice of the Wireless Network Provider (WNP) and wireless access network technology used. If the MBU would have a choice of WNP, it would choose one resulting in providing best-of best-effort service to MobiHealth system, rather than an arbitrary WNP.

Application-level QoS

End-users of health telemonitoring application are healthcare professionals and their patients. However, only the healthcare professionals can define the application QoS requirements posed on the application (Broens et al., 2007). These requirements encompass reliable, error-free data exchange between the MBU and BEsys without losses and at a minimum delay. The use of TCP/IP protocol in combination with MBU data storage ensures the application data recovery in case of data losses due to poor networks performance. This paper focuses on a minimum data delay requirement posed on the MobiHealth system; we focus on the extra-BAN communication network delay, having a major contribution to the application-data delay (Wac et al., 2005). The MobiHealth system performance is managed based on the Keep-Alive message Round Trip Time (*KA-RTT*). It is the time it takes for a Keep-Alive control message originated at the DS at the MBU (Pawar et al., 2007; SUN, 2001)), to be received by the SO at the BEsys and returned (without processing) to the DS at the MBU.

The *KA-RTT* reflects the delay induced by the underlying networks and the processing delays in the protocol stacks at the MBU and the BEsys. The (wireless) access network uplink (*i.e.* MBU to the BEsys) and downlink (*i.e.* BEsys to the MBU) contribute significantly to the *KA-RTT* (Bults et al., 2005b; Wac et al., 2005).

MobiHealth System Measurements

In this paper we focus on health telemonitoring application provided by MobiHealth system to COPD patients. We have collected *KA-RTT* measurement data for one patient living in Geneva (Switzerland), using the application along one month (17.11.07 - 15.12.07), while following his daily routines. The patient has spent 69.8% of time in two top geographical locations: home and office.

Application Flow

The BAN samples patient's pulse rate, oxygen saturation, plethysmogram and alarm button activity at a frequency of 128 Hz. A sample consists of 5 Bytes of application-data. An application aggregates a unit of 1 s of data, *i.e.*, 640 Bytes, and compresses it (losslessly) before sending it to BEsys. The reduction in size relative to the uncompressed size is 80–85%; it decreases as variability of the values of the measured vital signs. The MHIP adds 10 Bytes of protocol overhead per a compressed data unit. The overall data rate sent by the DS at the MBU to the SO at the BEsys is around 1.5 kbps.

MBU and BEsys Platforms

As a MBU we have used Qtek 9090 with Intel® PXA263 400 MHz processor, 128 MB RAM, running Windows Mobile 2003 SE PocketPC OS. The device has been dedicated for the executed measurements. The Qtek used GPRS (class 10) interface for extra-BAN communication; GPRS network was provided by Sunrise operator and WLAN by University of Geneva. The BEsys was a standard high performance server dedicated to MobiHealth telemonitoring services. The server was placed at University of Twente.

Measurements Instrumentation

The *KA-RTT* values (in milliseconds) were measured every 10 seconds continuously during the telemonitoring application execution. Moreover, the geographical location, time, network received signal strength indication, remaining battery level and data-rate sent by the DS at the MBU to the SO at the BEsys (in B/s), have been logged every second.

Predictions Tasks Definition

The goal of the research reported in this chapter is to assess the feasibility of predicting the *KA-RTT* values based on classification models build from the measurements data. We aimed to answer the following question: "Having collected 5 days of history of *KA-RTT* values as observed at different geographical locations, times, with use of different wireless networks and technologies, can we predict *KA-RTT* values for one of the next days?" For the purpose of this task we have taken Monday-Friday 26-30 November 2007 data as training (*i.e.* learning) dataset, and Monday 3 December 2007 data as testing (*i.e.* prediction) dataset.

Moreover, we have asked a question if data collected by one MBU device (emulating user 1 or 2 and denoted as D1 or D2) can be equally used for deriving predictions for this device or for the other device (cases denoted D1-2 if data collected by device 1 is used for predictions for device 2; D2-1 is the other way around). Furthermore we asked a question if joined set of data collected by both devices can be used for predictions provided to any of them (cases denoted D1,1-1 if predictions are made for D1 and D1,2-2 if for D2).

From our preliminary research we concluded that predicting the *KA-RTT* numerical value is practically impossible. Therefore, as the purpose of this study is explorative, we transformed *KA-RTT* numeric values (in milliseconds) into categorical values along nine different schemas (and not only along one schema, as it is done in many other studies). A categorization schema was derived twofold. Firstly, we have derived it based on healthcare practitioner's (*i.e.* MobiHealth user's) delay requirements posed for a vital sign data delivery. Secondly, we have derived it from by the *KA-RTT* numeric values distribution. Therefore we defined nine different *KA-RTT* classification tasks. For those tasks the *KA-RTT* value was classified in:

- Task 1: two intervals ('1', '0'): corresponding to *KA-RTT* values of [0, 750) and [750, ∞); task 1 is denoted further as c1-750,
- Task 2: two intervals ('1', '0'): corresponding to *KA-RTT* values of [0, 1000) and [1000, ∞); task 2 is denoted further as c1-1000,
- Task 3: two intervals ('1', '0'): corresponding to *KA-RTT* values of [0, 1500) and [1500, ∞); task 3 is denoted further as c1-1500,

- Task 4: two intervals ('1', '0'): corresponding to *KA-RTT* values of [0, 2500) and [2500, ∞); task 4 is denoted further as c1-2500,
- Task 5: two intervals ('1', '0'): corresponding to *KA-RTT* values of [0, 3000) and [3000, ∞); task 5 is denoted further as c1-3000,
- Task 6: four intervals derived from *KA-RTT* distribution ('1', '2', '3', '4'): corresponding to *KA-RTT* values of [0, 1811), [1811, 2221), [2221, 2609), and [2609, ∞); task 6 is denoted further as c2
- Task 7: five intervals ('1', '2', '3', '4', '5'): corresponding to *KA-RTT* values of [0, 500), [500, 1000), [1000, 1500), [1500, 2000) and [2000, ∞); task 7 is denoted further as c3-500,
- Task 8: five intervals ('1', '2', '3', '4', '5'): corresponding to *KA-RTT* values of [0, 750), [750, 1500), [1500, 2250), [2250, 3000), [3000, ∞); task 8 is denoted further as c3-750,
- Task 9: five intervals ('1', '2', '3', '4', '5'): corresponding to *KA-RTT* values of [0, 1000), [1000, 2000), [2000, 3000), [3000, 4000), [4000,∞); task 9 is denoted further as c3-1000.

Tasks 1-5 can be also called 'binary', as they aim to predict one of two possible values of the *KA-RTT* class.

Collected Data Summary

In the collected data, we distinguish nine fixed user's geographical locations and six trajectories (traversed while patient was mobile). We distinguish Sunrise-GPRS as wireless network provider - wireless technology 1 (provider-Tech1) combination and University of Geneva - WLAN as wireless provider 2 - technology 2 combination (provider-Tech2). We have collected in total 2'509'250 *KA-RTT* measurement seconds for two systems; 1'228'780 instances for D1 and 1'280'470 instances for D2.

Data Representation

Measurement of each *KA-RTT* instance has been associated with measurement of the MBU and BEsys platform parameters. These parameters we use now as discriminators for *KA-RTT* classification. These parameters are also called *features* (Witten & Frank, 2005) and *KA-RTT* itself is called a *target feature*.

Day of a week (DoW) represents a day of the week (1-7; 1 is Monday) and hour (hr) represented as an hour of a day (0-23). DoW and hr are the two features indicating time. We have collected data in time span of one month; if we would collect data for consecutive months and years, the month and year would be candidate features as well.

Geographical location (loc) represents the patient's geographical location, where 1 is home, 2 - office, 3 - shopping centre, etc. Network operator (op) represents the wireless network provider and wireless network technology used, where 1 is wireless network provider-Tech 1 (Sunrise-GPRS) and 2 is wireless network provider-Tech2 (UniGe-WLAN). Hr, DoW, loc and op are categorical data.

The network received signal strength indication (RSSI) at the MBU (sig) has been quantized into four values from 1 (none or a weak signal) to 4 (a maximum signal). Similarly, the MBU remaining battery level (bat) has value from 1 (none or a small fraction of battery left) to 4 (maximum battery level). The variables sig and bat are to be considered as ordinal data with a Likert scale, *i.e.* one cannot assume the intervals between values are the same but just that the values are ordered (DeVellis, 2003). It results the inherent way these values are derived from the MBU OS; they are not measured continuously, but in steps. Health telemonitoring data-rate (in Bytes/second) sent by the MBU to the BEsys is denoted as MBU-Rout and is a regular numerical value.

Training and Testing Accuracy

Accuracy of classification models build from the measurements data for a given classification task is derived in two phases: learning, called *training phase* and predictions called *testing phase*.

For a given model, its accuracy it is a percentage of correctly classified instances in training or testing phase and therefore called respectively training accuracy or testing accuracy. Accuracy has a value in range of 0…100%, where 0% means that none of instances were correctly classified, and 100% means all of them were classified correctly.

In a training phase, prediction models are derived (*i.e.* learned) by given data mining algorithm (with given parameters), based on the given training dataset. In a testing phase the derived models are tested on the given testing dataset. Training and testing datasets are non-overlapping subsets derived from the overall set of measurements data available for our research.

As a training procedure we choose 10 folds cross-validation (CV) executed on a training dataset (Witten & Frank, 2005). In each CV we obtain 10 models for an algorithm. The CV is repeated 10 times to obtain statistically sound training results, *i.e.* all 100 individual models are then used to estimate mean and variance of accuracy for the given algorithm on the given training dataset.

Testing procedure encompasses evaluation of the accuracy of the derived models on the testing dataset. The testing dataset is sometimes also denoted as a 'hold-out' data set, as it is held-out from the overall set of measurements data available for a given research. According to guidelines in data mining field, the accuracy obtained by an algorithm in a testing phase aims to represent a future (*i.e.*, any) predictions accuracy of this algorithm (Mitchell, 1999; Witten & Frank, 2005).

Accuracy of each classification model build from the training dataset for given classification task is to be compared with the task's baseline accuracy. Given our nine prediction tasks, where target feature is to be classified in one of two, four or five distinct categories, baseline accuracy we define based on a mode class (*i.e.* most probable class, which is more suitable than a mean class value) for this task (Witten & Frank, 2005). For a given task, firstly, the mode class is learned from the task's training dataset; then the task baseline accuracy is a probability of this mode class derived from task's training dataset.

A prediction algorithm, which uses the above logic and predicts the mode class value, we call an '*educated guess*' data mining method. It is implemented in WEKA software package as ZeroR (ZR) rule (Witten & Frank, 2005).

Prediction Results

In our predictions case study, we analyzed performance of diverse data mining techniques including Bayesian networks, rules and trees. The results are presented in six separate Tables 1-6, each table corresponds to a defined system case (D1, D2, D1-2, D2-1, D1,2-1 and D1,2-2) and rows correspond to nine classification tasks (c1-750, etc.). The table columns represent a) a user-case and its classification task; b) task's baseline accuracy in percentage; c) a result for a training phase: the most accurate algorithm and its parameters and its accuracy in percentage; d) a result for the testing phase: the most accurate algorithm and its accuracy in percentage; and e) the testing accuracy of the most accurate algorithm from a training phase (*i.e.* algorithm with parameters as given in the column c).

The names of the algorithms are abbreviated as follows: J48 for J48, PA for PART, JR for JRip, NB for Naïve Bayes, RF for Random Forest and ZR for ZeroR rule (*i.e.* performing along the task's baseline accuracy). For some tasks, no values exist in the tables ("-"). For these tasks, we assume that due to the distribution of target feature in the training and testing datasets at the level of 99-100%, the baseline accuracy is the only performance measure. For these tasks, there is no advantage

Table 1. System 1 data only used for training-validation and testing (Table 1-6. Predictions accuracy for the predictions experiment. Each row represents results for one prediction task and it gives its baseline accuracy, with an indication in brackets which is the majority class for a given task; the most accurate algorithm with its parameter setting and its accuracy found a training-validation phase; the most accurate algorithm and its accuracy found in testing phase; and the accuracy of the most accurate algorithm from a training-validation phase, in its testing phase)

D1	Baseline Acc (%)	Training-Validation		Testing		Training-Validation in Testing
class. task		alg. and param.	Acc (%)	algorithm	Acc (%)	Acc (%)
c1-750	100 ('0')	-	-	-	-	-
c1-1000	100 ('0')	-	-	-	-	-
c1-1500	100 ('0')	-	-	-	-	-
c1-2500	69 ('1')	J48 -C 15 B	70	JR	69	57
c1-3000	92 ('1')	J48 -C 15 B	79	ZR	92	88
c2	22 ('4')	J48 -C 15	51	NB	37	33
c3-500	81 ('5')	JR -N10 O2P	90	JR	81	81
c3-750	46 ('4')	J48 -C 15 B	54	NB	50	46
c3-1000	74 ('3')	J48 -C 15 B	68	JR	74	72

gained from data mining, which require training and testing phase hence would mean wasting of the computational resources.

For all algorithms, the training and testing times were in the order of one second or below.

Analyzing the results, we conclude that the advantage gained from data mining leaves no room for doubt. All results throughout all the six cases, show large advantage of algorithms over the baseline accuracy (*i.e.* accuracy of models is higher of up to 64% over the baseline accuracy).

Table 2. System 2 data only was used for training-validation and testing

D2	Baseline Acc (%)	Training-Validation		Testing		Training-Validation in Testing
class. task		alg. and param.	Acc (%)	algorithm	Acc (%)	Acc (%)
c1-750	72 ('0')	J48 -C 15 B	82	JR	73	70
c1-1000	49 ('0')	PA -C 25 B	89	JR	82	82
c1-1500	33 ('0')	PA -C 15	98	NB	98	97
c1-2500	86 ('1')	J48 -C 15 B	86	JR	87	86
c1-3000	96 ('1')	PA -C 35 B	95	JR	96	96
c2	69 ('1')	J48 -C 15 B	74	J48	80	79
c3-500	27 ('5')	J48 -C 15 B	78	JR	74	73
c3-750	40 ('2')	J48 -C 15 B	62	JR	56	53
c3-1000	51 ('1')	J48 -C 15 B	75	JR	75	73

Table 3. System 1 data was used for training-validation and system 2 data for testing

D1-2	**Baseline Acc (%)**	**Training-Validation**		**Testing**		**Training-Validation in Testing**
Class. task		**alg. and param.**	**Acc (%)**	**algorithm**	**Acc (%)**	**Acc (%)**
c1-750	100 ('0')	-	-	-	-	-
c1-1000	100 ('0')	-	-	-	-	-
c1-1500	100 ('0')	-	-	-	-	-
c1-2500	58 ('1')	PA -C 35 B	76	JR	59	45
c1-3000	89 ('1')	JR -N2 O2	79	JR	89	75
c2	33 ('4')	J48 -C 15 B	52	J48	36	36
c3-500	86 ('5')	JR -N10 O2P	90	NB	80	86
c3-750	53 ('4')	J48 -C 15 B	53	JR	53	41
c3-1000	76 ('3')	J48 -C 15 B	68	J48	76	75

This is especially visible for tasks when delay was classified in 2 categories (*i.e.*, tasks c1), where predictions accuracy reaches 80-100%. For tasks when delay was classified in 4-5 categories (*i.e.*, tasks c2, c3), the predictions accuracy increased up to even 4 times over the one of the baseline. This has been observed consistently for task c2.

The highest accuracy is mainly obtained by algorithms which test features sequentially (trees and rules: J48, Part and JRip). When examining the models produced by these algorithms, we observe that the geographical location, operator and time are the most predictive features, followed by application data-rate and device's signal and battery. NB models were in many tasks least accurate. This clear performance dichotomy between sequential (*e.g.* J48) and parallel (NB) algorithms suggests a weak interaction among all the seven features used for delay predictions.

However, in some tasks, NB models were most accurate. When examining the models produced by NB algorithms, we observed that

Table 4. System 2 data was used for training-validation and system 1 data for testing

D2-1	**Baseline Acc (%)**	**Training-Validation**		**Testing**		**Training-Validation in Testing**
class. Task		**alg. and param.**	**Acc (%)**	**algorithm**	**Acc (%)**	**Acc (%)**
c1-750	100 ('0')	-	-	-	-	-
c1-1000	100 ('0')	-	-	-	-	-
c1-1500	100 ('0')	-	-	-	-	-
c1-2500	74 ('1')	JR -N10 O5	76	JR	75	73
c1-3000	94 ('1')	PA -C 35 B	91	ZR	94	89
c2	17 ('4')	J48 -C 15 B	53	J48	41	41
c3-500	76 ('5')	NB -D	85	NB -D	76	76
c3-750	41 ('4')	J48 -C 15 B	64	J48	53	53
c3-1000	71 ('3')	J48 -C 15 B	72	ZR	71	70

Table 5. System 1 and 2 data was used for training-validation and system 1 data for testing

D1,2-1	Baseline Acc (%)	Training-Validation		Testing		Training-validation in Testing
class. Task		alg. and param.	Acc (%)	algorithm	Acc (%)	Acc (%)
c1-750	100 ('0')	-	-	-	-	-
c1-1000	100 ('0')	-	-	-	-	-
c1-1500	100 ('0')	-	-	-	-	-
c1-2500	69 ('1')	J48 -C 35	74	JR	74	64
c1-3000	92 ('1')	J48 -C 25 B	86	ZR	94	90
c2	22 ('4')	J48 -C 15 B	51	NBS	41	36
c3-500	81 ('5')	JR -N10 O2P	87	PA	76	76
c3-750	46 ('4')	J48 -C 15	58	NB -D	54	51
c3-1000	74 ('3')	J48 -C 15 B	70	ZR	71	69

the accuracy of these models was not related to strong interaction among all the seven features used for delay predictions, but it was related to the highly unbalanced distribution of delay (*i.e.* 90% probability of mode class).

We also analyzed the accuracy of best algorithms in a training phase, *i.e.*, algorithms with the highest accuracy in the training phase, against the accuracy of best algorithms in the testing phase. Without looking at results after applying the statistical significance tests, the difference between these accuracies ranges from 1-14% (c.f. column 'training' vs. 'training in testing (Tr in Te)'). Based on this observation alone we conclude that algorithms accuracy observed at training resulted not from over-fitting to training dataset, but from an effective generalization made by algorithms, which kept same level of accuracy on testing datasets. Moreover, after applying the Bonferroni adjustment for multiple comparisons and McNemar statistical significance tests, we saw that all differences vanish among the best

Table 6. System 1 and 2 data was used for training-validation and system 2 data for testing

D1,2-2	Baseline Acc (%)	Training-Validation		Testing		Training-Validation in Testing
class. Task		alg. and param.	Acc (%)	algorithm	Acc (%)	Acc (%)
c1-750	72 ('0')	NB	100	NB	100	100
c1-1000	49 ('0')	NB	100	NB	100	100
c1-1500	33 ('0')	NB -D	99	NB	100	100
c1-2500	86 ('1')	J48 -C 35	74	JR	59	52
c1-3000	96 ('1')	J48 -C 25 B	86	NB	89	88
c2	10 ('4')	J48 -C 15 B	51	J48	37	37
c3-500	27 ('5')	JR -N10 O2P	87	NB -D	86	86
c3-750	16 ('4')	J48 -C 15	58	JR	53	52
c3-1000	23 ('3')	J48 -C 15 B	70	JR	76	75

models in testing. That means that best training algorithms, were also amongst the best testing algorithms, *i.e.* they can be used further for deriving accurate predictions.

Overall, from prediction results we conclude that it is feasible to provide accurate predictions for a user based on own or the other user (*i.e.* system) data. It is equally feasible to provide predictions for a user based on own and other user data. In these cases (*i.e.*, D1,2-1 and D1,2-2) however the improvement (or loss) in accuracy is small (-5 to 7% for D1), and does not motivate the effort of including other user's training datasets. In other words, the data collected by D2 does not improve accuracy of predictions provided to user D1. For D2, this improvement is highly variable (-42 to 30%), and again, does not motivate the effort of including other user's training datasets. It is not clear why using first and second user datasets for the first user, provides more advantage than for the other user. For this experiment, we conclude that data collected by one user does not improve accuracy of predictions provided to other user.

Conclusion on Prediction Results

Analyzing the overall results, we conclude that the advantage gained from data mining leaves no room for doubt. We show however that accuracy of different algorithms varied, depending on what has been predicted. Best prediction results we obtained via tree-based data mining methods, which can learn tree-structured dependencies (like a random tree or a rule). This we explain by the structure of data which has been organized in a hierarchical way along user geographical location, network provider wireless technology and time variables. Similar conclusion has been already reported by (Nurmi et al., 2007) for user's geographical location and time-based activity prediction tasks. Therefore it is not surprising that when examining prediction models derived by trees and rules algorithms, we observe that user geographical location, network provider wireless technology and time (day of the week and hours) were most predictive variables. The mobile device received signal strength indication and battery fill level were not predictive, but we conclude it could be that because their distributions were very monotonic, *i.e.*, strength indication as well as battery fill level were almost all the time having a maximum value.

FUTURE RESEARCH DIRECTIONS

There are many possible scenarios, in which QoS-predictions service can evolve. Perhaps the biggest remaining challenge is that of bringing more discriminatory information to bear on the prediction tasks. Integrating more domain-specific information concerning *e.g.* cell-ID for a GPRS network or précising geographical location of the user further (*i.e.* to the meter of accuracy while using one of the techniques as proposed by (Hightower & Borriello, 2001)), or the current state of mobile device (*e.g.* memory, CPU usage) more accurately, is a feasible solution in the short term. We may also consider including domain-independent features (*e.g.* statistical measures like median) to the discriminatory power of the data mining techniques. We may ultimately have to mine the mobile computing literature to gather fresh insights on the part of delay values, currently defying classification.

Another possible future work direction is a research on a scalable solution for QoS-predictions service - limiting its scope of operation to a particular city, region or country, limiting the scope of wireless network providers or technologies for which predictions can be provided. Scoping can be dictated by the need of limiting the QoS information to be processed for deriving the QoS-predictions, or the need for higher accuracy of QoS-predictions provided for a restricted geographical location-area. Furthermore, the service can be limited in terms of network providers and wireless technologies for which the QoS-predictions service acquires QoS information.

A research direction for a long term includes a technical feasibility of the QoS-predictions service (*i.e.* delay-predictions service) to be evaluated in an operational MobiHealth system. So far we have proposed an architecture of such a solution, which we name QoS information system (QoSIS (Wac et al., 2009b). From implementation point of view we would like to research feasibility of deriving predictions on mobile device itself, without passing through the QoS-predictions provider on the Internet. We believe that it is duable given the current computing and storage capabilities of mobile devices. However, the efficiency and effectiveness of this solution needs to be compared to the one based on the QoS-predictions server on the Internet.

In our research we have already analyzed a business viability of an enterprise that based on a collaborative-sharing of the QoS information collected by mobile service users (Mobile Web 2.0 community), provides its QoS-predictions service back to these users. We name this enterprise QoSIS.net and we present our research outcomes in (Wac et al., 2009b).

We also need research on the following issue. If all mobile users will get prediction for an excellent QoS level provided by network X in geographical location A and will switch to this network at the same time, most probably they will experience low QoS (or even that they will be inhibited from using the network by the network admission control system). This low QoS will then be logged by QoS-predictions service provider in form of a new historical data. Based on this new historical data, future prediction for network X will not indicate its excellent QoS level. We believe that in a long term, system will find equilibrium based on what are the QoS required by the users and the QoS provided by networks. However, we admit that the load balancing in the network and QoS-predictions service usage is a future work area.

Future research directions aiming at QoS support for mobile users in 4G environments would include research of solutions that build upon best-effort networks to fulfil QoS requirements for any mobile application used by its user anywhere-anytime. We envision that QoS-predictions service will be used by other applications in *e.g.* entertainment (gaming), communication (voice) and infotainment (TV, web browsing) domains. Many of these applications are provided nowadays over fixed Internet infrastructures and soon they will be available 'on the move'. These applications provide strict QoS requirements. Namely, for example games of type first person shooter (Quake, Doom, Half-life, Counter Strike, American Army), where the fast step first person action counts, are so far most popular multiplayer games and as it has been proven, they have strict QoS requirements (Matsumoto, 2004) like the RTT value of 100 ms (Beigbeder et al., 2004), very low delay jitter and no (Quax et al., 2004). VoIP applications require RTT of 150 ms to maximum 400 ms, jitter of up to 50 ms and loss of 3% maximum (ITU-T, 2007a, 2007b). Internet Protocol Television (IPTV) services require a constant network capacity of around 2 Mbps in order to deliver its services to users at the satisfactory level (ITU-T, 2007c). Moreover, for web browsing (King, 2008), proved that service delay quality ratings show a drop-off at around 8 to 10 seconds; users required that the most important content on the web will be displayed to them with delay of maximum 2 s. These (and many others) applications provide strict QoS requirements; hence they will need a support to fulfil these requirements for a mobile user being anywhere-anytime. Using QoS-predictions service would be an option.

CONCLUSION

The overall conclusive remarks drawn upon the results obtained in our study on feasibility assessment for QoS-predictions indicate that the previously believed assumption that, it is not feasible to predict QoS provided by a network provider in a 'best-effort' networking environment, can be refuted. Namely, we show that it is feasible and sometimes straightforward to predict an accurately value of a QoS performance measure for a specific mobile application used at given geographical location-network provider-technology-time. The prediction accuracy is driven by the amount of available historical data. On one hand, we claim that mobile services need no longer be condemned to 'best-effort' service provided by the underlying heterogeneous networking environment; a mobile service can use QoS-predictions to proactively obtain 'best of best-effort' service for the end-to-end communication path. On the other hand, we envision that in order to match the mobile applications QoS requirements and QoE-expectations in plentiful 4G networking environments, mobile application and service providers need to take a necessary step and employ novel technological solutions for the QoS support in their service delivery processes. QoS-predictions service is just an example of such a novel support.

REFERENCES

Alpaydin, E. (2004). Introduction to machine learning. Cambridge, MA: MIT press.

Amirijoo, M., Tesanovic, A., Andersson, T., Hansson, J., & Son, S. H. (2006). *Finite Horizon QoS Prediction of Reconfigurable Firm Real-Time Systems.* Paper presented at the 12th IEEE Intl Conference on Embedded and Real-Time Computing Systems and Applications.

Andersen, D., Bansal, D., Curtis, D., Seshan, S., & Balakrishnan, H. (2000). *System Support for Bandwidth Management and Content Adaptation in Internet Applications.* Paper presented at the 4th Symposium on Operating Systems Design and Implementation (OSDI), San Diego, CA, USA.

Beigbeder, T., Coughlan, R., Lusher, C., Plunkett, J., & Agu, E. (2004). *The Effect of Loss and Latency on User Performance in Unreal Tournament 2003®.* Paper presented at the SIGCOMM'04 Workshops, Portland, Oregon, USA.

Bellavista, P., Küpper, A., & Helal, S. (2008). Location-Based Services: Back to the Future. *IEEE PERVASIVE COMPUTING*, *7*(2), 85–89. doi:10.1109/MPRV.2008.34

Bremler-Barr, A., Cohen, E., Kaplan, H., & Mansour, Y. (2003). Predicting and bypassing end-to-end Internet service degradations. *IEEE Journal on Selected Areas in Communications*, *21*(6), 961–978. doi:10.1109/JSAC.2003.814614

Broens, T., Huis in't Veld, R., Vollenbroek-Hutten, M., Hermens, H., van Halteren, A., & Nieuwenhuis, B. (2007). Determinants for successful telemedicine implementations: a literature study. *Journal of Telemedicine and Telecare*, *13*(6), 303–309. doi:10.1258/135763307781644951

Bults, R., Pawar, P., & van Halteren, A. (2005a). Context Aware Nomadic Mobile Service - A Case Study on M-Health Signal Processing. the Netherlands: University of Twente.

Bults, R., Wac, K., van Halteren, A., Konstantas, D., Jones, V., & Widya, I. (2004). *Body Area Networks for Ambulant Patient Monitoring Over Next Generation Public Wireless Networks.* Paper presented at the 13th IST Mobile and Wireless Communications Summit, Lyon, France.

Bults, R., Wac, K., van Halteren, A., Konstantas, D., & Nicola, V. (2005b). *Goodput Analysis of 3G wireless networks supporting m-health services.* Paper presented at the 8th International Conference on Telecommunications (ConTEL05), Zagreb, Croatia.

Chalmers, D., & Sloman, M. (1999a). *QoS and Context Awareness for Mobile Computing.* Paper presented at the 1st Intl. Symposium on Handheld and Ubiquitous Computing (HUC99), Karlsruhe, Germany.

Chalmers, D., & Sloman, M. (1999b). A survey of Quality of Service in mobile computing environments. *IEEE Communications Surveys and Tutorials*, *2*(2), 2–10. doi:10.1109/COMST.1999.5340514

Claffy, K., Miller, G., & Thompson, K. (1998). *The nature of the beast: recent traffic measurements from an Internet backbone.* Paper presented at the International Networking Conference (INET98), Geneva, Switzerland.

Cuevas, A., Moreno, J. I., Vidales, P., & Einsiedler, H. (2006). The IMS Platform: A Solution for Next Generation Network Operators to Be More Than Bit Pipes. *IEEE Commun. Mag. Advances in Service Platform Technologies*, *44*(8), 75–81.

De Vriendt, J., Laine, P., Lerouge, C., & Xu, X. (2002). Mobile Network Evolution: A Revolution on the Move. *IEEE Communications Magazine*, *40*(4), 104–111. doi:10.1109/35.995858

Dekleva, S., Shim, J. P., Varshney, U., & Knoerzer, G. (2007). Evolution and emerging issues in mobile wireless networks. *Communications of the ACM*, *50*(6), 38–43. doi:10.1145/1247001.1247003

DeVellis, R. F. (2003). Scale development: theory and applications. London: SAGE.

Digital Chocolate. (2008). *Seize the Minute.* Retrieved Sep 05, 2008, from www.digitalchocolate.com

Dokovsky, N., van Halteren, A., & Widya, I. (2004). *BANip: Enabling Remote Healthcare Monitoring with Body Area Networks.* Paper presented at the Intl Workshop on Scientific Engineering of Distributed Java Applications (FIJI03), Luxembourg.

Dood, A. (2005). The Essential Guide to Telecommunications. Upper Saddle River, NJ: Prentice Hall PTR.

Gao, Z., & Wu, G. (2005). *Combining QoS-based service selection with performance prediction.* Paper presented at the IEEE Intl Conf on e-Business Engineering (ICEBE05).

Gomez, G., & Sanchez, R. (2005). End-to-End Quality of Service over Cellular Networks: Data Services Performance Optimization in 2G/3G: John Wiley & Sons, Ltd.

Han, J., & Kamber, M. (2006). Data mining: concepts and techniques. San Francisco: Morgan Kaufmann.

Han, Q., & Venkatasubramanian, N. (2006). Information Collection Services for QoS-aware Mobile Applications. *IEEE Transactions on Mobile Computing*, *5*(5), 518–535. doi:10.1109/TMC.2006.65

Hansmann, U., Merk, L., Nicklous, M., & Stober, T. (2003). Pervasive Computing: The Mobile World. Berlin: Springer.

Hegge, S. (2007). Predicting end-to-end network characteristics on mobile devices. Unpublished MSc, University of Twente, Enschede, the Netherlands.

Hightower, J., & Borriello, G. (2001). Location systems for ubiquitous computing. *Computer*, *34*(8), 57–66. doi:10.1109/2.940014

Hutchison, D., Mauthe, A., & Yeadon, N. (1997). Quality-of-service architecture: Monitoring and control of multimedia communications. *Electronics & Communication Engineering Journal*, *9*(3), 100. doi:10.1049/ecej:19970303

Iannello, G., Palmieri, F., Pescapè, A., & Salvo Rossi, P. (2005). *End-to-End Packet-Channel Bayesian Model applied to Heterogeneous Wireless Networks.* Paper presented at the IEEE Globecom 2005, St. Louis (MO, USA).

ITU-T. (1993). General aspects of Quality of Service and Network Performance in Digital Networks, including ISDNs (Vol. I.350): ITU.

ITU-T. (2006). Framework for achieving end-to-end IP performance objectives (Vol. []: ITU.]. *Y (Dayton, Ohio)*, ▪▪▪, 1542.

ITU-T. (2007a). *Mean Opinion Score (MOS) terminology (Vol. P.800): ITU.*

ITU-T.(2007b). Vocabulary for performance and quality of service: Appendix I – Definition of Quality of Experience (QoE) (Vol. P.10/G.100): ITU-T.

ITU-T. (2007c). Working document: Quality of Experience Requirements for IPTV (Vol. FG IPTV-36): ITU.

ITU-T. (2008). Definitions of terms related to quality of service (Vol. []: ITU.]. *E (Norwalk, Conn.)*, ▪▪▪, 800.

King, A. (2008). Website Optimization: Speed, Search Engine & Conversion Rate Secrets: O'Reilly Media, Inc.

LaMarca, A., Chawathe, Y., Consolva, S., Hightower, J., Smith, I., Scott, J., et al. (2005). *Place Lab: Device Positioning Using Radio Beacons in the Wild.* Paper presented at the Pervasive Computing 2005, Munchen, Germany.

Lin, K. J. (2007). Building Web 2.0. *IEEE COMPUTER*, *40*(5), 101–102.

Matsumoto, C. (2004). Online Games Pose Carrier Conundrum. Light Reading - Networking the Telecom Industry.

Michaut, F., & Lepage, F. (2005). Application-oriented network metrology: Metrics and active measurement tools. *IEEE Communications Surveys & Tutorials*, *7*(2), 2–24. doi:10.1109/COMST.2005.1610543

Mirza, M., Sommers, J., Barford, P., & Zhu, X. (2007). *A machine learning approach to TCP throughput prediction.* Paper presented at the ACM SIGMETRICS07.

Mitchell, T. (1999). Machine Learning and Data Mining. *Communications of the ACM*, *42*(11), 30–36. doi:10.1145/319382.319388

MobiHealth. (2007). *Putting care in motion.* Retrieved Sep 05, 2008, from www.mobihealth.com

Mody, A. N., Blatt, S. R., Mills, D. G., McElwain, T. P., Thammakhoune, N. B., & Niedzwiecki, J. D. (2007). Recent advances in cognitive communications. *IEEE Communications Magazine*, *45*(10), 54–61. doi:10.1109/MCOM.2007.4342823

Nahrstedt, K., Xu, D., Wichadakul, D., & Li, B. (2001). QoS-aware middleware for ubiquitous and heterogeneous environments. *IEEE Communications Magazine*, *39*(11), 140–148. doi:10.1109/35.965372

Nurmi, P., Hassinen, M., & Lee, K. C. (2007). *A Comparative Analysis of Personalization Techniques for a Mobile Application.* Paper presented at the Intl Conf on Advanced Information Networking and Applications Workshops (AINAW07).

O'Reilly, T. (2005). *What is Web 2.0?* Retrieved Sep 05, 2008, from www.oreillynet.com/pub/a/oreilly/tim/news/2005/09/30/what-is-web-20.html

Ortiz, S. (2007). 4G Wireless Begins to Take Shape. *IEEE Computer*, *40*(11), 18–21.

Osterwalder, A., Ondrus, J., & Pigneur, Y. (2005). Skype's Disruptive Potential in the Telecom Market: A Systematic Comparison of Business Models. Lausanne, Switxerland: University of Lausanne.

Pawar, P., van Beijnum, B. J., Peddemors, A., & van Halteren, A. (2007). *Context-Aware Middleware Support for the Nomadic Mobile Services on Multi-homed Handheld Mobile Devices.* Paper presented at the 12th IEEE Symposium on Computers and Communications (ISCC07), Aveiro, Portugal.

Pawar, P., Wac, K., van Beijnum, B. J., Maret, P., van Halteren, A., & Hermens, H. (2008). *Context-Aware Middleware Architecture for Vertical Handover Support to Multi-homed Nomadic Mobile Services.* Paper presented at the 23rd Annual ACM Symposium on Applied Computing (ACMSAC08), Ceará, Brazil.

Quax, P., Monsieurs, P., Lamotte, W., De Vleeschauwer, D., & Degrande, N. (2004). *Objective and Subjective Evaluation of the Influence of Small Amounts of Delay and Jitter on a First Person Shooter Game.* Paper presented at the NetGames04, Portland, USA.

Quinlan, R. (1993). C4.5: Programs for Machine Learning. San Meteo, CA: Morgan Kaufmann.

Salvo Rossi, P., Romano, G., Palmieri, F., & Iannello, G. (2003a). *Bayesian Modelling for Packet Channels.* Paper presented at the Italian Workshop on Neural Nets (WIRN).

Salvo Rossi, P., Romano, G., Palmieri, F., & Iannello, G. (2003b). *A Hidden Markov Model for Internet Channels.* Paper presented at the IEEE Intl Symposium on Signal Processing and Information Technology (ISSPIT), Darmstadt, Germany.

Shepherd, D., Scott, A., & Rodden, T. (1996). Quality-of-Service Support for Multimedia Applications. *IEEE MultiMedia, 03*(3), 78–82. doi:10.1109/93.556542

Soh, W. S., & Kim, H. S. (2003). QoS Provisioning in Cellular Networks Based on Mobility Prediction Techniques. *IEEE Communications Magazine, 41*(1), 86–92. doi:10.1109/MCOM.2003.1166661

SUN. (2001). The Jini™Technology Surrogate Architecture Overview. Santa Clara, CA: SUN Microsystems Inc.

Sun, Y., Tsang, K. F., Tung, H. Y., Lam, K. L., & Ko, K. T. (2008). *QoS Prediction for High Speed Packet Access Networks.* Paper presented at the 5th IEEE Consumer Communications and Networking Conference (CCNC08).

Tachikawa, K. (2003). A Perspective on the Evolution of Mobile Communications. *IEEE Communications Magazine, 41*(10), 66–73. doi:10.1109/MCOM.2003.1235597

Tan, S. (2004). *Evolution of mobile technology and business models.* Tech. Rep., Center for Information and Communication Technologies, Lyngby, DK.

van Halteren, A., Bults, R., Wac, K., Dokovsky, N., Koprinkov, G., Widya, I., et al. (2004a). Wireless Body Area Networks for Healthcare: the MobiHealth Project. In G. Demiris (Ed.), Studies in Health Technology and Informatics (Vol. 108, pp. 181-193). Amsterdam, the Netherlands: IOS Press.

van Halteren, A., Bults, R., Wac, K., Konstantas, D., Widya, I., & Dokovsky, N. (2004b). Mobile Patient Monitoring: The MobiHealth System. *The Journal on Information Technology in Healthcare, 2*(5), 365–373.

van Halteren, A., & Pawar, P. (2006). *Mobile Service Platform: A Middleware for Nomadic Mobile Service Provisioning.* Paper presented at the 2nd Intl. IEEE Conference on Wireless and Mobile Computing, Networking and Communications (WiMob06), Montreal, Canada.

Wac, K. (2009). *Collaborative Sharing of Quality of Service-Information for Mobile Service Users.* Unpublished doctoral dissertation, University of Geneva, Geneva, Switzerland.

Wac, K., Bargh, M., van Beijnum, B. J., Bults, R., Pawar, P., & Peddemors, A. (2009a). Power- and Delay-Awareness of Health Telemonitoring Services: the MobiHealth System Case Study. *IEEE JSAC. Special Issue on Wireless and Pervasive Communications in Healthcare, 27*(4), 1–12.

Wac, K., Bults, R., Konstantas, D., Chen, H., & van Beijnum, B. J. (2009b). Toward Mobile Web 2.0-based business methods: Collaborative QoS-information Sharing for Mobile Service Users. In M. Head, (Ed.), Mobile and Ubiquitous Commerce: Advanced E-Business Methods (Vol. 4, pp. xx-xx). Hershey, PA: IGI Global.

Wac, K., Bults, R., van Halteren, A., Konstantas, D., & Nicola, V. (2005). *Measurements-based performance evaluation of 3G wireless networks supporting m-health services.* Paper presented at the 12th Multimedia Computing and Networking (MMCN05), San Jose, CA.

Want, R., Fishkin, K. P., Gujar, A., & Harrison, L. B. (1999). Bridging physical and virtual worlds with electronic tags. In *Intl Conference on Human Factors in Computing Systems (SIGCHI)*. Pittsburgh, PA: ACM.

Weiss, S. M., & Kulikowski, C. A. (1991). Computer systems that learn: classification and prediction methods from statistics, neural nets, machine learning, and expert systems. San Francisco, CA: Morgan Kaufmann Publishers Inc.

Witten, I., & Frank, E. (2005). Data Mining: Practical Machine Learning Tools and Techniques. San Francisco: Morgan Kaufmann.

Xiao, X., & Ni, L. M. (1999). Internet QoS: a big picture. *IEEE Network, 13*(2), 8–18. doi:10.1109/65.768484

Yang, L., Dai, Y., & Zhang, B. (2008). *Reliability Oriented QoS Driven Composite Service Selection Based on Performance Prediction.* Paper presented at the Intl Conference on Software Engineering Knowledge Engineering (SEKE'08), San Francisco, CA, USA, July 1-3, 2008.

ADDITIONAL READING

The recommended additional reading includes work on Quality of Service in mobile computing, for example a survey paper of (Chalmers & Sloman, 1999b) or diverse standardization documents of International Telecommunication Union including (ITU-T, 1993, 2006, 2008). State of the art on location determination technologies for mobile service users has been presented, amongst the others by (Bellavista et al., 2008; Hightower & Borriello, 2001; LaMarca et al., 2005). A field of machine learning and data mining techniques is extensively presented in (Alpaydin, 2004; J. Han & Kamber, 2006; Quinlan, 1993; Weiss & Kulikowski, 1991). The doctoral thesis of the author herself (Wac, 2009) presents the detailed aspects of QoS prediction service, including its detailed functional and non-functional requirements, proposed architectural design as well as more QoS-predictions case studies.

Chapter 16
Enhanced QoS through Cooperating Schemes in Next Generation Wireless Networks

Dimitris E. Charilas
National Technical University of Athens, Greece

Athanasios D. Panagopoulos
National Technical University of Athens, Greece

Philip Constantinou
National Technical University of Athens, Greece

ABSTRACT

This chapter addresses the critical issue of Quality of Service (QoS) provisioning in next generation wireless networks. While the QoS offered to users may be enhanced through innovative protocols and new technologies, future trends should take into account the efficiency of the resource allocation strategies and the network/terminal cooperation as well. 4G networks will be characterized by an heterogeneous environment where several access networks will be available. The purpose of this chapter is to summarize techniques that enable efficient distribution of resources exploiting the existing infrastructure. Such techniques may involve either smart selection mechanisms or cooperating schemes among network entities. Since decision-making processes are examined, the use of game theory is considered as a valuable asset in the authors' work. To this end, the chapter also collects applications of both non-cooperative and cooperative game theory applications in wireless networks. The main aspects of both game types are presented and several games are modeled.

INTRODUCTION

Future wireless networks will be heterogeneous. The heterogeneous wireless networks integrate different access networks, such as IEEE 802.15 Wireless Personal Area Networks (WPAN), IEEE 802.11 Wireless Local Area Networks (WLAN), IEEE 802.16 Wireless Metropolitan Area Networks (WMAN), General Packet Radio Service (GPRS), Enhanced Data rates for GSM Evolution (EDGE), Code Division Multiple Access (cdma) 2000, Wideband Code Division Multiple Access (WCDMA),

DOI: 10.4018/978-1-61520-771-8.ch016

satellite networks, etc. The proliferation of wireless access technologies, and the evolution of the end-user terminals (smart phones, PDAs, etc...) are leading fast towards a ubiquitous, pervasive and rich connectivity offer, such that the end users won't be only always connected, but also always covered by multiple access networks / technologies.

Users are expected to access personalized services with context-awareness: location, characteristics of the available networks, user preferences, application requirements and terminal capabilities. The vision is that users will not be tied down to a long-term contract with one single operator but will instead be able to dynamically choose access provision on a per call basis. The evolving competitive marketplace is expected to provide a choice of access networks in any given location, each offering different network technologies with varying characteristics to transport the user's communications application.

Selection of the most efficient and suitable access network to meet a specific application's QoS requirements has recently become a significant topic, the actual focus of which is maximizing the QoS experienced by the user. The main concept is that users will rely on intelligent network selection decision strategies to aid them in optimal network selection. The end-users can potentially take wise decisions on which access network to connect to on the basis of several merit functions including the current load of the network and the cost-for-connectivity. **The first section** of this chapter will collect modern network selection algorithms, focusing on the merits that may be used as selection criteria.

In general, the wide area access wireless networks have larger coverage and support better mobility but have lower data rates and require higher power consumption on mobile terminals, for instance cellular networks GPRS and Universal Mobile Telecommunications System (UMTS); the local area access wireless networks have higher data rates and consume much less power on the terminals but have smaller coverage with limited mobility, for instance nomadic wireless networks WLAN and Bluetooth. The tradeoff between coverage and data rate is due to the relation between radio signal attenuation and distance.

The second section of this chapter will expose new ideas and trends on cooperative communication networks, in which wireless nodes cooperate with each other by transmitting information. Such schemes promise significant gains in overall throughput and energy efficiency. A feasible solution to implement the envisioned wireless networks and sophisticated terminals is twofold. On the one side cellular and nomadic (short-range) heterogeneous networks should cooperate with each other and on the other side the terminals forming a cluster should also collaborate. The next generation cooperative networks architecture is based on cellular reception of data which is then forwarded or shared among mobile devices within each others' proximity over the short-range link. To implement such cooperative networks, it requires that all the coexisted heterogeneous (wide access/local access) wireless networks can be designed to synergize efficiently. In addition, the multi-modality terminals can exploit the highly cooperative heterogeneous networks to cooperate with peers to realize the envisioned network.

Since decision-making processes are also examined in this chapter, we consider the employment of game theory as a promising asset. **The third section** of this chapter is devoted to explain this fascinating tool. Game theory is a mathematical tool developed to understand competitive situations in which rational decision makers interact to achieve their objectives, aimed at modeling situations in which decision makers have to make specific actions that have mutual, possibly conflicting, consequences (Fudenberg & Tirole, 1991). It has been used primarily in economics, in order to model competition between companies. Game theory techniques have recently been applied to various engineering design problems in which the action of one component has impact on (and perhaps conflicts with) that of the other components. As a tool, it may be used for forming cooperation schemes among entities such as nodes, terminals or network providers. During the last years, game theory has widely been applied to networking, in most cases to solve routing and resource allocation problems in a competitive environment. Many references on the state of the art of application of game theory to wireless networks are collected by Altman (Altman, Boulogne, Azouzi, Jimenez & Wynter, 2006). The most fundamental concepts of non-cooperative game theory, as well as the modeling of several examples are given by Felegyhazi and Hubaux (2006).

NETWORK SELECTION MECHANISMS

Background

Utility-based functions are commonly used to describe user preference rating relationship for a number of metrics. There are several cases where users' perception of a service can be estimated through using end-to-end QoS, as explained by Jain (2004); there is however, as also stressed by Dohler (Dohler, Meddour, Senouci & Saadani, 2008), too much subjectivity which poses barriers to its use.

Ormond et al. (Ormond, Murphy & Muntean, 2006) propose a utility-based strategy for network selection in a multi-access network scenario. Their solution is a user-centric selection strategy based on maximizing consumer surplus subject to meeting user-defined constraints in terms of transfer completion time. Moreover, a low complexity, centralized network selection scheme, aiming to optimally distribute the end users to the networks of the heterogeneous wireless system, in the sense of maximizing the global spectrum efficiency is presented by Huiling (Huiling, Zhaoyang, Peng, Hsiao-Hwa & Shiju, 2006). Bari and Leung (2007a), on the other hand, based on the Euclidean distance, they calculate the measures of closeness/separation *S*, for the best and the worst cases. Then, for each network, the preference *P* is calculated, based on the relative closeness to the best and separation from the worst solutions. The access network with the highest *P* value is then selected. Finally, Xuejun et al. (Xuejun, Ling, Rut & Yanqi, 2007) adopt a Satisfaction Degree Function (SDF) to evaluate, according to user's predefined criteria, available networks and select the best one(s) according to such criteria. The criteria considered incorporate user policies and information, including dynamic network status and application requirements.

Song and Jamalipour (2005a, 2005b) propose a methodology to compare networks based on the level of end-to-end QoS provided. The methodology combines two mathematical methods, called Analytical Hierarchy Process (AHP) and Grey Relational Analysis (GRA). An extension of this methodology with additional analysis on weight specification is introduced by Charilas et al. (Charilas, Markaki, Nikitopoulos & Theologou, 2008a; Markaki, Charilas & Nikitopoulos, 2007). According to these approaches, network selection is based on an algorithm for the estimation of end user satisfaction.

Selection Criteria: Key Performance Indicators

Evaluating the performance of a wireless network presupposes the existence of adequate metrics that reflect the network's actual capacity to satisfy its users. Key Performance Indicators (KPIs) are a set of measurements used to keep track of a network status over the time. KPIs can be categorized in two types whether they describe the network's resources or the QoS provisioned. The main KPIs related to QoS can be measured in any type of packet-switched network; some of them are listed below:

- **Delay:** (also referred as latency for WLAN or Short Radio Range systems), is the necessary time for one data packet to get from one designed point to another. The round-trip delay is measured by the time needed for sending a packet that is returned to the sender. From this, the one-way delay can be calculated, being half of the round-trip delay. A measured delay much longer than the expected one indicates that there is traffic congestion in the network.
- **Jitter:** represents the delay variation of the received packets over time. Packets that are transmitted at a constant rate are not necessary received at a constant rate, due to network behaviour congestion. Jitter is, hence, the measure in time of the irregularity of the packets transmission. Several formulas for jitter calculation can be defined. Jitter can firstly be calculated as a raw spreading of the delay around the expected delay. If no expected delay is available, another reference must be chosen, for example the delay of the first received packet. Jitter may also be evaluated with reference to the mean delay of the previously received packet, using a recursive formula.
- **Peak user data throughput:** is the measure of the maximum rate achieved during the data transmission in the network. This KPI must refer to a single user.
- **Mean user data throughput:** is the measure of the average rate achieved during the data transmission in the network. This KPI must also refer to a single user. The calculation is usually made by comparing the size of the transmitted data with the time of transmission of these data, both for uplink and downlink.

Available bandwidth may be calculated the capacity minus utilization over a given time interval, where utilization is the percentage of capacity currently being consumed by aggregated traffic. Several approaches of network selection are based on utility functions and decisions are taken according to values of certain figures of merit from physical layer, such as SNR, SNIR and BER.

- **Signal-to-noise ratio**: is a term for the power ratio between the signal (useful information) and the noise power. SNRs are usually expressed in decibels.
- **Signal-to-noise + interference ratio**: the ratio, generally expressed in decibels, of the power of the wanted signal to the total power of interfering signals and noise, evaluated in specified conditions at a specified point of a transmission channel.
- **Bit Error Ratio (BER)**: some examples of BER are (a) transmission BER, *i.e.*, the number of erroneous bits received divided by the total number of bits transmitted; and (b) information BER, *i.e.*, the number of erroneous decoded (corrected) bits divided by the total number of decoded (corrected) bits.

The Importance of Pricing in Network Selection

Consider that the user has a choice of several available networks, dependent on the current location. Each network in the system employs a fixed price per byte pricing scheme but charges with different prices and is subjected to different background traffic patterns. Naturally every user wants to deliver his/her data timely at the lowest price. The more delay the user experiences, the less he/she is willing to pay. The user may employ different possible tactics to select the network which will maximise his/her satisfaction.

For example a user may choose to always select one designated network regardless of its current characteristics, or to always minimise his/her expenses by choosing the cheapest network, or decide to continually opt the strategy of random network selection. Based on the above, users may be classified as risk neutral, risk seeking or risk adverse according to their preferences.

Pricing schemes for network selection consider that the user's willingness to pay depends on the level of his satisfaction. In other words, the more satisfied the user is, the more he is willing to pay. Satisfaction can be modeled in many different ways. For example, in case of file downloading, the user perceives the required downloading time as the sole KPI (key performance indicator). On the other hand, when audio or video services are considered, the user requires low delay and jitter to feel satisfied.

For example, Ormond et al. (Ormond, Muntean & Murphy, 2005) presented an economic model for network selection. The authors validate their proposed strategy by comparing it with the "Always Cheapest network selection" and the "Consumer Surplus network selection" strategies.

Weights of Selection Criteria

So far we have collected a list of metrics that could aid the user in the selection process. Many selection schemes however require not only the values of selected criteria, but also the importance each one of them bears. In other words, assuming that n criteria may evaluate a network's performance is not sufficient; we also need to define the weights $w_1, \ldots, w_n$ that correspond to each one.

In most cases in bibliography these weights are mainly defined through the creation and analysis of questionnaires, which reflect the user's overall perception of a service. However, such approaches depend only on user feedback to determine relative weights and thus cannot be considered precise, since user's perception and opinion is subjective. Alternatively, other approaches consider the system's characteristics and adapt the values of weights according to each cell's performance (Charilas et al., 2008; Markaki et al., 2007). Due to their internal subjectiveness, weight specification can be implemented with the employment of tools such as fuzzy logic and multi-attribute decision making.

Fuzzy Logic-Based Schemes for Network Ranking

Modern approaches for network selection may additionally involve either Fuzzy Logic-based schemes or Multi Attribute Decision Making (MADM) schemes. In the first one a group of fuzzy logic rules in the form of linguistic IF-THEN have to be defined to model network selection (Kher, Somani & Gupta, 2005). However, such rules have to be configured by the user manually prior to selection and their complexity becomes overwhelming high as the number of attributes increases. Thus, the scalability of the fuzzy logic-based schemes is extremely low, which limits their usage in wireless networks selection. Despite the scalability issues, fuzzy logic-based schemes may be applied to the problem in question, since the objectivity of parameter weights and indicative values involves imprecision.

The fuzzy sets theory, introduced by Zadeh in1965 to deal with vague, imprecise and uncertain problems, has been used as a modeling tool for complex systems that can be controlled by humans but are hard to define precisely. The main characteristic of fuzziness is the grouping of individuals into classes that do not have sharply defined boundaries. The uncertain comparison judgment can be represented by the fuzzy number. For example, a triangular fuzzy number is the special class of fuzzy number whose membership is defined by three real numbers, expressed as (l, m, u), here l is the lower limit value, m is the most promising value and u is the upper limit value.

Applying Multi-Attribute Decision Making to Network Selection

Multi Attribute Decision Making (MADM) involves the selection of a series of criteria that influence a decision and the comparison of their importance, so that the best option among a set of alternatives can be reached. MADM methods focus on determining exactly how much importance each criterion yields concerning the final selection. Once this task has been completed, all alternatives can be evaluated regarding their performance on all criteria and the best alternative can be pointed out.

MADM provides a solid framework for network selection, since it constitutes a multi-criteria scenario. Researchers have considered the use of MADM algorithms (Bari & Leung, 2007b) to rank the candidate networks in a preference order. Numerous types of MADM algorithms exist. Several of them may be suitable for solving a decision problem so that the decision maker may encounter the task of selecting amongst a number of feasible methods the most appropriate one. The final objective of the problem is analyzed into a number of decision elements, which are afterwards further analyzed until the problem acquires a hierarchical structure, in the inferior level of which the alternative solutions of the problem are found. Common QoS parameters, such as Delay, Jitter, Packet Loss, Throughput etc can be employed as criteria in such schemes. Furthermore, additional criteria may involve Availability, Cost etc. MADM requires also an initial indication on each parameter's relative importance with regard to other parameters. As far as network selection is concerned, such an indication may be acquired from questionnaires or measurements. Criteria such as user satisfaction or cost are rather subjective and it is preferable that they are expressed in linguistic terms that correspond to fuzzy numbers.

The problem of network selection can therefore be modeled as $P = (A, C, w)$, where

- $A = \{1, \dots, N\}$ denotes the set of alternatives, in this case candidate networks
- $C = \{1, \dots, N\}$ denotes the set of criteria impacting the decision process of network selection,
- $w = \{1, \dots, M\}$ denotes the set of weights assigned to the selected criteria depending on the information about the specific service requested or the user QoS profile, so that $\sum_{i=1}^{M} w_i = 1$.

Based on the scores achieved for each one of the selected criteria (attributes), the i_{th} candidate network can be represented by a vector as follows:

$$DM_i = \begin{bmatrix} C1_i & C2_i & \cdots & CM_i \end{bmatrix} \tag{1}$$

For N alternative networks to be considered in the selection process, a matrix NW is formulated as follows:

$$DM = \begin{bmatrix} C1_1 & C2_1 & \cdots & CM_1 \\ C1_2 & C2_2 & \cdots & CM_2 \\ \vdots & \vdots & \vdots & \vdots \\ C1_N & C2_N & \cdots & CM_N \end{bmatrix} \tag{2}$$

Song and Jamalipour (2005b) as well as Charilas (2008) and Markaki (2007) demonstrate an example of how network selection can be modeled using two MADM methods, Analytical Hierarchy Process (AHP) and Grey Relational Analysis (GRA). AHP is used as a means of finding the optimal solution to a complex decision problem by synthesizing all problem-deciding factors. The elements of each level of the hierarchy are compared per two as far as the degree of preference of one against the other is concerned. This comparison is realized via matrices (the number of which equals the number of nodes in the structure), where the preferences for each pair of decision elements are declared. In the afore-mentioned approaches,

AHP is used to estimate the weights of all factors, which are then used in GRA.

On the other hand, GRA builds grey relationships between elements of two series to compare each member quantitatively. One of the series is composed of best quality entities, while the other series contains comparative entities. The less difference between the two series, the more preferable the comparative series. A Grey Relational Coefficient (GRC) is used to describe the relationship between them and is calculated according to the level of similarity and variability.

Bari and Leung (2007b) demonstrate how network selection can be modeled using another MADM method, called ELECTRE. This method also performs pair-wise comparisons among alternatives for each one of the attributes separately to establish outranking relationships between the alternatives.

Combining Fuzzy Logic and Multi-Attribute Decision Making

Several combinations of both scheme categories have also been proposed. MADM methods have been extended, so as to support fuzzy numbers. For example, Fuzzy AHP (FAHP), a fuzzy extension of AHP, was developed to solve the hierarchical fuzzy problems. In this approach, the pair wise comparisons in the judgment matrix are fuzzy numbers that are modified by the designer's focus. Based on a set of standardized answers (linguistic variables) provided through appropriate question forms, the corresponding triangular fuzzy values are defined and the pair wise comparisons matrix $\tilde{A}$ is constructed as

$$\tilde{A} = \left(\tilde{a}_{ij}\right)_{n\times n} = \begin{bmatrix} (1,1,1) & (l_{12},m_{12},u_{12}) & \cdots & (l_{1n},m_{1n},u_{1n}) \\ (l_{21},m_{21},u_{21}) & (1,1,1) & \cdots & (l_{2n},m_{2n},u_{2n}) \\ \vdots & \vdots & \vdots & \vdots \\ (l_{n1},m_{n1},u_{n1}) & (l_{n2},m_{n2},u_{n2}) & \cdots & (1,1,1) \end{bmatrix} \tag{3}$$

where $\tilde{a}_{ij}$ denotes a triangular fuzzy number depicting the relative strength of two elements. The necessity behind these new approaches lies behind the inability of conventional schemes to adequately handle the inherent uncertainty and imprecision associated with the mapping of the decision-maker's perception to exact numbers, since they cannot reflect the human thinking style. Kong and Liu (2007) presented a fuzzy madm algorithm to estimate parameter weights, as an integration of both subjective and objective weights.

COOPERATION AMONG USERS IN 4G NETWORKS

Background

The success of 4G depends on the combination of network and terminal heterogeneity (Frattasi, Fathi, Fitzek, Prasad & Katz, 2006; Katz & Fitzek, 2005). Network heterogeneity guarantees ubiquitous connection and provision of common services to the user, ensuring at least the same level of QoS when passing from one network's support to another one. Moreover, due to the simultaneous availability of different networks, heterogeneous services are also provided to the user (Frattasi, Fathi & Fitzek, 2006; Zhang, Fitzek & Katz, 2006).

Cooperation is divided into two categories (Zhang et al., 2006): *Macro cooperation* and *Micro cooperation*. In the macro cooperation the cooperating entities are wireless terminals, virtual access points, wireless routers and other macroscopic wireless system parts. The potential of macro cooperation is exploiting shorter radio propagation distance to reduce interference and to increase data rate. As a result, lower transmission and reception power is required. Additionally, cooperative diversity (or cooperation gain) can be exploited to form virtual antenna array (or virtual MIMO) to achieve network performance gain. In the micro cooperation the cooperating entities are microscopic components including processing units, functional parts and algorithms. The basic idea of micro cooperation is virtually sharing these microscopic components to obtain gain by designed cooperation mechanism.

By sharing energy resources, the cooperating terminals inherently share processing as well as spectrum resources. This makes cooperation a paradigm of gaining energy efficiency by sharing processing and spectrum resources among cooperating terminals. It is apparent that proper incentives, e.g. in terms of energy efficiency, have to be in place before each terminal agrees to participate in a cooperating network and to spend energy resources to serve the needs from other cooperating terminals.

Cooperative Services in 4G

The concept of node cooperation introduces a new form of diversity that results in an increased reliability of the communication, leading both to the extension of the coverage and the minimization of the power consumption (Frattasi, Can, Fitzek & Prasad, 2005; Perrucci, Fitzek, Boudali, Canovas, Nejsum & Strudstrap, 2007). In fact, mobile terminals are less susceptible to the channel variations and shadowing effects and can transmit at lower power levels in order to achieve a certain throughput, thus increasing their battery life. Furthermore, cooperative transmission strategies may increase the end-to-end capacity and hence the spectral efficiency of the system. In a cooperation scenario the terminals may communicate over a short range communication in parallel to the cellular communication. Such architecture offers virtual high data rate, lower energy consumption and new business models.

Frattasi (2005) and Perrucci (2007) show that terminals in the same multicast group with physical proximity can communicate with each other using high-speed wireless links. The basic idea behind this mechanism is illustrated in Figure 1. Users belonging to the same multicast group may download data and at the same time establishing Bluetooth connections to share downloaded data with less fortunate users. The decision about sharing resources involves a reputation matrix, Finally, resource sharing users are rewarded according to their contribution. Rewards may involve monetary value, points or even provision of higher QoS in future service requests.

In order to successfully deploy cooperative schemes in wireless networks, new network architectures have to be developed. Valavanis et al. (Valavanis, Ververidis, Vazirgianis, Polyzos & Norvag, 2003) describe a hybrid (e.g. between adhoc and infrastructure-based) architecture called MobiShare, that enables mobile devices to share their data encapsulated in services. Two approaches of network architectures may be considered so that cooperation among nodes can be established (Scaglione, Goeckel & Laneman, 2006):

- In the *centralized architecture* cooperative transmission is centrally activated and controlled by cluster access points (APs). All terminals communicate through a cluster AP, which handles routing to other clusters.
- In the *decentralized architecture* random cooperative clusters are formed. A random source transfers extra control information and link parameters in the message. The nodes can then infer their transmission schedule according to this information. They may be ignorant of the codes chosen by the other nodes, but the resulting cooperative gains are close to those of a centralized scenario in which codes are explicitly assigned to the nodes. A small cluster of nodes can act as a source and recruit additional nodes to form a larger cluster.

Figure 1. User cooperation

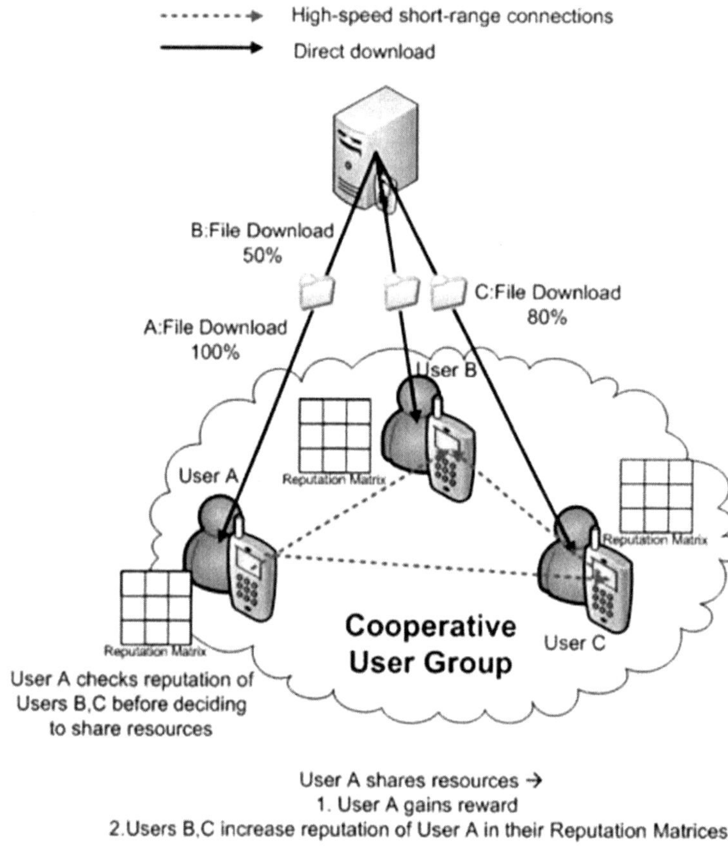

Incentive Mechanisms

Incentive mechanisms intend to provide a framework that urges players to cooperate for the best interest of all participants. In other words, they provide a *motive* so that each individual prefers to work along with others, sometimes sacrificing their own resources and sometimes benefiting from the resources of others. Two types of incentive mechanisms are distinguished:

- *Credit based systems*
- *Reputation based systems*

Chakravorty (Chakravorty, Banerjee, Agarwal & Pratt, 2003) developed a Mobile Bazaar (MoB) framework to allow collaborative networking between users based on incentives. MoB is an open market architecture which uses a payment system to manage resource trades and where mobile users opportunistically and flexibly trade various services. However, MoB supports only a single device per user since its main focus is on a user having only one device and getting all service through this device.

In some occasions some nodes may refuse to forward packets in order to conserve their limited resources (.e.g., energy), resulting in traffic disruption. Nodes exhibiting such behaviour are termed selfish. The basic idea for node punishment is that nodes should be rewarded or penalized based on their behaviour. Nodes that offer resources should be aided. On the other hand, selfish nodes should be gradually isolated from the network. This is bound to happen if nearby nodes refuse to forward packets deriving from the selfish ones. In addition, some systems may punish misbehaving nodes by isolating them from the network for a certain period of time in order to provide an incentive for users to cooperate.

A significant problem arising when dealing with node behavior issues is that sometimes, due to packet collisions and interference, cooperative nodes will be perceived as being selfish, which will trigger a retaliation situation. Such poor judgment may lead nodes to stop cooperating and thus degrade the overall network performance. In other words, disadvantaged nodes that are inherently selfish due to their precarious energy conditions shouldn't be excluded from the network using the same basis as for malicious nodes. Furthermore, users on the outskirts of a network are found at a disadvantage unrelated to their willingness to participate. Those users will not have as much traffic routed through them due to their location and furthermore will have lower congestion prices because of that. They will thus earn significantly less than a centralized node and be penalized for it resulting in low QoS

In several cases incentive mechanisms are required not only to motivate the user to cooperate but also to discourage him from cheating. For example, malicious users may refuse to pay. To this end, disincentives against cheating may also have to be considered.

Cooperation Based on Credits

In *credit-based systems*, the basic idea is to use notional credit, monetary or otherwise to pay off users for the congestion costs (transmission and battery costs) they incur from forwarding packets from other users. These credits can then be used to forward their own packets through other users, resulting in an incentive to act as relay points, especially where there is the greatest excess demand for traffic since this is when they earn the most. Users who do not cooperate will not be able to use the network themselves, having not earned any credits. Under the general token mechanism, a user's token counter is increased when it forwards, and decreased proportionally to the number of hops it needs when it sends. This inevitably means that a user needs to forward more than he sends and also limits the amount of information that any user can send at any given time, dependent on their store of tokens.

Zhang (Zhang, Ileri & Mandayam, 2008) proposes an incentive mechanism called bandwidth exchange, where a node can delegate a portion of its bandwidth to another node in exchange for relay cooperation. Buttayan and Hubaux (2003), as well as Zhong et al. (Zhong, Cheng & Yang, 2003) have developed pricing-based protocols where the amount charged per packet is determined exogenously and is the same for each node in the network. The drawback with the above approaches is the assumption of a simplified channel model - the energy required to forward a packet is assumed to be constant regardless of transmission distance. Stimulation mechanisms that take into account the fading channel have been developed during the last years. Results show that in this case cooperation is highly dependent on network geometry and configuration.

Incentives to cooperate have been proposed by Buttyan and Hubaux (2003) in the form of so-called *nuglets* that serve as a per-hop payment in every packet or in the form of counters to encourage forwarding. Both nuglets and counters reside in a secure module in each node, are incremented when nodes forward for others and decremented when they send packets for themselves. One of their findings is that, given such a module, increased cooperation is beneficial not only for the entire network but also for individual nodes.

Shastry and Adve (2006) focus their work on stimulating nodes to participate in cooperative diversity transmissions in wireless ad hoc networks. More specifically, they consider that the source and relay may interact through reimbursement prices, transmitter power control and forwarding/

protocol preferences such that their utilities are maximized.

Some systems propose using real money as credit, either directly or indirectly (to buy virtual credit). One problem that must be kept in mind is that the introduction of monetary rewards in the system may act not only as an incentive for collaboration, but also (for a poorly designed system) as an incentive for cheating: an adversary may attempt to corrupt the routing tables (both his own and those of others) in order to gain rewards. Such malicious behavior can be avoided by using a set of protocols that allow the operator to collect information based on which it can decide which accounts should be charged and which accounts should be credited. Such protocols may (Salem, Buttyan, Hubaux & Jakobsson, 2003):

a. Authenticate the source and the destination

The operator should be able to reliably identify the source and the destination of every packet received by the base stations in order to know which account to charge.

b. Authenticate the forwarding nodes

The operator must be able to reliably identify the forwarding nodes (both in the initiator and in the correspondent routes) during the session setup phase, and later verify that this exact set of nodes participated in the forwarding of each packet.

Cooperation Based on Reputation

In wireless networks, nodes can be thought of as members of a community that share a common resource. The key to solve problems related to node misbehavior derives from the strong binding between the utilization of a common resource and the cooperative behavior of the members of the community. Thus, all members of a community that share resources have to contribute to the community life in order to be entitled to use those resources. However, the members of a community are often unrelated to each other and have no information on one another's behavior. Generally, *reputation* is defined as the amount of trust inspired by a particular member of a community in a specific setting or domain of interest. Members that have a good reputation, because they helpfully contribute to the community life, can use the resources while members with a bad reputation, because they refused to cooperate, are gradually excluded from the community. Reputation is formed and updated along time through direct observations and through information provided by other members of the community. Michiardi and Molva (2002) classify reputation in three types:

- **Subjective reputation** is calculated directly from a subject's observation. A subjective reputation is calculated using a weighted mean of the observations' rating factors, giving more relevance to past observations. The reason why more relevance is given to past observations is that a sporadic misbehavior in recent observations should have a minimal influence on the evaluation of the final reputation value: as a result, it is possible to avoid false detections due to link breaks and to take into account the possibility of a localized misbehavior caused by disadvantaged nodes.
- **Indirect reputation** adds the information provided by other members of the community to the final value given to the reputation of a subject.
- **Functional reputation** allows for the possibility to calculate a global value of a subject's reputation that takes into account different observation/evaluation criteria.

Each type of reputation is obtained as a combination of different observations made by a subject over another subject with respect to a defined function. Furthermore, the above types of reputation information can be combined. When a node detects uncooperative behavior it disseminates this observation to other nodes which take action to avoid being affected by the node in question by changing traffic routes.

However, as with the token-based incentives system, trust management systems are subject to some significant problems. The first problem is that they take up considerable resources due to the constant transmission of observation data, which serves no purpose other than to monitor node behavior. This spends valuable bandwidth and battery power that could be used to send actual data. Trust management systems also suffer from vulnerabilities due to exchanging second hand information. Nodes may falsely accuse other nodes of misbehaving or collude with each other to cheat other users on the network.

Cooperative Relays and Diversity

Relay channels provide the fundamental concept of cooperative diversity. Cooperative relaying is a significant means to meet the demanding requirements of next generation wireless networks in terms of system coverage and capacity. The vision of providing coverage in cellular systems at reasonable cost by using mobile terminals as relay stations has recently led to tremendous research efforts and emerges as a valuable option for future generations of wireless networks. Radio Relays through Cooperative Diversity techniques create a virtual antenna array, or a multiple-input, multiple-output (MIMO) system (Tse & Viswanath, 2005).

While conventional relaying systems employ relays as pure forwarders that operate in a relay chain, it is the objective of cooperative relaying to go one step further. The idea is to exploit two fundamental features of the wireless medium: its broadcast nature, and its ability to achieve diversity through providing independent channels. While the broadcast nature is frequently considered to be a drawback as it leads to mutual interferences in a wireless network, the concepts of cooperative relaying aim at benefiting from the fact that a signal, once transmitted, can be received and usefully forwarded by multiple terminals. Moreover, by using truly intermediate relay stations one can benefit from a reduction of the end-to-end path loss. Relay cooperation has been recognized as an important mechanism to enhance connectivity and throughput in multi-hop wireless networks, especially under varying channel conditions. One major problem of relay cooperation is that relaying always incurs energy and possibly delay costs. To a rational and selfish node these costs are worth incurring only if it receives at least comparable returns in the long term.

The cooperative relay technique can further enhance the performance of a relay-based network by employing the antennas of different relay stations as a part of an antenna array, and applying the multi-antenna techniques among several relays. Coordinating the transmitted signals properly, the link quality, data rate and coverage of the relay-based system can be greatly improved. Here we are going to present the basics of cooperative relays.

We consider the fundamental cooperative scenario (Cover & Gamal, 1979; Zheng &Tse, 2003) as depicted in Figure 2, which include a single relay and where all nodes feature a single antenna. There can be developed various lower bounds based on the capacity (Laneman, Tse & Wornell, 2004), by employing three different coding schemes (Boyer, Falconer & Yanikomeroglu, 2004; Laneman et al., 2004)

a. *Facilitation*: in which the relay does not have an active role but eases the source by causing as little interference as possible,
b. *Cooperation*: in which the relay fully decodes the source message and retransmits, jointly with the data transmitted by the source,

Figure 2. The fundamental cooperative radio relaying scenario

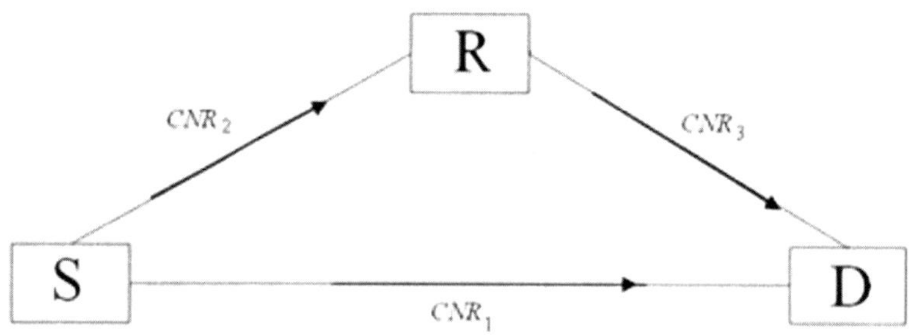

c. *Observation*, in which the radio relay encodes a quantized version of its received signal.

According to their forwarding strategy, the cooperative protocols can be categorized as (Zimmermann, Herhold & Fettweis, 2005):

a. *Amplify and Forward*, where the relay acts as an analogue repeater (transparent repeater, bent-pipe system),
b. *Decode and Forward*, where the relay fully decodes, encodes and retransmits the received signal (regenerative),
c. *Decode and Re-encode*, where the relay fully decodes the received message, but constructs a codeword that is different from the source codeword.

The transmission schemes based on protocol nature are further classified (Zimmermann et al., 2005):

a. *Fixed protocols*, where the relay always forwards a processed version of the received message,
b. *Adaptive protocols*, where the relay uses a threshold rule to make a decision autonomously, whether to forward or not and
c. *Feedback protocols*, where the relay has an assistant role on the transmission only when the destination requires extra information.

As a further classification, relays in non-regenerative systems can be categorized into (Hasna & Alouini, 2004): a) channel state information assisted relays, and b) blind relays. Non-regenerative systems with CSI assisted relays use instantaneous CSI of the first hop to control the gain introduced by the relay.

In the destination node of the simple cooperative relay of Figure 2, there may be various diversity combining techniques that can be employed and these are distinguished:

a. *Selection Combining (SC),* from the two received signals (from the relay and the direct link), the strongest signal is selected (or the one with the minimum attenuation),
b. *Switched combining*, where the receiver switches to another signal when the current SNR falls below a predefined threshold. This is a less efficient technique than selection combining,
c. *Equal Gain Combining (EGC),* where all the received signal are added coherently,
d. *Maximal Ratio Combing (MRC),* where the received signals are weighted with respect to their SNR and then summed.

There are many papers in the literature that are dealing with the performance of the cooperative relays under various fading channels (Safari & Uysal, 2008). Here we are going to present simply how the outage probability of an MRC and a SC receiver at the destination can be calculated. The source node **S** communicates with the destination node **D** through two different routes. The first signal is directly transmitted by the node **S** to the node **D** and the second signal is transmitted by the node **S** to the node **D** through the relay node **R** (dual-hop transmission). These two signal paths form two diversity branches which are combined by the node **D** using coherent combining to form the final received signal. The received Carrier-to-Noise (CNR) of link $j = 1, 2, 3$ is denoted (in linear scale) as CNR_j.

The relay **R** retransmission takes place in the form of the Decode-and-Forward technique (Regenerative relay) where the received signal is regenerated using the full receiver-transmitter processing chain containing the sequence of demodulation, channel decoding, encoding and modulation. Since the Relay Node R is a regenerative relay, the outage event of the dual-hop transmission is the event where CNR_2 or CNR_3 or both do not exceed a specified level. Thus, the outage event is the event where the minimum of CNR_2 and CNR_3 does not exceed a specified threshold, meaning that $\min\left(CNR_2, CNR_3\right)$ characterizes the dual-hop performance. Therefore, the dual-hop CNR_{23} is given by:

$$CNR_{23} = \min\left(CNR_2, CNR_3\right) \quad (4)$$

The above final result can also be verified in terms of spectral efficiency similarly to the methodology in (Boyer et al., 2004):

$$\left(\frac{C}{B}\right)_{23} = \min\left(\left(\frac{C}{B}\right)_2, \left(\frac{C}{B}\right)_3\right) \quad (5)$$

where $\left(C/B\right)_j = \log_2\left(1 + CNR_j\right)$.

The destination node **D** combines the direct-link Carrier-to-Noise Ratio CNR_1 and the dual-hop CNR CNR_{23} and the total CNR $TCNR_C$ is produced. Here, two different combining techniques are used: the Selection Combining (SC) and the Maximal Ratio Combining (MRC). If the SC technique is used, the destination node **D** selects the received signal with the greater CNR. Using (4) the final CNR $TCNR_C = TCNR_{SC}$ is given by:

$$TCNR_{SC} = \max\left(CNR_1, CNR_{23}\right) = \begin{cases} \max\left(CNR_1, CNR_2\right), CNR_2 < CNR_3 \\ \max\left(CNR_1, CNR_3\right), CNR_3 < CNR_2 \end{cases} \quad (6)$$

Adopting now MRC technique, node **D** adds the two received signals and using (4) the final CNR $\gamma_C = \gamma_{MRC}$ is given by:

$$CNR_{MRC} = CNR_1 + CNR_{23} = \begin{cases} CNR_1 + CNR_2, CNR_2 < CNR_3 \\ CNR_1 + CNR_3, CNR_3 < CNR_2 \end{cases} \quad (7)$$

From the above equations (6) and (7), the outage probability of the two combining techniques can be evaluated, after making some fundamental assumption for the fading channels.

Now we are going to discuss some issues regarding the access to the wireless medium of the participating nodes. The cooperation is facilitated through the use of either Time or Frequency Division Multiple Access (TDMA or FDMA) or a more bandwidth-efficient scheme such as Space Division Multiple Access (SDMA, smart antennas). In the TDMA or FDMA scheme, for the configuration of Figure 2, two degrees of freedom (DOF) are utilized: in the first time/frequency slot the source node **S** broadcasts a signal to both the relay node **R** and the destination node **D**, while in the second time/frequency slot the relay **R** retransmits the received signal to the destination D. In the SDMA scheme, only one DOF is used: **S** transmits and **R** receives while transmitting the previous transmission from **S**; the destination **D** simultaneously receives the transmission from **S** and the relayed transmission from **R** (interference is avoided through the use of spatial multiplexing such as smart or highly directive antennas). SDMA scheme can alternatively be viewed as a pipelining system, which explains the DOF superiority over TDMA/FDMA. This multiple access scheme has the same spectral efficiency as a direct link system but is less efficient in terms of energy as it consumes more energy per DOF slot.

The idea of cooperation has also gained interest and is utilized to the design of higher layer protocols e.g. MAC (Zhu & Cao, 2005). Khandani (Khandani, Abounadi, Modiano & Zhang, 2003) has employed the general concept of cooperation to the routing problem of ad hoc networks. Multiple nodes are cooperating in sending the information to a single receiver node, and they can precisely delay their transmitted signal to achieve perfect phase synchronization. The problem of finding the optimal cooperative route from a source node to a destination node can be formulated as a Dynamic Programming problem. Moreover, a new approach to cross layer design of multihop wireless networks with cooperative diversity has been proposed and succeeds in maximizing the gain in the physical layer and hitting off the interactions with higher layers (Long & Hossain, 2008). In this paper the authors have applied the nonlinear optimization techniques to develop the optimization frameworks. The first framework has to do with joint routing and cooperative resource allocation which minimizes the total power consumption, while the second framework uses congestion control through a utility function to break down a balance between maximizing the sum rate utility and minimizing total cost assumption.

Cross-Layer Optimization

Cross-layer optimization determines a general concept of useful interactions among different layers of the protocol stack with a view to improving wireless network performance. The parameters that should be exchanged among the layers are specific figure of merits at each layer, or particular information that can be used to estimate other quantities and determine accordingly the cross-layer optimization scenario. There are two approaches for the evaluation of the cross-layer design (Lin, Shroff & Srikant, 2006): the *evolutionary* approach also referred in the literature as layer-related approach that focuses mainly on the improvement of the interoperability among the layers with a view to optimizing the overall network performance. The second approach is the *revolutionary* that tries to optimize the network performance without binding to the general concept layering. There are other two design cross-layer approaches taking into account the exchange of signalling during the operation. These are (Aune, 2004; Giambene & Kota, 2006): the *implicit* cross-layer design, where there is no exchange of information among different layers during operation, but the design takes into consideration all the layer interactions, and the *explicit* cross-layer design, where signalling interactions among adjacent or not protocol layers are employed during the phase of operation.

For 4G networks, OFDM (Orthogonal Frequency Division Multiplexing) has been chosen to become the basic modulation scheme. OFDM is a combination of modulation and multiplexing. The general concept is to divide the available spectrum into several subcarriers. It has been shown that OFDM is an effective technique to combat frequency selective multipath fading. Furthermore, in an OFDM wireless network, different subcarriers can be allocated to provide a flexible multiuser access scheme and take advantage multiuser diversity. There are numerous papers in the literature that deal with the cross-layer optimization of OFDM systems for 4G Wireless Communications (Song & Li, 2005).

APPLICATIONS OF GAME THEORY IN WIRELESS NETWORKS

Background - Fundamentals of Game Theory

A game consists of a principal and a finite set of players $N = \{1,2,\ldots,N\}$, each of which selects a strategy $s_i \in S_i$ with the objective of maximizing his utility u_i. The utility function u_i (s): S→R characterizes each player's sensitivity to everyone's actions. In **non-cooperative** games, each player selects strategies without coordination with others. On the other hand, in a **cooperative game**, the players cooperatively try to come to an agreement, and the players have a choice to bargain with each other so that they can gain maximum benefit, which is higher than what they could have obtained by playing the game without cooperation.

When a player makes a decision, he can use either a pure or a mixed strategy. If the actions of the player are deterministic, he is said to use a pure strategy. If probability distributions are defined to describe the actions of the player, a mixed strategy is used. The strategy profile ***s*** is the vector containing the strategies of all players: $s = (s_i), i \in \mathbf{N} = (s_1, s_2, \ldots, s_N)$.

The equilibrium strategies are those that the players pick while trying to maximize their individual payoffs. In game theory, the Nash equilibrium is a solution concept of a game involving two or more players, in which no player has anything to gain by changing only his own strategy unilaterally. If each player has chosen a strategy and no player can benefit further by changing his strategy while the other players keep theirs unchanged, then the current set of strategy choices and the corresponding payoffs constitute a Nash equilibrium (Fudenberg & Tirole, 1991; MacKenzie & DaSilva, 2006).

In wireless networks, the players can be either wireless terminals striving for obtaining as much possible bandwidth from the share medium; or they can be wireless network providers aiming at increasing their market share or revenue. It is clear that in both cases the actions of one player can affect the others, possibly in a negative way.

Applications of Non-Cooperative Games

Medium Access Control

As a first example of a situation in which game theory is considered as an appropriate analysis tool, we choose to model random access to a communications channel, in this case slotted Aloha (MacKenzie & DaSilva, 2006; MacKenzie & Wicker, 2001a). Users who wish to transmit typically wish to do so as soon as possible. If multiple users try to transmit simultaneously, though, all accesses fail; in addition, unsuccessful attempts to transmit may cost to the network resources. The users trying to transmit have conflicting objectives. In slotted Aloha, time is divided into slots and via some method of synchronization, all users are presumed to know where the slot boundaries are located. When a user wishes to access the shared channel the user waits until the next slot boundary and then begins attempting to transmit. If two or more users try to transmit in the same slot, the users become "backlogged" and must attempt to transmit again in a future slot.

Let *G(n)* be the game in which there are currently *n* users. In each stage of *G(n)* each of the players must decide whether to transmit (T) or wait (W). If one player decides to transmit and the rest decide to wait, the player who transmits will receive a payoff of 1, and each of the other $(n-1)$ players will play *G(n–1)* in the next period. If either no users transmit or more than one user transmits, all players will play *G(n)* again in the next period. Players place a lower value on payoffs in later stages than on current payoffs. This is represented with a per-period discount factor $d < 1$. Let $u_{i,n}$ represent user *i*'s utility from playing *G(n)* and let *K* be a random variable denoting the number of other users who transmit in a given slot. We then come up with the following utility functions:

$$u_{i,n}(T) = \frac{P[K=0]}{1 - d \cdot P[K>0]}, \quad u_{i,n}(W) = \frac{d \cdot P[K=1]}{1 - d \cdot P[K \neq 1]} u_{i,n-1} \tag{8}$$

Pricing of Network Resources

Bennis and Lara (2008) address the problem of non-cooperative operators trying to maximize their profits by offering extra spectral resources to other operators starving for spectrum in an oligopoly market. The revenue and costs for primary operator *i* are calculated as follows:

$$\mathrm{Re}\, v(i) = c_1 M_i$$

$$Cost\left(q_i\right) = c_2 M_i \left(BW_i^{req} - a_i \frac{W_i - q_i}{M_i}\right)^2 \quad (9)$$

where c_1 and c_2 denote the weights for the revenue and cost functions respectively, $BW^{req}i$ denotes the bandwidth requirement for a primary operator and a_i is the spectral efficiency for primary operator *i*. M_i is the number of primary connections. Based on the aforementioned model, a game can be formulated where the players are the primary operators offering spectrum. Their strategy may be the price per unit of spectrum p_i and finally the payoff for every operator can be the profit (revenue-costs) after the spectrum transaction. The revenue for every operator can be calculated as:

$$prof_i(p_i) = q_i p_i + \mathrm{Re}\, v(i) - Cost\left(q_i\right) \quad (10)$$

where p_i denotes the set of prices offered by all players in the game and $p = \{p_1, \ldots, p_N\}$ is the set of prices. The best response function of operator *i*, given a set of prices offered by other primary operators p_i, is $B_i(p_i) = \arg\max_{p_i}(prof_i(p_i))$. Using the equality $\frac{\partial \Pr of_i(p)}{\partial p_i} = 0$ we get the Nash Equilibrium of the game.

Access Admission Control

The basic goal of an Admission Control algorithm in cellular networks is to control the admission of new sessions within the network with the purpose of maintaining the load of the network within some boundaries. When several radio technologies may at the same time meet the services demands, a decision is necessary to select the most suitable radio access technology on a per user basis. The decision about the target network can be based on either user or network/operator criteria (Charilas et al., 2008; Markaki et al., 2007; Ormond et al., 2006;). Admission control takes place each time a new session request is received and decides whether it should be allocated resources or be rejected due to lack of resources.

Network vs. Network

In this kind of games the networks constitute the players. As individual players in the game, the access networks will therefore try to maximize their own payoff by choosing the best available strategy in a rational manner, meaning that they will try to choose the request that best fits their characteristics. Such a game may be played in rounds. At each round of the game the networks should decide which request will maximize their payoff and then select it. Once a request is selected it is removed from the set of service requests and the game is repeated, until all requests have been selected (Charilas, Markaki & Tragos, 2008). The proposed game is *non-zero-sum* and *non-cooperative*, since a player is unable to bind and enforce agreements with other players.

User vs. Network

The main goal of such schemes is to maximize not only the QoS offered to customers, but also the provider's gain, therefore balancing the interests of both parties (Lin, Chatterjee, Das & Basu, 2005; Vlacheas, Charilas, Tragos & Markaki, 2008). It is assumed that each user has a contract with a specific service network setting the default network choice ("home" network provider); nevertheless, if case of insufficient resources, the customer is free to pursue higher QoS at another network, given that there is some kind of federation agreement between the visited and the home provider as in roaming (possibly under a small monetary penalty).

Suppose that there are N users and M service networks, which means that each user at any time can choose any network, giving a total of M^N possible states. Each user-network combination is considered as a two-player game G_j, $1 \leq j \leq M$. The proposed game is *non-cooperative* because, on the one hand, the service providers wish to maximize their revenue and, on the other hand, the users wish to maximize the quality of service received, keeping at the same time the expenses as low as possible. Since these two goals are obviously contradictory, the players do not have the slightest motivation to cooperate. The game is also *nonzero-sum*, since an increase in one player's payoff does not imply a decrease in the other player's payoff.

The user's revenue expresses in monetary value the quality of service offered to him, taking into account the cost, and may be modeled as $R = U \cdot q - C$, where U expresses the customer's utility function, q is a constant factor mapping the utility value to monetary value and C is the cost of the service from the customer's point of view (Lin et al., 2005; Vlacheas et al., 2008). It is also assumed that the provider's billing scheme takes into account the QoS percentage offered to customers, meaning that the customer pays an amount proportional to the level of QoS he receives.

In bibliography, in most cases, user satisfaction is monitored through utility. For instance, utility may be approximated by a sigmoid function as $U = \frac{1}{1 + e^{-a(b-Pb)}}$ where P_b is the packet blocking probability and a, b are constants which determine the steepness and the center of the curve (Lin et al., 2005). Exact user utility functions can be obtained through field tests and user surveys.

As far as the solution of the game is concerned, two cases are distinguished. Assuming the case where the system is not full, the user request will be accepted and the probability that a customer leaves is near to zero. In this case there is a Nash equilibrium, when the service provider accepts the request while the user remains with the network provider. Assuming now the case where the system is loaded to a certain extent or even overloaded, the user request may be not accepted and the probability that a customer leaves is non zero. Even in this case, there is also a pure strategy Nash equilibrium which highly depends on the relation between some terms in the payoffs. The new request is accepted, if the revenue generated from admitting the request is greater than the possible revenue loss if the user leaves. Otherwise, the network provider is better to reject the request.

Cell Selection

Cell selection is responsible for guaranteeing the required QoS by always keeping the mobile camped on a mobile station with good enough quality. The goal of cell selection procedures is to determine which base station is the optimal choice. Lin et al. (Lin, Xinbing, Youyun & Zheng, 2008) formulate the cell selection problem as the two-tier game. In the first tier, i.e., in inter-cell game, the mobile stations select the cell according to the optimal cell selection strategy derived from the expected payoff. In the second tier, i.e., in intra-cell game, the mobile stations choose the proper time-frequency resource in the serving cell to achieve the highest payoff.

Applications of Cooperative Games

Formation of Coalitions

A cooperative game is a game in which the players have the option of planning as a group in advance of choosing their actions. Let $N = \{1, 2, ..., N\}$ be a set of n players. Non-empty subsets of N, S, $T \subseteq N$ are called a coalition. The coalition form of an n-player game is given by the pair (N, u), where u is the characteristic function. A coalition that includes all of the players is called a grand coalition. The characteristic function assigns each coalition S its maximum gain, the expected total income of the coalition denoted $u(S)$. The core is the set of all feasible outcomes that no player or coalition can improve upon by acting for themselves. The objective is to allocate the resources so that the total utility of the coalition is maximized. In wireless networks the formation of coalitions involves the sharing of certain resources; however, as the costs of such resource sharing outweigh the benefits perceived by the nodes, users are less likely to participate, compromising overall network goals.

Bargaining

A very promising element of cooperative game theory is bargaining. Approaches to bargaining fall into two divisions: strategic and axiomatic bargaining. Strategic bargaining, such as the Rubinstein's model of bargaining, assumes that there is a bargaining process where the solution is achieved in a series of offers and counteroffers. The bargaining solution emerges as the equilibrium of a sequential game. The need for a bargaining process among the players is time-consuming and therefore unsuitable for WCDMA network with many users. On the other hand, Axiomatic bargaining ignores the bargaining process and assumes some desirable properties about the outcome of the bargaining process and then identifies process rules or axioms that guarantee this outcome. The operator serves as the arbitrator in the cooperative resource bargaining game.

The notion of axiomatic bargaining in cooperative game theory provides a good analytical framework to derive a desirable operative point that is fair and Pareto optimal. An allocation is Pareto optimal if there is no wasted utility, i.e. it is impossible to make any one party better off without making any other worse off. Some very well-known bargaining solutions are Nash, Raiffa-Kalai-Smorodinsky, utilitarian and modified Thomson and each one corresponds to a unique point on the Pareto optimal boundary. Yaiche et al. (Yaiche, Mazumdar & Rosenberg, 2000) consider the Nash bargaining solution for bandwidth allocation for elastic traffic in broadband networks. Furthermore, Siew-Lee and White (2008) propose a cooperative resource bargaining framework among the users and the wireless network operator, dividing the problem in two subcategories: the *symmetric problem* (where all players have the same bargaining power) and the *asymmetric problem* (where players may submit bids to the arbitrator as a form of tactics to alter the final result and benefit from the resource distribution).

Power Control

In the power control problem, each user's utility is increasing with his SNIR and decreasing with his power level (Goodman & Mandayam, 1999). If all other users' power levels were fixed, then increasing one's power would increase one's SINR. However, when a user raises her transmission power, this action increases the total system as sensed by the other users, driving their SNIRs down, inducing them to increase their own power levels. MacKenzie and Wicker (2001b) formulate a power-control game for a CDMA wireless system. Suppose that users transmit information at the rate R bits/second in L bit packets over a spread-spectrum bandwidth of W(Hz). Let p_j be the power transmitted by j user; we assume that users choose their power levels from the set of non-negative real numbers, $p_j \in [0,\infty)$. We can then define the signal-to-interference-and noise ratio of user j:

$$SNIR_j = \gamma_j = \frac{W}{R} \cdot \frac{h_j p_j}{\sum_{\forall i \neq j} h_i p_i + \sigma^2} \qquad (11)$$

where h_j is the path gain from user j to the base station and σ^2 is the power of the background noise at the receiver. We assume that the background noise is additive white Gaussian noise (AWGN). Given these preliminaries, the utility function of user j (measured in bits/J) is expressed by

$$u_j(p_j, \gamma_j) = \frac{R}{p_j}\left(1 - 2 \cdot BER\left(\gamma_j\right)\right)^L \tag{12}$$

where $BER(\gamma_j)$ is the bit error rate achieved by a given transmission scheme. If the user's transmit power is too high, precious battery power is squandered while having little impact on the bit error rate. The users will attempt to make the best possible choices, taking into account that the other users are doing the same thing. Another fundamental assumption is that the users have complete information about each other and that behave completely rationally. Then, according to game theory, the users will choose an operating point which is a Nash Equilibrium. Two new types of games may be introduced, according to MacKenzie and Wicker (2001b): the refereed and the repeated power control games. Gunturi and Paganini (2003) form a similar power control game, which is then expanded to a multi-cell case.

Spectrum Allocation

The spectrum sharing problem addresses the issue of how to allocate the limited available spectrum among multiple wireless devices. This problem has two important goals: efficiency and fairness. The allocation of spectrum should utilize as much of the resource as possible. However, when utilization is maximized, fairness should be compromised. A cooperative game for distributed spectrum sharing is discussed in the work of Suris (Suris, DaSilva, Han & MacKenzie, 2007). According to this approach, the available bandwidth is divided equally into multiple channels. Each node can transmit in any combination of channels at any time and can set it's transmit power on each channel. Receiver nodes are not considered game players, since they do not transmit.

Routing

In the routing problem, the source and the destination nodes can be viewed as the players of the game. The action set available to each player is the set of all possible paths from the source to the destination. In wireless ad hoc networks, for example, nodes communicate with far off destinations using intermediate nodes as relays. Since wireless nodes are energy constrained, it may not be in the best interest of a node to always accept relay requests. On the other hand, if all nodes decide not to expend energy in relaying, then network throughput will drop away dramatically. For this reason, ad hoc and peer-to-peer networks sometimes operate as voluntary resource sharing networks, relying on users' willingness to spend their own resources for the common good (DaSilva & Srivastava, 2004). The utility function of such a game may be modeled as

$$U_j(s) = a_j(s) + \beta_j(s) \tag{13}$$

where $a_j(s) = a_j\left(\sum_{i \in N, i \neq j} s_i\right)$ is the benefit accrued by a user from others' sharing of their resources and $\beta_j(s) = \beta_j(s_j)$ is the benefit (or cost) accrued by sharing one's own resources with others, s being the joint action taken by all players (s=0 stands for sharing, and s=1 for not sharing). The latter may be negative, since there may be a cost to participating in the network (such as faster depletion of a node's energy resources) or positive, if there exist financial incentives for participation or if the user derives satisfaction in doing so.

Resource Allocation in Shared Network Infrastructures

The high cost associated with the rollout of 3G services encourages operators to share network infrastructure. Operators are attracted to share network resources because of the lower expenses in infrastructure establishment, as well as reduced operation and maintenance expenses. Besides, operators can increase coverage by sharing or having complementary, geographically separated sites, especially in low-density suburban and rural areas where it is more cost-effective to share. Siew-Lee & White (2008) divide the resource allocation problem in a shared network into two sub-problems: a) resource sharing among the network operators, i.e. the co-owners of the network and b) resource bargaining among the users and Mobile Virtual Network Operators (MVNOs) of each network operator. For the latter case, the authors distinguish the symmetric and asymmetric bargaining games. According to this approach, for the case of WCDMA systems, the utility function of each player can be modeled, as the load factor assigned to him either for the uplink or downlink. An extended analysis is also provided for the comparison of Nash and Raiffa solutions, showing that the Raiffa solution takes into account the impact one's gain has over the loss of others, therefore providing fairer results.

FUTURE RESEARCH DIRECTIONS

Game-theoretic frameworks for wireless resource allocation are constantly being developed, exploring new aspects and cooperation trends. Nevertheless, cooperation cannot be taken for granted; even though in most cases players do obtain the optimal result by sharing resources with others, in certain cases it is not clear enough for them why they should not act selfishly or even not try to cheat. This assumption creates the need for more and more complex incentive mechanisms, which will encourage players to act in a cooperative manner. Such mechanisms are a valuable tool for making 4G networks work more efficiently, as the latter provide more opportunities for cooperation than traditional networks.

A fundamental description of how the authors envision future cooperative services and hybrid networks has been included in the chapter as well. It is expected that research will focus towards that direction in the near future, since next generation networks will offer unlimited interaction possibilities and thus the framework of this new reality needs to be defined in terms of communication protocols, security guarantees, incentives, pricing mechanisms etc. Once new hybrid network architectures have emerged, emphasis should be placed on how cooperative services can be implemented over such infrastructure.

Concerning the issue of network selection, the authors envision highly dynamic schemes that exploit efficiently the tools described in this chapter. Of particular interest would be the integration of prediction algorithms that will provide accurate estimations of the QoS level a user is expected to receive. Using Fuzzy Logic, one could also devise a set of IF-THEN rules based on certain QoS merits and weights, enabling the construction of a Fuzzy Inference System (FIS) that tailors membership functions to cell performance. Adaptive Fuzzy Inference Systems (ANFIS) may also be used in this context to provide self-learning decision systems and thus such frameworks should be further investigated.

CONCLUSION

The main goal of this chapter was to collect trends on efficient resource allocation through cooperation among entities in a wireless heterogeneous environment and selection of the best solution among a set of alternatives. In the first section, we presented the problem of network selection, including most common metrics and selection tools from other scientific fields that can be applied with success. In the second section, we described the basic trends and ideas on user cooperation in future networks, emphasizing on cooperation on both physical and application layers. In the third section, we briefly demonstrate how game theory can be applied to wireless networking. Using simple examples, we have shown how to capture wireless networking problems in corresponding games.

REFERENCES

Altman, E., Boulogne, T., El-Azouzi, R., Jimenez, T., & Wynter, L. (2006). A survey on networking games in telecommunications. *Computers & Operations Research*, *33*(2), 286–311. doi:10.1016/j.cor.2004.06.005

Anna, S., Goeckel, D. L., & Nicholas, L. J. (2006). Cooperative Communications in Mobile Ad Hoc Networks - Rethinking the link abstraction. *IEEE Signal Processing Magazine*, *23*(5).

Aune, F. (2004). Cross-Layer Design Tutorial. Norwegian University of Science and Technology, Department of Electronics and Telecommunications, Trondheim, Norway, published under Creative Common License.

Bari, F., & Leung, V. (2007b). Application of ELECTRE to Network Selection in a Heterogeneous Wireless Network Environment. *Wireless Communications and Networking Conference, 2007.*

Bari, F., & Leung, V. C. M. (2007a). Automated network selection in a heterogeneous wireless network environment. *IEEE Network*, *21*(Issue 1), 34–40. doi:10.1109/MNET.2007.314536

Bennis, M., & Juan, L. (2008). Non-Cooperative Operators in a Game-Theoretic Framework. *IEEE 19th International Symposium on Personal Indor Mobile Radio Communications (PIMRC).*

Boyer, J., Falconer, D. D., & Yanikomeroglu, H. (2004). Multihop diversity in wireless relaying channels. *IEEE Transactions on Communications*, *52*(10), 1820–1830. doi:10.1109/TCOMM.2004.836447

Buttyan, L., & Hubaux, J. P. (2003). Stimulating Cooperation in self-organizing mobile ad hoc networks. *MONET*, *8*, 579–592.

Cai, X., Chen, L., Sofia, R., & Wu, Y. (2007). Dynamic and User-Centric Network Selection in Heterogeneous Networks. *IEEE International Conference Performance, Computing, and Communications,* (pp. 538 – 544).

Chakravorty, R., Banerjee, S., Agarwal, S., & Pratt, I. (2005). MoB: A Mobile Bazaar for Wide-area Wireless Services. In ACM Mobicom.

Charilas, D., Markaki, O., Nikitopoulos, D., & Theologou, M. (2008). Packet-Switched Network Selection with the Highest QoS in 4G Networks. *Computer Networks*, *52*(1), 248–258. doi:10.1016/j.comnet.2007.09.005

Charilas, D., Markaki, O., & Tragos, E. (2008). A Theoretical Scheme for applying game theory and network selection mechanisms in access admission control. In *International Symposium on Wireless Pervasive Computing, Santorini.*

Cover, T. M., & El Gamal, A. A. (1979). Capacity theorems for the relay channel. *IEEE Transactions on Information Theory*, *25*, 572–584. doi:10.1109/TIT.1979.1056084

DaSilva Luiz, A. Srivastava Vivek (2004). Node Participation in Ad Hoc and Peer-to-Peer Networks: A Game-Theoretic Formulation. In *Workshop on Games and Emergent Behavior in Distributed Computing Environments, Birmingham, UK.*

Dohler, M., Meddour, D. E., Senouci, S.-M., & Saadani, A. (2008). Cooperation in 4G - Hype or Ripe? *IEEE Technology and Society Magazine*, *27*(1), 13–17. doi:10.1109/MTS.2008.918035

Felegyhazi, M., Hubaux, J.-P. (2006). Game Theory in Wireless Networks: A Tutorial. *EPFL technical report, LCA-REPORT-2006-002.*

Frattasi, S., & Can, B. Fitzek. F., & Prasad, R. (2005). Cooperative Services for 4G. In *Proceedings of the 14th IST Mobile & Wireless Communications.*

Frattasi, S., Fathi, H., & Fitzek, F.H.P. (2006). 4G: A User-Centric System. *Wireless Personal Communications Journal (WPC) - Special Issue on Advances in Wireless Communications: Enabling Technologies for 4G.*

Frattasi, S., Fathi, H., Fitzek, F. H. P., Prasad, R., & Katz, M. (2006). Defining 4G Technology from the User Perspective. *IEEE Network Magazine*, *20*(1), 35–41. doi:10.1109/MNET.2006.1580917

Fudenberg, D., & Tirole, J. (1991). Game Theory. MIT Press.

Gao, L., Wang, X., Xu, Y. & Li, Z. (2008). *A Game Approach for Cell Selection and Resource Allocation in Heterogeneous Wireless Networks.*

Giambene, G., & Kota, S. (2006). Cross-layer protocol optimization for satellite communication networks: A survey. *International Journal of Satellite Communications and Networking*, *24*, 323–341. doi:10.1002/sat.853

Goodman, D., & Mandayam, M. (1999). Power control for wireless data. *IEEE International Workshop on Mobile Multimedia Communications*, (pp. 55–63).

Gunturi, S., & Paganini, F. (2003). Game theoretic approach to power control in cellular CDMA. *Vehicular Technology Conference (VTC).*

Hasna, M., & Alouini, M.&M. (2004). A performance study of dual-hop transmissions with fixed gain relays. *IEEE Transactions on Wireless Communications*, *3*(6), 1963–1968. doi:10.1109/TWC.2004.837470

Huiling, J., Zhang, Z., Cheng, P., Chen, H.-H., & Li, S. (2006). Study on Network Selection for Next-Generation Heterogeneous Wireless Networks. In *IEEE 17th International Symposium on Personal, Indoor and Mobile Radio Communications,* (pp.1 – 5.)

Jain, R. (2004). Quality of experience. *IEEE MultiMedia*, *11*(1), 96–95. doi:10.1109/MMUL.2004.1261114

Katz, M., & Fitzek, F. H. P. (2005). Cooperative Techniques and Principles Enabling Future 4G Wireless Networks. In *The International Conference on EUROCON,* (pp. 21-24). Amir, K., Jinane, A., Eytan, M., & Lizhong, Z. (2003). Cooperative Routing in Wireless Networks. *Allerton Conference on Communications, Control and Computing.*

Kher, S., Somani, A. K., & Gupta, R. (2005). Network selection using fuzzy logic. In *2nd International Conference on Broadband Networks.* Feng, K. & Hong-yan, L. (2007). A New Fuzzy MADM Algorithm Based on Subjective and Objective Integrated Weights. In *International Conference on Service Systems and Service Management.*

Laneman, J. N., Tse, D. N. C., & Wornell, G. W. (2004). Cooperative Diversity in Wireless Networks: Efficient Protocols and Outage Behavior. *IEEE Transactions on Information Theory*, *50*(12), 3062–3080. doi:10.1109/TIT.2004.838089

Le, L., & Ekram, H. (2008). Cross-Layer Optimization Frameworks for Multihop Wireless Networks Using Cooperative Diversity. *IEEE Transactions on Wireless Communications*, *7*(7). doi:10.1109/TWC.2008.060962

Lin, H., Chaterjee, M., Das, S. K., & Basu, K. (2005). ARC: An Integrated Admission and Rate Control Framework for Competitive Wireless CDMA Data Networks Using Noncooperative Games. *IEEE Transactions on Mobile Computing*, *4*(3), 243–258. doi:10.1109/TMC.2005.35

Lin, X., Shroff, N. B., & Srikant, R. (2006). A Tutorial on Cross-Layer Optimization in Wireless Networsk. *IEEE Journal on Selected Areas in Communications*, *24*(8).

MacKenzie, A. B., & DaSilva, L. A. (2006). Game Theory for Wireless Communications. San Rafael, CA: Morgan and Claypool Publishers.

MacKenzie, A. B., & Wicker, S. B. (2001a). Game Theory and the Design of Self-Configuring, Adaptive Wireless Networks. *IEEE Communications Magazine*, *39*(11), 126–131. doi:10.1109/35.965370

MacKenzie, A. B., & Wicker, S. B. (2001b). Game Theory in Communications: Motivation, Explanation and Application to Power Control. *IEEE Global Telecommunications Conference, 2,* 821 – 826.

Markaki, O., Charilas, D., & Nikitopoulos, D. (2007). Enhancing Quality of Experience in Next Generation Networks through Network Selection Mechanisms. *Mobile Terminal Assisted Enhanced Services Provisioning in a B3G Environment Workshop, PIMRC 2007.*

Michiardi, P., & Molva, R. (2002). Core: A cooperative reputation mechanism to enforce node cooperation in mobile ad hoc networks. *Communications and Multimedia Security Conference.*

Ormond, O., Miro-Muntean, G., & Murphy, J. (2005). Economic Model for Cost Effective Network Selection Strategy in Service Oriented Heterogeneous Wireless Network Environment. In *Proc. of 10th IEEE/IFIP Network Operations and Management Symposium (NOMS 2006), Canada.*

Ormond, O., Murphy, J., & Muntean, G.-M. (2006). Utility-based Intelligent Network Selection in Beyond 3G Systems. *IEEE International Conference on Communications, 4,* 1831 – 1836.

Perrucci, G. P., Fitzek, F. H. P., Boudali, A., Canovas, M., Nejsum, P., & Studstrup, S. (2007). Cooperative Web Browsing for Mobile Phones. *International Symposium on Wireless Personal Multimedia Communications (WPMC'07), India.*

Qingyang, S., & Abbas, J. (2005a). A Network Selection Mechanism for Next Generation Networks. *IEEE International Conference on Communications, 16-20 May 2005, 2,* 1418- 1422.

Qingyang, S., & Abbas, J. (2005b). Network Selection in an integrated Wireless LAN and UMTS Environment Using Mathematical Modeling and Computing Techniques. *IEEE Personal Communications, 12*(3), 42–48.

Safari, M. & Uysal, M. (2008). Cooperative Diversity over Log-Normal Fading Channels: Performance Analysis and Optimization. *IEEE Transactions on Wireless Communications, 7*(5-2), 1963-1972.

Salem, N.B., Buttyan, L., Hubaux, J.-P., & Jakobsson, M. (2003). A charging and rewarding scheme for packet forwarding in multi-hop cellular networks. *ACM MobiHoc 03'.*

Shastry, N., & Adve, R. S. (2006). Stimulating Cooperative Diversity in Wireless Ad Hoc Networks through Pricing. *IEEE International Conference on Communications (ICC '06).*

Siew-Lee Hew, White L. (2008). Cooperative resource allocation games in shared networks: symmetric and asymmetric fair bargaining models. *IEEE Transactions on Wireless Communications, 7*(11-1), 4166-4175.

Song, G.-C., & Li, Y. (G.). (2005). Cross-layer optimization for OFDM wireless networks – Part I: theoretical framework. *IEEE Transactions on Wireless Communications*, *4*(2), 614–624. doi:10.1109/TWC.2004.843065

Suris, J. E., DaSilva, L. A., Han, Z., & MacKenzie, A. B. (2007). Cooperative Game Theory for Distributed Spectrum Sharing. *IEEE International Conference on Communications.*

Tse, D., & Viswanath, P. (2005). Fundamentals of Wireless Communication. Cambridge, UK: Cambridge University Press.

Valavanis, E., Ververidis, C., Vazirgianis, M., Polyzos, G. C., & Nørvåg, K. (2003). MobiShare: Sharing Context-Dependent Data and Services from Mobile Sources. In *Proc. IEEE/WIC Int'l Conference on Web Intelligence (WI 2003), Halifax, Canada.*

Vlacheas, P., Charilas, D., Tragos, E., & Markaki, O. (2008). Maximizing Quality of Service for Customers and Revenue for Service Providers through a Noncooperative Admission Control Game. In ICT Mobile Summit 2008, Stockholm.

Yaiche, H., Mazumdar, R. & Rosenberg, C. (2000). A game theoretic framework for bandwidth allocation and pricing in broadband networks. *IEEE/ACM Trans Networking, 8*(5), 667–678.

Zhang, D., Ileri, O., & Mandayam, N. (2008). Bandwidth exchange as an incentive for relaying. In *42nd Annual Conference on Information Sciences and Systems (CISS 2008).*

Zhang, Q., Fitzek, F. H. P., & Katz, M. (2006). Evolution of Heterogeneous Wireless Networks: Cooperative Networks. In *3rd International Conference of the Center for Information and Communication Technologies, Copenhagen, Denmark.*

Zheng, L., & Tse, D. N. C. (2003). Diversity and freedom: A fundamental trade off I multiple antenna channels. *IEEE Transactions on Information Theory, 49,* 1073–1096. doi:10.1109/TIT.2003.810646

Zhong, S., Chen, J., & Yang, Y. R. (2003). Sprite: A Simple, Cheat-Proof, Credit-Based System for Mobile Ad-Hoc Networks. In *Proc. of IEEE INFOCOM 2003.*

Zhu, H., & Cao, G. (2005). rDCF: a relay enabled medium access control protocol for wireless ad hoc networks. In *Proceedings of IEEE INFOCOM.*

Zimmermann, E., Herhold, P., & Fettweis, G. (2005). On the performance of cooperative relaying protocols in wireless networks. *European Transactions on Telecommunications, 16,* 5–16. doi:10.1002/ett.1028

ADDITIONAL READING

Arkko, J., Aboba, B., Korhonen, J. & Bari, F. (2008). Network Discovery and Selection Problem. *RFC5113.*

Bansal, S., & Baker, M. (2003). Observation-based Cooperation Enforcement in Ad hoc Networks. Stanford Technical Report.

Cesana, M. M. Ilaria & Antonio, Capone (2008). Modelling network selection and resource allocation in wireless access networks with non-cooperative games. In *5th IEEE International Conference on Mobile Ad Hoc and Sensor Systems, 2008(MASS 2008).*

Charilas, D., Markaki, O., Psarras, J., & Constantinou, P. (2009). Application of Fuzzy AHP and ELECTRE to Network Selection. In *the 1st International Conference on Mobile Lightweight Wireless Systems*, Athens, Greece.

Dang, M. S., Prakash, A., & Anvekar, D. K. (2000). Fuzzy logic based handoff in wireless networks. In *IEEE 51st Vehicular Technology Conference Proceedings, 3, 2375 - 2379 Tokyo.*

Economides, A., & Silvester, J. (1991). Multi-objective routing in integrated services networks: A game theory approach. In *IEEE INFOCOM Networking in the 90s/Proceedings of the 10th Annual Joint Conference of the IEEE and Communications Societies, 3, 1220–1227.*

Ferguson, T. S. Game Theory Text. *Mathematics Department, UCLA.*

Fitzek, F. H. P., & Katz, M. (2006). Cooperation in Wireless Networks: Principles and Applications – Real Egoistic Behavior is to Cooperate! Springer.

Kristensen, J. M., & Fitzek, F. H. P. (2006). The application of software defined radio in a cooperative wireless network. In *Software Defined Radio Technical Conference, Orlando, Florida, USA, SDR Forum.*

Liu, P. (2006). Cooperative wireless communications: a cross-layer approach. *IEEE Wireless Communications Magazine, 13*(4), 84–92. doi:10.1109/MWC.2006.1678169

Mikhailov, L., & Tsvetinov, P. (2004). Evaluation of services using a fuzzy analytic hierarchy process. *Applied Soft Computing, 5*, 23–33. doi:10.1016/j.asoc.2004.04.001

Militano, L., Fitzek, F. H. P., Iera, A., & Molinaro, A. (2007). On the beneficial effects of cooperative wireless peer to peer networking. In *Tyrrhenian International Workshop on Digital Communications (TIWDC'07), Ischia, Italy.*

Niyato, D., & Hossain, E. (2006). A cooperative game framework for bandwidth allocation in 4G heterogeneous wireless networks. In *Proc. IEEE ICC'06.*

Rasmusen, E. (2006). Games and Information, Fourth Edition: An Introduction to Game Theory. Wiley-Blackwell.

Renzo, M. D., Graziosi, F., & Santucci, F. (2008). Performance of cooperative multihop wireless systems over log-normal fading channels. In IEEE Globecom, 1-6.

Sendonaris, A., Erkip, E., & Aazhang, B. (1998). Increasing uplink capacity via user cooperation diversity. In *Proc. IEEE Int. Symp. Information Theory (ISIT), Cambridge, MA, 156.*

Shen, W., & Zeng, Q.-A. (2007). Resource Allocation Schemes in Integrated Heterogeneous Wireless and Mobile Networks. *ACM Journal of Networks, 2*, 5.

Srinivasan, V., Nuggehalli, P., & Chiasserini, C. F., Rao, & Ramesh R. (2003). Cooperation in Wireless Ad Hoc Networks. In IEEE INFOCOM.

Srivastava, V., Neel, J. A., MacKenzie, A. B., Hicks, J. E., DaSilva, L. A., Reed, J. H., & Gilles, R. P. (2005). Using Game Theory to Analyze Wireless Ad Hoc Networks. *IEEE Communications Surveys and Tutorials, 7*, 5. doi:10.1109/COMST.2005.1593279

Triantaphyllou, E. (2002). Multi-Criteria Decision Making Methods: A Comparative Study. Kluwer Academic Publishers.

Van der Meulen, E. C. (1971). Three-terminal communication channels. *Advances in Applied Probability, 3*, 120–154. doi:10.2307/1426331

Vassaki, S., Panagopoulos, A., & Constantinou, P. (2009). Game-theoretic approach of fair bandwidth allocation in DVB-RCS networks, In *International Workshop on Satellite and Space Communications* (IWSSC 2009), Siena, Italy.

Walid, Saad, Zhu, Han, M´erouane, Debbah, Are, Hjørungnes & Tamer, Basar (2009). Coalitional Game Theory for Communication Networks: A Tutorial, accepted to appear in the *IEEE Signal Processing Magazine, Special Issue on Game Theory.*

Yu, Y., Yong, B., & Lan, C. (2007). Utility dependent network selection using MADM in heterogeneous wireless networks. In *18th Annual IEEE International Symposium on Personal, Indoor and Mobile Radio Communications (PIMRC'07).*

Zadeh, L. A. (1965). Fuzzy sets. Inform. and Control, pp. 338-353.

Zhang, W. (2004). Handover Decision Using Fuzzy MADM in Heterogeneous Networks. In *IEEE Wireless Communications and Networking Conference*.

Chapter 17
Admission Control for QoS Provision in Mobile Wireless Networks

Georgios I. Tsiropoulos
National Technical University of Athens, Greece

Dimitrios G. Stratogiannis
National Technical University of Athens, Greece

John D. Kanellopoulos
National Technical University of Athens, Greece

Panayotis G. Cottis
National Technical University of Athens, Greece

ABSTRACT

Admission control is one of the key elements for ensuring the quality of service (QoS) in modern mobile wireless networks. Since such networks are resource constrained, supporting multimedia traffic guaranteeing its QoS levels is excessively challenging for call admission control (CAC) design. CAC is the most important radio resource management (RRM) function in wireless networks as its efficiency has a direct impact on network performance and QoS provision to end users. The goal of this chapter is to provide a thorough study of the basic concepts considering CAC design and a comprehensive analysis of the fundamental CAC schemes employed in wireless networks. The basic performance criterion considering CAC schemes is the probability of denying the access to the network for an arriving call, which is extensively studied in this chapter. Moreover, additional performance criteria are presented and discussed, which may help to provide an overall efficiency estimation of the available CAC schemes.

INTRODUCTION

Quality of Service (QoS) provision in wireless networks is closely related to the exploitation of network available resources and the maximization of the number of users. The increasing demand for multimedia services combined with the limited resources of wireless networks urge for efficient admission control schemes that achieve a compe-

DOI: 10.4018/978-1-61520-771-8.ch017

tent resource management along with appropriate QoS levels for end users. Call Admission Control (CAC) algorithms are employed to ensure that the admission of a new call into a resource limited network does not violate the Service Level Agreements (SLAs) concerning ongoing calls. Any service session supported by the network can be considered as call. An efficient admission control policy should accomplish the following:

- establish a robust priority assigning mechanism for handoff calls and different Service Class (SC) calls,
- exhibit a low Call Blocking Probability (CBP),
- manage a fair resource allocation,
- achieve a high network throughput and
- avoid congestion.

CAC schemes for wireless networks have been widely studied under different network architectures and network administrator policies. To apply a specific call admission policy, the various call types are classified into SCs with precise characteristics and QoS demands. Each SC call is treated differently depending on the criteria set according to the operating principles adopted for the admission procedure. The majority of CAC schemes base their admission criteria on an efficient resource management, accounted for either in terms of channels or of bandwidth units. The methods proposed usually set thresholds related to the desirable QoS for high priority SCs and handoff calls. The last trend in such schemes proposes the use of adjustable admission probabilities for low priority SC calls combined with thresholds for the dynamic handling of the input traffic.

Other CAC schemes examine Signal to Noise Ratio (SNR) levels to determine an admission criterion satisfying the QoS demands of end users. Such schemes have to deal with propagation and mobility issues. Under high traffic network conditions, an efficiency enhancing module may be incorporated into the CAC schemes employed renegotiating the resource allocation of ongoing calls. Through QoS re-negotiation and resource re-allocation, available resources can be retrieved to serve a high priority SC call request. The admission decision may also be controlled according to dynamic changes in call pricing, providing at the same time a stream flow management.

Finally, a critical issue in admission control is the performance evaluation of the proposed schemes to assess the provided QoS. The evaluation is performed under appropriate criteria. The metric studied most in the literature is CBP. Assigning priorities to certain call types causes a serious side effect, as certain SCs may monopolize the network resources leading to an unfair treatment of low priority SC calls. Thus, in shared resources networks, fairness among users of the same SC subjected to different channel conditions with different mobility characteristics and among users of different SCs should be investigated employing appropriate metrics. A call admission request represents potential revenue for the network operator. Thus, CAC schemes should be further evaluated employing performance metrics based on optimal pricing to maximize revenues. The above criteria constitute a thorough performance evaluation of admission control policies providing QoS in wireless multiservice networks.

BACKGROUND

One of the most fundamental objectives of the information society is the widespread implementation of communication and information technologies offering both opportunities and challenges to all users. Communication services may be offered to users either through a wired or a wireless infrastructure. Compared to wired networks, the superiority of wireless networks with regards to ease of installation and flexibility is evident. Wireless networks offer low installation cost combined with the advantage of overcoming even the most unfavorable geographic and climatic conditions of the region. On the contrary, wireless networks compared to wired ones suffer from lower bandwidth, higher delays and higher bit error rate. These limitations accompanied with the confined frequency range available for communication purposes result in a hard constrained network in terms of capability and attainable resources. Thus, providing QoS over such a network is quite challenging and complicated procedure for wireless networks.

The present telecommunication environment is an amalgamation of a large number of networks and administrative domains employing various access technologies. Considering this status combined with the increased user mobility makes it difficult to promise the same grade of service quality at all parts of the network. The problem of providing an adequate QoS level to end users becomes even more difficult when new traffic types, real-time (RT) and non-real-time (NRT) are entering into the wireless network list of services. In particular, in addition to requirements of traditional applications, such as email, web browsing, file transfer and instant messaging, new services introduced the recent years such as multimedia applications, video conferencing and voice over IP (VoIP) demand higher bandwidth, smaller delays and session continuity. As users are sensitive to the variance of network parameters, mainly to throughput and bit-error rate, maintaining them at a fixed level is extremely hard, especially for those characterized by high mobility.

Although QoS term is one of the most significant considering communication networks, there is no universal or common recognized definition of QoS. However, there are several definitions according to the communication level where the notion originated. Each one is based on the network layer characteristics considered. Thus, each technical community may perceive and interpret QoS in different ways, corresponding to various meanings and perspectives. Generally, QoS refers to a broad collection of networking mechanisms and techniques, the goal of which are to provide guarantees on the ability of the network to maintain acceptable and sustainable application levels. Concerning standard communication networks offering telephony services QoS was defined by ITU in standard X.902 as "A set of quality requirements on the collective behavior of one or more objects." (ITU-D, 2008). These objects correspond to all system parameters; therefore, this definition is too general to be helpful in system analysis. In the fields of ATM networks the Internet Engineering Task Force has defined QoS in RFC 1946 (1996): "As the demand for networked real time services grows, so does the need for shared networks to provide deterministic delivery services. Such deterministic delivery services demand that both the source application and the network infrastructure have the capabilities to request, setup, and enforce the delivery of the data. Collectively these services are referred to as bandwidth reservation and Quality of Service (QoS)".

Generally, it is more difficult to provide an acceptable QoS level to RT services rather than to NRT services, as the first need improved network performance compared to the latter. Moreover, multimedia traffic is more demanding as to standard communication services. Thus, considering both characteristics a more general definition of QoS is given as "The set of those quantitative and qualitative characteristics of a distributed multimedia system necessary to achieve the required functionality of an application." (Vogel, Kerherve, von Bochmann, & Gecsei, 1995). In this chapter QoS will be considered as a defined measure of performance in a wireless network. The specific network characteristic which will be mostly examined in this chapter is the availability of the system to serve a new or ongoing service request and will be defined in a subsequent section.

Apart from the QoS definition used in each analysis or system deployment, the main issue addressed is the way that QoS is conceived by either the network or the users. Therefore, QoS can be determined by the way that data are presented to end users and users satisfaction. To assess QoS offered by a system, a number of system parameters are examined. These parameters can be a subset of some system performance characteristics such as bandwidth/throughput, transit delay, error rate, outage probability, blocking and dropping probability, response time, availability (uptime), etc. Every parameter can range over a given set of values, each of which is related to a specific QoS level for the corresponding service offered. In a multiservice system, different services demand also different values for the relevant subset of QoS parameters.

QUALITY OF SERVICE IN ADMISSION CONTROL

Necessity for Radio Resource Management

The rapid increase of wireless communication networks users, combined with the demand for new multimedia services and the need for high speed communications are in contrast to the scarce spectrum resources allocated for wireless systems in international organizations. Thus, an efficient radio resource management (RRM) is of paramount importance, to allocate the available resources among contenting users in compliance with their needs and respective priorities, offering them the required QoS. In particular, RRM functionality is aimed to improve system performance by maximizing the overall system capacity in the wireless network maintaining at the same time the QoS of mobile user traffic. System capacity can be defined as the maximum traffic load that a system can accommodate, under the constraint that QoS is above a given level, and can be measured in bit/s/Hz/base station (BS) or Erlang/MHz/site. The conditions that limit the capacity of a wireless network are associated with the QoS requirements (such as latency, call duration, minimum acceptable throughput, etc.), the channel characteristics which can be further analyzed into acceptable noise levels and co-channel interference, the power control operation which is subjected to electromagnetic compatibility issues and the mobility characteristics of end users.

RRM can be thought as a set of mechanisms and techniques that control the usage of the limited radio resources, managing and allocating radio resource units (RRU) to call requests. Different definitions can be given to RRU according to the multiple access scheme implemented. Thus, in frequency division multiple access (FDMA) systems RRU is denoted by a certain bandwidth with a given carrier frequency, in time division multiple access (TDMA) schemes RRU is specified by a pair of frequency and time slot and in code division multiple access (CDMA) schemes RRU is defined by a pair of code sequence and power level. The basic RRM components can be classified, upon the system characteristics which are based on, into three main groups as follows (Ahmed, 2005). The mechanisms included into first group are driven by the physical characteristics of the channel and enclose frequency or time resource allocation schemes, such as load control, channel allocation and reservation, transmission rate control and packet scheduling. The techniques comprised into the second group are based on the adjustment of the transmitter power, either on the BS or the terminal. This technique is commonly referred as power control mechanism and it is subject to limitations considering inter- and intra-system interference and regulations concerning EMC standardization. The third group includes efficient network management schemes such as routing, call admission control (CAC), congestion control (CC), mobility management, handoff algorithms and BS assignment functions.

Operating Principles for Admission Control

An essential RRM mechanism fundamental for QoS provision used in wireless networks is CAC.

CAC algorithms are employed to ensure that the admittance of a new call into a resource constrained network does not violate the SLAs guaranteed by the network to ongoing users. When deciding to admit a new call into the network, there are many contradictory factors to be taken into consideration, such as resource usage optimization, revenue maximization, fairness provision, etc (Stratogiannis, Tsiropoulos, & Cottis, 2008). Thus, CAC constitutes a mechanism which is used to limit the number of call connections into the network as to provide different priorities among users with different QoS characteristics, increase network utilization and prevent congestion. Hence, when a call request from a mobile user arises, it may be either accepted or blocked. The latter is commonly known as call blocking. The corresponding probability, which is defined as the probability that a new call request is being denied service by the network, is called call blocking probability (CBP) and it is subjected to the decision taking part of CAC scheme employed. Evidently, an efficient CAC policy should achieve low CBP.

CAC is difficult to be implemented in practice, since the traffic flows in communication networks are inherently chaotic or "bursty" and traffic bursts are almost impossible to predict. Considering wireless networks, CAC schemes become more complicated, because of users mobility and variable link quality. In particular, an accepted call in a certain cell may have to be handed off to a neighboring cell owing to user mobility. The main consideration in handoff procedure it to maintain the continuity of the call while at the same time offering at least the minimum acceptable QoS level. During an ongoing call a mobile user may cross several cell boundaries; therefore, it may require a corresponding number of successful handoffs. According to this process, the new cell may not have any available resources to serve the handoff call, resulting in handoff failure, commonly known as call dropping. In literature, the probability of an ongoing call to be forced-terminated is defined as call dropping probability (CDP). It is widely accepted that users are more sensitive to CDP than to CBP, as dropping an ongoing call is annoying for end users; consequently, efficient CAC schemes should keep CDP as low as possible. A simple way to reduce CDP is to assign higher priorities to handoff calls compared to new ones. Therefore, the admission criteria for new and handoff calls are different.

Since CAC schemes operation is real-time, the CAC scheme algorithm used should be computationally quick. Moreover, the exact situation considering the available resources of the BS controller should be available for the CAC algorithm input data. The design and implementation of a CAC scheme should be performed very carefully as to minimize the false rejections and false admissions. As false rejection can be considered any call that can be rejected, even though there exist enough resources in the network to serve it. In this case, network resource optimization is not achieved, capacity is wasted and operator's revenue is not maximized. On the other hand, false admissions occur when a call request is accepted, albeit available resources are exhausted. In this case, the QoS level in not guaranteed and CDP is increased, resulting in degradation of users satisfaction.

Classification of Service Classes

Former generation wireless networks employed simple traffic shaping schemes, as all traffic was shaped uniformly by rate. This model was realistic since mostly one service (voice calls) was offered. In the view of modern wireless networks which offer a variety of services to end users, incoming traffic should be classified into different traffic types. Each traffic type is called service class (SC) and the procedure followed to determine in which SC a new call request falls into is called classification. Every SC has its own QoS characteristics, e.g. bitrate, packet delay, duration etc. Thus, each SC should be treated differently to differentiate the service implied for the user.

Although control mechanisms are more complex due to multiple SCs supported by the network, such networks are more flexible in resource allocation management and QoS provision. One of the main advantages of this policy, apart from different QoS characteristics for each SC, is the different priority levels applied among the SCs supported. This SC prioritization is usually based on QoS requirements, the pricing policy followed by the administrator and the SLAs with the users. This differentiation among incoming calls can be utilized by a network operator to treat different SC calls differently considering bandwidth allocation, admission decision process, pricing policy applied, etc.

A common classification mainly applied on wireless networks implies the differentiation among incoming calls into two general SCs, RT and NRT (Tsiropoulos, Stratogiannis, Kanellopoulos, & Cottis, 2008). This classification is primarily based on latency characteristic of each call. Generally, there is a deadline before a data packet should be delivered to its destination. If this requirement is strict or lenient for a given call then the call is characterized as RT or NRT call, respectively. In modern wireless networks supporting multimedia traffic a broader call classification is required. Apart from taking into consideration the latency parameter of each call, additional QoS requirements are considered such as bandwidth demand and call duration. Therefore, calls are categorized into multiple SCs (Tragos, Tsiropoulos, Karetsos, & Kyriazakos, 2008) such as voice, messaging, internet browsing, file transfer, teleconference etc.

CAC SCHEMES PROVIDING QOS

CAC Schemes Criteria

The basic operation of CAC schemes is to decide whether a call is accepted by the network or not. This decision is based on several criteria which are related to the specific QoS characteristics of the call request and the network parameters. Although QoS characteristics of the call are predetermined, the network parameters are variable and changeable in time. Thus, the CAC scheme employed should assure that QoS characteristics of ongoing calls will not be violated throughout their total duration. The most common criteria used in CAC schemes are listed below (Figure 1).

- Network load/resources: The limited network resources constitute the most essential factor in CAC design. Thus, CAC schemes based on this criterion should know the available resources in each cell before the decision is taken. In this case, the network load after the admission of the call request is computed; if the predicted network load remains under a given threshold the call request can be admitted, otherwise it is blocked. As handoff calls are treated differently by most CAC schemes, a set of channels may be reserved by each cell for handoff calls. Therefore the admission of a new call request is stricter, as the threshold employed in CAC scheme is lower, compared to the admission of handoff calls. These CAC schemes are known widely as Guard Channels (GC) schemes (Fang & Zhang, 2002).
- Connection/link quality: As BS and user terminals use the air interface as common medium for transmitting communication signals; link quality is an essential parameter in interference-limited wireless networks. Link quality refers to the quality of the radio link between the user terminal and the BS. To quantify this parameter, the signal strength received in mobile terminal and the interference caused to this link by other mobile terminals in the corresponding area is employed. Thus, CAC schemes admit a new call request, if they can maintain link quality for admitted calls over a

Figure 1. The main criteria of wireless networks which are employed in CAC schemes

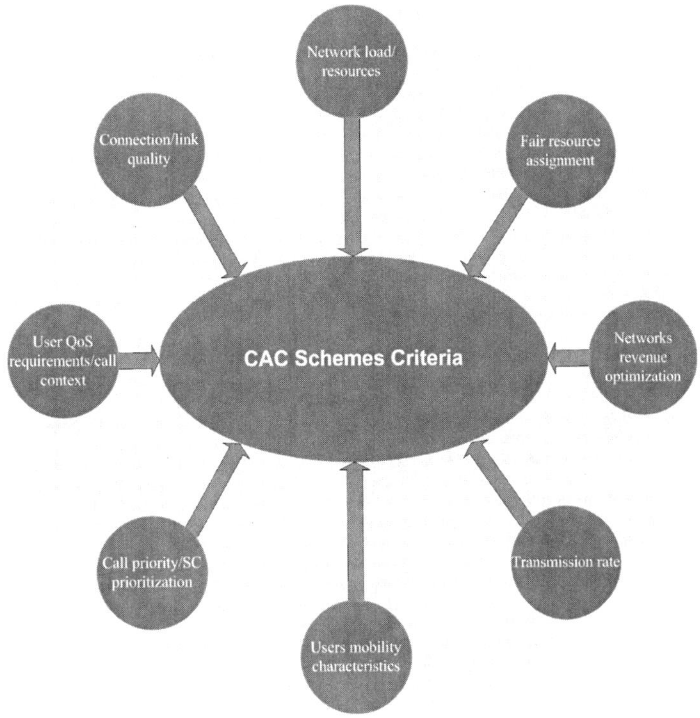

certain level. Otherwise, if the admission of the new call will result in a decrement of the links quality under a certain level, the call is rejected. CAC schemes based on link quality usually employ the signal to interference ratio (SIR) or the signal to noise plus interference ratio (SNIR) as an admission criterion, therefore, they are called in literature SIR/SNIR-based schemes (Liu & Zarki, 1994).

- User QoS requirements/call context: As users may request for services with different QoS parameters such as mean throughput, mean delay, bit error rate (BER) and bandwidth demands, the call requests are classified into different SCs. For each SC call request different admission criteria can be employed taking into consideration the specific QoS constraints of each SC and the available network resources. Thus, CAC schemes can be classified upon the number of SCs supported. Single SC CAC schemes constituted a simple and appropriate model for first and second generation (2G) wireless networks, as they were mainly based on voice service. The growing need for new services combined with the diffusion of new technologies, such as 2.5, 3G networks and also next generation networks (NGN), indicated the need for systems to support multiple SCs with multimedia traffic and enhanced QoS characteristics. Thus, advanced CAC schemes supporting multiple SCs were introduced during the last decade, categorizing stream flows and call requests into different types according to their QoS characteristics (Tragos, Tsiropoulos, Karetsos, & Kyriazakos, 2008). Multiple SC CAC design is more challenging since different CAC criteria are employed for the SCs supported, resulting to high complexity and difficulties considering the implementation in practice.

- Call priority/SC prioritization: The latter CAC criterion is strictly related to SC prioritization. Assigning certain SCs higher priority over the rest is a common technique in multiple SC CAC schemes. In particular, it is widely accepted that RT services have higher priority over NRT ones. Thus, a voice call is considered by the network as a high priority call compared to internet browsing. Moreover, different priorities can be given even in the same SC reflecting the differentiation between different user classes, which can be stemming from the application of different subscription fees policy. Furthermore, higher priorities are assigned to handoff calls and to calls related to emergency services. Different priority levels reflect different CAC criteria respectively, which are stricter for low priority SC calls and lenient for high priority ones (Stratogiannis, Tsiropoulos, & Cottis, 2008).

Prioritization schemes can be implemented mainly through three different techniques; channel borrowing, queuing and reservation schemes. In the first technique, if a cell has all its channels reserved, it can borrow available channels from the neighboring cells to serve high priority SC calls. In queuing schemes, if a cell has all resources occupied, a high priority call request is set into the queue until the available resources, capable of accommodating the call request into the cell, are released. Queuing systems can be applied to either high priority call requests or all incoming call requests (high and low priority). In the latter case their position into the queue can be adjusted according to call requests priority. On the other hand, the reservation schemes were first used to prioritize handoff calls by permanently reserving a portion of channels dedicated to handoff requests. These schemes have been extended to support multiple SCs and to provide different priority levels among them, by reserving channels for high priority SC calls.

- Users mobility characteristics: Users mobility is a critical factor in wireless networks as users travel along multiple cells, consequently, the traffic is changeable and it cannot be precisely predicted. Therefore, an active terminal may move from one cell to a neighboring one, resulting in calls handoff. If a handoff call cannot be served in the new cell, it is dropped resulting in increased CDP. Since users are more sensitive to CDP than CBP, CAC schemes are employed to control the handoff failure probability. Most schemes in literature are assigning higher priorities to handoff calls resulting in less stringent admission conditions for handoff calls. These schemes are the same with prioritization schemes mentioned above; with the only difference that they are designed to prioritize handoff calls (Ahmed, 2005).
- Transmission rate: CAC schemes are employed in wireless networks to guarantee the minimum bandwidth requirement for ongoing calls. On the other hand, each SC call may also have a maximum bandwidth requirement (Li & Chao, 2007). Based on the available resources, CAC schemes aim to provide each call with the highest bandwidth between the minimum and maximum requirement, while at the same time they try to reduce the CBP. Thus, some CAC schemes embed a mechanism, named QoS renegotiation technique. In this case, if the resources of network cell are insufficient, QoS renegotiation mechanism is activated to reduce the transmission rate of ongoing calls, up to an acceptable level, as much as it needs to admit a new call request. The transmission rate may be restored or fur-

ther increased when available resources are released from a terminated call.

- Networks revenue optimization: An efficient CAC scheme should provide high network revenue for the network operator by applying a high network utilization policy. On the other hand, there are strict limitations considering the total bandwidth constraints and the QoS guarantees. Any accepted call is not only accompanied with a reward coming from the revenue obtained, but also with a penalty if ongoing calls deteriorate their QoS. The reward may be the number of users or the portion of occupied bandwidth whereas the penalties may be defined by the probability of QoS deterioration. Finding in real time the optimum solution between reward and penalties constitutes a complicated problem (Hou, Yang, & Papavassiliou, 2002). CAC schemes based on these parameters are named revenue optimization CAC schemes.
- Fair resource assignment: The main problem of CAC schemes basing their admission criterion on the call priority is that often high priority calls monopolize network resources. This forces low priority calls to be severely blocked by the network resulting in high CBP values for low priority traffic flows (Tsiropoulos, Stratogiannis, Kanellopoulos, & Cottis, 2008). This fact is observed not only in networks supporting multiple SCs, where different priority levels are assigned to each SC, but also among different users in the same SC with different service agreements and mobility characteristics. Thus, there exist specific CAC schemes which take into consideration various fairness criteria based on different network parameters, such as network throughput, CBPs achieved etc., guaranteeing that no SC or user class dominates over network resources.

Classification of CAC Schemes

As CAC schemes are of most importance in wireless networks, a plethora of different techniques have been proposed and implemented on systems. Thus, CAC schemes can be classified into different general categories, each of which is based on either the criteria considered in the decision part of the CAC scheme or the specific design characteristic as stated at Table 1. The criteria employed by CAC schemes are usually related to the QoS parameters, which were mentioned above. Considering the specific design characteristics, each one has its own advantages and disadvantages. The selection between the different CAC design approaches should be made upon the wireless technology used, the number of SCs supported and the geographical peculiarities of the region where the network is installed.

The most common design option for CAC schemes is the grade of centralization applied in CAC, in other words the site into the network architecture where the decision making part of the CAC is embedded. Upon this parameter, CAC schemes can be classified into centralized, distributed or collaborative. In centralized schemes, CAC is realized in the mobile switching center (MSC), which is responsible for handling the services supported by the network. The information from each cell must be aggregated in MSC, where the admission decision is taken, and in turn the cell BS is informed to act accordingly. The main advantage of centralized CAC schemes is the increased efficiency level that they achieve. On the other hand, the increased complexity accompanied with the additional control data that they require, constitute them unrealistic in practice. Considering distributed CAC schemes, the decision making part is embedded in each cell BS and realizes CAC procedure independently from the rest cells. Therefore, they are more reliable and practical in implementation, however they are less efficient since they lack of global information for the network parameters; an information

Table 1. Different approaches for CAC classification in wireless networks

Approach to CAC Design	Design Option	Main Idea	Comments
Centralization	Centralized Distributed Collaborative	The site into the network architecture where the decision making part of the CAC is embedded.	Centralized design approach is unrealistic in practice. Distributed approach is less efficient.
Action type	Traffic descriptor (proactive) Measurement based (reactive)	Predictive estimation of QoS constraints or measurement during the transmission attempt.	Traffic descriptor based CAC schemes are very simple to be implemented, while measurement based need real time information.
Information availability level	Global Semi-local Local	Level of information span along the network.	Moving from global to local schemes the complexity is reduced along with their efficiency.
Grade of optimality	Optimal Near-optimal (sub-optimal)	Use of optimization techniques to apply the CAC policy maintaining QoS constraints.	Near-optimal policies are preferable compared to optimal, as the latter are more complex and lead to intractable solutions.
Traffic model	Incoming rate Service time	Different stochastic models can be employed for traffic model description.	If Poison arrivals and exponential holding times are mainly assumed, leading to a tractable analysis.
Spatial distribution of users	Uniform Non-uniform	The differentiation is based on position, direction and velocity of mobile terminals.	Uniform and non-homogeneous schemes correspond to homogeneous and non-homogeneous systems respectively.
Information size	Per cell Per call	The classification is based on the amount of information required for the admission procedure.	CAC schemes based on per call information are more complex that per cell schemes, but they are more efficient.
Link under consideration	Uplink Downlink	The asymmetry between uplink and downlink.	The uplink constraint is stricter due to power level consumption limitations.

available only in centralized CAC schemes. A hybrid design option, named collaborative approach (O'Callaghan, Gawley, Barry, & McGrath, 2004), constitutes a promising solution in CAC. In this case, information is exchanged between neighboring cells with respect to resource allocation and admission control, while the decision in taken locally in the BS of each cell. Hence, the advantages from both centralized and distributed CAC schemes are combined into a new powerful design architecture offering increased efficiency and reliability. The main disadvantage in this design option is its demand for high communication overhead.

CAC schemes can also be assorted into traffic descriptor based (alternatively called proactive) or measurement based (alternatively called reactive). In the former group of CAC schemes, admission decision is based on a given traffic pattern available to the algorithm. The CAC scheme examines whether the already reserved bandwidth increased by the bandwidth demand of the new call exceeds cell capacity. In this case, the call is blocked otherwise it is accepted. The most common traffic descriptor based CAC scheme is simple sum (Tragos, Tsiropoulos, Karetsos, & Kyriazakos, 2008) which simply ensures that the sum of requested resources does not exceed cell capacity. A new call (denoted as a) with maximum bandwidth demand r_{α} is admitted under the condition that the total existing bandwidth demand increased by r_{α} remains below maximum cell capacity μ. Thus, measured sum scheme admits a call α if the following test succeeds:

$$v + r_{\alpha} \leq \mu . \tag{1}$$

As multimedia traffic supported by modern networks is bursty in nature, traffic descriptor based CAC schemes overestimate bandwidth demands. This occurs since traffic descriptors specify the maximum bandwidth demand of each call, which is rarely used. On the other hand, traffic descriptor based CAC schemes are very simple; therefore, they are widely implemented mostly by switch and router vendors.

In measurement based CAC schemes the decision making part of the algorithm employs the actual network characteristics of the network, which may be the traffic load, the packet error rate etc. These network characteristics are measured through an appropriate mechanism, implying that they are more realistic compared to explicit traffic descriptors. Some interesting measurement based CAC schemes are considered in (Tragos, Tsiropoulos, Karetsos, & Kyriazakos, 2008) (Niyato & Hossain, 2005) which are based on actual traffic flow, occupied bandwidth, network load and packet loss accompanied with revenue award. The most fundamental parameter in measurement based CAC schemes is the measurement based mechanism itself, namely the way the parameter employed in CAC procedure is measured (Mase, 2004). This can be realized by either measuring a network parameter every sampling period according to a time-window policy (Jamin, Danzig, Shenker, & Zhang, 1997), or computing an average based on current and/or previous measurements (Floyd, 1996). Most CAC schemes embedded in CDMA systems are designed according to the measurement based technique (Stasiak, Wisniewski, & Zwierzykowski, 2005).

A further remarkable classification of CAC schemes can be achieved based on the amount of information available at the decision making part of the algorithm. This information may include the available or occupied channels of a cell, the total bandwidth allocated to ongoing users, the mean packet delay for each traffic flow, etc. If this information can span along the whole network, then the scheme can be characterized as global. It is reasonable that these schemes provide high efficiency, but they are exceptionally complex, since they need to exchange a huge amount of information among the cells of the network. In case that the area of information exchange is limited only among an area including at least the neighboring cells of the cell under consideration, then the CAC scheme is called semi-local. These schemes provide also high efficiency and they are less complex compared to global ones, but they still need a lot of information exchange. Apart from information exchanging schemes, local CAC schemes exist, which are based on the information concerning the cell under consideration. Local schemes are simple in implementation, however they are less efficient compared to global or semi-local schemes. This is mainly attributed to the fact that the load of a cell may be influenced by the load of neighboring cells due to users mobility, a characteristic which is not considered in local CAC schemes.

There are many CAC schemes available in literature proven to achieve the optimal solution in CAC problem, according to input parameters given for the admission decision process. However, the optimal CAC evokes a high computational complexity in design and implementation, due to a large number of states associated with the Markov decision problem (MDP). The large problem scale and multiple interdependent network parameters employed in optimal CAC schemes result in high complexity and increased processing time. Thus, optimal CAC schemes are not applicable in practice, as the admission decision is taken instantaneously upon the arrival of a call request. As an alternative approach to optimal CAC schemes, sub-optimal CAC schemes have been proposed which achieve significantly lower complexity in design and implementation and are capable of operating on-line. Sub-optimal CAC schemes are utilized to obtain a near-optimal solution to CAC problem, usually by employing intellectual techniques (heuristic functions, alternative approaches, etc) to simplify original problem complexity.

CAC schemes can also be separated based on information granularity (Jain & Knightly, 1999), which is dependent on the traffic model assumed, the spatial distribution of network users and the way that network information is obtained. CAC schemes may adopt a specific mobility pattern for users otherwise a simple resource policy for mobile users will be used. In the first case, the exact knowledge of the mobility characteristics of users, such as direction and velocity, help to precisely predict the handoff traffic load to each cell. The spatial distribution of users may be uniform or non-uniform, consequently the wireless network is assumed as homogeneous or non-homogeneous respectively. Information can be obtained by a cell for either each call or each SC stream flow. As the information for the network gets more analytical, the complexity of the CAC scheme increases along with its efficiency.

An additional taxonomy of CAC schemes can be performed based on differentiating the data rates between uplink and downlink. Unlike traditional voice services, the demand for bandwidth between uplink and downlink may be asymmetric in many multimedia services. In such systems, if CAC scheme employed allocates equal bandwidth for both uplink and downlink, then the overall system capacity could be limited by downlink (Yang, Feng, & Kheong, 2006). In this case, resources are deficiently exploited, bandwidth is wasted and CAC scheme efficiency remains low. Some CAC schemes consider a joint admission policy, taking into account both uplink and downlink. Thus, a new call request is accepted on condition that enough resources can be allocated to both uplink and downlink according to call QoS characteristics.

Network Model and System Analysis

The majority of the studies concerning CAC in wireless networks employ some standard assumptions to provide a tractable system analysis. Most assumptions and system models were obtained by classic traffic theory and applied to cellular network. The latter does not necessarily obey the same traffic model, as the mobility of the users and the emerging multimedia services indicate the need for new teletraffic theories and models, taking into account the new opportunities of wireless networks.

One of the most fundamental assumptions in system models for wireless networks concerning CAC is the call arrivals to the network which are considered to follow the Poisson distribution. Thus, new calls are assumed to arrive according to a Poisson distribution with rate $\lambda_{n,i}$ in cell i. If the network is assumed homogeneous, then the arrival rate is λ for every cell and the analysis is focused only to a single cell. The handoff call arrivals are also assumed to follow the Poisson distribution with rate $\lambda_{h,i}$. Poisson arrivals are not so obvious in the case of handoff calls, as the handoff traffic is strictly related to the mobility characteristics of the users. In (Chlebus & Ludwin, 1995) it has been proven that this assumption holds provided that there is no blocking in the network. Since this is an ideal case, in the same work a blocking scenario is assumed in order to examine how accurate the assumption of Poisson handoff arrivals is. The results indicated that this constitutes an approximation model and its performance is satisfactory. Moreover, in (Chlebus & Ludwin, 1995) the authors claim without providing the proof that in a blocking environment handoff traffic is a smooth process, which means that the variance is less than the mean value.

Channel holding time is defined as the amount of time that a call is assigned a channel in a certain cell. The channel may be released if the call is either completed or handed off to a neighboring cell. Another critical definition in wireless networks is call holding time (found in literature also as service time or requested call connection time, RCCT), which is defined as the total connection time initially requested by a call. Call holding time is subjected to the specific characteristics of the call, as calls belonging to different SCs may have also different duration. How long a call stays in a cell is another fundamental parameter in wireless networks and is widely called as cell residence time (CRT) or cell dwell time. CRT is mainly dependent on users mobility characteristics and on the geometry of the cells. The majority of the analysis existent in literature assume that channel holding time is modeled through an exponential distribution for both new and handoff calls. However, channel holding time does not necessarily obey the exponential distribution as there exist certain conditions to be held, which were extensively investigated in (Fang, Chlamtac, & Lin, 1998). Fang, Chlamtac and Lin (1998) have proven that channel holding time follows exponential distribution if and only if the cell dwell time is exponentially distributed. In the rest cases, the channel holding time cannot be modeled according the exponential distribution, handoff traffic cannot be modeled as Poisson distribution and the total incoming traffic to the cell cannot be assumed as Poisson arrivals. Some researchers adopt other distribution models for channel holding time such as lognormal (Jedrzycki & Leung, 1996) and general distribution (Rajaratmam & Takawira, 1999). Although the assumption of cell dwell time and channel holding time is complicated, most researchers model both parameters through exponential distribution as through this assumption the analysis is tractable and analytical formulas for CBP can be derived. A more rigorous analysis would require the analytical survey of network assumptions, but this is beyond the scope of this chapter. Thus, both (new and handoff) incoming traffic flows will be assumed as Poisson arrivals and channel holding time and cell dwell time will be modeled through exponential distribution with mean $1/\mu_i$.

In a complete resource sharing scheme (Lai, Misic, & Chanson, 1998) a call is admitted as long as there are sufficient network resources to accommodate it; otherwise it is rejected. The same policy is applied for new and handoff calls, making no segregation between them. By defining the state of a particular cell i at time t to be the number of occupied channels in that cell, represented by $\{c_i(t) \mid t \geq 0\}$, the cell can be modeled as a continuous-time Markov chain (CTMC). If the number of channels in a given cell i is C_i, then the system model is a typical M/M/C_i queue (Figure 2). Note that in order to adopt M/M/C_i model we should make the assumption that when the network operates under congestion, a call arrival (new or handoff) is blocked. This assumption relaxes the analysis from M/M/C_i/K, where K is maximum calls waiting to be admitted, into M/M/C_i, where no buffer is used. The truncated state space of cell i, represented by S_i, is given as

$$S_i = \{n_i; 0 \leq n_i \leq C_i\},$$

where, n_i denotes the number of occupied channels in a certain network state.

Let $\pi(n_i, n_i')$ define the transition rate from state n_i to state n_i', where n_i, $n_i' \in S_i$. Then, the transition probabilities for adjacent states are obtained as follow

$$\pi(n_i, n_{i+1}) = \lambda_{n,i} + \lambda_{h,i} \quad (2.a)$$

$$\pi(n_i, n_{i-1}) = n_i\mu_i . \quad (2.b)$$

Based on transition diagram stated in Figure 2 the following global balance equation is derived

$$(\lambda_{n,i} + \lambda_{h,i})\, p(n_i) = n_{i+1}\mu_i p(n_{i+1}), \quad (3)$$

Figure 2. Transition diagrams considering network state of the complete resource sharing scheme

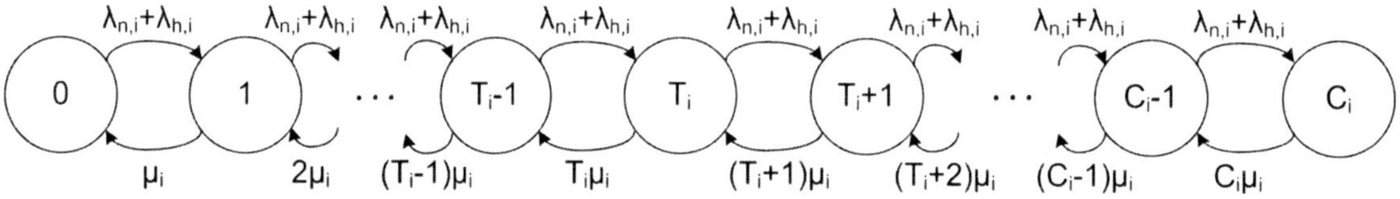

where, $p(n_i)$ denotes the steady state probability that the number of ongoing calls in cell i is n_i, defined as $\lim_{t\to\infty} \Pr ob(c_i(t) = n_i)$, where $n_i = 0, 1, \ldots, C_i$. From the global balance equation the steady state probability can be obtained as follows

$$p(n_i) = \frac{\rho^{n_i}}{n_i!} p(0), \tag{4}$$

where, ρ is the traffic intensity defined as $\rho = (\lambda_{n,i} + \lambda_{h,i})/\mu_i$ and $p(0)$ is the normalization factor defined as

$$p(0)^{-1} = \sum_{n_i=0}^{C_i} \frac{\rho^{n_i}}{n_i!}. \tag{5}$$

A new call is blocked on the grounds that all channels are occupied, thus, the new call blocking probability $P_n^b(i)$ in cell i is given by

$$P_n^b(i) = p(C_i). \tag{6}$$

Since no prioritization for handoff calls has been assumed for this general analysis, handoff failure probability $P_h^b(i)$ in cell i should be equal to $P_n^b(i)$, therefore we obtain

$$P_n^b(i) = P_h^b(i) = p(C_i). \tag{7}$$

This analysis can be carried out for multiple SCs and multiple cells, but it is complicated as the transition diagram has multiple dimensions (Fang & Zhang, 2002), therefore, the corresponding global balance equation is difficult to be solved. In most cases, different traffic flows, each of which corresponds to a specific SC, are considered to be independent; therefore, multiple one dimensional transition diagrams are obtained reducing the complexity of the problem. An interesting and mathematically robust analysis considering this problem is provided in (Li & Chao, 2007) where CBPs, handoff rates and QoS (named also as grade of service) are acquired in closed form expressions.

In the case of multiple independent SCs, the previous analysis is carried out separately for every SC. Suppose U SCs with the corresponding arrival and death rate for a cell i being

$$\lambda_{u,i}(n_{u,i}) = \lambda_{nu,i}(n_{u,i}) + \lambda_{hu,i}(n_{u,i}) \tag{8}$$

and $\mu_{u,i}$, respectively, where $u = 1, \ldots, U$, $\lambda_{nu,i}(n_{u,i})$ and $\lambda_{hu,i}(n_{u,i})$ are the respective call arrival rates for new and handoff u SC calls in the cell i, $\mu_{u,i}$ is the mean cell dwell time of the same u SC calls and $n_{u,i}$ are the total number of occupied channel by u SC calls in the cell i. The respective steady state probability of being $n_{u,i}$ channels occupied is given by

$$p_u(n_{u,i}) = p_u(0) \left(\frac{\lambda_{nu,i} + \lambda_{hu,i}}{\mu_{u,i}} \right)^{n_{u,i}} \frac{1}{n_{u,i}!}, \tag{9}$$

where, $p_u(0)$ is the normalization constant given by

$$p_u(0)^{-1} = \sum_{n_{u,i}=0}^{C_i} \left(\frac{\lambda_{nu,i} + \lambda_{hu,i}}{\mu_{u,i}} \right)^{n_{u,i}} \frac{1}{n_{u,i}!}. \tag{10}$$

Considering now the total number of SCs supported in cell i, the truncated state space is

$$S_i' = \left\{n_i = \left(n_{1,i}, n_{2,i}, \ldots, n_{U,i}\right);\ n_{1,i} + n_{2,i} + \ldots + n_{U,i} \leq C_i\right\}$$
.

The steady state probability of the network being at state n_i is given by

$$\mathbf{p}_{u,i}\left(\mathrm{n}_i\right) = \mathbf{p}_{u,i}\left(0\right)\prod\nolimits_{u=1}^{U}\left(\frac{\lambda_{nu,i} + \lambda_{hu,i}}{\mu_{u,i}}\right)^{n_{u,i}}\frac{1}{n_{u,i}!}, \quad (11)$$

where, $\mathbf{p}_{u,i}\left(0\right)$ being the normalization constant given by

$$\mathbf{p}^{-1}_{u,i}\left(0\right) = \sum_{\mathrm{n}_i \in S_i'}\prod\nolimits_{u=1}^{U}\left(\frac{\lambda_{nu,i} + \lambda_{hu,i}}{\mu_{u,i}}\right)^{n_{u,i}}\frac{1}{n_{u,i}!}. \quad (12)$$

For the complete resource sharing scheme the new CBP and handoff CDP for u SC in cell i are equal with the probability of the state where the network operates under congestion, e.g. all channels are occupied. Thus, the corresponding probability is given by

$$P_h^b\left(u,i\right) = P_n^b\left(u,i\right) = p\left(n_i^*\right), \quad (13)$$

where,
$$n_i^* = \left(n_{1,i}^*, n_{2,i}^*, \ldots, n_{U,i}^*\right);\ n_{1,i}^* + n_{2,i}^* + \ldots + n_{U,i}^* = C_i$$
.

Generally, there are two different approaches considering the whole network problem with J cells supported, $i = 1, 2, \ldots, J$. In the first case, the network can be assumed as homogeneous; therefore, one cell may be examined with its results representing the whole network state. Thus, the new CBP and handoff CDP determined above for i cell are the same for the whole network problem. In the second case, the network traffic is disproportionally distributed to network cells; consequently, the analysis should be further carried on to determine the admittance failure probabilities. This study is analytically presented in (Li & Chao, 2007) where additional QoS network parameters are examined.

Guard Channel / Fractional Guard Channel Policies (Thinning Schemes)

Complete resource sharing schemes cannot guarantee a certain level of QoS for handoff calls, especially during network congestion, as the whole number of channels are allotted indiscriminately. Therefore, GC schemes, called also reservation schemes, have been first proposed by Hong and Rappaport (1986) to prioritize handoff calls over new calls. The basic concept of these schemes is to permanently reserve a certain amount of channels or network resources only for serving handoff calls. The rest channels are available for both new and handoff calls. Thus, supposing that the total amount of channels is set C_i in the cell i and the channels available for common use are T_i ($T_i < C_i$), then the channels dedicated to handoff calls are $C_i - T_i$. Therefore, T_i operates as a threshold for new users, implying that a new call is admitted provided that at least T_{i+1} channels are available, while a handover call only needs one available channel to be accepted. Maintaining the same network assumptions stated above for the single SC complete resource sharing scheme the transition diagram for the CTMC is shown in Figure 3, represented by the same truncated state space S_i. The transition probabilities for adjacent states in GC schemes are obtained as follows

$$\pi\left(n_i, n_{i+1}\right) = \lambda_{n,i} + \lambda_{h,i} \text{ for}$$
$$\pi\left(n_i, n_{i+1}\right) = \lambda_{n,i} + \lambda_{h,i}\ 0 \leq n_i \leq T_i - 1 \text{ for}$$
$$0 \leq n_i \leq T_i - 1 \quad (14.a)$$

Figure 3. Transition diagrams considering network state of the Guard Channel scheme

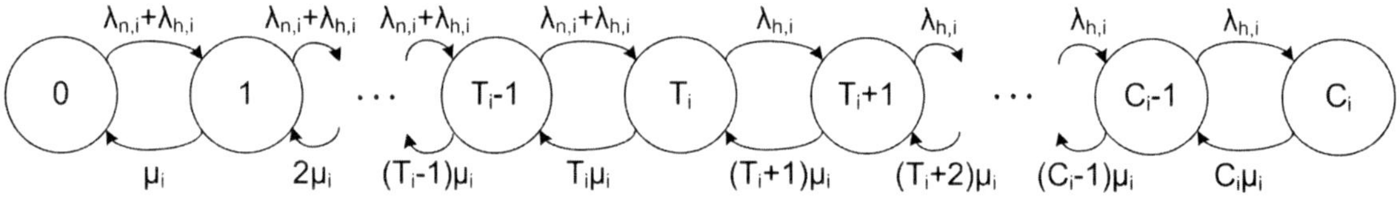

$$\pi\left(n_i, n_{i+1}\right) = \lambda_{h,i} \text{ for } \pi\left(n_i, n_{i+1}\right) = \lambda_{h,i}$$

$$T_i \le n_i \le C_i - 1 \text{ for } T_i \le n_i \le C_i - 1 \quad (14.b)$$

$$\pi\left(n_i, n_{i-1}\right) = n_i \mu_i \,. \quad (14.c)$$

Based on the global balance equations provided by CTMC on Figure 3, the following expression for the state space probability $p\left(n_i\right)$ is derived

$$p\left(n_i\right) = p\left(0\right)\left(\frac{\lambda_{n,i} + \lambda_{h,i}}{\mu_i}\right)^{n_i} \frac{1}{n_i!} \text{ for } p\left(n_i\right) = p\left(0\right)\left(\frac{\lambda_{n,i} + \lambda_{h,i}}{\mu_i}\right)^{n_i} \frac{1}{n_i!}$$

$$0 \le n_i \le T_i - 1 \text{ for } 0 \le n_i \le T_i - 1 \quad (15)$$

$$p\left(n_i\right) = p\left(0\right)\left(\frac{\lambda_{n,i} + \lambda_{h,i}}{\mu_i}\right)^{T_i}\left(\frac{\lambda_{h,i}}{\mu_i}\right)^{n_i - T_i} \frac{1}{n_i!} \text{ for } p\left(n_i\right) = p\left(0\right)\left(\frac{\lambda_{n,i} + \lambda_{h,i}}{\mu_i}\right)^{T_i}\left(\frac{\lambda_{h,i}}{\mu_i}\right)^{n_i - T_i} \frac{1}{n_i!}$$

$$T_i \le n_i \le C_i - 1 \text{ for } T_i \le n_i \le C_i - 1, \quad (16)$$

where, $p\left(0\right)$ is the normalization factor defined as

$$p\left(0\right)^{-1} = \sum_{n_i=0}^{T_i-1}\left(\frac{\lambda_{n,i} + \lambda_{h,i}}{\mu_i}\right)^{n_i} \frac{1}{n_i!} + \sum_{n_i=T_i}^{C_i}\left(\frac{\lambda_{n,i} + \lambda_{h,i}}{\mu_i}\right)^{T_i}\left(\frac{\lambda_{h,i}}{\mu_i}\right)^{n_i - T_i} \frac{1}{n_i!} \quad (17)$$

According to GC CAC scheme policy, a new call is blocked if the total occupied channels in the cell i are equal or above T_i. Thus, the new CBP is given by

$$P_n^b\left(i\right) = p\left(T_i\right) + p\left(T_{i+1}\right) + \ldots + p\left(C_i\right). \quad (18)$$

As GC schemes provide handoff prioritization, a handoff call is blocked if and only if all channels of i cell are occupied, consequently the handoff CDP is expressed as

$$P_h^b\left(i\right) = p\left(C_i\right). \quad (19)$$

If T_i becomes equal to C_i then the GC scheme become a complete resource sharing scheme. The main issue in GC schemes is the calculation of threshold value T_i that minimizes new CBP given the constraint for handoff CDP (Ramjee, Towsley, & Nagarajan, 1997) (Harine, Marie, Puigjaner, & Trivedi, 2001). The harder the constraint for handoff calls is, the better QoS is provided by the network. On the other hand, the lower values for handoff CDP imply higher values for new CBP. Thus, there is a tradeoff between new CDP and handoff CDP, which is reflected in threshold value calculation.

As modern wireless networks support multiples SCs with different QoS characteristics, sophisticated CAC schemes should be employed to guarantee the different QoS levels. Thus, GC schemes, which were employed to prioritize handoff calls over new ones, can be modified to utilize different thresholds for each SC (new and handoff) calls in the admission decision process. These schemes are commonly known in literature as multi-threshold bandwidth reservation schemes (Haung & Ho, 2002) (Chen, Yilmaz, & Yen, 2006). In such schemes two types of call prioritization can be applied, priority of SC and priority of call type. The former priority characterizes the type of the service and addresses the problem of different SC calls prioritization by employing different thresholds for each SC calls. The analysis is similar to the one presented above for complete resource sharing scheme supporting multiple SCs.

The different SCs supported by the network are usually RT and NRT SCs, while in other cases the SCs supported may be more than 2 (Tragos, Tsiropoulos, Karetsos, & Kyriazakos, 2008) each of which having its own QoS demands. The latter priority type characterizes the call type such as new or handoff call. The main problem of multiple thresholding schemes is the exact calculation of critical parameters such as threshold values. In the literature, there exist some algorithms for calculating thresholds mainly based on revenue optimization techniques, but most of them are computationally expensive and difficult to be solved (Chen, Yilmaz, & Yen, 2006). In particular, threshold estimation is a multi-constraint combinatorial problem, where multiple network limitations should be taken into account.

The fixed portion of channels reserved for handoff calls or high priority SC calls by GC or multi-threshold schemes respectively is determined mainly based on a priori knowledge of the traffic patterns. Therefore, they lack adaptability as they are static in nature, thereby being unable to cope with network dynamics. In realistic situations, call arrival rates do not remain static and they cannot be precisely described through traffic patterns. Thus, a network with time-varying traffic requires dynamic CAC schemes, which usually yield better performance. Dynamic (named also adaptive) GC schemes are based on either a mobility prediction technique employing the users mobility ratio, the direction and the location of the mobile terminal, or measurements of some network parameters such as available resources, CBP and CDP. Their basic objective is to adapt the thresholds employed by the CAC scheme as to optimize a given objective function. This function can be based on different algorithms, such as a penalty – revenue approach concerning the admittance or rejection of the incoming calls, a network utilization optimization policy, etc. Both mobility prediction and measurement based dynamic CAC schemes are strongly dependent on the efficiency of the mobility estimation algorithm or the measurement technique employed, respectively. The extended results presented in literature indicate that dynamic GC schemes outperform the static GC schemes (Yu & Leung, 1997) (Bartolini & Chlamtac, 2001). On the other hand, dynamic GC schemes are more complex compared to static ones, requiring increased communication between BSs and large processing load.

Another variation of the GC schemes presented above is the fractional guard channels (FGC) schemes. FGC scheme was first proposed by Ramjee (1996) and was proven to be more general than GC schemes. The basic idea of the FGC scheme lies on accepting an arriving new call under a certain probability $a(n_i)$, where $0 \le a(n_i) \le 1$ and $n_i = 0,1,\ldots,C_i$, which depends on the number of busy channels in cell i. Generally, this probability is variable among different states of the network and in the case that $a(0) \ge a(1) \ge \ldots \ge a(C_i)$ the new call stream becomes gradually thinner with the increase of occupied channels. Handoff calls are unhinderly admitted unless there are no channels available. Due to the fact that a certain number of channels were reserved in simple GC schemes, new CBP and handoff CDP vary greatly as the number of occupied channels change. On the other hand, in FGC schemes the application of the admission probabilities eliminates the use of threshold in admission decision providing a smoother change in new CBP and handoff CDP.

Assume a FGC scheme with the same network assumptions stated in the GC schemes of the present section. Network states can be modeled according the CTMC shown in Figure 4, which are represented by the truncated state space S_i. As observed in Figure 4, the transition probabilities

Figure 4. Transition diagrams considering network state of the Fractional Guard Channel scheme

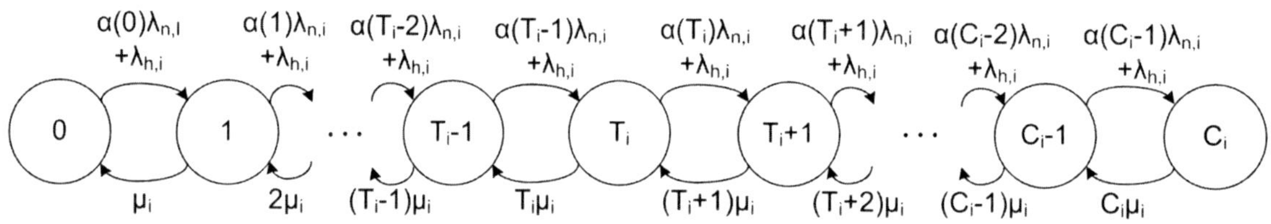

for adjacent states in FGC schemes are obtained are the following

$$\pi\left(n_i, n_{i+1}\right) = \alpha\left(n_i\right)\lambda_{n,i} + \lambda_{h,i} \tag{20.a}$$

$$\pi\left(n_i, n_{i-1}\right) = n_i \mu_i \,. \tag{20.b}$$

Consequently, the global balance equation can be obtained by applying the fundamental rule of birth-death process in Figure 4, "rate up equals rate down" (Cooper, 1981), as follows

$$\left(\alpha\left(n_i\right)\lambda_{n,i} + \lambda_{h,i}\right) p\left(n_i\right) = \left(n_{i+1}\right)\mu_i p\left(n_{i+1}\right). \tag{21}$$

By solving the global balance equation the stationary distribution for the FGC scheme is obtained

$$p\left(n_i\right) = \frac{\prod_{k=0}^{n_i} \alpha\left(k\right)\lambda_{n,i} + \lambda_{h,i}}{\mu_i^{n_i} n_i!} p\left(0\right) \text{ for}$$

$$p\left(n_i\right) = \frac{\prod_{k=0}^{n_i} \alpha\left(k\right)\lambda_{n,i} + \lambda_{h,i}}{\mu_i^{n_i} n_i!} p\left(0\right)$$

$$0 \le n_i \le C_i \text{ for } 0 \le n_i \le C_i, \tag{22}$$

where

$$p\left(0\right)^{-1} = \sum_{n_i=0}^{C_i} \frac{\prod_{k=0}^{n_i} \alpha\left(k\right)\lambda_{n,i} + \lambda_{h,i}}{\mu_i^{n_i} n_i!}. \tag{23}$$

The new CBP and handoff CDP can be obtained as

$$P_n^b\left(i\right) = \sum_{n_i=0}^{C_i} \left(1 - \alpha\left(n_i\right)\right) p\left(n_i\right) \tag{24}$$

and

$$P_h^b\left(i\right) = p\left(C_i\right) \tag{25}$$

respectively.

Obviously, when $a(0) = a(1) = ...a(T_i) = 1$ and $a\left(T_{i+1}\right) = ... = a\left(C_i\right) = 0$ the simple GC scheme is obtained with T_i operating as threshold for the new calls. Moreover, by setting all fractional probabilities equal to unity, e.g. $a\left(0\right) = a\left(1\right) = ... = a\left(C_i\right) = 1$, the complete resource sharing scheme is obtained. Thus, FGC schemes are proven to be more general with GC and resource sharing schemes being special cases. Due to the fractional probability $\alpha\left(n_i\right)$ embedded in FGC schemes, they are used to obtain new schemes for multimedia wireless networks, specially designed to comply with network administrator specifications and SCs QoS requirements.

FGC schemes are further employed to prioritize high priority SCs in multimedia wireless networks. These schemes are also known as thinning or probabilistic CAC schemes (Wang, Fan, & Pan, 2008)(Tsiropoulos, Stratogiannis, Kanellopoulos, & Cottis, 2008). Each SC call is assigned with its own probability being the same for all the calls of a certain SC. Higher SCs in terms of priority are assigned with respective higher fractional probabilities. In this case, the previous analysis for FGC schemes can be generalized according to the same analysis employed for multi-SCs in complete resource sharing and GC schemes (Wang, Fan, & Pan, 2008).

CAC Schemes Based on Signal Quality (SIR-Based)

In modern wireless access technologies, interference poses critical constraints concerning mainly the signal quality. This situation has an impact not only on network conditions but also on system capacity. Particularly, in CDMA wireless networks interference is the dominant factor affecting their performance in terms of capacity and QoS provision to end users. Thus, the SINR value is an adequate metric of the signal quality. CDMA-based air interfaces are mainly influenced by interference caused by other users from the same network instead of Gaussian noise, consequently the noise effect is usually neglected focusing mainly on SIR.

CAC schemes implemented for interference - limited networks employ as admission criterion either the interference levels caused by a new incoming call or the signal quality levels achieved. Hence, interference based CAC schemes admit new calls only if the SNR/SIR values can maintain a minimum signal quality level. The SNR/SIR levels correspond to predefined QoS levels for new and ongoing users. This simple approach offers a tool to reduce interference in wireless networks, while on the other hand is constitutes an efficient admission criterion.

Two simple SIR-based solutions were proposed by (Liu & Zarki, 1994) for controlling the signal quality. This is realized by considering the SIR value achieved by the new call. The call is admitted provided that this value is higher than the minimum SIR value. Both implemented schemes are based on the residual capacity of the cell formulating each time an appropriate admission criterion. In the first scheme the residual capacity of the network is defined as

$$R_k = \left[\frac{1}{SIR_{th}} - \frac{1}{SIR_k}\right], \tag{26}$$

where SIR_k is the uplink SIR value in a cell k and SIR_{th} is the threshold value that imposes whether a call is admitted or not. The residual capacity of the cell is calculated when a new user arrives and if is greater than zero the incoming call is admitted otherwise the call is rejected. The second proposed algorithm follows the same rationale taking also into account the impact of admitting one call on cell k itself and its adjacent cells $C(k)$ as well. This is done by encompassing an interference coupling parameter β in the above defined residual capacity leading to

$$R_{k,j} = \left[\frac{1}{\beta}\left(\frac{1}{SIR_{th}} - \frac{1}{SIR_j}\right)\right], \quad j \in C(k) \tag{27}$$

These simple algorithms were evolved taking into account inter-cell interference. In residual capacity estimation, the parameter $L_m(j,k)$ is employed, representing the predicted additional intercell interference. The use of this parameter has a great impact on CBPs achieved, which are significantly reduced. QoS guarantees are also provided by using certain SIR thresholds, which correspond to certain levels for blocking rate (Kim, Shin, & Lee, 2000). The residual capacity is given by

$$R_{k,j} = \left[\frac{1}{SIR_{th}} - \frac{1}{SIR_j} - \frac{1}{L_m(j,k)}\right], \quad j \in C(k) \tag{28}$$

Bandwidth Adaptation and QoS Re-Negotiation

Wireless networks support a variety of services which can be categorized into rate adaptive applications and constant bitrate (CBR) services. The calls of the latter category demand a constant bitrate throughout their duration. In such services, e.g. voice calls, a further increase of their bandwidth above the standard demand will not improve their QoS. On the other hand, in rate adaptive services users specify at their connection request the range of bandwidth required to be supported by the network. Thus, each user seeking for a rate adaptive service has a minimum and a maximum bandwidth request. Moreover, rate variations may originate from the dynamic nature of wireless environment combined with the mobility of the user terminals. Thus, bandwidth adaptation algorithms are employed in modern wireless networks to improve network utilization and guarantee the QoS requirements of ongoing calls, offering them at least their minimum bandwidth demand. When the network condition are favorable and enough resources are available, the existent free resources are allocated to ongoing rate adaptive users according two different strategies based on SCs priorities (Li & Chao, 2007). In the first scheme the available resources are fairly allocated to all ongoing users without taking into account their priority levels. In the second scheme, resources are first allocated to the high priority SC calls, till the resources are exhausted or the maximum bandwidth demand of these calls is met. If enough resources are still available, then the scheme assigns the rest available resources to the next high priority SC calls. The same procedure follows until all resources are exhausted or all calls are served with their maximum bandwidth demand. Apart from taking into account priority levels, resource assignment in rate adaptive services may be performed based on more complicated schemes. In (Sen, Jawanda, Basu, & Das, 1998) an optimal resource assignment strategy is proposed for maximizing a revenue strategy, while in (Sherif, Habib, Nagshineh, & Kermani, 2000) an adaptive resource allocation algorithm is proposed to maximize bandwidth utilization and attempt to provide fairness with a generic algorithm.

Generally, when a call arrives into a certain cell the network either may have enough resources to provide a bandwidth between its minimum and maximum demand or it may operate under congestion which means that it cannot provide even the minimum bandwidth of the new call request. In the first case the call is admitted as QoS requirements in terms of bandwidth demands are met. In the second case, bandwidth adaptation CAC algorithms, also known as rate adaptive schemes, are applied to find an optimal solution concerning the resource allocation serving as many users as possible while providing QoS guarantees to ongoing calls minimizing admittance failure probabilities. This is implemented by reducing the rate of some users which have higher rates than the declared minimum values as much as needed to accommodate the new call request. In some bandwidth adaptation CAC algorithms, this procedure is followed only for handoff or high priority SC call requests (Tragos, Tsiropoulos, Karetsos, & Kyriazakos, 2008) (Lindemann, Lohmann, & Thümmler, 2004). However, it must be mentioned that user rates cannot be reduced under the minimum rate values so as to ensure QoS provision, therefore, in case that all users operate at their lowest bandwidth requirement then a new call request is rejected. Rate degradation may be applied according to a prioritization or a non-prioritization scheme. In the former scheme, the rate degradation policy is first applied to the SC calls with the lowest priority. If the resources released are still not enough to admit the call request then the calls of the next priority level are examined. In the non-prioritization scheme, all calls served with higher rate than their minimum bandwidth demand decrease their rate so as to admit the call request. A useful QoS metric in QoS renegotiation CAC schemes is the degradation ratio which is defined as the proportion of the number of degraded calls to the number of ongoing calls (Kwon, Choi, Bisdikian, & Naghshineh, 1999). Moreover, though network measurements the degradation probability can be determined. Higher or lower degradation probability values denote a more or less aggressive CAC design approach respectively.

Except for rate degradation, the reverse procedure is followed in case where enough available resources exist so as to offer higher rates to ongoing calls. This rate upgrade policy can be applied in accordance with two ways. In the first one, a rate adaptive resource allocation scheme mentioned above is employed to exploit the available resources (Li & Chao, 2007). In the second one, the calls which had recently their rate decreased are the first ones to have their rate restored (Tragos, Tsiropoulos, Karetsos, & Kyriazakos, 2008). If there are still enough available resources, a resource allocation scheme is employed to allocate the current available resources to ongoing users.

QoS renegotiation schemes and especially rate degradation algorithms must be treated carefully and should constitute the last step of a CAC scheme in their effort to gain the needed resources for admitting a new call. There are many applications, such as voice calls or video streaming, the rate of which cannot be reduced without QoS degradation or being noticed by the user. Furthermore, since users are sensitive to multiple fluctuations in the rate of the applications used, it is preferable to employ some thresholds in rate upgrade procedure. This policy implies that a rate upgrade is accomplished only if the available resources after the upgrade are above the threshold value defined (Tragos, Tsiropoulos, Karetsos, & Kyriazakos, 2008).

Price Driven CAC Schemes

One of the most widely studied CAC approaches in literature are price driven CAC schemes. They are usually employed to apply a pricing policy in CAC procedure as to maximize the total revenue while guaranteeing at the same time the QoS requirements. Moreover, they are used in wireless networks as means to shape incoming traffic and prevent network congestion (Cocchi, Shenker, Estrin, & Zhang, 1993) (Evci & Fino, 2001). The main idea which they are based on is that users act independently and sometimes selfishly, since they seek to maximize their own utility, without taking into consideration the current state of the network. Thus, through the application of an appropriate pricing policy they can behave in a way that improves overall network utilization and system performance. Price driven CAC schemes can be further categorized into two main groups. The CAC schemes belonging to the first category embed a dynamic pricing module into call admittance procedure, while CAC schemes belonging to the second category adopt the pricing policy to determine the network parameters employed in admittance procedure.

A thorough analysis of a CAC scheme belonging to the fist category is examined in (Hou, Yang, & Papavassiliou, 2002). In this analysis, CAC scheme assigns a certain price so as to shape the incoming traffic of new calls. Handoff calls are admitted unhinderly by the network till all network resources are occupied. The price $p(t)$ determined by the network at a given time t is dynamic, aiming at maximizing the total utility of users. Its calculation is based on the optimal arrival rate λ^*. Thus, if the total incoming rate $\lambda(t)$ is below λ^* a normal price p_o is assigned to new calls, which is acceptable by all users. Therefore, the accepted traffic load of new calls $\lambda_{in}(t)$ is equal to the total incoming rate, $\lambda_{in}(t) = \lambda(t)$.

Otherwise, in the case that $\lambda(t)$ is above λ^*, an increased price $p(t)$ is determined based on a demand function defined as follows

$$D(t) = e^{-\left(\frac{p(t)}{p_o}-1\right)^2}, \quad D(t) = e^{-\left(\frac{p(t)}{p_o}-1\right)^2} \quad p(t) \geq p_o, \quad p(t) \geq p_o, \tag{29}$$

where, $D(t)$ denotes the fraction of the users which accept the price $p(t)$. The value of $p(t)$ should be suitable so as the accepted traffic load, which is lower than the total incoming rate as $\lambda_{in}(t) = D(t)\lambda(t)$, has to be below the optimal arrival rate λ^*. As $p(t)$ is continuous over the time and dynamically adjusted according the system conditions any, the condition

$$\lambda_{in}(t) \leq \lambda^* \tag{30}$$

is always satisfied.

Price driven CAC schemes belonging to the second category employ a revenue estimation function or mechanism, which is based on a certain network parameter (thresholds applied, SIR, etc.). Through an appropriate method they aim at maximizing the revenue obtained, consequently an optimal value for the network parameter is obtained. Moreover, the CAC scheme design has to be appropriate so as the network parameter value examined should be close or equal to the ideal value. Therefore, the revenue optimization is achieved and at the same time CAC scheme performance is satisfactory.

Such an analysis is presented in (Kwon, Kim, Choi, & Naghshineh, 2000) (Kwon, Choi, & Naghshineh, 1998), where U SCs are assumed. The CAC scheme employed is a multiple threshold one, where the values of the thresholds used are determined by the pricing policy applied and the QoS requirements of the SCs supported. Each u SC call is charged with price r_u, $u = 1, 2, \ldots, U$. Let a feasible user allocation $\mathbf{x} = (x_1, x_2, \ldots, x_U)$ exists, where x_u denotes the number of ongoing u SC calls in the cell. Hence, the total revenue is obtained as follows

$$\sum\nolimits_{\forall \mathbf{x}} (\mathbf{r} \cdot \mathbf{x})\pi(\mathbf{x}), \tag{31}$$

where, $\pi(\mathbf{x})$ is the probability of the network cell to be in state $\mathbf{x}$ and $\mathbf{r} = (r_1, r_2, \ldots, r_U)^T$ is the pricing vector. The main objective of the study is to obtain the threshold values t_u for each SC by solving the non linear programming (NLP) problem of maximizing the total revenue, while guaranteeing the QoS requirements. A method to transform the problem from a NLP one to a linear programming (LP) problem can be applied, consequently threshold values can be easily obtained.

Apart from obtaining the optimal values for thresholds employed in a multiple SC GC CAC scheme, pricing strategies are useful to calculate the "Bandwidth Market Price" (BMP) (Ibrahim, Chinneck, & Periyalwar, 2003), which is defined as the current price to transmit 1Gbit of traffic using 1kbps of network bandwidth. To determine the BMP the total revenue of each user is examined. This includes the basic service, which corresponds to *i*th customer bitrate (BR_i), and more advanced services such as a guaranteed end-to-end delay (ETED) and/or frame error rate (FER). The charge of BR_i, ETED and FER are denoted as T_B, T_E and T_F, respectively. Therefore, the overall charge for a given *i* call is given as

$$T_i = T_B + T_E + T_F. \tag{32}$$

Assuming that T_B is directly proportional to BR_i and BMP the BR charge is obtained as

$$T_B = BR_i \times BMP \tag{33}$$

Charges T_E and T_F are proportional to T_B, therefore they are given by

$$T_E = \alpha \times T_B = \alpha \times BR_i \times BMP \tag{34}$$

and

$$T_F = \beta \times T_B = \beta \times' R_i \times BMP \tag{35}$$

respectively, where parameters α, β are positive numbers usually below unity. From the above mentioned, *i*th user total charge T_i can be obtained as

$$T_i = \left(1+\alpha+\beta\right)' \ \mathrm{R}_i \times \mathrm{BMP} \quad (36)$$

The overall charge T in the cell under consideration is obtained as

$$T = \sum\nolimits_{\forall i \in cell} T_i = \sum\nolimits_{\forall i \in cell} \left(1+\alpha+\beta\right)' \ \mathrm{R}_i \times \mathrm{BMP} \quad (37)$$

The revenue based policy aims at maximizing the service provider revenue by dynamically adjusting the BMP and the resource allocation of ongoing users. Applying this objective to the above formula, the optimal BMP for maximizing total revenue and the resource allocation to ongoing users considering the optimal BMP are obtained.

EVALUATION CRITERIA AND PERFORMANCE METRICS

CBP Estimation

The most common criteria employed to evaluate the performance of all CAC schemes proposed are CBP and CDP. In cases where the assumptions made allow the application of Markov chain analysis, analytical formulas for CBP and CDP are derived (Li & Chao, 2007) (Fang & Zhang, 2002)(Tsiropoulos, Stratogiannis, Kanellopoulos, & Cottis, 2008). Therefore, the evaluation of CAC schemes employed can be based on these criteria. In measurement based CAC schemes CBP and CDP are estimated by measuring the calls blocked or dropped respectively into a predefined time window. The main aim of each CAC scheme proposed in literature is to reduce as much as possible both probabilities through the decision making process employed. Moreover, QoS requirements for ongoing calls should be met providing at the same time handoff prioritization. Both probabilities used for CAC schemes evaluation are mostly dependent on the input traffic load, the number of already accepted calls, the bandwidth requirements of each call and the policy applied for handoff calls (Tragos, Tsiropoulos, Karetsos, & Kyriazakos, 2008).

In the case of single SC wireless networks the evaluation of CAC schemes considering failure probabilities is mainly focused on handoff prioritization (Fang & Zhang, 2002) (Yavuz & Leung, 2006). The divergence between CBP and CDP becomes greater as the policy for handoff prioritization becomes stricter. In GC schemes this is realized by lowering threshold value T_i while in FGC probability $\alpha\left(n_i\right)$ becomes lower. To measure the prioritization achieved among new and handoff calls an appropriate priority index (PRIN) is defined as the fraction of new CBP to handoff CDP, as follows

$$\mathrm{PRIN} = \frac{\text{new CBP}}{\text{handoff CDP}} > 1 \quad (38)$$

Evidently, in order to achieve handoff prioritization this index should be above unity as new CBP should be greater than handoff CDP.

A similar analysis is applied in multiple SC wireless networks. Apart from handoff prioritization, different SCs should correspond to respective different priority levels. Thus, considering that u SC should have absolute priority over u+1 SC, where $u, u+1 \in U$, then new CBP_u and handoff CDP_u of u SC should be lower than new CBP_{u+1} and handoff CDP_{u+1} of u+1 SC respectively. Therefore, the divergence between failure probabilities among different SCs is more critical in multiple SC wireless networks. The above mentioned index can be modified to estimate the prioritization level among different SCs. In particular, $\mathrm{PRIN}\left(u, u'\right)$ measures the prioritization achieved among u and u' SC, $u, u' \in U$, which is defined as

$$\mathrm{PRIN}\left(u, u'\right) = \frac{\text{new CBP}_u + \text{handoff CDP}_u}{\text{new CBP}_{u'} + \text{handoff CDP}_{u'}} \quad (39)$$

where, $\text{PRIN}(u,u') > 1$ if $u < u'$ or $\text{PRIN}(u,u') < 1$ if $u > u'$.

Fairness and Resource Allocation

The handoff or SC prioritization is desirable up to a certain extent. When the network is heavily loaded, high priority calls monopolize the available resources resulting in high failure probabilities considering low priority calls. Thus, a low priority call may always be blocked while high priority calls are unhinderly admitted by the wireless network. Therefore, unfairness in resource allocation and admittance process occurs, which constitute a side effect of handoff or SC prioritization mechanism. Fairness is correlated with resource allocation. Hence, unfairness grows as the network resources are more disproportionally allocated. There are some CAC schemes which can adapt their admission procedure as to provide the desirable fairness among different priority level calls. On the other hand, there are different fairness indexes proposed in literature which aim at evaluating the existed CAC schemes in terms of fairness achieved.

Fairness can be straightforwardly embedded in admission decision process of measurement, GC and FGC CAC schemes. In measurement based CAC schemes unfairness is measured into a time window and admission decision process is adapted according to the QoS requirements and the already unfairness exhibited (Hwang & Noh, 2005) (Racunica, Menouni Hayar, & Bonnet, 2004). This is achieved by measuring the failure probabilities for each priority level calls in each time window. The unfairness is determined by an appropriate index; in many cases the divergence among the failure probabilities is used. To achieve an increased fairness performance during the next time window, CAC scheme determines the admission probabilities for each priority level calls. In GC and FGC CAC schemes fairness could be achieved by adjusting dynamically the thresholds for low priority calls employed or the admission probabilities used, respectively (Kim & Kim, 2005).

To apply a fair CAC policy either a measurement based or a GC CAC scheme is employed in most cases. The main task executed to apply the desired policy is to estimate fairness. A simple way to measure fairness in CAC procedure is to calculate the difference between the total resources allocated (Ahn & Ramakrishna, 2004) to different priority level calls or the corresponding admittance failure probabilities (Hwang & Noh, 2005) (Nasser & Hassanein, 2004). In particular, consider a wireless network supporting U SCs and let $\text{P}_{b,u}$ be the overall (new and handoff) blocking probability of u SC calls. Assuming u+1 SC calls have absolute priority over u SC calls, fairness is obtained by the difference between the blocking probabilities of consecutive SCs which should not exceed a predefined deviation value ε (Nasser & Hassanein, 2004), defined as

$$\left|\text{P}_{b,u} - \text{P}_{b,u+1}\right| \le \varepsilon \tag{40}$$

where, $u = 1, 2, \ldots, U-1$ and $0 \le \varepsilon \le 1$. Parameter value ε reflects the maximum tolerable deviation between the admittance failure probabilities of two consecutive SCs. A similar relation can be defined for single SC wireless networks based on new CBP and handoff CDP.

It should be noted that, the above mentioned relation is a QoS specification that should be met by a CAC scheme and not a way to measure unfairness emerged in the admission decision procedure. Thus, appropriate fairness indexes (*FIs*) have to be used, each of which should have the following desired properties (Jain, Chiu, & Hawe, 1984):

- The *FI* should be independent of the number of users sharing the resources or the

number of call priority levels supported by the network. Therefore, *FI* should be applicable to two or more either users or call priority levels.
- *FI* should be independent of scale. Thus, the unit of measurement should not matter enabling the comparison between different wireless systems or CAC schemes.
- The values of *FI* should be bounded between 0 and 1. In this case, value 0 corresponds to totally unfair and 1 to totally fair systems. This property facilitates the expression of fairness as a percentage value, which helps the intuitive understanding of *FI*.
- *FI* values should be continuous and any change made to allocated resources, even a slight one, should be reflected to *FI*.

In literature there are many *FIs* proposed, but Jain's Fairness Index (*JFI*) is the most widespread one (Jain, Chiu, & Hawe, 1984). Given a system with N contending users and a resource allocation $\mathbf{x} = \left\{x_1, x_2, \ldots, x_N;\ x_n \geq 0;\ n = 1, 2, \ldots, N\right\}$, $JFI(\mathbf{x})$ is determined as below

$$JFI\left(\mathbf{x}\right) = \frac{\left[\sum_{n=1}^{N} x_n\right]^2}{\sum_{n=1}^{N} x_n^2}. \tag{41}$$

It is proven that *JFI* satisfies the above mentioned properties for *FIs*, therefore in the case of total fairness in resource allocation *JFI* is equal to unity, while in total unfairness is equal to 0. Alternatively to *JFI*, any function of the form

$$FI\left(\mathbf{x}\right) = \frac{\left[\frac{1}{N}\sum_{n=1}^{N} x_n\right]^r}{\frac{1}{N}\sum_{n=1}^{N} x_n^r}, \qquad r \in \mathbb{N}^* - \left\{1\right\} \tag{42}$$

satisfies the *FI* requirements set, thus it can also be used as a *FI*.

Revenue Approach

An important efficiency metric for evaluating the performance of the CAC scheme applied in wireless networks is the total revenue obtained. The maximization of the total revenue along with the satisfaction of the QoS requirements constitutes one of the main targets of the CAC schemes applied. To measure the total revenue, the average income by each SC supported should be determined. This is further classified into the income obtained from new and handoff calls for a certain SC. Thus, the total average income of u SC denoted as R_u is given by

$$R_u = R_{n,u} + R_{h,u} \tag{43}$$

where, $R_{n,u}$ and $R_{h,u}$ correspond to the average revenue of new and handoff u SC calls (Yilmaz & Chen, 2006). The total average revenue R of the wireless network cell under consideration is determined as

$$R = \sum_{u=1}^{U} R_u = \sum_{u=1}^{U}\left(R_{n,u} + R_{h,u}\right), \tag{44}$$

where, U is the total number of SCs supported by the wireless network.

The average revenue of new and handoff u SC calls are dependent on the corresponding incoming rate $\lambda_{n,u}$, the average call duration τ_u of u SC and the charging price $p\left(t\right)$ determined by the demand function defined above (Hou, Yang, & Papavassiliou, 2002). Thus, $R_{n,u}$ and $R_{h,u}$ can be calculated as

$$R_{n,u} = \int_0^{T_o} \lambda_{n,u} p\left(t\right) \tau_u dt, \tag{45}$$

$$R_{h,u} = \int_0^{T_o} \lambda_{h,u} p\left(t\right) \tau_u dt, \tag{46}$$

respectively, where $T_o = 24h$. If revenue optimization is preferable, then the charging price for each SC calls is obtained, while specific network parameters and CAC design specifications are determined.

FUTURE RESEARCH DIRECTIONS

Due to the time varying characteristics of mobile wireless systems, the limited available resources and the users mobility, RRM has riveted scientists attention. New techniques based on cross layer design will enhance the existent CAC schemes providing a robust mechanism to exploit the potentials of emerging wireless networks. In the new wireless era, where different networks providers will exist supporting various radio access technologies, novel CAC schemes should be developed to handle the heterogeneous traffic. Moreover, the new multimedia services introduced the recent years and the increasing demand for high rate applications constitute the development of new CAC schemes more imperative than ever.

CAC schemes can be evolved into a powerful tool to ensure the QoS provision of mobile wireless networks, increase resources exploitation and prevent networks overload. Moreover, they should be cognizant of the various QoS parameters that should be satisfied, such as low CBP and CDP, fairness among users, session continuity during handoffs, etc. Each SC supported by the mobile wireless system should be treated in accordance with its QoS requirements and network capabilities. Novel CAC schemes can be based on economic models incorporating price bargaining and resource allocation according to game theoretic analysis. Although CAC is an old problem which has been thoroughly studied in literature worldwide, it still constitutes an open research field as the new trends emerged in wireless networks alter CAC objectives creating new perspectives and demands.

CONCLUSION

In this chapter the importance of CAC in wireless networks for providing QoS guarantees has been investigated. CAC algorithms are important for wireless networks not only for providing the expected QoS requirements to mobile users, but also to maintain network consistency and prevent congestion. To address the problem of CAC the main term of QoS has been firstly examined. Different QoS levels supported by the network correspond to the respective SCs offered to mobile users. Each SC has its own requirements and specifications which should be met to offer a satisfactory QoS to end users. Thus, various challenges arise in efficient CAC schemes design that have been determined and thoroughly investigated in the present chapter. To measure the appropriateness of a certain CAC scheme several criteria should be satisfied. Therefore, an analytical presentation of CAC criteria followed by a comprehensive classification of existing schemes has been performed. The main idea of CAC scheme classification is that different schemes apply individual criterion on admission procedure. Moreover, various system architectures exist which demand different CAC schemes, properly designed to adapt to system characteristics. Furthermore, the concerns of the network administrator should be taken into account, applying the policy needed for revenue optimization and maximum resource exploitation through CAC. Analytical models for the most common CAC schemes have been exhibited. An efficient CAC scheme should achieve low failure probabilities, high network resources exploitation, fairness in resource allocation among different users and revenue optimization. To evaluate the performance of CAC schemes studied according to these aspects, various efficiency criteria have been presented. The key idea of this chapter, apart from offering a comprehensive study of CAC process in wireless networks, is to lay emphasis on the CAC method itself as a powerful tool to provide the desirable QoS level to mobile users along with the maximization of network resource exploitation.

REFERENCES

Ahmed, M. H. (2005). Call Admission Control in Wireless Networks: A Comprehensive Survey. *IEEE Communications Surveys & Tutorials*, *7*(1), 50–66. doi:10.1109/COMST.2005.1423334

Ahn, C. W., & Ramakrishna, R. S. (2004). QoS provisioning dynamic connection-admission control for multimedia wireless networks using a Hopfield neural network. *IEEE Transactions on Vehicular Technology*, *53*(1), 106–117. doi:10.1109/TVT.2003.822000

Bartolini, N., & Chlamtac, I. (2001). Improving call admission control procedures by using hand-off rate information. *Wireless Communications and Mobile Computing*, *1*(3), 257–268. doi:10.1002/wcm.17

Chen, I. R., Yilmaz, O., & Yen, I. L. (2006). Admission Control Algorithms for Revenue Optimization With QoS Guarantees in Mobile Wireless Networks. *Wireless Personal Communications: An International Journal*, *38*(3), 357–376. doi:10.1007/s11277-006-9037-6

Chlebus, E., & Ludwin, W. (1995). Is Handoff Traffic Really Poissonian? *Fourth IEEE International Conference on Universal Personal Communications* (pp. 348-353). Tokyo: IEEE.

Cocchi, R., Shenker, S., Estrin, D., & Zhang, L. (1993). Pricing in computer networks: motivation, formulation, and example. *IEEE/ACM Transactions on Networking, 1* (6), 614 - 627.

Cooper, R. B. (1981). Introduction to Queueing Theory. New York: Elsevier North Holland Inc.

Evci, C., & Fino, B. (2001). Spectrum management, pricing, and efficiency control in broadbandwireless communications. *Proceedings of the IEEE*, *89*(1), 105–115. doi:10.1109/5.904509

Fang, Y., Chlamtac, I., & Lin, Y. B. (1998). Channel Occupancy Times and Handoff Rate for Mobile Computing and PCS Networks. *IEEE Transactions on Computers*, *47*(6), 679–692. doi:10.1109/12.689647

Fang, Y., & Zhang, Y. (2002). Call admission control schemes and performance analysis in wirelessmobile networks. *IEEE Transactions on Vehicular Technology*, *51*(2), 371–382. doi:10.1109/25.994812

Floyd, S. (1996). Comments on measurement-based admission control for controlled-load services. Berkeley: Lawrence Berkeley Laboratory.

Harine, G., Marie, R., Puigjaner, R., & Trivedi, K. (2001). Loss formulas and their application to optimization for cellular networks. *IEEE Transactions on Vehicular Technology*, *50*(3), 664–673. doi:10.1109/25.933303

Haung, Y. R., & Ho, J. M. (2002). Distributed call admission control for a heterogeneous PCS network. *IEEE Transactions on Computers*, *51*(12), 1400–1409. doi:10.1109/TC.2002.1146706

Hou, J., Yang, J., & Papavassiliou, S. (2002). Intergration of Pricing with Call Admission Control to Meet QoS Requirements in Cellular Networks. *IEEE Transactions on Parallel and Distributed Systems*, *13*(9), 898–910. doi:10.1109/TPDS.2002.1036064

Hwang, Y. H., & Noh, S. K. (2005). A call admission control scheme for heterogeneous service considering fairness in wireless networks. *Fourth Annual ACIS International Conference on Computer and Information Science 2005 (ICIS 2005)* (pp. 688- 692). Jeju Island, South Korea: IEEE Computer Society.

Ibrahim, W., Chinneck, J. W., & Periyalwar, S. (2003). A QoS-based charging and resource allocation framework for next generation wireless networks. *Wireless Communications and Mobile Computing*, *3*(7), 895–906. doi:10.1002/wcm.179

ITU-D. S. G. (2008, May). *Teletraffic Engineering Handbok.* Retrieved April 18, 2009, from http://oldwww.com.dtu.dk/teletraffic/handbook.html

Jain, R., & Knightly, E. W. (1999). A framework for design and evaluation of admission controlalgorithms in multi-service mobile networks. INFOCOM '99 (pp. 1027-1035). New York: IEEE.

Jain, R. K., Chiu, D. M., & Hawe, W. R. (1984). A quantitative measure of fairness and discrimination for resource allocation in shared computer systems. Maynard, Massachusetts: Digital Equipment Corporation.

Jamin, S., Danzig, P. B., Shenker, S. J., & Zhang, L. (1997). A Measurement-based Admission Control Algorithm for Integrated Services Packet Networks. *IEEE/ACM Transactions on Networking, 5* (1), 56-70.

Jedrzycki, C., & Leung, V. C. (1996). Probability distribution of channel holding time in cellulartelephony systems. *IEEE 46th Vehicular Technology Conference 1996 (VTC '96). 1*, pp. 247 - 251. Atlanta, Georgia: IEEE.

Kim, I. M., Shin, B. C., & Lee, D. J. (2000). SIR-based call admission control by intercell interference prediction for DS-CDMA systems. *IEEE Communications Letters*, *4*(1), 29–31. doi:10.1109/4234.823540

Kim, K. I., & Kim, S. H. (2005). A Light Call Admission Control with Inter-Cell and Inter-Service Fairness in Heterogeneous Packet Radio Networks. *IEICE Transactions*, *88-B*(10), 4064–4073. doi:10.1093/ietcom/e88-b.10.4064

Kwon, T., Choi, Y., Bisdikian, C., & Naghshineh, M. (1999). Measurement-based Call Admission Control for adaptive multimedia in wireless/mobile networks. *IEEE Wireless Communications and Networking Conference 1999, (WCNC'99). 2*, pp. 540-544. New Orleans: IEEE.

Kwon, T., Choi, Y., & Naghshineh, M. (1998). Optimal Distributed Call Admission Control for Multimedia Services in Mobile Cellular Network. *International Workshop on Mobile Multimedia Communication 1998 (MoMuC '98)* (pp. 477-482). Berlin: IEEE.

Kwon, T., Kim, S., Choi, Y., & Naghshineh, M. (2000). Threshold-type call admission control in wireless/mobile multimedia networks using prioritised adaptive framework. *IEEE Electronics Letters*, *36*(9), 852–854. doi:10.1049/el:20000639

Lai, F. S., Misic, J., & Chanson, S. T. (1998). Complete sharing versus partitioning: Quality of service managment for wireless multimedia networks. *7th International Conference on Computer Communications and Networks 1998* (pp. 584 - 593). Lafayette, Louisiana, USA: IEEE Computer Society.

Li, W., & Chao, X. (2007). Call Admission Control for an Adaptive Heterogeneous Multimedia Mobile Network. *IEEE Transactions on Wireless Communications*, *6*(2), 515–525. doi:10.1109/TWC.2006.05192

Lindemann, C., Lohmann, M., & Thümmler, A. (2004). Adaptive call admission control for QoS/revenue optimization in CDMA cellular networks. [WINET]. *ACM Journal on Wireless Networks*, *10*(4), 457–472. doi:10.1023/B:WINE.0000028549.11958.89

Liu, Z., & Zarki, M. E. (1994). SIR-based call admission control for DS-CDMA cellular systems. *IEEE Journal on Selected Areas in Communications*, *12*(4), 638–644. doi:10.1109/49.286670

Mase, K. (2004). Toward scalable admission control for VoIP networks. *IEEE Communications Magazine*, *42*(7), 42–47. doi:10.1109/MCOM.2004.1316529

Nasser, N., & Hassanein, H. (2004). Seamless QoS-Aware Fair Handoff in Multimedia Wireless Networks with Optimized Revenue. *IEEE Canadian Conference on Electrical and Computer Engineering (CCECE)* (pp. 1195-1198). Niagara Falls, Canada: IEEE.

Niyato, D., & Hossain, E. (2005). Call admission control for QoS provisioning in 4G wireless networks: issues and approaches. *IEEE Network*, *19*(5), 5–11. doi:10.1109/MNET.2005.1509946

O'Callaghan, M., Gawley, N., Barry, M., & McGrath, S. (2004). Admission Control for Heterogeneous Networks. 13th IST Mobile & Wireless Communications. Lyon.

Racunica, I., Menouni Hayar, A., & Bonnet, C. (2004). An opportunistic scheduling with fairness for NRT traffic in presence of RT traffic for UMTS/TDD. *7th International Symposium on Wireless Personal Multimedia Communications 2004 (WPMC'04)*. Abano Terme, Italy: EURECOM+1410.

Rajaratmam, M., & Takawira, F. (1999). Performance analysis of highway cellular networks using generalised arrival and generalised service time distributions. *16th International Teletraffic Congress (ITC-16)*. *3*, pp. 11-22. Edinburgh: Elsevier.

Ramjee, R., Towsley, D., & Nagarajan, R. (1997). On optimal call admission control in cellular networks. *Wireless Networks*, *3*(1), 29–41. doi:10.1023/A:1019172226345

Sen, S., Jawanda, J., Basu, K., & Das, S. (1998). Quality-of-Service degradation strategies in multimedia wireless networks. *48th IEEE Vehicular Technology Conference, 1998 (VTC 98)*, *3*, pp. 1884-1888. Ottawa.

Sherif, M. R., Habib, I. W., Nagshineh, M. N., & Kermani, P. K. (2000). Adaptive allocation of resources and call admission control for wireless ATM using generic algorithm. *IEEE Journal on Selected Areas in Communications*, *18*(2), 268–282. doi:10.1109/49.824812

Stasiak, M., Wisniewski, A., & Zwierzykowski, P. (2005). Uplink Blocking Probability for a Cell with WCDMA Radio Interface and Differently Loaded Neighbouring Cells. *Advanced Industrial Conference on Telecommunications/Service Assurance with Partial and Intermittent Resources Conference/E-Learning on Telecommunications Workshop (AICT/SAPIR/ELETE'05)* (pp. 402-407). Lisbon: IEEE Computer Society.

Stratogiannis, D. G., Tsiropoulos, G. I., & Cottis, P. G. (2008). Call Admission Control in Wireless Networks: Probabilistic Approach and Efficiency Evaluation. *International Conference on Wireless Communications and Mobile Computing Conference 2008 (IWCMC '08)* (pp. 712 - 717). Crete Island: IEEE.

Tragos, E., Tsiropoulos, G., Karetsos, G., & Kyriazakos, S. (2008). Admission Control for QoS support in Heterogeneous 4G Wireless Networks. *IEEE Network Magazine*, *22*(3), 30–37. doi:10.1109/MNET.2008.4519962

Tsiropoulos, G. I., Stratogiannis, D. G., Kanellopoulos, J. D., & Cottis, P. G. (2008). Efficiency evaluation of class-based call admission control schemes for wireless communications. *IEEE International Symposium on Wireless Communication Systems (ISWCS '08)* (pp. 69 - 73). Reykjavik: IEEE.

Vogel, A., Kerherve, B., von Bochmann, G., & Gecsei, J. (1995). Distributed Multimedia and QOS: A Survey. *IEEE MultiMedia*, *2*(2), 10–19. doi:10.1109/93.388195

Wang, X., Fan, F., & Pan, Y. (2008). A More Realistic Thinning Scheme for Call Admission Control in Multimedia Wireless Networks. *IEEE Transactions on Computers*, *57*(8), 1143–1147. doi:10.1109/TC.2008.29

Yang, X., Feng, G., & Kheong, D. S. (2006). Call admission control for multiservice wireless networks with bandwidth asymmetry between uplink and downlink. *IEEE Transactions on Vehicular Technology*, *55*(1), 360–368. doi:10.1109/TVT.2005.861202

Yavuz, E. A., & Leung, V. C. (2006). Computationally efficient method to evaluate the performance of guard-channel-based call admission control in cellular networks. *IEEE Transactions on Vehicular Technology*, *55*(4), 1412–1424. doi:10.1109/TVT.2006.877470

Yilmaz, O., & Chen, I. (2006). Utilizing Call Admission Control to Derive Optimal Pricing of Multiple Service Classes in Wireless Cellular Networks. *12th International Conference on Parallel and Distributed Systems (ICPADS'06)*. *1*, pp. 605-612. Minneapolis: IEEE Computer Society.

Yu, O. T., & Leung, V. C. (1997). Adaptive resource allocation for prioritized call admission over an ATM-based wireless PCN. *IEEE Journal on Selected Areas in Communications*, *15*(7), 1208–1225. doi:10.1109/49.622906

ADDITIONAL READING

Akyildiz, I., Xie, J., & Mohanty, S. (2004). A survey of mobility management in nextgeneration all-IP-based wireless systems. *IEEE Wireless Communications*, *11*(4), 16–28. doi:10.1109/MWC.2004.1325888

Ayyagari, D., & Ephremides, A. (1998). Admission Control with Priorities: Approaches for Multi-rate Wireless Systems. *IEEE International Conference on Universal Personal Communications 1998 (ICUPC'98)*. *1*, pp. 301 - 305. Florence: IEEE.

Bari, F., & Leung, V. C. (2007). Automated Network Selection in a Heterogeneous Wireless Network Environment. *IEEE Network*, *21*(1), 34–40. doi:10.1109/MNET.2007.314536

Bernet, Y., Ford, P., Yavatkar, R., Baker, F., Zhang, F., Speer, M., et al. (2000). A Framework for Integrated Services Operation over Diffserv Networks. RFC 2998, IETF.

Bharghavan, V., Lee, K., Lu, S., Ha, S., Li, J., & Dwyer, D. (1998). The TIMELY adaptive resource management architecture. *IEEE Personal Communications*, *5*(4), 20–31. doi:10.1109/98.709366

Blake, S., Black, D., Carlson, M., Davies, E., Wang, Z., & Weiss, W. (1998). An Architecture for Differentiated Services. RFC 2475, IETF.

Braden, R., Clark, D., & Shenker, S. (1994). Integrated Services in the Internet Architecture: an Overview. RFC 1633, IETF.

Buddhikot, M., Chandranmenon, G., Han, S., Lee, Y. W., Miller, S., & Salgareli, L. (2003). Integration of 802.11 and third-generation wireless data networks. IEEE INFOCOM 2003 (pp. 503-512). San Francisco: IEEE.

Casoni, M., Immovilli, G., & Merani, M. L. (2002). Admission control in T/CDMA systems supporting voice and dataapplications. *IEEE Transactions on Wireless Communications*, *1*(3), 540–548. doi:10.1109/TWC.2002.800552

Floroiu, J. W., Ruppelt, R., Sisalem, D., & Stephanopoli, J. V. (2003). Seamless handover in terrestrial radio access networks: a case study. *IEEE Communications Magazine*, *41*(11), 110–116. doi:10.1109/MCOM.2003.1244930

3GPP23.107. (2003). *Quality of Service (QoS) Concept and Architecture (Release 5)*. TS 23.107 V5.11.0.

Gupta, V. (2004). *Media independent handover service*. Draft technical requirements 21-04-0087-08-0000, IEEE 802.21.

IEEE Guide for Determining Fault Location on AC Transmission and Distribution Lines. (2005). 1-36.

Imre, S. (2007). Dynamic Call Admission Control for Uplink in 3G/4G CDMA-Based Systems. *IEEE Transactions on Vehicular Technology*, *56*(5), 2617–2629. doi:10.1109/TVT.2007.899932

Jiang, P., & Sekercioglu, A. (2008). A dynamic call admission control scheme for optimization with QoS provisioning in multiclass cellular networks. *Q2SWinet '08: Proceedings of the 4th ACM symposium on QoS and security for wireless and mobile networks* (pp. 19-26). Vancouver, British Columbia, Canada: ACM.

Koskie, S., & Gajic, Z. (2005). A Nash Game Algorithm for SIR-Based Power Control in 3G Wireless CDMA Networks. *Trasactions on Networking*, *13*(5), 1017–1026. doi:10.1109/TNET.2005.857068

Manzoni, P., Ghosal, D., & Serazzi, G. (1995). Impact on mobility on TCP/IP: An integrated performance study. *IEEE Journal on Selected Areas in Communications*, *13*(5), 858–867. doi:10.1109/49.391748

McNair, J., & Zhu, F. (2004). Vertical Handoffs in Fourth-Generation Multinetwork Environments. *IEEE Wireless Communications*, *11*(3), 8–15. doi:10.1109/MWC.2004.1308935

Naghshineh, M., & Schwartz, M. (1996). Distributed Call Admission Control in Mobile/Wireless Networks. *IEEE Journal on Selected Areas in Communications*, *14*(4), 711–717. doi:10.1109/49.490422

Niyato, D., & Hossain, E. (2008). Competitive Pricing in Heterogeneous Wireless Access Networks: Issues and Approaches. *IEEE Network*, *22*(6), 4–11. doi:10.1109/MNET.2008.4694168

Wu, D. (2005). QoS provisioning in wireless networks. *Wireless Communications and Mobile Computing*, *5*(8), 957–969. doi:10.1002/wcm.359

Zhang, Q., & Zhang, Y. Q. (2008). Cross-Layer Design for QoS Support in Multihop Wireless Networks. *Proceedings of the IEEE*, *96*(1), 64–76. doi:10.1109/JPROC.2007.909930

Chapter 18
Efficient Power Allocation in E-MBMS Enabled 4G Networks

Antonios Alexiou
University of Patras, Greece; & Research Academic Computer Technology Institute, Greece

Christos Bouras
University of Patras, Greece; & Research Academic Computer Technology Institute, Greece

Vasileios Kokkinos
University of Patras, Greece; & Research Academic Computer Technology Institute, Greece

ABSTRACT

The plethora of mobile multimedia services that are expected to face high penetration, poses the need for the deployment of a resource economic scheme in Long Term Evolution (LTE) networks. To this direction, the Evolved - Multimedia Broadcast / Multicast Service (E-MBMS) is envisaged to play an instrumental role for LTE proliferation and set the basis for a successful 4th Generation (4G) standardization process. One of the most critical aspects of E-MBMS performance is the selection of the most efficient radio bearer, in terms of power consumption. This chapter presents the prevailing radio bearer selection mechanisms and examines their performance in terms of power consumption. Furthermore, it discusses problems regarding the high power requirements for the realization of E-MBMS and evaluates the proposed techniques/solutions. Finally, this chapter presents a novel mechanism for efficient power control during E-MBMS transmissions that conforms to LTE requirements for simultaneous provision of multiple multimedia sessions.

INTRODUCTION

Nowadays, mobile industry rapidly evolves towards a multimedia-oriented model for providing rich services, such as mobile TV and mobile streaming. Long Term Evolution (LTE) networks address this emerging trend, by shaping the future mobile landscape in a more power and spectral efficient way than its predecessors.

LTE technology improves spectral efficiency and sector capacity and lowers the telecommunication costs for service provision, making use of new and reformed spectrum opportunities and better integration with other open standards. These enhancements compared to Universal Mobile Telecommunication System (UMTS) technology, give LTE networks the

DOI: 10.4018/978-1-61520-771-8.ch018

opportunity to offer high throughput, low latency, plug and play, improved end-user experience and simple architecture resulting in low operating expenditures.

More specific, LTE networks provide high peak rates of at least 100 Mbps in the downlink and 50 Mbps in the uplink. Contrary to UMTS networks that provide peak rates of 384 Kbps (or 21 Mbps with HSDPA) in the downlink, LTE networks may overcome the recent increase of mobile data usage and emergence of new applications such as mobile TV and streaming contents. However, the plethora of mobile multimedia services that are expected to face high penetration, poses the need for deploying complementary resource economic schemes.

To this direction, the Evolved - Multimedia Broadcast/Multicast Service (E-MBMS) is envisaged to play an instrumental role for the LTE proliferation in mobile market and set the basis for a successful 4th Generation (4G) standardization process. E-MBMS constitutes the evolutionary successor of MBMS, which was introduced in the Release 6 of UMTS (3rd Generation Partnership Project TR 23.846, 2003; 3rd Generation Partnership Project TS 22.146, 2008). It is a unidirectional service which targets at the resource economic delivery of multimedia data from a single source entity to multiple recipients. The main requirement during the provision of E-MBMS services is to make an efficient overall usage of radio and network resources and more importantly, to reduce the power requirements for the provision of such demanding services.

Power in mobile networks is the most limited resource and may lead to significant capacity decrease when misused. Providing multicast or broadcast services to a meaningful proportion of a cell coverage area may require significant amounts of power dedicated to the multicast or broadcast transmission. Several techniques, such as Dynamic Power Setting (DPS), Macro Diversity Combining (MDC) and Rate Splitting (RS) have been introduced in order to minimize the base station's total E-MBMS transmission power. This chapter examines the operation and performance of these techniques and demonstrates the amount of power that could be saved through their employment.

Furthermore, a critical aspect of E-MBMS performance is the selection of the most efficient radio bearer, in terms of power consumption, for the transmission of multimedia traffic. The system should conceive and adapt to continuous changes that occur in such dynamic wireless environments and optimally allocate resources. The selection of the most efficient radio bearer is an open issue in today's E-MBMS infrastructure and several mechanisms have been proposed to this direction. Nevertheless, the selection of the most appropriate mechanism is plagued with uncertainty, since each mechanism may provide specific advantages. In this chapter, the prevailing radio bearer selection mechanisms are presented and compared in terms of power consumption so as to highlight the advantages that each mechanism may provide.

Finally, this chapter presents a novel mechanism for efficient power control during E-MBMS transmissions that incorporates the advantages of each mechanism. The most remarkable advantage of the proposed mechanism, that actually differentiates it from the other approaches, is that it conforms to LTE requirements for the simultaneous provision of multiple multimedia sessions. This approach is compared with the aforementioned approaches in terms of both power consumption and complexity so as to highlight its enhancements and underline the necessity for its incorporation in E-MBMS specifications.

Main objective of this chapter is to present the main characteristics regarding the operation and performance of E-MBMS and moreover to highlight the significance of power control during E-MBMS transmissions. The reader will become familiar with the most crucial problems that have a direct impact on E-MBMS performance; and moreover, the reader will be introduced to the proposed techniques/solutions.

BACKGROUND

In MBMS rich wireless multimedia data is transmitted simultaneously to multiple recipients, by allowing resources to be shared in an economical way. MBMS efficiency is derived from the single transmission of identical data over a common channel without clogging up the air interface with multiple replications of the same data.

The major factor for integrating MBMS into UMTS networks was the rapid growth of mobile communications technology and the massive spread of wireless data and wireless applications. The increasing demand for communication between one sender and many receivers led to the need for point-to-multipoint (PTM) transmission. PTM transmission is opposed to the point-to-point (PTP) transmission, using the unicast technology, which is exclusively used in conventional UMTS networks (without the MBMS extension). Broadcast and multicast technologies constitute an efficient way to implement this type of communication and enable the delivery of a plethora of high-bandwidth multimedia services to a large users' popularity.

From the service and operators' point of view, the employment of MBMS framework involves both an improved network performance and a rational usage of radio resources, which in turns leads to extended coverage and service provision. In parallel, users are able to realize novel, high bit-rate services, experienced until today only by wired users. Such services include Mobile TV, weather or sports news as well as fast and reliable data downloading (Holma & Toskala, 2007).

MBMS Operation Modes

As the term MBMS indicates, there are two types of service mode: the broadcast mode and the multicast mode. Each mode has different characteristics in terms of complexity and packet delivery.

The broadcast service mode is a unidirectional PTM transmission type. Actually, with broadcast, the network simply floods data packets to all nodes within the network. In this service mode, content is delivered, using PTM transmission, to a specified area without knowing the receivers and whether there is any receiver in the area. As a consequence, the broadcast mode requires no subscription or activation from the users' point of view.

In the multicast operation mode, data is transmitted solely to users that explicitly request such a service. More specifically, the receivers have to signal their interest for the data reception to the network and then the network decides whether the user may receive the multicast data or not. Thus, in the multicast mode there is the possibility for the network to selectively transmit to cells, which contain members of a multicast group. Either PTP or PTM transmission can be configured in each cell for the multicast operation mode (3rd Generation Partnership Project TS 22.146, 2008).

Unlike the broadcast mode, the multicast mode generally requires a subscription to the multicast subscription group and then the user joining the corresponding multicast group. Moreover, due to the selective data transmission to the multicast group, it is expected that charging data for the end user will be generated for this mode, unlike the broadcast mode.

MBMS Architecture

The MBMS framework requires minimal modifications in the current UMTS architecture. As a consequence, this fact enables the fast and smooth upgrade from pure UMTS networks to MBMS-enhanced UMTS networks. Actually, MBMS consists of a MBMS bearer service and a MBMS user service. The latter represents applications, which offer for example multimedia content to the users, while the MBMS bearer service provides methods for user authorization, charging and Quality of Service (QoS) improvement to prevent unauthorized reception (3rd Generation Partnership Project TS 22.146, 2008).

Figure 1. UMTS and MBMS architecture

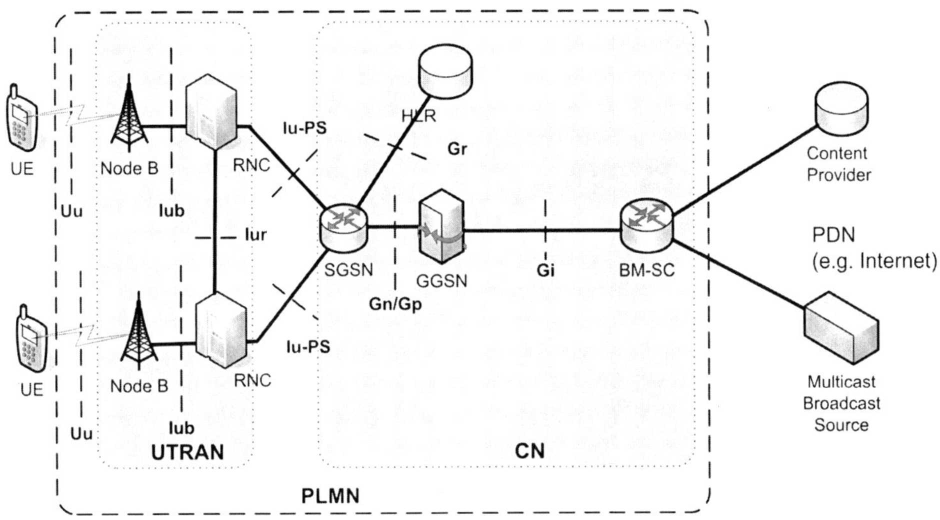

UMTS network is split in two main domains: the User Equipment (UE) domain and the Public Land Mobile Network (PLMN) domain. The UE domain consists of the equipment employed by the user to access the UMTS services. The PLMN domain consists of two land-based infrastructures: the Core Network (CN) and the UMTS Terrestrial Radio Access Network (UTRAN) (Figure 1). The CN is responsible for switching/routing voice and data connections, while the UTRAN handles all radio-related functionalities. The CN is logically divided into two service domains: the Circuit-Switched (CS) service domain and the Packet-Switched (PS) service domain (Holma & Toskala, 2007). The CS domain handles the voice-related traffic, while the PS domain handles the packet transfer. The remainder of this chapter will focus on the UMTS packet-switching mechanism.

The PS portion of the CN in UMTS consists of two kinds of General Packet Radio Service (GPRS) Support Nodes (GSNs), namely Gateway GSN (GGSN) and Serving GSN (SGSN) (Figure 1). SGSN is the centerpiece of the PS domain. It provides routing functionality, interacts with databases (like Home Location Register (HLR)) and manages many Radio Network Controllers (RNCs). SGSN is connected to GGSN via the Gn interface and to RNCs via the Iu interface. GGSN provides the interconnection of UMTS network (through the Broadcast Multicast–Service Center) with other Packet Data Networks (PDNs), like the Internet.

UTRAN consists of two kinds of nodes: the first is the RNC and the second is the Node B. Node B constitutes the base station and provides radio coverage to one or more cells (Figure 1). Node B is connected to the UE via the Uu interface (based on the Wideband Code Division Multiple Access, WCDMA technology) and to the RNC via the Iub interface. One RNC with all the connected to it Node Bs is called Radio Network Subsystem (RNS) (Holma & Toskala, 2007).

The major modification in the existing UMTS platform for the provision of the MBMS framework is the addition of a new entity called Broadcast Multicast–Service Center (BM-SC, see Figure 1). Actually, BM-SC acts as entry point for data delivery between the content providers and the UMTS network and is located in the PS domain of the CN. The BM-SC entity communicates with existing UMTS networks and external PDNs (3rd Generation Partnership Project TR 23.846, 2003; 3rd Generation Partnership Project TS 22.146, 2008).

Figure 2. UMTS multicast without MBMS enhancement

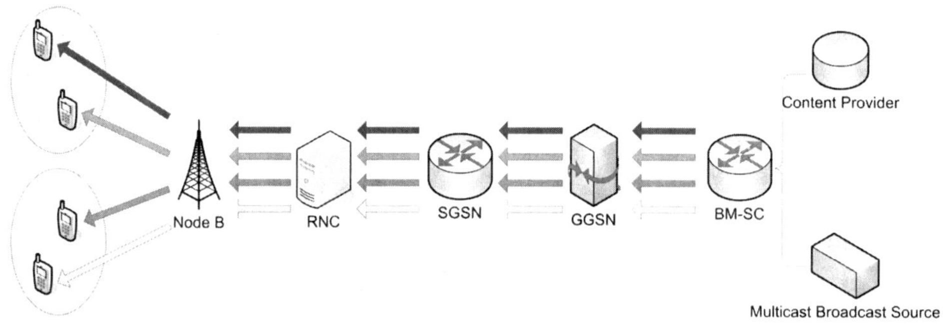

The BM-SC is responsible for both control and user planes of a MBMS service. More specifically, the function of the BM-SC can be separated into five categories: Membership, Session and Transmission, Proxy and Transport, Service Announcement and Security function. The BM-SC Membership function provides authorization to the UEs requesting to activate a MBMS service. According to the Session and Transmission function, the BM-SC can schedule MBMS session transmissions and shall be able to provide the GGSN with transport associated parameters, such as QoS and MBMS service area. As far as the Proxy and Transport function is concerned, the BM-SC is a proxy agent for signaling over Gmb reference point between GGSNs and other BM-SC functions. Moreover, the BM-SC Service Announcement function must be able to provide service announcements for multicast and broadcast MBMS user services and provide the UE with media descriptions specifying the media to be delivered as part of a MBMS user service. Finally, MBMS user services may use the Security functions for integrity or confidentiality protection of the MBMS data, while the specific function is used for distributing MBMS keys (Key Distribution Function) to authorized UEs.

Multicast Mode of MBMS

MBMS multicast efficiency improvement in UMTS networks can be derived from the following figures. More specifically, Figure 2 and Figure 3 present UMTS multicast functionality without and with MBMS enhancement respectively.

Without the MBMS enhancement, multicast data is replicated as many times as the total number of multicast users in all interfaces. Obviously, a bottleneck is placed when the number of users increases significantly. All interfaces are heavily

Figure 3. UMTS multicast with MBMS enhancement

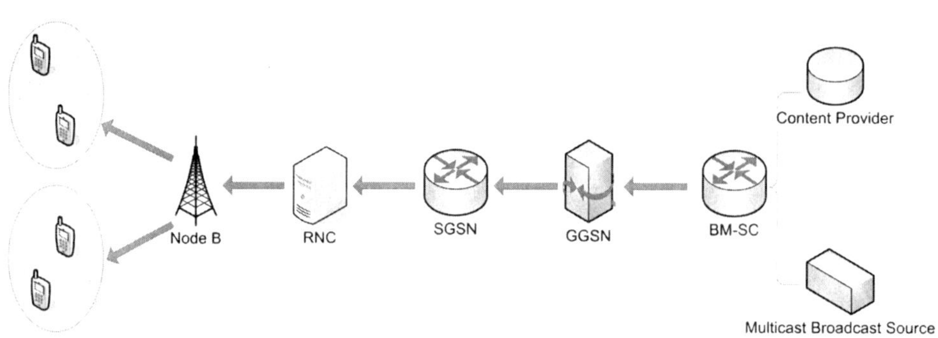

Figure 4. Packet delivery in MBMS multicast mode

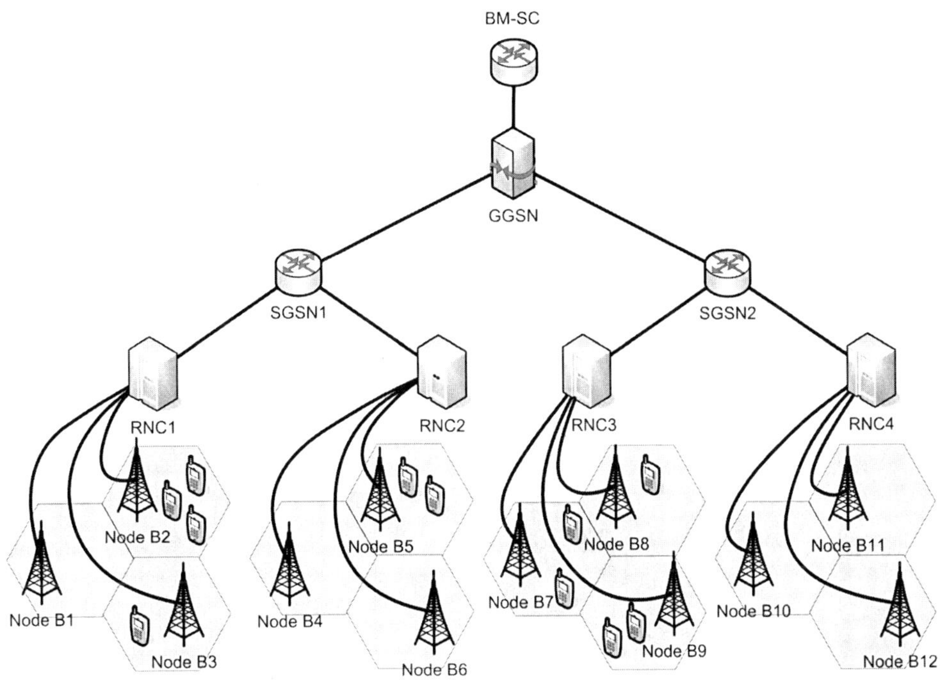

overloaded due to the multiple transmissions of the same data. On the other hand, MBMS multicast benefits UMTS networks through the radio and network resources' sharing. Only a single stream per MBMS service of identical data is essential for the delivery of the multicast content, thus saving expensive resources. Conclusively, MBMS multicast data distribution is optimally configured throughout the UMTS network.

Packet Delivery Process

An overview of the multicast data flow procedure during a MBMS service provision is presented in this paragraph. Figure 4 depicts a subset of a UMTS-MBMS network. In this architecture, there are two SGSNs connected to a GGSN, four RNCs, and twelve Node Bs. Furthermore, eleven members of a multicast group are located in six cells. The BM-SC acts as the interface towards external sources of traffic. The presented analysis assumes that a data stream that comes from an external PDN, through BM-SC, must be delivered to the eleven UEs as illustrated in Figure 4.

The analysis presented in the following paragraphs, covers the forwarding mechanism of the data packets between the BM-SC and the UEs. With multicast, the packets are forwarded only to those Node Bs that have multicast users. Therefore, in Figure 4, the Node Bs 2, 3, 5, 7, 8, 9 receive the multicast packets issued by the BM-SC. We briefly summarize the five steps occurred for the delivery of the multicast packets.

Initially, the BM-SC receives a multicast packet and forwards it to the GGSN that has registered to receive the multicast traffic. Then, the GGSN receives the multicast packet and by querying its multicast routing lists, it determines which SGSNs have multicast users residing in their respective service areas. In Figure 4, the GGSN duplicates the multicast packet and forwards it to the SGSN1 and the SGSN2 (Alexiou, Antonellis, Bouras & Papazois, 2006). Then, both destination SGSNs receive the multicast packets and, having queried their multicast routing lists, determine which RNCs are to receive the multicast packets. The destination RNCs receive the multicast packet and send it to the Node Bs that have established the appropriate radio bearers for the multicast application. In Figure 4, these are: Node B2, B3, B5, B7, B8, and B9. The multicast users receive the multicast packets on the appropriate radio bearers, by dedicated channels transmitted to individual users separately or by common channels transmitted to all members in the cell (Alexiou, Antonellis, Bouras & Papazois, 2006).

MBMS Multicast Mode Radio Bearers

According to current MBMS specifications, the transmission of the MBMS multicast packets over the Iub and Uu interfaces may be performed on common (Forward Access Channel - FACH), on dedicated (Dedicated Channel - DCH) channels or on the shared channel named High Speed - Downlink Shared Channel (HS-DSCH), introduced in Release 5. The main requirement is to make an efficient overall utilization of the radio resources: this makes a common channel the favorite choice, since many users can access the same resource at the same time.

More specifically, the transport channel that the 3rd Generation Partnership Project (3GPP) decided to use as the main transport channel for PTM MBMS data transmission is the FACH with turbo coding and Quadrature Phase Shift Keying (QPSK) modulation at a constant transmission power (3rd Generation Partnership Project TR 23.846, 2003). DCH is a PTP channel and hence, it suffers from the inefficiencies of requiring multiple DCHs to carry the data to a group of users. However, DCH can employ fast closed-loop power control and soft handover mechanisms and generally is a highly reliable channel (Boni, Launay, Mienville & Stuckmann, 2004; Holma & Toskala, 2007). The allocation of HS-DSCH as transport channel affects the obtained data rates and the remaining capacity to serve Release '99 users (users served by DCH). High Speed Downlink Packet Access (HSDPA) cell throughput increases when more HSDPA power is allocated, while DCH throughput simultaneously decreases (Holma & Toskala, 2006).

Radio Bearers' Selection

The importance of the selection of the most efficient transport channel in terms of power consumption is a key point for MBMS, since a wrong transport channel selection for the transmission of the MBMS data could result to a significant decrease in the total capacity of the system. Several studies and simulations have been carried out focusing on the threshold for switching between dedicated and common resources. The 3GPP MBMS Counting Mechanism was the prevailing approach mainly due to its simplicity of implementation and function (3rd Generation Partnership Project TS 25.346, 2009). On the other hand, Vriendt, Gomez Vinagre & Van Ewijk (2004) claim that for a FACH with transmission power set to 4 Watt, the threshold for switching from dedicated to common resources is around 7 UEs per cell.

However, only the information about the number of users in a cell may not be sufficient so as to select the appropriate radio bearer (PTP or PTM) for the specific cell. The inefficiencies of the above works and the power limitations that wireless network encounter motivated novel approaches, indicating that there is no need for a priori information and predefined switching thresholds; while, the assignment of the radio bearer should be performed in order to minimize the Node B's power requirements (3rd Generation Partnership Project TR 25.922, 2007). An interesting study under these assumptions is presented in B-BONE (2003), where the authors propose a switching point between PTP and PTM bearers, based on power consumption; while, Alexiou, Bouras & Rekkas (2007) propose a power control scheme for efficient radio bearer selection in MBMS enabled UMTS networks.

However, none of these works and approaches considers the power saving techniques that have been proposed (Alexiou, Bouras & Kokkinos, 2007). Techniques, such as FACH Dynamic Power Setting and Rate Splitting are proven to reduce the power requirements during MBMS transmissions and should be incorporated in the MBMS operation.

POWER CONTROL IN E-MBMS MULTICAST MODE

Power control is one of the most critical aspects in MBMS due to the fact that downlink transmission power in UMTS networks is a limited resource and must be shared efficiently among all MBMS users in a cell. Power control aims at minimizing the transmitted power, eliminating in this way the intercell interference. However, when misused, the use of power control may lead to a high level of wasted power and worse performance results.

On the PTP downlink transmissions, fast power control is used to maintain the quality of the link and thus to provide a reliable connection for the receiver to obtain the data with acceptable error rates. Transmitting with just enough power to maintain the required quality for the link also ensures that there is minimum interference affecting the neighboring cells. However, when a user consumes a high portion of power, more than actually is required, the remaining power, allocated for the rest of the users, is dramatically decreased, thus leading to a significant capacity loss in the system.

During PTM downlink transmissions, Node B transmits at a power level that is high enough to support the connection to the receiver with the highest power requirement among all receivers in the multicast group. This would still be efficient because the receiver with the highest power requirement would still need the same amount of power in a unicast link, and by satisfying that particular receiver's requirement, the transmission power will be enough for all the other receivers in the multicast group. Consequently, the transmitted power is kept at a relatively high level most of the time, which in turn, increases the signal quality at each receiver in the multicast group. On the other hand, a significant amount of power is wasted and moreover intercell interference is increased.

As a consequence, downlink transmission power plays a key role in MBMS planning and optimization. This section provides an analytical description of the HS-DSCH, DCH and FACH power profiles that are employed during PTP and PTM transmission. The following analysis refers to a macrocell environment with parameters described in Table 1 (3rd Generation Partnership Project TR 101.102, 2002; Holma & Toskala, 2007).

HS-DSCH Power Profile

HS-DSCH is a rate controlled rather than a power controlled transport channel. Although there are two basic modes for allocating HS-DSCH transmission power (Holma & Toskala, 2006), this chapter will focus on a dynamic method in order to provide only the required, marginal amount of power so as to satisfy all the serving multicast users and, in parallel, eliminate interference. Two major measures for HSDPA power planning are: the HS-DSCH Signal-to-Interference-plus-Noise Ratio (SINR) metric and the Geometry factor (*G*). SINR for a single-antenna Rake receiver is calculated as in (1) (Holma & Toskala, 2006):

$$SINR = SF_{16} \frac{P_{HS-DSCH}}{pP_{own} + P_{other} + P_{noise}} \quad (1)$$

where $P_{HS\text{-}DSCH}$ is the HS-DSCH transmission power, P_{own} is the own cell interference experienced by the mobile user, P_{other} the interference from neighboring cells and P_{noise} the Additive White Gaussian Noise. Parameter p is the orthogonality

Table 1. Macrocell simulation assumptions

Parameter	Value
Cellular layout	Hexagonal grid
Number of cells	18
Sectorization	3 sectors/cell
Site to site distance	1 Km
Cell radius	0.577 Km
Maximum BS Tx power	20 Watt (43 dBm)
Other BS Tx power	5 Watt (37 dBm)
Common channel power	1 Watt (30 dBm)
Propagation model	Okumura Hata
Multipath channel	Vehicular A (3km/h)
Orthogonality factor	0.5
E_b/N_0 target	5 dB

factor (p = 0: perfect orthogonality), while SF_{16} is the spreading factor of 16.

Geometry factor is another major measure that indicates the users' position throughout a cell. A lower G is expected when a user is located at the cell edge. G is calculated as in (2) (Holma & Toskala, 2007):

$$G = \frac{P_{own}}{P_{other} + P_{noise}} \tag{2}$$

There is a strong relationship between the HS-DSCH allocated power and the obtained MBMS cell throughput. This relationship can be disclosed in the three following steps. Initially, we have to define the target MBMS cell throughput. Once the target cell throughput is set, the next step is to define the way that this throughput relates to the SINR (Holma & Toskala, 2006). Finally, we can describe how the required HS-DSCH transmission power ($P_{HS\text{-}DSCH}$) can be expressed as a function of the SINR value and the user location (in terms of G) as in (3) (Holma & Toskala, 2006):

$$P_{HS-DSCH} \geq SINR[p - G^{-1}]\frac{P_{own}}{SF_{16}} \tag{3}$$

DCH Power Profile

The total downlink transmission power allocated for all MBMS users in a cell that are served by multiple DCHs is variable. It mainly depends on the number of serving users, their location in the cell, the bit rate of the MBMS session and the experienced signal quality E_b/N_0 for each user. Equation 4 calculates the Node B's total DCH transmission power required for the transmission of the data to n users in a specific cell (Perez-Romero, Sallent, Agusti & Diaz-Guerra, 2005).

$$P_T = \frac{P_P + \sum_{i=1}^{n} \frac{(P_N + x_i)}{\frac{W}{(E_b/N_0)_i R_{b,i}} + p} L_{p,i}}{1 - \sum_{i=1}^{n} \frac{p}{\frac{W}{(E_b/N_0)_i R_{b,i}} + p}} \tag{4}$$

where P_T is the base station's total transmitted power, P_p is the power devoted to common control channels $L_{p,i}$ is the path loss, $R_{b,i}$ the ith user transmission rate, W the bandwidth, P_N the background noise, p is the orthogonality factor (p = 0 for perfect

Figure 5. DCH transmission (Tx) power

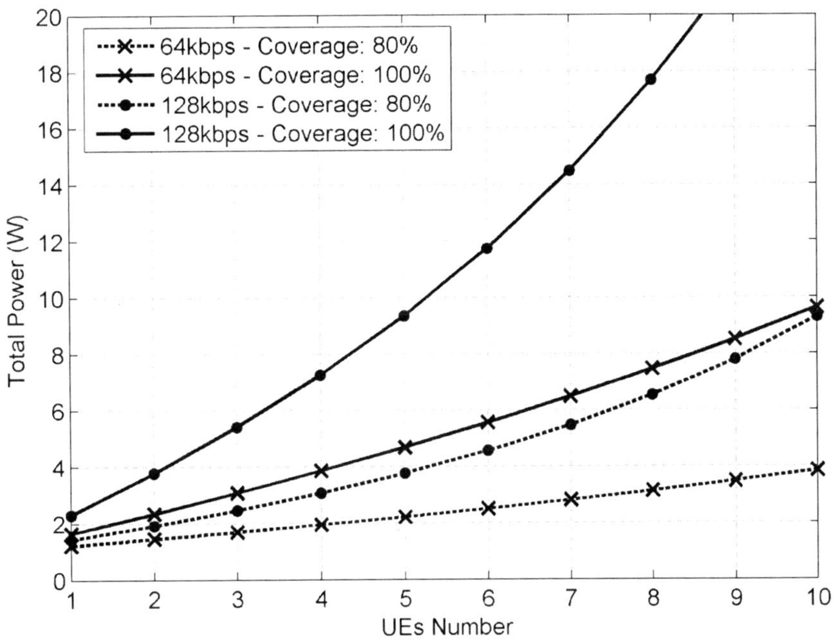

orthogonality) and x_i is the intercell interference observed by the ith user given as a function of the transmitted power by the neighboring cells P_{Tj}, $j=1,\ldots K$ and the path loss from this user to the jth cell L_{ij}. More specifically (Perez-Romero, Sallent, Agusti & Diaz-Guerra, 2005):

$$x_i = \sum_{j=1}^{K} \frac{P_{Tj}}{L_{ij}} \quad (5)$$

DCH may be used for the delivery of PTP MBMS services, while can not be used to serve large multicast populations since high downlink transmission power would be required. Figure 5 depicts the downlink transmission power when MBMS multicast data is delivered over multiple DCHs (one separate DCH per user). Obviously, higher power is required to deliver higher MBMS data rates. In addition, increased cell coverage area and larger user groups lead to higher power consumption.

FACH Power Profile

A FACH essentially transmits at a fixed power level since fast power control is not supported. FACH is a PTM channel and must be received by all users throughout the cell (or the part of the cell that the users reside in), thus, the fixed power should be high enough to ensure the requested QoS in the desired coverage area of the cell, irrespective of users' location. FACH power efficiency strongly depends on maximizing diversity as power resources are limited. Diversity can be obtained by the use of a longer Transmission Time Interval (TTI) in order to provide time diversity against fast fading (fortunately, MBMS services are not delay sensitive) and the use of combining transmissions from multiple cells to obtain macro diversity (3rd Generation Partnership Project TR 25.803, 2005; Parkvall, Englund, Lundevall & Torsner, 2006).

Table 2 presents some indicative FACH downlink transmission power levels obtained for various

Table 2. FACH Tx power levels

Cell Coverage	Service Bit Rate (Kbps)	Required Tx Power (Watt)
50%	32	1.8
	64	2.5
95%	32	4.0
	64	7.6

cell coverage areas and MBMS bit rates, without assuming diversity techniques (3rd Generation Partnership Project TR 25.803, 2005). A basic constraint is that the delivery of high data rate MBMS services over FACH is not feasible, since excessive downlink transmission power would be required (overcoming the maximum available power of 20 Watt). High bit rates can only be offered to users located very close to Node B.

POWER SAVING TECHNIQUES

In this section, the main problem during a MBMS session is highlighted and the proposed techniques to overcome this problems are presented. The analysis that follows will constitute the guide for our assumptions and simulation experiments.

The main problem during a MBMS session, in terms of power consumption, is the exceedingly high fixed power levels when allocating FACH as transport channel. As an example, we mention that in order to provide a 128 Kbps MBMS service with a FACH coverage set to the 95% of the cell, 16 Watt of power are required. If we contemplate that the maximum transmission power of the Node B is 20 Watt (which should be shared among all the users of the cell and among all the possible services), it becomes comprehensible that this level of power makes impossible the provision of services with such bit rates. The techniques which are stated in the remaining of this section partly overcome this problem, since they reduce the FACH transmission power levels.

Dynamic Power Setting (DPS)

DPS is the technique where the transmission power of the FACH can be determined based on the worst user's path loss. This way, the FACH transmission power is allocated dynamically; and the FACH transmission power will need to cover the whole cell only if one (or more) user is at the cell boundary. To perform DPS, the MBMS users need to turn on measurement report mechanism while they are on the Cell_FACH state. Based on such measurement reports, the Node B can adjust the transmission power of the FACH (Chuah, Hu & Luo, 2004).

This is presented in Figure 6, where the Node B sets its transmission power based on the worst user's path loss (*i.e.* distance). The information about the path loss is sent to the Node B via uplink channels. The examination of Figure 6 reveals that 4.0 Watt are required in order to provide a 32 Kbps service to the 95% of the cell. However, supposing that all the MBMS users are found near the Node B (10% coverage) only 0.9 Watt are required. In that case, 3.1 Watt (4.0 Watt minus 0.9 Watt) can be saved while delivering a 32 Kbps service, as with DPS the Node B will set its transmission power so as to cover only the 10% of the cell. The corresponding power gain increases to 6.2 Watt for a 64 Kbps service and to 13.4 Watt for a 128 Kbps service. These high sums of power underline the need for using this technique.

Figure 6. FACH Tx power with DPS (RL: Radio Link)

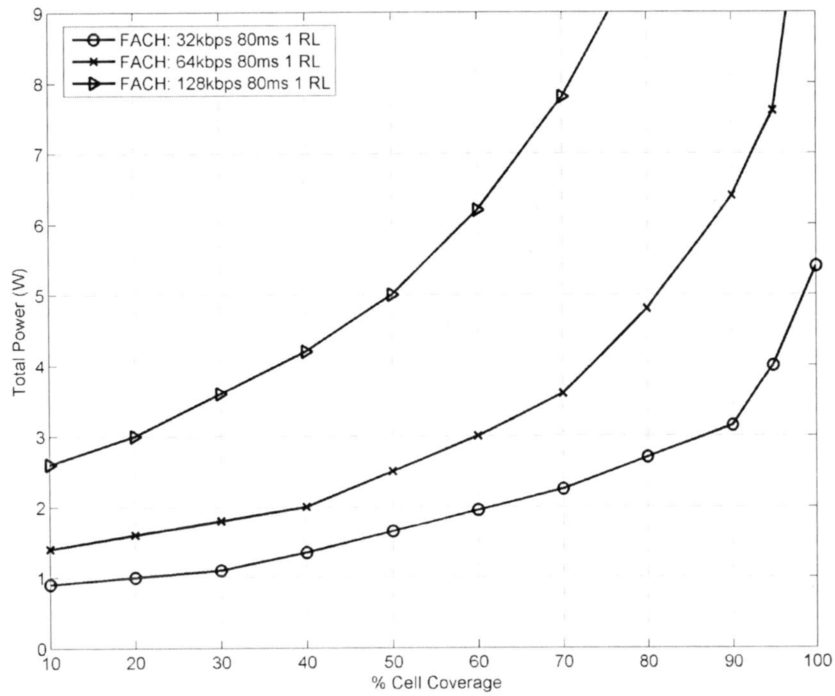

Macro Diversity Combining (MDC)

Diversity is a technique to combine several copies of the same message received over different channels. Macro Diversity is normally applied as diversity switching where two or more base stations serve the same area, and control over the mobile is switched among them. Basically, the Diversity Combining concept consists of receiving redundantly the same information bearing signal over two or more fading channels, and combine these multiple replicas at the receiver in order to increase the overall received Signal-to-Noise Ratio (SNR).

Figure 7 presents how the FACH transmission power level changes with cell coverage when MDC is applied. For the needs of the simulation we considered that a 64 Kbps service should delivered, using 1, 2 or 3 Node Bs (or radio links). TTI is assumed to be 80 ms. The main idea with regard to MDC is to decrease the power level from a Node B when it serves users near the cell edge. However, as we assume 3 sectors per cell (see Table 1), this technique can also be used for distances near the Node B, where each sector is considered as one radio link (RL). Succinctly, in Table 3 we mention some cases that reveal the power gains with this technique.

As the user receives data from two (or three) Node Bs, simultaneously the required power of each Node B is decreased; however, the total required power remains the same and sometimes it is higher. Nevertheless, this technique is particularly useful when the power level of a specific Node B is high, while respectively the power level of its neighboring Node B is low.

Rate Splitting (RS)

The RS technique assumes that the MBMS data stream is scalable, thus it can be split into several streams with different QoS. Only the most important stream is sent to all the users in the cell to provide the basic service. The less important

Figure 7. FACH Tx power with MDC (1Radio Link (RL), 2RLs and 3RLs)

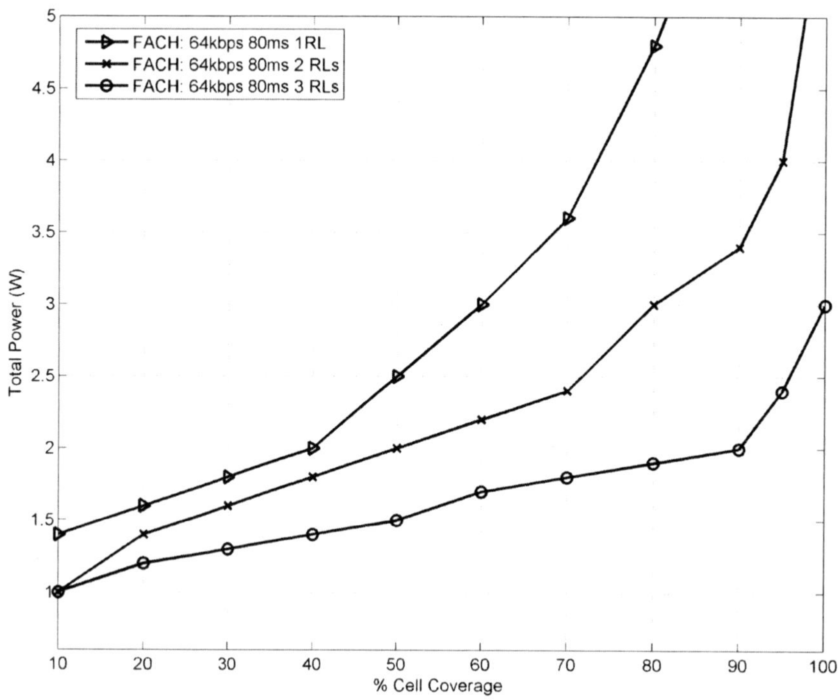

streams are sent with less amount of power or coding protection and only the users who have better channel conditions (*i.e.* the users close to Node B) can receive those to enhance the quality on top of the basic MBMS. This way, transmission power for the most important MBMS stream can be reduced because the data rate is reduced, and the transmission power for the less important streams can also be reduced because the coverage requirement is relaxed (3rd Generation Partnership Project R1-021239, 2002).

In the following scenario, we consider that a 64 Kbps service can be split in two streams of 32 Kbps. The first 32 Kbps stream (basic stream) is provided throughout the whole cell, as it is supposed to carry the important information of the MBMS service. On the contrary, the second 32 Kbps stream is sent only to the users who are close to the Node B (50% of the cell area) providing the users in the particular region the full 64 Kbps service. Figure 8 depicts the way the operation of

Table 3. Indicative FACH Tx power levels with MDC

Cell Coverage	Radio Links	Required Tx Power (Watt)
50%	1	2.5
	2	2.0
	3	1.5
95%	1	7.6
	2	4.0
	3	2.4

Figure 8. MBMS provision with RS

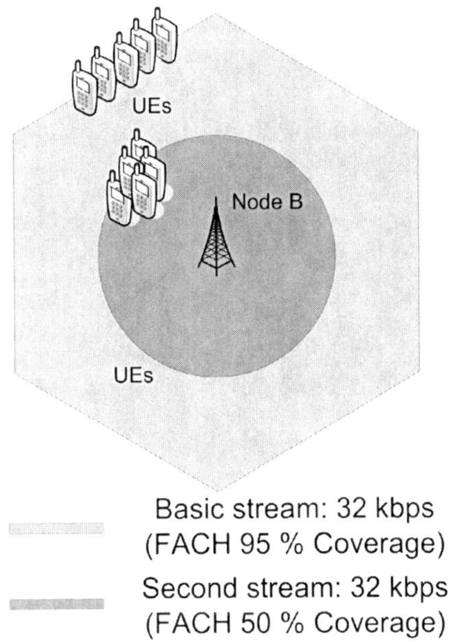

the RS technique, in terms of channel selection and cell coverage.

From Table 2 it can be seen that this technique requires 5.8 Watt (4.0 for the basic stream and 1.8 for the second). On the other hand, in order to deliver a 64 Kbps service using a FACH with 95% coverage the required power would be 7.6 Watt. Thus, 1.8 Watt can be saved through the RS technique. However, it is worth mentioning that this power gain involves certain negative results. Some of the users will not be fully satisfied, as they will only receive the 32 Kbps of the 64 Kbps service, even if these 32 Kbps have the important information. As the observed difference will be small, the Node B should weigh between the transmission power and the users' requirements.

Usage of Longer TTI and Space Diversity (LTTI)

These two methods can be employed in the physical layer to benefit every member of the MBMS group in a cell. Space-time processing techniques exploit diversity in both the spatial and temporal domains. On the one hand, an increment in TTI length (from 20 ms to 80 ms) can provide significant power gain; however, the use of longer TTI introduces more complexity and larger memory space requirement in the mobile station. On the other hand, space diversity assumes two transmit antennas and a single data stream in order to improve the signal quality and reduce the power requirements. The main benefit of using Space Time Transmit Diversity (STTD) is a reduction in the downlink E_b/N_0 requirement. These improvements in E_b/N_0 requirement impact upon both downlink system capacity and downlink service coverage (3rd Generation Partnership Project R1-021234, 2002; 3rd Generation Partnership Project TR 25.803, 2005). Fortunately, some MBMS services are not delay sensitive. In that case, diversity can be obtained by using the LTTI technique (Figure 9).

Table 4 demonstrates certain cases that reveal the sums of power that can be saved while delivering a 64 Kbps service, by increasing the TTI length and obtaining STTD. The above power levels are

Figure 9. FACH Tx power with LTTI

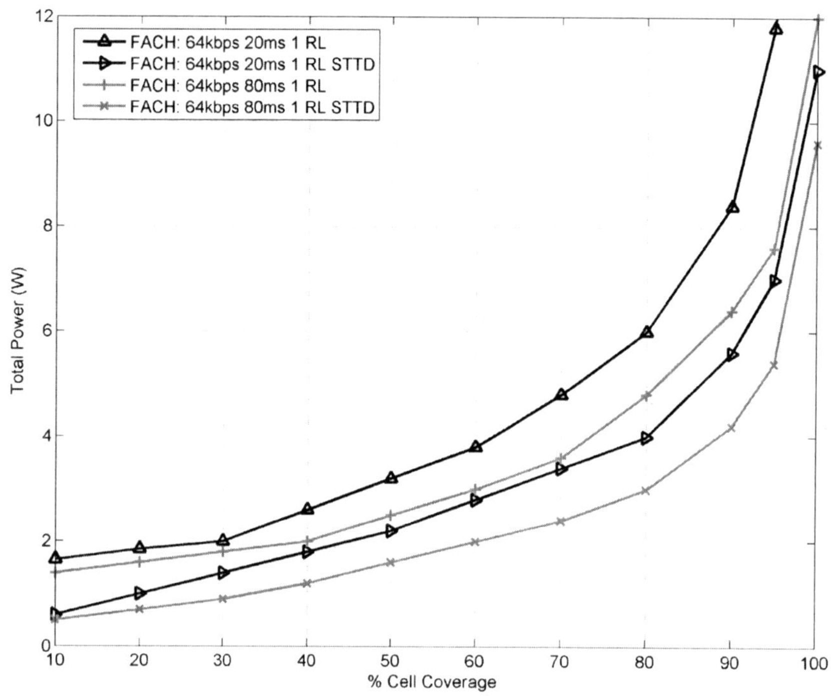

indicative of the sums of power that can be saved with the LTTI technique.

Mixed Usage of Multiple DCH channels and FACH (MDF)

The MDF can significantly decrease the Node B's transmission power, depending on the number and the location of the users that receive the MBMS service. In this approach, the FACH channel only covers the inner part of the sector (*e.g.* 50% of the sector area) and provides the MBMS service to the users that are found in this part ("inner part" users). The rest of the users are served using DCH to cover the remaining outer cell area ("outer part" users). Figure 10 represents the way of providing a MBMS service according to the MDF technique. The total downlink power consumption including

Table 4. Indicative FACH Tx power levels with LTTI

Cell Coverage	TTI (ms)	Required Tx Power (Watt)
50%	20 - no STTD	3.2
	20 - with STTD	2.2
	80 - no STTD	2.5
	80 - with STTD	1.6
95%	20 - no STTD	11.8
	20 - with STTD	7.0
	80 - no STTD	7.6
	80 - with STTD	5.4

Figure 10. MBMS provision with MDF

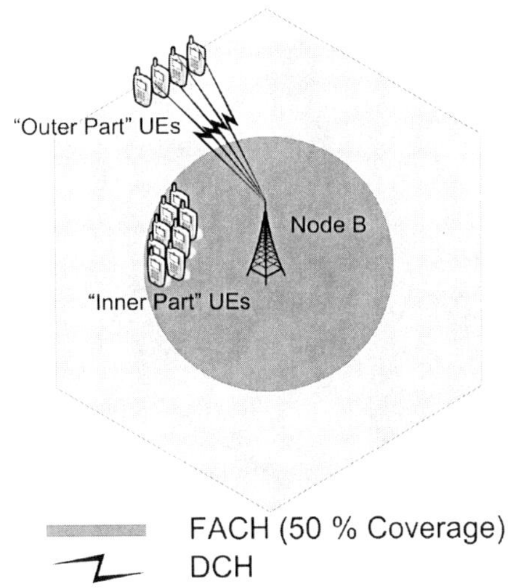

FACH and dedicated channels obviously depends on the number of users who are served by DCHs and their location (3rd Generation Partnership Project R1-021240, 2002).

The main goal is to examine how the transmission power is affected by the number of users. Figure 11 represents the Node B's total transmission power as a function of the number of the "outer part" users. The total power in Figure 11 includes the power that is required in order to cover the 50% of the cell with FACH (*i.e.* 2.5 Watt). Moreover, the number of the "inner part" users is assumed to be high enough, so as to justify the choice of FACH as the transport channel in the inner part.

Apart from the power required for the MDF technique, the power allocation level of a FACH with 95% cell coverage is depicted. This parallel plot intends to highlight the fact that a switch from MDF to a pure FACH has to be performed or vice versa. For instance, in Figure 11 when the "outer part" users exceeds 7, the total power (*i.e.* the power to cover the inner part with FACH plus the power to cover the outer part with DCHs) exceeds the power that is required in order to cover the whole cell with a single FACH. Thereby, it is more "power efficient" to use a FACH with 95% coverage.

Except for the power gain described above, the MDF technique ensures one more advantage. This advantage relies to the fact that DCH supports soft handover, while FACH does not. Since with this technique the users that are found near the cell edge are served with DCHs, their transition to another cell will be smoother, as the MBMS service will be provided uninterruptedly.

EXISTING RADIO BEARER SELECTION MECHANISMS

During the provision of MBMS multicast services the system should conceive and adapt to continuous changes that occur in dynamic wireless environments and optimally allocate resources. Under this prism, a critical aspect of MBMS performance is the selection of the most efficient radio bearer for the transmission of MBMS multicast data. It is worth mentioning that this is still an open issue in today's MBMS infrastructure mainly due to its catalytic role in Radio Resource Management (RRM).

Figure 11. Node B's Tx power with MDF

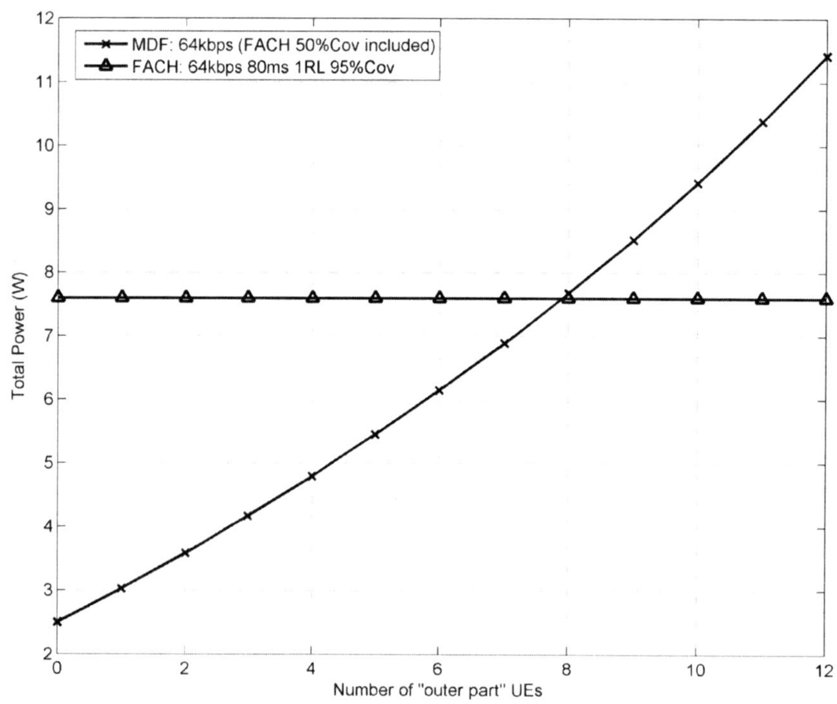

There exist two main research directions during the radio bearer selection procedure. According to the first approach, a single transport channel (PTP or PTM) can be deployed in a cell at any given time. In this case, a switching threshold is actually set that defines when each channel should be deployed. On the other hand, the second approach performs a simultaneous deployment of PTP and PTM modes. A combination of these modes is scheduled and both dedicated and common bearers are established in parallel in a cell. In the following paragraphs we present the main representative approaches of each of the two research directions.

MBMS Counting Mechanism (TS 25.346)

The 3GPP MBMS Counting Mechanism (or TS 25.346) constitutes the prevailing approach of switching between PTP (multiple DCHs) and PTM (FACH) radio bearers, mainly due to its simplicity of implementation and function (3rd Generation Partnership Project TS 25.346, 2009). According to this mechanism, a single transport channel (PTP or PTM) can be deployed in a cell at any given time. The decision on the threshold between PTP and PTM bearers is operator dependent, although it is proposed that it should be based on the number of serving MBMS users. In other words, a switch from PTP to PTM resources should occur, when the number of users in a cell exceeds a predefined threshold.

Figure 12 presents the operation of TS 25.346. Assuming that the threshold is 8 UEs (a mean value for the threshold proposed in the majority of research works), TS 25.346 will command Node B to switch from DCH to FACH when the number of users exceeds this predefined threshold, since HS-DSCH is not supported. Simultaneously, Figure 12 reveals the inefficiencies of TS 25.346. At simulation time 10 sec when the group that receives the MBMS service consists of 9 users (exceeding the switching threshold), this mecha-

Figure 12. 3GPP TS 25.346 Tx Power Levels

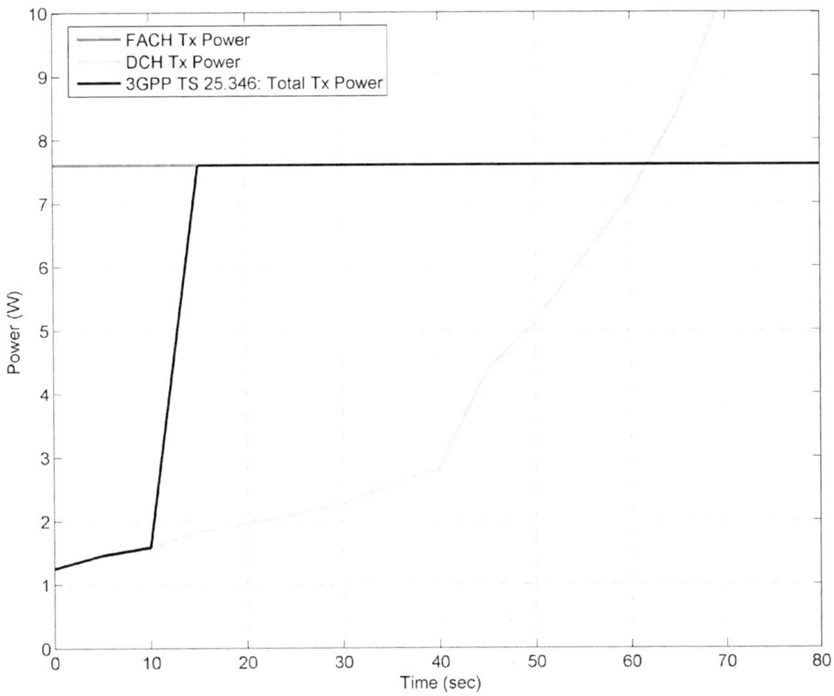

nism will switch from DCH to FACH, leading to a power wasting of 6 Watt.

Therefore, we could say that this mechanism provides a non realistic approach because mobility and current location of the mobile users are not taken into account. Moreover, this mechanism does not support FACH Dynamic Power Setting. Therefore, when employed, FACH has to cover the whole cell area, leading to power wasting. Finally, TS 25.346 does not support the HS-DSCH, a transport channel that could enrich MBMS with broadband characteristics.

MBMS PTP/PTM Switching Algorithm (TR 25.922)

3GPP MBMS PTP/PTM switching algorithm, or TR 25.922 (3rd Generation Partnership Project TR 25.922, 2007), assumes that a single transport channel can be deployed in a cell at any given time. However, contrary to TS 25.346, it follows a power based approach when selecting the appropriate radio bearer, aiming at minimizing the Node B's power requirements during MBMS transmissions.

In TR 25.922, instead of using solely DCHs, HS-DSCH can also be transmitted. However, the restricted usage of either DCH or HS-DSCH in PTP mode may result to significant power losses. In both cases, the PTP (DCH or HS-DSCH, since the switching between HS-DSCH and DCH is not supported in this mechanism) and the PTM power levels are compared and the case with the lowest power requirements is selected. In general, for small number of multicast UEs, PTP bearers are favored. As the number of users increases, the usage of PTM bearer is imperative.

An example of this mechanism's operation is presented in Figure 13. In this example, TR 25.922 supports only DCH transmissions in PTP mode. According to Figure 13, for small UE population (up to simulation time 60 sec) DCH is preferred compared to FACH since its power consumption is lower. As the number of users increases (from

Figure 13. 3GPP TR 25.922 (with DCH) Tx Power Levels

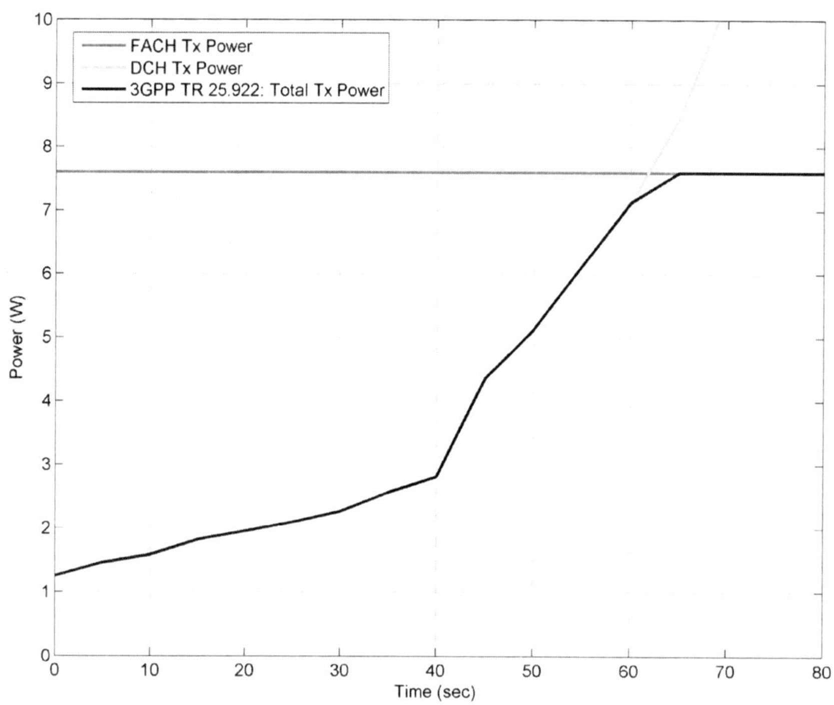

simulation time 60 sec) a switch from DCH to FACH is performed in order to ensure the lowest possible power consumption. However, even though TR 25.922 overcomes several inefficiencies of the TS 25.346 mechanism, still it does not support FACH Dynamic Power Setting, leading in turn, to increased power consumption in PTM transmissions.

Mechanism Proposed in 3GPP TSG RAN1 R1-02-1240

The above mechanisms allow a single PTP or PTM transport channel deployment at any given time. In 3rd Generation Partnership Project R1-021240 (2002), an alternative idea is presented, which is based on the simultaneous/combined usage of PTP and PTM bearers for MBMS transmissions. In particular, this approach considers the mixed usage of DCHs and FACH for the transmission of the MBMS data over the UTRAN interfaces. According to this approach, the FACH channel only covers an inner area of a cell/sector and provides the MBMS service to the users that are found in this part. The rest of the users are served using DCHs to cover the remaining outer cell area. The power for serving the outer part users is calculated as in equation (4). The total downlink power consumption, including FACH and dedicated channels, is the sum of these two power levels (Figure 14).

However, as clearly concluded in 3rd Generation Partnership Project R1-021240 (2002), this approach is only beneficial when the number of outer part users that use the DCHs is extremely small (less than 5). This suggests that the use of DCH in association with FACH for MBMS services is rather limited for real world traffic scenarios.

Novel Mechanism for PTP and PTM Bearers Combination

At this point, it should be mentioned that none of the above mechanisms takes into account the ability of the Node Bs to support many simulta-

Figure 14. 3GPP TSG RAN1 R1-02-1240 Tx Power Levels

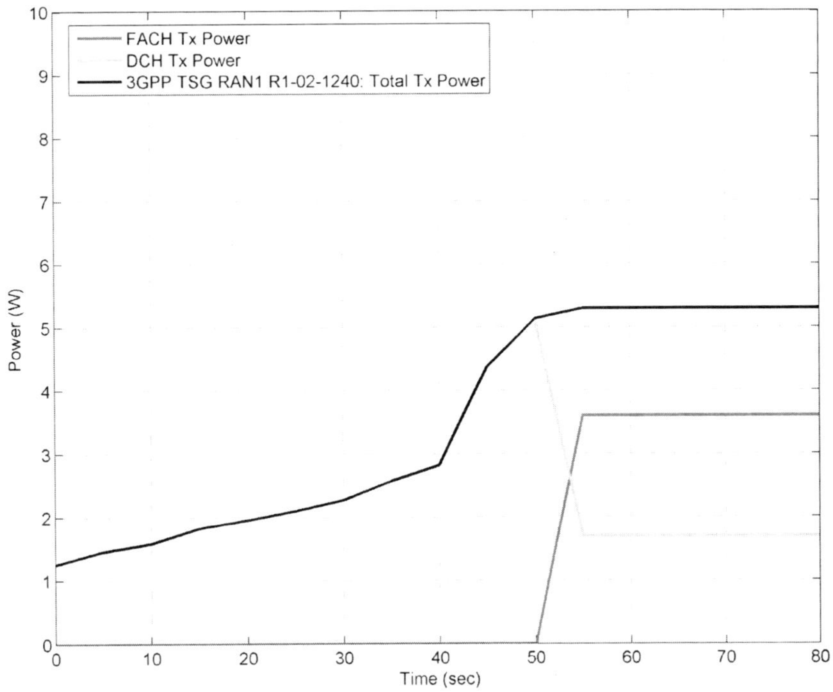

neous MBMS sessions. MBMS transmissions have increased power requirements and consume a large portion of the available power recourses. Consequently, the number of parallel MBMS sessions that a base station could support is limited; and the selection of the appropriate radio bearer for a MBMS service should be done with respect to other existing MBMS sessions in the corresponding cell.

To this direction, this section presents a power control mechanism that considers the Node B's transmission power level as the key criterion for the selection of the appropriate MBMS radio bearer. The goal achieved by mechanism work is threefold: At a first level, our mechanism proposes a more realistic and adaptive to dynamic wireless environments approach, by employing a power based switching criterion and allowing the combined usage of transport channels. At a second level, our mechanism contributes to Radio Resource Management (RRM) mechanisms of UMTS by presenting a novel framework for MBMS that optimally utilizes power resources. At a third level, a major advantage of our mechanism is its ability to ensure the service continuity in the system when multiple parallel MBMS services are delivered.

Parameters

The number of parallel MBMS sessions that a Node B could support depends on many parameters. We could classify these parameters in three categories:

- User related parameters,
- MBMS session related parameters and
- Provider related parameters.

User related parameters are parameters such as UEs' distances from the base stations and UEs' QoS parameters. The number of active MBMS sessions per cell, the number of UEs per MBMS session per cell and the bit rates of the MBMS services are some

Figure 15. Block diagram of the mechanism

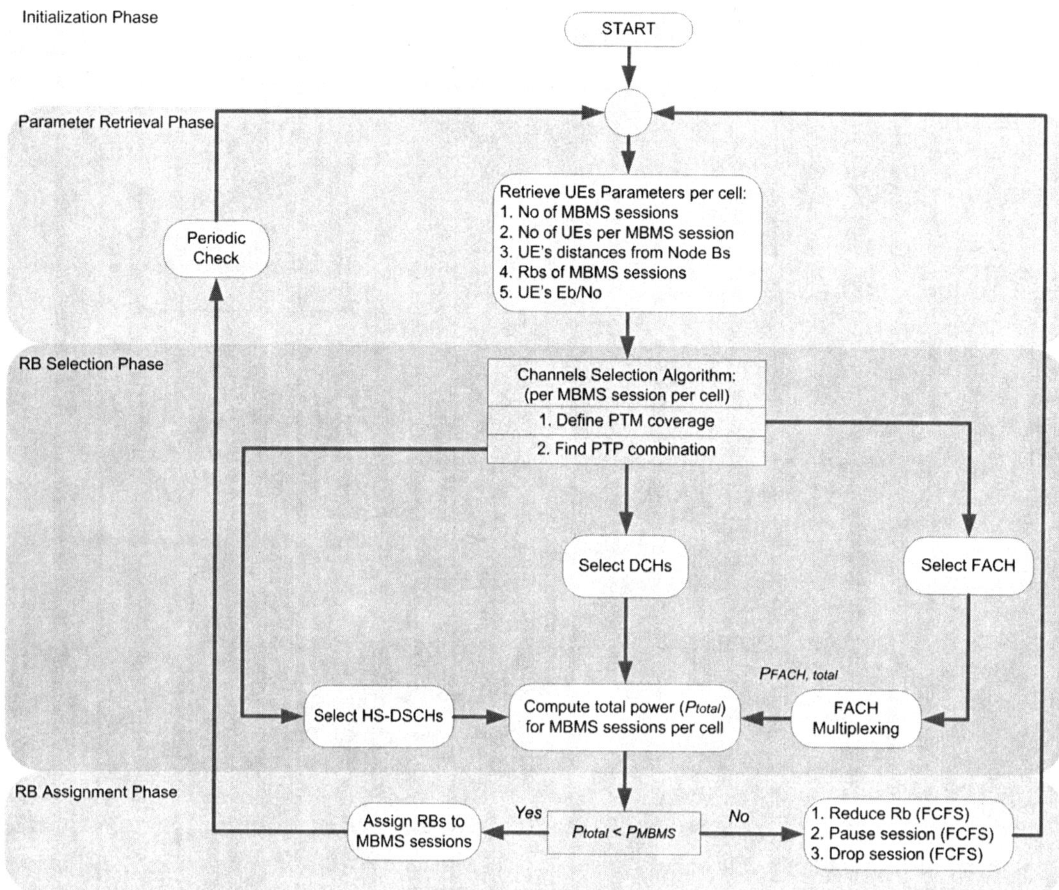

of the MBMS session related parameters. Finally, the portion of the available power recourses of base stations that could be used for MBMS transmissions is a provider related parameter. All these parameters should be considered in the RRM of MBMS so as to have efficient power control.

Architecture and Functionality

The remaining of this paragraph presents the architecture and functionality of the proposed mechanism that is used for the efficient data transmission of parallel MBMS services in LTE. The block diagram of the proposed mechanism is illustrated in Figure 15. According to Figure 15, the mechanism consists of four distinct operation phases: the initialization phase, the parameter retrieval phase, the radio bearer (RB) selection phase and the RB assignment phase.

Initialization Phase

The initialization phase (Figure 15) launches the mechanism when one user expresses his interest in receiving a MBMS service (*i.e.* the mechanism begins when the first user requests the first MBMS service).

Parameter Retrieval Phase

The parameter retrieval phase is responsible for retrieving the parameters of the existing MBMS users and services in each cell. In this phase, the mechanism requires the two of the three types of

Figure 16. Cell areas and zones

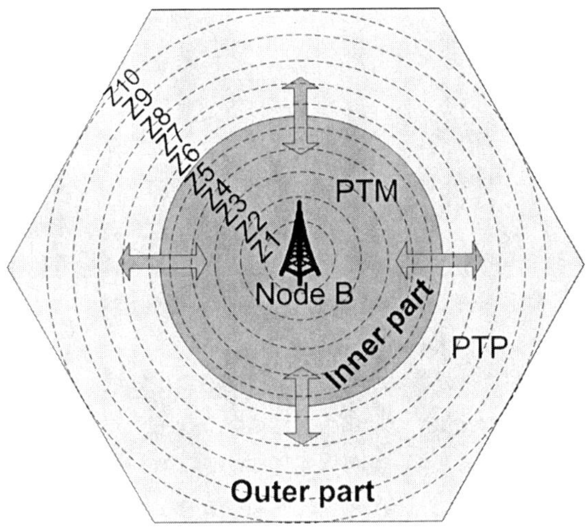

parameters, mentioned in the beginning of this paragraph: the user related parameters and the MBMS session related parameters. Regarding the latter type of parameters, the mechanism requires information about the number of active MBMS sessions per cell, the number of UEs per MBMS session per cell and the bit rates of the MBMS sessions. This information is retrieved from the BM-SC. On the other hand, the user related parameters are retrieved from the UEs through uplink channels.

Radio Bearer Selection Phase

The RB selection phase is dedicated to the selection of the transport channels for the MBMS sessions in any cell of the network. The most critical operations of the phase are executed by the Channels Selection Algorithm block (Figure 15). The algorithm executed in this block selects the combination of PTP and PTM bearers that minimizes the downlink base station's transmission power in any cell of the network that multicast users are residing. In particular, the algorithm is executed in two steps. In the first step (Define PTM coverage) the algorithm estimates the optimum coverage of FACH for the users' distribution of any MBMS session in the cell. This coverage area is called inner part of the cell as illustrated in Figure 16. In the second step (Find PTP combination), the mechanism decides which PTP bearer(s) will cover the rest part of the cell (outer part - Figure 16). It has to be mentioned that the above cell characterization is done for every MBMS session of the corresponding cell.

In order to estimate the optimum coverage of FACH (for any MBMS session in the cell) in Define PTM coverage step (Figure 15), the algorithm initially divides the cell in ten zones (Z1 to Z10). Each zone Zi refers to a circle with radius equal to 10i% of the cell radius. Afterwards, the algorithm scans all the zones and calculates the total base station's transmission power for the following 21 transport Channel Configurations (CC):

- CC1: No FACH used. All users of the specific MBMS session are covered by DCHs.
- CC2: No FACH used. All users of the specific MBMS session are covered by HS-DSCHs.
- CC3: FACH for UEs up to Z1. All the rest UEs covered by DCHs.
- CC4: FACH for UEs up to Z1. All the rest UEs covered by HS-DSCHs.
- ....

- CC19: FACH for UEs up to Z9. All the rest UEs covered by DCHs.
- CC20: FACH for UEs up to Z9. All the rest UEs covered by HS-DSCHs.
- CC21: FACH for all UEs (up to Z10). DCHs and HS-DSCHs are not used.

The CC that consumes less power indicates the coverage of the FACH and determines the inner part of the cell. The same procedure is executed simultaneously for any MBMS session in the cell. The output of the Define PTM coverage step is the coverage of the FACH for any MBMS session in the examined cell.

Once the appropriate FACH coverage is defined, the algorithm enters the Find PTP combination step (see Figure 15), which determines the appropriate PTP radio bearer(s) that will cover the MBMS users residing in the outer part of the cell. The procedure is similar to the procedure described in the Define PTM coverage step. The algorithm scans all the zones in the outer part of the cell and calculates the total base station's transmission power in order to cover all the outer part MBMS users only with PTP bearers. The first zone of the outer part is Z(inner part+1), therefore the algorithm will have to scan the following PTP transport Channel Configurations (PTP_CC):

- PTP_CC1: DCHs for outer part UEs up to Z(inner part+1). All the rest outer part UEs (up to Z10) covered by HS-DSCHs.
- PTP_CC2: DCHs for outer part UEs up to Z(inner part+2). All the rest outer part UEs (up to Z10) covered by HS-DSCHs.
- ....
- PTP_CC(10-inner part): All MBMS users in the outer part cell are covered by DCHs. HS-DSCHs are not used.
- PTP_CC(10-inner part+1): HS-DSCHs for outer part UEs up to Z(inner part+1). All the rest outer part UEs (up to Z10) covered by DCHs.
- PTP_CC(10-inner part+2): HS-DSCHs for outer part UEs up to Z(inner part+2). All the rest outer part UEs (up to Z10) covered by DCHs.
- ....
- PTP_CC(2*(10-inner part)): All MBMS users in the outer part cell for the specific session are covered by HS-DSCHs. DCHs are not used.

After these calculations, the different PTP_CCs are compared and the PTP_CC with the lowest power requirements determines the PTP transport channel configuration for the outer part MBMS UEs of the specific MBMS session in the cell. The procedure is recursively executed for any MBMS session in the examined cell.

Generally, the output of the Channels Selection Algorithm block is the combination of PTM and PTP transport channels that consumes the lowest power resources between all possible combinations in the corresponding cell for any MBMS session running in it.

In the case of FACH there is another block in the mechanism's block diagram named FACH Multiplexing. When the number of MBMS sessions requiring FACH in cell is greater than one, these FACHs should be multiplexed onto a Secondary Common Control Physical Channel (S-CCPCH) (3rd Generation Partnership Project TS 25.211, 2009; 3rd Generation Partnership Project TS 25.212, 2009). After the multiplexing procedure, the capacity of the S-CCPCH is calculated and based on this, the total power required for the common channels ($P_{FACH,total}$) in the corresponding base station is estimated. In this chapter we consider a one to one mapping between MBMS sessions (MBMS point-to-multipoint Traffic Channels - MTCHs) and FACHs.

The last action performed in the RB selection phase is the computation of the total base station's power (P_{total}) required so as to support all the MBMS sessions in each cell of the network.

Table 5. UE Number, Coverage per time period

Time (sec)	UEs Number	Coverage (%)	Best Performance
0-50	25	50	Our Mechanism
	7	80	
51-100	25	50	R1-02-1240 and Our Mechanism
	2	80	
101-150	17	50	TR 25.922 (HS-DSCH) and Our Mechanism
151-200	4	50	All except TR 25.922 (HS-DSCH)

However, at this point we have to mention that the selected radio bearers are not yet assigned to the MBMS sessions. This action is performed in the following phase.

Radio Bearer Assignment Phase

During the RB assignment phase, the P_{total} is compared with the available power assigned by the network provider to MBMS sessions in each base station (P_{MBMS}). Obviously, P_{MBMS} constitutes the third type of parameters, known as provider related parameter. If P_{total} is smaller than P_{MBMS} then the selected from the RB selection phase transport channels are assigned to MBMS sessions and the MBMS data transfer phase begins. In case when P_{total} is bigger than P_{MBMS}, a session reconfiguration procedure should occur due to the fact that the Node B has no available radio resources so as to serve all the MBMS sessions in the examined cell. Three possible reconfiguration events could be used in such a case. The first is the reduction of the transmission rate of a MBMS session, the second is the pause of a MBMS session for a short time period and the last is the cancellation of the service. The simplest policy that the mechanism could adopt in order to perform these reconfiguration events, is a First Come First Served (FCFS) policy. Following the FCFS policy and considering the available power, the mechanism performs the optimum event to the most recent sessions.

The above description refers to a dynamic model, in the sense that the UEs are assumed to be moving throughout the topology and the number of MBMS sessions varies. The parameter retrieval phase is triggered at regular time intervals so as to take into account the changes in user related parameters, MBMS session related parameters and operator related parameters. This periodic computation inserts a further complexity as this information is carried in through uplink channels. This entails that a certain bandwidth fraction must be allocated for the transmission of this information in uplink channels, thus resulting to a system's capacity reduction. A further complexity is inserted due to the fact that the mechanism is executed many times. In particular, if we suppose that N base stations are served with multicast users, while each of these base stations serves Mi ($i = 1...N$) parallel MBMS sessions, then the number of executions of the mechanism is computed as in (6):

$$K = \sum_{i=1}^{N} M_i \qquad (6)$$

Comparison with 3GPP Approaches

This scenario lasts for 200 sec and can be divided into four time periods, depending on the number of MBMS users. According to this scenario, a 64 Kbps service should be delivered to a group of users, whose initial position at each time period is presented in Table 5. For example, for the time period 0 to 50 sec, 25 UEs receive the 64 Kbps service at distance 50% of the cell radius and 7 UEs at distance 80% of the cell radius.

Figure 17. (a) Power consumption and (b) complexity comparison

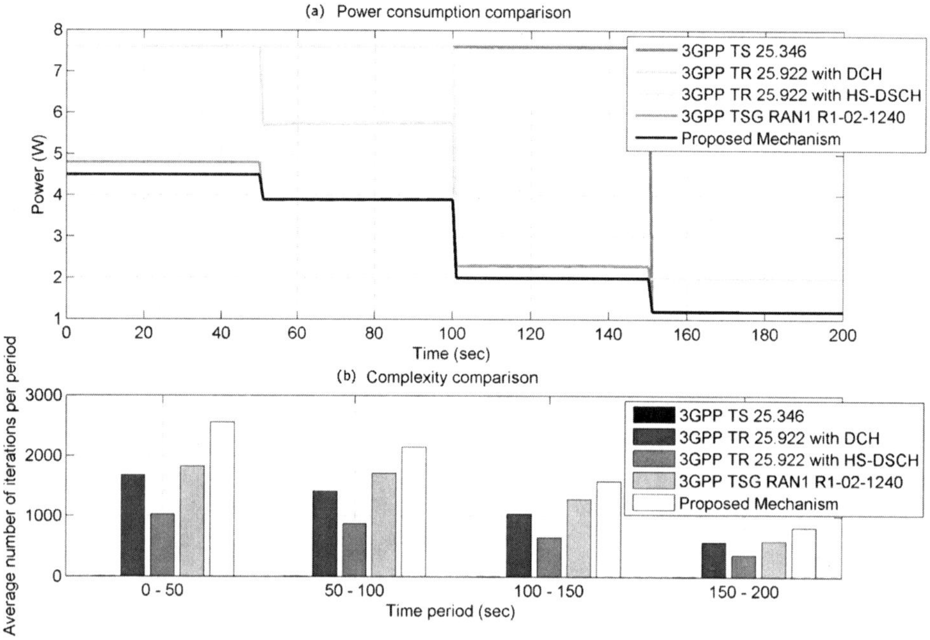

Figure 17a depicts the power levels of the examined radio bearer selection mechanisms. As it can be noticed from Figure 17, the proposed mechanism and the mechanism presented in 3GPP TSG RAN1 R1-02-1240 have the best performance in general.

For example, for the period 0-50 sec, the total number of users in the cell is 32. By assuming that the threshold for switching between DCH and FACH (HS-DSCH is not supported) in TS 25.346 is 8 UEs, TS 25.346 will deploy a FACH with 100% cell coverage (requiring 7.6 W).

The high initial users' population favors the deployment of FACH in order to serve all the UEs in 3GPP TR 25.922. However, as TS 25.346, TR 25.922 does not support FACH Dynamic Power Setting. This is the reason why TR 25.922 (with DCH or HS-DSCH) has the same power requirements with TS 25.346 (7.6 W) for the time period 0-50 sec.

The mechanism proposed in 3GPP TSG RAN1 R1-02-1240 allows the mixed usage of DCHs and FACH and supports FACH Dynamic Power Setting. As shown in Figure 17a, this mechanism requires 4.8 W in order to serve all the users in the cell, for the first time period. This derives from the fact that for the specific scenario, this mechanism will deploy only a FACH with 80% coverage, while the user with the worst path loss resides in the borders of zone Z8.

Finally, Figure 17a depicts the power requirements of the proposed mechanism for the examined scenario. For the time period 0-50 sec, the output of the Channels Selection Algorithm block (Figure 15) specifies that the users up to Z5 should be server by a FACH. Moreover, the most efficient combination of PTP bearers for the outer part MBMS users is to serve the remaining 7 users in zone Z8 with HS-DSCH. Therefore, 4.5 W in total are required in order to serve all the MBMS users with this mechanism. Obviously, the proposed mechanism ensures minimized power consumption. A significant power budget, ranging from 0.3 to 3.1 W, may be saved for the period 0-50 sec compared with the other approaches.

On the other hand, Figure 17b presents the computational overhead that each mechanisms inserts (number of iterations required to calculate

Table 6. Comparison of the mechanisms

Mechanism	Advantages	Disadvantages
TS 25.346	1) Low complexity 2) Easy to implement 3) 3GPP standardized	1) High power requirements 2) No mobility support 3) No HS-DSCH support 4) No dynamic FACH support
TR 25.922	1) Support all transport channels 2) 3GPP standardized	1) High power requirements 2) No switching between HS-DSCH and DCH support 3) No dynamic FACH support
3GPP R1-02-1240	1) Power efficient 2) Support combined usage of FACH and DCH 3) Support dynamic FACH	1) High complexity 2) No standardized 3) No HS-DSCH support
Proposed Mechanism	1) Power efficient 2) Support combined usage of all transport channels 3) Support dynamic FACH 4) Support multiple MBMS sessions	1) High complexity 2) No standardized (Novel Mechanism)

the power of the available transport channels and assign the ideal channel), based on the above scenario.

In general, TS 25.346 inserts the lowest computational overhead (number of iterations constant and equal to one), because TS 25.346 requires only the number of served MBMS users in order to assign the appropriate transport channel. On the other hand, the other approaches have higher computational overhead due to the fact that these mechanisms have to periodically retrieve the parameters of existing MBMS users. Moreover, these approaches have to calculate the power consumption of the transport channels that each mechanism supports; and based on this calculation to assign the ideal radio bearer. The fact that the proposed mechanism supports all the available transport channels and examines all possible transport channels configurations explains why the number of iterations in this case is higher than the other approaches.

To sum up, Table 6 presents a cumulative, direct comparison between all mechanisms analyzed in this chapter. The main conclusion is that the proposed mechanism outperforms the other approaches in terms of power consumption, since significant power budget is saved. It puts together the benefits of all mechanisms by providing a scheme that is based on the concept of transport channels combination; and performs an optimal power resource allocation in LTE base stations. And even if the complexity of the proposed mechanism is higher than the complexity of the other mechanisms, the benefits from the optimal power planning counterbalance the complexity issues raised. This fact is strongly enhanced mainly due to the power limited LTE networks, which entails that power strategies are of key importance in order to obtain high capacity.

Managing Parallel MBMS Sessions

The major advantage of the proposed mechanism is its ability to manage multiple parallel MBMS sessions. In order to evaluate this ability, we setup a simulation scenario where multiple MBMS services are transmitted in parallel to several user groups residing in a cell. In particular, we suppose that four user groups receive four distinct MBMS services with characteristics presented in Table 7.

Table 7. Scenario parameters

MBMS No.	Duration (sec)	Rb (Kbps)	UEs Number	Maximum Coverage	Channel
1	0 - 600	64	10	80%	HS-DSCH
2	50 - 600	64	22 + 6	20%+50%	FACH+DCH
3	100 - 150	64	2 to 13	60%	DCH
	151 - 300	64	14 to 19	60%	HS-DSCH
	301 - 600	64	20 to 27	60%	FACH
4	150 - 560	64	7	70%	DCH
	561 - 600	32	7	80%	DCH

Moreover, Table 7 presents the appropriate transport channel (with respect to power consumption as presented in previous sections) to serve each group at each time interval.

Figure 18 depicts the power consumption of each MBMS session as well as the total, aggregative power required to support the transmission of all services to the multicast users in the corresponding cell.

Users of the 1st MBMS session are served with a HS-DSCH channel, due to the small population, throughout the whole service time. At simulation time 50 sec, MBMS service 2 is initiated (Figure 18). At this time instant, the mechanism, through the RB selection phase, selects FACH (for the 22 inner part users) and DCHs (for the 6 outer part users) as the most efficient transport channel combination for the transmission of the MBMS traffic.

MBMS service 3 starts at simulation time 100 sec. At this time the 3rd multicast group consists of only two UEs; thus, the mechanism selects multiple DCHs for this MBMS service. The number of users receiving the service successively increases (join requests), reaching 13 UEs at simulation time 150 sec, 19 at simulation time 300 sec and 27 at the end of the simulation time. The increasing number of users in the group forces the mechanism to perform a channel switching from DCH to HS-DSCH at simulation time 151 sec and another one from HS-DSCH to FACH at simulation time 301 sec, securing, in this way, the efficient resource utilization.

At this point we have to mention that from simulation time 300 sec until the end of the simulation, MBMS services 2 and 3 employ FACHs for the transmission of the MBMS data (see Table 7). During this time interval, the deployment of two parallel FACHs forces the mechanism to perform a FACH multiplexing procedure in the RB selection phase. Consequently, a single S-CCPCH with bit rate of 128 Kbps is used to deliver MBMS services 2 and 3. Moreover, P_{total} is lower than P_{MBMS}, which translates into efficient provision of the three parallel MBMS sessions.

At simulation time 150 sec, the MBMS service 4 is initiated and is targeted to a multicast group consisting of seven members. Multiple DCHs are selected by the mechanism to deliver the MBMS content to the 4th multicast group. Additionally, at the same time instance, P_{total} still remains smaller than P_{MBMS}, which means that the MBMS service 4 is accepted for transmission in the system. From simulation time 150 until the end of the simulation, four parallel MBMS sessions running in the system and the proposed mechanism handles them in an efficient way.

Due to the fact that the users of the 4th multicast group are moving towards the cell edge an increase in P_{total} occurs and at simulation time 560 sec; P_{total} exceeds P_{MBMS} value (Figure 18). Thus, a session reconfiguration procedure is performed,

Figure 18. Power levels of the MBMS sessions

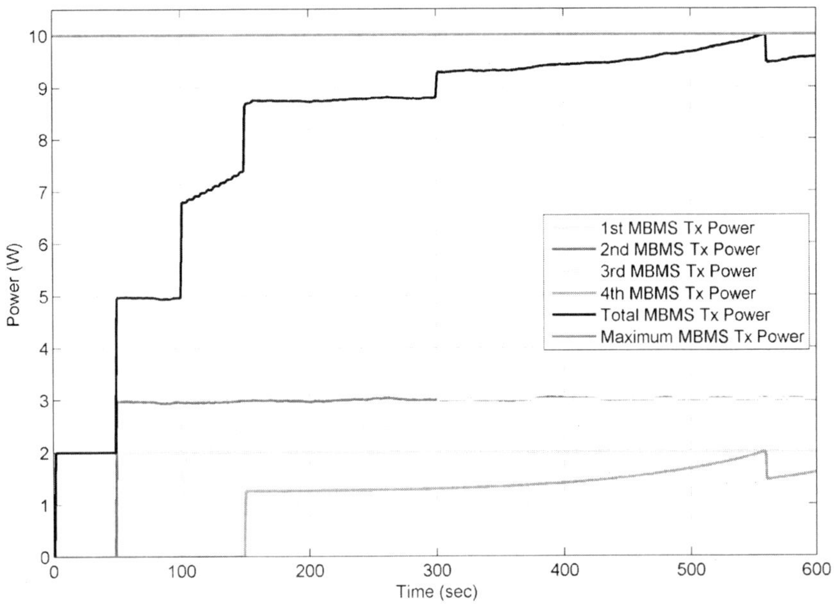

forcing the MBMS service 4 to reduce its bit rate from 64 Kbps to 32 Kbps in order to ensure the efficient service of four parallel MBMS sessions without any interruption.

FUTURE RESEARCH DIRECTIONS

Regarding the operation of the proposed mechanism, several enhancements could be incorporated in order to further improve the MBMS performance. The steps that follow this work could be at a first level the evaluation of the mechanism through additional simulation scenarios. The scenarios could be simulated in the ns-2 simulator, in which the proposed mechanism could be implemented. In that way, we could measure, except from the performance of our mechanism, other parameters such as delays in UTRAN interfaces during MBMS transmissions. Furthermore, several power saving techniques such as Rate Splitting and Macro Diversity Combining could be integrated in the proposed mechanism. The use of these techniques will further improve the overall performance of the proposed mechanism, which in turn means that a better utilization of radio and network resources could be achieved.

Additionally, the capacity and functionality of the proposed mechanism may be further improved by incorporating the enhancements that could be obtained from the use of Multiple Input Multiple Output (MIMO) antenna techniques in HSDPA. MIMO systems are a prerequisite for LTE networks. Early LTE requirements consider two transmit and receive antennas (MIMO 2x2) and approximately, double data rates are obtained with the same base station power compared with conventional HS-DSCH single antenna systems. Therefore, MIMO antenna techniques have the potential to address the unprecedented demand for wireless multimedia services and in particular for MBMS. Finally, it may be examined whether the Multicast Broadcast Single Frequency Network (MBSFN) transmission mode, included in the evolved UTRAN technologies of the LTE, could be used as an alternative PTM transmission mode for MBMS. MBSFN tries to overcome the cell edge problems of MBMS and to reduce the intercell interference. Therefore, MBSFN could be used in order to achieve very high receiver output SNR and significantly improve the overall spectral efficiency.

CONCLUSION

This chapter introduced the key concepts of MBMS services. The main target was to highlight the importance of power control and its commanding role during the delivery of MBMS multicast content, for the overall efficiency of next generation networks. To this direction, the power profiles of several transport channels which could be employed for the transmission of MBMS services to the mobile users were investigated. Moreover, the reader was introduced to certain problems that MBMS current specifications are facing and become familiar with techniques/solutions proposed to overcome such limitations.

Finally, this chapter underlined the significance of selecting the appropriate transport channel for the delivery of MBMS services so as to optimally allocate power resources. The reader was introduced to several radio bearer selection mechanisms that focus on conceiving and adapting to continuous changes that occur in dynamic wireless environments. These mechanisms could further improve the MBMS performance since they ensure the economic and rational usage of the expensive radio and network resources, enabling in this way the mass market delivery of multimedia services to mobile users.

REFERENCES

Alexiou, A., Antonellis, D., Bouras, C., & Papazois, A. (2006, October). *An Efficient Multicast Packet Delivery Scheme for UMTS*. Paper presented at the 9th ACM/IEEE International Symposium on Modeling, Analysis and Simulation of Wireless and Mobile Systems (MSWiM 2006), Torremolinos, Malaga, Spain.

Alexiou, A., Bouras, C., & Kokkinos, V. (2007, October). *Power Saving Techniques in MBMS Multicast Mode*. Paper presented at the 3rd ACM International Workshop on Wireless Multimedia Networking and Performance Modeling (WMuNeP 2007), Chania, Greece.

Alexiou, A., Bouras, C., & Rekkas, E. (2007, June). *A Power Control Scheme for Efficient Radio Bearer Selection in MBMS*. Paper presented at the 8th IEEE International Symposium on World of Wireless, Mobile and Multimedia Networks (WoWMoM 2007), Helsinki, Finland.

B-BONE. (2003). *Final Results with combined enhancements of the Air Interface*. Deliverable of the project (D2.5), IST-2003-507607.

Boni, A., Launay, E., Mienville, T., & Stuckmann, P. (2004, October). *Multimedia Broadcast Multicast Service – Technology Overview and Service Aspects*. Paper presented at the 5th IEE International Conference on 3G Mobile Communication Technologies (3G 2004), London, UK.

Chuah, P., Hu, T., & Luo, W. (2004, June). *UMTS Release 99/4 Airlink Enhancement for supporting MBMS Services*. Paper presented at the 2004 IEEE International Conference on Communications (ICC 2004), Paris, France.

Holma, H., & Toskala, A. (2006). HSDPA/HSUPA for UMTS: High Speed Radio Access for Mobile Communications. Chichester, UK: John Wiley & Sons.

Holma, H., & Toskala, A. (2007). WCDMA for UMTS: HSPA Evolution and LTE (4th edition). Chichester, UK: John Wiley & Sons.

Parkvall, S., Englund, E., Lundevall, M., & Torsner, J. (2006). Evolving 3G Mobile Systems: Broadband and Broadcast Services in WCDMA. *IEEE Communications Magazine*, *44*(2), 30–36. doi:10.1109/MCOM.2006.1593548

Perez-Romero, J., Sallent, O., Agusti, R., & Diaz-Guerra, M. (2005). Radio Resource Management Strategies in UMTS. Chichester, UK: John Wiley & Sons.

3rd Generation Partnership Project R1-021234. (2002). *Evaluation of combining gains for MBMS (incl. STTD)*. Qualcomm, TSG-RAN WG1#28.

3rd Generation Partnership Project R1-021239. (2002). *MBMS Power Usage, Lucent Technologies*. Lucent Technologies, TSG-RAN WG1#28.

3rd Generation Partnership Project R1-021240. (2002). *Power Usage for Mixed FACH and DCH for MBMS*. Lucent Technologies, TSG-RAN WG1#28.

3rd Generation Partnership Project TR 101.102. (2002). *Universal Mobile Telecommunications System (UMTS); Selection procedures for the choice of radio transmission technologies of the UMTS*. Version 3.2.0.

3rd Generation Partnership Project TR 23.846. (2003). *Technical Specification Group Services and System Aspects; Multimedia Broadcast/Multicast Service; Architecture and functional description (Release 6)*. Version 6.1.0.

3rd Generation Partnership Project TR 25.803. (2005). *Technical Specification Group Radio Access Network; S-CCPCH performance for MBMS, (Release 6)*. Version 6.0.0.

3rd Generation Partnership Project TR 25.922. (2007). *Technical Specification Group Radio Access Network; Radio resource management strategies (Release 7)*. Version 7.1.0.

3rd Generation Partnership Project TS 22.146. (2008). *Technical Specification Group Services and System Aspects; Multimedia Broadcast/Multicast Service; Stage 1 (Release 9)*. Version 9.0.0.

3rd Generation Partnership Project TS 25.211. (2009). *Physical channels and mapping of transport channels onto physical channels (FDD) (Release 8)*. Version 8.4.0.

3rd Generation Partnership Project TS 25.212. (2009). *Multiplexing and channel coding (FDD) (Release 8)*. Version 8.5.0.

3rd Generation Partnership Project TS 25.346. (2009). *Technical Specification Group Radio Access Network; Introduction of the Multimedia Broadcast Multicast Service (MBMS) in the Radio Access Network (RAN); Stage 2, (Release 8)*. Version 8.3.0.

Vriendt, J., Gomez Vinagre, I., & Van Ewijk, A. (2004). Multimedia Broadcast and Multicast Services in 3G Mobile Networks. Alcatel Telecommunication Review.

ADDITIONAL READING

Aho, K., Kurjenniemi, J., Haikola, V., Callender, C., & Ristaniemi, T. (2008). Multimedia Broadcast Multicast Service Performance and its Enhancements in WCDMA Networks. *Springer International Journal of Wireless Personal Communications*, *46*(2), 115–142. doi:10.1007/s11277-007-9413-x

Alexiou, A., Bouras, C., & Kokkinos, V. (2009, April). *An Efficient Mechanism for Power Control Optimization in MBMS Enabled UTRAN*. Paper presented at the IEEE Wireless Communications and Networking Conference 2009 (WCNC 2009), Budapest, Hungary.

Alexiou, A., Bouras, C., Kokkinos, V., & Rekkas, E. (2009). An improved mechanism for multiple MBMS sessions assignment in B3G cellular networks. Springer Wireless Networks.

Alexiou, A., Bouras, C., Kokkinos, V., & Rekkas, E. (2009, April). *Optimal MBMS Power Allocation Exploiting MIMO in LTE Networks.* Paper presented at the IEEE 69th Vehicular Technology Conference (VTC2009 Spring), Barcelona, Spain.

Christophorou, C., & Pitsillides, A. (2008, July). *A New Approach for Efficient MBMS Service Provision in UTRAN.* Paper presented at the IEEE Symposium on Computers and Communications 2008 (ISCC 2008), Marrakech, Morocco.

Dahlman, E., Parkvall, S., Sköld, J., & Beming, P. (2007). 3G EvolutionHSPA and LTE for Mobile Broadband. Linacre House, Jordan Hill, Oxford, United Kingdom: Elsevier.

Laiho, J., Wacker, A., & Novosad, T. (2006). Radio Network Planning and Optimisation for UMTS. The Atrium, Southern Gate, Chichester, England: John Wiley & Sons.

Lescuyer, P., & Lucidarme, T. (2008). Evolved Packet System (EPS) The LTE and SAE Evolution of 3G UMTS. The Atrium, Southern Gate, Chichester, England: John Wiley & Sons.

Nawrocki, M., Dohler, M., & Aghvami, H. (2006). Understanding UMTS Radio Network Modelling, Planning and Automated Optimisation. The Atrium, Southern Gate, Chichester, England: John Wiley & Sons.

Panayides, A., Christophorou, C., Antoniou, J., Pitsillides, A., & Vassiliou, V. (2008, July). *Power Counting for Optimized Capacity in MBMS Enabled UTRAN.* Paper presented at the IEEE Symposium on Computers and Communications 2008 (ISCC 2008), Marrakech, Morocco.

Rong, L., Haddada, O., & Elayoubi, S. (2008, December). *Analytical Analysis of the Coverage of a MBSFN OFDMA Network.* Paper presented at the IEEE Global Telecommunications Conference 2008 (GLOBECOM 2008), New Orleans, LA, USA.

Weng, J., Chen, J., & Cheng, K. (2008). Comparative study of broadcast and multicast in 3GPP and 3GPP2 networks. Elsevier Computer Communications.

Compilation of References

3GPP2. (2004). cdma2000 Evaluation Methodology Revision 0. TSG-C.R1002-0.

3GPP23.107. (2003). Quality of Service (QoS) Concept and Architecture (Release 5). TS 23.107 V5.11.0.

3rd Generation Partnership Project R1-021234. (2002). Evaluation of combining gains for MBMS (incl. STTD). Qualcomm, TSG-RAN WG1#28.

3rd Generation Partnership Project R1-021239. (2002). MBMS Power Usage, Lucent Technologies. Lucent Technologies, TSG-RAN WG1#28.

3rd Generation Partnership Project R1-021240. (2002). Power Usage for Mixed FACH and DCH for MBMS. Lucent Technologies, TSG-RAN WG1#28.

3rd Generation Partnership Project TR 101.102. (2002). Universal Mobile Telecommunications System (UMTS); Selection procedures for the choice of radio transmission technologies of the UMTS. Version 3.2.0.

3rd Generation Partnership Project TR 23.846. (2003). Technical Specification Group Services and System Aspects; Multimedia Broadcast/Multicast Service; Architecture and functional description (Release 6). Version 6.1.0.

3rd Generation Partnership Project TR 25.803. (2005). Technical Specification Group Radio Access Network; S-CCPCH performance for MBMS, (Release 6). Version 6.0.0.

3rd Generation Partnership Project TR 25.922. (2007). Technical Specification Group Radio Access Network; Radio resource management strategies (Release 7). Version 7.1.0.

3rd Generation Partnership Project TS 22.146. (2008). Technical Specification Group Services and System Aspects; Multimedia Broadcast/Multicast Service; Stage 1 (Release 9). Version 9.0.0.

3rd Generation Partnership Project TS 25.211. (2009). Physical channels and mapping of transport channels onto physical channels (FDD) (Release 8). Version 8.4.0.

3rd Generation Partnership Project TS 25.212. (2009). Multiplexing and channel coding (FDD) (Release 8). Version 8.5.0.

3rd Generation Partnership Project TS 25.346. (2009). Technical Specification Group Radio Access Network; Introduction of the Multimedia Broadcast Multicast Service (MBMS) in the Radio Access Network (RAN); Stage 2, (Release 8). Version 8.3.0.

Acanfora, S., Cacace, F., Iannello, G., & Vollero, L. (2006). Dynamic Configuration of MAC QoS Mechanisms in 802.11 Access Networks. In Next Generation Teletraffic and Wired/Wireless Advanced Networking, (LCNS 4003, pp. 542-553). Berlin: Springer-Verlag.

Adas, A. (1997). Traffic models in broadband networks. IEEE Communications Magazine, 35, 82–89. doi:10.1109/35.601746

Agarwal, S., Krishnamurthy, S., Katz, R., & Dao, S. (2001). Distributed Power Control in Ad-Hoc Wireless Networks. In 12th IEEE International Symposium on Personal, Indoor and Mobile Radio Communications (pp. 59-66), San Diego, CA. Alvarion BreezeAccess. (n.d.). VL datasheet. Retrieved from www.alvarion.com

Aggelou, G. (2009). Wireless Mesh Networking. New York: McGraw Hill Publishers. Capacity of Wireless Mesh Networks. (n.d.). White paper, BelAir Networks. Retrieved Aug 2006, from http://www.belairnetworks.com/resources/pdfs/ Mesh_Capacity_BDMC00040-C02.pd

Ahmad, A. M., & Abdul Manan Ahmad. (2004). Recurrent neural network with backpropagation through time algorithm for arabic recognition. In proceedings of 18th European Simulation Multiconference Graham Horton(c) SCS Europe.

Ahmed, M. H. (2005). Call Admission Control in Wireless Networks: A Comprehensive Survey. IEEE Communications Surveys & Tutorials, 7(1), 50–66. doi:10.1109/COMST.2005.1423334

Ahn, C. W., & Ramakrishna, R. S. (2004). QoS provisioning dynamic connection-admission control for multimedia wireless networks using a Hopfield neural network. IEEE Transactions on Vehicular Technology, 53(1), 106–117. doi:10.1109/TVT.2003.822000

Ahn, G. S., Campbell, A., Veres, A., & Sun, L. H. (2002). Supporting Service Differentiation for Real-Time and Best-Effort Traffic in Stateless Wireless Ad Hoc Networks (SWAN). IEEE Transactions on Mobile Computing, 1(3), 192–207. doi:10.1109/TMC.2002.1081755

Ahuja, R. K., Magnanti, T. L., & Orlin, J. B. (1993). Network Flows: Theory, Algorithms, and Applications. Englewood Cliffs, NJ: Prentice Hall.

Akhlaghi, Y., & Kompany-Zareh, M. (2005). Comparing radial basis function and feed-forward neural networks assisted by linear discriminant or principal component analysis for simultaneous spectrophotometric quantification of mercury and copper. Analytica Chimica Acta, 537, 331–338. doi:10.1016/j.aca.2004.12.079

Akyildiz, I. F., McNair, J., Carrasco, L., & Puigjaner, R. (1999). Medium Access Control Protocols for Multimedia Traffic in Wireless Networks. IEEE Network Magazine, 13(4), 39–47. doi:10.1109/65.777440

Akyildiz, I., Xie, J., & Mohanty, S. (2004). A survey of mobility management in nextgeneration all-IP-based wireless systems. IEEE Wireless Communications, 11(4), 16–28. doi:10.1109/MWC.2004.1325888

Alavi, H. S., Mojdeh, M., & Yazdani, N. (2005). A Quality of Service Architecture for IEEE 802.16 Standards. In Asia-Pacific Conference on Communications, (pp. 249 – 253).

Alenezi, A., Moses, S., & Trafalis, T. B. (2008). Real-time prediction of order flow times using support vector regression . Computers & Operations Research, 35, 3489–3503. doi:10.1016/j.cor.2007.01.026

Alexiou, A., Antonellis, D., Bouras, C., & Papazois, A. (2006, October). An Efficient Multicast Packet Delivery Scheme for UMTS. Paper presented at the 9th ACM/IEEE International Symposium on Modeling, Analysis and Simulation of Wireless and Mobile Systems (MSWiM 2006), Torremolinos, Malaga, Spain.

Alexiou, A., Bouras, C., & Kokkinos, V. (2007, October). Power Saving Techniques in MBMS Multicast Mode. Paper presented at the 3rd ACM International Workshop on Wireless Multimedia Networking and Performance Modeling (WMuNeP 2007), Chania, Greece.

Alexiou, A., Bouras, C., & Rekkas, E. (2007, June). A Power Control Scheme for Efficient Radio Bearer Selection in MBMS. Paper presented at the 8th IEEE International Symposium on World of Wireless, Mobile and Multimedia Networks (WoWMoM 2007), Helsinki, Finland.

Alonso, L., Agusti, R., & Sallent, O. (2000). A near-optimum MAC protocol based on the distributed queueing random access protocol (DQRAP) for a CDMA mobile communication system. Journal on Selected Areas in Communications, 18(9), 1701–1718. doi:10.1109/49.872957

Alonso-Zárate, J., Kartsakli, E., Skianis, C., Verikoukis, C., & Alonso, L. (2008). Saturation Throughput Analysis of a Cluster-based Medium Access Control Protocol for Ad Hoc Wireless Networks. Simulation: Transactions of the Society for Modeling and Simulation International, 84(12), 619–633. doi:10.1177/0037549708099232

Alonso-Zárate, J., Verikoukis, C., Kartsakli, E., Cateura, A., & Alonso, L. (2008). A near-optimum cross-layered distributed queuing protocol for wireless LAN [medium access control protocols for wireless LANs]. Wireless Communications, 15(1), 48–55. doi:10.1109/MWC.2008.4454704

Alpaydin, E. (2004). Introduction to machine learning. Cambridge, MA: MIT press.

Altman, E., Boulogne, T., El-Azouzi, R., Jimenez, T., & Wynter, L. (2006). A survey on networking games in telecommunications. Computers & Operations Research, 33(2), 286–311. doi:10.1016/j.cor.2004.06.005

Amirijoo, M., Tesanovic, A., Andersson, T., Hansson, J., & Son, S. H. (2006). Finite Horizon QoS Prediction of Reconfigurable Firm Real-Time Systems. Paper presented at the 12th IEEE Intl Conference on Embedded and Real-Time Computing Systems and Applications.

Andersen, D., Bansal, D., Curtis, D., Seshan, S., & Balakrishnan, H. (2000). System Support for Bandwidth Management and Content Adaptation in Internet Applications. Paper presented at the 4th Symposium on Operating Systems Design and Implementation (OSDI), San Diego, CA, USA.

Andrews, M., Kumaran, K., Ramanan, K., Stolyar, A., Vijayakumar, R., & Whiting, P. (2000). CDMA data QoS scheduling on the forward link with variable channel conditions. Tech. Rep., Bell Labs.

Anna, S., Goeckel, D. L., & Nicholas, L. J. (2006). Co-operative Communications in Mobile Ad Hoc Networks - Rethinking the link abstraction. IEEE Signal Processing Magazine, 23(5).

Ansel, P., Ni, Q., & Turletti, T. (2006). A simple and efficient scheduling scheme for IEEE 802.11e wireless LAN. Journal of Mobile Networks and Applications, 391-403.

ANSI/IEEE Std 802.11, 1999 Edition Part 11: Wireless LAN Medium Access Control and Physical Layer Specifications. (1999). Atheros AR6001X Radio On Chip. (n.d.). Retrieved from http://www.atheros.com

Antonio, G., Mario, M., & Mario, N. (2003). A scheduling algorithm for QoS support in IEEE802.11 networks. IEEE Wireless Communications, 36-43.

Arunesh, M., Shrivastava, V., Banerjee, S., & Arbaugh, W. (2006). Partially overlapped channels not considered harmful. In Proceedings of the joint international conference on Measurement and modeling of computer systems (pp. 63-74). New York: ACM Press.

Asokan, R., Natarajan, A. M., & Venkatesh, C. (2008). Quality-of-service routing using path and power aware techniques in mobile ad hoc networks. Journal of Computer Systems, Networks, and Communications, 2008(1).

Ata, B. (2006). Dynamic Control of a Multiclass Queue with Thin Arrival Streams. Operations Research, 54.

Aune, F. (2004). Cross-Layer Design Tutorial. Norwegian University of Science and Technology, Department of Electronics and Telecommunications, Trondheim, Norway, published under Creative Common License.

Awduche, D., Berger, L., Gan, D., Li, T., Srinivasan, V., & Swallow, G. (2001). RSVP-TE: Extensions to RSVP for LSP Tunnels. RFC 3209, IETF.

Awerbuch, B., Holmer, D., & Rubens, H. (2003). High throughput route selection in multi-rate ad hoc wireless networks. Tech. Rep., Johns Hopkins University, Computer Science Dept., Baltimore, MD.

Badis, H. & Al Agha, K. (2006). Quality of Service for Ad hoc Optimized Link State Routing Protocol (QOLSR). IETF Internet Draft (draft-badis-manet-qolsr-03.txt).

Baker, F., Iturralde, C., Faucheur, F. L., & Davie, B. (2001). RFC 3175: Aggregation of RSVP for IPv4 and IPv6 Reservations.

Balachandran, A., Voelker, G. M., Bahl, P., et al. (2002). Characterizing user behavior and network performance in a public wireless LAN. In proceedings of ACM SIGMETRICS Performance Evaluation Review, 195-205.

Balakrishnan, H., Lakshminarayanan, K., Ratnasamy, S., Shenker, S., Stoica, I. & Walfish, M. (2004). A Layered Naming Architecture for the Internet. SIGCOMM '04, 34, 343–352.

Banchs, A., & Perez, X. (2002). Providing Throughput Guarantees in IEEE 802.11 Wireless LAN. In [New York: IEEE Press.]. Proceedings of IEEE Wireless Conference on Networking Communication WCNC, 1, 130–138.

Banchs, A., & Vollero, L. (2006). Throughput analysis and optimal configuration of 802.11e EDCA. Computer Networks, 50(11),1749–1768. doi:10.1016/j.comnet.2005.07.008

Banchs, A., Iannello, G., Serrano, P., & Vollero, L. (2006). A CAPWAP Architecture for Dynamic Configuration of QoS MAC Parameters. In Proceedings 1st Workshop on Multimedia Applications over Wireless Networks (MediaWiN).

Banerjee, S., & Misra, A. (2002). Minimum energy paths for reliable communication in multi-hop wireless networks. In The 3rd ACM International Symposium on Mobile Ad Hoc Networking and Computing (pp. 146-156), Lausanne, Switzerland.

Bang, H. J., Ekman, T., & Gesbert, D. (2008). Channel predictive proportional fair scheduling. IEEE Transactions on Wireless Communications, 7(2), 482–487. doi:10.1109/TWC.2008.060729

Bari, F., & Leung, V. (2007). Application of ELECTRE to Network Selection in a Heterogeneous Wireless Network Environment. Wireless Communications and Networking Conference, 2007.

Bari, F., & Leung, V. C. (2007). Automated Network Selection in a Heterogeneous Wireless Network Environment. IEEE Network, 21(1), 34–40. doi:10.1109/MNET.2007.314536

Bartolini, N., & Chlamtac, I. (2001). Improving call admission control procedures by using hand-off rate information. Wireless Communications and Mobile Computing, 1(3), 257–268. doi:10.1002/wcm.17

Bauchot, F., Decrauzat, S., Marmigere, G., Merakos, L., & Passas, N. (1996). MASCARA, a MAC Protocol for Wireless ATM. In Proceedings of ACTS Mobile Summit 1996.

B-BONE. (2003). Final Results with combined enhancements of the Air Interface. Deliverable of the project (D2.5), IST-2003-507607.

Beigbeder, T., Coughlan, R., Lusher, C., Plunkett, J., & Agu, E. (2004). The Effect of Loss and Latency on User Performance in Unreal Tournament 2003®. Paper presented at the SIGCOMM'04 Workshops, Portland, Oregon, USA.

Bejerano, Y., & Han, S.-J. (2006). Cell Breathing Techniques for Balancing the Access Points Load in Wireless LANs. In Proceedings IEEE INFOCOM 2006, Twenty-fifth Annual Joint Conference of IEEE Computer and Communications Societies, (pp. 1-13). New York: IEEE Press.

Belghith, A., & Nuaymi, L. (2008). Performance evaluation for scheduling algorithms in WiMAX networks. In The 14th European Wireless conference 2008, (pp. 1-6).

Belhoul, A., Şekercioglu, Y. A., & Mani, N. (2008). Mobility protocols and RSVP performance in wireless IPv6 networks: Shortcomings and solutions. Wireless Communications and Mobile Computing, 8(9), 1183–1189. doi:10.1002/wcm.560

Bellalta, B., Cano, C., Oliver, M., & Meo, M. (2006). Modeling the IEEE 802.11e EDCA for MAC parameter optimization. In Proceedings 4th International Working Conference on performance modeling and evaluation of heterogeneous networks, Ilkley, UK.

Bellavista, P., Küpper, A., & Helal, S. (2008). Location-Based Services: Back to the Future. IEEE PERVASIVE COMPUTING, 7(2), 85–89. doi:10.1109/MPRV.2008.34

Bennis, M., & Juan, L. (2008). Non-Cooperative Operators in a Game-Theoretic Framework. IEEE 19th International Symposium on Personal Indor Mobile Radio Communications (PIMRC).

Beran, J. (1994). Statistics for long-Memory Processes. New York: Chapman and Hall.

Bermolen, P., & Rossi, D. (2008). Support vector regression for link load forecast. In Proceedings of Telecommunication Networking Workshop on QoS in Multiservice IP Networks, 268-273.

Bernaschi, M., Cacace, F., Iannello, G., Vellucci, M., & Vollero, L. (2009). OpenCAPWAP: An open source CAPWAP implementation for the management and configuration of WiFi hot-spots. Computer Networks, 53(2),217–230. doi:10.1016/j.comnet.2008.09.016

Bernaschi, M., Cacace, F., Iannello, G., Za, S., & Pescapé, A. (2005). Seamless internetworking performance of WLANs and cellular networks: architecture and performance issues in a Mobile Ipv6 scenario. IEEE Wireless Communications, 12(3),73–80. doi:10.1109/MWC.2005.1452857

Bernet, Y., Ford, P., Yavatkar, R., Baker, F., Zhang, F., Speer, M., et al. (2000). A Framework for Integrated Services Operation over Diffserv Networks. RFC 2998, IETF.

Beverly, R., Sollins, K., & Berger, A. (2006). SVM Learning of IP address structure for latency prediction. In Proceedings of SIGCOMM Workshop on Mining Network Data, 299-304.

Bhattacharya, A., Parlos, A. G., & Atiya, A. F. (2003). Prediction of MPEG-coded video source traffic using recurrent neural networks. IEEE Transactions on Signal Processing, 51(8), 2177–2190. doi:10.1109/TSP.2003.814470

Bianchi, G. (2000). Performance Analysis of the IEEE 802.11 Distributed Coordination Function. IEEE Journal on Selected Areas in Communications, 18(3), 535–547. doi:10.1109/49.840210

Bianchi, G., & Tinnirello, I. (2003). Kalman Filter Estimation of the Number of Competing Terminals in an IEEE 802.11 network. In IEEE INFOCOM 2003, Twenty-Second Annual Joint Conference of the IEEE Computer and Communications Societies: Vol. 2 (pp. 844-852). New York: IEEE Press.

Bianchi, G., Borgonovo, F., Fratta, L., Musumeci, L., & Zorzi, M. (1997). C-PRMA: A centralized packet multiple access for local wireless communications. IEEE Transactions on Vehicular Technology, 46(2), 422–436. doi:10.1109/25.580781

Bicket, J. (2005). Bit-rate Selection in Wireless Networks. Unpublished master's thesis, Massachusetts Institute of Technology, Cambridge, MA.

Blake, S., Black, D., Carlson, M., Davies, E., Wang, Z., & Weiss, W. (1998). An Architecture for Differentiated Services. RFC 2475, IETF.

Blazevic, L., Giordano, S., & Boudec, J.-Y. L. (2002). Self organized terminode routing. Cluster Computing, 5(2), 205–218. doi:10.1023/A:1013998030317

Bless, R. (2002). Dynamic Aggregation of Reservations for Internet Services. In International Conference on Telecommunication Systems – Modeling and Analysis, (ICTSM), (pp. 26–38).

Blough, D. M., Leoncini, M., Resta, G., & Santi, P. (2003). The k-Neighbors Approach to Interference Bounded and Symmetric Topology Control in Ad Hoc Networks. In ACM MobiHoc, Annapolis, MD.

Bollerslev, T. (1986). Generalized autoregressive conditional heteroscedasticity. Journal of Econometrics, 31, 307–327. doi:10.1016/0304-4076(86)90063-1

Bonald, T. (2005). Flow-level performance analysis of some opportunistic scheduling algorithms. European Transactions on Telecommunications 2005, (Vol.16, pp. 65- 75).

Boni, A., Launay, E., Mienville, T., & Stuckmann, P. (2004, October). Multimedia Broadcast Multicast Service – Technology Overview and Service Aspects. Paper presented at the 5th IEE International Conference on 3G Mobile Communication Technologies (3G 2004), London, UK.

Bose, P., Morin, P., Stojmenovic, I., & Urrutia, J. (2001). Routing with guaranteed delivery in ad hoc wireless networks. Wireless Networks, 7(6), 609–616. doi:10.1023/A:1012319418150

Bouam, S., & Othman, J. (2004). A 802.11 multiservices cross-layer approach for QoS management. In Vehicular Technology Conference, 2004. VTC2004-Fall. 2004 IEEE 60th, (Vol. 4, pp. 2698-2702).

Box, G., Jenkins, G. M., & Reinsel, G. (1994). Time series analysis: Forecasting and control, (3rd ed.). Upper Saddle River, NJ: Prentice-Hall.

Boyer, J., Falconer, D. D., & Yanikomeroglu, H. (2004). Multihop diversity in wireless relaying channels. IEEE Transactions on Communications, 52(10), 1820–1830. doi:10.1109/TCOMM.2004.836447

Braden, R. (Ed.). Zhang, L., Berson, S., Herzog, S., & Jamin, S. (1997). Resource ReSerVation Protocol (RSVP) -- Version 1 Functional Specification (RFC 2205). IETF Network Working Group.

Braden, R., Clark, D., & Shenker, S. (1994). Integrated Services in the Internet Architecture: an Overview. RFC 1633, IETF.

Brady, P. T. (1969). A model for generating on-off speech patterns in two-way conversation. The Bell System Technical Journal, 48(7), 2445–2472.

Bremler-Barr, A., Cohen, E., Kaplan, H., & Mansour, Y. (2003). Predicting and bypassing end-to-end Internet service degradations. IEEE Journal on Selected Areas in Communications, 21(6), 961–978. doi:10.1109/JSAC.2003.814614

Broens, T., Huis in't Veld, R., Vollenbroek-Hutten, M., Hermens, H., van Halteren, A., & Nieuwenhuis, B. (2007). Determinants for successful telemedicine implementations: a literature study. Journal of Telemedicine and Telecare, 13(6), 303–309. doi:10.1258/135763307781644951

Broomhead, D., & Lowe, D. (1988). Multivariable functional interpolation and adaptive Networks. Complex Systems, 2, 321–355.

Buddhikot, M., Chandranmenon, G., Han, S., Lee, Y. W., Miller, S., & Salgareli, L. (2003). Integration of 802.11 and third-generation wireless data networks. In IEEE INFOCOM 2003 (pp. 503-512). San Francisco: IEEE.

Bults, R., Pawar, P., & van Halteren, A. (2005a). Context Aware Nomadic Mobile Service - A Case Study on M-Health Signal Processing. the Netherlands: University of Twente.

Bults, R., Wac, K., van Halteren, A., Konstantas, D., & Nicola, V. (2005b). Goodput Analysis of 3G wireless networks supporting m-health services. Paper presented at the 8th International Conference on Telecommunications (ConTEL05), Zagreb, Croatia.

Bults, R., Wac, K., van Halteren, A., Konstantas, D., Jones, V., & Widya, I. (2004). Body Area Networks for Ambulant Patient Monitoring Over Next Generation Public Wireless Networks. Paper presented at the 13th IST Mobile and Wireless Communications Summit, Lyon, France.

Burkhart, M., Von Rickenbach, P., Wattenhofer, R., & Zollinger, A. (2004). Does topology control reduce interference? In ACM International Symposium on Mobile Ad Hoc Networking and Computing (pp. 9–19), Tokyo, Japan.

Buttyan, L., & Hubaux, J. P. (2003). Stimulating Co-operation in self-organizing mobile ad hoc networks. MONET, 8, 579–592.

Cacace, F., & Vollero, L. (2009). A Delay Monitoring Method for Up-link Flows in IEEE 802.11 Networks. In Proceedings of the 2nd International Conference on Simulation Tools and Techniques 2009 (pp. 1-8), ICST Brussels.

Cacace, F., Iannello, G., Vellucci, M., & Vollero, L. (2008). A Reactive Approach to QoS Provisioning in IEEE 802.11e WLANs. In Proceedings of the 2008 4th international telecommunication networking workshop on QoS in multiservice IP networks, 4th EURO-NGI Conference on Next Generation Internet Networks (pp. 253-260). New York: IEEE Press.

Cai, X., Chen, L., Sofia, R., & Wu, Y. (2007). Dynamic and User-Centric Network Selection in Heterogeneous Networks. IEEE International Conference Performance, Computing, and Communications, (pp. 538 – 544).

Cal, F., Conti, M., & Gregori, E. (1998) Ieee 802.11 wireless lan: Capacity analysis and protocol enhancement. In Fourth Annual ACM/IEEE International Conference on Mobile Computing and Networking (MobiCom '98).

Calhoun, P., Montemurro, & M., Stanley, D. (2008). CAPWAP Protocol Specification, version 15 (Internet draft). IETF Network Working Group.

Calhoun, P., Montemurro, & M., Stanley, D. (2008b). CAPWAP Protocol Binding for IEEE 802.11 version 12 (Internet draft). IETF Network Working Group.

Campbell, A.T., Gomez, J., Kim, S., & Turanyi, Z.L., Valko', A.G., & Wan, C.-Y. (2002). Internet micromobility. Journal of High Speed Networks, 11, 177–198.

Campbell, G., & Xu, W. (2001). Patent No. 6292493. USA.

Cao, L. J., & Tay, F. E. H. (2003). Support vector machine with adaptive parameters in financial time series forecasting. IEEE Transactions on Neural Networks, 14(6), 1506–1518. doi:10.1109/TNN.2003.820556

Casdagli, M., Jardins, D. D., Eubank, S., et al. (1992). Nonlinear modeling of chaotic time series: Theory and applications. In J. H. Kim & J. Stringer (Eds.), Applied Chaos (pp. 335-380). New York: John Wiley Sons, Inc.

Chakravorty, R., Banerjee, S., Agarwal, S., & Pratt, I. (2005). MoB: A Mobile Bazaar for Wide-area Wireless Services. In ACM Mobicom.

Chalise, B. K., & Czylwik, A. (2004). Robust Downlink Beamforming based upon Outage Probability Criterion. In IEEE Vehicular Technology Conference (VTC).

Chalmers, D., & Sloman, M. (1999a). QoS and Context Awareness for Mobile Computing. Paper presented at the 1st Intl. Symposium on Handheld and Ubiquitous Computing (HUC99), Karlsruhe, Germany.

Chalmers, D., & Sloman, M. (1999b). A survey of Quality of Service in mobile computing environments. IEEE Communications Surveys and Tutorials, 2(2), 2–10. doi:10.1109/COMST.1999.5340514

Chan, H., Fan, P., & Cao, Z. (2005). A Utility-Based Network Selection Scheme for Multiple Services in Heterogeneous Networks. In International Conference on Wireless Networks, Communications and Mobile Computing, 2005 (IWCMC'05) (pp. 1175-1180). Maui, Hawaii: IEEE.

Chandra, A., Gumalla, V., & Limb, J. O. (2000). Wireless Medium Access Control Protocols. IEEE Communications Surveys and Tutorials, 3(2), 2–15.

Chandra, R., Bahl, P., & Bahl, P. (2004). Multinet: Connecting to multiple IEEE 802.11 networks using a single wireless card". In Proceedings IEEE INFOCOM 2004, Twenty-third Annual Joint Conference of IEEE Computer and Communications Societies: Vol. 2 (pp. 882-893). New York: IEEE Press.

Chang, J.-H., & Tassiulas, L. (2000). Energy conserving routing in wireless ad-hoc networks. IEEE Conference on Computer Communications, (pp. 22-31), Tel-Aviv, Israel

Chang, J-H. & Tassiulas, L. (2004). Maximum lifetime routing in wireless sensor networks. IEEE/ACM Transactions on Networking, 12, 609-619.

Chang, M., Lee, M., & Lee, H. (2005). An Efficient Resource Reservation Scheme Based on Gray-Cell in Wireless Mobile Networks. In IEEE Wireless Communications and Networking Conference (WCNC), 3, 1311–1316.

Chang; P. R., & Hu, J, T, (1997). Optimal nonlinear adaptive prediction and modeling of MPEG video in ATM networks using pipelined recurrent neural networks, Selected Areas in Communications, 15(6), 1087-1100.

Charilas, D., Markaki, O., & Tragos, E. (2008). A Theoretical Scheme for applying game theory and network selection mechanisms in access admission control. In International Symposium on Wireless Pervasive Computing, Santorini.

Charilas, D., Markaki, O., Nikitopoulos, D., & Theologou, M. (2008). Packet-Switched Network Selection with the Highest QoS in 4G Networks. Computer Networks, 52(1), 248–258. doi:10.1016/j.comnet.2007.09.005

Chaskar, H. (Ed.). (2003). RFC 3583: Requirements of a Quality of Service (QoS) Solution for Mobile IP.

Chatzigiannakis, I., Dimitriou, T., Nikoletseas, S., & Spirakis, P. (2006). A probabilistic algorithm for efficient and robust data propagation in smart dust networks. Ad-Hoc Networks Journal, 4(5).

Chaudet, C., & Guérin Lassous, I. (2001). BRuIT: Bandwidth Reservation under Interferences Influence. (Research Rep. No. 4308), INRIA.

Chen, B., Jamieson, K., Balakrishnan, H., & Morris, R. (2001). Span: An energy-efficient coordination algorithm for topology maintenance in Ad Hoc wireless networks. In Mobile Computing and Networking (pp. 85-96).

Chen, D., & Varshney, P. (2007a). A survey of void handling techniques for geographic routing in wireless networks. In Communications Surveys and Tutorials, (pp. 50–67). Washington, DC: IEEE.

Chen, D., & Varshney, P. K. (2007b). On-demand geographic forwarding for data delivery in wireless sensor networks. Computer Communications, 30(14-15), 2954–2967. doi:10.1016/j.comcom.2007.05.022

Chen, D., Deng, J., & Varshney, P. (2005). On the forwarding area of contention-based geographic forwarding for ad hoc and sensor networks. Sensor and Ad Hoc Communications and Networks, (pp. 130 – 141).

Chen, I. R., Yilmaz, O., & Yen, I. L. (2006). Admission Control Algorithms for Revenue Optimization With QoS Guarantees in Mobile Wireless Networks. Wireless Personal Communications: An International Journal, 38(3), 357–376. doi:10.1007/s11277-006-9037-6

Chen, J., Jiao, W., & Guo, Q. (2005). An integrated QoS control architecture for IEEE 802.16 broadband wireless access systems. Global Telecommunications Conference, (Globecom) 2005: (Vol. 6, pp. 3330-3335).

Chen, J., Jiao, W., & Wang, H. (2005). A Service Flow Management Strategy for IEEE 802.16d Broadband Wireless Access Systems in TDD Mode. Proceedings of IEEE International Conference on Communications, ICC 2005, (Vol. 5, pp.3422 – 3426).

Chen, K.-C. (1994). Lee, & C.-Hua. (1994). Group randomly addressed polling for wireless data networks. Proceedings of the IEEE, ICC, 1713–1717.

Chen, S. G. F., & Cui, J. (2006). Avoid void in geographic routing for data aggregation in sensor networks. International Journal of Ad Hoc and Ubiquitous Computing (IJAHUC), Special Issue on Wireless Sensor Networks.

Chen, S., & Nahrstedt, K. (1998). On finding multi-constrained paths. IEEE International Conference on Communications, (pp. 874-879), Atlanta, USA. Cisco Aironet 1240AG Series 802.11A/B/G Access Point. (n.d). Retrieved from www.cisco.com.

Chen, W.-T., & Huang, L.-C. (2000). RSVP Mobility Support: A Signalling Protocol for Integrated Services Internet with Mobile Hosts. In IEEE/ACM INFOCOM (pp. 1283–1292), Tel Aviv, Israel.

Chen, X., Zhai, H., Tian, X., & Fang, Y. (2006). Supporting QoS in IEEE 802.11e Wireless LANs. IEEE Transaction on Wireless Communications, 2217-2227.

Cheng, Y., Jiang, H., Zhuang, W., Niu, Z., & Lin, C. (2005). Efficient Resource Allocation for China's 3G/4G Wireless Networks. IEEE Communications Magazine, 43(1), 76–83. doi:10.1109/MCOM.2005.1381878

Chenxi, Z., & Corson, M. S. (2002). QoS routing for mobile ad hoc networks. In Proceedings of Twenty-First Annual Joint Conference of the IEEE Computer and Communications Societies, IEEE INFOCOM 2002 (pp. 958-967). New York: IEEE Inc.

Chevillon, G., & Hendry, D. (2005). Non-parametric direct multi-step estimation for forecasting economic processes. International Journal of Forecasting, 21, 201–218. doi:10.1016/j.ijforecast.2004.08.004

Chisci, L., Pecorella, T., & Fantacci, R. (2006). Dynamic bandwidth allocation in GEO satellite networks: A predictive control approach. Control Engineering Practice, 14, 1057–1067. doi:10.1016/j.conengprac.2005.06.010

Chlebus, E., & Ludwin, W. (1995). Is Handoff Traffic Really Poissonian? Fourth IEEE International Conference on Universal Personal Communications (pp. 348-353). Tokyo: IEEE.

Choi, D.I., & Lee, Y. (2001). Performance Analysis of a Dynamic Priority Queue for Traffic Control of Bursty Traffic in ATM Networks. IEEE Communications Proceedings, 148.

Choi, H., Igm, Y., An, S., & Nam, C. (1999). Dynamic Queue Management Mechanism for Service Enhancement in Wireless Intelligent Network Environment. In IEEE Global Communications Conference (Globecom).

Chou, C. T., & Shin, K. S. (2004). Analysis of Adaptive Bandwidth Allocation in Wireless Networks with Multilevel Degradable Quality of Service. IEEE Transactions on Mobile Computing, 3(1), 5–17. doi:10.1109/TMC.2004.1261813

Chuah, P., Hu, T., & Luo, W. (2004, June). UMTS Release 99/4 Airlink Enhancement for supporting MBMS Services. Paper presented at the 2004 IEEE International Conference on Communications (ICC 2004), Paris, France.

Chui, K. (1992). An introduction to wavelets. New York: Academic Press.

Claffy, K., Miller, G., & Thompson, K. (1998). The nature of the beast: recent traffic measurements from an Internet backbone. Paper presented at the International Networking Conference (INET98), Geneva, Switzerland.

Clark, B. N., Colbourn, C. J., & Johnson, D. S. (1990). Unit disk graphs. Discrete Mathematics, 86(1-3), 165–177. doi:10.1016/0012-365X(90)90358-O

Clementi, A. E. F., Penna, P., & Silvestri, R. (1999). Hardness Results for the Power Range Assignment Problem in Packet Radio Networks. In G. Goos, J. Hartmanis, & J. V. Leeuwen, (Eds.), Randomization, Approximation, and Combinatorial Optimization. Algorithms and Techniques. Berlin: Springer.

Clements, M. P., & Hendry, D. F. (1996). Multi-step estimation for forecasting. Oxford Bulletin of Economics and Statistics, 58(4), 657–684.

Cocchi, R., Shenker, S., Estrin, D., & Zhang, L. (1993). Pricing in computer networks: motivation, formulation, and example. IEEE/ACM Transactions on Networking, 1 (6), 614 - 627.

Cooper, R. B. (1981). Introduction to Queueing Theory. New York: Elsevier North Holland Inc.

Corradi, M. Garroppo, R.G., Giordano, S. & Pagano, M. (2001). Analysis of f-ARIMA processes in the modeling of broadband traffic. Proceedings of IEEE International Conference on Communications, 3, 964-968.

Couto, D. D., & Morris, R. (2001). Location proxies and intermediate node forwarding for practical geographic forwarding. Technical report, MIT Laboratory for Computer Science.

Cover, T. M., & El Gamal, A. A. (1979). Capacity theorems for the relay channel. IEEE Transactions on Information Theory, 25, 572–584. doi:10.1109/TIT.1979.1056084

CRAWDAD. (n.d.). Retrieved from http://crawdad.cs.dartmouth.edu

Crisostomo, S., Sargento, S., Natkaniec, M., & Vicari, N. (2005). A QoS architecture integrating mobile Ad-Hoc and infrastructure networks. In Proceedings of the ACS/IEEE 2005 international Conference on Computer Systems and Applications (pp. 897-903). Washington, DC: IEEE Computer Society.

Cristianini, N., & Shawe-Taylor, J. (2000). An Introduction to Support Vector Machines and Other Kernel-Based Learning Methods. Cambridge, UK: Cambridge Univ. Press.

Crutchfield, J. P. (1993). The calculi of emergence: Computation, dynamics, and induction. Oji International Seminar Complex Systems — from Complex Dynamics to Artificial Reality. Numazu, Japan.

Crutchfield, J. P. (1994). Is anything ever new? Considering emergence. Santa Fe Institute.

Cuevas, A., Moreno, J. I., Vidales, P., & Einsiedler, H. (2006). The IMS Platform: A Solution for Next Generation Network Operators to Be More Than Bit Pipes. IEEE Commun. Mag. Advances in Service Platform Technologies, 44(8), 75–81.

Dangerfield, I., Malone, D., & Leith, D. (2006). Experimental Evaluation of 802.11e EDCA for Enhanced Voice over WLAN Performance. In Proceedings of the 4th International Symposium on Modeling and Optimization in Mobile, Ad Hoc and Wireless Networks, (pp. 1-7).

Das, S., & Sen, S., Basu, k., & Lin, H. (2003). A Framework for bandwidth degradation and call admission control schemes for multiclass traffic in next-generation wireless networks. IEEE Journal on Selected Areas in Communications, 21(10), 1790–1802. doi:10.1109/JSAC.2003.815928

DaSilva Luiz, A. Srivastava Vivek (2004). Node Participation in Ad Hoc and Peer-to-Peer Networks: A Game-Theoretic Formulation. In Workshop on Games and Emergent Behavior in Distributed Computing Environments, Birmingham, UK.

Datta, S., Stojmenovic, I., & Wu, J. (2002). Internal node and shortcut based routing with guaranteed delivery in wireless networks. Cluster Computing, 5(2), 169–178. doi:10.1023/A:1013985610753

Daubechies, I. (1992). Ten lectures on wavelets. SIAM, Philadelphia, PA.

David, G., Jose F., M., Daniel, C., & Javier, G. (2007). Policy-based channel access mechanism selection for QoS provision in IEEE 802.11e. IEEE Vehicular Technology Magazine, 29-34.

De Vriendt, J., Laine, P., Lerouge, C., & Xu, X. (2002). Mobile Network Evolution: A Revolution on the Move. IEEE Communications Magazine, 40(4), 104–111. doi:10.1109/35.995858

De Wolf, T., & Holvoet, T. (2005). Emergence Versus Self-Organization: Different Concepts but Promising When Combined. In S. G. D. M. S. Brueckner, A. Karageorgos & R. Nagpal, (Eds.), Engineering Self Organizing Systems: Methodologies and Applications

De, S., Das, S. K., Wu, H., & Qiao, C. (2002). A resource efficient RT-QoS routing protocol for mobile ad hoc networks. In The 5th International Symposium on Wireless Personal Multimedia Communications, (pp. 257-261).

Dekleva, S., Shim, J. P., Varshney, U., & Knoerzer, G. (2007). Evolution and emerging issues in mobile wireless networks. Communications of the ACM, 50(6), 38–43. doi:10.1145/1247001.1247003

Deng, D. J., & Chang, R. S. (1999). A priority scheme for IEEE 802.11 DCF access method. IEICE Transactions on Communications. E (Norwalk, Conn.), 82-B(1), 96–102.

Deng, D. J., & Yen, H. C. (2005). Quality-of-Service Provision System for Multimedia Transmission in IEEE 802.11 Wireless LANs. IEEE Journal on Selected Areas in Communications, 23(6), 1240–1252. doi:10.1109/JSAC.2005.845632

DeVellis, R. F. (2003). Scale development: theory and applications. London: SAGE.

Deyun, G., Jianfei, C., & Chen, C. W. (2008). Admission Control Based on Rate-Variance Envelop for VBR Traffic Over IEEE 802.11e HCCA WLANs. IEEE Transactions on Vehicular Technology, 1778–1788.

Deyun, G., Jianfei, C., & King Ngi, N. (2005). Admission control in IEEE 802.11e wireless LANs. IEEE Network, 19(4), 6–13. doi:10.1109/MNET.2005.1470677

Dharmaraju, D., & Roy-Chowdhury, A. (2002). INORA - A Unified Signaling and Routing Mechanism for QoS Support in Mobile Ad hoc Networks. In Proceedings of the 2002 International Conference on Parallel Processing Workshops (pp. 86-93). Washington, DC: IEEE Computer Society.

Dierks, T., & Rescorla, E. (2006). The Transport Layer Security (TLS) Protocol Version 1.1 (RFC 4346), IETF Network Working Group.

Digital Chocolate. (2008). Seize the Minute. Retrieved Sep 05, 2008, from www.digitalchocolate.com

Dimitriadis, G., & Pavlidou, F. N. (2004). Comparative performance evaluation of EDCF and EY-NPMA protocols. IEEE Communications Letters, 8(1), 42–44. doi:10.1109/LCOMM.2003.822501

D-Link AirPlusTM DWL-G630 Wireless Cardbus Adapter Manual. (n.d.). Retrieved from http://www.dlink.com

Dohler, M., Meddour, D. E., Senouci, S.-M., & Saadani, A. (2008). Cooperation in 4G - Hype or Ripe? IEEE Technology and Society Magazine, 27(1), 13–17. doi:10.1109/MTS.2008.918035

Dokovsky, N., van Halteren, A., & Widya, I. (2004). BANip: Enabling Remote Healthcare Monitoring with Body Area Networks. Paper presented at the Intl Workshop on Scientific Engineering of Distributed Java Applications (FIJI03), Luxembourg.

Domingo, M. C., & Remondo, D. (2004). A cooperation model between ad hoc networks and fixed networks for service differentiation. In 29th Annual IEEE International Conference on Local Computer Networks, (pp. 692-693). New York: IEEE Inc.

Dood, A. (2005). The Essential Guide to Telecommunications. Upper Saddle River, NJ: Prentice Hall PTR.

Doulamis, A. D., Doulamis, N. D., & Kollias, S. D. (2003). An adaptable neural-network model for recursive nonlinear traffic prediction and modeling of MPEG video sources. IEEE Transactions on Neural Networks, 14(1), 150–166. doi:10.1109/TNN.2002.806645

Dyson, D. A., & Haas, Z. J. (1999). A dynamic packet reservation multiple access scheme for wireless ATM. ACM/Baltzer Journal of Mobile Networks & Applications, 4, 87-99.

El Housseini, S., & Alnuweiri, H. (2005). Adaptive Contention-Window MAC Algorithms for QoS-Enabled Wireless LANs. In Proceedings of 2005 International Conference on Wireless Networks, Communications and Mobile Computing: Vol.1 (pp. 368-374).

ElBatt, T. A., Krishnamurthy, S. V., Connors, D., & Dao, S. (2000). Power Management for Throughput Enhancement in Wireless Ad-Hoc Networks. In IEEE International Conference on Communications (pp. 1506 - 1513), New Orleans, LA.

Elman, J. L. (1990). Finding structure in time. Cognitive Science, 14(2), 179–211.

Engelstad, P. E., & Østerbø, O. N. (2005). Non-Saturation and Saturation Analysis of IEEE 802.11e EDCA with Starvation Prediction. In Proceedings of the 8th ACM Symposium on Modeling, Analysis and Simulation of Wireless and Mobile Systems MSWIM 05 (pp. 224-233). New York: ACM Press.

Engle, R. F. (1982). Autoregressive conditional heteroscedasticity with estimates of the variance of united kingdom inflation. Econometrica, 50, 987–1008. doi:10.2307/1912773

Ephremides, A. (2002). Energy concerns in wireless networks. In IEEE Wireless Communications, (pp. 48 – 59)

Ether2. (2005). Ether 2 - Distributing the Queue. Retrieved from www.ether2.com

ETSI. (1998), EN 300 652 V1.2.1, Broadband Radio Access Network (BRAN); HIgh PErformance Radio Local Area Network (HIPERLAN) Type 1. Functional Specification.

Evci, C., & Fino, B. (2001). Spectrum management, pricing, and efficiency control in broadbandwireless communications. Proceedings of the IEEE, 89(1), 105–115. doi:10.1109/5.904509

Fan, Y., Huang, C., & Tseng, Y. (2006). Multimedia Services in IEEE 802.11e WLAN Systems. In Proceedings of the ACM International Wireless Communications and Mobile Computing Conference IWCMC 06 (pp. 401-405). New York: ACM Press.

Fang, Q., Gao, J., & Guibas, L. J. (2006). Locating and bypassing holes in sensor networks. Mobile Networks and Applications, 11(2), 187–200. doi:10.1007/s11036-006-4471-y

Fang, Y., & Zhang, Y. (2002). Call admission control schemes and performance analysis in wirelessmobile networks. IEEE Transactions on Vehicular Technology, 51(2), 371–382. doi:10.1109/25.994812

Fang, Y., Chlamtac, I., & Lin, Y. B. (1998). Channel Occupancy Times and Handoff Rate for Mobile Computing and PCS Networks. IEEE Transactions on Computers, 47(6), 679–692. doi:10.1109/12.689647

Feldmann, A., Gilbert, A. C., & Willinger, W. (1998). Data networks as cascades: investigating the multifractal nature of Internet WAN traffic. Computer Communication Review, 28(4), 42–55. doi:10.1145/285243.285256

Felegyhazi, M., Hubaux, J.-P. (2006). Game Theory in Wireless Networks: A Tutorial. EPFL technical report, LCA-REPORT-2006-002.

Feng Zhao, L. G. (July 2004). Wireless Sensor Networks: An Information Processing Approach. San Francisco: Morgan Kaufmann.

Feng, H. F., & Shu, Y. T. Wang, S. Y. & Ma, M. D. (2006). SVM-Based models for predicting WLAN traffic. In Proceedings of IEEE International Conference on Communications, (pp. 597-602).

Ferrari, G., Tonguz, O. K., & Bhatt, M. (2004). Impact of receiver sensitivity on the performance of sensor networks. In IEEE Vehicular Tech. Conf. VTC Spring, Milan, Italy.

Finn, G. G. (1987). Routing and addressing problems in large metropolitan-scale internetworks. Technical report, University of Southern California, Marina del Rey, Information Sciences Institut.

Floroiu, J. W., Ruppelt, R., Sisalem, D., & Stephanopoli, J. V. (2003). Seamless handover in terrestrial radio access networks: a case study. IEEE Communications Magazine, 41(11), 110–116. doi:10.1109/MCOM.2003.1244930

Floyd, S. (1996). Comments on measurement-based admission control for controlled-load services. Berkeley: Lawrence Berkeley Laboratory.

Frattasi, S., & Can, B. Fitzek. F., & Prasad, R. (2005). Cooperative Services for 4G. In Proceedings of the 14th IST Mobile & Wireless Communications.

Frattasi, S., Fathi, H., & Fitzek, F.H.P. (2006). 4G: A User-Centric System. Wireless Personal Communications Journal (WPC) - Special Issue on Advances in Wireless Communications: Enabling Technologies for 4G.

Frattasi, S., Fathi, H., Fitzek, F. H. P., Prasad, R., & Katz, M. (2006). Defining 4G Technology from the User Perspective. IEEE Network Magazine, 20(1), 35–41. doi:10.1109/MNET.2006.1580917

Freitag, J., da Fonseca, N. L. S., & de Rezende, J. F. (2006). Tuning of 802.11e network parameters. IEEE Communications Letters, 10(8).doi:10.1109/LCOMM.2006.1665127

Fu, H., & Kim, D. I. (2006). Analysis of Throughput and Fairness with Downlink Scheduling in WCDMA Networks. IEEE Transactions on Wireless Communications, 5.

Fudenberg, D., & Tirole, J. (1991). Game Theory. MIT Press.

Gabriel, K., & Sokal, R. R. (1969). A new statistical approach to geographic variation analysis.

Gao, D., Cai, J., & Ngan, K. N. (2005). Admission Control in IEEE 802.11e Wireless LANs. IEEE Network, 19(4),6–13. doi:10.1109/MNET.2005.1470677

Gao, L., Wang, X., Xu, Y. & Li, Z. (2008). A Game Approach for Cell Selection and Resource Allocation in Heterogeneous Wireless Networks.

Gao, X., Wu, G., & Miki, T. (2004). End-to-end QoS provisioning in mobile heterogeneous networks. IEEE Wireless Communications, 11(3), 24–34. doi:10.1109/MWC.2004.1308940

Gao, Z., & Wu, G. (2005). Combining QoS-based service selection with performance prediction. Paper presented at the IEEE Intl Conf on e-Business Engineering (ICEBE05).

Garcia-Macias, J. A., Rousseau, F., Berger-Sabbatel, G., Toumi, L., & Duda, A. (2003). Quality of Service and Mobility for the Wireless Internet. Wireless Networks, 9, 341–352. doi:10.1023/A:1023647311052

Garey, M., & Johnson, D. (1979). Computers and Intractability: A Guide to the Theory of NP-Completeness. New York: W.H. Freeman and Company.

Garroppo, R. G., Giordano, S., Lucetti, S., & Tavanti, L. (2007). Providing air-time usage fairness in IEEE 802.11 networks with the deficit transmission time (dtt) scheduler. Wireless Networks, 13(4), 481–495. doi:10.1007/s11276-006-9201-7

Ge, Y., & Kuo, G. (2006). An Efficient Admission Control Scheme for Adaptive Multimedia Services in IEEE 802.16e Networks. IEEE 64th Vehicular Technology Conference, VTC-2006, (pp.1–5).

Ge, Y., Hou, J. C., & Choi, S. (2007). An analytic study of tuning systems parameters in IEEE 802.11e enhanced distributed channel access. Elsevier Computer Networks, 51(8), 1955–1980. doi:10.1016/j.comnet.2006.07.018

Gerasimov, I., & Simon, R. (2002). Performance analysis for ad hoc QoS routing protocols. In International Mobility and Wireless Access Workshop, MobiWac'02, (pp. 87-94).

Ghaderi, M., Capka, J., & Boutaba, R. (2003). Prediction-based admission control for DiffServ wireless internet. Proceedings of IEEE Vehicular Technology Conference, 3, 1974–1978.

Giambene, G., & Kota, S. (2006). Cross-layer protocol optimization for satellite communication networks: A survey. International Journal of Satellite Communications and Networking, 24, 323–341. doi:10.1002/sat.853

Głowacz, A., Natkaniec, M., Kosek, K., & Szott, S. (2008). Guarantying QoS in MANETs using MAC Layer Measurements. In 17th ICT Mobile & Wireless Communications Summit. Dublin: IIMC International Information Management Corporation.

Gomez, G., & Sanchez, R. (2005). End-to-End Quality of Service over Cellular Networks: Data Services Performance Optimization in 2G/3G: John Wiley & Sons, Ltd.

Goodman, D., & Mandayam, M. (1999). Power control for wireless data. IEEE International Workshop on Mobile Multimedia Communications, (pp. 55–63).

Govindan, Y.-J. K. R., Karp, B., & Shenker, S. (2006). Lazy cross-link removal for geographic routing. In SenSys '06: Proceedings of the 4th international conference on Embedded networked sensor systems, (pp. 112–124). New York: ACM.

Gowrishankar & Satyanarayana, P.S. (2008). Neural network based traffic prediction for wireless data networks. International Journal of Computational Intelligence Systems, 1(4), 379–389. doi:10.2991/ijcis.2008.1.4.9

Gozdecki, J., Pacyna, P., Marques, V., Aguiar, R. L., Garcia, C., Moreno, J. I., et al. (2003). An IP QoS Architecture for 4G Networks. In International workshop on architectures for quality of service in the Internet 2003 (Art-QoS 2003). 2698, (pp. 18-28). Warsaw, POLOGNE: Springer.

Granger, C. W. J., Joyeux. R. (1980). An introduction to long memory time series models and fractional differencing. Journal of Time Series Analysis, 1, 15–39. doi:10.1111/j.1467-9892.1980.tb00297.x

Gu, D., & Zhang, J. (2003). A New Measurement-Based Admission Control Method for IEEE 802.11 Wireless Local Area Networks. In Proceedings IEEE International Symposium on Personal, Indoor, and Mobile Radio Communications: Vol. 3 (pp. 2009-2013). New York: IEEE Press.

Gundavelli, S. (Ed.). Devarapalli, V., Chowdhury, K., & Patil, B. (2008). RFC 5213: Proxy Mobile IPv6.

Gunturi, S., & Paganini, F. (2003). Game theoretic approach to power control in cellular CDMA. Vehicular Technology Conference (VTC).

Gupta, N., & Das, S. (2002). Energy-aware on-demand routing for mobile ad hoc networks. In 4th International Workshop on Distributed Computing (pp. 164-173), Calcutta, India.

Gupta, V. (2004). Media independent handover service. Draft technical requirements 21- 04-0087-08-0000, IEEE 802.21.

Gutierrez, F., Varvarigos, E., & Vassiliadis, S. (2000). Multicost routing in max-min fair share networks. In 38th Annual. Allerton Conference on Communication, Control and Computing, (pp. 1294—1304).

Haffner, P., Tur, G., & Wright, J. H. (2003). Optimizing SVMs for complex call classification. Proceedings of International Acoustics, Speech, and Signal Processing, 1, 632–635.

Hahne, E. L., & Choudhury, A. K. (2002). Dynamic Queue Length Thresholds for Multiple Loss Priorities. IEEE/ACM Transactions on Networking, 10.

Haken, H. (2000). Information and Self-Organization: A Macroscopic Approach to Complex Systems. Berlin: Springer Verlag.

Hamidian, A., & Korner, U. (2006). An Enhancement to the IEEE 802.11e EDCA Providing QoS Guarantees. Springer Telecommunication Systems, 31(2-3), 195–212. doi:10.1007/s11235-006-6520-z

Han, J., & Kamber, M. (2006). Data mining: concepts and techniques. San Francisco: Morgan Kaufmann.

Han, Q., & Venkatasubramanian, N. (2006). Information Collection Services for QoS-aware Mobile Applications. IEEE Transactions on Mobile Computing, 5(5), 518–535. doi:10.1109/TMC.2006.65

Hancock, R., Karagiannis, G., Loughney, J., & Van den Bosch, S. (2005). Next Steps in Signaling (NSIS): Framework. RFC 4080.

Hansmann, U., Merk, L., Nicklous, M., & Stober, T. (2003). Pervasive Computing: The Mobile World. Berlin: Springer.

Harine, G., Marie, R., Puigjaner, R., & Trivedi, K. (2001). Loss formulas and their application to optimization for cellular networks. IEEE Transactions on Vehicular Technology, 50(3), 664–673. doi:10.1109/25.933303

Harsha, S., Kumar, A., & Sharma, V. (2006). An Analytical Model for the Capacity Estimation of Combined VoIP and TCP File Transfers over EDCA in an IEEE 802.11e WLAN. In Proceedings 14th IEEE International Workshop on Quality of Service (pp. 178-187). New York: IEEE Press.

Hasna, M., & Alouini, M.&M. (2004). A performance study of dual-hop transmissions with fixed gain relays. IEEE Transactions on Wireless Communications, 3(6), 1963–1968. doi:10.1109/TWC.2004.837470

Haung, Y. R., & Ho, J. M. (2002). Distributed call admission control for a heterogeneous PCS network. IEEE Transactions on Computers, 51(12), 1400–1409. doi:10.1109/TC.2002.1146706

Haykin, S. (2001). Feedforward: Neural Networks: An Introduction. Retrieved from media.wiley.com/product_data/excerpt/19/04713491/0471349119.pdf

He, T., Stankovic, J., Lu, C., & Abdelzaher, T. (2003). Speed: A stateless protocol for real-time communication in sensor networks. In International Conference on Distributed Computing Systems (ICDCS).

He, Y., & Abdel Wahab, H. (2006). HQMM: A Hybrid QoS Model for Mobile Ad-hoc Networks. 11th IEEE Symposium on Computers and Communications, ISCC '06, (pp. 194-200). New York: IEEE Inc.

Hegge, S. (2007). Predicting end-to-end network characteristics on mobile devices. Unpublished MSc, University of Twente, Enschede, the Netherlands.

Heinanen, J., Baker, F., Weiss, W., & Wroclawski, J. (1999). Assured Forwarding PHB Group. RFC 2597, IETF.

Heissenbuttel, M., Braun, T., Bernoulli, T., & Walchli, M. (2003). BLR: Beacon-less routing algorithm for mobile ad-hoc networks.

Hightower, J., & Borriello, G. (2001). Location systems for ubiquitous computing. Computer, 34(8), 57–66. doi:10.1109/2.940014

Holland, G., Vaidya, N. H., & Bahl, P. (2001). A rate-adaptive MAC protocol for multi-hop wireless networks. International Conference on Mobile Computing and Networking (pp. 236-251). New York: ACM.

Holland, G., Vaidya, N., & Bahl, P. (2001). A rate-adaptive MAC protocol for multi-hop wireless networks. In Proceedings of the 7th annual international conference on Mobile computing and networking (MobiCom '01) (pp. 236–251). New York: ACM.

Holland, J. (1998). Emergence: from Chaos to Order. Reading, MA: Addison-Wesley.

Holma, H., & Toskala, A. (2006). HSDPA/HSUPA for UMTS: High Speed Radio Access for Mobile Communications. Chichester, UK: John Wiley & Sons.

Holma, H., & Toskala, A. (2007). WCDMA for UMTS: HSPA Evolution and LTE (4th edition). Chichester, UK: John Wiley & Sons.

Hong, D., & Rappaport, S. S. (1986). Traffic Model and Performance Analysis for Cellular Mobile Radio Telephone Systems with Prioritized and Nonprioritized Handoff Procedures. IEEE Transactions on Vehicular Technology, 35(3), 77–92. doi:10.1109/T-VT.1986.24076

Hosking, J. R. M. (1981). Fractional differencing. Biometrika, 68(1), 165–176. doi:10.1093/biomet/68.1.165

Hossain, E., & Leung, K. (Eds.). (2007). Wireless Mesh Networks: Architectures and Protocols. Berlin: Springer.

Hou, J., Yang, J., & Papavassiliou, S. (2002). Intergration of Pricing with Call Admission Control to Meet QoS Requirements in Cellular Networks. IEEE Transactions on Parallel and Distributed Systems, 13(9), 898–910. doi:10.1109/TPDS.2002.1036064

Hou, J., Yang, J., & Papavassiliou, S. (2002). Intergration of Pricing with Call Admission Control to Meet QoS Requirements in Cellular Networks. IEEE Transactions on Parallel and Distributed Systems, 13(9), 898–910. doi:10.1109/TPDS.2002.1036064

http://mosquitonet.stanford.edu/software.html. (n.d.).

http://ramp.ucsd.edu/pawn/sigcomm-trace/. (n.d.).

Hu, M., Zhang, J., & Sadowsky, J. (2004). Traffic aided opportunistic scheduling for wireless networks: algorithms and performance bounds. Computer Networks: The International Journal of Computer and Telecommunications Networking, 46(4), 505–518.

Hua, Z., Ming, L., Imrich, C., & Prabhakaran, B. (2004). A survey of quality of service in IEEE 802.11 networks. IEEE Wireless Communications, 6-14.

Huang, C., Lai, W., Luo, R., & Yan, Y. (2005). Application of support vector machines to bandwidth reservation in sectored cellular communications. Engineering Applications of Artificial Intelligence, 18, 585–594. doi:10.1016/j.engappai.2004.11.009

Huang, J., & Niu, Z. (2007). A Cross-Layer Proportional Fair Scheduling Algorithm with Packet Length Constraint in Multiuser OFDM Networks. In . Proceedings of IEEE International Conference on Global Telecommunications Conference, GLOBECOM, 07, 3489–3493. doi:10.1109/GLOCOM.2007.662

Huang, N.-F., & Chen, W.-N. (2003). RSVP Extensions for Real-Time Services in Hierarchical Mobile IPv6. [MONET]. Mobile Networking and Applications, 8, 625–634. doi:10.1023/A:1026022325876

Huang, X., & Fang, Y. (2006). Multi-constrained soft-QoS provisioning in wireless sensor networks. In 3rd international conference on Quality of service in heterogeneous wired/wireless networks, (pp. 465-478), Canada. IEEE 802.11b Wireless LAN PC Card User's Guide Version 5.0. (n.d.). Retrieved from http://www.ansel.com.mx

Huiling, J., Zhang, Z., Cheng, P., Chen, H.-H., & Li, S. (2006). Study on Network Selection for Next-Generation Heterogeneous Wireless Networks. In IEEE 17th International Symposium on Personal, Indoor and Mobile Radio Communications, (pp.1 – 5.)

Hutchison, D., Mauthe, A., & Yeadon, N. (1997). Quality-of-service architecture: Monitoring and control of multimedia communications. Electronics & Communication Engineering Journal, 9(3), 100. doi:10.1049/ecej:19970303

Hwang, Y. H., & Noh, S. K. (2005). A call admission control scheme for heterogeneous service considering fairness in wireless networks. Fourth Annual ACIS International Conference on Computer and Information Science 2005 (ICIS 2005) (pp. 688- 692). Jeju Island, South Korea: IEEE Computer Society.

I.S. 802.11 (2007). IEEE Standard for Information Technology - Telecommunications and Information Exchange between Systems - Local and Metropolitan Area Networks - Specific Requirements Part 11: Wireless LAN Medium Access Control (MAC) and Physical Layer (PHY) Specifications, IEEE802.11-REVma/D9.0, Revision of Std. 802.11-1999, August 2007.

Iannello, G., Palmieri, F., Pescapè, A., & Salvo Rossi, P. (2005). End-to-End Packet-Channel Bayesian Model applied to Heterogeneous Wireless Networks. Paper presented at the IEEE Globecom 2005, St. Louis (MO, USA).

Ibrahim, W., Chinneck, J. W., & Periyalwar, S. (2003). A QoS-based charging and resource allocation framework for next generation wireless networks. Wireless Communications and Mobile Computing, 3(7), 895–906. doi:10.1002/wcm.179

IEEE 802.11e WG. (2005). IEEE Standard for Information Technology--Telecommunications and Information Exchange Between Systems--LAN/MAN Specific Requirements--Part 11 Wireless Medium Access Control and Physical Layer specifications, Amendment 8: Medium Access Control Quality of Service Enhancements.

IEEE 802.11 WG. (1999). International Standard [for] Information Technology – Telecommunications and Information Exchange between systems-Local and metropolitan area networks-Specific Requirements - Part 11:Wireless LAN Medium Access Control (MAC) and Physical Layer (PHY) specifications, Reference number ISO/IEC 8802-11:1999(E).

IEEE 802.11g WG. (2003). International Standard for Information Technology – Telecommunications and Information Exchange between systems-Local and metropolitan area networks-Specific Requirements - Part 11:Wireless LAN Medium Access Control (MAC) and Physical Layer (PHY) specifications, Amendment 4: Further Higher Data Rate Extension in the 2.4GHz Band.

IEEE 802.11n/D11.0. (2009). Unapproved Draft Standard for Information Technology-Telecommunications and information exchange between systems-Local and metropolitan area networks-Specific requirements-Part 11: Wireless LAN Medium Access Control (MAC) and Physical Layer (PHY) specifications Amendment: Enhancements for Higher Throughput.

IEEE Draft Std. 802.11e/D8.0. (2004). Medium Access Control (MAC) Quality of Service (QoS) Enhancements.

IEEE Standard for Local and Metropolitan Area Networks Part 16: Air Interface for Fixed Broadband Wireless Access Systems. (2004). IEEE Std 802.16-2004 (Revision of IEEE Std 802.16-2001),(pp.1-857).

IEEE Std 802.11n. (2008). IEEE Draft Standard for Wireless LAN Medium Access Control (MAC) and Physical Layer (PHY) Specifications - Amendment 5: Enhancements for Higher Throughput. IEEE Unapproved Draft Std. P803.22n/D5.0.

IEEE Std. 802.11. (2007). IEEE Standard for Wireless LAN Medium Access Control (MAC) and Physical Layer (PHY) Specifications. IEEE Std 802.11-2007 (Revision of IEEE Std 802.11-1999).

IEEE Std. 802.11e. (2005). IEEE Standard for Wireless LAN Medium Access Control (MAC) and Physical Layer (PHY) Specifications - Amendment 8: Medium Access Control (MAC) Quality of Service Enhancements. IEEE Std. 802.11e.

IEEE Std. 802.11g. (2003). IEEE Standard for Wireless LAN Medium Access Control (MAC) and Physical Layer (PHY) Specifications - Amendment 4: Further Higher Data Rate Extension in the 2.4 GHz Band. IEEE Std. 802.11g.

IEEE. (1999). IEEE Std 802.11b-Wireless LAN Medium Access Control (MAC) and Physical layer (PHY) Specifications: High Speed Physical Layer (PHY) in the 2.4 GHZz Band, 1999.

IEEE. 802.11 Standard for Wireless LAN: Medium Access Control (MAC) and Physical Layer (PHY) Specification. (2007). New York: IEEE Inc.

Ismail, R., & Josang, A. (2002). The beta reputation system. In Proceedings of the 15th Bled Conference on Electronic Commerce.

Issariyakul, T., & Hossain, E. (2006). Channel-Quality-Based Opportunistic Scheduling with ARQ in Multi-Rate Wireless Networks: Modeling and Analysis. IEEE Transactions on Wireless Communications, 5.

Istook, E., & Martinez, T. (2002). Improved back-propagation learning in neural networks with windowed momentum. International Journal of Neural Systems, 12(3-4), 303–318. doi:10.1142/S0129065702001114

ITU-D. S. G. (2008, May). Teletraffic Engineering Handbok. Retrieved April 18, 2009, from http://oldwww.com.dtu.dk/teletraffic/handbook.html

ITU-T Recommendation G.711. (1998). Pulse code modulation (PCM) of voice frequencies.

ITU-T. (2006). Framework for achieving end-to-end IP performance objectives (Vol. []: ITU.]. Y (Dayton, Ohio), •••, 1542.

ITU-T. (2007a). Mean Opinion Score (MOS) terminology (Vol. P.800): ITU.

ITU-T.(2007b). Vocabulary for performance and quality of service: Appendix I – Definition of Quality of Experience (QoE) (Vol. P.10/G.100): ITU-T.

ITU-T. (2007c). Working document: Quality of Experience Requirements for IPTV (Vol. FG IPTV-36): ITU.

ITU-T. (2008). Definitions of terms related to quality of service (Vol. []: ITU.]. E (Norwalk, Conn.), •••, 800.

ITU-T.(1993). General aspects of Quality of Service and Network Performance in Digital Networks, including ISDNs (Vol. I.350): ITU.

Jacobson, V., Nichols, K., & Poduri, K. (1999). An Expedited Forwarding PHB. RFC 2598, IETF.

Jaffe, J. (1984). Algorithms for finding paths with multiple constraints. Networks Journal, 14, 95–116. doi:10.1002/net.3230140109

Jähnert, J., Zhou, J., Aguiar, R. L., Marques, V., Wetterwald, M., & Melin, E. (2005). The 'pure-IP' Moby Dick 4G architecture. Computer Communications, 28(9), 1014–1027. doi:10.1016/j.comcom.2004.11.013

Jain, K., Padhye, J., Padmanabhan, V. N., & Qiu, L. (2003). Impact of interference on multi-hop wireless network performance. MobiCom.

Jain, R. (2004). Quality of experience. IEEE MultiMedia, 11(1), 96–95. doi:10.1109/MMUL.2004.1261114

Jain, R. K., Chiu, D. M., & Hawe, W. R. (1984). A quantitative measure of fairness and discrimination for resource allocation in shared computer systems. Maynard, Massachusetts: Digital Equipment Corporation.

Jain, R., & Knightly, E. W. (1999). A framework for design and evaluation of admission controlalgorithms in multi-service mobile networks. INFOCOM '99 (pp. 1027-1035). New York: IEEE.

Jain, R., Puri, A., & Sengupta, R. (1999). Geographical routing using partial information for wireless ad hoc networks.

Jalali, A., Padovani, R., & Pankaj, R. (2000). Data throughput of CDMA HDR: a high efficiency-high data rate personal communication wireless system. In . Proceedings of IEEE VTC, 2000, 1854–1858.

Jamin, S., Danzig, P. B., Shenker, S. J., & Zhang, L. (1997). A Measurement-based Admission Control Algorithm for Integrated Services Packet Networks. IEEE/ACM Transactions on Networking, 5 (1), 56-70.

Jedrzycki, C., & Leung, V. C. (1996). Probability distribution of channel holding time in cellulartelephony systems. IEEE 46th Vehicular Technology Conference 1996 (VTC '96). 1, pp. 247 - 251. Atlanta, Georgia: IEEE.

Jiang, C., & Tsai, T. (2006). Token bucket based CAC and packet scheduling for IEEE 802.16 broadband wireless access networks. Consumer Communications and Networking Conference, CCNC 2006. (Vol.1, pp.183 – 187).

Jiang, M., Nikolic, M., Hardy, S., & Trajkovic, L. (2001). Impact of self-similarity on wireless data Network performance. In Proceedings of IEEE International Conference on Communications, (pp. 477-481).

Johnson, D., Perkins, C., & Arkko, J. (2004). Mobility Support in IPv6 (RFC 3775), IETF Network Working Group.

Jung, E.-S., & Vaidya, N. H. (2005). A Power Control MAC Protocol for Ad Hoc Networks. In The 11th Annual International Conference on Mobile Computing and Networking (pp. 55-66), Cologne, Germany.

Kaloxylos, A., Vali, D., Paskalis, S., Panagiotou, G., Gonianakis, G., & Zervas, E. (2006). Mobility Support for a QoS Aggregation Protocol. In International Symposium on Communication Systems, Networks and Digital System Processing (CSNDSP), Patra, Greece.

Kamal, G., Mounir, A., & Annie, G. (2006). Dynamic Resource Reservation in IEEE 802.16 Broadband Wireless Networks. In IWQoS 2006, (pp.140 – 148).

Kamerman, A., & Monteban, L. (1997). WaveLAN-II: A high-performance wireless LAN for the unlicensed band. Bell Labs Technical Journal, 2(3), 118–133. doi:10.1002/bltj.2069

Karagiorgas, N., Kokkinos, P., Papageorgiou, C., & Varvarigos, E. A. (2007). Routing in Wireless Ad-hoc Networks with Variable Transmission Power. IEEE 18th International Symposium on Personal, Indoor and Mobile Radio Communications (pp. 1-6). Athens, Greece

Karetsos, G. T., Kyriazakos, S. A., Groustiotis, E., Giandomenico, F. D., & Mura, I. (2005). A Hierarchical Radio Resource Management Framework for Integrating WLANs in Cellular Networking Environments. IEEE Wireless Communications, 12(6), 11–17. doi:10.1109/MWC.2005.1561940

Karol, M. J., Liu, Z., & Eng, K. Y. (1995). An Efficient Demand-Assignment Multiple Access Protocol for Wireless Packet (ATM) Networks. ACM/Baltzer Journal of Wireless Networks, 1(3), 267-279.

Karp, B., & Kung, H. T. (2000a). Gpsr: greedy perimeter stateless routing for wireless networks. In MobiCom '00: Proceedings of the 6th annual international conference on Mobile computing and networking, (pp. 243–254). New York: ACM Press.

Karp, B., & Kung, H. T. (2000b). GPSR: greedy perimeter stateless routing for wireless networks. In Mobile Computing and Networking, (pp. 243–254).

Kartsakli, E., Alonso-Zárate, J., Verikoukis, C., & Alonso, L. (2009). Cross-Layer Scheduling with QoS Support over a Distributed Queuing MAC for Wireless LANs. Mobile Networks and Applications (MONET) Special Issue on Recent Advances in IEEE 802.11 WLANs: Protocols, Solutions and Future Directions.

Kartsakli, E., Cateura, A., Alonso-Zárate, J., Verikoukis, C., & Alonso, L. (2007a). Cross-layer enhancement for wlan systems with heterogeneous traffic based on DQCA. In IEEE International Conference onCommunications, 2007. ICC '07., (pp. 5708–5713).

Kartsakli, E., Cateura, A., Alonso-Zárate, J., Verikoukis, C., & Alonso, L. (2007b). Opportunistic Scheduling using an Enhanced Channel State Information Update Scheme for WLAN Systems with DQCA. In Proc. of the IEEE Vehicular Technology Conference Spring (VTC Spring 2007). Dublin, Ireland.

Katz, M., & Fitzek, F. H. P. (2005). Cooperative Techniques and Principles Enabling Future 4G Wireless Networks. In The International Conference on EUROCON, (pp. 21-24). Amir, K., Jinane, A., Eytan, M., & Lizhong, Z. (2003). Cooperative Routing in Wireless Networks. Allerton Conference on Communications, Control and Computing.

Kawadia, V., & Kumar, P. R. (2003) Power Control and Clustering in Ad Hoc Networks. IEEE INFOCOM 2003, San Francisco, CA.

Keerthi, S., Shevade, S. K., Bhattacharyya, C., & Murthy, K. (2001). Improvements to Platt's SMO algorithm for SVM classifier design. Neural Computation, 13(3), 637–649. doi:10.1162/089976601300014493

Kempf, J. (Ed.). (2002). RFC 3374: Problem Description: Reasons For Performing Context Transfers Between Nodes in an IP Access Network.

Khan, L., Awad, M., & Thuraisingham, B. (2007). A new intrusion detection system using support vector machines and hierarchical clustering . The Very Large Data Bases Journal, 16(4), 507–521. doi:10.1007/s00778-006-0002-5

Kher, S., Somani, A. K., & Gupta, R. (2005). Network selection using fuzzy logic. In 2nd International Conference on Broadband Networks. Feng, K. & Hong-yan, L. (2007). A New Fuzzy MADM Algorithm Based on Subjective and Objective Integrated Weights. In International Conference on Service Systems and Service Management.

Khotanzad, A., & Sadek, N. (2003). Multi-scale high-speed network traffic prediction using combination of neural network. Proceedings of the International Joint Conference on Neural Networks, 2, 1071–1075.

Kim, I. M., Shin, B. C., & Lee, D. J. (2000). SIR-based call admission control by intercell interference prediction for DS-CDMA systems. IEEE Communications Letters, 4(1), 29–31. doi:10.1109/4234.823540

Kim, J. G., & Widjaja, I. (1996). PRMA/DA: A New Media Access Control Protocol for Wireless ATM. In Proceedings of ICC 1996, (pp. 1-19).

Kim, K. I., & Kim, S. H. (2005). A Light Call Admission Control with Inter-Cell and Inter-Service Fairness in Heterogeneous Packet Radio Networks. IEICE Transactions, 88-B(10), 4064–4073. doi:10.1093/ietcom/e88-b.10.4064

Kim, M., & Noble, B. (2001). Mobile network estimation. In Proceedings of the Seventh Annual International Conference on Mobile Computing and Networking, (pp. 298-309).

Kim, Y.-J., Govindan, R., Karp, B., & Shenker, S. (2005a). Geographic routing made practical. In NSDI'05: Proceedings of the 2nd conference on Symposium on Networked Systems Design & Implementation, (pp. 217–230). Berkeley, CA: USENIX Association.

Kim, Y.-J., Govindan, R., Karp, B., & Shenker, S. (2005b). On the pitfalls of geographic face routing. In DIALM-POMC '05: Proceedings of the 2005 joint workshop on Foundations of mobile computing, (pp. 34–43). New York: ACM.

King, A. (2008). Website Optimization: Speed, Search Engine & Conversion Rate Secrets: O'Reilly Media, Inc.

Kirn, S., Herzog, O., Lockermann, P., & Spaniol, O. (Eds.). (2006). Multiagent Engineering. Heidelberg: Springer.

Kirousis, L. M., Kranakis, E., Krizanc, D., & Pelc, A.(n.d.). Power consumption in packet radio networks. Theoretical Computer Science, 243, 41–64.

Knopp, R., & Humblet, P. (1995). Information Capacity and Power Control in Single Cell Multiuser Communications. IEEE International Conference on Communications (ICC).

Ko, B.-J., & Rubenstein, D. (2003). A Distributed, Self-stabilizing Protocol for Placement of Replicated Resources in Emerging Networks. In 11th IEEE International Conference on Network Protocols (ICNP), Atlanta, GA.

Koga, H., Kashihara, S., Fukuda, Y., Lida, K., & Oie, Y. (2006). A quality-aware VoWLAN architecture and its quantitative evaluations. IEEE Wireless Communications, 52-59.

Kokkinos, P., Papageorgiou, C., & Varvarigos, E. (2005). Energy aware routing in wireless ad hoc networks. IEEE 6th International Symposium on a World of Wireless, Mobile and Multimedia Networks, (pp. 306- 311), Naxos, Greece.

Kong, Z., Tsang, D. H. K., Bensaou, B., & Gao, D. (2004). Performance analysis of IEEE 802.11e contention-based channel access. IEEE Journal on Selected Areas in Communications, 22(10),2095–2106. doi:10.1109/JSAC.2004.836019

Konrad, A., Zhao, B. Y., Joseph, A. D., & Ludwig, R. (2001). A markov-based channel model algorithm for wireless networks. In Proceedings of the 4th ACM international workshop on Modeling, analysis and simulation of wireless and mobile systems (MSWIM '01) (pp. 28–36). New York: ACM.

Koodli, R. (2005). Fast Handovers for Mobile IPv6, RFC 4068, IETF.

Koodli, R. (2008). Mobile IPv6 Fast Handovers, RFC 5268, IETF.

Korkmaz, T., & Krunz, M. (2001). Multi-Constrained Optimal Path Selection. In 12th Annual Joint Conference of the IEEE Computer and Communications Societies, (pp. 834 - 843), Anchorage, AK.

Kountouris, M., & Gesbert, D. (2005). Robust Multi-user Opportunistic Beamforming for Sparse Networks. In IEEE Signal Processing Advances for Wireless Communications (SPAWC) workshop.

Kranakis, E., Singh, H., & Urrutia, J. (1999). Compass routing on geometric networks. In Proc. 11 th Canadian Conference on Computational Geometry, (pp. 51–54), Vancouver.

Krithikaivasan, B. Dekal K., & Medhi, D. (2004). Adaptive bandwidth provisioning envelope based on discrete temporal network measurements. In Proceedings of IEEE Computer and Communications Societies, (pp. 1786-1796).

Krunz, M., & Makowski, A. (1998). A source model for VBR video traffic based on M/G/ input processes. In Proceedings of IEEE Computer and Communications Societies, (pp. 1441-1449).

Kuhn, F., & Zollinger, A. (2003). Ad-hoc networks beyond unit disk graphs. In DIALM-POMC '03: Proceedings of the 2003 joint workshop on Foundations of mobile computing, (pp. 69–78). New York: ACM.

Kuhn, F., Moscibroda, T., & Wattenhofer, R. (2004). Unit disk graph approximation. In DIALM-POMC '04: Proceedings of the 2004 joint workshop on Foundations of mobile computing, (pp. 17–23). New York: ACM.

Kuhn, F., Wattenhofer, R., & Zollinger, A. (2002). Asymptotically optimal geometric mobile ad-hoc routing. In DIALM '02: Proceedings of the 6th international workshop on Discrete algorithms and methods for mobile computing and communications, (pp. 24–33). New York: ACM.

Kuhn, F., Wattenhofer, R., Zhang, Y., & Zollinger, A. (2003a). Geometric ad-hoc routing: Of theory and practice.

Kuhn, F., Wattenhofer, R., Zhang, Y., & Zollinger, A. (2003b). Geometric adhoc routing: of theory and practice. In PODC '03: Proceedings of the twenty second annual symposium on Principles of distributed computing, (pp. 63–72). New York: ACM Press.

Kuhn, F., Wattenhofer, R., & Zollinger, A. (2003c). Worst-Case Optimal and Average-Case Efficient Geometric Ad-Hoc Routing. In Proc. 4 th ACM Int. Symposium on Mobile Ad-Hoc Networking and Computing (MobiHoc).

Kulavaratharasah, M. D., & Aghvami, A. H. (1999). Teletraffic Performance Evaluation of Microcellular Personal Communication Networks (PCN's) with Prioritized Handoff Procedures. IEEE Trasactions on Vehicular Technology, 48 (1), 137-152.

Kuo, G.-S., & Ko, P.-C. (2000). Dynamic RSVP for Mobile IPv6 in Wireless Networks. In IEEE Vehicular Technology Conference (VTC) Spring, Tokyo, Japan, 1, 455–459.

Kwon, E., Jung, K., Lee, J., & Ryu, S. (2005). A Performance Model for Admission Control in IEEE 802.16. Proceedings of WWIC 2005, (Vol. 3510, pp.159–168).

Kwon, T., Choi, Y., & Naghshineh, M. (1998). Optimal Distributed Call Admission Control for Multimedia Services in Mobile Cellular Network. International Workshop on Mobile Multimedia Communication 1998 (MoMuC '98) (pp. 477-482). Berlin: IEEE.

Kwon, T., Choi, Y., Bisdikian, C., & Naghshineh, M. (1999). Measurement-based Call Admission Control for adaptive multimedia in wireless/mobile networks. IEEE Wireless Communications and Networking Conference 1999, (WCNC'99). 2, pp. 540-544. New Orleans: IEEE.

Kwon, T., Kim, S., Choi, Y., & Naghshineh, M. (2000). Threshold-type call admission control in wireless/mobile multimedia networks using prioritised adaptive framework. IEEE Electronics Letters, 36(9), 852–854. doi:10.1049/el:20000639

Kwon, T., Lee, H., Choi, S., Kim, J., & Cho, D. (2005). Design and Implementation of a Simulator Based on a Cross-Layer Protocol between MAC and PHY layers in a WiBro Compatible IEEE802.16e OFDMA System. IEEE Communications Magazine, 43(12), 136–146. doi:10.1109/MCOM.2005.1561931

Kyasanur, P., & Vaidya, N. H. (2005). Routing and Interface Assignment in multi-channel multi-interface Wireless networks. In Proc. WCNC, 4, (pp. 2051-2056), New Orleans, LA.

Lacage, M., Manshaei, M. H., & Turletti, T. (2004). IEEE 802.11 Rate Adaptation: A Practical Approach (Research Rep. No. 5208), INRIA.

Lagkas, T. D., Papadimitriou, G. I., & Pomportsis, A. S. (2006a). QAP: A QoS supportive Adaptive Polling Protocol for Wireless LANs. Elsevier Computer Communications, 29(5), 618–633.

Lagkas, T. D., Papadimitriou, G. I., & Pomportsis, A. S. (2006b). SQAP: A Simple QoS supportive Adaptive Polling Protocol for Wireless LANs. Elsevier Computer Communications, 29(8), 934–937.

Lagkas, T. D., Papadimitriou, G. I., Nicopolitidis, P., & Pomportsis, A. S. (2007a). Priority-Oriented Adaptive Control With QoS Guarantee for Wireless LANs. IEEE Transactions on Vehicular Technology, 56(4), 1761–1772. doi:10.1109/TVT.2007.897223

Lagkas, T. D., Papadimitriou, G. I., Nicopolitidis, P., & Pomportsis, A. S. (2007b). A New Hybrid Medium Access Mechanism with Integrated QoS Support for Wireless LANs. In Proceedings of IEEE/CVT '07, Delft, the Netherlands.

Lagkas, T. D., Papadimitriou, G. I., Nicopolitidis, P., & Pomportsis, A. S. (2008). A Novel Method of Serving Multimedia and Background Traffic in Wireless LANs. IEEE Transactions on Vehicular Technology, 57(5), 3263–3267. doi:10.1109/TVT.2007.912341

Lai, F. S., Misic, J., & Chanson, S. T. (1998). Complete sharing versus partitioning: Quality of service managment for wireless multimedia networks. 7th International Conference on Computer Communications and Networks 1998 (pp. 584 - 593). Lafayette, Louisiana, USA: IEEE Computer Society.

LaMarca, A., Chawathe, Y., Consolva, S., Hightower, J., Smith, I., Scott, J., et al. (2005). Place Lab: Device Positioning Using Radio Beacons in the Wild. Paper presented at the Pervasive Computing 2005, Munchen, Germany.

Laneman, J. N., Tse, D. N. C., & Wornell, G. W. (2004). Cooperative Diversity in Wireless Networks: Efficient Protocols and Outage Behavior. IEEE Transactions on Information Theory, 50(12), 3062–3080. doi:10.1109/TIT.2004.838089

Layuan, L., & Chunlin, L. (2007). A QoS multicast routing protocol for clustering mobile ad hoc networks. Computer Communications, 30(7), 1641–1654. doi:10.1016/j.comcom.2007.01.017

Le, L., & Ekram, H. (2008). Cross-Layer Optimization Frameworks for Multihop Wireless Networks Using Cooperative Diversity. IEEE Transactions on Wireless Communications, 7(7). doi:10.1109/TWC.2008.060962

Lee, K., Kim, M., Chanson, S., Yu, C., & Lee, J. (2003). CORP – A Method of Concatenation and Optimization for Resource Reservation in Mobile Internet. IEICE Transactions on Communications, Special Issue on Internet Technology . E (Norwalk, Conn.), 86-B(2), 479–489.

Lee, S., Ahn, G. S., Zhang, X., & Campbell, A. T. (2000). INSIGNIA: An IP-Based Quality of Service Framework for Mobile Ad Hoc Networks. Journal of Parallel and Distributed Computing, Special issue on Wireless and Mobile Computing and Communications, 60, 374-406.

Lee, Y. W., Kim, S., Park, J., & Kim, S. H. (2007). A Lightweight Implementation of RSVP-TE protocol for MPLS-TE Signaling. Computer Communications, 30(6), 1199–1204. doi:10.1016/j.comcom.2006.12.003

Lei, C., & Heinzelman, W. B. (2005). QoS-aware routing based on bandwidth estimation for mobile ad hoc networks. IEEE Journal on Selected Areas in Communications, 23(3), 561–572. doi:10.1109/JSAC.2004.842560

Leong, B., Liskov, B., & Morris, R. (2006). Geographic routing without planarization. In NSDI'06: Proceedings of the 3rd conference on 3rd Symposium on Networked Systems Design & Implementation, (pp. 25–25). Berkeley, CA: USENIX Association.

Leong, B., Mitra, S., & Liskov, B. (2005). Path vector face routing: Geographic routing with local face information. In ICNP '05: Proceedings of the 13TH IEEE International Conference on Network Protocols (ICNP'05), (pp. 147–158). Washington, DC: IEEE Computer Society.

Lera, A., Molinaro, A., & Pizzi, S. (2007). Channel-Aware Scheduling for QoS and Fairness Provisioning in IEEE 802.16/WiMAX Broadband Wireless Access Systems. IEEE Network, 21(5), 34–41. doi:10.1109/MNET.2007.4305171

Leung, B. J., & Kim, K. K. (2003). Frequency Assignment for IEEE 802.11 Wireless Networks. Presented at 58th IEEE Vehicular Technology Conference (VTC 2003 - Fall), (pp. 1422-1426), Orlando, FL.

Levanti, A. Giordano, F.; & Tinnirello (2007), I., A CAPWAP-Compliant Solution for Radio Resource Management in Large-Scale 802.11 WLAN. In Proceedings of IEEE Global Telecommunications Conference, 2007. (pp.3645-3650), 26-30 Nov. 2007.

Levine, D. A., Akyldiz, I. F., & Naghshineh, M. (1997). A Resource Estimation and Call Admission Algorithm for Wireless Multimedia Networks Using the Shadow Cluster Concept. IEEE/ACM Transactions on Networking, 5 (1), 1-12.

Li, C. H., Zhu, X. J., Cao, G. Y., Sheng, S., & Hu, M. R. (2008). Identification of the Hammerstein model of a PEMFC stack based on least squares support vector machines. Journal of Power Sources, (1): 303–316. doi:10.1016/j.jpowsour.2007.09.049

Li, W., & Chao, X. (2007). Call admission control for an adaptive heterogeneous multimedia mobile network. IEEE Transactions on Wireless Communications, 6(2), 515–524. doi:10.1109/TWC.2006.05192

Li, Z., & Garcia-Luna-Aceves, J. (2007). Finding multi-constrained feasible paths by using depth-first search. Wireless Networks, 13(3), 323–334. doi:10.1007/s11276-006-7528-8

Liang, J., & Du, R. (2007). Model-based fault detection and diagnosis of HVAC systems using support vector machine method. International Journal of Refrigeration, 30(6), 1104–1114. doi:10.1016/j.ijrefrig.2006.12.012

Liang, Q. (2002). Ad Hoc wireless network traffic—self-similarity and forecasting. IEEE Communications Letters, 6(7), 297–299. doi:10.1109/LCOMM.2002.801327

Lim, W.-S., Kim, D.-K., Suh, Y. J., & Won, J.-J. (2009). Implementation and Performance study of IEEE 802.21 in Intergrated IEEE 802.11/802.16e Networks. Computer Communications, 32(1), 134–143. doi:10.1016/j.comcom.2008.09.034

Lin, C. R., & Gerla, M. (1997). Asynchronous multimedia multihop wireless network. In Sixteenth Annual Joint Conference of the IEEE Computer and Communications Societies, INFOCOM '97 (pp. 118-125). New York: IEEE Inc.

Lin, C. R., & Liu, J. S. (1999). QoS routing in ad hoc wireless networks. Journal on Selected Areas in Communications, 17(8), 1426–1438. doi:10.1109/49.779924

Lin, H., Chaterjee, M., Das, S. K., & Basu, K. (2005). ARC: An Integrated Admission and Rate Control Framework for Competitive Wireless CDMA Data Networks Using Noncooperative Games. IEEE Transactions on Mobile Computing, 4(3), 243–258. doi:10.1109/TMC.2005.35

Lin, J., Chou, C., & Liu, C. (2008). Comparison of WiMAX scheduling algorithms and proposals for the rtPS QoS class. In The 22nd international Conference on advanced Information Networking and Applications-Workshops, (pp. 68-74).

Lin, K. J. (2007). Building Web 2.0. IEEE COMPUTER, 40(5), 101–102.

Lin, X., Lakshdisi, M., & Stojmenovic, I. (2001). Location based localized alternate, disjoint, multi-path and component routing schemes for wireless networks. In MobiHoc '01: Proceedings of the 2nd ACM international symposium on Mobile ad hoc networking & computing, (pp. 287–290). New York: ACM.

Lin, X., Shroff, N. B., & Srikant, R. (2006). A Tutorial on Cross-Layer Optimization in Wireless Networsk. IEEE Journal on Selected Areas in Communications, 24(8).

Lindemann, C., Lohmann, M., & Thümmler, A. (2004). Adaptive call admission control for QoS/revenue optimization in CDMA cellular networks. [WINET]. ACM Journal on Wireless Networks, 10(4), 457–472. doi:10.1023/B:WINE.0000028549.11958.89

Liu, H., Yu, H., Liu, X., Chuah, C., & Mohapatra, P. (2007). Scheduling Multiple Partially Overlapped Channels in Wireless Mesh Networks. In Proceedings IEEE International Conference on Communications, ICC '07, (pp. 3817-3822). New York: IEEE Press.

Liu, L., & Feng, G. (2007). Simulated Annealing Based Multi-constrained QoS Routing in Mobile ad hoc Networks. International Journal of Wireless Personal Communications, 41(3), 393–405. doi:10.1007/s11277-006-9149-z

Liu, Q., Wang, X., & Giannakis, G. B. (2006). A cross-layer scheduling algorithm with QoS support in wireless networks. IEEE Transactions on Vehicular Technology, 55(3), 839–847. doi:10.1109/TVT.2006.873832

Liu, X. (2003). Optimal Opportunistic Scheduling in Wireless Networks. In IEEE 58th Vehicular Technology Conference, (Vol.3, pp. 1417 -1421).

Liu, Z., & Zarki, M. E. (1994). SIR-based call admission control for DS-CDMA cellular systems. IEEE Journal on Selected Areas in Communications, 12(4), 638–644. doi:10.1109/49.286670

Ma, M. Lu, J.C., Sanjay, K., & Ng, B.C. (2007). A Three-Tier Framework and scheduling to Support QoS Service in WiMAX. In IEEE 6th International Conference on Information, Communications & Signal Processing, (pp.1-5).

Ma, M., & Ng, B. C. (2006). Supporting Differentiated Services in Wireless Access Networks. Proceedings of IEEE International Conference on Communication Systems, ICCS, 2006, 1–5.

MacKenzie, A. B., & DaSilva, L. A. (2006). Game Theory for Wireless Communications. San Rafael, CA: Morgan and Claypool Publishers.

MacKenzie, A. B., & Wicker, S. B. (2001a). Game Theory and the Design of Self-Configuring, Adaptive Wireless Networks. IEEE Communications Magazine, 39(11), 126–131. doi:10.1109/35.965370

MacKenzie, A. B., & Wicker, S. B. (2001b). Game Theory in Communications: Motivation, Explanation and Application to Power Control. IEEE Global Telecommunications Conference, 2, 821 – 826.

Mai, Y., Yang, C., & Lin, Y. (2007). Cross-Layer QoS Framework in the IEEE 802.16 Network. The 9th International Conference on Advanced Communication Technology, (Vol. 3, pp. 2090 – 2095).

Mallat, S. (1989). A theory for multiresolution signal decomposition: the wavelet representation. IEEE Transactions on Pattern Analysis and Machine Intelligence, 11, 674–693. doi:10.1109/34.192463

Malli, M., et al. (2004). Adaptive Fair Channel Allocation for QoS Enhancement in IEEE 802.11 Wireless LANs. In Proceedings 2004 IEEE International Conference on Communications (pp. 3470-3475). New York: IEEE Press.

Manner, ed., J., Karagiannis, G., & McDonald, A. (2009). NSLP for Quality of Service signalling. Internet Draft, work in progress. draft-ietf-nsis-qos-nslp-10.txt.

Manner, J., & Raatikainen, K. (2003). Localized QoS Management for Multimedia Applications in Wireless Access Networks. In IASTED Internet and Multimedia Systems and Applications (IMSA) (pp. 193–200), Honolulu, HI.

Manner, J., Toledo, A., Mihailovic, A., Munoz, H., Hepworth, E., & Khouaja, Y. (2002). Evaluation of mobility and quality of service interaction. Computer Networks, 38(2), 137–163. doi:10.1016/S1389-1286(01)00255-9

Manzoni, P., Ghosal, D., & Serazzi, G. (1995). Impact on mobility on TCP/IP: An integrated performance study. IEEE Journal on Selected Areas in Communications, 13(5), 858–867. doi:10.1109/49.391748

Marina, M. K., & Das, S. R. (2005). A Topology Control Approach for Utilizing Multiple Channels in Multi-Radio Wireless Mesh Networks. In IEEE Broadnets 2005, Boston.

Markaki, O., Charilas, D., & Nikitopoulos, D. (2007). Enhancing Quality of Experience in Next Generation Networks through Network Selection Mechanisms. Mobile Terminal Assisted Enhanced Services Provisioning in a B3G Environment Workshop, PIMRC 2007.

Mase, K. (2004). Toward scalable admission control for VoIP networks. IEEE Communications Magazine, 42(7), 42–47. doi:10.1109/MCOM.2004.1316529

Mathar, R., & Mattfeldt, J. (1996). Optimal transmission ranges for mobile communication in linear multihop packet radio networks. Wireless Networks, 2(4), 329–342. doi:10.1007/BF01262051

Matsumoto, C. (2004). Online Games Pose Carrier Conundrum. Light Reading - Networking the Telecom Industry.

McNair, J., & Zhu, F. (2004). Vertical Handoffs in Fourth-Generation Multinetwork Environments. IEEE Wireless Communications, 11(3), 8–15. doi:10.1109/MWC.2004.1308935

Meyer, Y. (1992). Wavelets and Operators. Cambridge, UK: Cambridge University Press.

Michaut, F., & Lepage, F. (2005). Application-oriented network metrology: Metrics and active measurement tools. IEEE Communications Surveys & Tutorials, 7(2), 2–24. doi:10.1109/COMST.2005.1610543

Michiardi, P., & Molva, R. (2002). Core: A cooperative reputation mechanism to enforce node cooperation in mobile ad hoc networks. Communications and Multimedia Security Conference.

Mieghem, P. & Kuipers, F. (2004). Concepts of Exact QoS Routing Algorithms. IEEE/ACM Transactions on Networking, 12(5), 851-864.

Mihaela, V. D. S., Yiannis, A., & Zhiping, H. (2006). Optimized scalable video streaming over IEEE 802.11 a/e HCCA wireless networks under delay constraints. IEEE Transactions on Mobile Computing, 755–768.

Mirchandani, V., Prodan, A., & Debenham, J. (Aug 2007) A Method and Study of Topology Control based Self-Organization in Mesh Networks. In Proc. IEEE AccessNets (BWIA Workshop), Ottawa, Canada.

Mirchandani, V., Prodan, A., & Debenham, J. (Oct 2007). Impact of Topology Control on the Performance of a Self-Organization Scheme for Wireless Mesh Networks. In Proc. ACM WICON, Austin, TX.

Mirza, M., Sommers, J., Bardford, P., & Zhu, X. (2007). A machine learning approach to TCP throughput prediction. In Proceedings of the 2007 ACM SIGMETRICS international conference on Measurement and modeling of computer systems, (pp. 97-108).

Mitchell, T. (1999). Machine Learning and Data Mining. Communications of the ACM, 42(11), 30–36. doi:10.1145/319382.319388

MobiHealth. (2007). Putting care in motion. Retrieved Sep 05, 2008, from www.mobihealth.com

MOBILEMAN. (2005). Mobile Metropolitan Ad hoc Networks, IST-2001-38113, Oct 2005.

Mody, A. N., Blatt, S. R., Mills, D. G., McElwain, T. P., Thammakhoune, N. B., & Niedzwiecki, J. D. (2007). Recent advances in cognitive communications. IEEE Communications Magazine, 45(10), 54–61. doi:10.1109/MCOM.2007.4342823

Mohammad, M. R., Ekram, H., & Vijay, K. B. (2008). Controlled Channel Access Scheduling for Guaranteed QoS in 802.11e-Based WLANs. IEEE Transactions on Wireless Communications, 1287–1297.

Monego, H. I., Bodanese, E. L., Nacamura, L., & Souza, R. D. (2005). A dynamic Resource Allocation Scheme for Providing QoS in Packet-Switched Cellular Networks. Second International Workshop Mobility Aware Technologies and Applications 2005 (MATA'05) (pp. 117-126). Montreal, Canada: Springer.

Moon, B., & Aghvami, H. (2001). RSVP Extensions for Real-Time Services in Wireless Mobile Networks. IEEE Communications Magazine, 39(12), 52–59. doi:10.1109/35.968812

Moon, B., & Aghvami, H. (2003). DiffServ extensions for QoS provisioning in IP Mobiltiy environments. IEEE Wireless Communications, 10(5), 38–44. doi:10.1109/MWC.2003.1241096

Moon, B., & Aghvami, H. (2004). Quality of Service Mechanisms in all-IP Wireless Access Networks. IEEE Journal on Selected Areas in Communications, 22(5), 873–888. doi:10.1109/JSAC.2004.826924

Moraru, L., Leone, P., Nikoletseas, S., & Rolim, J. (2008). Path Quality Detection Algorithms for Near Optimal Geographic Routing in Sensor Networks with Obstacles. Wireless Communications and Mobile Computing Journal, 1–13.

Moskowitz, R., & Nikander, P. (2006). RFC 4423: Host Identity Protocol (HIP) Architecture. Next Steps In Signaling IETF Working Group. (n.d.). Retrieved from http://www.ietf.org/html.charters/nsis-charter.html

Muqattash, A., & Krunz, M. (2003, April). A Distributed Transmission Power Control Protocol for Mobile Ad Hoc Networks. The 22nd Annual Joint Conference of the IEEE Computer and Communications Societies, (pp. 113 - 128), San Fransisco.

Mussabir, Q., Yao, W., Niu, Z., & Fu, X. (2007). Optimized FMIPv6 Using IEEE 802.21 MIH Services in Vehicular Networks. IEEE Transactions on Vehicular Technology, 56(6), 3397–3407. doi:10.1109/TVT.2007.906987

Mykoniati, E., Charalampous, C., Georgatsos, P., Damilatis, T., Goderis, D., & Trimintzios, G. (2003). Admission Control for Providing QoS in IP DiffServ Networks: the TEQUILA Approach. IEEE Communications, special issue on QoS Advances: the European IST Premium IP Projects, 41(1), 38-44.

Nadeem, T., & Agrawala, A. (2004). Ieee 802.11 fragmentation-aware energy efficient ad-hoc routing protocols. In 2004 IEEE International Conference on Mobile Ad-hoc and Sensor Systems, (pp. 90–103).

Naghshineh, M., & Schwartz, M. (1966). Distributed Call Admission Control in Mobile/Wireless Networks. IEEE Journal on Selected Areas in Communications, 14(4), 711–717. doi:10.1109/49.490422

Nahrstedt, K., Xu, D., Wichadakul, D., & Li, B. (2001). QoS-aware middleware for ubiquitous and heterogeneous environments. IEEE Communications Magazine, 39(11), 140–148. doi:10.1109/35.965372

Narayanaswami, S., Kawadia, V., Sreenivas, R. S., & Kumar, P. R. (2002). Power Control in Ad-Hoc Networks: Theory, Architecture, Algorithm and Implementation of the COMPOW Protocol. In European Wireless, Florence, Italy. NetLogo. (n.d.). Retrieved March 2007 from http://ccl.northwestern.edu/netlogo/

Nasser, N., & Hassanein, H. (2004). Seamless QoS-Aware Fair Handoff in Multimedia Wireless Networks with Optimized Revenue. IEEE Canadian Conference on Electrical and Computer Engineering (CCECE) (pp. 1195-1198). Niagara Falls, Canada: IEEE.

Natkaniec, M., & Pach, A. R. (2000). An Analysis of Modified Backoff Mechanism in IEEE 802.11 Networks. Polish-German Teletraffic Symposium (pp. 89-96).

Natkaniec, M., & Pach, A. R. (2002). PUMA - a new channel access protocol for wireless LANs. International Symposium on Wireless Personal Multimedia Communications, (vol. 3, pp. 1351- 1355).

Natkaniec, M., Gozdecki, J., & Sargento, S. (2007). A QoS Architecture Integrating Ad-Hoc and Infrastructure in Next Generation Networks. In B. Simak, R. Bestak, & E. Kozowska (Eds.), IFIP International Federation for Information Processing, Personal Wireless Communications (pp. 25–37). New York: Springer.

Natkaniec, M., Kosek-Szott, K., Szott, S., Gozdecki, J., Głowacz, A., & Sargento, S. (2009). Supporting QoS in Integrated Ad-Hoc Networks. Wireless Personal Communications, Springer, ISSN 1572-834X (Online).

Neely, M.J., Modiano, E., & Rohrs, C.E. (2005). Dynamic Power Allocation and Routing for Time Varying Wireless Networks. IEEE Journal Selected Area of Communications, 23.

Ni, Q., Romdhani, L., & Turletti, T. (2004). A Survey of QoS Enhancements for IEEE 802.11 Wireless LAN. Journal of Wireless Communication and Mobile Computing, 4(5), 547–566. doi:10.1002/wcm.196

Nikaen, N., Bonnet, C., Moret, Y., & Rai, I. A. (2002). 2LQoS - Two-layered QoS model for reactive routing protocols for mobile ad-hoc networks. In Proceedings of SCI - 6th World Multiconference on Systemics, Cybernetics and Informatics.

Nikkilä, R., Langutin, D., Asplund, H., & Ranta, M. (2004). Prototyping of Vertical Handover Enabled Services. Wireless World Research, (WWWRF11), Oslo, Norway.

Nikoletseas, S., & Powell, O. (2007). Simple and efficient geographic routing around obstacles for wireless sensor networks. In Proceedings of the 6th International Workshop on Efficient and Experimental Algorithms (WEA), (LNCS pp. 161–174). Berlin: Springer-Verlag.

Niyato, D., & Hossain, E. (2005). Call Admission Control for QoS Provisioning in 4G Wireless Networks: Issues and Approaches. IEEE Network, 19(5), 5–11. doi:10.1109/MNET.2005.1509946

Niyato, D., & Hossain, E. (2006). A Cooperative Game Framework for Bandwidth Allocation in 4G Heterogeneous Wireless Networks. International Conference on Communications 2006 (ICC'06). 2, 1418-1422. Instabul: IEEE.

Niyato, D., & Hossain, E. (2007). Radio Resource Management Games in Wireless Networks: An Approach to Bandwidth Allocation and Admission Control for Polling Service in IEEE 802.16. IEEE Wireless Communications, 14(1), 27–35. doi:10.1109/MWC.2007.314548

Niyato, D., & Hossain, E. (2008). Competitive Pricing in Heterogeneous Wireless Access Networks: Issues and Approaches. IEEE Network, 22(6), 4–11. doi:10.1109/MNET.2008.4694168

Nurmi, P., Hassinen, M., & Lee, K. C. (2007). A Comparative Analysis of Personalization Techniques for a Mobile Application. Paper presented at the Intl Conf on Advanced Information Networking and Applications Workshops (AINAW07).

O'Callaghan, M., Gawley, N., Barry, M., & McGrath, S. (2004). Admission Control for Heterogeneous Networks. 13th IST Mobile & Wireless Communications. Lyon.

O'Reilly, T. (2005). What is Web 2.0? Retrieved Sep 05, 2008, from www.oreillynet.com/pub/a/oreilly/tim/news/2005/09/30/what-is-web-20.html

Oliveira, C., Kim, J. B., & Suda, T. (1998). An Adaptive Bandwidth Reservation Scheme for High-Speed Multimedia Wireless Network. IEEE Journal on Selected Areas in Communications, 16(6), 858–874. doi:10.1109/49.709449

Ormond, O., Miro-Muntean, G., & Murphy, J. (2005). Economic Model for Cost Effective Network Selection Strategy in Service Oriented Heterogeneous Wireless Network Environment. In Proc. of 10th IEEE/IFIP Network Operations and Management Symposium (NOMS 2006), Canada.

Ormond, O., Murphy, J., & Muntean, G.-M. (2006). Utility-based Intelligent Network Selection in Beyond 3G Systems. IEEE International Conference on Communications, 4, 1831 – 1836.

Ortiz, S. (2007). 4G Wireless Begins to Take Shape. IEEE Computer, 40(11), 18–21.

Osterwalder, A., Ondrus, J., & Pigneur, Y. (2005). Skype's Disruptive Potential in the Telecom Market: A Systematic Comparison of Business Models. Lausanne, Switxerland: University of Lausanne.

Ostring, S., Sirisena, H., & Hudson, I. (2000). Designing and managing networks for self-similarity. In Proceedings of Engineering Mathematics and Applications, (pp. 235-238).

Otal, B., Alonso, L., & Verikoukis, C. (2009). Highly Reliable Energy-Saving MAC for Wireless Body Sensor Networks in healthcare systems. Journal Selected Area of Communications special issue on "Wireless and Pervasive Communications for Healthcare.

Pahlavan, K., Krishnamurthy, P., Hatami, A., Ylianttila, M., Mareka, J., & Pichna, R. (2000). Handoff in hybrid mobile data networks. IEEE Personal Communications, 7(2), 34–47. doi:10.1109/98.839330

Pal, A., Dogan, A., & Ozguner, F. (2002). MAC Layer Protocols for Real-time Traffic in Ad hoc Wireless Networks. IEEE International Conference on Parallel Processing (pp. 539-546). New York: IEEE Inc.

Pan, P., Schulzrinne, H., & Hahne, E. (2000). BGRP: A Tree-Based Aggregation Protocol for inter-Domain Reservations. Journal of Communications and Networks, 2(2), 157–167.

Pang, A., Chen, J., Chen, Y., & Agrawal, P. (2004). Mobility and session management: UMTS vs. CDMA 2000. IEEE Wireless Communications, 11(4), 30–43. doi:10.1109/MWC.2004.1325889

Papadimitriou, G. I., Lagkas, T. D., & Pomportsis, A. S. (2003). HIPERSIM: A Sense Range Distinctive Simulation Environment for HiperLAN Systems. Simulation. Transactions of The Society for Modelling and Simulation International, 79(8), 462–481. doi:10.1177/0037549703040235

Papadopouli, M., Shen, H., Raftopoulos, E., Ploumidis, M., & Campos, F. (2005). Short-term traffic forecasting in a campus-wide wireless network. In Proceedings of 16th Annual IEEE International Symposium on Personal Indoor and Mobile Radio Communications, (pp. 1446-1452).

Papageorgiou, C., Kokkinos, P., & Varvarigos, E. (2006). Multicost Routing over an Infinite Time Horizon in Energy and Capacity Constrained Wireless Ad-hoc Networks. In European Conference on Parallel Computing (pp. 931-940), Dresden, Germany.

Papageorgiou, C., Kokkinos, P., & Varvarigos, E. (2008). Implementing distributed multicost routing in mobile ad hoc networks using DSR. In 6th ACM international symposium on Mobility management and wireless access, (pp. 35-42), Vancouver, Canada.

Papagiannaki, K. Taft, N. Zhang, Z., & Diot, C. (2003). Long-term forecasting of internet backbone traffic: Observations and initial models. In Proceedings of IEEE Computer and Communications Societies, (pp. 1178-1188).

Park, E. C., Kim, D. Y., Choi, C. H., & So, J. M. (2007). Improving Quality of Service and Assuring Fairness in WLAN Access Networks. IEEE Transactions on Mobile Computing, 6(4), 337–350. doi:10.1109/TMC.2007.53

Park, E.-C., Kim, D.-Y., Choi, C.-H., & So, J. (2007). Improving Quality of Service and Assuring Fairness in WLAN Access Networks. IEEE Transactions on Mobile Computing, 6(4),337–350. doi:10.1109/TMC.2007.53

Park, K., & Willinger, W. (Eds.). (2000). Self-Similar Network Traffic and Performance Evaluation. New York: John Wiley & Sons, Inc.

Park, S.-J., & Sivakumar, R. (2002). Load-sensitive transmission power control in wireless ad-hoc networks. IEEE Globecom 2002, Taipei, Taiwan.

Parkvall, S., Englund, E., Lundevall, M., & Torsner, J. (2006). Evolving 3G Mobile Systems: Broadband and Broadcast Services in WCDMA. IEEE Communications Magazine, 44(2), 30–36. doi:10.1109/MCOM.2006.1593548

Paskalis, S., Kaloxylos, A., Zervas, E., & Merakos, L. (2003). An Efficient RSVP-Mobile Internetworking Scheme. Mobile Networks and Applications, 8, 197–207. doi:10.1023/A:1023385429587

Paul, F., & Geoff, H. (1998) Quality of Service. New York: John Wiley & Sons, Inc.

Pavon, J. d., & Choi, S. (2003). Link adaptation strategy for IEEE 802.11 WLAN via received signal strength measurement. In IEEE International Conference on Communications, 2003. ICC '03, (Vol. 2, pp. 1108–1113).

Pawar, P., van Beijnum, B. J., Peddemors, A., & van Halteren, A. (2007). Context-Aware Middleware Support for the Nomadic Mobile Services on Multi-homed Handheld Mobile Devices. Paper presented at the 12th IEEE Symposium on Computers and Communications (ISCC07), Aveiro, Portugal.

Pawar, P., Wac, K., van Beijnum, B. J., Maret, P., van Halteren, A., & Hermens, H. (2008). Context-Aware Middleware Architecture for Vertical Handover Support to Multi-homed Nomadic Mobile Services. Paper presented at the 23rd Annual ACM Symposium on Applied Computing (ACMSAC08), Ceará, Brazil.

Pawlikowski, K., Jeong, H. D. J., & Lee, J. S. R. (2002). On Credibility of Simulation Studies of Telecommunication Networks. IEEE Communications Magazine, 40(1), 132–139. doi:10.1109/35.978060

Perez-Romero, J., Sallent, O., Agusti, R., & Diaz-Guerra, M. (2005). Radio Resource Management Strategies in UMTS. Chichester, UK: John Wiley & Sons.

Perkins, C. (1996). RFC 2003: IP Encapsulation within IP.

Perkins, C. (2002). Mobile IP. IEEE Communications Magazine, 40(5), 66–82. doi:10.1109/MCOM.2002.1006976

Perkins, C. (2002). IP Mobility Support in IPv4 (RFC 3344). IETF Network Working Group.

Perkins, C., Royer, E., & Das, S. (2000). Quality of Service in Ad-hoc On-demand Distance Vector Routing. IETF Internet Draft, (draft-ietf-manet-qos-00.txt). POLARISTM TOTAL RADIOTM solution from RF Micro Devices. (n.d.). Retrieved from http://www.rfmd.com/

Perrucci, G. P., Fitzek, F. H. P., Boudali, A., Canovas, M., Nejsum, P., & Studstrup, S. (2007). Cooperative Web Browsing for Mobile Phones. International Symposium on Wireless Personal Multimedia Communications (WPMC'07), India.

Petras, D., & Kramling, A. (1996). MAC protocol with polling and fast collision resolution for an ATM air interface. In Proceedings of IEEE ATM Workshop 1996.

Platt, J. (1998). Sequential minimal optimization: a fast algorithm for training support vector machines, (Tech. Rep. MSR-TR-98-14). Microsoft Research; 1998.

Pong, D., & Moors, T. (2003). Call Admission Control for IEEE 802.11 Contention Access Mechanism. In Proceedings IEEE Global Telecommunications Conference, Globecom 2003: Vol. 1 (pp. 174-178). New York: IEEE Press.

Pong, D., & Moors, T. (2004). Fairness and capacity trade-off in IEEE 802.11 WLANs. In Proceedings of 29th Annual IEEE International Conference on Local Computer Networks 2004, (pp. 310-317).

Prodan, A. (2009). Topology Control on Multi-radio Infrastructure Wireless Mesh Networks, Unpublished doctoral dissertation, University of Technology, Sydney (UTS), Australia.

Prodan, A., Mirchandani, V., & Debenham, J. (2008). A Topology Control based Self-Organization in Wireless Mesh Networks. International Journal of Communication Networks and Distributed Systems (IJCNDS).

Pubill, D., & Perez-Neira, A. I. (2006). Handoff Optimization with Fuzzy Logic in 802.11 Networks. Information Processing and Management of Uncertainty (IPMU) Conference.

Qiao, Y. Skicewicz, J., & Dinda, P. (2004). An empirical study of the multiscale predictability of network traffic. In Proceedings of the 13th IEEE International Symposium on High Performance Distributed Computing, (pp. 66-76).

Qinglin, Z., & Tsang, H. K. (2008). An Equal-Spacing-Based Design for QoS Guarantee in IEEE 802.11e HCCA Wireless Networks. IEEE Transactions on Mobile Computing, 1474–1490.

Qingyang, S., & Abbas, J. (2005a). A Network Selection Mechanism for Next Generation Networks. IEEE International Conference on Communications, 16-20 May 2005, 2, 1418- 1422.

Qingyang, S., & Abbas, J. (2005b). Network Selection in an integrated Wireless LAN and UMTS Environment Using Mathematical Modeling and Computing Techniques. IEEE Personal Communications, 12(3), 42–48.

Qiu, D., & Shroff, N. B. (2004). A predictive flow control scheme for efficient network utilization and QoS. IEEE/ACM Transactions on Networking, 12(1), 161-172.

Quax, P., Monsieurs, P., Lamotte, W., De Vleeschauwer, D., & Degrande, N. (2004). Objective and Subjective Evaluation of the Influence of Small Amounts of Delay and Jitter on a First Person Shooter Game. Paper presented at the NetGames04, Portland, USA.

Quinlan, R. (1993). C4.5: Programs for Machine Learning. San Meteo, CA: Morgan Kaufmann.

Racunica, I., Menouni Hayar, A., & Bonnet, C. (2004). An opportunistic scheduling with fairness for NRT traffic in presence of RT traffic for UMTS/TDD. 7th International Symposium on Wireless Personal Multimedia Communications 2004 (WPMC'04). Abano Terme, Italy: EURECOM+1410.

Radhakrishnan, S., Rao, N., Racherla, G., Sekharan, C., & Batsell, S. (1999). Dst - a routing protocol for ad hoc networks using distributed spanning trees. In IEEE Wireless Communications and Networking Conference, (pp. 100– 104).

Rahimi, M., Shah, H., Sukhatme, G., Heidemann, J., & Estrin, D. (2003). Studying the Feasibility of Energy Harvesting in a Mobile Sensor Network. IEEE International Conference on Robotics and Automation (pp. 19- 24). Taipei, Taiwan.

Rajaratmam, M., & Takawira, F. (1999). Performance analysis of highway cellular networks using generalised arrival and generalised service time distributions. 16th International Teletraffic Congress (ITC-16). 3, pp. 11-22. Edinburgh: Elsevier.

Ramachandran, K. N., Belding, E. M., Almeroth, K. C., & Buddhikot, M. M. (2006). Interference-Aware Channel Assignment in Multi-Radio Wireless Mesh Networks. Infocom, (pp. 1-12), Barcelona, Spain.

Ramanathan, R., & Rosales-Hain, R. (2000). Topology Control of Multihop Wireless Networks Using Transmit Power Adjustment. In IEEE Conference on Computer Communications (pp. 404 - 413), Tel-Aviv, Israel.

Ramjee, R., Towsley, D., & Nagarajan, R. (1997). On optimal call admission control in cellular networks. Wireless Networks, 3(1), 29–41. doi:10.1023/A:1019172226345

Ramos, N., Panigrahi, D., & Dey, S. (2003). ChaPLeT: Channel-dependent Packet Level Tuning for Service Differentiation in IEEE 802.11e. In Proceedings International Symposium on Wireless Personal Multimedia Communications: Vol. 3 (pp. 89-90).

Raniwala, A., & Chiueh, T.-c. (2004). Architecting a High-Capacity Last-Mile Wireless Mesh Network. MobiCom, Philadelphia.

Raniwala, A., & Chiueh, T.-c. (2005). Architecture and Algorithms for an IEEE 802.11- based Multi-channel Wireless Mesh Network. In Proceedings IEEE Infocom, (vol. 3, pp. 2223-2234), Miami, FL.

Raniwala, A., Gopalan K., & Chiueh, T-c. (n.d.). Centralized Channel Assignment and Routing Algorithms for Multi-channel wireless mesh networks. Mobile Computing and Communications Review, 8(2).

Rath, H. K., Bhorkar, A., & Sharma, V. (2006). An Opportunistic DRR (O-DRR) Uplink Scheduling Scheme for IEEE 802.16-based Broadband Wireless Networks. In IETE, International Conference on Next Generation Networks (ICNGN).

Robinson, J. W., & Randhawa, T. S. (2004). Saturation Throughput Analysis of IEEE 802.11e Enhanced Distributed Coordination Function. IEEE Journal on Selected Areas in Communications, 22(5),917–928. doi:10.1109/JSAC.2004.826929

Rodoplu, V., & Meng, T. (1998). Minimum energy mobile wireless networks. IEEE International Conference on Communications, (pp. 1333 – 1344).

Romdhani, L., Ni, Q., & Turletti, T. (2003). Adaptive EDCF: Enhanced Service Differentiation for IEEE 802.11 Wireless Ad Hoc Networks. In Proceedings 2003 IEEE Wireless Communications and Network Conference: Vol. 3 (pp. 1373-1378). New York: IEEE Press.

Rong, B., Qian, Y., & Chen, Y. (2007). Adaptive power allocation and call admission control in multiservice WiMAX access networks. IEEE Wireless Communications, 14(1), 14–19. doi:10.1109/MWC.2007.314546

Rosen, E., Viswanathan, A., & Callon, R. (2001). RFC 3031: Multiprotocol Label Switching Architecture

Rumelhart, D. E., Hinton, G. E., & Williams, R. J. (1986). Learning internal representations by error propagation. In D. E. Rumelhart & J. L. McCleland (Eds.), Parallel Distributed Processing, (vol. 1). Cambridge, MA: MIT Press.

Ryu, S., Ryu, B., Seo, B., & Shi, M. (2005). Urgency and efficiency based wireless downlink packet scheduling algorithm in OFDMA system. In IEEE 61st Vehicular Technology Conference, (Vol. 3 pp. 1456 – 1462).

Sack, J.-R., & Urrutia, J. (2000). Handbook of computational geometry. Amsterdam: North- Holland Publishing Co.

Sadeghi, B., Kanodia, V., Sabharwal, A., & Knightly, E. (2002). Opportunistic media access for multirate ad hoc networks. In Proceedings of the 8th annual international conference on Mobile computing and networking (MobiCom '02) (pp. 24–35). New York: ACM.

Sadek, N., Khotanzad, A., & Chen, T. (2003). ATM dynamic bandwidth allocation using F-ARIMA prediction model, In Proceedings of International Conference on Computer Communications and Networks, (359-363).

Safari, M. & Uysal, M. (2008). Cooperative Diversity over Log-Normal Fading Channels: Performance Analysis and Optimization. IEEE Transactions on Wireless Communications, 7(5-2), 1963-1972.

Saghir, M., Wan, T. C., & Budiarto, R. (2006). A new cross-layer framework for QoS multicast applications in mobile ad hoc networks. International Journal of Computer Science and Network Security, 6(10), 142–151.

Saha, D. (2004, November-December). Mukherjee, A. Misra, I.S. Chakraborty, M. Subhash, N. (2004). Mobility support in IP: a survey of related protocol. IEEE Network, 18(6), 34–40. doi:10.1109/MNET.2004.1355033

Sakiyama, Y., Yuki, H., Moriya, T., Hattori, K., Suzuki, M., Shimada, K., & Honma, T. (2008). Predicting human liver microsomal stability with machine learning techniques. Journal of Molecular Graphics & Modelling, (6): 907–915. doi:10.1016/j.jmgm.2007.06.005

Salem, N.B., Buttyan, L., Hubaux, J.-P., & Jakobsson, M. (2003). A charging and rewarding scheme for packet forwarding in multi-hop cellular networks. ACM MobiHoc 03'.

Salkintzis, A. K., Fors, C., & Pazhyannur, R. (2002). WLAN-GPRS integration for next generation mobile data networks. IEEE Transactions on Wireless Communications, 9(5), 112–124. doi:10.1109/MWC.2002.1043861

Sallent, O., Perez-Romero, J., Agusti, R., Giupponi, L., Kloeck, C., Martoyo, I., et al. (2006). Resource Auctioning Mechanisms in Heterogeneous Wireless Access Networks. Vehicular Technology Conference 2006 (VTC '06), 1, 52-56. Melbourne: IEEE.

Salomon, J., King, S., & Osborne, M. (2002). Framewise phone classification using support vector machines. In Proceedings of International Spoken Language Processing, (2645-2648).

Salvo Rossi, P., Romano, G., Palmieri, F., & Iannello, G. (2003a). Bayesian Modelling for Packet Channels. Paper presented at the Italian Workshop on Neural Nets (WIRN).

Salvo Rossi, P., Romano, G., Palmieri, F., & Iannello, G. (2003b). A Hidden Markov Model for Internet Channels. Paper presented at the IEEE Intl Symposium on Signal Processing and Information Technology (ISSPIT), Darmstadt, Germany.

Sanda T., E., Fu, X., Jeong, S., Manner, J., & Tshofenig, H. (2009). Applicability Statement of NSIS Protocols in Mobile Environments. Internet Draft, work in progress. draft-ietf-nsis-applicability-mobility-signaling-12.txt.

Sang, A., & Li, S. (2002). A predictability analysis of network traffic. Computer Networks, 39, 329–345. doi:10.1016/S1389-1286(01)00304-8

Sang, J., Jeong, D., & Jeon, W. (2006). Cross-layer Design of Packet Scheduling and Resource Allocation in OFDMA Wireless Multimedia Networks. In IEEE 63rd Vehicular Technology Conference, (VTC) 2006, (pp. 309-313).

Santi, P. (2005). Topology Control in Wireless Ad Hoc and Sensor Networks. New York: Wiley.

Sarma, N., & Nandi, S. (2008). A priority based QoS-Aware MAC protocol (PQAMP) in mobile ad hoc networks. 4th ACM symposium on QoS and security for wireless and mobile networks, Q2SWinet (pp. 79-82), New York: ACM.

Schmid, S., & Wattenhofer, R. (2006). Algorithmic models for sensor networks. In 20th International Parallel and Distributed Processing Symposium, 2006. IPDPS 2006.

Schmidt, M. (1996). Identifying speaker with support vector networks. In Proceedings of Interface, (pp. 305-308).

Schmitz, R. T.-M. M. H. H., & Effelsberg, W. (2004). The impact of wireless radio fluctuations on ad hoc network performance. In Proceedings of the 29th Annual IEEE international Conference on Local Computer Networks, (pp. 594–601), Washington, DC.

Schulzrinne, H., & Hancock, R. (2009). GIST: General Internet Signalling Transport. Internet Draft, work in progress. draft-ietf-nsis-ntlp-19.txt.

Schulzrinne, H., Casner, S., Frederick, R., & Jacobson, V. (1996). RTP: A Transport Protocol for Real-Time Applications. RFC 1889, IETF Audio-Video Transport Working Group.

Sen, S., Jawanda, J., Basu, K., & Das, S. (1998). Quality-of-Service degradation strategies in multimedia wireless networks. 48th IEEE Vehicular Technology Conference, 1998 (VTC 98), 3, pp. 1884-1888. Ottawa.

Shah, K., Bohacek, S., & Jonckheere, E. (2003). On the predictability of data network traffic. In Proceedings of the 2003 American Control Conference, 2, 1619-1624.

Shah, R., & Rabaey, J. (2002). Energy aware routing for low energy ad hoc sensor networks. In IEEE Wireless Communications and Networking Conference (pp. 350-355), Florida.

Shah, S. H., & Nahrstedt, K. (2002). Predictive location-based QoS routing in mobile ad hoc networks. IEEE International Conference on Communications, ICC'02 (pp. 1022-1027). New York: IEEE Inc.

Shakkottai, G. S., & Stolyar, A. (2001). Scheduling algorithms for a mixture of real time and non-real-time data in HDR. International Teletraffic Conference (ITC).

Shakkottai, S., Rappaport, T., & Karlsson, P. (2003). Cross-layer design for wireless networks. Communications Magazine, IEEE, 41(10), 74–80. doi:10.1109/MCOM.2003.1235598

Shalizi, C. R. (2001). Causal Architecture, Complexity and Self-Organization in Time Series and Cellular Automata. University of Wisconsin at Madison.

Shankar, S., & van der Schaar, M. (2007). Performance analysis of video transmission over IEEE 802.11a/e WLANs. IEEE Transactions on Vehicular Technology, 56(4), 2346–2362. doi:10.1109/TVT.2007.897646

Sharif, M., & Hassibi, B. (2005). On the Capacity of MIMO Broadcast Channel with Partial Side Information. IEEE Transactions on Information Theory, 51.

Shastry, N., & Adve, R. S. (2006). Stimulating Cooperative Diversity in Wireless Ad Hoc Networks through Pricing. IEEE International Conference on Communications (ICC '06).

Shen, Q., Lo, A., Seah, W., & Ko, C.-C. (2000). On Providing Flow Transparent Mobility Support for IPv6-based Wireless Real-Time Services. In IEEE International Workshop on Mobile Multimedia Communications (MoMuC), (pp. 2B41–2b46), Tokyo, Japan.

Shenker, S., Patridge, C., & Guerin, R. (1997). Specification of Guaranteed Quality of Service. RFC 2212, IETF.

Shenoy, N., & Montalvo, R. (2005). A Framework for Seamless Roaming across Cellular and Wireless Local Area Networks. IEEE Wireless Communications, 12(3), 50–57. doi:10.1109/MWC.2005.1452854

Shepherd, D., Scott, A., & Rodden, T. (1996). Quality-of-Service Support for Multimedia Applications. IEEE MultiMedia, 03(3), 78–82. doi:10.1109/93.556542

Sherif, M. R., Habib, I. W., Nagshineh, M. N., & Kermani, P. K. (2000). Adaptive allocation of resources and call admission control for wireless ATM using generic algorithm. IEEE Journal on Selected Areas in Communications, 18(2), 268–282. doi:10.1109/49.824812

Shi, M., Shen, X., & Mark, J. (2004). IEEE 802.11 roaming and authentication in wireless LAN/cellular mobile networks. IEEE Wireless Communications, 11(4), 66–75. doi:10.1109/MWC.2004.1325893

Shigang, C., & Nahrstedt, K. (1999). Distributed quality-of-service routing in ad hoc networks. IEEE Journal on Selected Areas in Communications, 17(8), 1488–1505. doi:10.1109/49.780354

Shu, Y. T., Feng, H. F., Wang, H., & Ma, M. D. (2006). Dynamic admission control in hybrid QoS networks with WiFi access. Science in China Series F- . Information Sciences, 49(4), 409–420.

Shu, Y. T., Jin, Z. G., Wang, J. D., & Yang, O. W. (2000). Prediction-based admission control using FARIMA models. Proceedings of IEEE International Conference on Communications, 3, 1325–1329.

Shu, Y. T., Jin, Z. G., Zhang, L. F., et al. (1999). Traffic prediction using FARIMA models. In Proceedings of IEEE International Conference on Communications, (pp. 891-895).

Siew-Lee Hew, White L. (2008). Cooperative resource allocation games in shared networks: symmetric and asymmetric fair bargaining models. IEEE Transactions on Wireless Communications, 7(11-1), 4166-4175.

Singh, N., & Vives, X. (1984). Price and Quantity Competition in a Differentiated Duopoly. The Rand Journal of Economics, •••, 546–554. doi:10.2307/2555525

Singh, S., Woo, M., & Raghavendra, C. (1998). Power-Aware Routing in Mobile Ad Hoc Networks. In The Fourth Annual ACM/IEEE International Conference on Mobile Computing and Networking (pp. 181 - 190), Dallas, TX.

Sinha, P., Sivakumar, R., & Bharghavan, V. (1999). CEDAR: a core-extraction distributed ad hoc routing algorithm. In Proceedings of Eighteenth Annual Joint Conference of the IEEE Computer and Communications Societies, INFOCOM '99, (pp. 202-209). New York: IEEE Inc.

Sobrinho, J. L., & Krishnakumar, A. S. (1999). Quality-of-service in ad hoc carrier sense multiple access wireless networks. IEEE Journal on Selected Areas in Communications, 17(8), 1353–1368. doi:10.1109/49.779919

Sofia, R., Guerin, R., & Veiga, P. (2003). SICAP, A Shared-segment Inter-domain Control Aggregation Protocol. In Workshop on High Performance Switching and Routing (HPSR), (pp. 73–78).

Soh, W. S., & Kim, H. S. (2003). QoS Provisioning in Cellular Networks Based on Mobility Prediction Techniques. IEEE Communications Magazine, 41(1), 86–92. doi:10.1109/MCOM.2003.1166661

Soliman, H., Castelluccia, C., El Malki, K., & Bellier, L. (2005). Hierarchical Mobile IPv6 Mobility Management (HMIPv6), RFC 4140, IETF.

Song, G.-C., & Li, Y. (G.). (2005). Cross-layer optimization for OFDM wireless networks – Part I: theoretical framework. IEEE Transactions on Wireless Communications, 4(2), 614–624. doi:10.1109/TWC.2004.843065

Song, Q., & Jamalipour, A. (2005). A Network Selection Mechanism for Next Generation Network. IEEE International Conference on Communications 2005 (ICC'05), (pp. 1418-1422). Seoul: IEEE.

Song, W., & Zhuang, W. (2005). QoS provisioning via Admision Control in Cellular/Wireless LAN interworking. 2nd International Conference on Broadband Networks 2005 (BroadNets 2005). 1, (pp. 543-550). Boston: IEEE.

Song, W., Jiang, H., Zhuang, W., & Shen, X. (2005). Resource Management for QoS support in Cellular/WLAN Interworking. IEEE Network, 19(5), 12–18. doi:10.1109/MNET.2005.1509947

Spiegel, M. R. (1992). Theory and Problems of Probability and Statistics. New York: McGraw-Hill.

Sroka, S., & Karl, H. (2003). Performance Evaluation of a QoS-Aware Handover Mechanism. Proceedings 8th IEEE International Symposium on Computers and Communications: Vol. 1 (pp. 117- 124). New York: IEEE Press.

Stasiak, M., Wisniewski, A., & Zwierzykowski, P. (2005). Uplink Blocking Probability for a Cell with WCDMA Radio Interface and Differently Loaded Neighbouring Cells. Advanced Industrial Conference on Telecommunications/Service Assurance with Partial and Intermittent Resources Conference/E-Learning on Telecommunications Workshop (AICT/SAPIR/ELETE'05) (pp. 402-407). Lisbon: IEEE Computer Society.

Stefan, M., Sunghyun, C., & Gudio, R. H., Ole, K., & Bernhard, W. (2003). Analysis of IEEE 802.11e for QoS support in wireless LANs. IEEE Wireless Communications, 40-50. The Network Simulator ns-2. (n.d.). Retrieved from http://www.isi.edu/nsnam/ns/

Stemm, M., Katz, R., & Seshan, S. (2000). A network measurement architecture for adaptive applications. Proceedings of IEEE Computer and Communications Societies, 1, 285–294.

Stock, J. H., & Watson, M. W. (1999). A comparison of linear and nonlinear univariate models for forecasting macroeconomic time-series. In R. F. Engle & H. White, (Eds.), Cointegration, causality and forecasting: A Festschrift in honour of Clive W. J. Granger, (pp. 1-44). Oxford, UK: Oxford University Press.

Stojmenovic, I. (2005). Handbook of Sensor Networks: Algorithms and Architectures. Chichester, UK: John Wiley and Sons, Inc.

Stojmenovic, I., & Lin, X. (2001). Loop-free hybrid single-path/flooding routing algorithms with guaranteed delivery for wireless networks. IEEE Transactions on Parallel and Distributed Systems, 12(10), 1023–1032. doi:10.1109/71.963415

Stojmenovic, I., & Lin, X. (2001). Power-aware localized routing in wireless networks. IEEE Transactions on Parallel and Distributed Systems, 12(11), 1122–1133. doi:10.1109/71.969123

Stojmenovic, I., Nayak, A., Kuruvila, J., Ovalle-Martinez, F., & Villanueva-Pena, E. (2005). Physical layer impact on the design and performance of routing and broadcasting protocols in ad hoc and sensor networks. In Computer Communications, (Vol. 28).

Stojmenovic, I., Russell, M., & Vukojevic, B. (2000). Depth first search and location based localized routing and qos routing in wireless networks. In ICPP '00: Proceedings of the 2000 International Conference on Parallel Processing, (pp. 173), Washington, DC: IEEE Computer Society.

Stratogiannis, D. G., Tsiropoulos, G. I., & Cottis, P. G. (2008). Call Admission Control in Wireless Networks: Probabilistic Approach and Efficiency Evaluation. International Conference on Wireless Communications and Mobile Computing Conference 2008 (IWCMC '08) (pp. 712 - 717). Crete Island: IEEE.

Su, C. T., Tong, L. I., & Leou, C. M. (1997). Combination of time series and neural network for reliability forecasting modeling. Journal of Chinese Institute of Industrial Engineers, 14(4), 419–429.

Subramanian, A. P., Krishnan, R., Das, S. R., & Gupta, H. (2005). Minimum Interference Channel Assignment in Multi-Radio Wireless Mesh Networks. In 13th International Conference on Network Protocols (ICNP 2005), Boston, MA.

Sun, Y., Tsang, K. F., Tung, H. Y., Lam, K. L., & Ko, K. T. (2008). QoS Prediction for High Speed Packet Access Networks. Paper presented at the 5th IEEE Consumer Communications and Networking Conference (CCNC08).

SUN. (2001). The Jini™Technology Surrogate Architecture Overview. Santa Clara, CA: SUN Microsystems Inc.

Sung, A. H., & Mukkamala, S. (2003). Identifying important features for intrusion detection using SVM and neural networks. In Proceedings of the 2003 International Symposium on Applications and the Internet Technology, (pp. 209-216).

Suris, J. E., DaSilva, L. A., Han, Z., & MacKenzie, A. B. (2007). Cooperative Game Theory for Distributed Spectrum Sharing. IEEE International Conference on Communications.

Suykens, J. A. K., Gestel, T. V., Brabanter, J. D., Moor, B. D., & Vandewalle, J. (2002). Least squares support vector machines. Hoboken, NJ: World Scientific.

Tachikawa, K. (2003). A Perspective on the Evolution of Mobile Communications. IEEE Communications Magazine, 41(10), 66–73. doi:10.1109/MCOM.2003.1235597

Talukdar, A., Badrinath, B., & Acrarya, A. (2001). MRSVP: A Resource Reservation Protocol for an Integrated Services Network with Mobile Hosts. [WINET]. Wireless Networks, 7(1), 5–19. doi:10.1023/A:1009035929952

Tan, S. (2004). Evolution of mobile technology and business models. Tech. Rep., Center for Information and Communication Technologies, Lyngby, DK.

Tang, D., & Baker, M. (2000). Analysis of a local-area wireless network. In Proceedings of ACM Mobile Computing and Networking, (pp. 1-10).

Tanigawa, Y., Kim, J., Tode, H., & Murakami, K. (2006). Proportional Control and Deterministic Protection of QoS in IEEE 802.11e Wireless LAN. In Proceedings International Conference on Wireless Communications and Mobile Computing, IWCMC 06 (pp. 1147 – 1152). New York: ACM Press.

Tarraf, A. A., Habib, I. W., Saadawi, T. N., & Ahmed, S. A. (1993). ATM multimedia traffic prediction using neural networks. In Proceedings of IEEE Global Data Networking, (pp. 77–84).

Terzis, A., Wang, L., Ogawa, J., & Zhang, L. (1999). A Two-Tier Resource Management Model for the Internet. In Global Internet.

Thomas, M. (2002). Analysis of Mobile IP and RSVP Interactions. Internet Draft, work in progress. draft-thomas-nsis-rsvp-analysis-00.txt.

Toh, C.-K. (2001). Maximum Battery Life Routing to Support Ubiquitous Mobile Computing in Wireless Ad Hoc Networks. IEEE Communications Magazine, 39(6), 138–147. doi:10.1109/35.925682

Toussaint, G. (1991). Some unsolved problems on proximity graphs.

Tragos, E., Tsiropoulos, G., Karetsos, G., & Kyriazakos, S. (2008). Admission Control for QoS support in Heterogeneous 4G Wireless Networks. IEEE Network Magazine, 22(3), 30–37. doi:10.1109/MNET.2008.4519962

Tragos, E., Tsiropoulos, G., Karetsos, G., & Kyriazakos, S. (2008). Admission Control for QoS support in Heterogeneous 4G Wireless Networks. IEEE Network Magazine, 22(3), 30–37. doi:10.1109/MNET.2008.4519962

Tse, D., & Viswanath, P. (2005). Fundamentals of Wireless Communication. Cambridge, UK: Cambridge University Press.

Tseng, C.-C., Lee, G.-C., Liu, R.-S., & Wang, T.-P. (2003). HMRSVP: A Hierarchical Mobile RSVP Protocol. [WINET]. Kluwer Wireless Networks, 9(2), 95–102. doi:10.1023/A:1021833430898

Tsiropoulos, G. I., Stratogiannis, D. G., Kanellopoulos, J. D., & Cottis, P. G. (2008). Efficiency evaluation of class-based call admission control schemes for wireless communications. IEEE International Symposium on Wireless Communication Systems (ISWCS '08) (pp. 69 - 73). Reykjavik: IEEE.

Tumer, K., & Wolpert, D. H. (2004). Collectives and the design of complex systems. New York: Springer.

Urrutia, J. (2002). Routing with guaranteed delivery in geometric and wireless networks, (pp. 393–406).

Valavanis, E., Ververidis, C., Vazirgianis, M., Polyzos, G. C., & Nørvåg, K. (2003). MobiShare: Sharing Context-Dependent Data and Services from Mobile Sources. In Proc. IEEE/WIC Int'l Conference on Web Intelligence (WI 2003), Halifax, Canada.

Vali, D., Paskalis, S., Kaloxylos, A., & Merakos, L. (2004). A Survey on QoS Signaling. IEEE Communications Surveys and Tutorials, 6(4).

Van der Vegt, A. J. (2002). Auto Rate Fallback Algorithm for the IEEE 802.11a Standard, (Tech. Rep.), Utrecht University.

van Halteren, A., & Pawar, P. (2006). Mobile Service Platform: A Middleware for Nomadic Mobile Service Provisioning. Paper presented at the 2nd Intl. IEEE Conference on Wireless and Mobile Computing, Networking and Communications (WiMob06), Montreal, Canada.

van Halteren, A., Bults, R., Wac, K., Dokovsky, N., Koprinkov, G., Widya, I., et al. (2004a). Wireless Body Area Networks for Healthcare: the MobiHealth Project. In G. Demiris (Ed.), Studies in Health Technology and Informatics (Vol. 108, pp. 181-193). Amsterdam, the Netherlands: IOS Press.

van Halteren, A., Bults, R., Wac, K., Konstantas, D., Widya, I., & Dokovsky, N. (2004b). Mobile Patient Monitoring: The MobiHealth System. The Journal on Information Technology in Healthcare, 2(5), 365–373.

Vapnik, V. N. (1995). The Nature of Statistical Learning Theory. New York: Springer-Verlag.

Varshney, U., & Jain, R. (2001). Issues in Emerging 4G Wireless Networks. IEEE Computer, 34(6), 94–96.

Vegard, H. (2006). Tutorial: Opportunistic Multiuser Scheduling for Wireless Networks. Department of Electronics and Telecommunications, Norwegian University of Science and Technology (NTNU), Jun. 2006. Retrieved May, 10, 2009, from www.unik.no/personer/porten/MoPSAR/MOPSARTutorial_Hassel.pdf

Vidales, P., Baliosian, J., Serrat, J., Mapp, G., Stajano, F., & Hopper, A. (2005). Autonomic System for Mobility Support in 4G Networks. Journal on Selected Areas in Communications, 23(12), 2288–2304. doi:10.1109/JSAC.2005.857198

Vlacheas, P., Charilas, D., Tragos, E., & Markaki, O. (2008). Maximizing Quality of Service for Customers and Revenue for Service Providers through a Noncooperative Admission Control Game. In ICT Mobile Summit 2008, Stockholm.

Vogel, A., Kerherve, B., von Bochmann, G., & Gecsei, J. (1995). Distributed Multimedia and QOS: A Survey. IEEE MultiMedia, 2(2), 10–19. doi:10.1109/93.388195

Vriendt, J., Gomez Vinagre, I., & Van Ewijk, A. (2004). Multimedia Broadcast and Multicast Services in 3G Mobile Networks. Alcatel Telecommunication Review.

Wac, K. (2009). Collaborative Sharing of Quality of Service-Information for Mobile Service Users. Unpublished doctoral dissertation, University of Geneva, Geneva, Switzerland.

Wac, K., Bargh, M., van Beijnum, B. J., Bults, R., Pawar, P., & Peddemors, A. (2009a). Power- and Delay-Awareness of Health Telemonitoring Services: the MobiHealth System Case Study. IEEE JSAC . Special Issue on Wireless and Pervasive Communications in Healthcare, 27(4), 1–12.

Wac, K., Bults, R., Konstantas, D., Chen, H., & van Beijnum, B. J. (2009b). Toward Mobile Web 2.0-based business methods: Collaborative QoS-information Sharing for Mobile Service Users. In M. Head, (Ed.), Mobile and Ubiquitous Commerce: Advanced E-Business Methods (Vol. 4, pp. xx-xx). Hershey, PA: IGI Global.

Wac, K., Bults, R., van Halteren, A., Konstantas, D., & Nicola, V. (2005). Measurements-based performance evaluation of 3G wireless networks supporting m-health services. Paper presented at the 12th Multimedia Computing and Networking (MMCN05), San Jose, CA.

Wan, L., Ma, W., & Guo, Z. (2007). A cross-layer packet scheduling and subchannel allocation scheme in 802.16e OFDMA System. In Wireless Communications and Networking Conference, (pp. 1865 - 1870).

Wang, D., & Liu, K. (2007). A Novel QoS MAC Protocol for Multi-hop Ad Hoc Networks. International Conference on Wireless Communications, Networking and Mobile Computing, WiCom (pp. 1484-1487).

Wang, H. J., Katz, R. H., & Giese, J. (1999). Policy-Enabled Handoffs Across Heterogeneous Wireless Networks. In 2nd IEEE Workshops on Mobile Computing and Applications (WMCSA '99) (pp. 51-60). New Orleans: IEEE.

Wang, H., Pi, D., & Sun, Y. (2007). Online SVM regression algorithm-based adaptive inverse control. Neurocomputing, 70, 952–959.

Wang, J., Zhai, H., & Fang, Y. (2004). Opportunistic packet scheduling and media access control for wireless LANs and multi-hop ad hoc networks. Wireless Communications and Networking Conference, 2004. WCNC. 2004 IEEE, (pp. 1234–1239 Vol.2).

Wang, P., Jiang, H. & Zhuang, W. (2006) IEEE 802.11e enhancements for voice service. IEEE Journal for Wireless Communications, 30-35.

Wang, S. C., & Helmy, A. (2006). Performance Limits and Analysis of Contention-based IEEE 802.11 MAC. In Proceedings of 31st Annual IEEE Conference on Local Computer Networks 2006, (pp. 418-425).

Wang, W., Liew, S. C., & Li, V. O. K. (2005). Solutions to performance problems in VoIP over a 802.11 wireless LAN. IEEE Transactions on Vehicular Technology, 54(1), 366–384. doi:10.1109/TVT.2004.838890

Wang, W., Liu, F., Ji, Y., & Nararat, R. (2007). Admission Control for Non-preprovisioned Service Flow in Wireless Metropolitan Area Networks. In Fourth European Conference on Universal Multiservice Networks, ECUMN '07, (pp. 243–249).

Wang, W., Men, C., & Lu, W. (2008). Online prediction model based on support vector machine. Neurocomputing, 71, 550–558. doi:10.1016/j.neucom.2007.07.020

Wang, X., Eun, D., & Wang, W. (2007). A Dynamic TCP-Aware Call Admission Control Scheme for Generic Next Generation Packet-Switched Wireless Network. IEEE Transactions on Wireless Communications, 6(9), 3344–3352. doi:10.1109/TWC.2007.06064

Wang, X., Fan, F., & Pan, Y. (2008). A More Realistic Thinning Scheme for Call Admission Control in Multimedia Wireless Networks. IEEE Transactions on Computers, 57(8), 1143–1147. doi:10.1109/TC.2008.29

Wang, X., Mellor, J., & Al-Begain, K. (2003). Towards Providing QoS for Intergrated Cellular and WLAN Networks. In 4th Annual Postgraduate Symposium on the Convergence of Telecommunications (PGNET'03), (pp. 207-211). Liverpool, UK: JMU.

Wang, Z., & Crowcroft, J. (1996). Quality-of-Service Routing for Supporting Multimedia Applications. Journal of Selected Areas in Communications, 14(7), 1228–1234. doi:10.1109/49.536364

Want, R., Fishkin, K. P., Gujar, A., & Harrison, L. B. (1999). Bridging physical and virtual worlds with electronic tags. In Intl Conference on Human Factors in Computing Systems (SIGCHI). Pittsburgh, PA: ACM.

Wattenhoffer, R., & Zollinger, A. (2004). XTC: A Practical Topology Control Algorithm for Ad-Hoc Networks. In IEEE Parallel and Distributed Processing Symposium, Santa Fe, NM.

Weiss, G. (Ed.). (1999). Multiagent Systems A Modern Approach to Distributed Artificial Intelligence.Cambridge, MA: Massachusetts Institute of Technology.

Weiss, S. M., & Kulikowski, C. A. (1991). Computer systems that learn: classification and prediction methods from statistics, neural nets, machine learning, and expert systems. San Francisco, CA: Morgan Kaufmann Publishers Inc.

Werbos, P. (1990). Backpropagation trough time, what it does and how do it. Proceedings of IEEE on Special Issue on Neural Networks, 78, 1550–1560.

Werbos, P. J. (1974). Beyond regression: New tools for prediction and analysis in the behavioral sciences. Ph.D. Thesis, Harvard University, Cambridge, MA.

Wireless LAN Medium Access Control (MAC) and Physical Layer(PHY) specifications. (1999). ANSI/IEEE Std 802.11: 1999 (E) Part 11, ISO/IEC 8802-11.

Wisely, D., Aghvami, H., Gwyn, S. L., Zahariadis, T., Manner, J., & Gazi, V. (2003). Transparent IP Radio Access for Next-Generation Mobile Networks. IEEE Wireless Communication, 10(4), 26–35. doi:10.1109/MWC.2003.1224976

Witten, I., & Frank, E. (2005). Data Mining: Practical Machine Learning Tools and Techniques. San Francisco: Morgan Kaufmann.

Wongthavarawat, K., & Aura, G. (2003). Packet scheduling for QoS support in IEEE 802.16 broadband wireless access systems. International Journal of Communication Systems, 16(1), 81–96. doi:10.1002/dac.581

Woo, K., Yu, C., Youn, H. Y., & Lee, B. (2001). Non-Blocking, Localized Routing Algorithm for Balanced Energy Consumption in Mobile Ad Hoc Network. In 9th IEEE International Symposium on Modeling, Analysis and Simulation of Computer and Telecommunication Systems (pp. 117-124), Cincinnati, OH.

Wooldridge, M. (2002). An Introduction to Multiagent Systems. Chichester,UK: John Wiley & Sons.

Wroclawski, J. (1997). Specification of the Controlled-Load Network Element Service. RFC 2211, IETF.

Wu, D. (2005). QoS provisioning in wireless networks. Wireless Communications and Mobile Computing, 5(8), 957–969. doi:10.1002/wcm.359

Xergias, S. A., Passas, N., & Marekos, L. (2005). Flexible Resource Allocation in IEEE 802.16 Wireless Metropolitan Area Networks. In The 14th IEEE Workshop on Local and Metropolitan Area Networks, LANMAN 2005, (pp.-6).

Xiao, H., Seah, W. K. G., Lo, A., & Chua, K. C. (2000). A Flexible Quality of Service Model for Mobile Ad hoc Networks. In IEEE Vehicular Technology Conference Proceedings, (pp. 445-449). New York: IEEE Inc.

Xiao, X., & Ni, L. M. (1999). Internet QoS: a big picture. IEEE Network, 13(2), 8–18. doi:10.1109/65.768484

Xiao, Y., & Li, H. (2004). Voice and Video Transmissions with Global Data Parameter Control for the IEEE 802.11e Enhanced Distributed Channel Access. IEEE Transactions on Parallel and Distributed Systems, 15(11),1041–1053. doi:10.1109/TPDS.2004.72

Xiao, Y., Li, H., & Li, B. (2007). Bandwidth Sharing Schemes for Multimedia Traffic in the IEEE 802.11e Contention-Based WLANs. IEEE Transactions on Mobile Computing, 6(7),815–831. doi:10.1109/TMC.2007.1054

Xiong, L., & Mao, G. (2007). Saturated Throughput Analysis of IEEE 802.11e EDCA. Computer Networks, 51(1),3047–3068. doi:10.1016/j.comnet.2007.01.002

Xu, W., & Campbell, G. (1992). A Near Perfect Stable Random Access Protocol for a Broadcast Channel. In Proc. of the IEEE International Conference on Communications 1992 (ICC'92), (Vol. 1, pp. 370-374).

Xue, Q., & Ganz, A. (2003). Ad hoc QoS on-demand routing (AQOR) in mobile ad hoc networks. Special Issue on Routing in Mobile and Wireless Ad Hoc Networks, 63(2), 154–165.

Xue, Y., Nguyen, H., & Nahrstedt, K. (2007). CA-AQM: Channel-Aware Active Queue Management for Wireless Networks. IEEE International Conference on Communications (ICC).

Yaiche, H., Mazumdar, R. & Rosenberg, C. (2000). A game theoretic framework for bandwidth allocation and pricing in broadband networks. IEEE/ACM Trans Networking, 8(5), 667–678.

Yang, L., Dai, Y., & Zhang, B. (2008). Reliability Oriented QoS Driven Composite Service Selection Based on Performance Prediction. Paper presented at the Intl Conference on Software Engineering Knowledge Engineering (SEKE'08), San Francisco, CA, USA, July 1-3, 2008.

Yang, L., Zerfos, P., & Sadot, E. (2005). Architecture Taxonomy for Control and Provisioning of Wireless Access Points (CAPWAP), (RFC 4118). IETF Network Working Group.

Yang, X., Feng, G., & Kheong, D. S. (2006). Call admission control for multiservice wireless networks with bandwidth asymmetry between uplink and downlink. IEEE Transactions on Vehicular Technology, 55(1), 360–368. doi:10.1109/TVT.2005.861202

Yavuz, E. A., & Leung, V. C. (2006). Computationally efficient method to evaluate the performance of guard-channel-based call admission control in cellular networks. IEEE Transactions on Vehicular Technology, 55(4), 1412–1424. doi:10.1109/TVT.2006.877470

Ye, F., Chen, A., Liu, S., & Zhang, L. (2001). A scalable solution to minimum cost forwarding in large sensor networks. In Proceedings of Tenth International Conference on Computer Communications and Networks, (pp. 304 –309).

Ye, J., Hou, J., & Papavassiliou, S. (2002). A Comprehensive Resource Management Framework for Next Generation Wireless networks. IEEE Transactions on Mobile Computing, 1(4), 249–264. doi:10.1109/TMC.2002.1175539

Ye, W., Heidemann, J. & Estrin, D. (2003). Medium Access Control with Coordinated, Adaptive Sleeping for Wireless Sensor Networks. IEEE/ACM Transactions on Networking, 12, 493-506.

Yihai, Z., & Gulliver, T. A. (2005). Quality of service for ad hoc on-demand distance vector routing. IEEE International Conference on Wireless And Mobile Computing, Networking And Communications, WiMob'05 (pp. 192-196).

Yilmaz, O., & Chen, I. (2006). Utilizing Call Admission Control to Derive Optimal Pricing of Multiple Service Classes in Wireless Cellular Networks. 12th International Conference on Parallel and Distributed Systems (ICPADS'06). 1, pp. 605-612. Minneapolis: IEEE Computer Society.

Ying, G., Kunz, T., & Lamont, L. (2003). Quality of service routing in ad-hoc networks using OLSR. In Proceedings of the 36th Annual Hawaii International Conference on System Sciences.

Ying, Z., Anand, A. L., & Jacob, L. (2003). A QoS enabled MAC protocol for multi-hop ad hoc wireless networks. IEEE International Performance, Computing, and Communications Conference, IPCCC'03 (pp. 149- 156). New York: IEEE Inc.

Ylitalo, J., Jokikyyny, T., Kauppinen, T., Tuominen, A. J., & Laine, J. (2003). Dynamic Network Interface Selection in Multihomed Mobile Hosts. Thirty-Sixth Hawai'i International conference on System Sciences (HICSS-36). 9, (pp. 315). Big Island, Hawai'i: IEEE Computer Society.

You, C., & Chandra, K. (1999). Time series models for Internet data traffic. In Proceedings of the 24th Conference on Local Computer Networks, (pp. 164-171).

Youngki, H., & Varshney, P. (2003). An adaptive QoS routing protocol with dispersity for ad-hoc networks. In Proceedings of the 36th Annual Hawaii International Conference on System Sciences.

Youngseok, L., Kyoungae, K., & Choi, Y. (2002). Optimization of AP placement and channel assignment in wireless lans. In Proceedings of the 27th Annual IEEE Conference on Local Computer Networks, LCN 02 (pp. 831-836). Los Alamitos: IEEE Computer Society Press.

Younis, M., Youssef, M., & Arisha, K. (2003). Energy-aware management for cluster-based sensor networks. Computer Networks, 43(5), 649–668. doi:10.1016/S1389-1286(03)00305-0

Yu, O. T., & Leung, V. C. (1997). Adaptive resource allocation for prioritized call admission over an ATM-based wireless PCN. IEEE Journal on Selected Areas in Communications, 15(7), 1208–1225. doi:10.1109/49.622906

Yuan, X. & Liu, X. (2002). Heuristic algorithms for multi constrained quality-of service routing. IEEE/ACM Transactions on Networking, 10(2), 244-256.

Zhai, H., Chen, X., & Fang, Y. (2006). A call admission and rate control scheme for multimedia support over IEEE 802.11e wireless LANs. ACM Wireless Networks, 12(4),451–463. doi:10.1007/s11276-006-6545-y

Zhang, D., Ileri, O., & Mandayam, N. (2008). Bandwidth exchange as an incentive for relaying. In 42nd Annual Conference on Information Sciences and Systems (CISS 2008).

Zhang, L., & Zeadally, S. (2004). HARMONICA: Enhanced QoS Support with Admission Control for IEEE 802.11 Contention-based Access. In Proceedings 10th IEEE Real-Time and Embedded Technology and Applications Symposium (pp. 64–71). Los Alamitos: IEEE Computer Society Press.

Zhang, Q., Fitzek, F. H. P., & Katz, M. (2006). Evolution of Heterogeneous Wireless Networks: Cooperative Networks. In 3rd International Conference of the Center for Information and Communication Technologies, Copenhagen, Denmark.

Zhang, Q., Guo, C., Guo, Z., & Zhu, W. (2003). Efficient mobility management for efficient handoff between WWAN and WLAN. IEEE Communications Magazine, 41(11), 102–108. doi:10.1109/MCOM.2003.1244929

Zhang, W. (2005). Bearer Service Allocation and Pricing in Heterogeneous Wireless Networks. International Conference on Communications 2005 (ICC 2005) (pp. 1367-1371). Seoul: IEEE.

Zhang, Z., & Shen, H. (2005). Application of online-training SVMs for real-time intrusion detection with different considerations. Computer Communications, 28, 1428–1442. doi:10.1016/j.comcom.2005.01.014

Zhao, Q., & Tong, L. (2000). A Dynamic Queue MAC Protocol for Random Access Channels with Multipacket Reception. Asilomar Conference.

Zheng, L., & Tse, D. N. C. (2003). Diversity and freedom: A fundamental trade off I multiple antenna channels. IEEE Transactions on Information Theory, 49, 1073–1096. doi:10.1109/TIT.2003.810646

Zhong, S., Chen, J., & Yang, Y. R. (2003). Sprite: A Simple, Cheat-Proof, Credit-Based System for Mobile Ad-Hoc Networks. In Proc. of IEEE INFOCOM 2003.

Zhou, B. He, D., & Sun, Z. (2006). Traffic modeling and prediction using ARIMA/GARCH model. In Proceedings of 2nd EuroNGI Conference on Next Generation Internet Design and Engineering.

Zhu, H., & Cao, G. (2005). rDCF: a relay enabled medium access control protocol for wireless ad hoc networks. In Proceedings of IEEE INFOCOM.

Zhuang, W., Gan, Y., Loh, K., & Chua, K. (2003). Policy-based QoS management architecture in an integrated UMTS and WLAN environment. IEEE Communications Magazine, 41(11), 118–125. doi:10.1109/MCOM.2003.1244931

Zimmermann, E., Herhold, P., & Fettweis, G. (2005). On the performance of cooperative relaying protocols in wireless networks. European Transactions on Telecommunications, 16, 5–16. doi:10.1002/ett.1028

Zorba, N., & Perez-Neira, A. I. (2007). CAC for Multibeam Opportunistic Schemes in Heterogeneous WiMax Systems under QoS Constraints. In Global Telecommunications Conference, GLOBECOM apos07, (pp. 4296 – 4300).

Zorzi, M., & Rao, R. R. (2003). Geographic random forwarding (geraf) for ad hoc and sensor networks: multihop performance. IEEE Transactions on Mobile Computing, 2(4), 337–348. doi:10.1109/TMC.2003.1255648

Zorzi, M., Rao, R. R., & Milstein, L. B. (1995). On the accuracy of a first-order Markov model for data transmission on fading channels. In . Proceedings of ICUPC, 1995, 211–215.

Zou, L. Z. X., & Lu, M. (2004). Pager: A distributed algorithm for the deadend problem of location-based routing in sensor networks. In International Conference on Computer Communications and Networks, Chicago, IL.

About the Contributors

Thomas Lagkas received the B.S. degree (with honors) in computer science from the Department of Informatics, Aristotle University, Thessaloniki, Greece. He received the PhD degree on "Wireless Communication Networks" from the same department, in 2006. During his PhD studies, he was awarded a PhD candidates' scholarship by the Research Committee of the Aristotle University. He has been an adjunct Lecturer at the Department of Informatics and Telecommunications Engineering, University of Western Macedonia, Greece, since 2007. He has been a Laboratory Associate at the Technological Educational Institute of Thessaloniki, since 2004, and a Scientific Associate, since 2008. He has been awarded a postdoctoral research scholarship by the State Scholarship Foundation. His interests are in the areas of wireless communication networks, medium access control, QoS provision and mobile multimedia communications with relevant publications at a number of widely recognized international scientific journals and conferences.

Pantelis Angelidis received his diploma in Electrical Engineering and his Ph.D. in Telecommunications Engineering from the School of Electrical Engineering, Aristotle University of Thessaloniki in 1989 and 1993 respectively. He is currently with the University of Western Macedonia, Greece, Ass. Prof. on Health Telematics. He has worked as a technology expert in Telecommunications for the past 20 years. He has served as a project manager in over twenty international research projects. He has presented more than 50 papers in international scientific conferences and has published more than 20 original articles in international research journals. He serves as an evaluator of IST projects for the European Commission and the Greek Government. Dr Angelidis is the leader of the CEN eEHIC PT, a member of the EC M/403 eHealth-INTEROP PT and the ITU Steering Committee of the Centre of Excellence Programme for Eastern Europe and CIS countries and board secretary of HL7 Hellas. Dr. Angelidis is the founder and president of VIDAVO, a health telematics company. His research interests include WSN, BAN, protocols and services in ehealth.

Loukas Georgiadis is an assistant professor of Informatics and Telecommunications Engineering at the University of Western Macedonia,Greece. His research interests include combinatorial algorithms (mainly for graph and network problems), data structures, and optimization. Loukas obtained his PhD in Computer Science from Princeton University under the supervision of Professor Robert E. Tarjan. Prior to joining the University of Western Macedonia he held research positions at the University of Aarhus, Denmark, and Hewlett-Packard Laboratories, USA.

* * *

Antonios Alexiou obtained his Diploma from the Department of Electrical and Computer Engineering of the Aristotle University of Thessaloniki (Greece). Furthermore, he obtained his Master Degree and his PhD from the Computer Engineering and Informatics Department of Patras University (Greece). His research interests include data networks, mobile telecommunications networks, multicast routing and group management, radio resource management, radio network planning and optimization, wireless networks. He has published over 40 papers in Journals, International Conferences and Book Chapters.

Luis Alonso reached the PhD degree in the Department of Signal Theory and Communications of the Universitat Politècnica de Catalunya in 2001. He currently holds a permanent tenured position in the University, as he is an Associate Professor of the Radio Communications Research Group. He has participated in several research programs, networks of excellence, COST actions and integrated research projects. He has been the Project Coordinator of a Marie Curie ToK Action (funded by the European Union). He is currently the Scientific in Charge of a three-year Research Project funded by the Spanish Ministry of Science and Technology. His research interests are within the field of medium access protocols, radio resource management, cross-layer optimization and QoS features for all kind of wireless communications systems.

Jesús Alonso-Zárate received his M.Sc. and Ph.D. degrees in Telecommunications Engineering from the Universitat Politècnica de Catalunya (UPC), Barcelona, Spain, in March 2004 and February 2009, respectively. From 2004 to 2005, he worked as an Information Technology consultant at Everis former DMR Consulting). In 2005, he was granted by the Centre Tecnològic de Telecomunicacions de Catalunya (CTTC) to obtain the Ph. D. on Signal Theory and Communications at the UPC. While completing his Ph. D. degree, he was granted in 2006 by the European Space Agency (ESA) and by the Generalitat de Catalunya to attend the Space Studies Program (SSP'06) at the International Space University (ISU) in Strasbourg, France. His main research interests are focused on Wireless Communications and include the design and modeling of Wireless Networks, Medium Access Control Protocols, Cross-Layer Design, Scheduling Algorithms, Radio Resource Management, Cooperative Networks, and Cognitive Radio Networks, for ad hoc, sensor, and infrastructure-based networks.

Bert-Jan van Beijnum received his MSc and PhD in Electrical Engineering from the University of Twente, the Netherlands. He is an assistant professor in the Remote Monitoring and Treatment section of the Bio Signals and Systems research group at the University of Twente, the Netherlands. His research is embedded in the projects of the Centre for Telematics and Information Technology and the Institute of Biomedical Technology. His research interests include Autonomic Computing, Mobile Virtual Communities, Telemedicine, Information Systems, ICT Management, Task Assignment Systems and algorithms, Application layer mobility handover mechanisms and QoS.

Carolina Blanch Perez del Notario was born in Pamplona, Spain. She received her M.S. degree in telecommunications engineering in January 2000 from the University of Navarra, Spain. In 2001 she obtained an M.S. in artificial intelligence from the Katholieke Universiteit Leuven. Since then she has been working as a research engineer in the digital components technology department at IMEC, Leuven, Belgium. Her research interests include video coding and transmission with focus on energy-efficiency, quality-of-service support and cross-layer design for optimization of wireless video transmission. Currently she is working on run-time resource management techniques.

Anthony Boucouvalas has worked at GEC Hirst Research Centre, and became a Group Leader and a Divisional Chief Scientist until 1987 when he joined Hewlett Packard (HP) Laboratories as a Project Manager, where he worked in the areas of optical communication systems, optical networks, and instrumentation until 1994, when he joined Bournemouth University. In 1996, he became a Professor of multimedia communications and in 1999, he became the Director of the Microelectronics and Multimedia Research Centre. In 1996, he joined the University of Peloponnese as a Professor and now the Head of the department. His current research interests span the fields of wireless communications, optical fibre communications and components, multimedia communications, and human-computer interfaces, where he has published over 200 papers. He has contributed to the formation of IrDA as an industry standard and he is now a Member of the IrDA Architectures Council. He is a Fellow of the Royal Society for the Encouragement of Arts, Manufacturers and Commerce (FRSA) and a Fellow of IEE (FIEE). In 2002, he became a Fellow of the Institute of Electrical and Electronic Engineers (FIEEE) for contributions to optical fibre components and optical wireless communications. He is a Member of the New York Academy of Sciences, and the Association for Computing Machinery (ACM). He is an Editor of numerous journals and in the organising committees of many conferences.

Christos Bouras obtained his Diploma and PhD from the Department Of Computer Engineering and Informatics of Patras University (Greece). He is currently a Professor in the above department. Also he is a scientific advisor of Research Unit 6 in Research Academic Computer Technology Institute (CTI), Patras, Greece. His research interests include Analysis of Performance of Networking and Computer Systems, Computer Networks and Protocols, Telematics and New Services, QoS and Pricing for Networks and Services, e-Learning Networked Virtual Environments and WWW Issues. He has extended professional experience in Design and Analysis of Networks, Protocols, Telematics and New Services. He has published 250 papers in various well-known refereed conferences and journals. He is a co-author of 8 books in Greek. He has been a PC member and referee in various international journals and conferences. He has participated in R&D projects such as RACE, ESPRIT, TELEMATICS, EDUCATIONAL MULTIMEDIA, ISPO, EMPLOYMENT, ADAPT, STRIDE, EUROFORM, IST, GROWTH and others. Also he is member of experts in the Greek Research and Technology Network (GRNET), Advisory Committee Member to the World Wide Web Consortium (W3C), IEEE - CS Technical Committee on Learning Technologies, IEEE ComSoc Radio Communications Committee, IASTED Technical Committee on Education WG6.4 Internet Applications Engineering of IFIP, ACM, IEEE, EDEN, AACE, New York Academy of Sciences and Technical Chamber of Greece.

Richard Bults holds a Masters degree (cum laude) in Telematics and a Bachelor degree in Technical Computer Science. He is a researcher in the Remote Monitoring and Treatment section of the Bio Signals and Systems research group at the University of Twente, the Netherlands. His research interests are design of mobile telemedicine systems and QoS evaluation and control of these systems. Richard is also one of the Mobihealth BV founding fathers and the CTO of this privately owned company. MobiHealth's mission is to give patients full mobility during remote health monitoring sessions while staying in touch with their care professional. He is responsible for the company's product portfolio and tele-monitoring solutions consultancy.

Filippo Cacace graduated in electronic engineering at Politecnico di Milano in 1988, where he received the PhD degree in Computer Science in 1991. Since 2006 he is an Assistant Professor at the University Campus Bio-Medico di Roma. He co-authored more than 50 research papers on data models and database programming languages, wireless and heterogeneous computer networks, mobile applications, control theory and systems biology. He participated in the Italian and European projects: E-NEXT, CONTENT, NETQoS, OneLAB, Quasar, WebMinds.

Dimitris E. Charilas is a Ph.D. Candidate in the Department of Electrical and Computer Engineering of the National Technical University of Athens (NTUA). He received the Diploma in Electrical and Computer Engineering from NTUA in 2006 and an MBA in Techno-Economic Systems from NTUA in 2008. He has worked as a research assistant in the Telecommunications Laboratory of the Institute of Communication and Computer Systems (ICCS-NTUA). He is currently with the Mobile Radiocommunications Laboratory of ICCS-NTUA. His research interests include mobile services, quality of service, wireless communications, resource allocation and game theory applications.

Hsiao-Hwa Chen is currently a full Professor at National Sun Yat-Sen University, Taiwan. He has authored or co-authored over 200 technical papers in major journals and conferences and five books in the areas of communications. He served as symposium co-chair of major international conferences, including IEEE VTC, IEEE ICC, IEEE Globecom, IEEE WCNC, etc. He served or is serving as an Editorial Board member or/and Guest Editor of IEEE COMMUNICATIONS MAGAZINE, IEEE JSAC, IEEE Wireless Communications Magazine, IEEE Networks Magazine, IEEE TRANS. ONWIRELESS COMMUNICATIONS, IEEE COMMUNICATIONS LETTERS, IEEE TRANSACTIONS ON VEHICULAR TECHNOLOGY, IEEE Vehicular Technology Magazine, Wireless Communications and Mobile Computing Journal and the International Journal of Communication Systems, etc. He is a Guest Professor at Zhejiang University and Shanghai Jiao Tung University, China.

Philip Constantinou received the Diploma in Physics from the National University of Athens in 1972, the Master of Applied Science in Electrical Engineering from the University of Ottawa, Ontario, Canada in 1976 and the PhD degree in Electrical Engineering in 1983 from Carleton University, Ottawa, Ontario, Canada. From 1976 to 1979 was with Telesat Canada. In 1980 he joined the Ministry of Communications in Ottawa, Canada. From 1984 to 1989 he was with the National Research Centre Demokritos in Athens, Greece where he was involved on several research projects in the area of Mobile Communications. In 1989 he joined NTUA where he is currently Professor. His current research interests include Mobile Satellite Communications and Interference Problems on Digital Communications Systems.

Panayotis G. Cottis was born in Thessaloniki, Greece, in 1956. He received the Diploma in Mechanical and Electrical Eng. and the Dr. Eng. Degree from the National Technical University of Athens (NTUA), Greece, in 1979 and 1984, respectively, and the M.Sc. Degree from the University of Manchester,(UMIST), Manchester, U.K., in 1980. In 1986, he joined the School of Electrical and Computer Eng. NTUA, where he has been a Professor since 1996. He has published more than 130 papers in international technical journals and conference proceedings. His current research interests include wave propagation, wireless networks and satellite communications. From September 2003 to September 2006, Professor Cottis served as the Vice-Rector of NTUA.

Antoine Dejonghe was born in Ottignies (Belgium) in 1977. He received the Electrical Engineering degree and the Ph. D. degree from the Université catholique de Louvain (UCL), Louvain-la-Neuve, Belgium, in 2000 and 2004, respectively. From September 2000 until November 2004, he was pursuing a Ph.D. degree within the Communications and Remote Sensing Laboratory (TELE), as a Research Fellow of the Belgian National Fund for Scientific Research (FNRS). Since December 2004, he has been with the Wireless Group of IMEC, Leuven, where he's currently Program Manager for Cognitive Reconfigurable Baseband activities.

Der-Jiunn Deng received the B.S. degree in computer science from Tung-Hai University, the M.S. degree in information management from National Taiwan University of Science and Technology, and the Ph.D. degree in electrical engineering from National Taiwan University in 1997, 1999, and 2005, respectively. He joined National Changhua University of Education as an assistant professor in the Department of Computer Science and Information Engineering in August 2005 and then became an associate professor in February 2009. Dr. Deng visited Iowa State University, USA, in 2006. His research interests include multimedia communication, quality-of-service, and wireless networks. Dr. Deng is the guest editor of the special issue on Ubiquitous Multimedia Computing: Systems, Networking, and Applications for the International Journal of Ad Hoc and Ubiquitous Computing (IJAHUC). He is now serving as a program chair for the 2nd International Conference on Computer Science and its Applications (CSA-09). Dr. Deng is a member of the IEEE.

Claude Desset (S'94--M'01) was born in Bastogne, Belgium, in 1974. Graduated (summa cum laude) as an electrical engineer from the Université catholique de Louvain (UCL), Louvain-la-Neuve, in 1997, he then achieved a PhD in the same university until 2001, funded by the Belgian national fund for scientific research (FNRS). His doctoral research mainly included joint source-channel coding for image transmissions, focusing on unequal error protection, global optimization of a transmission chain, and image reconstruction from incomplete data. He also worked in channel coding, especially bit error rate approximation of error-correcting codes and code selection for specific applications. In 2001 he joined IMEC, Leuven, Belgium, to work as a senior researcher in the design of ultra-low-power wireless communication systems. He focused on body area networks, developing ultra-wideband solutions and optimizing the power at system level by considering both air interface and front-end architecture. He also worked in MIMO communications, link adaptation, and turbo coding/processing. Since 2006, he works in a cross-disciplinary team tackling power-performance optimization of wireless terminals by tuning all the components to the current conditions and requirements in order to achieve the system-wide optimum.

Huifang Feng is an Assistant Professor of Computer Science at Northwest Normal University. She received the Ph.D. degree in computer science from Tianjin University in 2006, the M.S. degree in computer science from Lanzhou University in 1996, and the B.A. degree in Mathematical Sciences from Northwest Normal University in 1993. She is a member of IEICE of Japan. Her research interests are in computer communication networks, wireless networks, modeling and simulation.

Tong Gan was born in Nanjing, P. R. China in 1978. He received his bachelor degree in 2000 from the school of communication engineering, South East University of Nanjing. In 2005 he achieved a PhD in information engineering from Nanyang Technological University, Singapore. His doctoral research focuses on error-resilient transmission of image/video over error-prone networks. Since 2006 he worked at IMEC, Leuven, Belgium as a research engineer in digital components department. His research area includes H.264/AVC codec optimization and performance modeling, and cross-layer design and optimization of wireless video streaming system.

Vasileios Gkamas was born in Ioannina, Greece in 1981. He obtained his Diploma of Information and Communication Systems Engineering from the University of Aegean, Greece, in 2004 and his MSc on Integrated Hardware and Software Systems from the Department of Computer Engineering and Informatics, University of Patras, in 2007. Now, he is a Ph.D. candidate at the department of Computer Engineering and Informatics, University of Patras. Since 2006, he is with the Networking Technology Sector of the Research Academic Computer Technology Institute (RA-CTI), Patras, Greece, participating in numerous National and European projects. Also from 2007, he is laboratorial staff at the Department of Informatics and Means of Mass Communication of the Technological Educational Institute of Patras. His research activities are in the areas of energy-aware protocols for wireless ad hoc and wireline networks, grid computing technologies and design and implementation of data networks.

Melanie Hilario has a Ph.D. in computer science from the University of Paris VI and currently works at the University of Geneva's Artificial Intelligence Laboratory. She has initiated and participated in several European research projects on neuro-symbolic integration, meta-learning, and biological text mining. She has served on the program committees of many conferences and workshops in machine learning, data mining, and artificial intelligence. She is currently an Associate Editor of the International Journal on Artificial Intelligence Tools and a member of the Editorial Board of the Intelligent Data Analysis journal.

Yueh-Min Huang is a Distinguished Professor in the Department of Engineering Science, National Cheng-Kung University, Taiwan, R.O.C. His research interests include wireless networks, artificial intelligence, and e-Learning. He received his MS and Ph.D. degrees in Electrical Engineering from the University of Arizona in 1988 and 1991 respectively. He has co-authored 3 books and has published over 140 refereed journal research papers. Dr. Huang has received many research awards, such as the Best Paper Award of 2007 IEA/AIE Conference, Best Paper Award of the Computer Society of the Republic of China in 2003, the Awards of Acer Long-Term Prize in 1996, 1998, and 1999, Excellent Research Awards of National Microcomputer and Communication Contests in 2006. Dr. Huang is in the editorial board of the Journal of Wireless Communications and Mobile Computing, Journal of Security and Communication Networks and International Journal of Communication Systems. Huang is a member of the IEEE as well as IEEE Communication, Computer, and Circuits and Systems Societies.

Giulio Iannello graduated in Electronic Engineering at Politecnico di Milano in 1981 and received a Ph.D in Computer Science and Computer Engineering from the University of Napoli Federico II in 1987. Currently he is Full Professor of Computer Science and Computer Engineering at the University Campus Bio-Medico di Roma. His current research interests include wireless and high performance computer networks, multimedia data processing, biomedical data and image processing, design and analysis of parallel algorithms, performance evaluation of parallel and distributed systems. He has published over 100 journal and conference papers in these and related areas. He is member of several Program Committees of international conferences. Dr. Iannello is a member of the IEEE and ACM.

Victor LU Jinchang received his B.E. in Information and Communication Engineering from Zhejiang University, China, in 1989. He received his M.S. in Information Studies from Nanyang Technological University, Singapore, in 2005. He is currently a Ph.D. candidate working toward his Ph.D. degree in the School of Electrical and Electronic Engineering, Nanyang Technological University, Singapore. His research interests include QoS support mechanisms and algorithms in WiMAX networks, cognitive radio networks & cognitive Digital TV over WiMAX and WRAN networks and bioinformatics Self-organization & Self-optimization Autonomic Networks.

Alexandros Kaloxylos received the B.Sc. degree in Computer Science from the University of Crete, Greece, in 1993, the M.Phil. degree in Computing and Electrical Engineering from the Heriot-Watt University, Scotland, in 1994 and the Ph.D. degree in Informatics and Telecommunications from the University of Athens in 1999. From 1990 to 1993 he was a staff member of the Computer Centre of the University of Crete, and a researcher in the Foundation of Research and Technology Hellas (FORTH). From 1994 to 1995 he was a research associate at the University of Wales. From 1995 to 2001 he was a staff member of the Communications Network Laboratory of the University of Athens. From 2001 to 2003 he was a member of the Greek Broadband Task Force. In 2002 he joined the faculty of the University of Peloponnese, where he is presently a Lecturer in the Department of Telecommunications Science and Technology. He has participated in numerous projects realized in the context of EU programs as well as National Initiatives. His research interests are in the design and performance analysis of broadband and wireless/mobile networks and services.

John D. Kanellopoulos (SM'90) was born in Athens, Greece, on December 12, 1948. He received the Diploma of Mechanical and Electrical Engineering and the Dr. Eng. degree from National Technical University of Athens (NTUA), in 1971 and 1979, respectively. He also received the DIC and Ph.D. degrees from the Imperial College of Science, Technology and Medicine, University of London, London, U.K., in 1979. In November 1979, he joined the Department of Electrical Engineering, NTUA, where he is now a Professor. He has published more than 100 papers in international journals and transactions. His areas of interest are microwave propagation through rain media, satellite communication systems, nonlinear optics, rough surface scattering and waveguide propagation.

Elli Kartsakli received the Diploma in Electrical and Computer Engineering from the National Technical University of Athens, Greece, in 2003 and a MSc. in Mobile and Satellite Communications from the University of Surrey, UK, in 2004. She is currently a PhD student in the Department of Signal Theory and Communications (TSC) of the Technical University of Catalunya (UPC). Her research interests include MAC layer optimization and cross-layer design for WLANs. Vasileios Kokkinos obtained his diploma from the Physics Department of the University of Patras, Greece in October 2003. Next, he was accepted in the postgraduate program "Electronics and Information Processing" in the same department and in March 2006 he obtained his Master Degree. Currently, he is a PhD student in the Computer Engineering and Informatics Department. He works in the Research Unit 6 of Research Academic Computer Technology Institute since September 2006. His research interests include wireless networks, third generation mobile telecommunications networks, multicast routing and group management and radio resource management. He has published eight papers in well-known refereed conferences.

Panagiotis Kokkinos received a Diploma in Computer Engineering and Informatics in 2003 and an MS degree in Integrated Software and Hardware Systems in 2006, both from the University of Patras, Greece. He has also worked in the private sector. Currently, he is a PhD student at the Department of Computer Engineering and Informatics of the University of Patras. His research activities are in the areas of wireless networks and grid computing.

Dimitri Konstantas is Professor and Vice-Dean at the Faculty of Social and Economic Sciences of the University of Geneva (CH), department of Information Systems, heading the Advanced Systems Group (ASG). He was previously professor and Chair of the APS group at the University of Twente, The Netherlands. For the last 20 years, prof. Konstantas, is active in research in the areas of Object Oriented systems, agent technologies, Multimedia applications and e-commerce services, with numerous publications in international conferences, journals, books and book chapters, a long participation and leadership in many European projects and with many nominations as consultant and scientific expert for several international companies and governments. Since 2002 his main research areas are mobile and wireless multimedia services and applications, with special interest in mobile health and location based services.

Katarzyna Kosek-Szott received her M.Sc. degree in telecommunications from the AGH University of Science and Technology, Krakow, Poland in 2006. Now, she is working towards her Ph.D. degree at the Department of Telecommunications, AGH UST. Her general research interests are focused on wireless networking. The major topics include wireless LANs (especially ad-hoc networks) and quality of service provisioning within these networks. She is a reviewer of international journals and conferences. She is/was involved in several European projects: DAIDALOS II, CONTENT, CARMEN as well as grants supported by the Ministry of Science and Higher Education. She has also co-authored a number of research papers.

Pierre Leone is Assistant Professor at the Department of Computer Science of the University of Geneva. He was a scientific researcher at the Engineering School of Geneva where he was involved in various research projects in the fields of collaborative optimization algorithms on distributed systems and hardware development. He received his Ph.D. degree in Mathematics from the University of Geneva and has a background of electrical engineer with orientation in Computer Science. He spent a post-doctoral year as a Visiting Lecturer of the Mathematics, Department of the Auckland University in New-Zealand.

Maode Ma received his Ph.D. degree in computer science from Hong Kong University of Science and Technology in 1999. Now, Dr. Ma is a tenured Associate Professor in the School of Electrical and Electronic Engineering at Nanyang Technological University in Singapore. He has extensive research interests including wireless networking and wireless network security. He has been a member of the technical program committee for more than 90 international conferences. He has been a technical track chair, tutorial chair, publication chair, and session chair for more than 40 international conferences. Dr. Ma has published more than 130 international academic research papers on wireless networks. He currently serves as an Associate Editor for IEEE Communications Letters, an Editor for IEEE Communications Surveys and Tutorials, and an Associate Editor for other 6 International Journals.

Olivier Marcé joined former Alcatel R&I in 1999 after 2 years in French Research Institute INRIA. Since, he worked in the field of IP network and IP routing, as well as wireless networking. He actively worked on the subject of Active Networking. His main subjects of interests are inter-domain both in wired and wireless, as well as user-defined networks. He participated and contributed to the definition of several other research projects in national or European funding framework, IST, Eureka, RNRT or RNTL. He is author or co-authors of more than thirty international patents, mainly in IP network domain. In Bell-Labs France, he is manager of international research projects focused on the integration of the IP technologies with the mobile radio networks architecture.

Bart Masschelein received his degree of Industrial Engineer in '98 from the Katholieke Hogeschool Brugge-Oostende, in Belgium. That year he started a part-time MSc-course in Electronic System Design, in collaboration with Leeds Metropolitan University. In '98, he also joined the Multi-Media group at the Interuniversity Micro Electronics Center (IMEC) in Leuven, Belgium. His first assignment was the implementation of the FlexWave-II architecture, a wavelet- based still-image compression engine with MPEG-4 functionality. This architecture uses the Local Wavelet Transform, which is an implementation of the Discrete Wavelet Transform focusing on high performance and low memory requirements. Following this, he worked on the Scalable Video Codec, where he collaborated with the MPEG community to establish the final standard. His task focused on the optimization of a real-time SVC decoder for mobile applications, together with the implementation of a Media Aware Network Element for real-time adaption of a video stream based on external parameters. Currently he is working on 3DTV applications, and more specifically multiple-camera setups.

Vinod Mirchandani has earned Ph.D (Australia), MS (Canada), BE (IISc-India) and BSc (India) degrees. He is currently working as a private telecommunications consultant and a lecturer. He was a Senior Research Associate at the University of Technology, Sydney, Australia from 2006-2008, where he worked on Alcatel-Lucent (Bell labs, Paris) projects. He conducted lecturing and research duties at the University of Sydney from 2004-2006. From 1999-2003 he worked as a Senior Research Engineer at Motorola Labs, where he also won a BRAVO award. Throughout his career, he has won numerous awards and prestigious scholarships. He has also authored several publications and has pending patent(s) in the Wireless LAN area. His current research interests are in the area of wireless mesh networks, mobile networks, performance evaluation, grid computing, protocol design and quality of service. He has also been a reviewer for IEEE Communications, IEEE Networks magazines and several reputed IEEE conferences.

Luminita Moraru is currently a Ph.D. candidate in the TCS-sensor lab of the Computer Science Department of the University of Geneva. She received a B.S. degree in Electrical Engineering and Computer Science from the Polytechnic University of Bucharest, in 2004, and a M.S. degree in Computer Science (Embedded Systems) from the University of Science and Technology of Lille, in 2005. Her research interests are in sensor networks, mobile ad-hoc networks, security and reputation based trust. Her current research focuses on security and QoS of routing protocols for sensor networks.

Marek Natkaniec received his M.Sc. and Ph.D. degrees in telecommunications from the AGH University of Science and Technology, Cracow, Poland in 1997 and 2002, respectively. In 1997 he joined AGH University of Science and Technology. Now, he works as an assistant professor at the Department of

Telecommunications. He is responsible for lectures, projects and laboratories of computer networks. His general research interests are in wireless networks. Particular topics include wireless LANs, designing of protocols, formal specification of communication protocols, modeling and performance evaluation of communication networks, quality of service, multimedia service provisioning and cooperation of networks. He is a reviewer for international journals and conferences. He has/had actively participated in European projects: MOCOMTEL, PRO-ACCESS, DAIDALOS I, DAIDALOS II, CONTENT, CARMEN, MEDUSA as well as grants supported by the Ministry of Science and Higher Education. Marek Natkaniec co-authored five books and over 80 research papers.

Sotiris Nikoletseas is a Professor at the Computer Engineering and Informatics Department of Patras University, Greece and Director of the SensorsLab at CTI. His research interests include Algorithmic Techniques in Distributed Computing (focus on sensor and mobile networks), Probabilistic Techniques and Random Graphs, and Algorithmic Engineering. He has coauthored over 120 publications in Journals and refereed Conferences, several Book Chapters and two Books (one on Sensor Networks, and another on the Probabilistic Method), while he has delivered several invited talks and tutorials. He has served as the Program Committee Chair of many Conferences, and as Editorial Board Member of major Journals. He has co-initiated international conferences on sensor networking. He has coordinated several externally funded European Union R&D Projects related to fundamental aspects of modern networks.

Athanasios D. Panagopoulos received the Diploma Degree in Electrical and Computer Engineering and the Dr. Engineering Degree from National Technical University of Athens (NTUA) in July 1997 and in April 2002. From September 2003 to December 2008, he gave lectures at School of Pedagogical and Technological Education, as part-time Assistant Professor. From January 2005 to May 2008, he was head of the Satellite Division of Hellenic Authority for the Information and Communication Security and Privacy. Since May 2008, he is Lecturer in the Department of Electrical and Computer Engineering of NTUA. He has published more than 135 papers in International Journals and Transactions and Conference proceedings. He is the recipient of URSI General Assembly Young Scientist Award in 2002 and 2005. His research interests include mobile radio communication systems design, wireless and satellite communications networks, propagation effects on multiple access systems and on communication protocols, MIMO systems channel modeling, interference mitigation techniques, and cognitive communications. He participates to ITU-R and ETSI Study Groups. He is Senior Member of IEEE and member of Technical Chamber of Greece. He serves on the editorial boards of the Hindawi International Journal of Antennas and Propagation, International Journal of Vehicular Technology and Elsevier Physical Communication and since October 2008 as an Associate Editor of IEEE Transactions on Antennas and Propagation.

Sarantis Paskalis received his B.Sc. degree (with honors), M.Sc. and Ph.D degrees in Computer Science from the Department of Informatics and Telecommunications at the University of Athens, Greece in 1995, 1998, and 2004 respectively. He is a staff member of the Communication Networks Laboratory of the University of Athens. His research interests include network performance analysis, signaling, protocol design and evaluation for QoS in mobile networks.

Sofie Pollin received a degree in electrical engineering in 2002 and a Ph.D. degree in 2006 (with honors) from the Katholieke Universiteit Leuven, Belgium. Since October 2002 she has been a researcher at the Wireless Research group of the Interuniversity Microelectronics Center (IMEC). In the summer of 2004 she was a visiting scholar at National Semiconductor, Santa Clara, California. In the summer of 2005 she was a visitor at UC Berkeley. In 2006, she started as a post-doctoral researcher at UC Berkeley working on coexistence issues in wireless communication networks. Since 2008, she is a senior researcher at IMEC working on cognitive radio.

Ante Prodan is a Ph.D candidate at the Faculty of Engineering and IT, University of Technology Sydney. He is recipient of Alcatel-Lucent scholarship and the faculty innovation award as well as the author of over 10 peer reviewed papers. Ante's main interests lie in the areas of telecommunications, wireless mesh networks and applied Artificial Intelligence in particular multiagent systems. His Ph.D work is focused on the development of a multiagent system for topology control on multi-radio wireless mesh networks. Before commencing his Ph.D, Ante has worked as an IT consultant for a variety of businesses in recruitment, insurance, tourism and computer-aided manufacturing.

Jose Rolim is Full Professor at the Department of Computer Science of the University of Geneva where he heads the Theoretical Computer Science and Sensor Lab (TCSensor Lab). He received his Ph.D. degree in Computer Science from the University of California, Los Angeles, in the area of formal languages and complexity. He has published many articles in the areas of theoretical computer science, distributed systems, randomization and computational complexity and leads two major national projects in the areas of Power Aware Computing and Games and Complexity. He also participates as a partner in two European Projects in the areas of Dynamic Systems and Foundations of Sensor Networks. Prof. Rolim participates in the editorial board of several journals and conferences and he is the Steering Committee Chair and General Chair of the IEEE Distributed Computing Conference in Sensor Systems. He has been Program Committee Chair of major conferences such as ICALP, IPDPS, RANDOM and served as Program Committee Member of all major conferences in theoretical computer science.

Dimitrios G. Stratogiannis was born in Larissa, Greece on January 6, 1982. He received the Diploma Degree in Electrical and Computer Engineering from the Democritus University of Thrace (DUTH), Xanthi Greece in July 2005. In October 2008 he received his Master in Business Administration (MBA) from Athens University of Economics and Business (AUEB), Athens, Greece. In October 2005 he joined the wireless and satellite communications group of National Technical University of Athens (NTUA), where he is a Research Associate. For his academic progress and excellence he has been awarded by the State Scholarship Foundation and the Technical Chamber of Greece. He participated and he has served or is serving as chair in international conferences. His research interests include wireless, mobile and satellite communications networks, mobile computing and radio communication systems design. He is member of IEEE, IEEE Communications society and of the Technical Chamber of Greece.

Szymon Szott received his M.Sc. degree in telecommunications from the AGH University of Science and Technology, Krakow, Poland in 2006. Currently he is working towards his Ph.D. degree at the Department of Telecommunications, AGH UST. His professional interests are related to wireless networks (in particular: QoS provisioning and security of ad-hoc networks). He is a reviewer of international journals and conferences. He is/was involved in several European projects: DAIDALOS II, CONTENT, CARMEN as well as grants supported by the Ministry of Science and Higher Education. He is the co-author of several research papers.

Georgios I. Tsiropoulos was born in Athens, Greece in 1982. He received his Diploma in Electrical and Computer Engineering (summa cum laude) from National Technical University (NTUA) of Athens in 2005 and the M.Sc. in techno-economics in 2009 from the same institute. Since 2006, he is technical consultant in the Ministry of Transport and Communications of the Hellenic Republic and Research Associate in the Wireless & Satellite Communications Group of NTUA. Mr. Tsiropoulos is member of IEEE and Technical Chamber of Greece. He served or is serving as the Chair or TPC member of many international conferences and he has published more than 20 technical papers in leading journals, conferences and books. His research interests are in the areas of wireless, satellite, mobile and heterogeneous networks including radio resource management, quality of service and mobile ad hoc networks as well as physical and link layer issues for BPLC networks (Broadband over Power Line Communications).

Dimitra Vali received her M.Sc. degree in Communication Systems and Data Networks in 1998 and her B.Sc. degree in Computer Science and Telecommunications in 1995, from the Department of Informatics and Telecommunications, University of Athens. In 1995 she became a staff member of the Communication Networks Laboratory of the University of Athens. Since 1999 she has joined the Hellenic Telecommunications Organisation of Greece (OTE), working for the Mobile Systems and Radiocommunications Laboratories of OTE Research. She has participated in various research projects, including the ACTS WAND project and the IST MOBIVAS project. She has also been actively involved in a number of EURESCOM projects as well as several national research activities. Her main areas of research interest include QoS in backbone networks, wireless networking, next generation networks, fixed and mobile networks convergence and voice over IP.

Emmanouel (Manos) Varvarigos was born in Athens, Greece, in 1965. He received a Diploma in Electrical and Computer Engineering from the National Technical University of Athens in 1988, and the M.S. and Ph.D. degrees in Electrical Engineering and Computer Science from the Massachusetts Institute of Technology in 1990 and 1992, respectively. He has held faculty positions at the University of California, Santa Barbara (1992-1998, as an Assistant and later an Associate Professor) and Delft University of Technology, the Netherlands (1998-2000, as an Associate Professor). In 2000 he became a Professor at the department of Computer Engineering and Informatics at the University of Patras, Greece, where he heads the Communication Networks Lab. He is also the Director of the Network Technologies Sector at the Research Academic Computer Technology Institute (RA-CTI), which through its involvement in pioneering research and development projects, has a major role in the development of network technologies and telematic services in Greece. Professor Varvarigos has served in the organizing and program committees of several international conferences, primarily in the networking area, and in national committees. He has also worked as a researcher at Bell Communications Research, and has consulted with several companies in the US and in Europe. His research activities are in the areas of network protocols, optical networks, QoS routing, wireless networks, network services, parallel and grid computing.

Christos Verikoukis is Senior Researcher with CTTC. He obtained his degree in Physics and his M.Sc. in Telecommunications Engineering from the Aristotle University of Thessaloniki (AUTh) in 1994 and 1997, respectively. He got his PhD from the Technical University of Catalonia (UPC), in 2000, in the area of Broadband Indoor Wireless Communications. From 2003 to February 2004 he was researcher in the Southeastern Europe Telecommunications & Informatics Research Institute (INA),

Thessaloniki, Greece. Dr. Verikoukis has coordinated and participated several research projects funded by the Spanish and the Greek governments and by the EC (INTERREG, Marie-Curie, PEOPLE, IST, MEDEA+, CELTIC, ICT and eTEN). He has published more than 60 articles in international journals and conferences, he is co-editor in one book and co-author in 7 book chapters.

Luca Vollero received the M.Sc. and Ph.D. degrees in Telecommunications and Computer Science from the University of Napoli "Federico II" respectively in 2001 and 2005. In 2001 he joined the DIS - Dipartimento di Informatica e Sistemistica - of the University "Federico II" of Napoli. In 2005 Luca worked at the Consorzio Nazionale CINI in Napoli as a researcher. In 2005-2006 Luca worked at the NEC Network Labs in Heidelberg (Germany) as a researcher. From 2006, Luca is an Assistant Professor at University Campus Bio-Medico of Rome. His general research interests are wireless networks and image processing. He served and is serving as reviewer for international journals and conferences. He has/had actively participated in the Italian and European projects: CADENUS, DAIDALOS, E NEXT, CONTENT, NETQoS, OneLAB, Quasar, WebMinds . Luca co-authored over 25 research papers.

Katarzyna Wac is a PhD candidate at University of Geneva, Switzerland conducting research in area of Quality of Service (QoS) in mobile applications. She is also a research staff member at the University of Twente, the Netherlands, where she is affiliated with the MobiHealth BV spin-off company. She has received her BSc and MSc degrees (cum laude) in Computer Science from Wroclaw University of Technology, Poland, and her MSc in Telematics (cum laude) from University of Twente. Her current research interests include mobile applications and services with special emphasis on supporting adaptive multimedia protocols and end-to-end QoS mechanisms, especially in a mobile healthcare domain.

Nizar Zorba holds a BSc. in Electrical Engineering by JUST University (2002, Jordan), a MPhil in Mobile Communications by University of Zaragoza (2002-2004, Spain), a MPhil in Economics by University of Zaragoza (2003-2005, Spain), a MPhil in Signal Processing by Universitat Politecnica de Catalunya-UPC (2004-2006, Spain), and a PhD in Signal Processing for Communications by UPC (December 2007, Spain). During and after his PhD studies, he was been working at CTTC where he participated in 8 European research projects for ICT, Celtic Eureka and the European Space Agency. Since Fall of 2008, he works as an assistant professor at the University of Jordan and as a consultant for European research projects Celtic Eureka ICARUS and R2D2. He is author of more than 40 peer-reviewed journals and international conferences, 3 book chapters and 5 patents. His research interests are MIMO Multiuser, Cross Layer QoS designs and resource optimization in WLAN, WiMax and LTE.

Index

C

D

M

N

O

P

T

U

V

W